Joseph

Masamichi Takesaki

Theory of Operator Algebras I

Springer-Verlag
New York Heidelberg Berlin

Masamichi Takesaki
Department of Mathematics
University of California at Los Angeles
Los Angeles, California 90024
USA

AMS Subject Classification: 22D25, 46LXX, 47CXX, 47DXX

With 2 Figures.

Library of Congress Cataloging in Publication Data

Takesaki, Masamichi, 1933–
Theory of operator algebras I.

Bibliography: p.
Includes index.
1. Operator algebras. I. Title.
QA326.T34 512'.55 79-13655

Printed in the United States of America.

9 8 7 6 5 4 3 2 1

ISBN 0-387-90391-7 Springer-Verlag New York
ISBN 3-540-90391-7 Springer-Verlag Berlin Heidelberg

Contents

Introduction v

Chapter I
Fundamentals of Banach Algebras and C^*-Algebras 1

0. Introduction 1
1. Banach Algebras 2
2. Spectrum and Functional Calculus 6
3. Gelfand Representation of Abelian Banach Algebras 13
4. Spectrum and Functional Calculus in C^*-Algebras 17
5. Continuity of Homomorphisms 21
6. Positive Cones of C^*-Algebras 23
7. Approximate Identities in C^*-Algebras 25
8. Quotient Algebras of C^*-Algebras 31
9. Representations and Positive Linear Functionals 35
10. Extreme Points of the Unit Ball of a C^*-Algebra 47
11. Finite Dimensional C^*-Algebras 50
Notes 54
Exercises 55

Chapter II
Topologies and Density Theorems in Operator Algebras 58

0. Introduction 58
1. Banach Spaces of Operators on a Hilbert Space 59
2. Locally Convex Topologies in $\mathscr{L}(\mathfrak{H})$ 67
3. The Double Commutation Theorem of J. von Neumann 71
4. Density Theorems 79
Notes 99

Chapter III

Conjugate Spaces 101

0. Introduction 101
1. Abelian Operator Algebras 102
2. The Universal Enveloping von Neumann Algebra of a C^*-Algebra 120
3. W^*-Algebras 130
4. The Polar Decomposition and the Absolute Value of Functionals 139
5. Topological Properties of the Conjugate Space 147
6. Semicontinuity in the Universal Enveloping von Neumann Algebra* 157

Notes 179

Chapter IV

Tensor Products of Operator Algebras and Direct Integrals 181

0. Introduction 181
1. Tensor Product of Hilbert Spaces and Operators 182
2. Tensor Products of Banach Spaces 188
3. Completely Positive Maps 192
4. Tensor Products of C^*-Algebras 203
5. Tensor Products of W^*-Algebras 220

Notes 229

6. Integral Representations of States 230
7. Representation of $L^2(\Gamma,\mu)\otimes\mathfrak{H}$, $L^1(\Gamma,\mu)\otimes_\gamma\mathcal{M}_*$, and $L(\Gamma,\mu)\bar{\otimes}\mathcal{M}$ 253
8. Direct Integral of Hilbert Spaces, Representations, and von Neumann Algebras 264

Notes 287

Chapter V

Types of von Neumann Algebras and Traces 289

0. Introduction 289
1. Projections and Types of von Neumann Algebras 290
2. Traces on von Neumann Algebras 309

Notes 335

3. Multiplicity of a von Neumann Algebra on a Hilbert Space 336
4. Ergodic Type Theorem for von Neumann Algebras* 344
5. Normality of Separable Representations* 352
6. The Borel Spaces of von Neumann Algebras 359
7. Construction of Factors of Type II and Type III 362

Notes 374

Appendix

Polish Spaces and Standard Borel Spaces 375

Bibliography 387

Monographs 387
Papers 389

Notation Index 409

Subject Index 411

Introduction

Mathematics for infinite dimensional objects is becoming more and more important today both in theory and application. *Rings of operators*, renamed *von Neumann algebras* by J. Dixmier, were first introduced by J. von Neumann fifty years ago, 1929, in [254] with his grand aim of giving a sound foundation to mathematical sciences of infinite nature. J. von Neumann and his collaborator F. J. Murray laid down the foundation for this new field of mathematics, *operator algebras*, in a series of papers, [240], [241], [242], [257] and [259], during the period of the 1930s and early in the 1940s. In the introduction to this series of investigations, they stated *Their solution* (to the problems of understanding rings of operators)[1] *seems to be essential for the further advance of abstract operator theory in Hilbert space under several aspects. First, the formal calculus with operator-rings leads to them. Second, our attempts to generalize the theory of unitary group-representations essentially beyond their classical frame have always been blocked by the unsolved questions connected with these problems. Third, various aspects of the quantum mechanical formalism suggest strongly the elucidation of this subject. Fourth, the knowledge obtained in these investigations gives an approach to a class of abstract algebras without a finite basis, which seems to differ essentially from all types hitherto investigated.* Since then there has appeared a large volume of literature, and a great deal of progress has been achieved by many mathematicians. The motivations of Murray and von Neumann seem to have been fully verified. Many important results and powerful techniques were added to the theory. Various related fields of mathematics have emerged, and a number of topics in this subject have branched out to independent fields.

[1] Added by the author.

The main characteristic of this subject can be stated as a complex of analysis and algebra: the results are phrased in algebraic terms, while the techniques are highly analytic. Sometimes, one might run into problems directly related to the foundation of mathematics such as the continuum hypothesis. One might be amazed to realize the possibility of such an elaborated algebraic structure in this wild area involving high degrees of infinity.

The theory of operator algebras is concerned with self-adjoint algebras of bounded linear operators on a Hilbert space closed under the norm topology, C^*-algebras, or the weak operator topology, von Neumann algebras. C^*-algebras are characterized as a special class of Banach algebras by means of a simple system of axioms. A concrete realization of a C^*-algebra as an algebra of operators on a Hilbert space is regarded as a representation of the algebra. Thus, the study of C^*-algebras consists of two parts: one is concerned with the intrinsic structure of algebras and the other deals with the representations of a C^*-algebra. Needless to say, these two parts are closely related, and indeed the algebraic structure of a C^*-algebra is studied through various representations of the algebra. Thus, this division of the theory stays at a formal level. Nevertheless, the separation of problems has positive effects: for instance, a systematic usage of inequivalent representations of a C^*-algebra provides flexible techniques even if it is given as a concrete algebra of operators on a specially chosen Hilbert space. Indeed, this freedom in choosing an appropriate representation is one of the main merits of the axiomatic approach to operator algebras.

Being infinite dimensional, our problems require careful investigation of approximation process; thus the study of topological structures is inevitable. For this reason, the topological, analytical aspect of operator algebras receives more of our attention than the algebraic aspect in this first volume. After establishing the basic foundation in Chapter I, the Banach space duality for operator algebras will be studied throughout the text. The reader will find a strong similarity between our theory and measure theory on locally compact spaces. In fact, the study of abelian C^*-algebras will be reduced to that of locally compact spaces, and a substantial part of our theory is called noncommutative integration theory.

Each chapter begins with an introduction to its basic facts. Sections and paragraphs with * sign are somewhat technical; the reader who wants to get rather a quick grasp of the theory may postpone these parts. The sign ** indicates the end of the technical paragraph. Comments and historic background are placed at the end of each chapter and some sections as notes. Complements to a section or a chapter and some results of special interest are stated as exercises with † sign and references.

In the succeeding volume, the author will discuss further, among other topics, noncommutative integration theory, the so-called Tomita–Takesaki theory, automorphism groups of operator algebras, crossed products, infinite tensor products, the structure of von Neumann algebras of type III, approximately finite dimensional von Neumann algebras, and the existence of a continuum of nonisomorphic factors.

The author would like to express here his sincere gratitude to Professors H. A. Dye, R. V. Kadison, D. Kastler, M. Nakamura, Y. Misonou and J. Tomiyama from whom he received scientific as well as moral support at several stages of the work. A major part of the preparation was done at the University of Aix-Marseille-Luminy, ZiF, the University of Bielefeld, while the author was on leave from the University of California, Los Angeles. He acknowledges gratefully a generous support extended to him, for a part of the preparation, from the Guggenheim Foundation. The author is very grateful to Mrs. L. Beerman for typing the manuscript skillfully with great patience.

Chapter I
Fundamentals of Banach Algebras and C^*-Algebras

0. Introduction

In this this first chapter, we lay the foundation for later discussion, giving elementary results in Banach algebras and C^*-algebras. The first three sections are devoted to the general Banach algebras. The most important results in these sections are Theorem 2.5, Corollary 2.6, and Theorem 3.11, which are really fundamental in the theory of Banach algebras. Discussion of C^*-algebras starts from Section 4. As an object of the theory of operator algebras, a C^*-algebra is a uniformly closed self-adjoint algebra A of bounded linear operators on a Hilbert space $\mathfrak{H}$. The major task of the theory of operator algebras is to find descriptions of the structure of $\{A,\mathfrak{H}\}$. This problem splits into two problems:

(a) Find descriptions of the algebraic structure of A alone;
(b) Given an algebra A, find all possible pairs $\{B,\mathfrak{K}\}$ such that B is isomorphic to A as an abstract algebra.

The first approach to problem (a) is to characterize a uniformly closed self-adjoint algebra of bounded linear operators on a Hilbert space as an abstract algebra, i.e., without using a Hilbert space. A solution to this question is given by postulates (i)–(vi) in Section 1, for a C^*-algebra, and is proved in Theorem 9.18. Problem (b) leads us to the representation theory of C^*-algebras. Namely, an action of a C^*-algebra A is viewed as a representation on a Hilbert space, and problem (b) is translated in this terminology as follows:

(b′) Find descriptions of all representations of a given C^*-algebra.

The obvious question after the postulates were once laid down is the existence of representations, which is answered, as mentioned, by Theorem 9.18. It turns out (Theorem 9.14) that there is a strong link between positive linear functionals and representations. Section 9 is the highlight of the chapter. A characterization of extreme points of the unit ball of a C^*-algebra is given in Section 10, which will be used in Chapter III to show that a W^*-algebra is unital. Section 11 is devoted to a sketch of finite dimensional C^*-algebras and their representations.

1. Banach Algebras

Let **R** and **C** denote always the real number field and the complex number field, respectively.

Definition 1.1. Let A be a Banach space over **C**. If A is an algebra over **C** in which the multiplication satisfies the inequality

$$\|xy\| \leq \|x\| \|y\|,$$

then A is called a *Banach algebra.*

The inequality

$$\|x_1 y_1 - x_2 y_2\| \leq \|x_1\| \|y_1 - y_2\| + \|x_1 - x_2\| \|y_2\|$$

shows that the product xy is a continuous function of two variables x and y.

If E is a Banach space over **C**, then the set $\mathscr{L}(E)$ of all bounded operators on E is a Banach algebra with the natural algebraic operations and norm.

Definition 1.2. If a Banach algebra A admits a map: $x \mapsto x^* \in A$ with the following properties:

(i) $(x^*)^* = x$;

(ii) $(x + y)^* = x^* + y^*$;

(iii) $(\alpha x)^* = \bar{\alpha} x^*$;

(iv) $(xy)^* = y^* x^*$;

(v) $\|x^*\| = \|x\|$;

for every $x, y \in A$ and $\alpha \in \mathbf{C}$, then A is called an *involutive Banach algebra* and the map: $x \mapsto x^*$ the *involution* (or **-operation*) of A. If the involution of A satisfies the following additional condition:

(vi) $\|x^* x\| = \|x^*\| \|x\|$, $x \in A$;

then A is called a *C^*-algebra.*

Let Ω be a locally compact space. The set $C_\infty(\Omega)$ of all continuous functions on Ω vanishing at infinity is a C^*-algebra with the following structure:

$$(\lambda x + \mu y)(\omega) = \lambda x(\omega) + \mu y(\omega);$$
$$(xy)(\omega) = x(\omega)y(\omega);$$
$$x^*(\omega) = \overline{x(\omega)};$$
$$\|x\| = \sup\{|x(\omega)|:\omega \in \Omega\};$$

for every $x,y \in C_\infty(\Omega)$, $\lambda,\mu \in \mathbf{C}$ and $\omega \in \Omega$. The C^*-algebra $C_\infty(\Omega)$ is abelian. The algebra $C_\infty(\Omega)$ has an identity if and only if Ω is compact. In this case, $C_\infty(\Omega)$ is denoted simply by $C(\Omega)$.

If $\mathfrak{H}$ is a Hilbert space, then the Banach algebra $\mathscr{L}(\mathfrak{H})$ of all bounded operators on $\mathfrak{H}$ is a C^*-algebra with the involution: $x \mapsto x^*$ defined as the adjoint operator x^* of x. If the dimension of $\mathfrak{H}$ is greater than one, then $\mathscr{L}(\mathfrak{H})$ is not abelian.

Proposition 1.3. *If A is a Banach algebra with an identity* 1, *then there exists a norm $\|\cdot\|_0$ on A such that*: (i) *the new norm $\|\cdot\|_0$ is equivalent to the original norm $\|\cdot\|$*; (ii) $(A,\|\cdot\|_0)$ *is a Banach algebra*; (iii) $\|1\|_0 = 1$.

If $\|I\| = 1$ done $\leftarrow \|I\| \le \|I\| \cdot \|I\| \Rightarrow \|I\| \ge 1$ Assume $\|I\| > 1$

PROOF. For each $x \in A$, let L_x denote the operator: $y \in A \mapsto xy \in A$. The map: $x \mapsto L_x$ is then injective because $L_x 1 = x$. Put $\|x\|_0 = \|L_x\|$, $x \in A$. By the inequality $\|xy\| \le \|x\|\|y\|$, we have $\|x\|_0 \le \|x\|$. On the other hand, we have

$$\|x\|_0 = \|L_x\| = \sup\{\|xy\|:\|y\| \le 1\} \ge \|x\|/\|1\|.$$

Hence the norm $\|\cdot\|_0$ is equivalent to the original norm. Assertions (ii) and (iii) are almost automatic now.   Q.E.D.

By this proposition, we assume always that the norm of the identity is one if it exists. A Banach algebra with an identity is said to be *unital*.

$1^* = 1 \cdot 1^* = (1^* \cdot 1)^* = (1^* 1^*)^* = 1$

Remark 1.4. If A is a unital involutive Banach algebra, then we have $1^* = 1$. Furthermore, if A is a unital C^*-algebra, then the condition $\|1\| = 1$ follows automatically from postulate (vi).

If a given involutive Banach algebra A is not unital, then we can imbed A into a unital involutive Banach algebra A_1 as an ideal in the following way: We take the direct sum $A \oplus \mathbf{C}$ as a linear space A_1, in which we define a Banach algebra structure by

$$(x,\lambda)(y,\mu) = (xy + \mu x + \lambda y, \lambda\mu);$$
$$(x,\lambda)^* = (x^*,\bar{\lambda});$$
$$\|(x,\lambda)\| = \|x\| + |\lambda|;$$

for every (x,λ), $(y,\mu) \in A \oplus \mathbf{C}$. The map: $x \in A \mapsto (x,0) \in A_{\mathrm{I}}$ is an isometric isomorphism and the element $(0,1)$ is the identity of A_{I}. Identifying each $x \in A$ and $(x,0) \in A_{\mathrm{I}}$, we write $(x,\lambda) = x + \lambda 1 \in A_{\mathrm{I}}$. Under the identification, A is an ideal of A_{I}.

However, in the case when A is a C^*-algebra, A_{I} is not a C^*-algebra in general. Therefore, we should adjust the norm in A_{I} as follows:

Proposition 1.5. *If a C^*-algebra A is not unital, then there exists a norm in A_{I} which makes A_{I} a C^*-algebra.*

PROOF. Noticing that A is an ideal of A_{I}, we put $L_x y = xy$ for $x \in A_{\mathrm{I}}$ and $y \in A$. By postulate (vi) for C^*-algebras, we have $\|x\| = \|L_x\|$ for each $x \in A$, so that we can put $\|x\| = \|L_x\|$ for each $x \in A_{\mathrm{I}}$ without confusion. Suppose $\|L_x\| = 0$ for some $x = x' + \lambda 1$, $x' \in A$, $\lambda \in \mathbf{C}$, $\lambda \neq 0$. For every $y \in A$, we have

$$0 = \frac{1}{\lambda} xy = \frac{1}{\lambda} x'y + y,$$

so that $-(1/\lambda)x'$ is a left identity of A; hence $[-(1/\lambda)x']^*$ is a right identity of A. Therefore, we have

$$-\frac{1}{\lambda}x' = \left(-\frac{1}{\lambda}x'\right)\left(-\frac{1}{\lambda}x'\right)^* = \left(-\frac{1}{\lambda}x'\right)^*,$$

which means that $-(1/\lambda)x'$ must be the identity of A. This contradicts the assumption on A. Thus, the function: $x \in A_{\mathrm{I}} \mapsto \|x\|$ is indeed a norm in A_{I}. Since A is complete and of codimension one in A_{I}, A_{I} is complete. For any $x \in A_{\mathrm{I}}$ and $\varepsilon > 0$, there exists a $y \in A$ with

$$\|xy\| \geq (1 - \varepsilon)\|x\| \quad \text{and} \quad \|y\| \leq 1.$$

Recalling that xy is in A, we have

$$\begin{aligned} \|x^*x\| &\geq \|y^*(x^*x)y\| = \|(xy)^*(xy)\| \\ &= \|xy\|^2 \geq (1-\varepsilon)^2\|x\|^2. \end{aligned}$$

Therefore, we have $\|x\|^2 \leq \|x^*x\| \leq \|x^*\|\|x\|$ for every $x \in A_{\mathrm{I}}$, which means that $\|x^*x\| = \|x\|^2$. Q.E.D.

Therefore, whenever we consider a C^*-algebra A which is not unital, then A_{I} will denote the unital C^*-algebra obtained by the above procedure.

In a unital Banach algebra A, an element x of A is said to be *invertible* or *regular* if there exists an $x' \in A$ with $x'x = xx' = 1$. The element x' is called the *inverse* of x and denoted by x^{-1}. The inverse of an invertible element is unique. The set $G(A)$ of all invertible elements in A is a group with respect to the multiplication, which will be called the *general linear group* of A.

Proposition 1.6. *If an element $x \in A$ is in open unit ball $\|x - 1\| < 1$, then x is invertible and the inverse x^{-1} is written as*

$$x^{-1} = \sum_{n=0}^{\infty} (1 - x)^n,$$

where $a^0 = 1$ for any $a \in A$.

PROOF. Since $\sum_{n=0}^{\infty} \|1 - x\|^n < +\infty$, the series $\sum_{n=0}^{\infty} (1 - x)^n$ converges in norm. Put $x' = \sum_{n=0}^{\infty} (1 - x)^n$. We have then

$$\begin{aligned} x'x = xx' &= (1 - (1 - x))x' = x' - (1 - x)x' \\ &= 1 + \sum_{n=1}^{\infty} (1 - x)^n - \sum_{n=1}^{\infty} (1 - x)^n \\ &= 1. \end{aligned}$$

Q.E.D.

Proposition 1.7. *The group $G(A)$ is an open subset of A. More precisely, if $\|x - x_0\| < 1/\|x_0^{-1}\|$ for an $x_0 \in G(A)$, then x is invertible and x^{-1} is represented by*

$$x^{-1} = \left(\sum_{n=0}^{\infty} [x_0^{-1}(x_0 - x)]^n\right) x_0^{-1}.$$

PROOF. Let $x_0 \in G(A)$ and $\|x - x_0\| < 1/\|x_0^{-1}\|$. We have then

$$\begin{aligned} \|1 - x_0^{-1}x\| &= \|x_0^{-1}(x_0 - x)\| \\ &\leq \|x_0^{-1}\|\|x_0 - x\| < 1, \end{aligned}$$

so that $x_0^{-1}x$ has the inverse written by

$$\begin{aligned} (x_0^{-1}x)^{-1} &= \sum_{n=0}^{\infty} (1 - x_0^{-1}x)^n \\ &= \sum_{n=0}^{\infty} [x_0^{-1}(x_0 - x)]^n. \end{aligned}$$

Hence it follows that

$$x^{-1} = x_0^{-1} + \sum_{n=0}^{\infty} [x_0^{-1}(x_0 - x)]^n x_0^{-1}.$$

Q.E.D.

Corollary 1.8. *The inverse x^{-1} is a continuous function of x in $G(A)$.*

PROOF. With the notations as in Proposition 1.7, we have

$$\begin{aligned} \|x^{-1} - x_0^{-1}\| &= \left\|\sum_{n=1}^{\infty} [x_0^{-1}(x_0 - x)]^n x_0^{-1}\right\| \\ &\leq \|x_0^{-1}\| \sum_{n=1}^{\infty} \|x_0^{-1}\|^n \|x_0 - x\|^n \\ &= \|x_0^{-1}\|^2 \|x_0 - x\|(1 - \|x_0^{-1}\|\|x - x_0\|)^{-1}. \end{aligned}$$

Q.E.D.

EXERCISES

1. Let G be a locally compact topological group with left invariant Haar measure ds and modular function δ_G. Let $L^1(G)$ denote the Banach space of all integrable functions on G with respect to the Haar measure equipped with norm

$$\|x\|_1 = \int_G |x(s)|\, ds, \qquad x \in L^1(G).$$

Define a multiplication (convolution) and an involution in $L^1(G)$ as follows:

$$(xy)(t) = \int_G x(s)y(s^{-1}t)\, ds, \qquad x,y \in L^1(G);$$

$$x^*(t) = \delta_G(t)^{-1}\overline{x(t^{-1})}.$$

(a) Show that $L^1(G)$ is an involutive Banach algebra.
(b) Show that $L^1(G)$ is unital if and only if G is discrete.
(c) Show that $L^1(G)$ is abelian if and only if G is also.
The algebra $L^1(G)$ is called the L^1 *group algebra* of G.

2. Let Ω be a locally compact space and G a locally compact topological group with left invariant Haar measure ds. Suppose G acts on Ω from the right in the following sense: for each $s \in G$, the map: $\omega \in \Omega \mapsto \omega s \in \Omega$ is a homeomorphism; the unit $e \in G$ acts on Ω as the identity map, i.e., $\omega e = \omega$ for every $\omega \in \Omega$; for each $s,t \in G$, $\omega(st) = (\omega s)t$, $\omega \in \Omega$; the map: $(\omega,s) \in \Omega \times G \mapsto \omega s \in \Omega$ is continuous. The pair (Ω,G) is called a *topological transformation group*, or G itself is called a topological transformation group of Ω. Let $\mathscr{K}(\Omega \times G)$ be the space of all continuous functions on $\Omega \times G$ with compact support. Define an algebraic structure in $\mathscr{K}(\Omega \times G)$ as follows:

$$\begin{cases} (\lambda x + \mu y)(\omega,s) = \lambda x(\omega,s) + \mu y(\omega,s); \\ (xy)(\omega,s) = \int_G x(\omega,t)y(\omega t,t^{-1}s)\, dt; \\ x^*(\omega,s) = \delta_G(s)^{-1}\overline{x(\omega s,s^{-1})}. \end{cases}$$

(a) Show that the completion A of $\mathscr{K}(\Omega \times G)$ with respect to the norm defined by

$$\|x\| = \int_G \sup_{\omega \in \Omega} |x(\omega,s)|\, ds$$

is an involutive Banach algebra.
(b) Show that A is unital if and only if Ω is compact and G is discrete.

2. Spectrum and Functional Calculus

Suppose that A is a unital algebra over $\mathbf{C}$. For each $x \in A$, the set

$$\mathrm{Sp}_A(x) = \{\lambda \in \mathbf{C} : (x - \lambda) \quad \text{is not invertible}\}^1$$

is called the *spectrum* of x in A. The complement of $\mathrm{Sp}_A(x)$ in $\mathbf{C}$ is called the *resolvent* of x.

[1] In a unital algebra, we often identify a scalar $\lambda \in \mathbf{C}$ and the element $\lambda 1$ in the algebra.

If A is not unital, then the *quasi-spectrum* $\mathrm{Sp}'_A(x)$ of $x \in A$ is the spectrum $\mathrm{Sp}_{A_\mathrm{I}}(x)$ of x in A_I, where A_I is the algebra obtained by adjunction of an identity to A. The quasi-spectrum $\mathrm{Sp}'_A(x)$ always contains zero.

Proposition 2.1. *If A is a unital algebra over* $\mathbf{C}$, *then for any $x, y \in A$, we have*

$$\mathrm{Sp}_A(xy) \cup \{0\} = \mathrm{Sp}_A(yx) \cup \{0\}.$$

PROOF. Suppose $\lambda \notin \mathrm{Sp}_A(xy) \cup \{0\}$. Then the inverse $(xy - \lambda)^{-1} = u$ exists. Since $xyu = uxy = 1 + \lambda u$, we have

$$(yx - \lambda)(yux - 1) = \lambda,$$
$$(yux - 1)(yx - \lambda) = \lambda;$$

hence $yx - \lambda$ is invertible. Hence $\lambda \notin \mathrm{Sp}_A(yx) \cup \{0\}$. By symmetry, $\mathrm{Sp}_A(xy) \cup \{0\} = \mathrm{Sp}_A(yx) \cup \{0\}$. Q.E.D.

Definition 2.2. For each element x of an algebra A over $\mathbf{C}$, the quantity

$$\|x\|_{\mathrm{sp}} = \sup\{|\lambda| : \lambda \in \mathrm{Sp}'_A(x)\}$$

is called the *spectral radius* of x.

By Proposition 2.1, we have $\|xy\|_{\mathrm{sp}} = \|yx\|_{\mathrm{sp}}$ for every $x, y \in A$.

Proposition 2.3. *If A is a Banach algebra, then the spectrum of any element of A is compact; hence the spectral radius is always finite.*

PROOF. Considering A_I if necessary, we may assume that A is unital. Let $x \in A$. Consider a map $f : \lambda \in \mathbf{C} \mapsto x - \lambda \in A$. The resolvent of x is then nothing else but the inverse image $f^{-1}(G(A))$ of the general linear group $G(A)$ of A under the continuous map f; hence it is open by Proposition 1.7. Hence $\mathrm{Sp}_A(x)$ is closed. Suppose now $|\lambda| > \|x\|$. The inequality $\|(1/\lambda)x\| < 1$ implies by Proposition 1.6 that the series $\sum_{n=0}^{\infty} [(1/\lambda)x]^n$ converges to $[(1/\lambda)x - 1]^{-1}$. Hence the inverse

$$(x - \lambda 1)^{-1} = \frac{1}{\lambda}\left(\frac{1}{\lambda} - x\right)^{-1}$$

exists, so that λ is in the resolvent of x. Therefore, $\mathrm{Sp}_A(x)$ is contained in the disk $|\lambda| \le \|x\|$. Hence it is compact. Q.E.D.

The above proof shows that $\|x\|_{\mathrm{sp}} \le \|x\|$, $x \in A$. Now, let

$$r(x) = \liminf_{n \to \infty} \|x^n\|^{1/n}, \qquad x \in A.$$

The power series $\sum_{n=0}^{\infty} (1/\lambda^{n+1})x^n$ converges for $|\lambda| > r(x)$, and the sum of this series is indeed the inverse $(x - \lambda)^{-1}$. Hence we have

$$\liminf_{n \to \infty} \|x^n\|^{1/n} = r(x) \ge \sup\{|\lambda| : \lambda \in \mathrm{Sp}_A(x)\} = \|x\|_{\mathrm{sp}}.$$

Let $f(\lambda) = (x - \lambda)^{-1}$ for $\lambda \notin \mathrm{Sp}_A(x)$. By Proposition 1.7, if $|\lambda - \lambda_0| < \|f(\lambda_0)\|^{-1}$ for $\lambda_0 \notin \mathrm{Sp}_A(x)$ and $\lambda \in \mathbf{C}$, then $\lambda \notin \mathrm{Sp}_A(x)$ and $f(\lambda)$ has the expression

$$f(\lambda) = \sum_{n=0}^{\infty} (\lambda - \lambda_0)^n f(\lambda_0)^{n+1}.$$

Hence $f(\lambda)$ is a holomorphic function on the resolvent of x. The power series $\sum_{n=0}^{\infty} x^n/\lambda^{n+1}$ is the Laurent expansion of $f(\lambda)$ at infinity. Let φ be a bounded linear functional of A. The function

$$f_\varphi(\lambda) = \langle f(\lambda),\varphi \rangle$$

is a holomorphic function on the resolvent of x, where $\langle a,\varphi \rangle$ denotes the value of φ at $a \in A$. The function $f_\varphi(\lambda)$ is holomorphic for $|\lambda| > \|x\|_{\mathrm{sp}}$; hence its Laurent expansion $\sum_{n=0}^{\infty} (1/\lambda^{n+1})\langle x^n,\varphi \rangle$ converges for $|\lambda| > \|x\|_{\mathrm{sp}}$. Hence we have

$$\lim_{n\to\infty} \frac{1}{\lambda^{n+1}} \langle x^n,\varphi \rangle = 0, \qquad \varphi \in A^*.$$ [2]

By the uniform boundedness theorem, the sequence $\{(1/\lambda^{n+1})x^n\}$ is bounded for $|\lambda| > \|x\|_{\mathrm{sp}}$, so that for any $|\lambda| > \|x\|_{\mathrm{sp}}$ there exists $\alpha > 0$ such that $\|x^n\| \leq \alpha|\lambda|^{n+1}$ $n = 1,2,\ldots$. Therefore, we have

$$\limsup_{n\to\infty} \|x^n\|^{1/n} \leq \lim_{n\to\infty} \alpha^{1/n}|\lambda| = |\lambda|.$$

Thus, we get

$$\limsup_{n\to\infty} \|x^n\|^{1/n} \leq \|x\|_{\mathrm{sp}}.$$

Combining this with the earlier inequality, we obtain the following result:

Proposition 2.4. *For any element x of a Banach algebra A, we have*

$$\|x\|_{\mathrm{sp}} = \lim_{n\to\infty} \|x^n\|^{1/n}.$$

Theorem 2.5. *The spectrum $\mathrm{Sp}_A(x)$ of any element x of a unital Banach algebra A is always nonempty.*

PROOF. Suppose $\mathrm{Sp}_A(x) = \varnothing$ for some $x \in A$. With the same notation as above, the function $f_\varphi(\lambda)$ is holomorphic everywhere; hence it is an entire function. But we have $\lim_{\lambda\to\infty} f_\varphi(\lambda) = 0$, so that f_φ is identically zero by Liouville's theorem, which means that $\langle f(\lambda),\varphi \rangle \equiv 0$ for every $\varphi \in A^*$. Hence $f(\lambda) \equiv 0$. But this is impossible by definition. Therefore, $\mathrm{Sp}_A(x) \neq \varnothing$. Q.E.D.

Corollary 2.6. *If a Banach algebra A is a division ring, then A is isomorphic to the complex field.*

[2] A^* denotes the conjugate space of the Banach space A.

PROOF. Let x be an element of A. By Theorem 2.5, $\mathrm{Sp}_A(x)$ contains at least one point $\lambda \in \mathbf{C}$. By definition, the element $x - \lambda$ is not invertible, which means that $x = \lambda$ by assumption. Q.E.D.

Let x be a fixed element of a unital Banach algebra A. Let f be a holomorphic function in an open neighborhood U_f of $\mathrm{Sp}_A(x)$, and C be a smooth simple closed curve in U_f enclosing $\mathrm{Sp}_A(x)$. We assign the positive orientation to C as in complex analysis. For each $\varphi \in A^*$, we consider a continuous function: $\lambda \mapsto f(\lambda)\langle(\lambda - x)^{-1},\varphi\rangle \in \mathbf{C}$ on the curve C. Put

$$F(\varphi) = \frac{1}{2\pi i}\int_C f(\lambda)\langle(\lambda - x)^{-1},\varphi\rangle\, d\lambda.$$

The map: $\varphi \in A^* \mapsto F(\varphi) \in \mathbf{C}$ is a linear functional of A^* and

$$|F(\varphi)| \leq \frac{1}{2\pi} l\|\varphi\|\sup\{|f(\lambda)|\|(\lambda - x)^{-1}\| : \lambda \in C\},$$

where l is the length of the curve C. Hence there exists an element F in the second conjugate space A^{**} of A such that

$$\langle F,\varphi\rangle = F(\varphi) = \frac{1}{2\pi i}\int_C f(\lambda)\langle(\lambda - x)^{-1},\varphi\rangle\, d\lambda.$$

On the other hand, the A-valued function: $\lambda \mapsto f(\lambda)(\lambda - x)^{-1} \in A$ is continuous, so that the limit y of

$$\sum_{i=0}^{n} f(\lambda_i)(\lambda_i - x)^{-1}(\lambda_i - \lambda_{i+1}),$$

as $\lim \max|\lambda_i - \lambda_{i+1}| = 0$, exists, where $\{\lambda_0,\lambda_1,\ldots,\lambda_n,\lambda_{n+1} = \lambda_0\}$ are dividing points of the curve C. It is clear then that $\langle y,\varphi\rangle = \langle F,\varphi\rangle$, $\varphi \in A^*$. Hence F lies in A. By Cauchy's theorem for contour integrals, this F does not depend on the choice of the curve C, but only on the function f. Therefore, we denote F by $f(x)$. Symbolically, $f(x)$ is given by

$$f(x) = \frac{1}{2\pi i}\int_C f(\lambda)(\lambda - x)^{-1}\, d\lambda.$$

Proposition 2.7. *In the above situation, the map*: $f \mapsto f(x)$ *is a homomorphism of the algebra* $A(\mathrm{Sp}_A(x))$ *of all functions holomorphic in a neighborhood of* $\mathrm{Sp}_A(x)$ *into* A, *which sends the constant function* 1 *to the identity* 1 *of* A *and the function*: $\lambda \in \mathbf{C} \mapsto \lambda \in \mathbf{C}$ *into* x.

PROOF. The linearity of the correspondence is clear. Let f and g be two functions holomorphic in neighborhoods U_f and U_g of $\mathrm{Sp}_A(x)$. Let $U = U_f \cap U_g$, and C_i, $i = 1,2$, be a smooth simple closed curve in U enclosing $\mathrm{Sp}_A(x)$ such that the curve C_2 lies completely inside the curve C_1. We have then

$$
\begin{aligned}
f(x)g(x) &= \left(\frac{1}{2\pi i}\int_{C_1} f(\lambda)(\lambda - x)^{-1}\, d\lambda\right)\left(\frac{1}{2\pi i}\int_{C_2} g(\mu)(\mu - x)^{-1}\, d\mu\right) \\
&= -\frac{1}{4\pi^2}\iint_{C_1 \times C_2} f(\lambda)g(\mu)(\lambda - x)^{-1}(\mu - x)^{-1}\, d\lambda\, d\mu \\
&= -\frac{1}{4\pi^2}\iint_{C_1 \times C_2} f(\lambda)g(\mu)\frac{1}{\lambda - \mu}\{(\mu - x)^{-1} - (\lambda - x)^{-1}\}\, d\lambda\, d\mu \\
&= -\frac{1}{4\pi^2}\iint_{C_1 \times C_2} \frac{f(\lambda)g(\mu)}{\lambda - \mu}(\mu - x)^{-1}\, d\lambda\, d\mu \\
&\quad + \frac{1}{4\pi^2}\iint_{C_1 \times C_2} \frac{f(\lambda)g(\mu)}{\lambda - \mu}(\lambda - x)^{-1}\, d\lambda\, d\mu.
\end{aligned}
$$

The second term is

$$
\frac{1}{4\pi^2}\int_{C_1}\left(\int_{C_2}\frac{g(\mu)}{\lambda - \mu}\, d\mu\right) f(\lambda)(\lambda - x)^{-1}\, d\lambda = 0,
$$

because $g(\mu)/(\lambda - \mu)$ is holomorphic inside the curve C_2 if λ lies on C_1. Hence we get

$$
\begin{aligned}
f(x)g(x) &= \frac{1}{2\pi i}\int_{C_2}\left(\frac{1}{2\pi i}\int_{C_1}\frac{f(\lambda)}{\lambda - \mu}\, d\lambda\right) g(\mu)(\mu - x)^{-1}\, d\mu \\
&= \frac{1}{2\pi i}\int_{C_2} f(\mu)g(\mu)(\mu - x)^{-1}\, d\mu \\
&= (fg)(x).
\end{aligned}
$$

Suppose $f(\lambda) \equiv 1$, $\lambda \in \mathbf{C}$. We have then

$$
f(x) = \frac{1}{2\pi i}\int_C (\lambda - x)^{-1}\, d\lambda,
$$

choosing a suitable smooth simple closed curve C. Let C be the circle $\{\lambda : |\lambda| = \|x\| + \varepsilon\}$ for some $\varepsilon > 0$. We have then

$$
(\lambda - x)^{-1} = \sum_{n=0}^{\infty}\frac{x^n}{\lambda^{n+1}}
$$

uniformly for every $\lambda \in C$. Hence we get

$$
\begin{aligned}
f(x) &= \frac{1}{2\pi i}\int_C \sum_{n=0}^{\infty}\frac{x^n}{\lambda^{n+1}}\, d\lambda \\
&= \sum_{n=0}^{\infty}\frac{1}{2\pi i}\left(\int_C \frac{1}{\lambda^{n+1}}\, d\lambda\right) x^n = 1.
\end{aligned}
$$

It is now clear that if $f(\lambda) = \lambda$, $\lambda \in \mathbf{C}$, then $f(x) = x$. Q.E.D.

Proposition 2.8. (Spectral Mapping Theorem). *Let x be a fixed element of a unital Banach algebra A. If f is a holomorphic function on a neighborhood of* $\mathrm{Sp}_A(x)$, *then we have*

$$\mathrm{Sp}_A(f(x)) = f(\mathrm{Sp}_A(x)).$$

Furthermore, if g is a holomorphic function on a neighborhood of $f(\mathrm{Sp}_A(x))$, then we get

$$(g \circ f)(x) = g(f(x)).$$

PROOF. Let $y = f(x)$. If $\mu \notin f(\mathrm{Sp}_A(x))$, then $h(\lambda) = 1/(f(\lambda) - \mu)$ is holomorphic on a neighborhood of $\mathrm{Sp}_A(x)$. Put $z = h(x)$. We have then

$$(y - \mu)z = z(y - \mu) = h(x)(f(x) - \mu) = (h(f - \mu))(x) = 1$$

by Proposition 2.7. Hence μ is in the resolvent of y. If $\mu \in f(\mathrm{Sp}_A(x))$, then $\mu = f(\lambda_0)$ for some $\lambda_0 \in \mathrm{Sp}_A(x)$. There exists a holomorphic function h on a neighborhood of $\mathrm{Sp}_A(x)$ such that

$$f(\lambda) - \mu = (\lambda - \lambda_0)h(\lambda).$$

Hence we have, again by Proposition 2.7,

$$y - \mu = (x - \lambda_0)h(x) = h(x)(x - \lambda_0).$$

Since $x - \lambda_0$ is not invertible, neither is $y - \mu$. Thus, μ falls in $\mathrm{Sp}_A(y)$.

Now choose smooth simple closed curves C_1 and C_2 in such a way that C_1 encloses $f(\mathrm{Sp}_A(x))$ and is contained in the domain of g, and C_2 encloses the inverse image of C_1 under f and is contained in the domain of f. We then compute:

$$\begin{aligned}
(g \circ f)(x) &= \frac{1}{2\pi i}\int_{C_1} (g \circ f)(\lambda)(\lambda - x)^{-1}\,d\lambda \\
&= \frac{-1}{4\pi^2}\int_{C_1}\left(\int_{C_2} g(\mu)(\mu - f(\lambda))^{-1}\,d\mu\right)(\lambda - x)^{-1}\,d\lambda \\
&= -\frac{1}{4\pi^2}\int_{C_2} g(\mu)\left\{\int_{C_1}(\mu - f(\lambda))^{-1}(\lambda - x)^{-1}\,d\lambda\right\}d\mu \\
&= \frac{1}{2\pi i}\int_{C_2} g(\mu)(\mu - f(x))^{-1}\,d\mu \\
&= g(f(x)).
\end{aligned}$$

Q.E.D.

Proposition 2.9. *Let A be a unital Banach algebra. If U is an open subset of* $\mathbf{C}$, *then the set $E_U = \{x \in A : \mathrm{Sp}_A(x) \subset U\}$ is open in A.*

PROOF. For each $x \in E_U$, the function: $\lambda \in U^c \mapsto (x-\lambda)^{-1} \in A$ is continuous and $\lim_{\lambda \to \infty} \|(x-\lambda)^{-1}\| = 0$. Hence for any $x_0 \in E_U$, $\sup_{\lambda \in U} \|(x_0-\lambda)^{-1}\| < +\infty$. Let $\delta = \inf\{1/\|(x_0-\lambda)^{-1}\| : \lambda \in U^c\} > 0$. If $\|x - x_0\| < \delta$, then for any $\lambda \in U^c$, we have

$$\|(x_0-\lambda)-(x-\lambda)\| < \frac{1}{\|(x_0-\lambda)^{-1}\|};$$

hence $(x-\lambda)^{-1}$ exists by Proposition 1.7, so that $\mathrm{Sp}_A(x) \subset U$. Thus, we get $x \in E_U$. Q.E.D.

Now, we define two very important functions, the exponential and the logarithmic functions, of elements of a unital Banach algebra A. Since the exponential function: $\lambda \in \mathbf{C} \mapsto \exp \lambda \in \mathbf{C}$ is an entire function, we can define $\exp x$ for all elements x of a unital Banach algebra A. The exponential of x, $\exp x$, is also given by the absolutely convergent power series

$$\exp x = \sum_{n=0}^{\infty} \frac{1}{n!} x^n,$$

where $x^0 = 1$ as usual. It is straightforward to see that if x and y commute then

$$\exp(x+y) = (\exp x)(\exp y).$$

In particular, $\exp(-x) = (\exp x)^{-1}$; hence the exponential function maps A into the general linear group $G(A)$. Since $\exp tx$, $t \in \mathbf{R}$, is a continuous one parameter subgroup of $G(A)$ in the sense that

$$\exp(s+t)x = (\exp sx)(\exp tx), \qquad s,t \in \mathbf{R}$$
$$\lim_{t \to t_0} \|\exp tx - \exp t_0 x\| = 0,$$

$\exp x$ is connected to the identity 1 by an arc, $\exp tx$, $0 \le t \le 1$, so that $\exp x$ lies in the connected component $G_0(A)$ of $G(A)$ containing the identity. The connected component $G_0(A)$ of $G(A)$ is called the *principal* component of $G(A)$. By Proposition 1.7, $G(A)$ is open in the Banach space A, so that it is locally arcwise connected; hence $G_0(A)$ is an open and closed subgroup of $G(A)$. Since the map: $x \in G(A) \mapsto axa^{-1} \in G(A)$ is an automorphism of $G(A)$ for every $a \in G(A)$ leaving the identity fixed, $G_0(A)$ is a normal subgroup of $G(A)$. The quotient group $G(A)/G_0(A)$ is called the *index group* of A.

The logarithm of an element $x \in A$ is not always defined because the logarithm is not entire. If $\mathrm{Sp}_A(x)$, $x \in A$, is contained in the domain of the principal logarithm $\mathrm{Log}\,\lambda$, then $\log x$ is defined to be $\mathrm{Log}\, x$ in the sense of Proposition 2.7. By Proposition 2.8, we have

$$\exp \log x = x$$

for each element x with $\mathrm{Sp}_A(x)$ contained in the domain of $\mathrm{Log}\,\lambda$. If $\|x-1\|_{\mathrm{sp}} < 1$, then $\log x$ is written by the absolutely convergent power series

$$\log x = -\sum_{n=1}^{\infty} \frac{1}{n}(1-x)^n.$$

Concerning the inversion problem for the exponential function in a Banach algebra, we have the following:

Proposition 2.10. *Let x be an element of a unital Banach algebra A. If* $\mathrm{Sp}_A(x)$ *is contained in the open strip:* $-\pi < \mathrm{Im}\,\lambda < \pi$, *then* $\log \exp x$ *is defined and*

$$\log \exp x = x.$$

We leave the proof to the reader. Apply Proposition 2.8.

EXERCISE

1. Let K be a compact subset of **C**. For a unital Banach algebra A, let A_K denote the subset $\{x \in A : \mathrm{Sp}_A(x) \subset K\}$. Show that if f is a holomorphic function on a neighborhood of K, then the functional calculus: $x \in A_K \mapsto f(x) \in A$ is continuous.

3. Gelfand Representation of Abelian Banach Algebras

In this section, we always assume that a Banach algebra A under consideration is *abelian*, i.e., commutative.

Definition 3.1. Let $\mathfrak{m}$ be an ideal of A. If there exists an element $e \in A$ such that $ex - x \in \mathfrak{m}$ for every $x \in A$, then $\mathfrak{m}$ is said to be *regular*. In other words, $\mathfrak{m}$ is regular if the quotient algebra $A/\mathfrak{m}$ admits an identity. The element $e \in A$ is called an *identity modulo* $\mathfrak{m}$.

Trivially, every ideal of a unital Banach algebra A is regular.

Proposition 3.2. *Let $\mathfrak{m}$ be a proper regular ideal of an abelian Banach algebra A. If e is an identity modulo $\mathfrak{m}$, then we have*

$$\inf\{\|e - x\| : x \in \mathfrak{m}\} \geq 1.$$

PROOF. Suppose $\|e - x\| < 1$ for some $x \in \mathfrak{m}$. Then the power series $y = \sum_{n=1}^{\infty} (e-x)^n$ converges. Since $(e-x)y = \sum_{n=2}^{\infty} (e-x)^n$, we have

$$\begin{aligned} y &= (e-x)y + e - x \\ &= ey - xy + e - x; \end{aligned}$$

hence $e = y - ey + xy + x \in \mathfrak{m}$. For any $a \in A$, we have $ea - a \in \mathfrak{m}$, so that $a \in \mathfrak{m}$. Thus $\mathfrak{m} = A$. This is a contradiction. Q.E.D.

Definition 3.3. A proper ideal $\mathfrak{m}$ of a Banach algebra A is said to be *maximal* if it is not contained in any ideal except $\mathfrak{m}$ itself and the entire algebra A.

Corollary 3.4. *The closure of any proper regular ideal of an abelian Banach algebra A is proper and, of course, regular. In particular, any maximal regular ideal of A is closed.*

Proposition 3.5. *Any proper regular ideal of an abelian Banach algebra A is contained in a maximal regular ideal.*

PROOF. Let e be an identity modulo a proper ideal $\mathfrak{m}$. We note first that any ideal of A containing $\mathfrak{m}$ is regular. It is easy to see that the family $\mathscr{F}$ of all ideals of A containing $\mathfrak{m}$ but not e is an inductive set under the inclusion ordering. Hence $\mathscr{F}$ has a maximal element by Zorn's lemma. Q.E.D.

Proposition 3.6. *Let $\mathfrak{m}$ be a closed ideal of a Banach algebra A (not necessarily abelian). The quotient Banach space $A/\mathfrak{m}$ is a Banach algebra with respect to the quotient algebra structure.*

PROOF. Let π be the canonical map of A onto $A/\mathfrak{m}$. We then have to show that $\|\pi(x)\pi(y)\| \leq \|\pi(x)\|\|\pi(y)\|$ for every $x, y \in A$. Suppose $\varepsilon > 0$. By the definition of the norm in $A/\mathfrak{m}$ (which is given by $\|\pi(x)\| = \inf\{\|x + m\| : m \in \mathfrak{m}\}$), we can find $m, n \in \mathfrak{m}$ such that

$$\|x + m\| \leq \|\pi(x)\| + \varepsilon, \qquad \|y + n\| \leq \|\pi(y)\| + \varepsilon.$$

We then compute:

$$\begin{aligned} \|\pi(x)\pi(y)\| &= \|\pi(x + m)\pi(y + n)\| = \|\pi((x + m)(y + n))\| \\ &\leq \|(x + m)(y + n)\| \leq \|x + m\|\|y + n\| \\ &\leq (\|\pi(x)\| + \varepsilon)(\|\pi(y)\| + \varepsilon). \end{aligned}$$

Letting $\varepsilon \to 0$, we get the required inequality, $\|\pi(x)\pi(y)\| \leq \|\pi(x)\|\|\pi(y)\|$. Q.E.D.

Proposition 3.7. *Let A be a unital abelian Banach algebra. If an element $x \in A$ is not invertible, then x is contained in some maximal ideal $\mathfrak{m}$ of A.*

PROOF. By assumption, the set Ax does not contain the identity 1; hence it is a proper ideal. Since A is unital, Ax is regular. By Proposition 3.5, Ax is contained in some maximal ideal $\mathfrak{m}$, which means that x belongs to $\mathfrak{m}$. Q.E.D.

Let $\mathfrak{m}$ be a maximal regular ideal of an abelian Banach algebra A. The quotient algebra $A/\mathfrak{m}$ is a unital Banach algebra which has no proper ideal except the trivial ideal $\{0\}$. Hence by Proposition 3.7, every nonzero element of $A/\mathfrak{m}$ is invertible, which means that $A/\mathfrak{m}$ is a field. Therefore, $A/\mathfrak{m}$ is isomorphic to the complex number field $\mathbf{C}$, by Corollary 2.6. Since any linear automorphism of $\mathbf{C}$ is the identity map of $\mathbf{C}$, an isomorphism of $A/\mathfrak{m}$ onto

C is unique, so that we can identify $A/\mathfrak{m}$ and **C** by this unique isomorphism. Therefore, each maximal regular ideal $\mathfrak{m}$ of A gives rise to a homomorphism $\omega_{\mathfrak{m}}$ of A onto the complex number field **C**.

Proposition 3.8. *Let A be an abelian Banach algebra. Let $\mathscr{M}(A)$ denote the set of all maximal regular ideals of A, and $\Omega(A)$ be the set of all nonzero homomorphisms of A onto* **C**. *Then the map: $\mathfrak{m} \in \mathscr{M}(A) \mapsto \omega_{\mathfrak{m}} \in \Omega(A)$ is a bijection, and the inverse map is given by: $\omega \in \Omega(A) \mapsto \omega^{-1}(0) \in \mathscr{M}(A)$.*

PROOF. We have shown that every $\mathfrak{m} \in \mathscr{M}(A)$ gives rise to an $\omega_{\mathfrak{m}} \in \Omega(A)$ such that $\mathfrak{m} = \omega_{\mathfrak{m}}^{-1}(0)$. If $\omega \in \Omega(A)$, then $\omega^{-1}(0) = \mathfrak{m}_\omega$ is certainly a maximal regular ideal of A because $A/\omega^{-1}(0)$ is isomorphic to **C**, and **C** has no nontrivial ideal. Furthermore, the isomorphism of $A/\omega^{-1}(0)$ to **C** is unique; hence $\omega = \omega_{\mathfrak{m}_\omega}$. Q.E.D.

If $\omega \in \Omega(A)$, then the kernel $\omega^{-1}(0) = \mathfrak{m}_\omega$ is a maximal regular ideal of A, so that it is closed; hence ω is continuous. (In general, a linear functional on a normed space is continuous if and only if its kernel is closed.) Moreover, we have the following:

Proposition 3.9. *In the above situation, $\Omega(A)$ is contained in the unit ball S^* of the conjugate space A^* of A.*

PROOF. If $\omega \in \Omega(A)$, then ω is bounded as mentioned above. For each $x \in A$, we have

$$|\omega(x)| = |\omega(x^n)|^{1/n} \leq \|\omega\|^{1/n}\|x^n\|^{1/n}, \quad n = 1,2,\ldots;$$

hence $|\omega(x)| \leq \lim_{n\to\infty} \|\omega\|^n\|x^n\|^{1/n} = \|x\|_{sp} \leq \|x\|$. Q.E.D.

Proposition 3.10. *The set $\Omega(A)$ is locally compact with respect to the $\sigma(A^*,A)$-topology. It is compact if A is unital.*

PROOF. Let $\Omega'(A) = \Omega(A) \cup \{0\}$. Then $\Omega'(A)$ is a subset of the unit ball S^* of A^*. Let $\{\omega_i\}$ be a net in $\Omega'(A)$ converging to ω_0 in the $\sigma(A^*,A)$-topology. For any $x,y \in A$, we have

$$\begin{aligned}\omega_0(xy) &= \lim \omega_i(xy) = \lim \omega_i(x)\omega_i(y)\\ &= \omega_0(x)\omega_0(y);\end{aligned}$$

hence $\omega_0 \in \Omega'(A)$. Thus $\Omega'(A)$ is a $\sigma(A^*,A)$-closed subset of S^*, so that it is compact. Of course, $\{0\}$ is closed in $\Omega'(A)$. Hence $\Omega(A)$ is an open subset of the compact space $\Omega'(A)$, so that it is locally compact.

Suppose that A is unital. The point 0 is isolated in $\Omega'(A)$ because $\omega(1) = 1$ for every $\omega \in \Omega(A)$, so that $\Omega(A)$ is closed in $\Omega'(A)$; hence it is compact. Q.E.D.

For each $x \in A$, we define a function $\hat{x}$ on the space $\Omega(A)$ by $\hat{x}(\omega) = \omega(x)$, $\omega \in \Omega(A)$. Clearly $\hat{x}$ is a continuous function on $\Omega(A)$. For each $\varepsilon > 0$, the set $\{\omega \in \Omega(A): |\hat{x}(\omega)| \geq \varepsilon\}$ is closed in $\Omega'(A)$, so that it is compact. Therefore, $\hat{x}$ vanishes at infinity.

Theorem 3.11. *If A is an abelian Banach algebra, then the map $\mathscr{F}: x \in A \mapsto \hat{x} \in C_\infty(\Omega(A))$ is a homomorphism of A into the abelian C^*-algebra $C_\infty(\Omega(A))$ of all continuous functions on $\Omega(A)$ vanishing at infinity. If A is unital, then $\Omega(A)$ is compact and $\mathrm{Sp}_A(x) = \hat{x}(\Omega(A))$. If A is not unital, then $\mathrm{Sp}'_A(x) = \hat{x}(\Omega(A)) \cup \{0\}$. Hence in any case,*

$$\|\hat{x}\| = \|x\|_{\mathrm{sp}}, \qquad x \in A.$$

Proof. It is clear that $\mathscr{F}$ is linear and multiplicative, hence a homomorphism. Suppose that A is unital. Let $x \in A$. If $\lambda \in \mathrm{Sp}_A(x)$, then $x - \lambda$ is not invertible, so that $x - \lambda$ falls in a regular maximal ideal $\mathfrak{m}$ of A. Hence $\omega_\mathfrak{m}(x - \lambda) = 0$, i.e., $\omega_\mathfrak{m}(x) = \lambda$. Conversely, if $\lambda = \omega(x)$ for some $\omega \in \Omega(A)$, then $x - \lambda \in \omega^{-1}(0)$; hence $x - \lambda$ is not invertible. Therefore, λ belongs to $\mathrm{Sp}_A(x)$ by definition. The assertion for nonunital A follows from the above. The assertion on the norm is now trivial. Q.E.D.

Definition 3.12. The map $\mathscr{F}$ is called the *Gelfand representation* and $\Omega(A)$ the *spectrum* of A. Each member of $\Omega(A)$ is called a *character* of A, i.e., a character is a homomorphism of A onto $\mathbf{C}$. The kernel $\mathscr{F}^{-1}(0)$ of $\mathscr{F}$ is called the *radical* of A. If $\mathscr{F}^{-1}(0) = \{0\}$, then A is said to be *semisimple*.

In other words, an abelian semisimple Banach algebra A is isomorphic to a subalgebra of the abelian C^*-algebra $C_\infty(\Omega)$ of all continuous functions on a locally compact space Ω vanishing at infinity.

Exercise

1. For a continuous function f on the unit interval $[0,1]$ with $f(0) = f(1)$, we define the Fourier coefficients $\{a_n\}$ by

$$a_n = \int_0^1 e^{-2\pi i n s} f(s)\, ds, \qquad n \in \mathbf{Z}.$$

Show that if $\{a_n\}$ is absolutely summable and if f does not vanish at any point, then the Fourier coefficients $\{b_n\}$ of $1/f$ are also absolutely summable.

(a) Let A be the set of all continuous functions f on $[0,1]$ with $f(0) = f(1)$ whose Fourier coefficients are absolutely summable. Trivially, A is a vector space over $\mathbf{C}$. Define a norm in A by

$$\|f\| = \sum_{n \in \mathbf{Z}} |a_n|.$$

Show that A is a Banach algebra under the multiplication as function, by proving $f(s) = \sum_{n \in \mathbf{Z}} a_n e^{2\pi i n s}$, $f \in A$.

(b) For each $t \in [0,1]$, define a character ω_t of A by $\omega_t(f) = f(t)$, $f \in A$.

(c) Let $f_1(s) = e^{2\pi i s}$. Then $f_1 \in A$ is invertible and $\|f_1\| = \|f_1^{-1}\| = 1$. For any $\omega \in \Omega(A)$, show that $\omega(f_1) = e^{2\pi i t}$ for some $t = t_\omega \in \Omega(A)$.

(d) Show that $\omega(f) = f(t_\omega)$ for every $f \in A$.

(e) Show that $\Omega(A)$ is identified with $[0,1]$ by the mapping $\omega \to t_\omega$.

(f) To get the original assertion, apply Proposition 3.7 to the Banach algebra A.

4. Spectrum and Functional Calculus in C^*-Algebras

Definition 4.1. Let A be an involutive Banach algebra. An element $x \in A$ is said to be *self-adjoint* or *hermitian* if $x = x^*$; *normal* if $x^*x = xx^*$; *unitary* if $x^*x = xx^* = 1$ when A is unital; a *projection* if $x^2 = x$ and $x^* = x$.

Every element x of an involutive Banach algebra A is represented uniquely in the form $x = x_1 + ix_2$ with two self-adjoint elements $x_1, x_2 \in A$, where x_1 and x_2 are obtained by

$$x_1 = \frac{1}{2}(x + x^*) \quad \text{and} \quad x_2 = \frac{1}{2i}(x - x^*).$$

Since

$$x^*x = x_1^2 + x_2^2 + i(x_1x_2 - x_2x_1);$$
$$xx^* = x_1^2 + x_2^2 - i(x_1x_2 - x_2x_1),$$

x is normal if and only if x_1 and x_2 commute. The elements x_1 and x_2 are called the *real part* and the *imaginary part* of x, respectively. We denote by A_h the set of all self-adjoint elements of A and by $U(A)$ or A_u the set of all unitary elements of A when A is unital. It is clear that A_h is a real Banach space and $A = A_h + iA_h$; $U(A)$ is a closed subgroup of the general linear group $G(A)$ of A, which is called the *unitary group* of A.

Proposition 4.2. *If A is a C^*-algebra, then $\|x\| = \|x\|_{sp}$ for every normal $x \in A$.*

PROOF. We compute as follows:

$$\begin{aligned}\|x^{2n}\|^2 &= \|(x^{2n})^*x^{2n}\| = \|(x^*x)^{2n}\| \\ &= \|(x^*x)^n(x^*x)^n\| = \|(x^*x)^n\|^2;\end{aligned}$$

hence

$$\begin{aligned}\|x^{2^n}\| &= \|(x^*x)^{2^{n-1}}\|^2 = \cdots = \|x^*x\|^{2^{n-1}} \\ &= \|x\|^{2^n}.\end{aligned}$$

Therefore, we get, by Proposition 2.4,

$$\|x\| = \lim_{n \to \infty} \|x^{2^n}\|^{2^{-n}} = \|x\|_{sp}. \qquad \text{Q.E.D.}$$

Proposition 4.3. *In a unital C^*-algebra A, we have the following:*

(i) $\mathrm{Sp}_A(u) \subset \{\lambda \in \mathbf{C} : |\lambda| = 1\}$ *for every* $u \in U(A)$;
(ii) $\mathrm{Sp}_A(h) \subset \mathbf{R}$ *for every* $h \in A_h$.

PROOF. Let $u \in U(A)$. We have then

$$1 = \|1\| = \|u^*u\| = \|u\|^2$$

so that $\|u\| = \|u^*\| = 1$. Since u is normal, $\|u\|_{sp} = \|u^*\|_{sp} = 1$. But $u^* = u^{-1}$, so that $\mathrm{Sp}_A(u^{-1}) = \{\lambda^{-1} : \lambda \in \mathrm{Sp}_A(u)\} = \{\bar{\lambda} : \lambda \in \mathrm{Sp}_A(u)\}$ is contained in the unit disk; hence $\mathrm{Sp}_A(u)$ must be contained in the unit circle of $\mathbf{C}$.

Let $h \in A_h$ and put $u = \exp(ih)$. We then have $u^* = \exp(-ih)$, so that u is in $U(A)$. Hence $\mathrm{Sp}_A(u)$ is contained in the unit circle. By Proposition 2.8, we have

$$\{\exp i\lambda : \lambda \in \mathrm{Sp}_A(h)\} = \mathrm{Sp}_A(u) \subset \{\lambda \in \mathbf{C} : |\lambda| = 1\}.$$

This is possible only if $\mathrm{Sp}_A(h) \subset \mathbf{R}$. Q.E.D.

Theorem 4.4. *An abelian C^*-algebra A is semisimple. If Ω is its spectrum, then the Gelfand representation $\mathscr{F}$ is indeed an isometric isomorphism of A onto $C_\infty(\Omega)$ preserving the *-operation.*

PROOF. Since A is abelian, every element $x \in A$ is normal, so that $\|x\| = \|x\|_{\mathrm{sp}}$ by Proposition 4.2. Hence the Gelfand representation $\mathscr{F}$ is isometric, so that A is semisimple. Since the image $\mathscr{F}(A)$ of A is complete, it is closed. Take an $\omega \in \Omega$. For each $h \in A_h$, we have $\omega(h) \in \mathrm{Sp}'_A(h) \subset \mathbf{R}$ by Theorem 3.11 and Proposition 4.3. For every $x \in A$, we write $x = h + ik$ with $h,k \in A_h$, and have

$$\begin{aligned}\omega(x^*) &= \omega(h - ik) = \omega(h) - i\omega(k)\\ &= \overline{\omega(x)}.\end{aligned}$$

Therefore, $\mathscr{F}: x \mapsto \hat{x}$ preserves the *-operation. Thus, $\mathscr{F}(A)$ is a self-adjoint closed subalgebra of $C_\infty(\Omega)$, and separates the points of Ω in the sense that for any distinct $\omega_1, \omega_2 \in \Omega$ there exists an $x \in A$ such that $\hat{x}(\omega_1) \neq \hat{x}(\omega_2)$. By the Stone–Weierstrass theorem, $\mathscr{F}(A)$ is dense in $C_\infty(\Omega)$; hence $\mathscr{F}(A) = C_\infty(\Omega)$ by the closedness. Q.E.D.

Proposition 4.5. *Let Ω be a locally compact space and $A = C_\infty(\Omega)$. The map: $\omega \in \Omega \mapsto \hat{\omega} \in \Omega(A)$ given by $\hat{\omega}(x) = x(\omega)$, $x \in A$, is a homeomorphism of Ω onto $\Omega(A)$.*

PROOF. Since Ω is a completely regular topological space, the map: $\omega \in \Omega \mapsto \hat{\omega} \in \Omega(A)$ is a homeomorphism of Ω into $\Omega(A)$. Let $\rho \in \Omega(A)$. Then ρ is a linear functional on $C_\infty(\Omega)$ such that $\rho(x) \geq 0$ if $x \geq 0$ as a function. Hence there exists a positive Radon measure μ on Ω such that $\rho(x) = \int_\Omega x(\omega)\,d\mu(\omega)$ and $\mu(\Omega) = 1$. We have then

$$\begin{aligned}\int_\Omega |x(\omega) - \rho(x)|^2\,d\mu(\omega) &= \rho((x - \rho(x))^*(x - \rho(x)))\\ &= 0,\end{aligned}$$

using the multiplicativity of ρ. Hence every $x \in C_\infty(\Omega)$ is μ-almost everywhere constant, which means that μ is concentrated at a single point $\omega \in \Omega$. Hence $\rho(x) = x(\omega)$. Thus $\rho = \hat{\omega}$. Q.E.D.

Let A be a C^*-algebra. A closed subalgebra B of A is called a *C^*-subalgebra* of A if $x \in B$ implies $x^* \in B$. When A is unital, a C^*-subalgebra B is said to be *unital in A* if B contains the identity of A. It is clear that the intersection of C^*-subalgebras of A is again a C^*-subalgebra of A. Hence for any subset E of A, there exists the smallest C^*-subalgebra B of A containing E. This algebra B is called the *C^*-subalgebra of A generated by E.*

Proposition 4.6. *Let A be a unital C^*-algebra. If $x \in A$ is normal, then there exists a unique isomorphism φ of $C(\mathrm{Sp}_A(x))$ onto the C^*-subalgebra B of A generated by x and 1 such that $\varphi(1) = 1$ and $\varphi(\iota) = x$ where ι denotes the function $\iota(\lambda) = \lambda$, $\lambda \in \mathrm{Sp}_A(x)$.*

PROOF. Let $S = \mathrm{Sp}_A(x)$. The normality of x implies the commutativity of B. Let Ω be the spectrum of B. Since the polynomials of x and x^* form a dense subalgebra of B, the map $\psi: \omega \in \Omega \mapsto \omega(x) \in S$ is a homeomorphism. It is easy to see that the composed map $\varphi: f \in C(S) \mapsto \mathscr{F}^{-1}(f \circ \psi) \in B$ is the desired isomorphism. Q.E.D.

In the above situation, if $\lambda \notin \mathrm{Sp}_A(x)$, then $x - \lambda$ is invertible in B because $(\lambda - \iota)$ is invertible in $C(S)$. Let f_1 be a holomorphic function on a neighborhood of S and $f = f_1|_S$. We then have, for any $\omega \in \Omega$,

$$\begin{aligned}\langle f_1(x), \omega\rangle &= \frac{1}{2\pi i}\int_C \langle(\lambda - x)^{-1}, \omega\rangle f_1(\lambda)\, d\lambda \\ &= \frac{1}{2\pi i}\int_C (\lambda - \omega(x))^{-1} f_1(\lambda)\, d\lambda \\ &= f_1(\omega(x)) = f(\psi(\omega)),\end{aligned}$$

where C means of course an appropriate smooth simple closed curve enclosing S. Hence we get $f_1(x) = \varphi(f)$. Therefore, the following definition is coherent with the functional calculus.

Definition 4.7. In the above situation, if f is a continuous function on $\mathrm{Sp}_A(x)$, then we denote by $f(x)$ the image $\varphi(f)$ of f under the isomorphism φ in the previous proposition.

We have the following:

$$\begin{aligned}(\lambda f + \mu g)(x) &= \lambda f(x) + \mu g(x); \\ (fg)(x) &= f(x)g(x); \\ \bar{f}(x) &= f(x)^*; \\ \mathrm{Sp}_A f(x) &= \{f(\lambda) : \lambda \in \mathrm{Sp}_A(x)\}; \\ \|f(x)\| &= \sup\{|f(\lambda)| : \lambda \in \mathrm{Sp}_A(x)\}.\end{aligned}$$

If $g \in C(f(\mathrm{Sp}_A(x))$, then we have

$$(g \circ f)(x) = g(f(x)).$$

If A is not unital, then $\mathrm{Sp}'_A(x)$ contains zero. Let A_{I} be the unital C^*-algebra obtained by adjunction of an identity to A. For a normal $x \in A$, and $f \in C(\mathrm{Sp}'_A(x))$, we consider $f(x)$ in A_{I}. If $f(0) = 0$, then $f(x)$ belongs to A because if ω_0 is a homomorphism of A_{I} with kernel A, then $\omega_0(f(x)) = f(\omega_0(x)) = f(0) = 0$. Therefore, $f(x)$ makes sense provided $f(0) = 0$.

For each $h \in A_{\mathrm{h}}$, we write:

$$|h| = (h^2)^{1/2}, \qquad h_+ = \tfrac{1}{2}(|h| + h), \qquad h_- = \tfrac{1}{2}(|h| - h).$$

The elements h_+, h_-, and $|h|$ are respectively called the *positive* part, the *negative* part, and the *absolute value* of h. The decomposition $h = h_+ - h_-$ is called the *Jordan decomposition* of h. Of course, $\mathrm{Sp}'_A(|h|)$, $\mathrm{Sp}'_A(h_+)$, and $\mathrm{Sp}'_A(h_-)$ are nonnegative real numbers. It is easy to check that h_+ and h_- are characterized as elements in A_h such that

$$h = h_+ - h_-, \qquad h_+h_- = 0, \quad \text{and} \quad \mathrm{Sp}'_A(h_+) \cup \mathrm{Sp}'_A(h_-) \subset \mathbf{R}_+.$$

Proposition 4.8. *Let B be a C^*-subalgebra of a C^*-algebra A and $x \in B$. We have the following*:

(i) $\mathrm{Sp}'_A(x) = \mathrm{Sp}'_B(x)$;
(ii) *If A is unital and B is a unital subalgebra, then* $\mathrm{Sp}_A(x) = \mathrm{Sp}_B(x)$.

Proof. By adjunction of an identity, (i) follows from (ii), so that we have only to prove (ii). In general, it is always true that $\mathrm{Sp}_A(x) \subset \mathrm{Sp}_B(x)$. Suppose that x is self-adjoint. By Proposition 4.3, we have $\mathrm{Sp}_B(x) \subset \mathbf{R}$. Take a scalar $\lambda \notin \mathrm{Sp}_A(x)$. We have to show $\lambda \notin \mathrm{Sp}_B(x)$. We may assume then that λ is real. For any $\varepsilon > 0$, $\lambda_\varepsilon = \lambda + i\varepsilon$ is not in $\mathrm{Sp}_B(x)$, so that $(x - \lambda_\varepsilon)^{-1}$ exists in B. By the continuity of the inverse operation in $G(A)$, $(x - \lambda_\varepsilon)^{-1}$ converges to $(x - \lambda)^{-1}$ in $G(A)$ as $\varepsilon \to 0$ because $(x - \lambda_\varepsilon) - (x - \lambda) = (\lambda - \lambda_\varepsilon) = -i\varepsilon$. Since B is closed, $(x - \lambda)^{-1}$ belongs to B. Hence $\lambda \notin \mathrm{Sp}_B(x)$. Thus $\mathrm{Sp}_A(x) = \mathrm{Sp}_B(x)$ for a self-adjoint $x \in B$. In the general case, we proceed as follows. If $x \in B$ is invertible in A, then x^*x is invertible in A, and so is in B by the self-adjointness of x^*x. Hence x is left invertible in B. Considering xx^*, we see that x is right invertible in B also. Therefore, x is invertible in B. Applying this argument to $x - \lambda$, $\lambda \in \mathbf{C}$, we conclude assertion (ii). Q.E.D.

Proposition 4.9. *If A is a unital C^*-algebra, then every $x \in A$ is a linear combination of four unitary elements.*

Proof. Let $h \in A_h$ with $\|h\| \leq 1$. Put

$$u = h + i(1 - h^2)^{1/2}.$$

Here we use a functional calculus for h, i.e., Proposition 4.6. It follows then that u is unitary. It is obvious that $h = \frac{1}{2}(u + u^*)$. For a general $x \in A$, we can obtain the conclusion by considering the real and the imaginary parts of $x/\|x\|$. Q.E.D.

Proposition 4.10. *Let K be a compact subset of* $\mathbf{C}$. *Let A_K denote the set of all normal elements x with $\mathrm{Sp}_A(x) \subset K$. If f is a continuous function on K, then the functional calculus*: $x \in A_K \mapsto f(x) \in A$ *is continuous.*

Proof. By the Stone–Weierstrass theorem, for any $\varepsilon > 0$ there exists a polynomial p of λ and $\bar{\lambda}$ such that $\sup_{\lambda \in K} |p(\lambda,\bar{\lambda}) - f(\lambda)| < \varepsilon$. Since p is a polynomial, there exists a $\delta > 0$ such that $\|p(x,x^*) - p(y,y^*)\| < \varepsilon$ if $\|x - y\| < \delta$ and $\|x\|,\|y\| \leq M$ with a preassigned constant $M > 0$. Let $M = \sup\{|\lambda| : \lambda \in K\}$.

If $\|x - y\| < \delta$ and $x,y \in A_K$, then we have

$$\|f(x) - f(y)\| \le \|f(x) - p(x,x^*)\| + \|p(x,x^*) - p(y,y^*)\| + \|p(y,y^*) - f(y)\| < 3\varepsilon. \qquad \text{Q.E.D.}$$

EXERCISES

1. Let u be a unitary element in a unital C^*-algebra A. Show that if $\mathrm{Sp}_A(u) \neq \{\lambda \in \mathbf{C}: |\lambda| = 1\}$ then there exists $h \in A_h$ such that $u = \exp ih$.

2. Let Γ be the unit circle $\{\lambda \in \mathbf{C}: |\lambda| = 1\}$. Let u be the unitary element in $C(\Gamma)$ defined by $u(\lambda) = \lambda$. Show that there is no element h in $C(\Gamma)$ such that $u = \exp ih$.

3. Let A be a unital C^*-algebra. Show that an element x of A is self-adjoint if and only if $\lim_{t\to 0} (1/t)(\|1 + itx\| - 1) = 0$. (*Hint*: If $h \in A$ is self-adjoint, then $\exp ith = 1 + ith + O(t)$ is unitary for every $t \in \mathbf{R}$. If k is another nonzero self-adjoint element of A, then $\|1 + ith - tk\| \ge \|1 - tk\| \neq 1 + O(t)$.)

4. Let Γ be a completely regular topological space. Let $C_b(\Gamma) = A$ be the set of all bounded complex valued continuous functions on Γ. Define an algebraic structure in A by the following:

$$(\lambda x + \mu y)(\gamma) = \lambda x(\gamma) + \mu y(\gamma);$$
$$(xy)(\gamma) = x(\gamma)y(\gamma), \qquad x^*(\gamma) = \overline{x(\gamma)};$$
$$\|x\| = \sup|x(\gamma)|.$$

   (a) Show that A is a unital C^*-algebra.
   (b) Let $\omega_\gamma(x) = x(\gamma)$ for each $\gamma \in \Gamma$ and $x \in A$. Show that the map: $\gamma \in \Gamma \mapsto \omega_\gamma \in \Omega(A)$ is an injective homeomorphism of Γ onto a dense subset of $\Omega(A)$.
   (c) Show that the image of Γ under the above map in (b) is open in $\Omega(A)$ if and only if Γ is locally compact.
   (d) Show that if f is a continuous map of Γ into a compact space K, then there exists a unique continuous map g of $\Omega(A)$ into K such that $f(\gamma) = g(\omega_\gamma)$, $\gamma \in \Gamma$. (This space $\Omega(A)$ is characterized by this universal property. $\Omega(A)$ is called the *Stone–Čech compactification* of Γ.)

5. Show that an abelian C^*-algebra is separable if and only if its spectrum is separable in the sense of the existence of a countable basis for open sets.

5. Continuity of Homomorphisms

Definition 5.1. If a homomorphism of an involutive Banach algebra into another preserves the *-operation, then it is called a **-homomorphism*.

Proposition 5.2. *If π is a *-homomorphism of an involutive Banach algebra A into a C^*-algebra B, then*

$$\|\pi(x)\| \le \|x\|, \qquad x \in A.$$

PROOF. By Proposition 4.2, we have $\|y\| = \|y\|_{\text{sp}}$ for every $y \in B_{\text{h}}$. For every $x \in A$, we have

$$\text{Sp}'_A(x) \supset \text{Sp}'_B(\pi(x)),$$

so that $\|\pi(x)\|_{\text{sp}} \leq \|x\|_{\text{sp}} \leq \|x\|$. Hence we get

$$\begin{aligned}\|\pi(x)\|^2 &= \|\pi(x)^*\pi(x)\| = \|\pi(x^*x)\| \\ &= \|\pi(x^*x)\|_{\text{sp}} \leq \|x^*x\| \\ &\leq \|x^*\|\|x\| = \|x\|^2;\end{aligned}$$

so that we have $\|\pi(x)\| \leq \|x\|$, $x \in A$. Q.E.D.

Proposition 5.3. *Let A be a C^*-algebra and B an involutive Banach algebra. If π is a $*$-isomorphism of A into B, then*

$$\|\pi(x)\| \geq \|x\|, \qquad x \in A.$$

Remark. If we know that $\pi(A)$ is closed in B, then the proposition follows automatically from the previous one by considering π^{-1}. Since the closedness of $\pi(A)$ is not a trivial matter, this proposition supplements the previous result.

PROOF. Let $h \in A_h$ and $k = \pi(h)$. In order to show the inequality $\|h\| \leq \|k\|$, we may assume, considering the C^*-subalgebras generated by h and k, respectively, that A and B are both abelian. By adjunction of an identity, we also assume that A and B are both unital and π preserves the identities. It follows that the spectrums $\Omega(A)$ and $\Omega(B)$ are both compact, and the transpose ${}^t\pi$ of π induces a continuous map of $\Omega(B)$ into $\Omega(A)$. By the compactness of $\Omega(B)$, ${}^t\pi(\Omega(B))$ is compact, hence closed in $\Omega(A)$. Suppose ${}^t\pi(\Omega(B)) \neq \Omega(A)$. Considering the Gelfand representation of A, we can find two nonzero elements $a,b \in A$ such that $ab = 0$ and $\omega(b) = 1$ for every $\omega \in {}^t\pi(\Omega(B))$. It follows then that $\pi(a)\pi(b) = 0$ and $\langle \pi(b),\omega \rangle = 1$ for every $\omega \in \Omega(B)$, so that $\pi(b)$ does not belong to any maximal ideal of B; hence $\pi(b)$ must be invertible by Proposition 3.7. But this contradicts the fact that $\pi(a)\pi(b) = 0$ and $\pi(a) \neq 0$. Thus, we have $\Omega(A) = {}^t\pi(\Omega(B))$, so that

$$\begin{aligned}\|k\| \geq \|k\|_{\text{sp}} &= \sup\{|\langle \pi(h),\omega \rangle| : \omega \in \Omega(B)\} \\ &= \sup\{|\langle h, {}^t\pi(\omega) \rangle| : \omega \in \Omega(B)\} \\ &= \sup\{|\langle h,\omega \rangle| : \omega \in \Omega(A)\} = \|h\|.\end{aligned}$$

For a general $x \in A$, we have

$$\begin{aligned}\|x\| &= \|x^*x\|^{1/2} \leq \|\pi(x)^*\pi(x)\|^{1/2} \\ &\leq \|\pi(x)\|.\end{aligned}$$

Q.E.D.

Corollary 5.4. *A $*$-isomorphism of a C^*-algebra into another C^*-algebra is an isometry.*

Proposition 5.5. *Let A be a Banach algebra and B a C^*-algebra. An isomorphism π of A into B with self-adjoint range $\pi(A)$ is continuous.*

PROOF. By the self-adjointness of $\pi(A)$, we can introduce an involution in A such a way that $\pi(x^*) = \pi(x)^*$, $x \in A$. As in the proof of Proposition 5.2, we have

$$\|\pi(x)\|^2 \leq \|x^*\|\,\|x\|, \qquad x \in A.$$

Hence we have only to prove the continuity of the *-operation in A. To do that, we shall appeal to the closed graph theorem. Let $\{x_n\}$ be a sequence in A such that $\lim\|x_n - x\| = 0$ and $\lim\|x_n^* - y\| = 0$ for some $x, y \in A$. We have then

$$\|\pi(x) - \pi(x_n)\|^2 \leq \|x^* - x_n^*\|\,\|x - x_n\|,$$
$$\|\pi(y) - \pi(x_n^*)\|^2 \leq \|y^* - x_n\|\,\|y - x_n^*\|.$$

Since $\{x_n^*\}$ and $\{x_n\}$ are both bounded, we have

$$\lim_{n\to\infty} \|\pi(x) - \pi(x_n)\| = 0, \qquad \lim_{n\to\infty} \|\pi(y) - \pi(x_n^*)\| = 0,$$

so that $\pi(x)^* = \pi(y)$; hence $x^* = y$. Thus, the *-operation in A is continuous by the closed graph theorem. Hence there exists $k > 0$ such that $\|x^*\| \leq k\|x\|$, $x \in A$, and so

$$\|\pi(x)\|^2 \leq \|x^*\|\,\|x\| \leq k\|x\|^2, \qquad x \in A. \qquad \text{Q.E.D.}$$

Corollary 5.6. *An isomorphism (not necessarily *-preserving) of a C^*-algebra onto another C^*-algebra is continuous.*

6. Positive Cones of C^*-Algebras

The next result allows us to represent a C^*-algebra on a Hilbert space as a uniformly closed self-adjoint algebra of operators; see Theorem 9.18.

Theorem 6.1. *For a self-adjoint element x of a C^*-algebra A, the following three conditions are equivalent:*

(i) $\mathrm{Sp}'_A(x) \subset [0,\infty)$;
(ii) $x = y^*y$ *for some* $y \in A$;
(iii) $x = h^2$ *for some* $h \in A_h$.

The set P of all elements $x \in A_h$ satisfying any of the above conditions is a closed convex cone in A with $P \cap (-P) = \{0\}$.

PROOF. (i) $\Rightarrow$ (iii): If $\mathrm{Sp}'_A(x) \subset [0,\infty)$, then we can form $h = x^{1/2}$ by Definition 4.7. We then have $h = h^*$ and $h^2 = x$.

(iii) $\Rightarrow$ (ii): It is trivial.

(iii) $\Rightarrow$ (i): If $x = h^2$ with $h \in A_h$, then we have

$$\mathrm{Sp}'_A(x) = \{\lambda^2 : \lambda \in \mathrm{Sp}'_A(h)\} \subset [0,\infty),$$

as $\mathrm{Sp}'_A(h)$ is real.

(ii) $\Rightarrow$ (iii): Let $P = \{x \in A_h : \mathrm{Sp}'_A(x) \subset [0,\infty)\}$, and S be the unit ball of A. We assume, for the moment, that A is unital. We then have

$$P \cap S = \{x \in A_h \cap S : \|x - 1\| \leq 1\}$$

by Theorem 4.4. Considering the normalization $\|x\|^{-1}x$ of a nonzero $x \in A_h$, we have

$$P = \{x \in A_h : \| \|x\| - x\| \leq \|x\|\}.$$

Hence P is closed in A_h. For any $x, y \in P \cap S$, we have

$$\begin{aligned} \|1 - \tfrac{1}{2}(x + y)\| &= \tfrac{1}{2}\|(1 - x) + (1 - y)\| \\ &\leq \tfrac{1}{2}(\|1 - x\| + \|1 - y\|) \leq 1, \end{aligned}$$

so that $\frac{1}{2}(x + y) \in P \cap S$. Thus $P \cap S$ is convex. As $P = \lambda P$ for any $\lambda > 0$, P is a convex cone. If $x \in P \cap (-P)$, then $\mathrm{Sp}_A(x) = \{0\}$, so that $\|x\|_{\mathrm{sp}} = 0$. Hence $x = 0$. Therefore, $P \cap (-P) = \{0\}$. In the general case, we consider the C^*-algebra A_{I} obtained by adjunction of an identity to A. Let P_{I} be the corresponding set in A_{I}. Then we have $P = P_{\mathrm{I}} \cap A$. Hence P is a closed convex cone with $P \cap (-P) = \{0\}$ by the proceeding arguments.

Now, let $h = y^*y$ for some $y \in A$. Considering $h_+^{1/2}$ and $h_-^{1/2}$, we can find elements $u,v \in A_h$ such that $h = u^2 - v^2$ and $uv = 0$. We have then

$$(yv)^*(yv) = vy^*yv = v(u^2 - v^2)v = -v^4 \in (-P),$$

by the already established implication (iii) $\Rightarrow$ (i). Let $yv = k_1 + ik_2$ with $k_1,k_2 \in A_h$. We then have

$$\begin{aligned} (yv)(yv)^* &= -(yv)^*(yv) + (k_1 - ik_2)(k_1 + ik_2) + (k_1 + ik_2)(k_1 - ik_2) \\ &= -(yv)^*(yv) + 2k_1^2 + 2k_2^2 \in P, \end{aligned}$$

as $-(yv)^*(yv)$, k_1^2 and k_2^2 all belong to P. By Proposition 2.1, we have

$$\mathrm{Sp}'_A((yv)(yv)^*) = \mathrm{Sp}'_A((yv)^*(yv)),$$

so that $(yv)^*(yv) \in P \cap (-P) = \{0\}$, which implies $v^4 = 0$. Hence $v = 0$. Therefore, $y^*y = u^2$ as desired. Q.E.D.

Definition 6.2. An element $x \in A_h$ is said to be *positive* and written $x \geq 0$ if x satisfies any of the three conditions in the theorem. We denote the set of all positive elements of A by A_+.

Since A_+ is a convex cone with $A_+ \cap (-A_+) = \{0\}$, A_+ induces an order structure in the real Banach space A_h. We write $x \geq y$ for $x,y \in A_h$ such that $x - y \in A_+$. For each $x \in A$, $(x^*x)^{1/2}$ is called the *absolute value* of x and denoted by $|x|$.

Proposition 6.3. *Let A be a C^*-algebra. If x and y are positive elements of A such that $x \geq y$, then $x^\alpha \geq y^\alpha$ for any $0 \leq \alpha \leq 1$.*

PROOF. By adjunction of an identity, we may assume that A is unital. Let $E = \{\alpha \in \mathbf{R}: x^\alpha \geq y^\alpha\}$. It follows that E is closed and contains 0 and 1. We then prove that E is convex. Suppose, for a moment, that x and y are invertible. Let $\alpha, \beta \in E$. Then

$$x^{-\alpha/2} y^\alpha x^{-\alpha/2} \leq 1;$$
$$x^{-\beta/2} y^\beta x^{-\beta/2} \leq 1.$$

Hence we have $\|x^{-\alpha/2} y^\alpha x^{-\alpha/2}\| \leq 1$ and $\|x^{-\beta/2} y^\beta x^{-\beta/2}\| \leq 1$, so that $\|y^{\alpha/2} x^{-\alpha/2}\|^2 \leq 1$ and $\|y^{\beta/2} x^{-\beta/2}\|^2 \leq 1$. Therefore, we get

$$\begin{aligned} 1 &\geq \|(x^{-\beta/2} y^{\beta/2})(y^{\alpha/2} x^{-\alpha/2})\| = \|x^{-\beta/2} y^{(\alpha+\beta)/2} x^{-\alpha/2}\| \\ &\geq \|x^{-\beta/2} y^{(\alpha+\beta)/2} x^{-\alpha/2}\|_{\mathrm{sp}} \\ &= \|x^{-(\alpha+\beta)/4} y^{(\alpha+\beta)/2} x^{-(\alpha+\beta)/4}\|_{\mathrm{sp}} \quad (\text{since } \|ab\|_{\mathrm{sp}} = \|ba\|_{\mathrm{sp}}), \\ &= \|x^{-(\alpha+\beta)/4} y^{(\alpha+\beta)/2} x^{-(\alpha+\beta)/4}\|. \end{aligned}$$

Therefore, we have $x^{-(\alpha+\beta)/4} y^{(\alpha+\beta)/2} x^{(\alpha+\beta)/4} \leq 1$; hence $y^{(\alpha+\beta)/2} \leq x^{(\alpha+\beta)/2}$. Thus E contains $\frac{1}{2}(\alpha + \beta)$. Hence $E \supset [0,1]$.

Now, let us drop the assumption for the invertibility of x and y. For any $\varepsilon > 0$, we have

$$(y + \varepsilon)^\alpha \leq (x + \varepsilon)^\alpha, \qquad 0 \leq \alpha \leq 1.$$

Hence in the limit $\varepsilon \to 0$, we get $y^\alpha \leq x^\alpha$. Q.E.D.

EXERCISES

†1. Let A be a C^*-algebra. Show that the following conditions are equivalent: (i) A is abelian; (ii) A_h is a lattice; (iii) A_h^* is a lattice [131,330].

†2. Let A be a C^*-algebra. Show that if the condition $0 \leq x \leq y$, $x,y \in A$, implies $x^2 \leq y^2$, then A is abelian [264].

7. Approximate Identities in C^*-Algebras

In general, the presence of an identity in a given algebra eases the discussion greatly. As we have done often already, we enlarge the algebra by adjunction of an identity to a given algebra. But this approach does not work always. For example, if we consider an ideal of the algebra in question, then the enlarged algebra by adjunction of an identity to the ideal is no longer an ideal even in the case that the entire algebra is unital. Therefore, the technique of adjunction of an identity cannot cover all the interesting cases. In this respect, the notion of an approximate identity is very useful, and we shall show that every C^*-algebra has an approximate identity.

Definition 7.1. Let A be a Banach algebra. An *approximate identity* of A is a net $\{u_i\}$ of elements of A with the properties

$$\lim\|u_i x - x\| = 0, \tag{1}$$

$$\lim\|x u_i - x\| = 0, \qquad x \in A. \tag{2}$$

If only condition (1) (resp. (2)) holds, then $\{u_i\}$ is called a *left* (resp. *right*) approximate identity. If $\{u_i\}$ is bounded, then $\{u_i\}$ is called a *bounded* approximate identity. If A is a C^*-algebra, then we require two more properties for an approximate identity $\{u_i\}$:

$$0 \le u_i \le u_j \quad \text{if} \quad i \le j; \tag{3}$$

$$\|u_i\| \le 1. \tag{4}$$

Lemma 7.2. *Let A be a C^*-algebra. For $\varepsilon > 0$, put $f_\varepsilon(t) = t/(t + \varepsilon)$ for each $t \in [0,\infty)$.*

(i) *If h and k are positive elements in A such that $h \le k$, then $f_\varepsilon(h) \le f_\varepsilon(k)$ for every $\varepsilon > 0$.*
(ii) *If h is a positive element of A, then*

$$\lim_{\varepsilon \to 0} \|f_\varepsilon(h)x - x\| = 0$$

for every x in the closed right ideal of A generated by h. Symmetrically,

$$\lim_{\varepsilon \to 0} \|x f_\varepsilon(h) - x\| = 0$$

for every x in the closed left ideal of A generated by k.

PROOF. Let A_I denote the C^*-algebra obtained by adjunction of an identity to A. We have then $h + \varepsilon \le k + \varepsilon$; so $(k + \varepsilon)^{-1/2}(h + \varepsilon)(k + \varepsilon)^{-1/2} \le 1$; hence, taking the inverse,

$$(k + \varepsilon)^{1/2}(h + \varepsilon)^{-1}(k + \varepsilon)^{1/2} \ge 1;$$

so we get $(h + \varepsilon)^{-1} \ge (k + \varepsilon)^{-1}$. Hence we obtain

$$\begin{aligned} f_\varepsilon(h) &= (h + \varepsilon)^{-1}h = 1 - \varepsilon(h + \varepsilon)^{-1} \\ &\le 1 - \varepsilon(k + \varepsilon)^{-1} = (k + \varepsilon)^{-1}k = f_\varepsilon(k). \end{aligned}$$

Thus, assertion (i) follows.

Suppose h is a fixed element of A_+. Let $u_\varepsilon = f_\varepsilon(h)$. Then $\|u_\varepsilon\| \le 1$. Set

$$\mathfrak{m} = \left\{x \in A_I : \lim_{\varepsilon \to 0} \|u_\varepsilon x - x\| = 0\right\}.$$

Clearly, $\mathfrak{m}$ is a right ideal of A_I. If $\{x_n\}$ is a sequence in $\mathfrak{m}$ with $x = \lim_{n\to\infty} x_n \in A_I$, then for any $\varepsilon > 0$ there exists n with $\|x - x_n\| < \varepsilon/3$. Choose

$\delta > 0$ so small that $\|u_\lambda x_n - x_n\| < \varepsilon/3$ for $0 < \lambda < \delta$. We have then, for any $0 < \lambda < \delta$,

$$\begin{aligned}\|u_\lambda x - x\| &\le \|u_\lambda x - u_\lambda x_n\| + \|u_\lambda x_n - x_n\| + \|x_n - x\| \\ &< \|u_\lambda\|\|x - x_n\| + \|u_\lambda x_n - x_n\| + \|x_n - x\| \\ &< \frac{\varepsilon}{3} + \frac{\varepsilon}{3} + \frac{\varepsilon}{3} = \varepsilon.\end{aligned}$$

Hence $\mathfrak{m}$ is a closed right ideal. Now, we have

$$\begin{aligned}h - u_\varepsilon h &= h - (h + \varepsilon)^{-1} h^2 \\ &= \varepsilon(h + \varepsilon)^{-1} h = \varepsilon u_\varepsilon,\end{aligned}$$

so that $\|h - u_\varepsilon h\| \le \varepsilon$. Hence $h \in \mathfrak{m}$; so $\mathfrak{m}$ contains the closed right ideal of A generated by h. Q.E.D.

Corollary 7.3. *Under the same notations as above, we have the following:*

(i) $\lim_{\varepsilon \to 0} \|f_\varepsilon(h)x - x\| = 0$ (*resp.* $\lim_{\varepsilon \to 0} \|x f_\varepsilon(h) - x\| = 0$) *if and only if* x *belongs to the closed right (resp. left) ideal* $\mathfrak{m}$ *of* A *generated by* h.
(ii) *If* $xx^* \le \lambda h^2$ (*resp.* $x^*x \le \lambda h^2$) *for some* $\lambda > 0$, *then* x *belongs to* $\mathfrak{m}$.

PROOF. Just note that $hA_{\mathrm{I}} = hA + \mathbf{C}h$ and $\|ax\|^2 = \|axx^*a\| \le \lambda\|ah^2a^*\| = \lambda\|ah\|^2$ if $xx^* \le \lambda h^2$. Q.E.D.

Theorem 7.4. *Let S_0 be the open unit ball of a C^*-algebra A. If $\mathfrak{m}$ is a left ideal (resp. right ideal) of A, then the open unit ball $S_0 \cap \mathfrak{m}_+$ of the positive part $\mathfrak{m}_+$ of $\mathfrak{m}$ is upward directed and forms a right (resp. left) approximate identity for the closure $\bar{\mathfrak{m}}$ of $\mathfrak{m}$.*

PROOF. Let A_{I} be the unital C^*-algebra obtained by adjunction of an identity to A. Since $A_{\mathrm{I}} = A + \mathbf{C}$, $\mathfrak{m}$ is a left ideal of A_{I}. Take any $u_1, u_2 \in \mathfrak{m}_+ \cap S_0$. Put $h_i = (1 - u_i)^{-1} u_i \in \mathfrak{m}$, $i = 1,2$. It follows that $u_i = (1 + h_i)^{-1} h_i$, $i = 1,2$. Put $h = h_1 + h_2$ and $u = (1 + h)^{-1} h \in \mathfrak{m}_+$. By Lemma 7.2, $u \ge u_i$, $i = 1,2$ and

$$\|u\| = \|(1 + h)^{-1} h\| \le \frac{\|h\|}{1 + \|h\|} < 1.$$

Hence u belongs to $\mathfrak{m}_+ \cap S_0$. This means that $\mathfrak{m}_+ \cap S_0$ is upward directed.

Let $\varepsilon > 0$. For an $x \in \mathfrak{m}$, put $h = x^*x \in \mathfrak{m}_+$ and $u_\varepsilon = (h + \varepsilon)^{-1} h \in \mathfrak{m} \cap S_0$. We have then

$$\begin{aligned}\|x(1 - u_\varepsilon)^{1/2}\| &= \|\varepsilon x(h + \varepsilon)^{-1/2}\| \\ &= \varepsilon\|(h + \varepsilon)^{-1/2} x^*x (h + \varepsilon)^{-1/2}\|^{1/2} \\ &= \varepsilon\|(h + \varepsilon)^{-1} h\| \le \varepsilon.\end{aligned}$$

Hence, for any $x \in \mathfrak{m}$, there exists $u_\varepsilon \in \mathfrak{m}_+ \cap S_0$ such that $\|x(1-u_\varepsilon)^{1/2}\| \leq \varepsilon$. Suppose $v \geq u_\varepsilon$ and $v \in \mathfrak{m}_+ \cap S_0$. We have

$$\begin{aligned}\|x - xv\| &= \|x(1-v)\| \leq \|x(1-v)^{1/2}\|\|(1-v)^{1/2}\| \\ &\leq \|x(1-v)x^*\|^{1/2} = \|x(1-v)x^*\|_{\mathrm{sp}}^{1/2} \\ &\leq \|x(1-u_\varepsilon)x^*\|_{\mathrm{sp}}^{1/2} = \|x(1-u_\varepsilon)x^*\|^{1/2} \\ &= \|x(1-u_\varepsilon)^{1/2}\| \leq \varepsilon.\end{aligned}$$

Therefore, $\mathfrak{m}_+ \cap S_0$ is a right approximate identity for $\mathfrak{m}$ with respect to its own ordering.

Set $\mathfrak{n} = \{x \in A : \lim xu = x,\ u \in \mathfrak{m}_+ \cap S_0\}$. We claim that $\mathfrak{n}$ is closed. If $x \in \bar{\mathfrak{n}}$, then for any $\varepsilon > 0$, there exists $y \in \mathfrak{n}$ with $\|x - y\| < \varepsilon/3$. Then there exists $u_0 \in \mathfrak{m}_+ \cap S_0$ such that $\|y - yu\| < \varepsilon/3$ for any $u \geq u_0$ with $u \in \mathfrak{m}_+ \cap S_0$. We then have, for $u \geq u_0$, $u \in \mathfrak{m}_+ \cap S_0$,

$$\begin{aligned}\|x - xu\| &\leq \|x - y\| + \|y - yu\| + \|yu - xu\| \\ &\leq \frac{\varepsilon}{3} + \frac{\varepsilon}{3} + \frac{\varepsilon}{3} = \varepsilon.\end{aligned}$$

Since $\mathfrak{n} \supset \mathfrak{m}$, $\mathfrak{n}$ contains $\bar{\mathfrak{m}}$. Therefore, $\mathfrak{m}_+ \cap S_0$ is a right approximate identity for $\bar{\mathfrak{m}}$. Q.E.D.

Corollary 7.5. *Any C^*-algebra A admits an approximate identity. If A is separable, then A admits an approximate identity consisting of a sequence.*

PROOF. The positive part of the open unit ball $S_0 \cap A_+$ is an approximate identity by the previous theorem. Suppose A is separable. Let $\{v_n\}$ be a dense sequence in $S_0 \cap A_+$. By induction, we choose $\{u_n\}$ in $S_0 \cap A_+$. If u_n were chosen, then u_{n+1} would be an element in $S_0 \cap A_+$ such that

$$u_{n+1} \geq v_i, \qquad i = 1,2,\ldots,n, \quad u_{n+1} \geq u_n.$$

This is possible because $S_0 \cap A_+$ is upward directed. Since the square root is continuous in $S_0 \cap A_+$ by Proposition 4.10, for any $\varepsilon > 0$ and $x \in A$, there exists an n such that $\|x(1-v_n)^{1/2}\| < \varepsilon$, which implies that $\|x - xu_k\| < \varepsilon$ for $k \geq n+1$ as seen in the proof of the theorem. Therefore, $\{u_n\}$ is a right approximate identity of A. But $\{u_n\} = \{u_n^*\}$ is a left approximate identity of A. Hence it is an approximate identity. Q.E.D.

Remark 7.6. If A is a separable C^*-algebra, then A admits an approximate identity consisting of a mutually commuting sequence. See Exercise 7.3.

If A is an abelian C^*-algebra, then the self-adjoint part A_h of A is a vector lattice, which is checked by the Gelfand representation, Theorem 4.4. Therefore, the ordered real Banach space A_h enjoys the Riesz decomposition property, that is, if $0 \leq x \leq y_1 + y_2$ with $x, y_1, y_2 \in A_+$, then there exists x_1 and x_2 in A such that $0 \leq x_1 \leq y_1$, $0 \leq x_2 \leq y_2$ and $x = x_1 + x_2$. But, in general, this decomposition property does not hold for a nonabelian C^*-algebra. In fact, the Riesz decomposition property characterizes the commutativity of a C^*-algebra. But the following decomposition theorem holds for any C^*-algebra.

Theorem 7.7 (Asymmetric Riesz Decomposition Theorem). *Let A be a C^*-algebra. If $\{x_1, \ldots, x_m\}$ and $\{y_1, \ldots, y_n\}$ in A satisfying the equation*

$$\sum_{i=1}^{m} x_i^* x_i = \sum_{j=1}^{n} y_j^* y_j,$$

then there exists a family $\{z_{i,j}: 1 \leq i \leq m, 1 \leq j \leq n\}$ such that

$$x_i x_i^* = \sum_{j=1}^{n} z_{i,j}^* z_{i,j}, \qquad i = 1,2, \ldots, m,$$

$$y_j y_j^* = \sum_{i=1}^{m} z_{i,j} z_{i,j}^*, \qquad j = 1,2, \ldots, n.\text{[3]}$$

Proof. As usual, we may assume that A is unital. Set $a = \sum_{i=1}^{m} x_i^* x_i = \sum_{j=1}^{n} y_j^* y_j$ and $u_t = (a+t)^{-1} a$, $t > 0$. Since $a^{1/2}$ is approximated by polynomials of a, for any $z \in A$ with $z^* z \leq a$, we have

$$\lim_{t \to 0} \|z(1 - u_t)\| = 0$$

by Lemma 7.2. Put

$$z_{i,j,t} = y_j (a+t)^{-1} a^{1/2} x_i^*,$$

for $t > 0$; $i = 1,2, \ldots, m$; $j = 1,2, \ldots, n$. For any $s,t > 0$, we have

$$\begin{aligned}(z_{i,j,s} - z_{i,j,t})&(z_{i,j,s} - z_{i,j,t})^* \\ &= y_j[(a+s)^{-1} - (a+t)^{-1}] a^{1/2} x_i^* x_i a^{1/2} [(a+s)^{-1} - (a+t)^{-1}] y_j^* \\ &\leq y_j[(a+s)^{-1} - (a+t)^{-1}] a^2 [(a+s)^{-1} - (a+t)^{-1}] y_j^* \\ &= [y_j(u_s - u_t)][y_j(u_s - u_t)]^*.\end{aligned}$$

Hence we have $\lim_{s,t \to 0} \|z_{i,j,s} - z_{i,j,t}\| = 0$. Let

$$z_{i,j} = \lim_{t \to 0} z_{i,j,t} \in A.$$

[3] This result for operators on a Hilbert space can be proven more directly by arguments similar to those involved in polar decomposition, even if infinitely many x_i and y_j are involved. See Exercise II.3.4.

We have then

$$\begin{aligned} y_j y_j^* &= \lim_{t\to 0} (y_j u_t)(y_j u_t)^* \\ &= \lim_{t\to 0} y_j(a+t)^{-1} a^{1/2} \left(\sum_{i=1}^m x_i^* x_i \right) a^{1/2}(a+t)^{-1} y_j^* \\ &= \sum_{i=1}^m \left(\lim_{t\to 0} y_j(a+t)^{-1} a^{1/2} x_i^* x_i a^{1/2} (a+t)^{-1} y_j^* \right) \\ &= \sum_{i=1}^m z_{i,j} z_{i,j}^*, \end{aligned}$$

$$\begin{aligned} \sum_{j=1}^n z_{i,j}^* z_{i,j} &= \sum_{j=1}^m (\lim_{t\to 0} x_i a^{1/2}(a+t)^{-1} y_j^* y_j a^{1/2}(a+t)^{-1} x_i^*) \\ &= \lim_{t\to 0} x_i a^{1/2}(a+t)^{-1} \left(\sum_{j=1}^m y_j^* y_j \right) (a+t)^{-1} a^{1/2} x_i^* \\ &= \lim_{t\to 0} x_i (a+t)^{-1} a^2 (a+t)^{-1} x_i^* \\ &= \lim_{t\to 0} (x_i u_t)(x_i u_t)^* = x_i x_i^*. \end{aligned}$$

Q.E.D.

Corollary 7.8. *Under the same conditions, if* $\sum_{j=1}^m y_j^* y_j \le \sum_{i=1}^n x_i^* x_i$, *then there exists a family* $\{z_{i,j}: 1 \le i \le n, 1 \le j \le m\}$ *such that*

$$y_j y_j^* = \sum_{i=1}^n z_{i,j} z_{i,j}^*$$

$$x_i x_i^* \ge \sum_{j=1}^m z_{i,j}^* z_{i,j}.$$

PROOF. Put $y_{m+1} = (\sum_{i=1}^n x_i^* x_i - \sum_{j=1}^m y_j^* y_j)$, and then apply the theorem. Q.E.D.

EXERCISES

1. Let $L^1(G)$ be as in Exercise 1.1. For each neighborhood V of the unit e of G, let u_V be a positive function such that $u_V(V^c) = 0$ and $\int_G u_V(t)\,dt = 1$. Show that $\{u_V : V$ runs all neighborhoods of $e\}$ is an approximate identity of $L^1(G)$ with norm ≤ 1. (*Hint*: Use the fact that the space $\mathscr{K}(G)$ of all continuous functions on G with compact support is dense in $L^1(G)$.)

2. Let Ω, G, and A be as in Exercise 1.2. For each compact subset K of Ω, let f_K be a positive continuous function in $\mathscr{K}(\Omega)$ such that $f_K(\omega) = 1$ on K and $0 \le f_K \le 1$; and for each neighborhood V of the unit $e \in G$, let u_V be a continuous positive function on G such that $u_V(V^c) = 0$ and $\int_G u_V(s)\,ds = 1$. Set $u_{K,V}(\omega,s) = f_K(\omega)u_V(s)$. Show that $\{u_{K,V}\}$ is an approximate identity in A.

3. A positive element a of a C^*-algebra A is said to be *strictly* positively if $\varphi(a) > 0$ for any nonzero positive linear functional φ on A. (See section 9 for this problem and the next one.)
   (a) Show that the strict positivity of a positive element a in a unital C^*-algebra A is equivalent to the invertibility of a.
   (b) Show that every separable C^*-algebra admits a strictly positive element.
   (c) Let a be a strictly positive element of a C^*-algebra A. Show that for any non-degenerate representation $\{\pi,\mathfrak{H}\}$ of A, $\lim_{t\to 0} \|\pi(f_t(a))\xi - \xi\| = 0$ with $f_t(\lambda) = \lambda/(\lambda + t)$, $t > 0$.
   (d) Show that $\lim_{t\to 0} \|f_t(a)x - x\| = 0$ for every $x \in A$. (*Hint*: Use Dini's theorem for the functions: $\varphi \mapsto \varphi(x^* f_t(a)^2 x)$ on the space of $\varphi \in A^*_+$ with $\|\varphi\| \leq 1$.)
   (e) Show that a C^*-algebra with a strictly positive element admits a commuting approximate identity.

4. Let $\{\Gamma,\mu\}$ be a nonatomic σ-finite measure space with $\mu(\Gamma) = +\infty$. Let A be the C^*-algebra generated by projections e in $L^\infty(\Gamma,\mu)$ with $\mu(e) < +\infty$. Show that for any $h \in A_+$ there exists a state φ on A such that $\varphi(h) = 0$. (*Hint*: A is not unital. So there exists an orthogonal sequence $\{p_n\}$ of projections in A such that $hp_n \leq 1/n$. Let $\{e_n\}$ be a sequence of projections in A such that $e_n \leq p_n$ and $\mu(e_n) < 1/2^n$. Put $e = \sum_{n=1}^\infty e_n \in A$. If $\{\varphi_n\}$ is a sequence of states on A such that $\varphi_n(e_n) = 1$, then $\varphi(e) = 1$ and $\varphi(h) = 0$ for any accumulation point φ of $\{\varphi_n\}$ in A^*.)

5. (a) Show that if $x,y \in \mathscr{L}(\mathfrak{H})$ satisfy the inequality $x^*x \leq y^*y$, then there exists $a \in \mathscr{L}(\mathfrak{H})$ such that $x = ay$ and $\|a\| \leq 1$. (*Hint*: Apply arguments similar to those involved in polar decomposition.)
   (b) Give a straightforward proof of Theorem 7.7 for $A = \mathscr{L}(\mathfrak{H})$ as an application of (a).

6. Let A be a C^*-algebra and $\mathfrak{m}$ a closed ideal of A. Show that any positive element of A majorized by a positive element of $\mathfrak{m}$ belongs to $\mathfrak{m}$. (*Hint*: Use an approximate identity of $\mathfrak{m}$.)

8. Quotient Algebras of C^*-Algebras

Theorem 8.1. *Any closed ideal $\mathfrak{m}$ of a C^*-algebra A is self-adjoint in the sense that $x \in \mathfrak{m}$ implies $x^* \in \mathfrak{m}$, and the quotient involutive Banach algebra $A/\mathfrak{m}$ is a C^*-algebra.*

PROOF. Let $\{u_i\}$ be an approximate identity of $\mathfrak{m}$. If $x \in \mathfrak{m}$, then

$$\lim\|x^* - x^*u_i\| = \lim\|u_ix - x\| = 0;$$

hence x^* belongs to $\mathfrak{m}$ since $\mathfrak{m}$ is closed and $x^*u_i \in \mathfrak{m}$.

It is now clear that the quotient algebra $A/\mathfrak{m}$ is an involutive Banach algebra. For each $x \in A$, let $\dot{x}$ denote the element in $A/\mathfrak{m}$ corresponding to x. We shall verify the inequality

$$\|\dot{x}^*\dot{x}\| \geq \|\dot{x}\|^2, \qquad \dot{x} \in A/\mathfrak{m}.$$

Since $yu_i - y \to 0$ for every $y \in \mathfrak{m}$, we have

$$\begin{aligned}\limsup\|x - xu_i\| &= \limsup\|x - xu_i + yu_i - y\| \\ &= \limsup\|(x + y)(1 - u_i)\| \\ &\leq \|x + y\|.\end{aligned}$$

Hence we get

$$\begin{aligned}\|\dot{x}\| &= \inf\{\|x + y\| : y \in \mathfrak{m}\} \\ &\geq \limsup\|x - xu_i\| \geq \liminf\|x - xu_i\| \\ &\geq \inf\{\|x + y\| : y \in \mathfrak{m}\} = \|\dot{x}\|,\end{aligned}$$

so that

$$\|\dot{x}\| = \lim\|x - xu_i\|, \qquad x \in A.$$

Therefore, we have, for every $z \in \mathfrak{m}$,

$$\begin{aligned}\|\dot{x}\|^2 &= \lim\|x - xu_i\|^2 = \lim\|(x - xu_i)^*(x - xu_i)\| \\ &= \lim\|(1 - u_i)x^*x(1 - u_i)\| \\ &= \lim\|(1 - u_i)x^*x(1 - u_i) + (1 - u_i)z^*z(1 - u_i)\| \\ &= \lim\|(1 - u_i)(x^*x + z)(1 - u_i)\| \\ &\leq \|x^*x + z\|,\end{aligned}$$

so that

$$\|\dot{x}\|^2 \leq \inf\{\|x^*x + z\| : z \in \mathfrak{m}\} = \|\dot{x}^*\dot{x}\|,$$

as we wanted to show. Q.E.D.

Corollary 8.2. *If π is a *-homomorphism of a C^*-algebra A onto another C^*-algebra B, then π induces canonically a *-isomorphism $\dot{\pi}$ of the quotient C^*-algebra $A/\pi^{-1}(0)$ of A by the kernel $\pi^{-1}(0)$ onto B.*

PROOF. By Proposition 5.2, π is continuous, so that the kernel $\pi^{-1}(0)$ is a closed ideal of A. Hence by Theorem 8.1, the quotient algebra $A/\pi^{-1}(0)$ is a C^*-algebra. It is obvious that π induces a *-isomorphism $\dot{\pi}$ of $A/\pi^{-1}(0)$ onto B by $\dot{\pi}(x + \pi^{-1}(0)) = \pi(x)$, $x \in A$. Q.E.D.

Proposition 8.3. *Let A be an abelian C^*-algebra with spectrum Ω.*

(i) *There exists a bijective inclusion reversing correspondence between closed ideals $\mathfrak{m}$ of A and closed subsets Γ of Ω which is given by*

$$\begin{aligned}\mathfrak{m}_\Gamma &= \{x \in A : \omega(x) = 0 \text{ for every } \omega \in \Gamma\}, \\ \Gamma_\mathfrak{m} &= \{\omega \in \Omega : \omega(x) = 0 \text{ for every } x \in \mathfrak{m}\}.\end{aligned}$$

(ii) *The spectrum of the quotient abelian C^*-algebra $A/\mathfrak{m}$ is homeomorphic to the closed subset $\Gamma_\mathfrak{m}$, and the spectrum of $\mathfrak{m}$ is homeomorphic to $\Gamma_\mathfrak{m}^c$.*

PROOF. Let Γ be a closed subset of Ω. Put $\mathfrak{m}_\Gamma = \{x \in A : \omega(x) = 0$ for every $\omega \in \Gamma\}$. Clearly $\mathfrak{m}_\Gamma$ is a closed ideal of A. For any closed ideal $\mathfrak{m}$ of A, put $\Gamma_\mathfrak{m} = \{\omega \in \Omega : \omega(x) = 0$ for every $x \in \mathfrak{m}\}$. It is also clear that $\Gamma_\mathfrak{m}$ is a closed subset of Ω. By definition, we know that

$$\mathfrak{m}_{\Gamma_{\mathfrak{m}}} \supset \mathfrak{m}_{\Gamma} \quad \text{and} \quad \Gamma_{\mathfrak{m}_\Gamma} \supset \Gamma.$$

Suppose $\mathfrak{m}$ is given first. Let π be the canonical homomorphism of A onto the quotient algebra $A/\mathfrak{m}$. If $x_0 \notin \mathfrak{m}$, then $\pi(x_0) \neq 0$. Let ω' be a character of $A/\mathfrak{m}$ with $\omega'(\pi(x_0)) \neq 0$. Put $\omega(x) = \omega'(\pi(x))$, $x \in A$. It follows that ω is a character, $\omega(x_0) \neq 0$, and $\omega(x) = 0$ for every $x \in \mathfrak{m}$. Hence $\omega \in \Gamma_{\mathfrak{m}}$ and $\omega(x_0) \neq 0$. Hence $x_0 \notin \mathfrak{m}_{\Gamma_{\mathfrak{m}}}$. Therefore, we obtain the reversed inclusion $\mathfrak{m}_{\Gamma_{\mathfrak{m}}} \subset \mathfrak{m}$.

Now, suppose Γ is given first. Let $\omega_0 \in \Gamma^c$. Let U be a compact neighborhood of ω_0 with $U \cap \Gamma = \varnothing$. By Urysohn's lemma, there exists a continuous function f on Ω such that $f(U^c) = 0$ and $f(\omega_0) = 1$. It follows that $f(\Gamma) = 0$. Let x be the element of A corresponding to f by the Gelfand representation. We have then $\omega(x) = 0$ for every $\omega \in \Gamma$, so that $x \in \mathfrak{m}_\Gamma$; but $\omega_0(x) \neq 0$. Therefore, we get $\Gamma \supset \Gamma_{\mathfrak{m}_\Gamma}$. Thus $\Gamma = \Gamma_{\mathfrak{m}_\Gamma}$.

To show assertion (ii), we consider the transpose ${}^t\pi$ of A which maps isometrically the conjugate space $(A/\mathfrak{m})^*$ of $A/\mathfrak{m}$ onto the annihilator space $\mathfrak{m}^0$ of $\mathfrak{m}$ in A^*. Clearly, ${}^t\pi$ is a homeomorphism with respect to the weak* topology. Furthermore, we know ${}^t\pi(\Omega(A/\mathfrak{m})) \subset \Omega$. It is easy to check that ${}^t\pi(\omega')$, $\omega' \in (A/\mathfrak{m})^*$, is a character if and only if ω' is. Therefore, noticing that $\Omega \cap \mathfrak{m}^0 = \Gamma_{\mathfrak{m}}$, we conclude that ${}^t\pi$ sets up a homeomorphism of the spectrum of $A/\mathfrak{m}$ onto $\Gamma_{\mathfrak{m}}$.

For each $\omega \in \Omega$, consider the restriction $\omega|_{\mathfrak{m}} = I(\omega)$, which is a character of $\mathfrak{m}$ or zero. By definition, $\omega|_{\mathfrak{m}} \neq 0$ if and only if $\omega \in \Gamma_{\mathfrak{m}}^c$. Hence the map I is a continuous map of $\Gamma_{\mathfrak{m}}^c$ into the spectrum $\Omega(\mathfrak{m})$ of $\mathfrak{m}$. If ω_1 and ω_2 are two distinct points of $\Gamma_{\mathfrak{m}}^c$, then there exists an $f \in C_\infty(\Omega)$ such that $f(\omega_1) \neq f(\omega_2)$ and $f(\Gamma_{\mathfrak{m}}) = 0$ because $\Gamma_{\mathfrak{m}}^c$ is open in Ω. The element $x \in A$ corresponding to f belongs to $\mathfrak{m}$ by assertion (i) and separates ω_1 and ω_2. Therefore, the map I is injective. Let ω be a character of $\mathfrak{m}$. Take an element $e \in \mathfrak{m}$ with $\omega(e) = 1$. Define a functional $\bar{\omega}$ on A by $\bar{\omega}(x) = \omega(ex)$. This makes sense because $eA \subset \mathfrak{m}$. Clearly $\bar{\omega}$ is linear. For any $x, y \in A$, we have

$$\begin{aligned}\bar{\omega}(xy) &= \omega(e(xy)) = \omega(e(xy))\omega(e) = \omega(e(xy)e) \\ &= \omega((ex)(ey)) = \omega(ex)\omega(ey) = \bar{\omega}(x)\bar{\omega}(y).\end{aligned}$$

Hence $\bar{\omega}$ is a character of A. For any $x \in \mathfrak{m}$, we have

$$\bar{\omega}(x) = \omega(ex) = \omega(e)\omega(x) = \omega(x),$$

so that $I(\bar{\omega}) = \omega$. Thus, the map I is surjective. Let U be a compact neighborhood of an $\omega_0 \in \Gamma_{\mathfrak{m}}^c$ with $U \subset \Gamma_{\mathfrak{m}}^c$. By the Gelfand representation, there exists an element $x \in \mathfrak{m}$ such that $\omega_0(x) = 1$ and $\omega(x) = 0$ for every $\omega \in U^c$, which means that $I(U)$ is a compact neighborhood of $I(\omega_0)$ in $\Omega(\mathfrak{m})$. Thus, I is a homeomorphism of $\Gamma_{\mathfrak{m}}^c$ onto $\Omega(\mathfrak{m})$. Q.E.D.

The ideal $\mathfrak{m}_\Gamma$ is called the *kernel* of Γ and $\Gamma_{\mathfrak{m}}$ is called the *hull* of $\mathfrak{m}$.

Corollary 8.4. *Let π be a *-homomorphism of a C^*-algebra A into another C^*-algebra B. If a is a normal element of A and f is a continuous function on $\mathrm{Sp}'_A(a)$ with $f(0) = 0$, then*

(i) $\mathrm{Sp}'_A(a) \supset \mathrm{Sp}'_B(\pi(a))$;
(ii) $f(\pi(a)) = \pi(f(a))$.

PROOF. Considering the restriction of π to the abelian C^*-algebra generated by a, we may assume that A is abelian. It follows that $\pi(A)$ is an abelian C^*-subalgebra of B. By Proposition 4.8, the spectrum of an element and the functional calculus do not depend on C^*-subalgebras, so we may assume that $\pi(A) = B$; hence B is abelian. Let Ω be the spectrum of A. We then identify A and $C_\infty(\Omega)$ by the Gelfand representation, Theorem 4.4. By the previous proposition, B is considered as $C_\infty(\Gamma)$ where Γ is a closed subset of Ω and π as the restriction map: $x \in C_\infty(\Omega) \mapsto x|_\Gamma \in C_\infty(\Gamma)$. Since $\mathrm{Sp}'_A(x) = x(\Omega) \cup \{0\}$, $x \in C_\infty(\Omega)$, and $\mathrm{Sp}'_B(y) = y(\Gamma) \cup \{0\}$, $y \in C_\infty(\Gamma)$, we have $\mathrm{Sp}'_B(\pi(a)) \subset \mathrm{Sp}'_A(a)$. Since $f(a)$ is nothing else but the function in $C_\infty(\Omega)$: $\omega \in \Omega \mapsto f \circ a(\omega) \in \mathbf{C}$, it is obvious that $\pi(f(a)) = f(\pi(a))$. Q.E.D.

Proposition 8.5. *If π is a *-homomorphism of a C^*-algebra A onto another C^*-algebra B, then for any fixed $a \in A_+$, we have*

$$\pi(\{x \in A : x^*x \leq a\}) = \{y \in B : y^*y \leq \pi(a)\}$$

PROOF. Put $b = \pi(a)$. Suppose y is an element of B with $y^*y \leq b$. Take an $h \in A$ with $\pi(h) = y^*y$. Replacing h by $\frac{1}{2}(h + h^*)$, we assume that h is self-adjoint. Let $a - h = (a - h)_+ - (a - h)_-$ be the Jordan decomposition of $a - h$. We then have

$$0 \leq b - y^*y = \pi(a - h) = \pi((a - h)_+) - \pi((a - h)_-),$$
$$\pi((a - h)_+)\pi((a - h)_-) = \pi((a - h)_+(a - h)_-) = 0.$$

Hence $\pi((a - h)_-) = 0$, that is, $b - y^*y = \pi((a - h)_+)$. Replacing h again by $a - (a - h)_+$, we assume that $h \leq a$ and $\pi(h) = y^*y$. Choose $z \in A$ with $\pi(z) = y$, and put $k = z^*z - h$. It follows that $\pi(k) = \pi(z^*z) - \pi(h) = y^*y - y^*y = 0$. Hence $\pi(k) = 0$, that is, k belongs to $\pi^{-1}(0)$; thus its positive and negative parts k_+ and k_- belong to $\pi^{-1}(0)$, respectively. Thus $\pi(k_+) = \pi(k_-) = 0$. For each $t > 0$, put

$$u_t = (a + k_+ + t)^{-1}(a + k_+) \quad \text{and} \quad x_t = z(a + k_+ + t)^{-1}(a + k_+)^{1/2}a^{1/2}.$$

We then have

$$z^*z = h + k \leq a + k \leq a + k_+;$$

and for any $s,t > 0$,

$$\begin{aligned}(x_s - x_t)^*(x_s - x_t) &= a^{1/2}(a + k_+)^{1/2}[(s + a + k_+)^{-1} - (t + a + k_+)^{-1}]z^*z \\ &\quad \times [(s + a + k_+)^{-1} - (t + a + k_+)^{-1}](a + k_+)^{1/2}a^{1/2} \\ &\leq [a^{1/2}(u_s - u_t)][a^{1/2}(u_s - u_t)]^*.\end{aligned}$$

But $\{u_s : s > 0\}$ is a right approximate identity for the left ideal of A generated by $a + k_+$ by Lemma 7.2. Since $a \leq a + k_+$, $\{u_s\}$ is also a right approximate identity for $a^{1/2}$. Hence we have

$$\|x_s - x_t\|^2 \le \|a^{1/2}(u_s - u_t)\|^2 \to 0 \quad \text{as} \quad s,t \to 0.$$

Put $x = \lim_{s\to 0} x_s$. We have

$$x_s^* x_s \le a^{1/2} u_s^2 a^{1/2} \le a, \qquad s > 0,$$

so that $x^*x \le a$. On the other hand, we have

$$\pi(x) = \lim_{s\to 0} \pi(x_s) = \lim_{s\to 0} y(s + b)^{-1} b = y$$

by Corollary 7.3. Q.E.D.

EXERCISES

1. Let A be a unital C^*-algebra. Let $\mathfrak{m}$ be a closed ideal of A and π denote the canonical homomorphism of A onto $A/\mathfrak{m}$. Show that if u is a unitary element in $A/\mathfrak{m}$ with $\mathrm{Sp}_{A/\mathfrak{m}}(\dot{u}) \neq \{\lambda \in \mathbf{C} : |\lambda| = 1\}$, then there exists a unitary element $u \in A$ such that $\dot{u} = \pi(u)$.

2. Let A be a C^*-algebra and $\mathfrak{m}$ a closed ideal. Show that if B is a C^*-subalgebra of A then $B + \mathfrak{m}$ is a C^*-subalgebra of A. (Of course, the problem is to show the closedness of $B + \mathfrak{m}$.)

3. Let Ω be the unit disk $\{\lambda \in \mathbf{C} : |\lambda| \le 1\}$ and Γ be the unit circle $\{\lambda \in \mathbf{C} : |\lambda| = 1\}$. Let A be the C^*-algebra $C(\Omega)$ and $\mathfrak{m}$ the closed ideal of A corresponding to Γ by Proposition 8.3. Show that there exists a unitary element $\dot{u}$ in $A/\mathfrak{m}$ which is not the image of any unitary element in A under π.

4. Let $\mathfrak{m}$ and $\mathfrak{n}$ be closed ideals of a C^*-algebra A. Show that $\mathfrak{m} + \mathfrak{n}$ is a closed ideal of A and that $(\mathfrak{m} + \mathfrak{n})_+ = \mathfrak{m}_+ + \mathfrak{n}_+$. (*Hint*: Apply Proposition 8.5.)

5. Let A be an abelian C^*-algebra and $\mathfrak{m}$ a closed ideal of A. Let π be the canonical homomorphism of A onto $A/\mathfrak{m}$.
   (a) Show that if A is separable then every self-adjoint element k in $A/\mathfrak{m}$ is of the form $k = \pi(h)$ for some self-adjoint element h such that $\|k\| = \|h\|$. Show further that if $k \ge 0$, then h can be so chosen that $h \ge 0$. (*Hint*: Use the Tietze–Urysohn extension theorem for a metric space. The spectrum $\Omega(A)$ of A is metrizable on account of separability.)
   (b) Show that the separability assumption for A in (a) can be lifted. (*Hint*: Choose an element $h' \in A$ with $k = \pi(h')$, then consider the C^*-subalgebra B generated by h' which is separable. Then apply (a) to $k \in \pi(B)$ and B.)

9. Representations and Positive Linear Functionals

Definition 9.1. Let A be an involutive Banach algebra. A *representation* of A is a $*$-homomorphism π of A into the C^*-algebra $\mathscr{L}(\mathfrak{H})$ of all bounded operators on a Hilbert space $\mathfrak{H}$. The Hilbert space $\mathfrak{H}$ is called the *representation space* of π. In order to specify the representation space together with a representation, we write $\{\pi,\mathfrak{H}\}$ or $\mathfrak{H}_\pi$. Two representations $\{\pi_1,\mathfrak{H}_1\}$ and $\{\pi_2,\mathfrak{H}_2\}$ of A are said to be *unitarily equivalent* if there exists an isometry U of $\mathfrak{H}_1$ onto $\mathfrak{H}_2$ such that $U\pi_1(x)U^* = \pi_2(x)$, $x \in A$; we write this fact as $\{\pi_1,\mathfrak{H}_1\} \cong \{\pi_2,\mathfrak{H}_2\}$ or $\pi_1 \cong \pi_2$. If $\pi(x) \neq 0$ for every nonzero $x \in A$, then π is called *faithful.*

Proposition 9.2. *Let $\{\pi,\mathfrak{H}\}$ be a representation of an involutive Banach algebra A. The following statements are then equivalent:*

(i) *The closed subspace $[\pi(A)\mathfrak{H}]$ spanned by $\pi(a)\xi$, $a \in A$, $\xi \in \mathfrak{H}$, coincides with the whole space $\mathfrak{H}$.*
(ii) *For any nonzero $\xi \in \mathfrak{H}$, there exists an element $a \in A$ with $\pi(a)\xi \neq 0$.*

Proof. Suppose that (i) holds, and that $\pi(a)\xi = 0$ for every $a \in A$. For any $\eta \in \mathfrak{H}$, we have

$$(\pi(a)\eta|\xi) = (\eta|\pi(a)^*\xi) = (\eta|\pi(a^*)\xi) = 0.$$

Hence ξ is orthogonal to $[\pi(A)\mathfrak{H}]$. By assumption, this means $\xi = 0$. Hence (ii) follows.

Conversely, suppose that (ii) holds. Let ξ be a vector of $\mathfrak{H}$ orthogonal to $[\pi(A)\mathfrak{H}]$. We then have

$$0 = (\xi|\pi(a^*a)\xi) = (\xi|\pi(a^*)\pi(a)\xi) = (\pi(a)\xi|\pi(a)\xi), \qquad a \in A,$$

so that $\|\pi(a)\xi\|^2 = 0$ for every $a \in A$. By assumption, $\xi = 0$. Thus (i) follows.
Q.E.D.

Definition 9.3. A representation $\{\pi,\mathfrak{H}\}$ of A is said to be *proper* or *nondegenerate* if either (i) or (ii) of the previous proposition holds. Otherwise, the closed subspace $[\pi(A)\mathfrak{H}]$ is called the *essential* space of π and denoted by $\mathfrak{H}(\pi)$.

We mainly consider nondegenerate representations, so we mean by a representation a nondegenerate one unless there is danger of confusion.

Definition 9.4. A linear functional ω on an involutive Banach algebra A is called *positive* if $\omega(x^*x) \geq 0$ for every $x \in A$. A positive linear functional of norm one is called a *state.* If $\omega(x^*x) \neq 0$ for every nonzero $x \in A$, then ω is said to be *faithful.*

In general, for a linear functional f on an involutive Banach algebra A, the *adjoint* functional f^* of f is defined by

$$f^*(x) = \overline{f(x^*)}, \qquad x \in A.$$

If $f = f^*$, then f is said to be *self-adjoint* or *hermitian.* For a pair of hermitian linear functionals f, g, we write $f \geq g$ if $f - g$ is positive.

Let $\{\pi,\mathfrak{H}\}$ be a representation of A. For any pair ξ, η in $\mathfrak{H}$, we define a functional $\omega(\pi;\xi,\eta)$ by

$$\langle x,\omega(\pi;\xi,\eta)\rangle = (\pi(x)\xi|\eta), \qquad x \in A.$$

It is then obvious that

$$\omega(\pi;\xi,\eta)^* = \omega(\pi;\eta,\xi),$$
$$\omega(\pi;\xi,\xi) \geq 0.$$

The positive linear functional $\omega(\pi,\xi,\xi)$ is often abbreviated as $\omega(\pi;\xi)$.

Proposition 9.5. *If ω is a positive linear functional on an involutive Banach algebra A, then*

$$\omega(y^*x) = \overline{\omega(x^*y)}, \tag{1}$$
$$|\omega(y^*x)|^2 \leq \omega(x^*x)\omega(y^*y), \qquad x,y \in A. \tag{2}$$

Proof. Equation (1) follows from the polarizations

$$4y^*x = \sum_{n=0}^{3} i^n(x + i^n y)^*(x + i^n y),$$

$$4xy^* = \sum_{n=0}^{3} i^n(x + i^n y)(x + i^n y)^*.$$

Inequality (2) follows from the fact that

$$\begin{aligned} 0 &\leq \omega((\lambda x + \mu y)^*(\lambda x + \mu y)) \\ &= |\lambda|^2\omega(x^*x) + 2\mathrm{Re}\ \lambda\mu\omega(x^*y) + |\mu|^2\omega(y^*y) \end{aligned}$$

for every pair λ,μ in $\mathbf{C}$. Q.E.D.

Inequality (2) is called the *Cauchy–Schwarz inequality*. If A is unital, then (1) and (2) imply the following immediately:

$$\omega(x^*) = \overline{\omega(x)}, \tag{3}$$
$$|\omega(x)|^2 \leq \omega(1)\omega(x^*x), \qquad x \in A. \tag{4}$$

Lemma 9.6. *If ω is a positive linear functional on A, then the set $N_\omega = \{x \in A : \omega(x^*x) = 0\}$ is a left ideal.*

Proof. Let x and y be elements of N_ω. We then have

$$\begin{aligned} \omega((x + y)^*(x + y)) &= \omega(x^*x + x^*y + y^*x + y^*y) \\ &= 2\mathrm{Re}\ \omega(x^*y) \leq 2\omega(x^*x)^{1/2}\omega(y^*y)^{1/2} = 0; \end{aligned}$$

hence $x + y$ falls in N_ω. If a is an element of A, then

$$\begin{aligned} \omega((ax)^*(ax)) &= \omega(x^*a^*ax) \\ &\leq \omega(x^*x)^{1/2}\omega((a^*ax)^*(a^*ax))^{1/2} = 0; \end{aligned}$$

so ax falls in N_ω. The replacement of a by a scalar in the above computation shows that $\mathbf{C}N_\omega = N_\omega$. Hence N_ω is a left ideal. Q.E.D.

Definition 9.7. The left ideal N_ω is called the *left kernel* of ω. Similarly, we can define the *right kernel* of ω.

Suppose now a positive linear functional ω on an involutive Banach algebra A is given. For each $x \in A$, let $\eta_\omega(x)$ denote the coset $x + N_\omega$ in the quotient space A/N_ω. We equip the complex vector space A/N_ω with the inner product defined by

$$(\eta_\omega(x)|\eta_\omega(y)) = \omega(y^*x), \qquad x, y \in A. \tag{5}$$

We denote by $\mathfrak{H}_\omega$ the Hilbert space obtained as the completion of A/N_ω. We shall see soon that the linear operator: $\eta_\omega(x) \in A/N_\omega \mapsto \eta_\omega(ax) \in A/N_\omega$ for each $a \in A$ is extended to a bounded operator $\pi_\omega(a)$ on the Hilbert space $\mathfrak{H}_\omega$, and that the map $\pi_\omega : a \in A \mapsto \pi_\omega(a) \in \mathscr{L}(\mathfrak{H}_\omega)$ is indeed a representation of A. To do this, we need a few preparations.

Lemma 9.8. *Let A be a unital Banach algebra. If a is an element of A with $\|1 - a\|_{\mathrm{sp}} < 1$, then there exists $b \in A$ with $b^2 = a$. Furthermore, if A is an involutive Banach algebra and if a is hermitian, then a self-adjoint element can be chosen as the above b.*

Proof. By assumption, $\mathrm{Sp}_A(a)$ is contained in the open disk $D = \{\lambda \in \mathbf{C} : |\lambda - 1| < 1\}$. Let $f(\lambda)$ be the analytic continuation of the function $\sqrt{\lambda}$ from the open interval (0,1) to the disk D. Put $f(a) = b$. By Proposition 2.7, we get $a = b^2$. Since $f(\lambda)$ has a Taylor series expansion around 1 with real coefficients converging in the disk D, $b = f(a)$ is self-adjoint if a is. Q.E.D.

Lemma 9.9. *If A is a unital involutive Banach algebra, then every positive linear functional ω of A is continuous and $\|\omega\| = \omega(1)$.*

Proof. If $x \in A$ is self-adjoint and $\|x\| < 1$, then $1 - x$ is of the form y^*y by Lemma 9.8, so that

$$\omega(1) - \omega(x) = \omega(1 - x) = \omega(y^*y) \geq 0.$$

If $\|x\| < 1$, then $\|x^*x\| \leq \|x\|^2 < 1$; hence

$$|\omega(x)|^2 \leq \omega(1)\omega(x^*x) \leq \omega(1)^2.$$

Therefore, we get $\|\omega\| \leq \omega(1)$. Clearly, $\|\omega\| \geq \omega(1)$ since $\|1\| = 1$. Q.E.D.

Lemma 9.10. *Let A be an involutive Banach algebra and ω a positive linear functional on A. For each $a \in A$, we set $\omega_a(x) = \omega(axa^*)$, $x \in A$. Then ω_a is a continuous positive linear functional and $\|\omega_a\| \leq \omega(aa^*)$.*

Proof. Let A_{I} be the unital involutive Banach algebra obtained by adjunction of an identity. Putting $\tilde{\omega}_a(x) = \omega(axa^*)$ for each $x \in \tilde{A}$, we get a linear functional $\tilde{\omega}_a$ on A_{I}, where we should note that A is an ideal of A_{I}, so that the definition of $\tilde{\omega}_a$ makes sense. Since $\tilde{\omega}_a(x^*x) = \omega(ax^*xa^*) = \omega((xa^*)^*(xa)) \geq 0$ for every $x \in A_{\mathrm{I}}$, $\tilde{\omega}_a$ is positive on A_{I}; hence it is continuous by Lemma 9.9 and $\|\tilde{\omega}_a\| = \tilde{\omega}_a(1) = \omega(aa^*)$. Thus, we get $\|\omega_a\| \leq \|\tilde{\omega}_a\| = \omega(aa^*)$. Q.E.D.

Lemma 9.11. *Let A be an involutive Banach algebra with a bounded approximate identity $\{u_i\}$ of norm $\leq \gamma$. If ω is a continuous positive linear functional on A, then*

(i) $\omega(x^*) = \overline{\omega(x)}$,
(ii) $|\omega(x)|^2 \leq \gamma^2\|\omega\|\omega(x^*x), \qquad x \in A.$

PROOF. For each $x \in A$, we have

$$\begin{aligned}\omega(x^*) &= \lim \omega(x^*u_i) = \lim \omega(u_i^*x)^- \\ &= \lim \omega((x^*u_i)^*)^- = \omega(x^*)^- \\ &= \overline{\omega(x)},\end{aligned}$$

$$\begin{aligned}|\omega(x)|^2 &= \lim |\omega(u_i^*x)|^2 \leq \limsup \omega(u_i^*u_i)\omega(x^*x) \\ &\leq \gamma^2\|\omega\|\omega(x^*x).\end{aligned}$$

Q.E.D.

Proposition 9.12. *A positive linear functional on a C^*-algebra is continuous.*

PROOF. Let ω be a positive linear functional on a C^*-algebra A. Let S denote the unit ball of A. Let $\{x_n\}$ be a sequence in $A_+ \cap S$. We shall show that $\{\omega(x_n)\}$ is bounded. To this end, take an arbitrary summable sequence $\{\lambda_n\}$ with $\lambda_n \geq 0$. It follows that the series $\sum_{k=1}^{\infty} \lambda_k x_k$ converges to an element $x \in A_+$. Since $\sum_{k=1}^{n} \lambda_k x_k \leq x$, we have

$$\sum_{k=1}^{n} \lambda_k \omega(x_k) = \omega\left(\sum_{k=1}^{n} \lambda_k x_k\right) \leq \omega(x).$$

Hence $\sum_{k=1}^{\infty} \lambda_k \omega(x_k) < +\infty$. This is true for every positive summable sequence $\{\lambda_k\}$, so that the sequence $\{\omega(x_k)\}$ is bounded. It follows that $M = \sup\{\omega(x) : x \in A_+ \cap S\} < +\infty$. If $x \in A_h \cap S$, then we have

$$|\omega(x)| \leq \omega(x_+) + \omega(x_-) \leq 2M,$$

where $x = x_+ - x_-$ is the Jordan decomposition of x. Therefore, we get, for every $x \in S$,

$$|\omega(x)| \leq |\omega(\tfrac{1}{2}(x + x^*)) + \left|\omega\left(\frac{1}{2i}(x - x^*)\right)\right| \leq 4M.$$

Hence $\|\omega\| \leq 4M$. Q.E.D.

Remark 9.13. More generally, a positive linear functional on an involutive Banach algebra with a bounded approximate identity is continuous. But the proof of this fact requires further argument; see [3, 396]. We shall instead assume the continuity of positive linear functionals in question.

Theorem 9.14. *Let A be an involutive Banach algebra with a bounded approximate identity. To any (continuous) positive linear functional ω, there corresponds uniquely, within unitary equivalence, a representation $\{\pi_\omega, \mathfrak{H}_\omega\}$ of A*

with a vector ξ_ω *such that*

(i) $[\pi_\omega(A)\xi_\omega] = \mathfrak{H}_\omega$,

(ii) $\omega(x) = (\pi_\omega(x)\xi_\omega|\xi_\omega)$, $x \in A$.

PROOF. *Uniqueness*: Let $\{\pi'_\omega, \mathfrak{H}'_\omega, \xi'_\omega\}$ be another representation of A with properties (i) and (ii). Define a map U_0 of $\pi'_\omega(A)\xi'_\omega$ onto $\pi_\omega(A)\xi_\omega$ by

$$U_0\pi'_\omega(x)\xi'_\omega = \pi_\omega(x)\xi_\omega, \qquad x \in A.$$

We then have

$$\begin{aligned}(U_0\pi'_\omega(x)\xi'_\omega|U_0\pi'_\omega(y)\xi'_\omega) &= (\pi_\omega(x)\xi_\omega|\pi_\omega(y)\xi_\omega)\\ &= (\pi_\omega(y)^*\pi_\omega(x)\xi_\omega|\xi_\omega) = (\pi_\omega(y^*x)\xi_\omega|\xi_\omega)\\ &= \omega(y^*x) = (\pi'_\omega(y^*x)\xi'_\omega|\xi'_\omega)\\ &= (\pi'_\omega(x)\xi'_\omega|\pi'_\omega(y)\xi'_\omega).\end{aligned}$$

Hence U_0 is well defined and an isometry of $\pi'(A)\xi'$ onto $\pi(A)\xi$, so that it is extended to an isometry U of $\mathfrak{H}'_\omega$ onto $\mathfrak{H}_\omega$ since the range and the domain of U_0 are both dense in $\mathfrak{H}_\omega$ and $\mathfrak{H}'_\omega$, respectively. For any pair $x,y \in A$, we have

$$\begin{aligned}\pi_\omega(x)U_0\pi'_\omega(y)\xi'_\omega &= \pi_\omega(x)\pi_\omega(y)\xi_\omega = \pi_\omega(xy)\xi_\omega\\ &= U_0\pi'_\omega(xy)\xi'_\omega = U_0\pi'_\omega(x)\pi'_\omega(y)\xi'_\omega;\end{aligned}$$

hence $\pi_\omega(x)U = U\pi'_\omega(x)$, $x \in A$. Therefore, U sets up the unitary equivalence of π_ω and π'_ω.

Existence: Let N_ω be the left kernel of ω and $\mathfrak{H}_\omega$ the completion of the pre-Hilbert space A/N_ω with respect to the inner product defined by (5). For each $a \in A$, set

$$\pi^0_\omega(a)\eta_\omega(x) = \eta_\omega(ax), \qquad x \in A. \tag{6}$$

For any a, x and $y \in A$, we have, by Lemma 9.10,

$$\begin{aligned}|(\pi^0_\omega(a)\eta_\omega(x)|\eta_\omega(y))| &= |\omega(y^*ax)|\\ &\leq \omega(y^*y)^{1/2}\omega(x^*a^*ax)^{1/2}\\ &= \|\eta_\omega(y)\|\omega_{x^*}(a^*a)^{1/2} \leq \|\eta_\omega(y)\|\,\|a^*a\|^{1/2}\omega(x^*x)^{1/2}\\ &\leq \|a\|\,\|\eta_\omega(x)\|\,\|\eta_\omega(y)\|,\end{aligned}$$

so that $\pi^0_\omega(a)$ is extended to a bounded operator $\pi_\omega(a)$ on $\mathfrak{H}_\omega$. It is now routine to check that the map $\pi_\omega : a \in A \mapsto \pi_\omega(a) \in \mathscr{L}(\mathfrak{H}_\omega)$ is a *-homomorphism.

By Lemma 9.11(ii), we have

$$|\omega(x)| \leq \gamma\|\omega\|^{1/2}\|\eta_\omega(x)\|, \qquad x \in A,$$

where γ means the bound of an approximate identity of A as in the lemma. Hence ω is extended to a bounded linear functional on the Hilbert space $\mathfrak{H}_\omega$. By the Riesz theorem, there exists a unique vector $\xi_\omega \in \mathfrak{H}_\omega$ such that

$$\omega(x) = (\eta_\omega(x)|\xi_\omega), \qquad x \in A. \tag{7}$$

We then get, for any $x,y \in A$,

$$(\eta_\omega(x)|\eta_\omega(y)) = \omega(y^*x) = (\eta_\omega(y^*x)|\xi_\omega)$$
$$= (\pi_\omega(y^*)\eta_\omega(x)|\xi_\omega) = (\eta_\omega(x)|\pi_\omega(y)\xi_\omega);$$

hence

$$\eta_\omega(x) = \pi_\omega(x)\xi_\omega,$$
$$\omega(x) = (\eta_\omega(x)|\xi_\omega) = (\pi_\omega(x)\xi_\omega|\xi_\omega), \qquad x \in A. \qquad \text{Q.E.D.}$$

Definition 9.15. The representation $\{\pi_\omega,\mathfrak{H}_\omega\}$ constructed in the above theorem is called *the cyclic representation* of A *induced* by ω. Sometimes, it is denoted by $\{\pi_\omega,\mathfrak{H}_\omega,\xi_\omega\}$ to indicate the vector corresponding to ω. The construction of $\{\pi_\omega,\mathfrak{H}_\omega,\xi_\omega\}$ employed above is called *the Gelfand–Naimark–Segal construction.* In general, if a representation $\{\pi,\mathfrak{H}\}$ of A admits a vector ξ such that $[\pi(A)\xi] = \mathfrak{H}$, where $[\mathfrak{M}]$ denotes the closed subspace of $\mathfrak{H}$ spanned by $\mathfrak{M}$ for any subset $\mathfrak{M}$ of $\mathfrak{H}$, then $\{\pi,\mathfrak{H}\}$ is said to be *cyclic* and ξ is called a *cyclic* vector for π.

Let $\{\{\pi_i,\mathfrak{H}_i\}: i \in I\}$ be a family of representations of A. Let $\mathfrak{H}$ be the direct sum Hilbert space $\sum_{i\in I}^\oplus \mathfrak{H}_i$. For each vector $\xi = \sum_{i\in I}^\oplus \xi_i \in \mathfrak{H}$ and $x \in A$, put

$$\pi(x)\xi = \sum_{i\in I}^\oplus \pi_i(x)\xi_i.$$

By virtue of Proposition 5.2, $\pi(x)\xi$ is a vector of $\mathfrak{H}$ and $\pi(x)$ is a bounded operator on $\mathfrak{H}$. It is easy to see that $\{\pi,\mathfrak{H}\}$ is a representation of A. The representation $\{\pi,\mathfrak{H}\}$ is called the *direct sum* of $\{\{\pi_i,\mathfrak{H}_i\}: i \in I\}$ and denoted $\sum_{i\in I}^\oplus \{\pi_i,\mathfrak{H}_i\}$. Each $\{\pi_i,\mathfrak{H}_i\}$ is called a *component* of $\{\pi,\mathfrak{H}\}$. It is obvious that $\{\pi,\mathfrak{H}\}$ is nondegenerate if and only if every $\{\pi_i,\mathfrak{H}_i\}$, $i \in I$, is also.

Definition 9.16. Given a representation $\{\pi,\mathfrak{H}\}$ of A, a closed subspace $\mathfrak{M}$ of $\mathfrak{H}$ is called an *invariant* subspace of $\{\pi,\mathfrak{H}\}$ if $\pi(x)\mathfrak{M} \subset \mathfrak{M}$ for every $x \in A$. In this case, the restriction $\pi(x)|_\mathfrak{M}$ of $\pi(x)$ to $\mathfrak{M}$ gives rise to a new representation of A on $\mathfrak{M}$, which will be denoted by $\pi_\mathfrak{M}$ and called a *subrepresentation* of π.

It is routine to show that the orthogonal complement $\mathfrak{M}^\perp$ of any invariant subspace $\mathfrak{M}$ of $\{\pi,\mathfrak{H}\}$ is also invariant and that $\{\pi,\mathfrak{H}\} \cong \{\pi_\mathfrak{M},\mathfrak{M}\} \oplus \{\pi_\mathfrak{M},\mathfrak{M}^\perp\}$. If $\{\pi,\mathfrak{H}\}$ has no invariant subspace other than $\mathfrak{H}$ and $\{0\}$, then it is said to be *irreducible* (or, more precisely, *topologically irreducible*).

Proposition 9.17. *Every nondegenerate representation $\{\pi,\mathfrak{H}\}$ of an involutive Banach algebra A is a direct sum of cyclic representations.*

PROOF. Let $\mathscr{F}$ denote the family of all subsets F of $\mathfrak{H}$ such that $[\pi(A)\xi]$ and $[\pi(A)\eta]$ are orthogonal for every distinct pair ξ,η in F. It follows that $\mathscr{F}$ is an inductive set under the inclusion ordering. By Zorn's lemma, there exists a maximal member $F = \{\xi_i\}_{i\in I}$ of $\mathscr{F}$. For each $i \in I$, the subspace $\mathfrak{H}_i = [\pi(A)\xi_i]$ is invariant for π, so that $\{\pi_{\mathfrak{H}_i},\mathfrak{H}_i\}$ is a subrepresentation of π. By definition, $\{\pi_{\mathfrak{H}_i},\mathfrak{H}_i\}$ is a cyclic representation of A with cyclic vector ξ_i. The maximality of F yields that $\mathfrak{H} = \sum_{i \in I}^{\oplus} \mathfrak{H}_i$; therefore,

$$\{\pi, \mathfrak{H}\} = \sum_{i \in I}^{\oplus} \{\pi_{\mathfrak{H}_i}, \mathfrak{H}_i\}. \qquad \text{Q.E.D.}$$

Theorem 9.18. *A C^*-algebra admits a faithful representation. Hence it is isometrically isomorphic to a uniformly closed self-adjoint algebra of operators on a Hilbert space.*

This theorem means that postulates (i)–(vi) in Section 1 characterize a uniformly closed self-adjoint algebra of operators on a Hilbert space without referring to a Hilbert space on which the algebra acts. The proof is an easy application of Theorems 6.1 and 9.14.

Proof. Let A be a C^*-algebra. If a is a nonzero element of A, then $-a^*a \notin A_+$. Since A_+ is a closed convex cone in the real Banach space A_{h}, there exists, by the Hahn–Banach theorem, a real linear functional f_a on A_{h} such that $f_a(y) \geq 0$ for every $y \in A_+$ and $f_a(-a^*a) < 0$. We extend f_a to the whole algebra A as follows:

$$f_a(x + iy) = f_a(x) + if_a(y), \qquad x, y \in A_{\mathrm{h}}.$$

It follows that f_a is a positive linear functional with $f_a(a^*a) > 0$. Let $\{\pi_a, \mathfrak{H}_a, \xi_a\}$ be the cyclic representation of A induced by f_a. We then have

$$\|\pi_a(a)\xi_a\|^2 = (\pi_a(a^*a)\xi_a | \xi_a) = f_a(a^*a) > 0,$$

so that $\pi_a(a) \neq 0$. Put

$$\{\pi, \mathfrak{H}\} = \sum_{a \in A - \{0\}}^{\oplus} \{\pi_a, \mathfrak{H}_a\}.$$

It follows that π is faithful. Hence π is a $*$-isomorphism of A onto $\pi(A)$; therefore it is an isometry by Corollary 5.4. Q.E.D.

Definition 9.19. An involutive Banach algebra is called an *A^*-algebra* if it admits a faithful representation.

Of course, a C^*-algebra is an A^*-algebra just as shown. Let A be an A^*-algebra. We define a new norm $\|\cdot\|_*$ in A by $\|x\|_* = \sup\{\|\pi(x)\| : \pi$ runs over all representations of $A\}$, $x \in A$. By virtue of Proposition 5.2, $\|x\|_*$ makes sense and $\|x\|_* \leq \|x\|$, $x \in A$. The completion of $\{A, \|\cdot\|_*\}$ is clearly a C^*-algebra, which will be called the *enveloping* C^*-algebra of A and denoted by $C^*(A)$.

Not having any possibility of a decomposition into a direct sum of subrepresentations, irreducible representations are clearly most fundamental among representations. The rest of this section is devoted to the existence of irreducible representations and their characterization in terms of positive linear functionals.

Proposition 9.20. *If $\{\pi,\mathfrak{H}\}$ is a representation of an involutive Banach algebra A, then the following two conditions are equivalent:*

(i) *$\{\pi,\mathfrak{H}\}$ is irreducible.*
(ii) *Only scalar multiplication operators commute with $\pi(A)$.*

PROOF. (i)$\Rightarrow$(ii): Suppose $x \in \mathscr{L}(\mathfrak{H})$ commutes with $\pi(A)$. We must show that $x = \lambda 1$ for some $\lambda \in \mathbf{C}$. Considering the real part and the imaginary part of x separately, we may assume that x is self-adjoint. All the spectral projections of x commute with $\pi(A)$, so that the range of any spectral projection of x is invariant for the representation π. Hence the irreducibility of π implies that all the spectral projections of x are either 0 or 1. Hence the spectrum of x must be a single point $\{\lambda\}$; therefore, $x = \lambda 1$.

(ii)$\Rightarrow$(i): Suppose $\mathfrak{K}$ is an invariant closed subspace of $\mathfrak{H}$. Let p be the projection of $\mathfrak{H}$ onto $\mathfrak{K}$. The invariance of $\mathfrak{K}$ yields that p commutes with every operator in $\pi(A)$. Hence p is a scalar multiple $\lambda 1$ of the identity. Since p is a projection, i.e., $p^2 = p$, λ must be either 0 or 1. Hence $\mathfrak{K} = \{0\}$ or $\mathfrak{H}$. This means that $\{\pi,\mathfrak{H}\}$ is irreducible. Q.E.D.

Definition 9.21. A positive linear functional φ of an involutive Banach algebra A is called *pure* if every positive linear functional ψ on A, majorized by φ in the sense that $\psi(x^*x) \leq \varphi(x^*x)$, is of the form $\lambda\varphi$, $0 \leq \lambda \leq 1$. We denote the set of all pure states by $P(A)$.

Theorem 9.22. *If φ is a (continuous) positive linear functional on an involutive Banach algebra A with a bounded approximate identity, then the following two statements are equivalent:*

(i) *φ is pure.*
(ii) *The cyclic representation $\{\pi_\varphi,\mathfrak{H}_\varphi,\xi_\varphi\}$ induced by φ is irreducible.*

PROOF. (i)$\Rightarrow$(ii): Let $\mathfrak{K}$ be an invariant closed subspace of $\mathfrak{H}_\varphi$ and p be the projection of $\mathfrak{H}_\varphi$ onto $\mathfrak{K}$. It follows that p commutes with every $\pi_\varphi(x)$, $x \in A$. Putting

$$\omega(x) = (\pi_\varphi(x)p\xi_\varphi | p\xi_\varphi), \qquad x \in A,$$

we obtain a continuous positive linear functional ω on A. Since we have

$$\begin{aligned}\omega(x^*x) &= \|\pi_\varphi(x)p\xi_\varphi\|^2 = \|p\pi_\varphi(x)\xi_\varphi\|^2 \\ &\leq \|\pi_\varphi(x)\xi_\varphi\|^2 = \varphi(x^*x), \qquad x \in A,\end{aligned}$$

ω is majorized by φ, so that $\omega = \lambda\varphi$, $0 \leq \lambda \leq 1$, by assumption. Hence we have, for every $x,y \in A$,

$$\begin{aligned}(\lambda\pi_\varphi(x)\xi_\varphi | \pi_\varphi(y)\xi_\varphi) &= \lambda\varphi(y^*x) = \omega(y^*x) \\ &= (\pi_\varphi(x)p\xi_\varphi | \pi_\varphi(y)p\xi_\varphi) = (p\pi_\varphi(x)\xi_\varphi | p\pi_\varphi(y)\xi_\varphi) \\ &= (p\pi_\varphi(x)\xi_\varphi | \pi_\varphi(y)\xi_\varphi),\end{aligned}$$

which implies that $p = \lambda 1$. Thus $p = 0$ or 1. Therefore π_φ is irreducible.

(ii) ⇒ (i): Suppose ω is a continuous positive linear functional on A majorized by φ. On the dense subspace $\pi_\varphi(A)\xi_\varphi$ of $\mathfrak{H}_\varphi$, define a new inner product by

$$\langle \pi_\varphi(x)\xi_\varphi | \pi_\varphi(y)\xi_\varphi \rangle = \omega(y^*x), \qquad x,y \in A.$$

It follows that the new inner product is majorized by the original one in $\mathfrak{H}_\varphi$, so that the new inner product makes sense, and there exists a bounded positive operator a of norm ≤ 1 on $\mathfrak{H}_\varphi$ such that

$$\langle \xi | \eta \rangle = (a\xi | \eta), \qquad \xi,\eta \in \mathfrak{H}_\varphi.$$

For every $x,y,z \in A$, we have

$$\begin{aligned}(a\pi_\varphi(x)\pi_\varphi(y)\xi_\varphi | \pi_\varphi(z)\xi_\varphi) &= \langle \pi_\varphi(x)\pi_\varphi(y)\xi_\varphi | \pi_\varphi(z)\xi_\varphi \rangle \\ &= \omega(z^*xy) = \omega((x^*z)^*y) \\ &= \langle \pi_\varphi(y)\xi_\varphi | \pi_\varphi(x^*)\pi_\varphi(z)\xi_\varphi \rangle \\ &= (a\pi_\varphi(y)\xi_\varphi | \pi_\varphi(x)^*\pi_\varphi(z)\xi_\varphi) = (\pi_\varphi(x)a\pi_\varphi(y)\xi_\varphi | \pi_\varphi(z)\xi_\varphi).\end{aligned}$$

Hence a commutes with $\pi_\varphi(x)$, $x \in A$. By Proposition 9.20(ii), a must be of the form $\lambda 1$. Therefore, we get

$$\begin{aligned}\omega(y^*x) &= (a\pi_\varphi(x)\xi_\varphi | \pi_\varphi(y)\xi_\varphi) \\ &= \lambda(\pi_\varphi(x)\xi_\varphi | \pi_\varphi(y)\xi_\varphi) = \lambda\varphi(y^*x), \qquad x,y \in A.\end{aligned}$$

Since the set of all y^*x is dense in A by the existence of an approximate identity, we have $\omega = \lambda\varphi$. The inequality $0 \leq \lambda \leq 1$ follows from the fact that $0 \leq a \leq 1$. Q.E.D.

Theorem 9.23. *An A^*-algebra A admits sufficiently many irreducible representations,* i.e., *for any nonzero $x \in A$, there exists an irreducible representation π of A with $\pi(x) \neq 0$. In particular, every C^*-algebra admits sufficiently many irreducible representations.*

Proof. Let B denote the enveloping C^*-algebra $C^*(A)$ of A. Since A is dense in B, the restriction of any irreducible representation of B to A is irreducible. Therefore, it suffices to prove the existence of sufficiently many irreducible representations of B. Let $\mathfrak{S}$ denote the set of all positive linear functionals on B of norm ≤ 1. It follows that $\mathfrak{S}$ is a $\sigma(B^*,B)$-compact convex subset of the conjugate space B^* of B as a Banach space. Extreme points of $\mathfrak{S}$ are either pure states or zero, so that the Krein–Milman theorem says that $\mathfrak{S}$ is the $\sigma(B^*,B)$-closed convex closure of zero and pure states. Therefore, if $\omega(x) = 0$, $x \in B$ for every pure state ω of B, then $\varphi(x) = 0$ for every $\varphi \in \mathfrak{S}$; hence $(\pi(x)\xi | \xi) = 0$ for every representation π and every unit vector $\xi \in \mathfrak{H}_\pi$; so $\pi(x) = 0$ for every π. By Theorem 9.18, $x = 0$. Hence for any nonzero $x \in B$, there exists a pure state ω of B with $\omega(x) \neq 0$. Thus, if π_ω is the cyclic representation of B induced by ω, then $\pi_\omega(x) \neq 0$ and π_ω is irreducible by Theorem 9.22. Therefore, B admits sufficiently many irreducible representations. Q.E.D.

The next obvious question is how one can relate a given representation to irreducible representations. This question leads us to the theory of disintegration of representations and states. We shall explore this later in Chapter IV.

Exercises

1. Let A be a C^*-algebra and $\mathscr{I}$ a closed ideal of A. Show that if $\{\pi,\mathfrak{H}\}$ is an irreducible representation of A, then the restriction of π to $\mathscr{I}$ is either the zero representation or irreducible.

2. Let A be a C^*-algebra and $\{\pi,\mathfrak{H}\}$ be an irreducible representation of A. Show that either $\pi(A) \supset \mathscr{L}\mathscr{C}(\mathfrak{H})$ or $\pi(A) \cap \mathscr{L}\mathscr{C}(\mathfrak{H}) = \{0\}$.

3. Let A be a separable C^*-algebra.
   (a) Show that any cyclic representation of A is separable, where we say that a representation is separable if the Hilbert space, on which the representation acts, is separable.
   (b) Show that A admits a faithful positive linear functional.
   (c) Show that A admits a faithful representation on a separable Hilbert space.

4. Let $L^1(G)$ be as in Exercise 1.1. On the Hilbert space $L^2(G)$ of all square integrable functions on G with respect to the Haar measure, set

$$(\lambda(x)\xi)(t) = \int_G x(s)\xi(s^{-1}t)\,ds, \qquad \xi \in L^2(G), \qquad x \in L^1(G).$$

   (a) Show that the map: $x \in L^1(G) \mapsto \lambda(x) \in \mathscr{L}(L^2(G))$ is a faithful nondegenerate representation of $L^1(G)$. Hence $L^1(G)$ is an A^*-algebra. This representation is called the *left regular representation.* The enveloping C^*-algebra of $L^1(G)$ is called the *group C^*-algebra of* G and denoted by $C^*(G)$. The norm closure of $\lambda(L^1(G))$ in $\mathscr{L}(L^2(G))$ is called the *restricted group C^*-algebra* of G and denoted by $C_r^*(G)$.
   (b) Let $\{U,\mathfrak{H}\}$ be a unitary representation of G, i.e., U is a homomorphism of G into the unitary group on $\mathfrak{H}$ such that for each fixed $\xi \in \mathfrak{H}$ the map: $s \in G \mapsto U(s)\xi \in \mathfrak{H}$ is continuous. Show that there exists a unique representation π_U of $L^1(G)$ on $\mathfrak{H}$ such that

$$(\pi_U(x)\xi|\eta) = \int_G x(s)(U(s)\xi|\eta)\,ds, \qquad x \in L^1(G), \qquad \xi,\eta \in \mathfrak{H}.$$

   (c) Let $\{\pi,\mathfrak{H}\}$ be a representation of $L^1(G)$. Show that there exists a unitary representation U of G on $\mathfrak{H}$ such that $\pi = \pi_U$. (*Hint*: Setting $(\lambda_s x)(t) = x(s^{-1}t)$, $x \in L^1(G)$, define U on the dense subspace $\pi(L^1(G))\mathfrak{H}$ by $U(s)\pi(x)\xi = \pi(\lambda_s x)\xi$; then prove that this definition makes sense and that U is extended to a representation with the required property.)
   (d) Show that the correspondence: $U \leftarrow \pi_U$ preserves the unitary equivalence of representations. For this reason, the representations U and π_U are often identified.

5. Let Ω, G, and A be as in Exercise 1.2. Let $\{\rho,\mathfrak{H}\}$ and $\{U,\mathfrak{H}\}$ be a representation of $C_\infty(\Omega)$ and a unitary representation of G on the same Hilbert space. Suppose that

ρ and U satisfy the equation

$$U(s)\rho(f)U(s)^* = \rho(\alpha_s f), \qquad f \in C_\infty(\Omega), \qquad s \in G, \tag{$*$}$$

where $(\alpha_s f)(\omega) = f(\omega s)$.

(a) Show that there exists a unique representation $\pi_{\rho,U}$ of A on $\mathfrak{H}$ such that

$$(\pi_{\rho,U}(x)\xi|\eta) = \int_G (\rho(x(\cdot,s))U(s)\xi|\eta)\,ds, \qquad x \in \mathscr{K}(\Omega \times G),$$

where $x(\cdot,s)$ means the function in $C_\infty(\Omega)$ given by: $\omega \in \Omega \mapsto x(\omega,s) \in \mathbf{C}$.

(b) Define actions $\bar{\rho}$ of $C_\infty(\Omega)$ and $\bar{\alpha}$ of G on $\mathscr{K}(\Omega \times G)$ by the following:

$$(\bar{\rho}(f)x)(\omega,s) = f(\omega s^{-1})x(\omega,s), \qquad f \in C_\infty(\Omega);$$
$$(\bar{\alpha}_t x)(\omega,s) = x(\omega,t^{-1}s), \qquad t \in G, \qquad x \in \mathscr{K}(\Omega \times G).$$

Let $\{\pi,\mathfrak{H}\}$ be a representation of G. Show that there exists a unique pair (ρ,U) of a representation ρ of $C_\infty(\Omega)$ and a unitary representation U of G on $\mathfrak{H}$ given by

$$\begin{cases} \rho(f)\pi(x)\xi = \pi(\bar{\rho}(f)x)\xi, & f \in C_\infty(G), \quad x \in \mathscr{K}(\Omega \times G), \\ U(s)\pi(x)\xi = \pi(\bar{\alpha}_s x)\xi, & s \in G, \quad \xi \in \mathfrak{H}, \end{cases}$$

which satisfies the Equations $(*)$ and $\pi = \pi_{\rho,U}$.

(c) Let $\{\rho,\mathfrak{H}\}$ be a representation of $C_\infty(\Omega)$. Consider the vector space $\mathscr{K}(\mathfrak{H},G)$ of all $\mathfrak{H}$-valued continuous functions on G with compact support. Define an inner product in $\mathscr{K}(\mathfrak{H},G)$ by

$$(\xi|\eta) = \int_G (\xi(s)|\eta(s))\,ds, \qquad \xi,\eta \in \mathscr{K}(\mathfrak{H},G).$$

Let $L^2(\mathfrak{H},G)$ be the completion of $\mathscr{K}(\mathfrak{H},G)$ with respect to the above inner product. Define actions $\tilde{\rho}$ of $C_\infty(G)$ and U of G on $\mathscr{K}(\mathfrak{H},G)$ by

$$\begin{cases} (\tilde{\rho}(f)\xi)(s) = \rho(\alpha_s^{-1}f)\xi(s), & f \in C_\infty(\Omega), \quad \xi \in \mathscr{K}(\mathfrak{H},G), \\ (U(t)\xi)(s) = \xi(t^{-1}s), & t \in G. \end{cases}$$

Show that $\tilde{\rho}$ and U give rise to a pair of representations of $C_\infty(\Omega)$ and G on $L^2(\mathfrak{H},G)$ satisfying Equation $(*)$, and that $\pi = \pi_{\rho,U}$ is a faithful representation of A if ρ is faithful. Hence A is an A^*-algebra.

(d) The enveloping C^*-algebra $C^*(A)$ of A is called the *covariance C^*-algebra* of $\{\Omega,G\}$ and denoted by $C^*(\Omega,G)$. A pair $\{\rho,U\}$ of a representation ρ of $C_\infty(\Omega)$ on $\mathfrak{H}$ and a unitary representation $\{U,\mathfrak{H}\}$ is said to be *covariant* if condition $(*)$ holds.

6. Consider the action of $\mathbf{R}$ on $\mathbf{R}$ by translation: $s \mapsto s + t$. Define a covariant representation $\{\rho,U\}$ of $C_\infty(\mathbf{R})$ and $\mathbf{R}$ on $L^2(\mathbf{R})$ by

$$\begin{cases} \rho(f)\xi(s) = f(s)\xi(s), & f \in C_\infty(\mathbf{R}), \\ U(t)\xi(s) = \xi(s+t), & s,t \in \mathbf{R}, \quad \xi \in L^2(\mathbf{R}). \end{cases}$$

Show that $\pi_{\rho,U}(x)$ is a compact operator, indeed of Hilbert–Schmidt class, on $L^2(\mathbf{R})$ for every $x \in \mathscr{K}(\mathbf{R} \times \mathbf{R})$, and that $\pi_{\rho,U}$ is irreducible.

7. Let A be a C^*-algebra and G a locally compact group. A homomorphism $\alpha: s \in G \mapsto \alpha_s \in \operatorname{Aut}(A)$ of G into the group $\operatorname{Aut}(A)$ of all automorphisms of A is called an *action* of G on A.

(a) Show that if φ is a state on A such that $\varphi \circ \alpha_s = \varphi$, $s \in G$, and for every $x,y \in A$, the function: $s \in G \mapsto \varphi(y^*\alpha(x))$ is continuous, then there exists a unique, up to unitary equivalence, continuous unitary representation U_φ of G on the Hilbert space $\mathfrak{H}_\varphi$ such that

$$U_\varphi(s)\pi_\varphi(x)U_\varphi(s)^* = \pi_\varphi \circ \alpha_s(s), \qquad x \in A, \qquad s \in G,$$
$$U_\varphi(s)\xi_\varphi = \xi_\varphi,$$

where $\{\pi_\varphi, \mathfrak{H}_\varphi, \xi_\varphi\}$ is the cyclic representation of A induced by the state φ. (*Hint*: Define $U_\varphi(s)$ by $U_\varphi(s)\pi_\varphi(x)\xi_\varphi = \pi_\varphi \circ \alpha_s(x)\xi_\varphi$, $x \in A$, $s \in G$.)

(b) Show that under the same condition as in (a), $\pi_\varphi(A)$ and $U_\varphi(G)$ are irreducible in the sense there is no closed nontrivial invariant subspace of $\mathfrak{H}_\varphi$ in common if and only if φ is extreme in the convex set $\mathfrak{S}_\alpha$ of all states of A satisfying the condition in (a) for φ. A state φ is called *ergodic* if the above condition holds. (*Hint*: Use the similar arguments as in the proof of Theorem 9.22.)

(c) Show that if A is unital and $\lim_{s \to e} \|\alpha_s(x) - x\| = 0$ for every $x \in A$, where e is the unit of G, then $\mathfrak{S}_\alpha$ is compact; thus $\mathfrak{S}_\alpha$ has extreme points whenever $\mathfrak{S}_\alpha \neq \varnothing$.

8. Let ω be a positive linear functional on a C^*-algebra A.

(a) Given n elements $x_1, x_2, \ldots, x_n \in A$, show that $\eta_\omega(x_1), \eta_\omega(x_2), \ldots, \eta_\omega(x_n)$ are linearly independent if and only if

$$\det(\omega(x_j^* x_i))_{\substack{1 \le i \le n \\ 1 \le j \le n}} \neq 0.$$

(b) Let $d(\omega) = \dim \mathfrak{H}_\omega$. Show that the map: $\omega \mapsto d(\omega)$ is lower semicontinuous on A_+^* with respect to the $\sigma(A^*,A)$-topology, where $d(\omega) = +\infty$ if $\dim \mathfrak{H}_\omega \geq \aleph_0$.

10. Extreme Points of the Unit Ball of a C^*-Algebra

Lemma 10.1. *Let A be an abelian C^*-algebra and S its closed unit ball.*

(i) *The set of all extreme points of S is exactly the unitary group $U(A)$ of A. Therefore, the existence of extreme points of S implies that A is unital.*

(ii) *The set of all extreme points of $S \cap A_+$ is exactly the set of all projections of A. If $h \in S \cap A_+$ is not extreme, then there exists an element $a \in A_+ \cap S$ such that*

$$ha \neq 0, \qquad \|h(1 + a)\| \leq 1, \quad \text{and} \quad \|h(1 - a)\| \leq 1.$$

Proof. Denoting the spectrum of A by Ω, we identify A with $C_\infty(\Omega)$. If $f \in S$ is not unitary, then there exists an $\omega_0 \in \Omega$ with $|f(\omega_0)| < 1$. By the continuity of f, we can choose a compact neighborhood of ω_0 such that $|f(\omega)| < 1$ for every $\omega \in U$. Put $\alpha = \sup\{|f(\omega)| : \omega \in U\} < 1$. Take a continuous function g on Ω such that $g(\omega_0) = 1$, $g(\omega) = 0$ for every $\omega \notin U$ and $0 \leq g(\omega) \leq 1$. Putting $h = (1 - \alpha)g$, we have $f = \frac{1}{2}\{(f + h) + (f - h)\}$ and $f \pm h \in S$, so that f is not extreme in S. Conversely, if f is unitary, then $|f(\omega)| = 1$ for every $\omega \in \Omega$. Every point on the unit circle in $\mathbf{C}$ is extreme in the unit disk, so that f must be extreme in S.

Let p be a projection of A. If $p = \frac{1}{2}(h + k)$ for some $h,k \in S \cap A_+$, then h and k both belong to $pA \cap S$ because $0 \leq h(1 - p) \leq p(1 - p) = 0$ and $0 \leq k(1 - p) \leq p(1 - p) = 0$. Since p is the identity of the C^*-algebra pA, it is extreme in the unit ball of pA; hence $h = k = p$. Therefore, p is extreme in $S \cap A_+$. Conversely, suppose $h \in S \cap A_+$ is not a projection. Then there exists $\omega_0 \in \Omega$ with $0 < h(\omega_0) < 1$. Take a compact neighborhood U of ω_0 such that $0 < h(\omega) < 1$ for every $\omega \in U$, and then choose a continuous function g on Ω such that $g(\omega_0) = 1$, $g(U^c) = 0$, and $0 \leq g(\omega) \leq 1$. For a small $\varepsilon > 0$, we have $h(1 \pm \varepsilon g) \in S \cap A_+$, and $hg \neq 0$. Hence $h = \frac{1}{2}\{h(1 + \varepsilon g) + h(1 - \varepsilon g)\}$ is not extreme. Putting $a = \varepsilon g$, we get an element a as required. Q.E.D.

Theorem 10.2. *Let A be a C^*-algebra and S its closed unit ball.*

(i) *There exists an extreme point in S if and only if A is unital.*
(ii) *When A is unital, $x \in S$ is extreme if and only if x^*x, hence xx^*, is a projection such that*

$$(1 - x^*x)A(1 - xx^*) = \{0\}.$$

Proof. Let A_I be the C^*-algebra obtained by adjunction of an identity to A. Let x be an element of S such that x^*x is not a projection. Applying Lemma 10.1 to the absolute value $|x|$ of x and the C^*-subalgebra B of A_I generated by $|x|$ and 1, we can find an $a \in B_+ \cap S$ such that

$$\|(1 \pm a)|x|\| \leq 1 \quad \text{and} \quad a|x| \neq 0.$$

We then have

$$\begin{aligned} x &= \tfrac{1}{2}\{x(1 + a) + x(1 - a)\}, \\ \|x(1 \pm a)\| &= \|(1 \pm a)x^*x(1 \pm a)\|^{1/2} \\ &= \|(1 \pm a)^2|x|^2\|^{1/2} = \|(1 \pm a)|x|\| \leq 1, \\ \|xa\| &= \|ax^*xa\|^{1/2} = \|a^2|x|^2\|^{1/2} = \|a|x|\| \neq 0. \end{aligned}$$

Hence x is not extreme in S. Therefore, if x is an extreme point of S, then x^*x is a projection. Let $p = x^*x$ and $q = xx^*$. Since $\{x(1 - p)\}^*x(1 - p) = (1 - p)x^*x(1 - p) = 0$, we have $x = xp$, so $q = xpx^*$ and $q^2 = xpx^*xpx^* = xp^3x^* = xpx^* = q$; thus q is a projection. We have $x = qx$ by the same reasoning as for $x = xp$. For any $a \in (1 - q)S(1 - p) \subset S$, we have

$$\begin{aligned} \|x \pm a\|^2 &= \|(x \pm a)^*(x \pm a)\| \\ &= \|x^*x \pm a^*x \pm x^*a + a^*a\| \\ &= \|p + (1 - p)a^*a(1 - p)\| = 1, \end{aligned}$$

where the last equality follows from the commutativity of p and

$$(1 - p)a^*a(1 - p).$$

Hence we have $(1-q)S(1-p) = \{0\}$, equivalently $(1-q)A(1-p) = \{0\}$. If $\{u_i\}$ is an approximate identity of A, then we have

$$0 = (1-q)u_i(1-p) = u_i - qu_i - u_ip + qu_ip,$$

hence $u_i = u_ip + qu_i - qu_ip$. Therefore, u_i converges to $p + q - qp \in A$. By definition, the limit of an approximate identity must be the identity of A if it exists. Therefore, if S has an extreme point, then A is unital.

Conversely, suppose A is unital. If $1 = \frac{1}{2}(a+b)$ for some $a,b \in S$, then we have $1 = \frac{1}{2}\{\frac{1}{2}(a+a^*) + \frac{1}{2}(b+b^*)\}$; hence $\frac{1}{2}(a+a^*) = \frac{1}{2}(b+b^*) = 1$ by Lemma 10.1. Therefore a and a^* commute, so that $a = a^* = 1$ by Lemma 10.1 again. Similarly, $b = b^* = 1$. Hence 1 is extreme in S. Suppose $x \in S$ satisfies the condition in (ii). Put $p = x^*x$ and $q = xx^*$. Suppose $x = \frac{1}{2}(x_1 + x_2)$ for some $x_1,x_2 \in S$. Put $y_1 = qx_1p$ and $y_2 = qx_2p$. Since $x = qxp$, we have

$$p = \tfrac{1}{4}(y_1^*y_1 + y_1^*y_2 + y_2^*y_1 + y_2^*y_2),$$
$$q = \tfrac{1}{4}(y_1y_1^* + y_1y_2^* + y_2y_1^* + y_2y_2^*).$$

Since p is the identity of the C^*-algebra pAp and $y_1^*y_1, \ldots, y_2^*y_2$ are all in pSp, we have

$$p = y_1^*y_1 = y_1^*y_2 = y_2^*y_1 = y_2^*y_2.$$

Similarly, we get

$$q = y_1y_1^* = y_1y_2^* = y_2y_1^* = y_2y_2^*.$$

Therefore, we have

$$\begin{aligned} y_1 &= qy_1 = (y_1y_1^*)y_1 = y_1(y_1^*y_1) = y_1(y_1^*y_2) \\ &= (y_1y_1^*)y_2 = qy_2 = y_2, \end{aligned}$$

so that $qx_1p = qx_2p = x$. If $(1-q)x_1p \neq 0$, then we have

$$\begin{aligned} (x_1p)^*(x_1p) &= p(qx_1 + (1-q)x_1)^*(qx_1 + (1-q)x_1)p \\ &= px_1^*qx_1p + px_1^*(1-q)x_1p \\ &= p + px_1^*(1-q)x_1p. \end{aligned}$$

Since $px_1^*(1-q)x_1p$ is positive and nonzero, we have

$$\|x_1p\|^2 = \|(x_1p)^*(x_1p)\| = \|p + px_1^*(1-q)x_1p\| > 1,$$

which contradicts the choice of x_1. Hence $(1-q)x_1p = 0$. Similarly, we have

$$(1-q)x_2p = qx_1(1-p) = qx_2(1-p) = 0.$$

By the assumption on p and q, we have

$$0 = (1-q)x_1(1-p) = (1-q)x_2(1-p).$$

Thus, we conclude $x_1 = y_1$ and $x_2 = y_2$; hence $x = x_1 = x_2$, so that x is extreme. Q.E.D.

11. Finite Dimensional C^*-Algebras

In this section, we examine the structure of a finite dimensional C^*-algebra. Let A be a finite dimensional C^*-algebra. The unit ball S of A is then compact; therefore it has an extreme point by the Krein–Milman theorem. Hence A is unital by Theorem 10.2. Since this is true for any *-subalgebra of A, we observe the following:

Lemma 11.1. *If A is a finite dimensional C^*-algebra then* (i) *A is unital and* (ii) *every ideal I of A is of the form $I = Ap$ for some central projection $p \in A$.*

PROOF. We have only to prove assertion (ii). Let p be the identity of I. For any $x \in A$, we have $xp \in I$; hence $p(xp) = xp$; so $px^* = px^*p = x^*p$. Thus p commutes with every $x \in A$. Q.E.D.

By definition, the center C of a C^*-algebra A is the set of all $a \in A$ such that $ax = xa$ for every $x \in A$. Obviously, C is an abelian C^*-algebra. If A is finite dimensional, then so is C; hence the spectrum Ω of C is a finite set, say $\{\omega_1,\omega_2,\ldots,\omega_m\}$. For each k, $1 \le k \le m$, let p_k be the element of C such that $p_k(\omega_j) = \delta_{k,j}$, where $\delta_{k,j}$ means, of course, the Kronecker symbol. It follows then that each p_k is a projection and $\sum_{k=1}^m p_k = 1$. Hence the algebra A is decomposed into the direct sum

$$A = \sum_{k=1}^{m} Ap_k.$$

Each Ap_k has a trivial center, i.e., the scalar multiples of the identity p_k. Hence, by Lemma 11.1, Ap_k has no nontrivial ideal, that is, Ap_k is simple. Now we state the structure of a finite dimensional C^*-algebra as follows:

Theorem 11.2. *If A is a finite dimensional C^*-algebra, then A is decomposed into the direct sum*

$$A = \sum_{k=1}^{m}{}^{\oplus} A_k,$$

where each A_k is isomorphic to the algebra of $n_k \times n_k$-matrices. The sequence $\{n_1,n_2,\ldots,n_m\}$ of positive integers is uniquely determined by A, up to permutations, and a complete invariant for the algebraic structure of A in the sense that if B is another finite dimensional C^*-algebra with the associated sequence $\{\bar{n}_1,\bar{n}_2,\ldots,\bar{n}_{\bar{m}}\}$, then A and B are isomorphic if and only if $m = \bar{m}$ and there exists a permutation σ of $\{1,\ldots,m\}$ such that $\bar{n}_k = n_{\sigma(k)}$, $k = 1,2,\ldots,m$.

PROOF. By the previous arguments, we may assume that A is simple and finite dimensional. We first note that $aAb \neq \{0\}$ for any nonzero a and b in A. This follows from the observation that the set $AaA = \{\sum_{i=1}^n x_i a y_i : x_i, y_i \in A\}$ is an ideal of A which must be the entire algebra A, being nonzero. Let B be a maximal abelian self-adjoint subalgebra of A. Being finite dimensional, the spectrum Γ of B is a finite set, say $\{\gamma_1,\gamma_2,\dots,\gamma_n\}$. For each i, $1 \leq i \leq n$, let e_i denote the projection of B corresponding to the characteristic function of the one point set $\{\gamma_i\}$. We have then that the $\{e_i\}$ are orthogonal and $\sum_{i=1}^n e_i = 1$, and B is isomorphic to $\mathbf{C}e_1 \oplus \mathbf{C}e_2 \oplus \cdots \oplus \mathbf{C}e_n$. It follows then that e_iAe_i, $1 \leq i \leq n$, commutes with every e_j, $1 \leq j \leq n$. Since B is maximal abelian, e_iAe_i is contained in B because the $\{e_j\}$ generate B, which means that $e_iAe_i = \mathbf{C}e_i$ for each $i = 1,2,\dots,n$. For fixed i and j, $e_iAe_j \neq \{0\}$ by the first remark. Choose a nonzero $x = e_ixe_j$. We have then $x^*x = e_jx^*xe_j = \lambda e_j$ and $xx^* = e_ixx^*e_i = \mu e_i$ for some λ and $\mu > 0$. But the equality $\lambda = \|x^*x\| = \|x\|^2 = \|xx^*\| = \mu$ shows indeed that $x^*x = \lambda e_j$ and $xx^* = \lambda e_i$. Put $u = \lambda^{-1/2}x$. We have then $u^*u = e_j$ and $uu^* = e_i$. Therefore, for any i and j, there exists $u \in A$ such that $u^*u = e_j$ and $uu^* = e_i$. For each i, $1 \leq i \leq n$, let u_i be an element of A such that $u_i^*u_i = e_1$ and $u_iu_i^* = e_i$. Put $u_{i,j} = u_iu_j^*$, $1 \leq i,j \leq n$. We then have

$$\begin{cases} u_{i,j}^* = u_{j,i}, \qquad \sum_{i=1}^n u_{i,i} = 1, \\ u_{i,j}u_{k,l} = \delta_{j,k}u_{i,l}. \end{cases} \tag{1}$$

We claim that $e_iAe_j = \mathbf{C}u_{i,j}$ for $i,j = 1,2,\dots,n$. In fact, if x is in e_iAe_j, then $xu_{j,i}$ falls in e_iAe_i, so that $xu_{j,i} = \lambda e_i$ for some $\lambda \in \mathbf{C}$; hence we get

$$x = xe_j = xu_{j,i}u_{i,j} = \lambda e_iu_{i,j} = \lambda u_{i,j}.$$

For each $x \in A$, let $\lambda_{i,j}(x)$ be a scalar such that $e_ixe_j = \lambda_{i,j}(x)u_{i,j}$. It follows then that

$$x = \sum_{i,j=1}^n e_ixe_j = \sum_{i,j=1}^n \lambda_{i,j}(x)u_{i,j}.$$

It is now straightforward to check that the map: $x \in A \mapsto (\lambda_{i,j}(x)) \in M_n(\mathbf{C})$ is a *-isomorphism of A onto the algebra $M_n(\mathbf{C})$ of all $n \times n$-matrices with complex coefficients. The number n is determined by the equality $n^2 = \dim A$, so that it is uniquely determined by A itself.

The other part of our assertion in the theorem is not hard to show. We leave it to the reader. Q.E.D.

Definition 11.3. A system $\{u_{i,j}: 1 \leq i,j \leq n\}$ of elements of a C^*-algebra A satisfying (1) is called a *matrix unit* of A.

Definition 11.4. A projection e in a C^*-algebra A is said to be *minimal* if $eAe = \mathbf{C}e$, because this means that e majorizes no other nonzero projection in A.

Lemma 11.5. *Let A be a finite dimensional simple C^*-algebra, hence isomorphic to $M_n(\mathbf{C})$ for some n, and e be a minimal projection of A. If $\{\pi,\mathfrak{H}\}$ is a* *representation of A, then the (cardinal) number* $m_\pi = \dim \pi(e)\mathfrak{H}$ *does not depend on the choice of the minimal projection e, and* $\dim \mathfrak{H} = n \cdot m_\pi$. *Furthermore, the representation* π *is uniquely determined by* m_π *up to unitary equivalence.*

PROOF. Let f be another minimal projection of A. As seen in the proof of Theorem 11.2, $fAe \neq \{0\}$. Choose a nonzero $x \in fAe$. We have then $x^*x = \lambda e$ and $xx^* = \lambda f$ for some $\lambda > 0$. Put $u = \lambda^{-1/2}x$. We then get $e = u^*u$ and $f = uu^*$. This means that $\pi(u)$ isometrically maps $\pi(e)\mathfrak{H}$ onto $\pi(f)\mathfrak{H}$, so that $m_\pi = \dim \pi(e)\mathfrak{H} = \dim \pi(f)\mathfrak{H}$. Let $\{e_1, \dots, e_n\}$ be orthogonal minimal projections with $\sum_{i=1}^n e_i = 1$. The existence of such a family was proven in the proof of Theorem 11.2. We then have $\dim \pi(e_i)\mathfrak{H} = m_\pi$, $i = 1,2, \dots, n$, so that $\dim \mathfrak{H} = n \cdot m_\pi$.

Suppose $\{\pi_1, \mathfrak{H}_1\}$ and $\{\pi_2, \mathfrak{H}\}$ are representations of A such that $m_{\pi_1} = m_{\pi_2}$. Let V be an isometry of $\pi_1(e_1)\mathfrak{H}_1$ onto $\pi_2(e_1)\mathfrak{H}_2$. Let $\{u_{i,j}\}$ be a matrix unit in A such that $u_{i,i} = e_i$, $i = 1,2, \dots, n$. Put

$$U = \sum_{i=1}^n \pi_2(u_{i,1})V\pi_1(u_{1,i}).$$

It is straightforward to check, based on (1), that U is an isometry of $\mathfrak{H}_1$ onto $\mathfrak{H}_2$. Furthermore, we have

$$\begin{aligned} U\pi_1(u_{i,j}) &= \sum_{k=1}^n \pi_2(u_{k,1})V\pi_1(u_{1,k})\pi_1(u_{i,j}) \\ &= \pi_2(u_{i,1})V\pi_1(u_{1,j}), \\ \pi_2(u_{i,j})U &= \pi_2(u_{i,j}) \sum_{k=1}^n \pi_2(u_{k,1})V\pi_1(u_{1,k}) \\ &= \pi_2(u_{i,1})V\pi_1(u_{1,j}), \end{aligned}$$

so that $U\pi_1(u_{i,j}) = \pi_2(u_{i,j})U$, $i,j = 1,2, \dots, n$. Since $\{u_{i,j}\}$ spans A linearly, we have $U\pi_1(x) = \pi_2(x)U$ for every $x \in A$. Q.E.D.

The number m_π is called the *multiplicity* of π.

Corollary 11.6. *A representation* $\{\pi, \mathfrak{H}\}$ *of a finite dimensional simple* C^*-*algebra A is irreducible if and only if* $m_\pi = 1$, *and such a representation is unique up to unitary equivalence.*

Lemma 11.7. *For any cardinal number m, and a finite dimensional simple* C^*-*algebra A, there exists a representation* $\{\pi, \mathfrak{H}\}$ *of A with* $m_\pi = m$.

PROOF. Let $\{\pi_0, \mathfrak{H}_0\}$ be an irreducible representation of A, and $\{\pi_i, \mathfrak{H}_i\}_{i \in I}$ be replicas of $\{\pi_0, \mathfrak{H}_0\}$ where $\operatorname{card} I = m$. Let $\{\pi, \mathfrak{H}\} = \sum_{i \in I}^{\oplus} \{\pi_i, \mathfrak{H}_i\}$. If e is a minimal projection of A, then $\dim \pi_0(e)\mathfrak{H}_0 = 1$, so that $\dim \pi(e)\mathfrak{H} = \dim(\sum_{i \in I}^{\oplus} \pi_i(e)\mathfrak{H}_i) = m$. Hence $m_\pi = m$. Q.E.D.

Lemma 11.8. *If $\{\pi,\mathfrak{H}\}$ is a finite dimensional representation of a finite dimensional simple C^*-algebra A, then the algebra $\pi(A)'$ of all operators on $\mathfrak{H}$ commuting with $\pi(A)$ is simple and isomorphic to the $m \times m$-matrix algebra $M_m(\mathbf{C})$ where $m = m_\pi$, and the multiplicity of the identity representation of $\pi(A)'$ is precisely n where $\dim A = n^2$.*

PROOF. Let $\{u_{i,j}\}$ be a matrix unit in A with $1 \le i,j \le n$ and $A \cong M_n(\mathbf{C})$. Let $B = \pi(A)'$ and $e = \pi(u_{1,1})$. Since B and e commute, each operators in B leaves $e\mathfrak{H}$ invariant, so that the map $\rho: x \in B \mapsto x|_{e\mathfrak{H}} \in \mathscr{L}(e\mathfrak{H})$ is a representation of B. Since we have, for every $x \in B$,

$$\begin{aligned} x = x \sum_{i=1}^{n} \pi(u_{i,i}) &= x \sum_{i=1}^{n} \pi(u_{i,1})\pi(u_{1,i}) \\ &= \sum_{i=1}^{n} \pi(u_{i,1})x\pi(u_{1,i}) \\ &= \sum_{i=1}^{n} \pi(u_{i,1})\rho(x)\pi(u_{1,i}), \end{aligned}$$

ρ is an isomorphism of B into $\mathscr{L}(e\mathfrak{H})$. Let y be an arbitrary element in $\mathscr{L}(e\mathfrak{H})$. Put $x = \sum_{i=1}^{n} \pi(u_{i,1})y\pi(u_{1,i})$. It is straightforward to check that $x \in B$ and $\rho(x) = y$. Hence ρ is an isomorphism of B onto $\mathscr{L}(e\mathfrak{H})$. Hence we get $B \cong M_m(\mathbf{C})$ since $\dim e\mathfrak{H} = m$. Since $\dim \mathfrak{H} = nm$, we conclude from Lemma 11.5 that the multiplicity of the identity representation of B is n. Q.E.D.

The algebra $\pi(A)'$ is called the *commutant* of $\pi(A)$, which will be discussed in greater detail in the rest of this book.

Let A be now a general finite dimensional C^*-algebra with center C. Let $N = \dim C$ and identify the spectrum of C with $\{1,2,\ldots,N\}$, and then decompose A into the direct sum

$$A = \sum_{i=1}^{N} Ap_i = \sum_{i=1}^{N}{}^{\oplus} A_i$$

of simple algebras according to Theorem 11.2. Each A_i is isomorphic to the $n_i \times n_i$-matrix algebra $M_{n_i}(\mathbf{C})$. Let $\{\pi,\mathfrak{H}\}$ be a representation of A. For each i, we put $\mathfrak{H}_i = \pi(p_i)\mathfrak{H}$, possibly trivial, and consider the representation $\pi_i: x \in A_i \mapsto \pi(x)|_{\mathfrak{H}_i}$. We then have the multiplicity m_i of the representation π_i of A_i, where we put $m_i = 0$ if $\mathfrak{H}_i = \{0\}$. In this way, we associate a sequence $m_\pi = \{m_1,m_2,\ldots,m_N\}$ of cardinal numbers with each representation π. We call this m_π the *multiplicity* of π. Summarizing the preceding lemmas, we obtain the following:

Theorem 11.9. *If A is a finite dimensional C^*-algebra, then the multiplicity $m_\pi = \{m_1,\ldots,m_N\}$ of a representation $\{\pi,\mathfrak{H}\}$ of A determines π up to unitary equivalence, and any sequence $\{m_1,\ldots,m_N\}$ appears as the multiplicity of some representation π. The commutant $B = \pi(A)'$ of $\pi(A)$ is isomorphic to the* *direct sum* $\sum_{i=1}^{N\oplus} M_{m_i}(\mathbf{C})$ *of the* $m_i \times m_i$*-matrix algebras provided* $\dim \pi < +\infty$, *and the multiplicity of the identity representation of* B *is* $\{n_1, n_2, \ldots, n_N\}$, *where*

$$A \cong \sum_{i=1}^{N} {}^{\oplus} M_{n_i}(\mathbf{C}),$$

with trivial interpretation for i *such that* $m_i = 0$.

EXERCISES

1. Show that a C^*-algebra is finite dimensional if and only if it admits a finite dimensional maximal abelian self-adjoint subalgebra.
2. Show that if a C^*-algebra A is reflexive as a Banach space, then A must be finite dimensional.
3. Show that a unital simple C^*-algebra A is finite dimensional if it contains a nonzero minimal projection. (*Hint*: Use the fact that $AeA = \{\sum_{i=1}^{n} x_i e y_i : x_1, \ldots, x_n, y_1, \ldots, y_n \in A\}$ is an ideal, hence $AeA = A$ for every $e \neq 0$, and that if e is a minimal projection, then $eAe = \mathbf{C}e$. Then show that $1 = \sum_{i=1}^{n} u_i e v_i$, and $x = \sum_{i,j=1}^{n} u_i e v_i x v_j^* e u_j^* = \sum_{i,j=1}^{n} \lambda_{i,j} u_i e u_j^*$; hence $\{u_i u_j^*\}$ spans A linearly.)

Notes

The theory presented in this chapter is a preliminary account of the general theory of Banach algebras and C^*-algebras. For those readers interested in more detailed theory for general Banach algebras, we refer to a recent book of F. F. Bonsall and J. Duncan, [3], where one finds an up-to-date treatise on the theory.

The aim of the theory of C^*-algebras is to understand a uniformly closed self-adjoint algebra of operators on a Hilbert space. Besides the completeness of the theory, the major advantage of the axiomatic approach to the subject of operator algebras over the spatial approach, i.e., studying the algebra on a preassigned Hilbert space, lies in the freedom of choosing an appropriate representation of the algebra in question. Although the majority of examples of C^*-algebras are given as algebras of operators on Hilbert spaces, the quotient C^*-algebras are given directly by appealing to the axioms. Therefore, the C^*-algebras constructed via the quotient process require the abstract treatment. A typical problem is, for example, the study of a property of an operator invariant under perturbation by compact operators, which involves the quotient C^*-algebra $\mathscr{L}(\mathfrak{H})/\mathscr{LC}(\mathfrak{H})$ of $\mathscr{L}(\mathfrak{H})$ by the ideal of compact operators on $\mathfrak{H}$. One never finds a concrete construction of a representation of $\mathscr{L}(\mathfrak{H})/\mathscr{LC}(\mathfrak{H})$ because it always involves Zorn's lemma or the axiom of choice.

The notion of an abstract Banach algebra was apparently introduced first by M. Nagumo in 1936 under the name "linear metric ring" [243], in connection with Hilbert's fifth problem concerning a topological group of invertible elements in a Banach algebra. The noble feature of the field was established by Gelfand, Naimark, and their collaborators in the 1940s. Gelfand [133] proved the fundamental theorem of Banach algebras, Theorem 2.5. Corollary 2.6 is called the Gelfand–Mazur theorem, which was announced by Mazur [232] and proved by Gelfand [133]. The postulates for C^*-algebras were given by Gelfand and Naimark [134] with an additional condition which assumes the invertibility of $1 + x^*x$ for every x. A C^*-algebra was defined to be a uniformly closed self-adjoint algebra of operators on a Hilbert space. An involutive Banach algebra satisfying the axioms in Definition 1.2 used to be called and is still sometimes called a *B^*-algebra.* The redundancy of the above extra condition was conjectured by Gelfand and Naimark, and indeed proved by Fukamiya and Kaplansky [131]; thus a B^*-algebra is a C^*-algebra. Further analysis of the postulates for a C^*-algebra is still going on. We refer to Bonsall and Duncan [3] for details. A great deal of work on the continuity of a homomorphism between Banach algebras has been accomplished. In Section 6, we touched very lightly the subject. For detail, we refer to a recent book of Sinclair [32]. Theorem 6.1 is essentially due to Fukamiya [131] and Kelley and Vaught [206]. Proposition 6.3 is known as Löwner–Heinz theorem [165], [220] and the proof here is due to Pedersen [285]. Segal [324] showed the existence of an approximate identity in a C^*-algebra. Theorem 7.7 is due to Pedersen [278]. Theorem 8.1 is due to Kaplansky [194] and Segal [325]. Theorem 9.14 is due to Gelfand–Naimark [134] and Segal [324]. Theorem 9.18 was shown by Gelfand and Naimark [134] with the additional hypothesis mentioned above. Theorem 9.22 is due to Segal [324] and Gelfand and Raikov [135]. Most of the material in Section 10 is due to Kadison [172].

Exercises[4]

1. Let $\mathfrak{H}$ be a separable infinite dimensional Hilbert space. The quotient C^*-algebra $\mathscr{A}(\mathfrak{H}) = \mathscr{L}(\mathfrak{H})/\mathscr{LC}(\mathfrak{H})$ is called the *Calkin algebra* on $\mathfrak{H}$. Let π be the canonical homomorphism of $\mathscr{L}(\mathfrak{H})$ onto $\mathscr{A}(\mathfrak{H})$. We call $x \in \mathscr{L}(\mathfrak{H})$ a *Fredholm operator* if $\pi(x)$ is invertible in $\mathscr{A}(\mathfrak{H})$. Let Fred($\mathfrak{H}$) denote the set of all Fredholm operators on $\mathfrak{H}$.
   (a) Show that if $x = uh$ is the polar decomposition of $x \in$ Fred($\mathfrak{H}$) then u and h are both in Fred ($\mathfrak{H}$).

[4] Exercises for the entire Chapter I.

(b) Show that if $h \in \mathrm{Fred}(\mathfrak{H})$ is self-adjoint then the range of h is closed. (*Hint*: If not, there exists a normalized orthogonal system $\{\xi_n\}$ in $\mathfrak{H}$ such that $h\xi_n \to 0$. If $kh = 1 + a$ with $a \in \mathscr{LC}(\mathfrak{H})$, then one can choose a subsequence $\{\xi_{n_j}\}$ such that $\lim_j a\xi_{n_j} = \eta$ exists. Hence $\xi_{n_j} = kh\xi_{n_j} - a\xi_{n_j} \to -\eta$, which contradicts the fact that $\{\xi_{n_j}\}$ is orthogonal and normalized.)

(c) Show that an $x \in \mathscr{L}(\mathfrak{H})$ is a Fredholm operator if and only if $\dim \ker(x) < +\infty$, $\dim \ker(x^*) < +\infty$ and x has a closed range, where $\ker(x)$ means the null space of x.

(d) For each $x \in \mathrm{Fred}(\mathfrak{H})$, let $\mathrm{index}(x) = \dim \ker(x^*) - \dim \ker(x)$. Show that $\mathrm{index}: x \in \mathrm{Fred}(\mathfrak{H}) \to \mathrm{index}\ x \in \mathbf{Z}$ is a homomorphism with respect to the multiplicative structure of $\mathrm{Fred}(\mathfrak{H})$ and the additive structure of $\mathbf{Z}$. (*Hint*: For any $x, y \in \mathscr{L}(\mathfrak{H})$, $\dim \ker(xy) = \dim \ker(y) + \dim(\ker(x) \cap \mathrm{range}(y))$. For any closed subspace $\mathfrak{M}$ and $\mathfrak{N}$ of $\mathfrak{H}$, $\dim(\mathfrak{M} \ominus (\mathfrak{M} \cap \mathfrak{N}^\perp)) = \dim(\mathfrak{N} \ominus \mathfrak{N} \cap \mathfrak{M}^\perp))$.)

(e) Show that if $p,q \in \mathscr{L}(\mathfrak{H})$ are projections with $\|p - q\| < 1$, then $\dim p\mathfrak{H} = \dim q\mathfrak{H}$.(*Hint*: Show that pq maps $q\mathfrak{H}$ onto $p\mathfrak{H}$ injectively.)

(f) Show that the map: $x \in \mathrm{Fred}(\mathfrak{H}) \mapsto \dim \ker(x) \in \mathbf{Z}$ is continuous. Hence the index is a continuous function of $\mathrm{Fred}(\mathfrak{H})$. (*Hint*: Show that the polar decomposition in $\mathrm{Fred}(\mathfrak{H})$ is continuous and use (e).)

(g) Show that $\mathrm{index}(a + x) = \mathrm{index}(a)$ for any $a \in \mathrm{Fred}(\mathfrak{H})$ and $x \in \mathscr{LC}(\mathfrak{H})$.

(h) Show that $\mathrm{index}(a) = \mathrm{index}(u)$ with $a = uh \in \mathrm{Fred}(\mathfrak{H})$ the polar decomposition.

(i) Show that for a partial isometry $u \in \mathrm{Fred}(\mathfrak{H})$, $u \in \mathscr{U}(\mathfrak{H}) + \mathscr{LC}(\mathfrak{H})$ if and only if $\mathrm{index}(u) = 0$.

(j) Show that for an $x \in \mathrm{Fred}(\mathfrak{H})$, $\pi(x)$ belongs to $G_0(\mathscr{A}(\mathfrak{H}))$ if and only if $\mathrm{index}(x) = 0$.

(k) Show that $G(\mathscr{A}(\mathfrak{H}))/G_0(\mathscr{A}(\mathfrak{H})) \cong \mathbf{Z}$.

2. Let A be a unital C^*-algebra. Let $U_0(A)$ denote the principal component of the unitary group $U(A)$ of A, i.e., the connected component containing the identity.

(a) For each $x \in G(A)$, set $|x| = (x^*x)^{1/2}$ and $u(x) = x|x|^{-1}$. Show that this definition makes sense and that $u(x)$ is a unitary element.

(b) Show that $G(A) = U(A)G_0(A)$.

(c) Show that $G_0(A) \cap U(A) = U_0(A)$. (*Hint*: Use the continuity of the map: $x \in G(A) \mapsto |x| \in A$.)

(d) Show that $\mathrm{index}(A) \cong U(A)/U_0(A)$.

3. Let A be the C^*-algebra $C(S^1)$ of the unit circle $S^1 = \{\lambda \in \mathbf{C}: |\lambda| = 1\}$. Show that $\mathrm{index}(A) \cong \mathbf{Z}$ following the following step.

(a) Show that if g is a continuous S^1-valued function on $\mathbf{R}$, then there exists an $\mathbf{R}$-valued continuous function f on $\mathbf{R}$ such that $g(t) = e^{2\pi i f(t)}$, $t \in \mathbf{R}$.

(b) Show that if $u \in U(A)$, then there exists an $\mathbf{R}$-valued continuous function f_u on $\mathbf{R}$ such that $u(e^{2\pi i t}) = e^{2\pi i f_u(t)}$, $t \in \mathbf{R}$.

(c) Show that the map $\text{Ind}: u \mapsto f_u(1) - f_u(0) \in \mathbf{Z}$ is a homomorphism of $U(A)$ onto $\mathbf{Z}$.

(d) Show that the kernel of the homomorphism Ind is precisely $U_0(A)$, and conclude that $\text{index}(A) \cong \mathbf{Z}$.

4. Let A be a unital Banach algebra. A continuous homomorphism: $t \in \mathbf{R} \mapsto x(t) \in G(A)$ from $\mathbf{R}$ into $G(A)$ is called a *one parameter group.*

(a) Show that $a = \lim_{t \to 0}(1/t)[x(t) - 1]$ exists and that $x(t) = \exp(ta)$. We say that a is the generator of $x(\cdot)$.

(b) Show that the principal component $G_0(A)$ is generated by $\exp a$, $a \in A$.

(c) $\exp(x + y) = \lim_{n \to \infty} [\exp(x/n) \cdot \exp(y/n)]^n$.

(d) $\exp(xy - yx) = \lim_{n \to \infty} [\exp(-x/n)\exp(-y/n)\exp(x/n)\exp(y/n)]^{n^2}$.

Chapter II
Topologies and Density Theorems in Operator Algebras

0. Introduction

By nature, our objects in this book are infinite dimensional, which makes topological and approximation arguments indispensable. In Section 1, we first study the Banach spaces of operators on a Hilbert space $\mathfrak{H}$. It is proved that the second conjugate space $\mathscr{L}\mathscr{C}(\mathfrak{H})^{**}$ of the C^*-algebra of all compact operators on $\mathfrak{H}$ as a Banach space is naturally identified with the Banach space $\mathscr{L}(\mathfrak{H})$ of all bounded operators on $\mathfrak{H}$. This result allows us to introduce, in Section 2, various kinds of locally convex topologies in $\mathscr{L}(\mathfrak{H})$ related to the duality of $\mathscr{L}(\mathfrak{H})$ and $\mathscr{L}\mathscr{C}(\mathfrak{H})^*$ as well as to the algebra structure of $\mathscr{L}(\mathfrak{H})$. In Section 3, the fundamental theorem of operator algebras (the double commutation theorem), due to J. von Neumann, is proved and a few of its immediate consequences are drawn. Section 4 is devoted to various approximation theorems. Among them, Theorem 4.8 is most important. It may be called the fundamental approximation theorem. In this section, the strong continuity of functional calculus is also shown. A striking consequence of this section, Theorem 4.18, is the algebraic irreducibility of an irreducible representation of a C^*-algebra. The proof presented here is not the most economic, it is drawn as a consequence from a more powerful result, the noncommutative Lusin's theorem, Theorem 4.15, which is somewhat technical. The reader, who does not want to be too technical at this stage is advised to skip the discussion after Theorem 4.11 and go on to the next chapter, where an easy proof of a part of Theorem 4.18 is presented, Proposition III.2.16.

1. Banach Spaces of Operators on a Hilbert Space

Definition 1.1. A *sesquilinear* form B on a Hilbert space $\mathfrak{H}$ is a map: $\mathfrak{H} \times \mathfrak{H} \mapsto \mathbf{C}$ with the following properties:

$$\begin{cases} B(\alpha\xi,\eta) = \alpha B(\xi,\eta) = B(\xi,\bar{\alpha}\eta), & \xi,\eta \in \mathfrak{H}, \quad \alpha \in \mathbf{C}; \\ B(\xi_1 + \xi_2,\eta) = B(\xi_1,\eta) + B(\xi_2,\eta), & \xi_1,\xi_2,\eta \in \mathfrak{H}; \\ B(\xi,\eta_1 + \eta_2) = B(\xi,\eta_1) + B(\xi,\eta_2), & \xi,\eta_1,\eta_2 \in \mathfrak{H}. \end{cases} \tag{1}$$

In general, a map B of $\mathfrak{H} \times \mathfrak{H}$ into a complex vector space satisfying the above conditions is said to be sesquilinear. If a sesquilinear form B satisfies the condition

$$B(\xi,\eta) = \overline{B(\eta,\xi)}, \qquad \xi,\eta \in \mathfrak{H}, \tag{2}$$

then it is called *hermitian* or *symmetric*.

If B is a sesquilinear form on $\mathfrak{H}$, then we have

$$B(\xi,\eta) = \tfrac{1}{4}\sum_{n=0}^{3} i^n B(\xi + i^n\eta, \xi + i^n\eta), \qquad \xi,\eta \in \mathfrak{H}. \tag{3}$$

This formula is called the *polarization identity*. Due to this equality, (2) is equivalent to the following:

$$B(\xi,\xi) \text{ is real for every } \xi \in \mathfrak{H}. \tag{2'}$$

Definition 1.2. A sesquilinear form B is said to be *positive* if

$$B(\xi,\xi) \geq 0, \qquad \xi \in \mathfrak{H}. \tag{4}$$

A positive sesquilinear form is, of course, hermitian.

For each sesquilinear form B, the *norm* $\|B\|$ of B is defined by

$$\|B\| = \sup\{|B(\xi,\eta)| : \|\xi\| \leq 1, \|\eta\| \leq 1\}. \tag{5}$$

If $\|B\| < +\infty$, then B is called *bounded.* The set $\mathscr{B}(\mathfrak{H})$ of all bounded sesquilinear forms becomes a Banach space in the natural fashion:

$$(\alpha B)(\xi,\eta) = \alpha B(\xi,\eta), \qquad B \in \mathscr{B}(\mathfrak{H}), \qquad \xi,\eta \in \mathfrak{H}, \qquad \alpha \in \mathbf{C};$$
$$(B_1 + B_2)(\xi,\eta) = B_1(\xi,\eta) + B_2(\xi,\eta), \qquad B_1,B_2 \in \mathscr{B}(\mathfrak{H}).$$

The following result is then an immediate consequence of the Riesz representation theorem for bounded linear functionals on a Hilbert space; we have already used it in Section I.9 and it will play a vital role in the sequel.

Theorem 1.3. *There is an isometric linear correspondence*: $t \leftrightarrow B$ *between* $\mathscr{L}(\mathfrak{H})$ *and* $\mathscr{B}(\mathfrak{H})$ *determined by*

$$(t\xi|\eta) = B(\xi,\eta).$$

Furthermore, this correspondence has the following properties:

(i) *t is hermitian if and only if B is;*
(ii) *t is positive if and only if B is.*

To each pair ξ, η in $\mathfrak{H}$, there corresponds an operator $t_{\xi,\eta}$ of rank one by

$$t_{\xi,\eta}\zeta = (\zeta|\eta)\xi, \qquad \xi \in \mathfrak{H}. \tag{6}$$

It is easily shown that every operator of rank one is of this form.

To proceed further discussion, we need notations for Banach spaces of sequences. The Banach space of all sequences $\{\lambda_n\}$ converging to zero with norm $\|\{\lambda_n\}\|_\infty = \sup|\lambda_n|$ is denoted by (c_0); the Banach space of all summable sequences $\{\lambda_n\}$ with norm $\|\{\lambda_n\}\|_1 = \sum_{n=1}^\infty |\lambda_n|$ is denoted by (l^1); the Banach space of all bounded sequences $\{\lambda_n\}$ with norm $\|\{\lambda_n\}\|_\infty = \sup|\lambda_n|$ is denoted by (l^∞). We have then the following dualities among these three Banach spaces:

$$(c_0)^* = (l^1), \qquad (l^1)^* = (l^\infty).$$

Proposition 1.4. *For every compact operator x on an infinite dimensional Hilbert space $\mathfrak{H}$, there exist two normalized orthogonal systems $\{\xi_n\}$, $\{\eta_n\}$ in $\mathfrak{H}$ and a positive sequence $\{\alpha_n\} \in (c_0)$ such that*

$$x = \sum_{n=1}^\infty \alpha_n t_{\eta_n,\xi_n}, \qquad \|x\| = \sup \alpha_n.$$

If x is self-adjoint in addition, then x is represented in the form

$$x = \sum_{n=1}^\infty \alpha_n t_{\xi_n,\xi_n} \quad \textit{and} \quad \mathrm{Sp}(x) = \{\alpha_n\} \cup \{0\}$$

be a normalized orthogonal system $\{\xi_n\}$ and a real sequence $\{\alpha_n\} \in (c_0)$, where Sp(x) *means the spectrum of x as an operator.*

Proof. At first, suppose x is self-adjoint. Let

$$x = \int_{-\|x\|}^{\|x\|} \lambda \, de(\lambda)$$

be the spectral decomposition of x. Put

$$e_n = \int_{|\lambda| \geq 1/n} de(\lambda), \qquad n = 1,2,\ldots.$$

It follows then that e_n is a projection of finite rank, since the restriction x_n of x onto the range $e_n\mathfrak{H}$ of e_n is invertible and compact. Observing that x_n is self-adjoint, we choose an orthogonal basis $\{\xi_{n,k}: 1 \leq k \leq m_n\}$ of $e_n\mathfrak{H}$ and real numbers $\{\alpha_{n,k}: 1 \leq k \leq m_n\}$ such that

$$x_n = \sum_{k=1}^{m_n} \alpha_{n,k} t_{\xi_{n,k},\xi_{n,k}}; \qquad |\alpha_{n,k}| \geq \frac{1}{n},$$

where $m_n = \dim e_n\mathfrak{H}$. Noticing that $e_n \leq e_{n+1}$, $n = 1,2,\ldots$, and that $\lim_{n\to\infty} \|x - xe_n\| = 0$, we can rearrange the systems $\{\xi_{n,k}: k = 1,2,\ldots,m_n;$

$n = 1,2, \ldots\}$ and $\{\alpha_{n,k} : k = 1,2, \ldots, m_n; n = 1,2, \ldots\}$ as a normalized orthogonal system $\{\xi_n\}$ and a real sequence $\{\alpha_n\}$ such that

$$x = \sum_{n=1}^{\infty} \alpha_n t_{\xi_n, \xi_n}.$$

It is now easily seen that $\mathrm{Sp}(x) = \{\alpha_n\} \cup \{0\}$.

Returning to the general case, let $x = uh$ be the polar decomposition of x. Applying the above arguments to h, we represent h in the form

$$h = \sum_{n=1}^{\infty} \alpha_n t_{\xi_n, \xi_n}, \qquad \alpha_n \geq 0.$$

Hence we get

$$x = uh = \sum_{n=1}^{\infty} \alpha_n u t_{\xi_n, \xi_n} = \sum_{n=1}^{\infty} \alpha_n t_{u\xi_n, \xi_n}.$$

Putting $\eta_n = u\xi_n$, $n = 1,2, \ldots$, we obtain the desired expression for x.

Q.E.D.

In the following, we shall denote by $\mathscr{L}\mathscr{C}(\mathfrak{H})$ the set of all compact operators on $\mathfrak{H}$. It follows from the general theory of linear operators on a Banach space, cf. [38], that $\mathscr{L}\mathscr{C}(\mathfrak{H})$ is a closed ideal of $\mathscr{L}(\mathfrak{H})$. We are now going to show that $\mathscr{L}(\mathfrak{H})$ is identified with the second conjugate space $\mathscr{L}\mathscr{C}(\mathfrak{H})^{**}$ of $\mathscr{L}\mathscr{C}(\mathfrak{H})$ in the canonical fashion as a Banach space.

For a pair $(\xi,\eta) \in \mathfrak{H} \times \mathfrak{H}$, we define a continuous linear functional $\omega_{\xi,\eta}$ on $\mathscr{L}(\mathfrak{H})$:

$$\omega_{\xi,\eta}(x) = (x\xi|\eta), \qquad x \in \mathscr{L}(\mathfrak{H}). \tag{7}$$

Obviously, $\omega_{\xi,\eta}$ induces a continuous linear functional on $\mathscr{L}\mathscr{C}(\mathfrak{H})$ by restriction. For a little while, we shall regard $\omega_{\xi,\eta}$ as an element of $\mathscr{L}\mathscr{C}(\mathfrak{H})^*$.

For each $\omega \in \mathscr{L}\mathscr{C}(\mathfrak{H})^*$, we define a sesquilinear form B_ω on $\mathfrak{H}$ by

$$B_\omega(\xi,\eta) = \langle t_{\xi,\eta},\omega\rangle, \qquad \xi,\eta \in \mathfrak{H}.$$

It follows then that B_ω is bounded; hence there exists an operator $t(\omega) \in \mathscr{L}(\mathfrak{H})$ such that

$$(t(\omega)\xi|\eta) = \langle t_{\xi,\eta},\omega\rangle, \qquad \xi,\eta \in \mathfrak{H}. \tag{8}$$

It is easy to check that $t(\omega_{\xi,\eta}) = t_{\xi,\eta}$.

Lemma 1.5. *For any $\omega \in \mathscr{L}\mathscr{C}(\mathfrak{H})^*$, $t(\omega)$ is a compact operator and*

$$\sum_{i \in I} |(t(\omega)\xi_i|\xi_i)| < +\infty \tag{9}$$

for every normalized orthogonal system $\{\xi_i\}_{i \in I}$ in $\mathfrak{H}$.

PROOF. Let $\{\xi_i\}_{i\in I}$ be a normalized orthogonal system in $\mathfrak{H}$. For each $i \in I$, let α_i be a scalar of modulus one such that

$$|(t(\omega)\xi_i|\xi_i)| = \alpha_i(t(\omega)\xi_i|\xi_i).$$

Let J be a finite subset of I. Then, since $\|\sum_{i\in J} t_{\xi_i,\alpha_i\xi_i}\| = 1$, we have

$$\sum_{i\in J} |(t(\omega)\xi_i|\xi_i)| = \sum_{i\in J} \alpha_i(t(\omega)\xi_i|\xi_i)$$
$$= \sum_{i\in J} \langle t_{\xi_i,\alpha_i\xi_i},\omega\rangle$$
$$= \left\langle \sum_{i\in J} t_{\xi_i,\alpha_i\xi_i},\omega \right\rangle \le \|\omega\|.$$

Therefore, the set $\{i\in I:(t(\omega)\xi_i|\xi_i)\ne 0\}$ is countable and series (9) converges.

Now, let $t(\omega) = uh$ be the polar decomposition of $t(\omega)$. We define a continuous functional φ on $\mathscr{LC}(\mathfrak{H})$ by $\varphi(x) = \omega(xu^*)$. Then we have

$$(t(\varphi)\xi|\eta) = \langle t_{\xi,\eta},\varphi\rangle = \langle t_{\xi,\eta}u^*,\omega\rangle$$
$$= \langle t_{\xi,u\eta},\omega\rangle = (t(\omega)\xi|u\eta)$$
$$= (u^*t(\omega)\xi|\eta) = (h\xi|\eta).$$

Hence we get $t(\varphi) = h$. Therefore, to prove the compactness of $t(\omega)$, we may assume that $t(\omega)$ is positive. Let

$$t(\omega) = \int_0^{\|t(\omega)\|} \lambda\, de(\lambda)$$

be the spectral decomposition of $t(\omega)$. If a projection

$$1 - e(\varepsilon) = \int_\varepsilon^{\|t(\omega)\|} de(\lambda)$$

is of infinite rank for some $\varepsilon > 0$, then the inequality

$$(t(\omega)\xi|\xi) = \int_\varepsilon^\infty \lambda\, d\|e(\lambda)\xi\|^2 \ge \varepsilon\|\xi\|^2$$

for each $\xi \in [1 - e(\varepsilon)]\mathfrak{H}$, yields that

$$\sum_{n=1}^\infty (t(\omega)\xi_n|\xi_n) = \infty$$

for an infinite normalized orthogonal system $\{\xi_n\}$ in $(1 - e(\varepsilon))\mathfrak{H}$. This contradicts the conclusion of our arguments above. Hence, the projection $1 - e(\varepsilon)$ is of finite rank for every $\varepsilon > 0$. By the inequality

$$\|t(\omega) - (1 - e(\varepsilon))t(\omega)\| = \|e(\varepsilon)t(\omega)\| \le \varepsilon,$$

$t(\omega)$ is uniformly well approximated by operators of finite rank, so that $t(\omega)$ is compact. Q.E.D.

Theorem 1.6. *Every $\omega \in \mathscr{LC}(\mathfrak{H})^*$ is of the form*

$$\omega = \sum_{n=1}^\infty \alpha_n\omega_{\xi_n,\eta_n} \quad \text{and} \quad \|\omega\| = \sum_{n=1}^\infty \alpha_n \tag{10}$$

for some normalized orthogonal systems $\{\xi_n\}$, $\{\eta_n\}$ and some positive sequence $\{\alpha_n\}$ in (l^1). Conversely, to any two normalized orthogonal systems $\{\xi_n\}$, $\{\eta_n\}$

and any positive sequence $\{\alpha_n\}$ *in* (l^1), *there corresponds a unique* $\omega \in \mathscr{LC}(\mathfrak{H})^*$ *by* (10).

PROOF. Let ω be an element of $\mathscr{LC}(\mathfrak{H})^*$. By Lemma 1.5, a compact operator $t(\omega)$ corresponds to ω. By Proposition 1.4, $t(\omega)$ is of the form

$$t(\omega) = \sum_{n=1}^{\infty} \alpha_n t_{\eta_n,\xi_n}$$

for some normalized orthogonal systems $\{\xi_n\}$, $\{\eta_n\}$ and some positive sequence $\{\alpha_n\} = \alpha$ in (c_0). For every $\beta = \{\beta_n\} \in (c_0)$, the operator

$$t_\beta = \sum_{n=1}^{\infty} \beta_n t_{\xi_n,\eta_n}$$

is a compact operator, so that we can define

$$\langle \alpha,\beta \rangle = \omega(t_\beta) = \sum_{n=1}^{\infty} \alpha_n \beta_n.$$

Since $\|t_\beta\| = \sup|\beta_n| = \|\beta\|_\infty$, we have

$$|\langle \alpha,\beta \rangle| \leq \|\omega\|\,\|t_\beta\| = \|\omega\|\,\|\beta\|_\infty;$$

therefore we conclude that α is in (l^1) and $\|\alpha\|_1 \leq \|\omega\|$.

Conversely, given two normalized orthogonal systems $\{\xi_n\}$ and $\{\eta_n\}$ in $\mathfrak{H}$, we put

$$\omega_\alpha = \sum_{n=1}^{\infty} \alpha_n \omega_{\xi_n,\eta_n}$$

for each sequence $\alpha = \{\alpha_n\} \in (l^1)$. Then we have, for every $x \in \mathscr{C}(\mathfrak{H})$,

$$\begin{aligned} |\langle x,\omega_\alpha \rangle| &= \left| \sum_{n=1}^{\infty} \alpha_n (x\xi_n|\eta_n) \right| \leq \sum_{n=1}^{\infty} |\alpha_n|\,|(x\xi_n|\eta_n)| \\ &\leq \|x\| \sum_{n=1}^{\infty} |\alpha_n| = \|x\|\,\|\alpha\|_1; \end{aligned}$$

hence we have

$$\|\omega_\alpha\| \leq \|\alpha\|_1, \qquad \alpha \in (l^1).$$ Q.E.D.

Recalling that $\mathscr{LC}(\mathfrak{H})$ is an ideal of $\mathscr{L}(\mathfrak{H})$, we define actions of $a \in \mathscr{L}(\mathfrak{H})$ on $\omega \in \mathscr{LC}(\mathfrak{H})^*$ by

$$\langle x,a\omega \rangle = \langle xa,\omega \rangle \quad \text{and} \quad \langle x,\omega a \rangle = \langle ax,\omega \rangle \tag{11}$$

for every $x \in \mathscr{C}(\mathfrak{H})$. A simple calculation shows that

$$t(a\omega) = at(\omega) \quad \text{and} \quad t(\omega a) = t(\omega)a. \tag{12}$$

Therefore, the set $\mathscr{LT}(\mathfrak{H})$ of all $t(\omega)$, $\omega \in \mathscr{LC}(\mathfrak{H})^*$, is an ideal of $\mathscr{L}(\mathfrak{H})$.

Definition 1.7. Each operator in $\mathscr{LT}(\mathfrak{H})$ is called a *nuclear* operator or an operator of *trace class*.

Theorem 1.8. *There is a one-to-one isometric correspondence $f \leftrightarrow x$ between the second conjugate space $\mathscr{LC}(\mathfrak{H})^{**}$ and $\mathscr{L}(\mathfrak{H})$ determined by*

$$\langle \omega_{\xi,\eta}, f \rangle = (x\xi|\eta), \qquad \xi,\eta \in \mathfrak{H}. \tag{13}$$

Proof. Suppose x is a bounded operator on $\mathfrak{H}$. Since every $\omega \in \mathscr{LC}(\mathfrak{H})^*$ is of the form

$$\omega = \sum_{n=1}^{\infty} \alpha_n \omega_{\xi_n,\eta_n},$$

we can define a linear function f_x on $\mathscr{LC}(\mathfrak{H})^*$ by

$$\langle \omega, f_x \rangle = \sum_{n=1}^{\infty} \alpha_n (x\xi_n|\eta_n). \tag{14}$$

By equality (10), we have

$$\begin{aligned} |\langle \omega, f_x \rangle| &\leq \sum_{n=1}^{\infty} |\alpha_n| |(x\xi_n|\eta_n)| \\ &\leq \|x\| \sum_{n=1}^{\infty} |\alpha_n| = \|x\| \|\omega\|, \end{aligned}$$

so that f_x is continuous on $\mathscr{LC}(\mathfrak{H})^*$, and

$$\|f_x\| \leq \|x\|. \tag{15}$$

Conversely, take an $f \in \mathscr{LC}(\mathfrak{H})^{**}$. Putting

$$B_f(\xi,\eta) = \langle \omega_{\xi,\eta}, f \rangle, \qquad \xi,\eta \in \mathfrak{H},$$

we get a sesquilinear form B_f on $\mathfrak{H}$. The inequality

$$\begin{aligned} |B_f(\xi,\eta)| &= |\langle \omega_{\xi,\eta}, f \rangle| \leq \|f\| \|\omega_{\xi,\eta}\| \\ &\leq \|f\| \|\xi\| \|\eta\| \end{aligned}$$

shows that B_f is bounded. Thus Theorem 1.3 yields that there exists a unique $x_f \in \mathscr{L}(\mathfrak{H})$ such that

$$(x_f\xi|\eta) = B_f(\xi,\eta) = \langle \omega_{\xi,\eta}, f \rangle. \tag{16}$$

Since every $\omega \in \mathscr{LC}(\mathfrak{H})^*$ is well approximated by finite linear combinations of the $\omega_{\xi,\eta}$ with respect to the norm topology in $\mathscr{LC}(\mathfrak{H})^*$, equality (16) and (14) yield that

$$\langle \omega, f \rangle = \langle \omega, f_{x_f} \rangle, \qquad \omega \in \mathscr{LC}(\mathfrak{H})^*. \tag{17}$$

As seen above, we have $\|x_f\| \leq \|f\|$. Hence we get, by (15)

$$\|x_f\| = \|f\|.$$

Therefore, the map: $x \mapsto f_x$ is a surjective isometry. Q.E.D.

Thus, $\mathscr{L}(\mathfrak{H})$ is regarded as the second conjugate space $\mathscr{L}\mathscr{C}(\mathfrak{H})^{**}$ of $\mathscr{L}\mathscr{C}(\mathfrak{H})$. In the following, we shall write

$$\langle \omega, f_x \rangle = \langle x, \omega \rangle, \qquad x \in \mathscr{L}(\mathfrak{H}), \qquad \omega \in \mathscr{L}\mathscr{C}(\mathfrak{H})^*.$$

Regarding $\mathscr{L}\mathscr{C}(\mathfrak{H})^*$ as a space of linear functionals on $\mathscr{L}(\mathfrak{H})$, we shall denote it by $\mathscr{L}_*(\mathfrak{H})$. For each nuclear operator $t(\omega)$, $\omega \in \mathscr{L}_*(\mathfrak{H})$, the value $\langle 1, \omega \rangle$ is called the *trace* of $t(\omega)$ and denoted by $\mathrm{Tr}(t(\omega))$. Then we have

$$\langle x, \omega \rangle = \mathrm{Tr}(xt(\omega)) = \mathrm{Tr}(t(\omega)x). \tag{18}$$

As seen already, there is a remarkable similarity between the duality in Banach spaces $\{(c_0),(l^1),(l^\infty)\}$ and that in $\{\mathscr{L}\mathscr{C}(\mathfrak{H}),\mathscr{L}\mathscr{C}(\mathfrak{H})^*,\mathscr{L}(\mathfrak{H})\}$. We shall discuss this similarity again later.

Exercises

1. Let $\mathfrak{H}$ be a Hilbert space and $x = u|x|$ be the polar decomposition of $x \in \mathscr{L}(\mathfrak{H})$.

   (a) Show that $x \in \mathscr{L}\mathscr{T}(\mathfrak{H})$ if and only if $|x|$ is nuclear.
   (b) Show that $|x|$ is nuclear if and only if $\sum_{i \in I} (|x|\xi_i|\xi_i) < +\infty$ for a normalized orthogonal basis $\{\xi_i\}_{i \in I}$ of $\mathfrak{H}$, and in this case that

   $$\mathrm{Tr}(x) = \sum_{i \in I} (x\xi_i|\xi_i).$$

   (*Hint*: Prove first that $|x|$ is compact.)
   (c) Show that every $x \in \mathscr{L}\mathscr{T}(\mathfrak{H})$ is a linear combination of four positive nuclear operators.
   (d) Show that if $x = t(\omega)$ for an $\omega \in \mathscr{L}_*(\mathfrak{H})$, then $\|\omega\| = \mathrm{Tr}(|x|)$.

2. Let $\mathfrak{H}$ be a separable infinite dimensional Hilbert space. Let $\mathscr{I}$ be a nontrivial ideal (not necessarily closed) of $\mathscr{L}(\mathfrak{H})$.

   (a) Show that $\mathscr{I}$ contains a nontrivial projection.
   (b) Show that $\mathscr{I}$ contains all projections of rank 1.
   (c) Show that $\mathscr{I}$ contains the ideal $\mathscr{L}\mathscr{F}(\mathfrak{H})$ of operators of finite rank. Hence $\mathscr{L}\mathscr{F}(\mathfrak{H})$ is the smallest ideal of $\mathscr{L}(\mathfrak{H})$.
   (d) Show that if h is a noncompact positive operator, then there exist $\lambda > 0$ and a projection e of infinite rank such that $\lambda e \le h$.
   (e) Show that if $a \in \mathscr{L}(\mathfrak{H})$ is noncompact, then there exist two operators b and c such that $bac = 1$. Hence $\mathscr{I}$ cannot contain a noncompact operator, i.e., $\mathscr{L}\mathscr{C}(\mathfrak{H})$ is the greatest ideal of $\mathscr{L}(\mathfrak{H})$.
   (f) Let $E(\mathscr{I})$ denote the set of sequences $\{\lambda_n\}$ such that $\sum_{n=1}^\infty \lambda_n t_{\xi_n,\xi_n} \in \mathscr{I}$ for some normalized orthogonal system $\{\xi_n\}$ in $\mathfrak{H}$. Show that $E(\mathscr{I})$ is an ideal of (c_0).
   (g) Show that $E(\mathscr{I})$ is invariant under the following operations: (i) permutations of indices of $\{\lambda_n\}$; (ii) forming a subsequence from a sequence in $E(\mathscr{I})$; (iii) adding a finitely many new terms in front of a sequence in $E(\mathscr{I})$.
   (h) Show that if E is an ideal of (c_0) invariant under the operations described in (g), then there exists an ideal $\mathscr{I}(E)$ of $\mathscr{L}(\mathfrak{H})$ such that $E = E(\mathscr{I}(E))$.
   (i) Show that $\mathscr{I}(E(\mathscr{I})) = \mathscr{I}$.

3. An operator $x \in \mathscr{L}(\mathfrak{H})$ is said to be of *Hilbert–Schmidt class* if x^*x is nuclear.
   (a) Show that the set of all operators of Hilbert–Schmidt class forms an ideal of $\mathscr{L}(\mathfrak{H})$. We denote it by $\mathscr{L}\mathscr{S}(\mathfrak{H})$.
   (b) Show that $\mathscr{L}\mathscr{S}(\mathfrak{H})$ is a Hilbert space under the inner product given by $(x|y) = \mathrm{Tr}(y^*x)$, $x,y \in \mathscr{L}\mathscr{S}(\mathfrak{H})$.
   (c) Show that if $\mathfrak{H} = L^2(0,1)$, then for any $x \in \mathscr{L}\mathscr{S}(\mathfrak{H})$ there exists a function $K \in L^2([0,1] \times [0,1])$ such that $(xf)(s) = \int_0^1 K(s,t)f(t)\,dt$, $f \in \mathfrak{H}$.
   (d) Show that the product of any operators of Hilbert–Schmidt class is nuclear.
   (e) Show that every nuclear operator is a product of operators of Hilbert–Schmidt class.

4. Let h be a self-adjoint operator on a separable Hilbert space $\mathfrak{H}$. Following the steps described below, show that for any $\varepsilon > 0$, there exists a self-adjoint operator $a \in \mathscr{L}\mathscr{S}(\mathfrak{H})$ with $\|a\|_2 = \mathrm{Tr}(a^*a)^{1/2} < \varepsilon$ such that the difference $k = h - a$ admits a normalized orthogonal basis of $\mathfrak{H}$ consisting of eigenvectors.
   (a) Fix a vector $\xi_1 \in \mathfrak{H}$ and $n = 1,2,\ldots$, set $\delta_n = 2\|h\|/n$, $\lambda_0 = -\|h\|$, and $\lambda_j = \lambda_0 + j\delta_n$, $j = 1,2,\ldots,n$. Let e_j be the spectral projection of h corresponding to the interval $[\lambda_{j-1},\lambda_j)$, and p_j be the projection of $\mathfrak{H}$ onto $\mathbf{C}e_j\xi_1$. Putting $p = \sum_{j=1}^n p_j$, show that $\|(1-p)hp\| \le \delta_n$. (*Hint*: $(1-p)hp_j\mathfrak{H} \subset e_j\mathfrak{H}$; hence the ranges of $(1-p)hp_j$, $j = 1,2,\ldots,n$, are orthogonal and $(1-p)hp_j = (1-p)(h-\lambda_j)p_j$.)
   (b) Show that $\|(1-p)hp\|_2 \le \sqrt{n}\delta_n \to 0$ as $n \to \infty$. (*Hint*: $\|x\|_2 \le \|x\|\sqrt{(\text{rank of } x)}$.)
   (c) Set $a_1 = -(1-p)hp - ph(1-p)$ and $k_1 = h - a_1$ with n so large that $\|a_1\|_2 \le \sqrt{n}\delta_n < \varepsilon/2$. Show that k_1 is self-adjoint and commutes with p.
   (d) Note that $p\xi_1 = \xi_1$. Set $q_1 = p$.
   (e) Let $\{\xi_n\}$ be a dense sequence of $\mathfrak{H}$. Apply the above procedure (a)–(d) to $(1-q_1)\xi_1$ and $k_1(1-q_1)$ on the Hilbert space $(1-q_1)\mathfrak{H}$ with ε replaced by $\varepsilon/2$ to obtain a_2, k_2 and q_2 such that $\|a_2\|_2 \le \varepsilon/4$, $k_2 = h - a_2$ commutes with q_2, and the rank of q_2 is finite.
   (f) Construct, by induction, $\{a_n\}$, $\{k_n\}$, and $\{q_n\}$ such that (i) $\{q_n\}$ is orthogonal family of projections of finite rank, (ii) $(\sum_{i=1}^n q_i)\xi_j = \xi_j$, $j = 1,2,\ldots,n$, (iii) $\|a_n\|_2 \le \varepsilon/2^n$, (iv) $h - \sum_{j=1}^n a_j = k_n = q_1hq_1 + \cdots + q_nhq_n + (1 - \sum_{i=1}^n q_i)h(1 - \sum_{i=1}^n q_i)$.
   (g) Setting $a = \sum_{n=1}^\infty a_n$ and $k = \lim_{n\to\infty} k_n = h - a$, conclude the original claim.
   (h) Note that the crucial point in the above procedure is the fact that $\lim_{n\to 0}\sqrt{n}\delta_n = 0$ in (b).

5. Let h be a self-adjoint operator on a separable Hilbert space $\mathfrak{H}$. The spectrum of h is said to be *absolutely continuous* if the function: $\lambda \in \mathbf{R} \mapsto \|e(\lambda)\xi\|^2 \in \mathbf{R}$ is absolutely continuous for every $\xi \in \mathfrak{H}$ where $h = \int \lambda\, de(\lambda)$ is the spectral decomposition of h.
   (a) Show that for any h, $\mathfrak{H}$ is the direct sum of $\mathfrak{H}_{a.c.}$ and $\mathfrak{H}_s$ such that (i) $\mathfrak{H}_{a.c.}$ (resp. $\mathfrak{H}_s$) of h to $\mathfrak{H}_{a.c.}$ (resp. $\mathfrak{H}_s$) is absolutely continuous (resp. singular in the sense that the spectral measure of h_s is singular with respect to the Lebesgue measure).
   (b) Determine when a multiplication operator on $L^2(\alpha,\beta)$ has an absolutely continuous spectrum.
   (c) Show that the spectrum of h is absolutely continuous if and only if there exists a direct sum decomposition $\{h,\mathfrak{H}\} \cong \sum^\oplus \{h_n,\mathfrak{H}_n\}$ such that $\mathfrak{H}_n = L^2(S_n,\mu_n)$ where S_n is a Borel subset of $\mathbf{R}$ and μ_n is the restriction of the Lebesgue measure, and $h_n\xi(t) = t\xi(t)$, $\xi \in \mathfrak{H}_n$.
   (d) Setting $u(t) = \exp(ith)$, show that the spectrum of h is absolutely continuous if and only if $\{\xi \in \mathfrak{H} : \int_{-\infty}^\infty |(u(t)\xi|\xi)|^2\,dt < +\infty\}$ is dense in $\mathfrak{H}$. (*Hint*: If $\varphi_\xi(t) = (u(t)\xi|\xi)$, $\xi \in \mathfrak{H}$, then the function φ_ξ is the Fourier–Stieltjes transform of the measure μ_ξ given by $d\mu_\xi(\lambda) = \|de(\lambda)\xi\|^2$.)

6. Show that if h and k are self-adjoint operators on a separable Hilbert space $\mathfrak{H}$, with $h - k \in \mathscr{L}\mathscr{T}(\mathfrak{H})$, then $h_{a.c.}$ and $k_{a.c.}$ are unitarily equivalent. Cf. [18, X.4.3].
7. Given an infinite dimensional Hilbert space $\mathfrak{H}$, let $\mathscr{A}(\mathfrak{H})$ denote the quotient C^*-algebra $\mathscr{L}(\mathfrak{H})/\mathscr{L}\mathscr{C}(\mathfrak{H})$, the Calkin algebra.
   (a) Show that there exists an isomorphism of the quotient abelian C^*-algebra $(l^\infty)/(c_0)$ into $\mathscr{A}(\mathfrak{H})$. (*Hint*: Let $\{\xi_n\}$ be a normalized orthogonal sequence of vectors in $\mathfrak{H}$. For each $\{\lambda_n\} \in (l^\infty)$, $\sum_{n=1}^\infty \lambda_n t_{\xi_n,\xi_n}$ is compact if and only if $\{\lambda_n\} \in (c_0)$.)
   (b) Show that $\mathscr{A}(\mathfrak{H})$ has no nontrivial representation on a separable Hilbert space.

2. Locally Convex Topologies in $\mathscr{L}(\mathfrak{H})$

Let $\mathfrak{H}$ be a Hilbert space. In this section, we introduce several useful locally convex topologies in $\mathscr{L}(\mathfrak{H})$, based on the duality between the Banach spaces $\mathscr{L}(\mathfrak{H})$ and $\mathscr{L}_*(\mathfrak{H})$ established in the previous section.

Since $\mathscr{L}(\mathfrak{H})$ is the conjugate space of the Banach space $\mathscr{L}_*(\mathfrak{H})$, we can naturally define the $\sigma(\mathscr{L}(\mathfrak{H}),\mathscr{L}_*(\mathfrak{H}))$-topology. This topology is determined by the family of seminorms

$$x \in \mathscr{L}(\mathfrak{H}) \mapsto |\langle x,\omega\rangle|, \qquad \omega \in \mathscr{L}_*(\mathfrak{H}).$$

But each $\omega \in \mathscr{L}_*(\mathfrak{H})$ has, by Theorem 1.6, the form

$$\omega = \sum_{n=1}^\infty \alpha_n \omega_{\xi_n,\eta_n}$$

for some normalized orthogonal systems $\{\xi_n\}$, $\{\eta_n\}$ and some positive sequence $\{\alpha_n\}$ in (l^1). Replacing ξ_n and η_n by $\sqrt{\alpha_n}\xi_n$ and $\sqrt{\alpha_n}\eta_n$, respectively, we conclude that every $\omega \in \mathscr{L}_*(\mathfrak{H})$ is represented in the form

$$\omega = \sum_{n=1}^\infty \omega_{\xi_n,\eta_n}, \qquad \sum_{n=1}^\infty \|\xi_n\|^2 < +\infty, \qquad \sum_{n=1}^\infty \|\eta_n\|^2 < +\infty. \tag{1}$$

Of course, any sequences $\{\xi_n\}$ and $\{\eta_n\}$ with $\sum_{n=1}^\infty \|\xi_n\|^2 < +\infty$ and $\sum_{n=1}^\infty \|\eta_n\|^2 < +\infty$ define an element $\omega \in \mathscr{L}_*(\mathfrak{H})$ by (1). Therefore, the $\sigma(\mathscr{L}(\mathfrak{H}),\mathscr{L}_*(\mathfrak{H}))$-topology is given by the family of seminorms

$$p(x) = \left|\sum_{n=1}^\infty (x\xi_n|\eta_n)\right|, \qquad \sum_{n=1}^\infty \|\xi_n\|^2 < +\infty, \qquad \sum_{n=1}^\infty \|\eta_n\|^2 < +\infty. \tag{2}$$

Definition 2.1. The locally convex topology described above is called the *σ-weak (operator) topology* in $\mathscr{L}(\mathfrak{H})$.

If an $\omega \in \mathscr{L}_*(\mathfrak{H})$ is positive, that is, $\omega(x^*x) \geq 0$, $x \in \mathscr{L}(\mathfrak{H})$, then the function

$$p_\omega(x) = \omega(x^*x)^{1/2}, \qquad x \in \mathscr{L}(\mathfrak{H}), \tag{3}$$

is a seminorm in $\mathscr{L}(\mathfrak{H})$.

Definition 2.2. The locally convex topology determined by the family $\{p_\omega\}$ of seminorms is called the *σ-strong* (*operator*) topology in $\mathscr{L}(\mathfrak{H})$.

By the preceeding consideration, the σ-strong operator topology is determined by the family of seminorms

$$x \in \mathscr{L}(\mathfrak{H}) \mapsto \left(\sum_{n=1}^{\infty} \|x\xi_n\|^2 \right)^{1/2}, \qquad \sum_{n=1}^{\infty} \|\xi_n\|^2 < +\infty. \tag{4}$$

If $\mathfrak{H}$ is of infinite dimension, then the *-operation in $\mathscr{L}(\mathfrak{H})$ is discontinuous with respect to the σ-strong operator topology. In fact, letting $\{\xi_n\}$ be an infinite normalized orthogonal sequence in $\mathfrak{H}, t_{\xi_1,\xi_n}$ converges σ-strongly to zero because

$$\left(\sum_{k=1}^{\infty} \|t_{\xi_1,\xi_n}\eta_k\|^2 \right)^{1/2} = \left(\sum_{k=1}^{\infty} |(\eta_k|\xi_n)|^2 \right)^{1/2}$$

converges to zero as $n \to \infty$ whenever $\sum_{k=1}^{\infty} \|\eta_k\|^2 < +\infty$. But $t^*_{\xi_1,\xi_n} = t_{\xi_n,\xi_1}$ does not converge σ-strongly to zero because

$$\|t^*_{\xi_1,\xi_n}\xi_1\| = \|\xi_n\| = 1, \qquad n = 1,2,\dots.$$

Therefore, the following definition makes sense.

Definition 2.3. The locally convex topology determined by the family of seminorms: $x \in \mathscr{L}(\mathfrak{H}) \to \{p_\omega(x)^2 + p_\omega(x^*)^2\}^{1/2}$, where ω runs over all positive elements in $\mathscr{L}_*(\mathfrak{H})$, is called the *σ-strong** (*operator*) *topology*.

Since the seminorms which define the σ-strong* topology are invariant under the *-operation, the *-operation is continuous with respect to the σ-strong* topology. Hence the σ-strong* topology is strictly finer than the σ-strong topology.

We introduce three more locally convex topologies. The locally convex topology determined by the seminorms: $x \in \mathscr{L}(\mathfrak{H}) \mapsto |(x\xi|\eta)|$, $\xi,\eta \in \mathfrak{H}$, is called the *weak* (*operator*) topology. It is nothing but the $\sigma(\mathscr{L}(\mathfrak{H}), \mathscr{L}\mathscr{F}(\mathfrak{H}))$-topology, where $\mathscr{L}\mathscr{F}(\mathfrak{H})$ is the space of all finite linear combinations of the $\omega_{\xi,\eta}$. The *strong* (*operator*) topology is the locally convex topology determined by the seminorms: $x \in \mathscr{L}(\mathfrak{H}) \mapsto \|x\xi\|$, $\xi \in \mathfrak{H}$. Also the *strong** (*operator*) topology is defined as the locally convex topology induced by the seminorms: $x \in \mathscr{L}(\mathfrak{H}) \mapsto (\|x\xi\|^2 + \|x^*\xi\|^2)^{1/2}$. The topology given by the norm $\|x\|$ is called the *uniform* (*operator*) topology.

The relation between these various topologies is as follows:

$$\begin{array}{ccccccc} \text{Uniform} \prec & \sigma\text{-strong*} & \prec & \sigma\text{-strong} & \prec & \sigma\text{-weak} \\ & \curlywedge & & \curlywedge & & \curlywedge \\ & \text{strong*} & \prec & \text{strong} & \prec & \text{weak,} \end{array}$$

where "$\prec$" means that the left-hand side is finer than the right-hand side.

Lemma 2.4. *Every σ-strongly* continuous linear functional on* $\mathscr{L}(\mathfrak{H})$ *is σ-weakly continuous; hence it is in* $\mathscr{L}_*(\mathfrak{H})$.

PROOF. Suppose ω is a σ-strongly* continuous linear functional on $\mathscr{L}(\mathfrak{H})$. Then there exists a sequence $\{\xi_n\}$ in $\mathfrak{H}$ with $\sum_{n=1}^{\infty} \|\xi_n\|^2 < +\infty$ such that $|\omega(x)| \leq 1$, $x \in \mathscr{L}(\mathfrak{H})$, whenever

$$p(x) = \left\{ \sum_{n=1}^{\infty} (\|x\xi_n\|^2 + \|x^*\xi_n\|^2) \right\}^{1/2} \leq 1.$$

It follows easily that

$$|\omega(x)| \leq p(x), \qquad x \in \mathscr{L}(\mathfrak{H}). \tag{5}$$

Let $\tilde{\mathfrak{H}} = \sum_{n=-\infty}^{\infty} \mathfrak{H}_n$, where $\mathfrak{H}_n$ is a replica of $\mathfrak{H}$ for $n = 1,2, \ldots$ and is a replica of the conjugate Hilbert space $\bar{\mathfrak{H}}$ of $\mathfrak{H}$ for $n = -1,-2, \ldots$. We denote the vector in $\bar{\mathfrak{H}}$ corresponding to $\xi \in \mathfrak{H}$ by $\bar{\xi}$. Of course, $\tilde{\xi} = \{\ldots, \bar{\xi}_2, \bar{\xi}_1, \xi_1, \xi_2, \ldots\}$ is an element of $\tilde{\mathfrak{H}}$. For each $\tilde{\eta} = \{\bar{\eta}_{-m}, \eta_m : m = 1,2, \ldots\} \in \tilde{\mathfrak{H}}$ and $x \in \mathscr{L}(\mathfrak{H})$, we put

$$\tilde{x}\tilde{\eta} = \{\overline{x^*\eta_{-m}}, x\eta_m : m = 1,2, \ldots\}.$$

Then $\tilde{x}$ is a bounded operator on $\tilde{\mathfrak{H}}$. We note that the map: $x \in \mathscr{L}(\mathfrak{H}) \mapsto \tilde{x} \in \mathscr{L}(\tilde{\mathfrak{H}})$ is linear but not multiplicative. By inequality (5), the map: $\tilde{x}\tilde{\xi} \mapsto \omega(x)$, $x \in \mathscr{L}(\mathfrak{H})$, is a bounded linear functional on the linear manifold $\{\tilde{x}\tilde{\xi} : x \in \mathscr{L}(\mathfrak{H})\}$. The Riesz theorem assures us that we can find a vector $\tilde{\eta} \in \tilde{\mathfrak{H}}$ such that

$$\omega(x) = (x\tilde{\xi}|\tilde{\eta}), \qquad x \in \mathscr{L}(\mathfrak{H}).$$

Rewriting this, we obtain

$$\omega = \sum_{n=1}^{\infty} (\omega_{\xi_n,\eta_n} + \omega_{\eta_{-n},\xi_{-n}}),$$

so that ω is σ-weakly continuous. Q.E.D.

Lemma 2.5. *Let S denote the closed unit ball of $\mathscr{L}(\mathfrak{H})$. Then we conclude*

(i) *σ-weak topology on S = weak topology on S;*
(ii) *σ-strong topology on S = strong topology on S;*
(iii) *σ-strong* topology on S = strong* topology on S.*

PROOF. By Theorem 1.6, the space $\mathscr{L}\mathscr{F}(\mathfrak{H})$ of all finite linear combinations of the $\omega_{\xi,\eta}$ is dense in the Banach space $\mathscr{L}_*(\mathfrak{H})$ with respect to the norm topology in $\mathscr{L}_*(\mathfrak{H})$. As has been pointed out once, the weak topology is the $\sigma(\mathscr{L},\mathscr{F})$-topology, so that the σ-weak topology and the weak topology are identical on every bounded subset of $\mathscr{L}(\mathfrak{H})$. Assertions (ii) and (iii) follow immediately from the fact that for a net $\{x_i\}$ in $\mathscr{L}(\mathfrak{H})$

$$x_i \to 0 \text{ (strongly)} \Leftrightarrow x_i^*x_i \to 0 \text{ (weakly)},$$
$$x_i \to 0 \text{ (σ-strongly)} \Leftrightarrow x_i^*x_i \to 0 \text{ (σ-weakly)},$$
$$x_i \to 0 \text{ (strongly*)} \Leftrightarrow x_i^*x_i + x_ix_i^* \to 0 \text{ (weakly)},$$
$$x_i \to 0 \text{ (σ-strongly*)} \Leftrightarrow x_i^*x_i + x_ix_i^* \to 0 \text{ (σ-weakly)}.$$

Q.E.D.

The standard results in the duality of Banach spaces and Lemmas 2.4, 2.5 yield the following result which will be used repeatedly.

Theorem 2.6. *Let $\mathscr{M}$ be a σ-weakly closed subspace of $\mathscr{L}(\mathfrak{H})$ and ω a bounded linear functional on $\mathscr{M}$.*

(i) *The following four statements are equivalent:*
  (i.1) *ω is weakly continuous;*
  (i.2) *ω is strongly continuous;*
  (i.3) *ω is strongly* continuous;*
  (i.4) $\omega = \sum_{i=1}^{n} \omega_{\xi_i,\eta_i}$.
(ii) *The following seven statements are equivalent:*
  (ii.1) *ω is σ-weakly continuous;*
  (ii.2) *ω is σ-strongly continuous;*
  (ii.3) *ω is σ-strongly* continuous;*
  (ii.4) $\omega = \sum_{n=1}^{\infty} \omega_{\xi_n,\eta_n}$, *where* $\sum_{n=1}^{\infty} \|\xi_n\|^2 < +\infty$ *and* $\sum_{n=1}^{\infty} \|\eta_n\|^2 < +\infty$;
  (ii.5) *ω is weakly continuous on $\mathscr{M} \cap S$;*
  (ii.6) *ω is strongly continuous on $\mathscr{M} \cap S$;*
  (ii.7) *ω is strongly* continuous on $\mathscr{M} \cap S$.*
(iii) *If $\mathscr{M}_*$ (resp. $\mathscr{M}_\sim$) denotes the space of all σ-weakly (resp. weakly) continuous linear functional on $\mathscr{M}$, then $\mathscr{M}_*$ is a closed subspace of the conjugate space $\mathscr{M}^*$ of $\mathscr{M}$ and $\mathscr{M}_\sim$ is dense in $\mathscr{M}_*$ with respect to the norm topology. Furthermore, $\mathscr{M}$ is isometrically isomorphic to the conjugate space of the Banach space $\mathscr{M}_*$ under the natural correspondence.*
(iv) *For a convex subset K of $\mathscr{M}$, the following six statements are equivalent:*
  (iv.1) *K is σ-weakly closed;*
  (iv.2) *K is σ-strongly closed;*
  (iv.3) *K is σ-strongly* closed;*
  (iv.4) *$K \cap rS$ is weakly (therefore σ-weakly) closed for every $r > 0$;*
  (iv.5) *$K \cap rS$ is strongly (therefore σ-strongly) closed for every $r > 0$;*
  (iv.6) *$K \cap rS$ is strongly* (therefore, σ-strongly*) closed for every $r > 0$.*

Proof. The implications (i.4) $\Rightarrow$ (i.1) $\Rightarrow$ (i.2) $\Rightarrow$ (i.3) are obvious. Now, suppose ω is strongly* continuous. Then there exists a finite subset $\{\xi_1, \dots, \xi_n\}$ in $\mathfrak{H}$ such that

$$|\omega(x)| \leq \left\{ \sum_{k=1}^{n} (\|x\xi_k\|^2 + \|x^*\xi_k\|^2) \right\}^{1/2}, \qquad x \in \mathscr{M}.$$

Using the direct sum Hilbert space

$$\tilde{\mathfrak{H}} = \bar{\mathfrak{H}}_{-n} \oplus \bar{\mathfrak{H}}_{-n+1} \oplus \cdots \oplus \bar{\mathfrak{H}}_{-1} \oplus \mathfrak{H}_1 \oplus \cdots \oplus \mathfrak{H}_n$$

and pretty much the same arguments as in the proof of Lemma 2.4, we can represent ω in the desired form (i.4). Assertion (ii) follows from Lemmas 2.4 and 2.5 directly. Assertions (iii) and (iv) follow from the general duality theory of Banach spaces. Q.E.D.

Proposition 2.7. *If a Hilbert space $\mathfrak{H}$ is separable, then the locally convex operator topologies in $\mathscr{L}(\mathfrak{H})$ defined above are all metrizable on bounded parts of $\mathscr{L}(\mathfrak{H})$.*

PROOF. We show here only the metrizability of the weak topology on the unit ball S. The metrizability of the other topologies is proven similarly. Let $\{\xi_n\}$ be a dense sequence in the unit ball of $\mathfrak{H}$. We define a distance function on S by

$$d(x,y) = \sum_{n,m=1}^{\infty} 2^{-(n+m)}|((x-y)\xi_n|\xi_m|$$

for each pair x,y in S. It is not hard to see that the distance function d gives rise to the weak topology on S. Q.E.D.

EXERCISES

1. Let $\mathfrak{H}$ be a Hilbert space with $\dim \mathfrak{H} = \infty$, and $\{e_n\}$ be an orthogonal sequence of projections in $\mathscr{L}(\mathfrak{H})$.
   (a) Show that $\{ne_n : n = 1,2,\ldots\}$ has 0 as an accumulation point in the σ-strong* topology. (*Hint*: If $\omega \in \mathscr{L}_*(\mathfrak{H})$, then $\{\omega(e_n)\} \in (l^1)$.)
   (b) Show that $\{ne_n : n = 1,2,\ldots\}$ does not admit a subsequence converging to zero weakly. (*Hint*: Any subsequence of $\{ne_n\}$ is unbounded; hence the uniform boundedness theorem applies.)
   (c) Show that all the locally convex topologies in $\mathscr{L}(\mathfrak{H})$ introduced in this section except the uniform topology are nonmetrizable.

2. Let $\mathfrak{H}$ be a Hilbert space with $\dim \mathfrak{H} = \infty$, and $\mathfrak{S}_0$ be the set of all states of $\mathscr{L}(\mathfrak{H})$ vanishing on $\mathscr{L}\mathscr{C}(\mathfrak{H})$. Let V be the set of all vector states of $\mathscr{L}(\mathfrak{H})$, i.e., $V = \{\omega_\xi : \xi \in \mathfrak{H}, \|\xi\| = 1\}$, and $\tilde{V}$ = the $\sigma(\mathscr{L}(\mathfrak{H})^*, \mathscr{L}(\mathfrak{H}))$-closure of V.
   (a) Show that if $f \in \tilde{V} \cap \mathfrak{S}_0$, then for any finite dimensional subspace $\mathfrak{M}$ of $\mathfrak{H}$, f is in the $\sigma(\mathscr{L}(\mathfrak{H})^*, \mathscr{L}(\mathfrak{H}))$-closure of $V_{\mathfrak{M}^\perp} = \{\omega_\xi : \xi \in \mathfrak{M}^\perp, \|\xi\| = 1\}$.
   (b) Show that $\tilde{V} \cap \mathfrak{S}_0$ is convex. (*Hint*: If $|f(x_i) - \omega_\xi(x_i)| < \varepsilon$, $i = 1,2,\ldots,n$, with $x_1 = 1$ and if $g \in \tilde{V} \cap \mathfrak{S}_0$, then find a vector $\eta \in [x_1\xi, \ldots, x_n\xi, x_1^*\xi, \ldots, x_n^*\xi]^\perp$ such that $|g(x_i) - \omega_\eta(x_i)| < \varepsilon$ using (a). Then $\lambda f + \mu g$ is approximated by $\omega_{\sqrt{\lambda}\xi + \sqrt{\mu}\eta}$ on $x_1, x_2, \ldots, x_n$ within ε.)
   (c) Show that $\tilde{V}$ contains all pure states on $\mathscr{L}(\mathfrak{H})$.
   (d) Show that $\tilde{V}$ contains $\mathfrak{S}_0$.

3. The Double Commutation Theorem of J. von Neumann

Let $\mathfrak{H}$ be a Hilbert space. For each subset $\mathscr{M}$ of $\mathscr{L}(\mathfrak{H})$, let $\mathscr{M}'$ denote the set of all bounded operators on $\mathfrak{H}$ commuting with every operator in $\mathscr{M}$. Clearly, $\mathscr{M}'$ is a Banach algebra of operators containing the identity operator 1. If $\mathscr{M}$

is invariant under the *-operation, that is, if $x \in \mathscr{M}$ implies $x^* \in \mathscr{M}$, then $\mathscr{M}'$ is a C^*-algebra acting on $\mathfrak{H}$, which is closed with respect to all the locally convex topologies defined in the previous section. Clearly, we have

$$\mathscr{M} \subset \mathscr{M}'' = \mathscr{M}^{(\text{iv})} = \mathscr{M}^{(\text{vi})} = \cdots,$$
$$\mathscr{M}' = \mathscr{M}''' = \mathscr{M}^{(\text{v})} = \cdots.$$

In this section, we shall demonstrate the fundamental theorem, Theorem 3.9, in the theory of operators, which is called the double commutation theorem of J. von Neumann.

Definition 3.1. If a subalgebra of $\mathscr{L}(\mathfrak{H})$ is invariant under the *-operation, then it is called a **-subalgebra* of $\mathscr{L}(\mathfrak{H})$ or a **-algebra of operators* on $\mathfrak{H}$. A *-subalgebra $\mathscr{M}$ of $\mathscr{L}(\mathfrak{H})$ is said to be *nondegenerate* if $[\mathscr{M}\mathfrak{H}] = \mathfrak{H}$. A *$C^*$-algebra of operators* on $\mathfrak{H}$ means a nondegenerate *-subalgebra of $\mathscr{L}(\mathfrak{H})$ which is closed under the uniform topology.

Definition 3.2. A *von Neumann algebra* on $\mathfrak{H}$ is a **-subalgebra* $\mathscr{M}$ of $\mathscr{L}(\mathfrak{H})$ such that

$$\mathscr{M} = \mathscr{M}''. \tag{1}$$

A *factor* is a von Neumann algebra $\mathscr{M}$ with trivial center $\mathscr{M} \cap \mathscr{M}' = \mathbf{C}1$.

If a subset $\mathscr{S}$ of $\mathscr{L}(\mathfrak{H})$ is invariant under the *-operation, then $\mathscr{S}''$, the double commutant of $\mathscr{S}$, is the smallest von Neumann algebra containing $\mathscr{S}$, and it is called the von Neumann algebra *generated* by $\mathscr{S}$. To specify the Hilbert space upon which a von Neumann algebra $\mathscr{M}$ acts, we shall often use the notation $\{\mathscr{M},\mathfrak{H}\}$.

If $\{\mathscr{M}_1,\mathfrak{H}_1\}$ and $\{\mathscr{M}_2,\mathfrak{H}_2\}$ are von Neumann algebras and if there exists an isometry U of $\mathfrak{H}_1$ onto $\mathfrak{H}_2$ such that

$$U\mathscr{M}_1U^* = \mathscr{M}_2, \tag{2}$$

then $\{\mathscr{M}_1,\mathfrak{H}_1\}$ and $\{\mathscr{M}_2,\mathfrak{H}_2\}$ are said to be *spatially isomorphic.* The map π of $\mathscr{M}_1$ onto $\mathscr{M}_2$ defined by

$$\pi(x) = UxU^*, x \in \mathscr{M}_1,$$

is called a *spatial isomorphism.* We denote this fact by $\{\mathscr{M}_1,\mathfrak{H}_1\} \cong \{\mathscr{M}_2,\mathfrak{H}_2\}$. If there exists an isomorphism π of $\mathscr{M}_1$ onto $\mathscr{M}_2$ which preserves the *-operation, that is, $\pi(x^*) = \pi(x)^*$, $x \in \mathscr{M}_1$, then $\mathscr{M}_1$ and $\mathscr{M}_2$ are said to be *isomorphic.* But we should note that an isomorphism is, in general, not spatial.

Let $\{\mathscr{M}_i,\mathfrak{H}_i\}_{i\in I}$ be a family of von Neumann algebras. Let $\mathfrak{H}$ denote the direct sum $\sum_{i\in I}^{\oplus} \mathfrak{H}_i$ of Hilbert spaces $\{\mathfrak{H}_i\}_{i\in I}$. Each vector $\xi = \{\xi_i\}_{i\in I}$ in $\mathfrak{H}$ is denoted by $\sum_{i\in I}^{\oplus} \xi_i$. For each bounded sequence $\{x_i\}_{i\in I}$ in $\prod_{i\in I} \mathscr{M}_i$, we define an operator x on $\mathfrak{H}$ by

$$x \sum_{i\in I}^{\oplus} \xi_i = \sum_{i\in I}^{\oplus} x_i\xi_i.$$

Then x is a bounded operator on $\mathfrak{H}$. We denote it by $\sum_{i\in I}^{\oplus} x_i$. Let $\mathscr{M}$ be the set of all such x. Then the following is easily verified:

Proposition 3.3. *The subset $\mathscr{M}$ of $\mathscr{L}(\mathfrak{H})$ is a von Neumann algebra on $\mathfrak{H}$.*

Definition 3.4. The algebra $\mathscr{M}$ is called the *direct sum* of $\{\mathscr{M}_i\}_{i\in I}$ and denoted by $\{\mathscr{M},\mathfrak{H}\} = \sum_{i\in I}^{\oplus} \{\mathscr{M}_i,\mathfrak{H}_i\}$ or simply by $\mathscr{M} = \sum_{i\in I}^{\oplus} \mathscr{M}_i$.

Let $\mathfrak{H}$ be a fixed Hilbert space and $\{\mathfrak{H}_i\}_{i\in I}$ be a family of replicas of $\mathfrak{H}$. Let $\tilde{\mathfrak{H}}$ denote the direct sum $\sum_{i\in I}^{\oplus} \mathfrak{H}_i$. Let U_i be the isometry of $\mathfrak{H}$ onto $\mathfrak{H}_i$. For any operator $x \in \mathscr{L}(\tilde{\mathfrak{H}})$, putting

$$x_{i,j} = U_i^* x U_j, \qquad i,j \in I, \tag{3}$$

we obtain a matrix $(x_{i,j})$ of bounded operators on $\mathfrak{H}$. Clearly, the map: $x \in \mathscr{L}(\tilde{\mathfrak{H}}) \mapsto (x_{i,j})$ is injective, so we may write $x = (x_{i,j})$. Then we have, for $x = (x_{i,j})$ and $y = (y_{i,j})$,

$$\begin{cases} \alpha x + \beta y = (\alpha x_{i,j} + \beta y_{i,j}), \\ x^* = (x_{j,i}^*), \\ xy = (z_{i,j}),\ z_{i,j} = \sum_{k\in I} x_{i,k} y_{k,j}, \end{cases} \tag{4}$$

where the last summation in $z_{i,j}$ is taken in the strong operator topology. For each $x \in \mathscr{L}(\mathfrak{H})$ we define an operator $\tilde{x}$ in $\mathscr{L}(\tilde{\mathfrak{H}})$ by $\tilde{x} \sum_{i\in I}^{\oplus} \xi_i = \sum_{i\in I}^{\oplus} x\xi_i$. Then the map $\pi: x \in \mathscr{L}(\mathfrak{H}) \to \tilde{x} \in \mathscr{L}(\tilde{\mathfrak{H}})$ is an isomorphism of $\mathscr{L}(\mathfrak{H})$ *into* $\mathscr{L}(\tilde{\mathfrak{H}})$.

Proposition 3.5. *For an operator $\tilde{x} \in \mathscr{L}(\tilde{\mathfrak{H}})$ to be of the form $\tilde{x} = \pi(x)$ for some $x \in \mathscr{L}(\mathfrak{H})$ it is necessary and sufficient that $\tilde{x}$ commutes with all $U_i U_j^*$.*

Proof. The necessity is trivial, so we have only to prove the sufficiency. Suppose $\tilde{x} \in \mathscr{L}(\tilde{\mathfrak{H}})$ commutes with every $U_i U_j^*$. Then we have

$$U_i^* \tilde{x} U_j = 0, \qquad i \neq j,$$
$$U_i^* \tilde{x} U_i = U_j^* x U_j.$$

Put $x = U_i^* \tilde{x} U_i$, which does not depend on $i \in I$. Then x is in $\mathscr{L}(\mathfrak{H})$ and $\tilde{x} = \pi(x)$. Q.E.D.

Corollary 3.6. *In the same situation as above, for any subset $\mathscr{M}$ of $\mathscr{L}(\mathfrak{H})$, we have*

$$\pi(\mathscr{M})'' = \pi(\mathscr{M}''). \tag{5}$$

Proof. Since $U_i U_j^* \in \pi(\mathscr{L}(\mathfrak{H}))'$, it is in $\pi(\mathscr{M})'$. Hence $\pi(\mathscr{M})''$ is contained in $\pi(\mathscr{L}(\mathfrak{H}))$. But π is an isomorphism of $\mathscr{L}(\mathfrak{H})$ onto $\pi(\mathscr{L}(\mathfrak{H}))$, so that $\pi(\mathscr{M})'' = \pi(\mathscr{M})'' \cap \pi(\mathscr{L}(\mathfrak{H})) = \pi(\mathscr{M}'')$. Q.E.D.

Lemma 3.7. *If $\mathscr{M}$ is a nondegenerate *-algebra of operators on a Hilbert space $\mathfrak{H}$, then ξ belongs to $[\mathscr{M}\xi]$ for every $\xi \in \mathfrak{H}$.*

Proof. Let p be the projection of $\mathfrak{H}$ onto $[\mathscr{M}\xi] = \mathfrak{K}$. Putting $\xi' = p\xi$ and $\xi'' = (1-p)\xi$, we have $\xi = \xi' + \xi''$. Since $\mathfrak{K}$ is invariant under the *-algebra $\mathscr{M}$, so is the orthogonal complement $\mathfrak{K}^{\perp}$ of $\mathfrak{K}$. Hence the equality $x\xi' + x\xi'' = x\xi \in \mathfrak{K}$, $x \in \mathscr{M}$, yields that $x\xi'' = 0$. Take an arbitrary $\eta \in \mathfrak{H}$ and $\varepsilon > 0$. We can choose $\{x_i\}_{1<i<n} \subset \mathscr{M}$ and $\{\xi_i\}_{1<i<n} \subset \mathfrak{H}$ such that

$$\left\| \eta - \sum_{i=1}^{n} x_i \xi_i \right\| < \varepsilon.$$

Therefore, we have

$$\begin{aligned}(\xi''|\eta) &= \lim \left(\xi'' \middle| \sum_{i=1}^{n} x_i \xi_i \right) \\ &= \lim \sum_{i=1}^{n} (x_i^* \xi'' | \xi_i) = 0,\end{aligned}$$

so that ξ'' must be zero. Thus, $\xi = \xi'$ falls in $\mathfrak{K} = [\mathscr{M}\xi]$. Q.E.D.

Lemma 3.8. *Under the same assumption for $\mathscr{M}$ as in the previous lemma, for every $\xi \in \mathfrak{H}$, $a \in \mathscr{M}''$, and $\varepsilon > 0$, there exists an element $b \in \mathscr{M}$ such that*

$$\|(a-b)\xi\| < \varepsilon.$$

Proof. It suffices to show that $[\mathscr{M}\xi] = [\mathscr{M}''\xi]$. Let p be the projection of $\mathfrak{H}$ onto $[\mathscr{M}\xi]$. Then p is in $\mathscr{M}'$, so that p commutes with every operator in $\mathscr{M}''$; hence $[\mathscr{M}\xi]$ is invariant under $\mathscr{M}''$. Hence, by Lemma 3.7, $[\mathscr{M}''\xi] \subset [\mathscr{M}\xi]$. Q.E.D.

Now, we can prove the so-called double commutation theorem of J. von Neumann, which is fundamental in the whole theory of operator algebras.

Theorem 3.9. *Suppose $\mathscr{M}$ is a *-algebra of operators on a Hilbert space $\mathfrak{H}$. If $\mathscr{M}$ is σ-strongly* closed, then:*

(i) *There exists a greatest projection e in $\mathscr{M}$, and it is the projection of $\mathfrak{H}$ onto $[\mathscr{M}\mathfrak{H}]$. For every $x \in \mathscr{M}$, we have*

$$x = xe = ex,$$

in other words, e is the identity element of $\mathscr{M}$.

(ii) *The double commutant $\mathscr{M}''$ of $\mathscr{M}$ is the set of all the operators of the form $x + \alpha 1$, $x \in \mathscr{M}$, $\alpha \in \mathbf{C}$. Therefore, if $\mathscr{M}$ is nondegenerate, then $\mathscr{M}$ is a von Neumann algebra, that is,*

$$\mathscr{M} = \mathscr{M}''. \tag{6}$$

Proof. At first, suppose $\mathscr{M}$ is nondegenerate. Take an arbitrary $a \in \mathscr{M}''$. We shall show that a is well approximated by elements of $\mathscr{M}$ in the σ-strong topology. Let $\tilde{\mathfrak{H}}$ be the direct sum $\sum_{n=1}^{\infty\oplus} \mathfrak{H}_n$ of countably infinitely many replicas $\mathfrak{H}_n$ of $\mathfrak{H}$. If $\{\xi_n\}$ is a sequence in $\mathfrak{H}$ with $\sum_{n=1}^{\infty} \|\xi_n\|^2 < +\infty$, then $\tilde{\xi} = \sum_{n=1}^{\oplus} \xi_n$ is a vector in $\tilde{\mathfrak{H}}$. Let π denote the isomorphism of $\mathscr{L}(\mathfrak{H})$ into

$\mathscr{L}(\tilde{\mathfrak{H}})$ defined in Proposition 3.5. Then by Corollary 3.6, $\pi(a) = \tilde{a}$ is in $\pi(\mathscr{M})''$. Since $\pi(\mathscr{M}) = \tilde{\mathscr{M}}$ is a nondegenerate *-algebra of operators on $\tilde{\mathfrak{H}}$, it follows from Lemma 3.8 that for any $\varepsilon > 0$, there exists an element $\tilde{b} = \pi(b)$ in $\tilde{\mathscr{M}}$ such that $\|(\tilde{a} - \tilde{b})\tilde{\xi}\| < \varepsilon$, which means that

$$\left(\sum_{n=1}^{\infty} \|(a - b)\xi_n\|^2\right)^{1/2} < \varepsilon.$$

Therefore, a belongs to the σ-strong closure of $\mathscr{M}$. By Theorem 2.6, the σ-strong* closure of $\mathscr{M}$ is the σ-strong closure of $\mathscr{M}$. Hence the double commutation relation (6) holds for a nondegenerate σ-strongly* closed *-algebra $\mathscr{M}$ of operators on $\mathfrak{H}$.

Returning to the general situation, we can conclude by Theorem 2.6 again that $\mathscr{M}$ is σ-weakly closed. Hence the unit ball $\mathscr{M} \cap S$ is σ-weakly compact; so by the Krein–Milman theorem, $\mathscr{M} \cap S$ has an extremal point. Therefore, by Theorem I.10.2, $\mathscr{M}$ admits an identity, say e. Of course, the identity of $\mathscr{M}$ is the greatest projection in $\mathscr{M}$. Let $\mathfrak{K}$ be the range of e. Let $\mathscr{M}_{\mathfrak{K}}$ be the set of all the restriction operators $x_{\mathfrak{K}}$ of $x \in \mathscr{M}$ onto $\mathfrak{K}$. Since the map: $x \in \mathscr{M} \mapsto x_{\mathfrak{K}} \in \mathscr{M}_{\mathfrak{K}}$ is a homeomorphism with respect to the σ-weak topology, the unit ball of $\mathscr{M}_{\mathfrak{K}}$ is σ-weakly compact, being the image of the unit ball of $\mathscr{M}$ under the homeomorphism. Hence $\mathscr{M}_{\mathfrak{K}}$ is σ-weakly closed by Theorem 2.6(iv). Therefore, it follows from the above arguments that $\mathscr{M}''_{\mathfrak{K}} = \mathscr{M}_{\mathfrak{K}}$. Every operator in $\mathscr{M}''$ leaves $\mathfrak{K}$ invariant as well as its orthogonal complement $\mathfrak{K}^\perp$. The restriction $\mathscr{M}_{\mathfrak{K}^\perp}$ of $\mathscr{M}$ to $\mathfrak{K}^\perp$ is reduced to the trivial algebra $\{0\}$, so that we have $(\mathscr{M}_{\mathfrak{K}^\perp})' = \mathscr{L}(\mathfrak{K}^\perp)$. Hence the restriction $a_{\mathfrak{K}^\perp}$ of each $a \in \mathscr{M}''$ is a scalar multiple $\lambda 1_{\mathfrak{K}^\perp}$ of the identity $1_{\mathfrak{K}^\perp}$ on $\mathfrak{K}^\perp$, so that $a - \lambda 1$ vanishes on $\mathfrak{K}^\perp$ and belongs to $\mathscr{M}''$. Therefore, $(a - \lambda 1)_{\mathfrak{K}}$ falls in $\mathscr{M}_{\mathfrak{K}}$; hence there exists an operator $b \in \mathscr{M}$ with $b_{\mathfrak{K}} = (a - \lambda 1)_{\mathfrak{K}}$. Both $a - \lambda 1$ and b vanish on $\mathfrak{K}^\perp$, so that $a - \lambda 1 = b$; hence $a = b + \lambda 1$. Q.E.D.

Suppose now $\mathscr{M}$ is a nondegenerate *-algebra of operators on a Hilbert space $\mathfrak{H}$. Take a projection $e \in \mathscr{M}$ with range $\mathfrak{K}$. Then each operator in $e\mathscr{M}e$ leaves $\mathfrak{K}$ invariant, so that the set of the restriction operator $x_{\mathfrak{K}}$ of $x \in e\mathscr{M}e$ onto $\mathfrak{K}$ is a nondegenerate *-algebra. We denote this algebra on $\mathfrak{K}$ by $\mathscr{M}_e$ or $\mathscr{M}_{\mathfrak{K}}$. On the other hand, every operator in $\mathscr{M}'$ leaves $\mathfrak{K}$ invariant; hence the map: $x \in \mathscr{M}' \mapsto x_{\mathfrak{K}} \in \mathscr{L}(\mathfrak{K})$ is a *-homomorphism. The image of $\mathscr{M}'$ under this map is also a nondegenerate *-algebra of operators on $\mathfrak{K}$. We denote it by $\mathscr{M}'_e$ or $\mathscr{M}'_{\mathfrak{K}}$.

Proposition 3.10. *If $\{\mathscr{M},\mathfrak{H}\}$ is a von Neumann algebra and e is a projection in $\mathscr{M}$ with range $\mathfrak{K}$, then $\mathscr{M}_e$ and $\mathscr{M}'_e$ are both von Neumann algebras on $\mathfrak{K}$ and*

$$(\mathscr{M}_e)' = \mathscr{M}'_e. \tag{7}$$

Proof. Let S denote the unit ball of $\mathscr{L}(\mathfrak{H})$. Then S is weakly compact, and so are both $S \cap \mathscr{M}$ and $S \cap \mathscr{M}'$. Hence $e(S \cap \mathscr{M})e$ and $(S \cap \mathscr{M}')e$ are both weakly compact, since the map: $x \in \mathscr{L}(\mathfrak{H}) \mapsto exe$ is weakly continuous and the map: $x \in e\mathscr{L}(\mathfrak{H})e \mapsto x_{\mathfrak{K}} \in \mathscr{L}(\mathfrak{K})$ is a weakly continuous isometry. Therefore, Theorem 2.6(iv) yields the first half of our assertion.

The inclusion $(\mathscr{M}_e)' \supset \mathscr{M}'_e$ is clear. We shall show the reverse inclusion. Take an arbitrary unitary operator $u' \in (\mathscr{M}_e)'$. Let $\mathfrak{M}$ denote the closed subspace $[\mathscr{M}\mathfrak{K}]$ spanned by $x\xi$, $x \in \mathscr{M}$, $\xi \in \mathfrak{K}$. Define an operator u'_0 on the finite linear combinations of $x\xi$'s, $x \in \mathscr{M}$, $\xi \in \mathfrak{K}$, by

$$u_0 \sum_{i=1}^{n} x_i\xi_i = \sum_{i=1}^{n} x_i u'\xi_i.$$

Then we have

$$\begin{aligned}\left\| u'_0 \sum_{i=1}^{n} x_i\xi_i \right\|^2 &= \sum_{i,j=1}^{n} (x_i u'\xi_i | x_j u'\xi_j) \\ &= \sum_{i,j=1}^{n} (ex_j^* x_i e u'\xi_i | u'\xi_j) \\ &= \sum_{i,j=1}^{n} (u' e x_j^* x_i e \xi_i | u'\xi_j) \\ &= \sum_{i,j=1}^{n} (e x_j^* x_i e \xi_i | \xi_j) \\ &= \sum_{i,j=1}^{n} (x_j^* x_i \xi_i | \xi_j) = \left\| \sum_{i=1}^{n} x_i\xi_i \right\|^2.\end{aligned}$$

Hence u'_0 can be extended to an isometry of $\mathfrak{M}$ onto $\mathfrak{M}$, which is also denoted by u'_0. Putting $u'_0\mathfrak{M}^\perp = 0$, where $\mathfrak{M}^\perp$ means the orthogonal complement of $\mathfrak{M}$, we obtain a partial isometry u'_0 on $\mathfrak{H}$. Noticing that the projection p of $\mathfrak{H}$ onto $\mathfrak{M}$ belongs to the center $\mathscr{M} \cap \mathscr{M}'$, we can easily conclude that u'_0 belongs to $\mathscr{M}'$. It is obvious by the construction of u'_0 that $(u'_0)_{\mathfrak{K}} = u'$. By Proposition I.4.9, every element in $(\mathscr{M}_e)'$ is a linear combination of four unitary operators in $(\mathscr{M}_e)'$. Therefore, the map: $x \in \mathscr{M}' \mapsto x_{\mathfrak{K}} \in (\mathscr{M}_e)'$ is surjective. Q.E.D.

Definition 3.11. The von Neumann algebras $\mathscr{M}_e$ and $\mathscr{M}'_e$ on $\mathfrak{K}$ obtained in the preceding proposition are respectively called the *reduced* von Neumann algebra of $\mathscr{M}$ on $\mathfrak{K}$ and the *induced* von Neumann algebra of $\mathscr{M}'$ on $\mathfrak{K}$. The map: $x \in \mathscr{M}' \mapsto x_{\mathfrak{K}} \in \mathscr{M}'_e$ is called the *induction* of $\mathscr{M}'$ onto $\mathscr{M}'_e$.

Proposition 3.12. *If $\{\mathscr{M},\mathfrak{H}\}$ is a von Neumann algebra, then every σ-weakly closed left (resp. right) ideal $\mathfrak{m}$ contains a unique projection e such that $\mathfrak{m} = \mathscr{M}e$ (resp. $\mathfrak{m} = e\mathscr{M}$). If $\mathfrak{m}$ is a two-sided ideal, then e belongs to the center $\mathscr{M} \cap \mathscr{M}'$. Therefore, if $\mathscr{M}$ is a factor, then every nonzero two-sided ideal is σ-weakly dense in it.*

Proof. Noticing that $\mathfrak{n} = \mathfrak{m} \cap \mathfrak{m}^*$ is a σ-weakly closed *-algebra of operators on $\mathfrak{H}$, we can find the greatest projection e in $\mathfrak{n}$ by Theorem 3.9. Then we have $\mathscr{M}e \subset \mathfrak{m}$, since $e \in \mathfrak{m}$ and $\mathfrak{m}$ is a left ideal. Conversely, take an arbitrary $x \in \mathfrak{m}$. Since x^*x is in $\mathfrak{m}$ and self-adjoint, x^*x falls in $\mathfrak{m} \cap \mathfrak{m}^* = \mathfrak{n}$. Hence $|x| = (x^*x)^{1/2}$ is in $\mathfrak{n}$, since $\mathfrak{n}$ is a C^*-algebra of operators on $\mathfrak{H}$; hence we have $|x|e = |x|$. Therefore, considering the polar decomposition $x = u|x|$ of x, we get $x = xe$; hence $x \in \mathscr{M}e$. Thus $\mathfrak{m} \subset \mathscr{M}e$.

If $\mathfrak{m}$ is a two-sided ideal, then we have, for every $x \in \mathscr{M}$, $ex \in \mathfrak{m} = \mathscr{M}e$; hence $ex = (ex)e = exe$. Considering x^*, we have $ex^* = ex^*e$, so that $ex = exe = (ex^*e)^* = (ex^*)^* = xe$. Therefore, e is central. Q.E.D.

Proposition 3.13. *Let $\{\mathscr{M},\mathfrak{H}\}$ be a von Neumann algebra and $\mathfrak{m}$ a two-sided ideal of $\mathscr{M}$. Let $\bar{\mathfrak{m}}$ be the σ-weak closure of $\mathfrak{m}$. Then, for every $x \in (\bar{\mathfrak{m}})_+$, there exists an increasing net $\{x_i\} \subset \mathfrak{m}_+$ convergent strongly to x.*

Proof. Let $\{u_i\}$ be an approximate identity in $\mathfrak{m}$. Let e be the greatest projection in $\bar{\mathfrak{m}}$. Then $\bar{\mathfrak{m}} = \mathscr{M}e$. From the proof of Theorem 3.9, it follows that e is the projection of $\mathfrak{H}$ onto $[\mathfrak{m}\mathfrak{H}]$. For every $a \in \mathfrak{m}$, we have $a\xi = \lim u_i a\xi$, $\xi \in \mathfrak{H}$, so that $\{u_i\}$ converges strongly to e, being bounded. Hence we have

$$x = x^{1/2}ex^{1/2} = \text{s-lim}\, x^{1/2}u_i x^{1/2};$$

so putting $x_i = x^{1/2}u_i x^{1/2} \in \mathfrak{m}$, we get a desired increasing net in $\mathfrak{m}_+$. Q.E.D.

Proposition 3.14. *Let $\{\mathscr{M},\mathfrak{H}\}$ be a von Neumann algebra. If $x = u|x|$ is the polar decomposition of an operator x in $\mathscr{M}$, then the partial isometry u belongs to $\mathscr{M}$ as well as the absolute value $|x|$.*

Proof. Let v be an arbitrary unitary operator in $\mathscr{M}'$. Then we have

$$x = vxv^* = vuv^*v|x|v^*.$$

Hence, by the uniqueness of the polar decomposition of x, we get $vuv^* = u$ and $v|x|v^* = |x|$, so that u and $|x|$ both commute with every unitary operator in $\mathscr{M}'$; thus u and $|x|$ are both in $\mathscr{M}$ because $\mathscr{M}'$ is spanned linearly by its unitary operators. Q.E.D.

Proposition 3.15. *Every two-sided ideal of a von Neumann algebra is self-adjoint.*

Proof. Let $\mathfrak{m}$ be a two-sided ideal of a von Neumann algebra $\{\mathscr{M},\mathfrak{H}\}$. Let x be an element of $\mathfrak{m}$ and $x = u|x|$ be the polar decomposition of x. Proposition 3.14, u is in $\mathscr{M}$, so that $|x| = u^*x$ is in $\mathfrak{m}$. Therefore $x^* = |x|u^*$ is in $\mathfrak{m}$. Q.E.D.

Definition 3.16. Let $\{\mathscr{M},\mathfrak{H}\}$ be a von Neumann algebra. A subset $\mathfrak{A}$ of $\mathfrak{H}$ is called *separating* (resp. *cyclic*) for $\mathscr{M}$ if $a\xi = 0$, $a \in \mathscr{M}$, for every $\xi \in \mathfrak{A}$ implies $a = 0$ (resp. the smallest invariant subspace $[\mathscr{M}\mathfrak{A}]$ under $\mathscr{M}$ containing $\mathfrak{A}$ is the whole space $\mathfrak{H}$).

Proposition 3.17. *Let $\{\mathscr{M},\mathfrak{H}\}$ be a von Neumann algebra. For a subset $\mathfrak{A}$ of $\mathfrak{H}$, the following two statements are equivalent:*

(i) *$\mathfrak{A}$ is cyclic for $\mathscr{M}$;*
(ii) *$\mathfrak{A}$ is separating for $\mathscr{M}'$.*

PROOF. (i) $\Rightarrow$ (ii): Suppose $\mathfrak{A}$ is cyclic for $\mathscr{M}$. Suppose that $a\mathfrak{A} = \{0\}$ for some $a \in \mathscr{M}'$. Let p be the projection of $\mathfrak{H}$ onto the null space of a. Then p is in $\mathscr{M}'$ and $p\mathfrak{H} \supset \mathfrak{A}$. Hence $p\mathfrak{H} \supset [\mathscr{M}\mathfrak{A}] = \mathfrak{H}$, so that $p = 1$, which means that $a = 0$.

(ii) $\Rightarrow$ (i): Suppose $\mathfrak{A}$ is separating for $\mathscr{M}'$. Let p denote the projection of $\mathfrak{H}$ onto $[\mathscr{M}\mathfrak{A}]$. Then p is in $\mathscr{M}'$ and $(1 - p)\mathfrak{A} = \{0\}$. Hence $1 - p = 0$ by assumption. Therefore, $\mathfrak{A}$ is cyclic for $\mathscr{M}$. Q.E.D.

Definition 3.18. A von Neumann algebra is said to be *σ-finite* if it admits at most countably many orthogonal projections.

Proposition 3.19. *Let $\{\mathscr{M},\mathfrak{H}\}$ be a von Neumann algebra. Then the following three statements are equivalent:*

(i) *$\mathscr{M}$ is σ-finite.*
(ii) *There exists a countable separating subset of $\mathfrak{H}$ for $\mathscr{M}$.*
(iii) *There exists a faithful positive linear functional in $\mathscr{M}_*$.*

PROOF. (i) $\Rightarrow$ (ii). Let $\{\xi_i\}_{i\in I}$ be a maximal family of vectors in $\mathfrak{H}$ such that for distinct $i,j \in I$, $[\mathscr{M}'\xi_i]$ and $[\mathscr{M}'\xi_j]$ are orthogonal. Since the projections e_i of $\mathfrak{H}$ onto $[\mathscr{M}'\xi_i]$ are in $\mathscr{M}$ and orthogonal, $\{\xi_i\}_{i\in I}$ must be countable. By the maximality of $\{\xi_i\}_{i\in I}$, we have $\sum_{i\in I}^{\oplus} [\mathscr{M}'\xi_i] = \mathfrak{H}$, i.e., $\sum_{i\in I} e_i = 1$. Hence $\{\xi_i\}_{i\in I}$ is cyclic for $\mathscr{M}'$, so that it is separating for $\mathscr{M}$ by Proposition 3.17.

(ii) $\Rightarrow$ (iii): Let $\{\xi_n\}$ be a countable separating family of vectors in $\mathfrak{H}$ for $\mathscr{M}$. Let φ be a positive linear functional in $\mathscr{M}_*$ defined by

$$\varphi(x) = \sum_{n=1}^{\infty} \frac{1}{2^n} \frac{1}{\|\xi_n\|^2} (x\xi_n|\xi_n), \qquad x \in \mathscr{M}.$$

Since $\varphi(x^*x) = 0$ if and only if $x\xi_n = 0$ for all n, φ is faithful.

(iii) $\Rightarrow$ (i): Let φ be a faithful positive linear functional on $\mathscr{M}$. Let $\{e_i\}_{i\in I}$ be an orthogonal family of projections in $\mathscr{M}$. Let I_n be the set of indices i with $\varphi(e_i) \geq 1/n$, $n = 1,2,\ldots$. Since φ is faithful, we have $I = \bigcup_{n=1}^{\infty} I_n$. But the inequality

$$\varphi(1) \geq \varphi\left(\sum_{i\in I_n} e_i\right) \geq \sum_{i\in I_n} \varphi(e_i) \geq \frac{1}{n} \cdot \text{card } I_n,$$

shows that I_n must be finite, where card I_n means the cardinal number of I_n. Hence I is countable. Q.E.D.

Proposition 3.20. *Let $\{\mathscr{M},\mathfrak{H}\}$ be a von Neumann algebra. If $\varphi \in \mathscr{M}_*$ is positive, then there exists a sequence $\{\xi_n\}$ in $\mathfrak{H}$ such that $\varphi = \sum_{n=1}^{\infty} \omega_{\xi_n}$ and $\sum_{n=1}^{\infty} \|\xi_n\|^2 < +\infty$, where ω_ξ means the functional in $\mathscr{M}_*$ given by $\omega_\xi(x) = (x\xi|\xi)$, $x \in \mathscr{M}$, for each $\xi \in \mathfrak{H}$.*

PROOF. We use the notations in the proof of Theorem 3.9. It follows from Theorem 2.6 that there exists a pair $\tilde{\eta}$, $\tilde{\zeta}$ in $\tilde{\mathfrak{H}}$ such that $\varphi(x) = (\pi(x)\tilde{\eta}|\tilde{\zeta})$, $x \in \mathcal{M}$, i.e., $\varphi = {}^t\pi(\omega_{\tilde{\eta},\tilde{\zeta}})$. Since $\varphi(x^*) = \varphi(x)$, we have $(\pi(x)\tilde{\eta}|\tilde{\zeta}) = (\pi(x)\tilde{\zeta}|\tilde{\eta})$, $x \in \mathcal{M}$, so that

$$4\varphi(x) = (\pi(x)(\tilde{\eta} + \tilde{\zeta})|\tilde{\eta} + \tilde{\zeta}) - (\pi(x)(\tilde{\eta} - \tilde{\zeta})|\tilde{\eta} - \tilde{\zeta}).$$

Therefore, we have $\varphi \leq {}^t\pi(\omega_{\tilde{\eta}+\tilde{\zeta}})$. We then define a sesquilinear form B_φ on $\pi(\mathcal{M})(\tilde{\eta} + \tilde{\zeta})$ by

$$B_\varphi(\pi(x)(\tilde{\eta} + \tilde{\zeta}), \pi(y)(\tilde{\eta} + \tilde{\zeta})) = \varphi(y^*x), \qquad x, y \in \mathcal{M}.$$

The inequality, $\varphi \leq {}^t\pi(\omega_{\tilde{\eta}+\tilde{\zeta}})$, guarantees that B_φ is well defined as well as bounded. Since B_φ is positive, there exists a positive bounded operator h on $[\pi(\mathcal{M})(\tilde{\eta} + \tilde{\zeta})]$ such that

$$\varphi(y^*x) = (h\pi(x)(\tilde{\eta} + \tilde{\zeta})|\pi(y)(\tilde{\eta} + \tilde{\zeta})), \qquad x, y \in \mathcal{M}.$$

It then follows that h commutes with $\pi(\mathcal{M})$. Set $\tilde{\xi} = h^{1/2}(\tilde{\eta} + \tilde{\zeta})$. We then have $\varphi(x) = (\pi(x)\tilde{\xi}|\tilde{\xi})$, which means that $\varphi = \sum_{n=1}^{\infty} \omega_{\xi_n}$, where $\tilde{\xi} = \sum_{n=1}^{\infty \oplus} \xi_n$. Q.E.D.

EXERCISES

1. Let A be a C^*-algebra of operators on a Hilbert space $\mathfrak{H}$. Show that if $A \subset \mathcal{LC}(\mathfrak{H})$, then $\{A' \cap \mathcal{LC}(\mathfrak{H})\}' \cap \mathcal{LC}(\mathfrak{H}) = A$.

2. Let A be a C^*-algebra with a nonzero minimal projection e.
   (a) Show that if $\{\pi, \mathfrak{H}\}$ is an irreducible representation of A such that $\pi(e) \neq 0$, then $\pi(e)$ is a projection of rank 1.
   (b) Show that if A is simple, then $A \cong \mathcal{LC}(\mathfrak{H})$ for some Hilbert space $\mathfrak{H}$.

3. Let $\{\Gamma_i, \mu_i\}_{i \in I}$ be a family of probability measure spaces and suppose I is uncountable. Let $\{\Gamma, \mu\} = \prod_{i \in I} \{\Gamma_i, \mu_i\}$ be the product measure space. Set $\mathfrak{H} = L^2(\Gamma, \mu)$. Let $\mathcal{M}$ be the von Neumann algebra on $\mathfrak{H}$ generated by multiplication operators by bounded measurable functions. Show that the constant function 1 is a separating and cyclic vector for $\mathcal{M}$, yet $\mathfrak{H}$ is not separable.

4. (Asymmetry Riesz Decomposition) Let $\mathcal{M}$ be a von Neumann algebra. Show that if $\sum_{i \in I} x_i^* x_i = \sum_{j \in J} y_j^* y_j$ in the σ-strong topology of $\mathcal{M}$, then there exists a family $\{z_{i,j}\}$ in $\mathcal{M}$ such that $\sum_j z_{i,j}^* z_{i,j} = x_i x_i^*$ and $\sum_i z_{i,j} z_{i,j}^* = y_j y_j^*$. (*Hint*: Set $a = \sum_{i \in I} x_i^* x_i = \sum_{j \in J} y_j^* y_j$. Show that there exists $\{s_i\}$ and $\{t_j\}$ such that $x_i = s_i a^{1/2}$, $y_j = t_j a^{1/2}$ and $\sum_{i \in I} s_i^* s_i = \sum_{j \in J} t_j^* t_j = p$ is the projection to the closure of the range of a. Set $z_{i,j} = t_j a s_i^*$.)

4. Density Theorems

In this section, we shall establish various density theorems in a von Neumann algebra that provide powerful tools for later study.

First, we begin with a discussion of the continuity of functional calculus.

Proposition 4.1. *The adjoint operation is strongly continuous in the set of all normal operators on a Hilbert space* $\mathfrak{H}$. *In other words, the strong operator topology and the strong* operator topology coincide in the set of normal operators.*

PROOF. The assertion is an immediate consequence of the following inequality for normal operators a and b:

$$\begin{aligned}\|(a^* - b^*)\xi\|^2 &= (aa^*\xi|\xi) + (bb^*\xi|\xi) - (ab^*\xi|\xi) - (ba^*\xi|\xi)\\ &= \|a\xi\|^2 - \|b\xi\|^2 + ((bb^* - ab^*)\xi|\xi) + ((bb^* - ba^*)\xi|\xi)\\ &= \|a\xi\|^2 - \|b\xi\|^2 + ((b-a)b^*\xi|\xi) + (\xi|(b-a)b^*\xi)\\ &\leq \|(a-b)\xi\|(\|a\xi\| + \|b\xi\|) + 2\|(b-a)b^*\xi\|\,\|\xi\|.\end{aligned}$$

Q.E.D.

Remark 4.2. The strong convergence of a net $\{a_i\}$ of normal operators to some operator does not guarantee the strong convergence of $\{a_i^*\}$. Namely, the normality of the limit operator is essential in the above result.

Let $\mathfrak{H}$ be a Hilbert space. Suppose that $\{a_1, \dots, a_n\}$ are a commutative family of normal operators on $\mathfrak{H}$. Let A be the C^*-algebra generated by $a_1, \dots, a_n$ and the identity 1. Obviously, A is abelian. So it is identified with the algebra $C(\Omega)$ of all continuous functions on the spectrum Ω of A, which is a compact space. Consider the map: $\omega \in \Omega \mapsto (a_1(\omega), \dots, a_n(\omega)) \in \mathbf{C}^n$. This map is of course continuous. Since $\{a_1, \dots, a_n, 1\}$ generates A, this map is injective, hence a homeomorphism by the compactness of Ω. Therefore, we may identify Ω and its image in $\mathbf{C}^n$. This subset Ω of $\mathbf{C}^n$ is called the *joint spectrum* of $\{a_1, \dots, a_n\}$. With this identification, each polynomial $p(\lambda_1, \dots, \lambda_n)$ of n variables corresponds to $p(a_1, a_2, \dots, a_n) \in A$. For example, if

$$\Lambda_j(\lambda_1, \dots, \lambda_n) = \lambda_j,$$

then $\Lambda_j(a_1, \dots, a_n) = a_j$. For each $f \in C(\Omega)$, the corresponding element in A is denoted by $f(a_1, \dots, a_n)$.

Given a closed subset G of $\mathbf{C}^n$, let $\mathscr{L}_G^n$ denote the set of all commuting n tuples $(a_1, \dots, a_n)$ of normal operators on $\mathfrak{H}$ whose joint spectrum is contained in G. In particular, $\mathscr{L}_{\mathbf{C}}$ denotes the set of all normal operators on $\mathfrak{H}$. If f is a continuous function on G, then the restriction $f|_\Omega$ of f to the joint spectrum Ω of $(a_1, \dots, a_n) \in \mathscr{L}_G^n$ defines an element $f|_\Omega(a_1, \dots, a_n)$ of the C^*-algebra generated by $\{a_1, a_2, \dots, a_n, 1\}$. We denote this element by $f(a_1, a_2, \dots, a_n)$.

Lemma 4.3. *If G is a compact subset of* $\mathbf{C}^n$, *then for any $f \in C(G)$, the functional calculus $f:(a_1, \dots, a_n) \in \mathscr{L}_G^n \mapsto f(a_1, \dots, a_n) \in \mathscr{L}_{\mathbf{C}}$ is strongly continuous.*

PROOF. We first recall that every algebraic operation is strongly* continuous on bounded subsets of $\mathscr{L}(\mathfrak{H})$ as a function of two variables. Hence, if p is a polynomial of n variables, then the map: $(a_1, \dots, a_n) \in \mathscr{L}_{\mathbf{C}^n}^n \mapsto p(a_1, \dots, a_n) \in \mathscr{L}_{\mathbf{C}}$ is strongly* continuous on bounded parts, hence on $\mathscr{L}_G^n$. By the Stone–

Weierstrass theorem, for any $f \in C(G)$, there exists a sequence $\{p_k(\lambda_1,\bar{\lambda}_1,\ldots,\lambda_n,\bar{\lambda}_n)\}$ of polynomials of $2n$ variables $(\lambda_1,\bar{\lambda}_1,\ldots,\lambda_n,\bar{\lambda}_n)$ such that

$$\lim_{k\to\infty} \sup\{|f(\lambda_1,\ldots,\lambda_n) - p_k(\lambda_1,\bar{\lambda}_1,\ldots,\lambda_n,\bar{\lambda}_n)|:(\lambda_1,\ldots,\lambda_n) \in G\} = 0.$$

Therefore, we have

$$\lim_{k\to\infty} \sup\{\|f(a_1,\ldots,a_n) - p_k(a_1,a_1^*,\ldots,a_n,a_n^*)\|:(a_1,\ldots,a_n) \in \mathscr{L}_G^n\} = 0$$

because $\|g(a_1,\ldots,a_n)\| = \sup\{|g(\lambda_1,\ldots,\lambda_n)|:(\lambda_1,\ldots,\lambda_n) \in \Omega\}$ for any $g \in C(\Omega)$ with Ω the joint spectrum of $(a_1,\ldots,a_n) \in \mathscr{L}_G^n$. As we have noted first, the map:

$$(a_1,\ldots,a_n) \in \mathscr{L}_G^n \mapsto p_k(a_1,a_1^*,\ldots,a_n,a_n^*)\xi \in \mathfrak{H}$$

is continuous for every $\xi \in \mathfrak{H}$ and $k = 1,2,\ldots$. Being a uniform limit of continuous maps, the map:

$$(a_1,\ldots,a_n) \in \mathscr{L}_G^n \mapsto f(a_1,\ldots,a_n)\xi \in \mathfrak{H}$$

is also continuous, which means that the functional calculus $f:(a_1,\ldots,a_n) \in \mathscr{L}_G^n \mapsto f(a_1,\ldots,a_n) \in \mathscr{L}_{\mathbf{C}}^n$ is strongly continuous. Q.E.D.

By definition, the Cayley transform is the map:

$$h \in \mathscr{L}_h = \mathscr{L}_{\mathbf{R}} \mapsto (h - i)(h + i)^{-1},$$

which maps a self-adjoint operator h into a unitary operator.

Lemma 4.4. *The Cayley transform is strongly continuous on* $\mathscr{L}_h$.

Proof. For any $h,k \in \mathscr{L}_h$, we have

$$(h - i)(h + i)^{-1} - (k - i)(k + i)^{-1} = 2i(h + i)^{-1}(h - k)(k + i)^{-1}.$$

Since $\|(h + i)^{-1}\| \le 1$, we have, for any $\xi \in \mathfrak{H}$,

$$\|(h - i)(h + i)^{-1}\xi - (k - i)(k + i)\xi\| \le 2\|(h - k)(k + i)^{-1}\xi\|.$$

Thus, the Cayley transform is strongly continuous. Q.E.D.

Lemma 4.5. *Let G be a closed subset of* **C**. *If f is a continuous function on G vanishing at infinity, then the map $f: a \in \mathscr{L}_G \mapsto f(a) \in \mathscr{L}_{\mathbf{C}}$ is strongly continuous.*

Proof. We first extend f to a continuous function on **C** vanishing at infinity by the Tietze extension theorem. We then define a function g on the two-dimensional torus $\mathbf{T}^2$ by

$$g(z,w) = \begin{cases} f\left(\dfrac{z+1}{i(z-1)} + \dfrac{w+1}{w-1}\right) & \text{for } |z| = |w| = 1, \quad z \neq 1 \text{ and } w \neq 1; \\ 0 & \text{otherwise.} \end{cases}$$

It follows that g is continuous on $\mathbf{T}^2$, so that the map: $(u,v) \in \mathscr{L}^2_{\mathbf{T}^2} \mapsto g(u,v) \in \mathscr{L}_{\mathbf{C}}$ is strongly continuous by the previous lemma. For each $a \in \mathscr{L}_{\mathbf{C}}$ with $a = h + ik$, $h,k \in \mathscr{L}_h$, we define

$$u(a) = (h - i)(h + i)^{-1} \quad \text{and} \quad v(a) = (k - i)(k + i)^{-1}.$$

We then have $(u(a),v(a)) \in \mathscr{L}^2_{\mathbf{T}^2}$ and $f(a) = g(u(a),v(a))$. By Lemma 4.4, the map: $a \in \mathscr{L}_{\mathbf{C}} \mapsto (u(a),v(a)) \in \mathscr{L}^2_{\mathbf{T}^2}$ is strongly continuous. Hence the functional calculus $f: a \in \mathscr{L}_G \mapsto f(a) \in \mathscr{L}_{\mathbf{C}}$ is strongly continuous, being the composition of three strongly continuous maps: $a \in \mathscr{L}_G \mapsto a \in \mathscr{L}_{\mathbf{C}}$; $a \in \mathscr{L}_{\mathbf{C}} \mapsto (u(a),v(a)) \in \mathscr{L}^2_{\mathbf{T}^2}$; and $(u,v) \in \mathscr{L}^2_{\mathbf{T}^2} \mapsto g(u,v) \in \mathscr{L}_{\mathbf{C}}$. Q.E.D.

Lemma 4.6. *If f is a bounded continuous function on a closed subset G of $\mathbf{C}$, then the functional calculus $f: a \in \mathscr{L}_G \mapsto f(a) \in \mathscr{L}_{\mathbf{C}}$ is strongly continuous.*

Proof. Let $D = \{\lambda \in \mathbf{C}: |\lambda| < 1\}$ be the open unit disk. We consider the function $f(\lambda)/\lambda$ on $G \cap D^c$, and extend it to a continuous function g on G vanishing at infinity by the Tietze extension theorem. Put $h(\lambda) = f(\lambda) - \lambda g(\lambda)$. It follows then that h is a continuous function on G with compact support, and that

$$f(a) = ag(a) + h(a), \qquad a \in \mathscr{L}_G.$$

By the previous lemma, the maps: $a \in \mathscr{L}_G \mapsto g(a) \in \mathscr{L}_{\mathbf{C}}$ and $a \in \mathscr{L}_G \mapsto h(a) \in \mathscr{L}_{\mathbf{C}}$ are both strongly continuous. For any $\xi \in \mathfrak{H}$, a and $b \in \mathscr{L}_G$, we have

$$\begin{aligned}\|(ag(a) - bg(b))\xi\| &\le \|g(a)(a - b)\xi\| + \|(g(a) - g(b))b\xi\| \\ &\le \|g(a)\|\,\|(a - b)\xi\| + \|(g(a) - g(b))b\xi\| \\ &\le \|g\|_\infty \|(a - b)\xi\| + \|(g(a) - g(b))b\xi\|,\end{aligned}$$

where $\|g\|_\infty = \sup\{|g(\lambda)|: \lambda \in G\}$. Hence $a \in \mathscr{L}_G \mapsto ag(a) \in \mathscr{L}_{\mathbf{C}}$ is strongly continuous; therefore so is the map: $a \in \mathscr{L}_G \mapsto f(a) \in \mathscr{L}_{\mathbf{C}}$. Q.E.D.

Theorem 4.7. *If f is a continuous function on a closed subset G of $\mathbf{C}$ such that $(1 + |\lambda|)^{-1} f(\lambda)$ is bounded on G, then the functional calculus $f: a \in \mathscr{L}_G \mapsto f(a) \in \mathscr{L}_{\mathbf{C}}$ is strongly continuous.*

Proof. As in the previous lemma, we write f as

$$f(\lambda) = \lambda g(\lambda) + h(\lambda), \qquad \lambda \in G,$$

where g is a bounded continuous function on G and h is a continuous function on G vanishing outside the open unit disk. We then have $f(a) = ag(a) + h(a)$, $a \in \mathscr{L}_G$. By the previous lemma, the maps: $a \in \mathscr{L}_G \mapsto g(a) \in \mathscr{L}_{\mathbf{C}}$ and $a \in \mathscr{L}_G \mapsto h(a) \in \mathscr{L}_{\mathbf{C}}$ are both strongly continuous. Applying the arguments in the last part of the proof of the previous lemma, we conclude the strong continuity of the functional calculus $f: a \in \mathscr{L}_G \mapsto f(a) \in \mathscr{L}_{\mathbf{C}}$. Q.E.D.

Theorem 4.8. *If A is a *-algebra of operators on a Hilbert space $\mathfrak{H}$, then the unit ball of A is strongly* dense in the unit ball of the weak closure of A.*

PROOF. Considering the uniform closure of A, we may assume that A is a uniformly closed *-algebra of operators (C^*-algebra) on $\mathfrak{H}$. Furthermore, replacing $\mathfrak{H}$ by the essential subspace $[A\mathfrak{H}]$, we may also assume that A is nondegenerate. Hence, by Theorem 3.9, the weak closure $\mathscr{M}$ of A is a von Neumann algebra on $\mathfrak{H}$. Let $\{u_i\}$ be an approximate identity of A. For every $x \in A$ and $\xi \in \mathfrak{H}$, we have $\|x\xi - u_i x\xi\| \leq \|x - u_i x\| \|\xi\| \to 0$. Since A is nondegenerate and $\{u_i\}$ is bounded, $\{u_i\}$ converges strongly, hence strongly*, to the identity 1. Therefore, the strong* closure of the unit ball of A contains the identity 1. Hence we may assume that A contains the identity.

Let S be the unit ball of $\mathscr{M}$. Define a map f of $\mathscr{M}$ into itself by

$$f(x) = 2(1 + xx^*)^{-1}x. \tag{1}$$

Let $x = uh = ku$ be the left and right polar decompositions of x, where $h = (x^*x)^{1/2}$ and $k = (xx^*)^{1/2}$. Since $uhu^* = k$ and $f(x) = 2(1 + k^2)^{-1}ku$, we have

$$f(x) = 2x(1 + x^*x)^{-1}. \tag{1'}$$

Since $|2t(1 + t^2)^{-1}| \leq 1$ for every $t \in \mathbf{R}$, the image of $\mathscr{M}$ under f is contained in S. Next, we define a map g S into $\mathscr{M}$ by:

$$g(x) = (1 + (1 - xx^*)^{1/2})^{-1}x, \qquad x \in S. \tag{2}$$

Then, as above, $g(x)$ may also be written in the form

$$g(x) = x(1 + (1 - x^*x)^{1/2})^{-1}, \qquad x \in S. \tag{2'}$$

Considering the equation $t = 2s(1 + s^2)^{-1}$, $-1 \leq t \leq 1$, and its solution $s = t(1 + (1 - t^2)^{1/2})^{-1}$, we have

$$f \circ g(x) = x = g \circ f(x), \qquad x \in S.$$

Thus f maps $\mathscr{M}$ onto S; furthermore, it maps A onto $A \cap S$. Thus if f is strongly* continuous, then $A \cap S$ is strongly* dense in S. Take arbitrary elements $x, y \in \mathscr{M}$. Then we have

$$\begin{aligned}
\tfrac{1}{2}[f(x) - f(y)] &= (1 + xx^*)^{-1}x - (1 + yy^*)^{-1}y \\
&= (1 + xx^*)^{-1}(x - y) + \{(1 + xx^*)^{-1} - (1 + yy^*)^{-1}\}y \\
&= (1 + xx^*)^{-1}(x - y) + (1 + xx^*)^{-1}(yy^* - xx^*)(1 + yy^*)^{-1}y \\
&= (1 + xx^*)^{-1}(x - y) + (1 + xx^*)^{-1}\{(y - x)y^* \\
&\quad + x(y^* - x^*)\}(1 + yy^*)^{-1}y.
\end{aligned}$$

Hence we get, for any $\xi \in \mathfrak{H}$,

$$\begin{aligned}
\|\tfrac{1}{2}(f(x) - f(y))\xi\| &\leq \|(x - y)\xi\| + \|(y - x)y^*(1 + yy^*)^{-1}y\xi\| \\
&\quad + \|(1 + xx^*)^{-1}x\| \|(y^* - x^*)(1 + yy^*)^{-1}y\xi\| \\
&\leq \|(x - y)\xi\| + \|(y - x)y^*(1 + yy^*)^{-1}y\xi\| \\
&\quad + \|(y^* - x^*)(1 + yy^*)^{-1}y\xi\|.
\end{aligned}$$

Similarly, we get, using equality (1′),

$$\|\tfrac{1}{2}[f(x) - f(y)]^*\xi\| \leq \|(x^* - y^*)\xi\| + \|(y^* - x^*)y(1 + y^*y)^{-1}y^*\xi\| + \|(y - x)(1 + y^*y)^{-1}y^*\xi\|.$$

Hence if a net $\{x_i\}$ in A converges strongly* to $y \in \mathcal{M}$, then $\{f(x_i)\}$ converges strongly* to $f(y)$. Hence f is strongly* continuous. Q.E.D.

By Theorem 2.6, the unit ball of A is dense in the unit ball of $\mathcal{M}$ with respect to all the locally convex topologies defined in Section 2 except the uniform topology.

Proposition 4.9. *The group of all unitary operators on a Hilbert space is strongly* closed.*

Proof. Let $\mathcal{U}$ denote the unitary group on a Hilbert space $\mathfrak{H}$. Let $\{u_i\}$ be a net in $\mathcal{U}$ convergent strongly* to $u \in \mathcal{L}(\mathfrak{H})$. Then, we have, for any $\xi \in \mathfrak{H}$,

$$\|u\xi\| = \lim\|u_i\xi\| = \|\xi\|,$$
$$\|u^*\xi\| = \lim\|u_i^*\xi\| = \|\xi\|.$$

Hence u and u^* are both isometries on $\mathfrak{H}$, which means that u is unitary. Q.E.D.

Remark 4.10. On the unitary group $\mathcal{U}(\mathfrak{H})$ of a Hilbert space $\mathfrak{H}$, the strong* topology, the strong topology, and the weak topology coincide. But, if $\mathfrak{H}$ is infinite dimensional, then $\mathcal{U}(\mathfrak{H})$ is not strongly closed in $\mathcal{L}(\mathfrak{H})$. In fact, let $\{\xi_n\}$ be a normalized orthogonal infinite sequence in $\mathfrak{H}$. For each $n = 1,2,\dots,$ let u_n be the unitary operator on $\mathfrak{H}$ defined by $u_n\xi_k = \xi_{k+1}$ for $k = 1,2,\dots,$ $n-1$, $u_n\xi_n = \xi_1$, and $u_n = 1$ on the orthogonal complement of $[\xi_1,\dots,\xi_n]$. Then the sequence $\{u_n\}$ converges strongly to the isometry u given by $u\xi_k = \xi_{k+1}$, $k = 1,2,\dots$ and $u = 1$ on the orthogonal complement of $[\xi_1,\xi_2,\dots]$. The strong limit u of $\{u_n\}$ is not unitary, because $u^*\xi_1 = 0$. In general, the strong limit of unitary operators is an isometry, but not necessarily unitary. At a glance, one might feel strange about the above facts. But this phenomena tells us that the strong* topology, the strong topology, and the weak topology are the same on the unitary group, but that their uniform structures are different. Actually, $\mathcal{U}(\mathfrak{H})$ is complete in the strong* topology and it is relatively compact but not compact in the weak topology.

For each subset A of $\mathcal{L}(\mathfrak{H})$, let $\bar{A}$ denote the strong* closure of A. For a unital C^*-algebra A, let $U(A)$ denote the group of unitary elements in A and let

$$U(A,\lambda) = \{u \in U(A): \|u - 1\| \leq \lambda\}, \qquad \lambda > 0.$$

Theorem 4.11. *If A is a C^*-algebra of operators on a Hilbert space $\mathfrak{H}$ and if it contains the identity* 1, *then we have*

$$\overline{U(A,\lambda)} = U(\bar{A},\lambda).$$

In particular, the unitary group $U(A)$ of A is strongly dense in the unitary group $U(\bar{A})$ of the von Neumann algebra $\bar{A}$ generated by A.*

PROOF. Since the ball $\{x \in \bar{A}: \|x - 1\| \leq \lambda\}$ is strongly* closed, $U(\bar{A},\lambda)$ is strongly* closed by Proposition 4.9. Hence $U(\bar{A},\lambda) \supset \overline{U(A,\lambda)}$. For a unitary operator u in $\bar{A}$, the inequality $\|u - 1\| \leq \lambda$ is equivalent to saying that

$$\mathrm{Sp}(u) \subset \{z \in \mathbf{C}: |z - 1| \leq \lambda\} = S_\lambda.$$

Let $u = \int_0^{2\pi} e^{i\theta}\, d\theta$ be the spectral decomposition of u. For $n = 1,2,\ldots$, put

$$u_n = \int_0^{\pi-(1/n)} \exp(i\theta)\, d\theta + \int_{\pi+(1/n)}^{2\pi} \exp(i\theta)\, d\theta$$
$$+ \exp\left(i\left(\pi - \frac{1}{n}\right)\right) \int_{\pi-(1/n)}^{\pi+(1/n)} de(\theta).$$

Then we have $\lim\|u_n - u\| = 0$. By the construction, $\mathrm{Sp}(u_n)$ does not contain -1. Hence we may assume $\|u - 1\| < 2$, that is, $\lambda < 2$. Then $\arg(z)$, $z \in S_\lambda$, is a continuous one-to-one function with range $[-\alpha,\alpha]$, where $\alpha = 2\sin^{-1}(\lambda/2)$, and we have

$$z = \exp(i\arg(z)), \qquad z \in S_\lambda.$$

Since $S_\lambda \supset \mathrm{Sp}(u)$ for every $u \in U(\bar{A},\lambda)$, the map: $u \in U(\bar{A},\lambda) \mapsto \arg(u) \in \bar{A}_h \cap \alpha S$ is strongly continuous by Theorem 4.7, where S denote the unit ball of $\bar{A}$. By the spectral mapping theorem, Proposition I.2.8, we have

$$u = \exp(i\arg(u)), \qquad u \in U(\bar{A},\lambda).$$

If $u \in U(\bar{A},\lambda)$, then $\arg(u) \in \bar{A}_h \cap \alpha S$. By Theorem 4.8, there exists a net $\{h_j\} \subset A_h \cap \alpha S$ converging strongly to $\arg(u)$, so that the net $u_j = \exp(ih_j)$ converges strongly to u by Theorem 4.7. By the spectral mapping theorem again, u_j belongs to $U(A,\lambda)$. Q.E.D.

***Lemma 4.12.** *Let $\{\mathcal{M},\mathfrak{H}\}$ be a von Neumann algebra. If a bounded net $\{x_i\}_{i\in I}$ in $\mathcal{M}$ converges strongly to x_0, then for any $\varepsilon > 0$, there exists a net $\{e_i\}_{i\in I}$ of projections in $\mathcal{M}$ converging strongly to the identity 1 such that $\|(x_i - x_0)e_i\| \leq \varepsilon$ for every $i \in I$.*

PROOF. We may assume $\|x_i\| \leq 1$ and $x_0 = 0$ by considering the net $\{x_i - x_0\}$. Put $y_i = x_i^*x_i$. Then, for every $\xi \in \mathfrak{H}$, we have $\|y_i\xi\| = \|x_i^*x_i\xi\| \leq \|x_i^*\|\|x_i\xi\| \leq \|x_i\xi\|$; hence $\{y_i\}_{i\in I}$ converges strongly to zero. Let χ_ε be the characteristic function of the interval $[0,\varepsilon^2]$. Putting $e_i = \chi_\varepsilon(y_i)$, $i \in I$, we obtain a net $\{e_i\}$ of projections in $\mathcal{M}$. The inequality $\varepsilon^2(1 - e_i) \leq y_i$ implies that $\{e_i\}$ converges strongly to 1. The inequality $\|e_i y_i\| \leq \varepsilon^2$ implies $\|x_i e_i\| < \varepsilon$. Q.E.D.

Theorem 4.13 (Noncommutative Egoroff's Theorem). *Let $\{\mathcal{M},\mathfrak{H}\}$ be a von Neumann algebra. Let A be a bounded subset of $\mathcal{M}$ and $\bar{A}$ be its strong closure. Take an arbitrary element $a \in \bar{A}$. Then, for any positive $\varphi \in \mathcal{M}_*$, any projection e, and any $\varepsilon > 0$, there exist a projection $e_0 \leq e$ in $\mathcal{M}$ and a sequence $\{a_n\}$ in A*

such that

$$\lim_{n\to\infty} \|(a_n - a)e_0\| = 0, \qquad \varphi(e - e_0) < \varepsilon.$$

PROOF. As usual, we may assume $e = 1$ and $a = 0$. There exists a net $\{a_i\}_{i\in I}$ in A with s-lim $a_i = 0$, where "s-lim" means the strong limit. By Lemma 4.12, there exists a net $\{e_i\}_{i\in I}$ of projections in $\mathcal{M}$ such that

$$\|a_i e_i\| \leq \tfrac{1}{2} \quad \text{and} \quad \text{s-lim}\, e_i = 1.$$

We choose an index i_1 such that

$$\varphi(1 - e_i) < \frac{\varepsilon}{2} \quad \text{for} \quad i \geq i_1.$$

Put $e_1 = e_{i_1}$. Then, of course, s-lim $a_i e_1 = 0$. Putting $b_i^{(1)} = e_1 a_i^* a_i e_1$, $i \geq i_1$, we get a bounded net $\{b_i^{(1)}\}$ in $e_1\mathcal{M}e_1$ with s-lim $b_i^{(1)} = 0$. By the same reasoning as for $\{a_i\}$, there exists a net $\{e_i^{(1)}\}$ of projections in $e_1\mathcal{M}e_1$ such that

$$\|a_i e_i^{(1)}\| < (\tfrac{1}{2})^2 \quad \text{and} \quad \text{s-lim}\, e_i^{(1)} = e_1.$$

We choose an index $i_2 \geq i_1$ such that

$$\varphi(e_1 - e_i^{(1)}) < (\tfrac{1}{2})^2\varepsilon \quad \text{for} \quad i \geq i_2.$$

Putting $e_2 = e_{i_2}^{(1)}$, we have

$$e_1 \geq e_2, \qquad \|a_i e_2\| < (\tfrac{1}{2})^2 \quad \text{for} \quad i \geq i_2,$$
$$\varphi(e_1 - e_2) < (\tfrac{1}{2})^2\varepsilon.$$

By induction, we obtain a decreasing sequence $\{e_n\}$ of projections in $\mathcal{M}$ and a sequence $\{i_n\}$ of indices with the properties

$$\|a_i e_n\| < (\tfrac{1}{2})^n \quad \text{for} \quad i \geq i_n,$$
$$\varphi(e_n - e_{n+1}) < (\tfrac{1}{2})^{n+1}\varepsilon, \qquad n = 1,2,\ldots.$$

Let $e_0 = $ s-lim e_n. Then we get

$$\varphi(1 - e_0) \leq \varepsilon \quad \text{and} \quad \|a_i e_0\| < (\tfrac{1}{2})^n \quad \text{for} \quad i \geq i_n, \qquad n = 1,2,\ldots,$$

so that the sequence $\{a_{i_n}\}$ and the projection e_0 are the ones desired. Q.E.D.

Corollary 4.14. *Let A be a C^*-algebra of operators on a Hilbert space and $\mathcal{M}$ be its weak closure. For every element $a \in \mathcal{M}$, positive $\varphi \in \mathcal{M}_*$, projection $e \in \mathcal{M}$, $\varepsilon > 0$, and $\delta > 0$, there exist a projection $e_1 \leq e$ in $\mathcal{M}$ and $a_1 \in A$ such that*

$$\|(a - a_1)e_1\| < \delta, \qquad \|a_1\| \leq \|ae\| \quad \textit{and} \quad \varphi(e - e_1) < \varepsilon.$$

PROOF. Let S denote the unit ball of $\mathcal{M}$. Applying Theorem 4.8 to $A \cap \|ae\|S$ and replacing a by ae in Theorem 4.13, we can easily verify the assertion.

Q.E.D.

Theorem 4.15 (Noncommutative Lusin's Theorem). *Let A be a C^*-algebra of operators on a Hilbert space and $\mathscr{M}$ be its weak closure. Take an arbitrary nonzero positive $\varphi \in \mathscr{M}_*$, nonzero projection $e \in \mathscr{M}$, $\varepsilon > 0$, and $\delta > 0$. Then the following statements hold:*

(i) *For every $a \in \mathscr{M}$, there exist a projection $e_0 \leq e$ in $\mathscr{M}$ and $a_0 \in A$ such that*

$$ae_0 = a_0 e_0, \qquad \varphi(e - e_0) < \varepsilon \quad \text{and} \quad \|a_0\| \leq (1 + \delta)\|ae_0\|.$$

(ii) *If $a \in \mathscr{M}_h$, then we can choose $a_0 \in A_h$ and a projection $e_0 \leq e$ in $\mathscr{M}$ so that*

$$ae_0 = a_0 e_0, \qquad \varphi(e - e_0) < \varepsilon,$$
$$\|a_0\| \leq \min\{2(1 + \delta)\|ae_0\|, \|a\| + \delta\}.$$

(iii) *If $1 \in A$ and $a \in U(\mathscr{M})$, then we can choose $a_0 \in U(A)$ and a projection $e_0 \leq e$ in $\mathscr{M}$ so that*

$$ae_0 = a_0 e_0, \qquad \varphi(e - e_0) < \varepsilon, \qquad \|a_0 - 1\| \leq \|a - 1\| + \delta.$$

PROOF. Case (i): We may assume $\|ae\| = 1$. We choose a positive $\varphi_0 \in \mathscr{M}_*$ with $\varphi_0(1) = 1$ and $\varphi_0((ae)^*(ae)) \geq 1 - \delta$. Put $\psi = \varphi + \varphi_0$ and $\varepsilon' = \min\{\varepsilon, \delta^2\}$. By Corollary 4.14, we can find $a_1 \in A$ and a projection $e_1 \leq e$ in $\mathscr{M}$ such that

$$\|a_1\| \leq \|ae\| = 1, \qquad \|(a - a_1)e_1\| < \frac{\delta}{2} \quad \text{and} \quad \psi(e - e_1) < \varepsilon'.$$

Applying Corollary 4.14 to $(a - a_1)e_1 \in \mathscr{M}$ again, we can find $a_2 \in A$ and a projection $e_2 \leq e_1$ in $\mathscr{M}$ so that

$$\|a_2\| \leq \|(a - a_1)e_1\|, \qquad \|(a - a_1 - a_2)e_2\| < (\tfrac{1}{2})^2\delta,$$
$$\psi(e_1 - e_2) < (\tfrac{1}{2})^2\varepsilon'.$$

By induction, we choose a sequence $\{a_n\}$ in A and a decreasing sequence $\{e_n\}$ of projections in $\mathscr{M}$ such that

$$\left\| \left(a - \sum_{k=1}^{n} a_k \right) e_n \right\| < (\tfrac{1}{2})^n\delta, \tag{3}$$

$$\|a_n\| \leq \left\| \left(a - \sum_{k=1}^{n-1} a_k \right) e_{n-1} \right\|, \tag{4}$$

$$\psi(e_{n-1} - e_n) < (\tfrac{1}{2})^n\varepsilon'. \tag{5}$$

By (3) and (4), $\sum_{k=1}^{\infty} a_k$ converges uniformly to $a_0 \in A$, and we have

$$\|a_0\| \leq \sum_{n=1}^{\infty} (\tfrac{1}{2})^n\delta + 1 = 1 + \delta.$$

Putting $e_0 = \text{s-lim}\, e_n$, we have, by (5),

$$\psi(e - e_0) < \sum_{n=1}^{\infty} (\tfrac{1}{2})^n\varepsilon' = \varepsilon'. \tag{6}$$

By (3), we have $ae_0 = a_0e_0$. Furthermore,

$$\begin{aligned}\|ae_0\| &= \|(ae_0)^*(ae_0)\|^{1/2} \geq \varphi_0((ae_0)^*(ae_0))^{1/2}\\ &\geq \varphi_0((ae)^*(ae))^{1/2} - \varphi_0((a(e-e_0))^*a(e-e_0))^{1/2}.\end{aligned}$$

On the other hand,

$$\varphi_0((a(e-e_0))^*a(e-e_0))^{1/2} \leq \psi(e-e_0) < \delta$$

since $\|ae\| = 1$, so that

$$\|ae_0\| \geq 1 - 2\delta \geq (1-3\delta)\|a_0\|$$

because $\|a_0\| \leq 1 + \delta$. Thus, if we replace δ by another sufficiently small positive number at the beginning, then we can conclude that $\|a_0\| \leq (1+\delta)\|ae_0\|$.

Case (ii): Let a be a self-adjoint element in $\mathscr{M}$. Let ε' and ψ be as in case (i). Let S denote the unit ball of $\mathscr{M}$. First, we will show that there exist a self-adjoint element $a_1 \in A$ and a projection $e_1 \leq e$ in $\mathscr{M}$ such that

$$\|(a-a_1)e_1\| < (\tfrac{1}{2})^2\delta, \qquad \|a_1\| \leq \min\{\|a\|, 2\|ae\|\},$$
$$\psi(e-e_1) < \tfrac{1}{2}\varepsilon'.$$

Suppose $2\|ae\| \geq \|a\|$. The strong closure of $A_h \cap \|a\|S$ contains a by Theorem 4.8, so that there exists a net $\{a_i\}$ in $A_h \cap \|a\|S$ converging strongly to a. Hence s-lim $a_ie = ae_1$ so that there exist, by Theorem 4.13, a projection $e_1 \leq e$ in $\mathscr{M}$ and $a_1 \in A_h$ such that

$$\|(a-a_1)e_1\| < (\tfrac{1}{2})^2\delta, \qquad \|a_1\| \leq \|a\| \leq 2\|ae\|,$$
$$\psi(e-e_1) < \tfrac{1}{2}\varepsilon'.$$

Suppose next $2\|ae\| \leq \|a\|$. Put $b_0 = eae + (1-e)ae + ea(1-e)$. Then b_0 belongs to the strong closure of $A_h \cap \|b_0\|S$ by Theorem 4.8. Hence we can find a projection $e_1 \leq e$ in $\mathscr{M}$ and $a_1 \in A_h$ such that

$$\|(a-a_1)e_1\| < (\tfrac{1}{2})^2\delta, \qquad \|a_1\| \leq \|b_0\| \leq 2\|ae\| \leq \|a\|,$$
$$\psi(e-e_1) < \tfrac{1}{2}\varepsilon',$$

noticing that $b_0e = ae$. Thus, in any case we can find the desired $a_1 \in A_h$ and a projection $e_1 \leq e$ in $\mathscr{M}$.

Putting

$$b_1 = e_1(a-a_1)e_1 + (1-e_1)(a-a_1)e_1 + e_1(a-a_1)(1-e_1),$$

we get a self-adjoint element $b_1 \in \mathscr{M}$ such that

$$(a-a_1)e_1 = b_1e_1, \qquad \|b_1\| \leq 2\|(a-a_1)e_1\|.$$

Applying the above arguments to b_1 instead of a, we find $a_2 \in A_h$ and a projection $e_2 \leq e_1$ such that

$$\|a_2\| \leq \|b_1\|, \qquad \|(b_1-a_2)e_2\| < (\tfrac{1}{2})^3\delta, \qquad \psi(e_1-e_2) < (\tfrac{1}{2})^2\varepsilon'.$$

By the equality

$$(b_1 - a_2)e_2 = b_1e_2 - a_2e_2 = (a - a_1)e_2 - a_2e_2$$
$$= (a - a_1 - a_2)e_2,$$

we have

$$\|(a - a_1 - a_2)e_2\| < (\tfrac{1}{2})^3\delta.$$

Hence, by induction, we can choose a decreasing sequence $\{e_n\}$ of projections in $\mathcal{M}$, a sequence $\{a_n\}$ in A_h, and a sequence $\{b_n\}$ in $\mathcal{M}_h$ such that

$$\|(b_{n-1} - a_n)e_n\| < (\tfrac{1}{2})^{n+1}, \tag{7}$$

$$\|a_n\| \leq \|b_{n-1}\| \leq 2\left\|\left(a - \sum_{k=1}^{n-1} a_k\right) e_{n-1}\right\|, \tag{8}$$

$$b_{n-1} = a - \sum_{k=1}^{n-1} a_k - (1 - e_{n-1})\left(a - \sum_{k=1}^{n-1} a_k\right)(1 - e_{n-1}), \tag{9}$$

$$\psi(e_{n-1} - e_n) < (\tfrac{1}{2})^n\varepsilon', \qquad n = 1,2,\ldots. \tag{10}$$

By the inequality

$$\|a_n\| \leq 2\left\|\left(a - \sum_{k=1}^{n-1} a_k\right) e_{n-1}\right\| = 2\|(b_{n-2} - a_{n-1})e_{n-1}\|$$
$$< (\tfrac{1}{2})^{n-1}\delta,$$

the series $\sum_{n=1}^{\infty} a_n$ converges uniformly to $a_0 \in A_h$, and we have

$$\|a_0\| \leq \min\{\|a\|, 2\|ae\|\} + \delta.$$

Put $e_0 =$ s-lim e_n. Then by the equality

$$(b_{n-1} - a_n)e_n = \left(a - \sum_{k=1}^{n} a_k\right) e_n,$$

we have

$$\left\|\left(a - \sum_{k=1}^{n} a_h\right) e_0\right\| < (\tfrac{1}{2})^{n+1}\delta, \qquad n = 1,2,\ldots;$$

hence we get $ae_0 = a_0e_0$. The inequality $\psi(e - e_0) < \varepsilon'$ follows from (10). By the same arguments as in case (i), we have

$$\|ae_0\| \geq \|ae\|(1 - 2\delta).$$

Noticing that $\|a_0\| \leq 2\|ae\| + \delta$, we have

$$\begin{aligned}\|ae_0\| &\geq \|ae\|(1 - 2\delta)\\ &\geq \tfrac{1}{2}(\|a_0\| - \delta)(1 - 2\delta)\\ &= \tfrac{1}{2}\|a_0\|\left(1 - \frac{\delta}{\|a_0\|}\right)(1 - 2\delta)\\ &\geq \tfrac{1}{2}\|a_0\|\left(1 - \frac{\delta}{\|ae_0\|}\right)(1 - 2\delta).\end{aligned}$$

Hence we have

$$\begin{aligned}\|a_0\| &\le 2(1-2\delta)^{-1}\left(1-\frac{\delta}{\|ae_0\|}\right)^{-1}\|ae_0\|\\ &\le 2(1-2\delta)^{-1}\left(1-\frac{\delta}{\|ae\|(1-2\delta)}\right)^{-1}\|ae_0\|.\end{aligned}$$

Therefore, if another sufficiently small δ' is chosen to start with, then we have

$$\|a_0\| \le 2(1+\delta)\|ae_0\|.$$

We need some preparation to prove case (iii), so that it will be proved after a few lemmas. Q.E.D.

Lemma 4.16. *Let $\{\mathcal{M},\mathfrak{H}\}$ be a von Neumann algebra and w a unitary operator in $\mathcal{M}$. If e is a projection in $\mathcal{M}$ with $\|(1-w)e\| < 1/8$, then there exists a unitary operator v in $\mathcal{M}$ such that*

$$ve = we \quad \text{and} \quad \|1-v\| \le 7\|(1-w)e\|.$$

Proof. Let $f = wew^*$. Then we have

$$\begin{aligned}\|e-f\| = \|e - wew^*\| &= \|e - we + we - wew^*\|\\ &\le \|(1-w)e\| + \|we(1-w^*)\|\\ &\le 2\|(1-w)e\| < \tfrac{1}{4}.\end{aligned}$$

Put $a = 1 + (1-e)(e-f)(1-e)$, and $u = (1-f)a^{-1/2}(1-e)$. Note here that a is positive and invertible, because $\|(1-e)(e-f)(1-e)\| < 1$. Since a and $(1-e)$ commute, we have

$$\begin{aligned}u^*u &= (1-e)a^{-1/2}(1-f)^2a^{-1/2}(1-e)\\ &= a^{-1/2}(1-e)(1-f)(1-e)a^{-1/2}\\ &= a^{-1/2}\{(1-e) + (1-e)(e-f)(1-e)\}a^{-1/2}\\ &= a^{-1/2}(1-e)aa^{-1/2} = 1-e.\end{aligned}$$

Hence uu^* is a projection and majorized by $(1-f)$. Suppose that $uu^* \ne (1-f)$. Then there exists a unit vector $\xi \in (1-f)\mathfrak{H}$ such that $(\xi|uu^*\eta) = 0$ for every $\eta \in \mathfrak{H}$, hence $u^*\xi = 0$. Therefore,

$$\begin{aligned}(1-e)(1-f)\xi &= a^{1/2}a^{-1/2}(1-e)(1-f)\xi\\ &= a^{1/2}(1-e)a^{-1/2}(1-f)\xi\\ &= a^{1/2}u^*\xi = 0,\end{aligned}$$

so that $e(1-f)\xi = (1-f)\xi$, which implies that

$$\begin{aligned}1 = \|e(1-f)\| &= \|e - ef\| = \|e(e-f)\|\\ &\le \|e-f\| < 1.\end{aligned}$$

This is a contradiction. Thus $uu^* = 1 - f$. Now, we have

$$\begin{aligned}\|(1-u)(1-e)\| &= \|(1-e)-(1-f)a^{-1/2}(1-e)\|\\ &= \|(1-e)-(e-f)a^{-1/2}(1-e)-(1-e)a^{-1/2}(1-e)\|\\ &= \|(1-e)(1-a^{-1/2})(1-e)-(e-f)a^{-1/2}(1-e)\|\\ &\le \|1-a^{-1/2}\| + \|a^{-1/2}\|\,\|e-f\|.\end{aligned}$$

Noticing that $|(1+t)^{-1/2}-1| \le |t|$ and $(1+t)^{-1/2} < 2$ for any real number t with $|t| < \frac{1}{4}$, we have

$$\begin{aligned}\|(1-u)(1-e)\| &\le \|e-f\| + 2\|e-f\|\\ &\le 6\|(1-w)e\|.\end{aligned}$$

Putting $v = we + u(1-e)$, we obtain a unitary operator in $\mathcal{M}$ with $ve = we$ such that

$$\begin{aligned}\|1-v\| &= \|1-we-u(1-e)\|\\ &= \|e+1-e-we-u(1-e)\|\\ &= \|(1-w)e+(1-u)(1-e)\| \le 7\|(1-w)e\|. \qquad \text{Q.E.D.}\end{aligned}$$

Lemma 4.17. *Let A be a unital C^*-algebra of operators on a Hilbert space $\mathfrak{H}$ and $\mathcal{M}$ be its weak closure. Let φ be an arbitrary nonzero positive linear functional in $\mathcal{M}_*$, u a unitary operator in $\mathcal{M}$, e a projection in $\mathcal{M}$, and $\varepsilon > 0$, $\delta > 0$. Then we can choose a unitary operator v in A and a projection $e_1 \le e$ in $\mathcal{M}$ such that*

$$\|(u-v)e_1\| < \delta, \qquad \|1-v\| \le \|1-u\|,$$
$$\varphi(e-e_1) < \varepsilon.$$

Proof. By Theorem 4.11, $\{v \in U(A): \|1-v\| \le \|1-u\|\}$ is strongly dense in $\{w \in U(\mathcal{M}): \|1-w\| \le \|1-u\|\}$. Therefore, our assertion follows from Theorem 4.6. Q.E.D.

Proof of Theorem 4.15. Case (iii): We will choose, by induction, a sequence $\{u_n\}$ in $U(A)$, a sequence $\{v_n\}$ in $U(\mathcal{M})$, and a decreasing sequence $\{e_n\}$ of projections in $\mathcal{M}$ such that

$$u_1 = 1, \qquad v_1 = u \quad \text{and} \quad e_1 = e, \tag{11}$$

$$ue_n = u_1u_2 \cdots u_nv_ne_n, \tag{12}$$

$$\|(1-u_n^*v_{n-1})e_n\| = \|(u_{n-1}-v_{n-1})e_n\| < (\tfrac{1}{2})^{n+2}\delta, \tag{13}$$

$$\varphi(e_{n-1}-e_n) < (\tfrac{1}{2})^n\varepsilon, \tag{14}$$

$$\|1-u_n\| \le \|1-v_{n-1}\| \le 7\|(1-u_{n-1}^*v_{n-2})e_{n-1}\|, \tag{15}$$

for $n = 2,3,\ldots$, where $v_0 = u$.

In fact, by Lemma 4.17, there exists $u_2 \in U(A)$ and a projection $e_1 \leq e$ in $\mathscr{M}$ such that

$$\|(u_2 - v_1)e_2\| < \frac{\delta}{8}, \qquad \|1 - u_2\| \leq \|1 - v_1\|, \qquad \varphi(e - e_1) < \frac{\varepsilon}{2}.$$

Since $u_2^* v_1 \in U(\mathscr{M})$, there exists, by Lemma 4.16, $v_2 \in U(\mathscr{M})$ such that

$$v_2 e_2 = u_2^* v_1 e_2 \quad \text{and} \quad \|1 - v_2\| \leq 7\|(1 - u_2^* v_1)e_2\| < \frac{7}{8}\delta < \delta.$$

Now, suppose that we have found $e_1, \dots, e_{n-1}, u_1, \dots, u_{n-1}$ and $v_1, \dots, v_{n-1}$ so that conditions (12)–(15) hold. Applying Lemma 4.17 to v_{n-1} and e_{n-1}, we find $u_n \in U(A)$ and a projection $e_n \leq e_{n-1}$ in $\mathscr{M}$ such that

$$\begin{gathered} \|(u_n - v_{n-1})e_n\| < (\tfrac{1}{2})^{n+2}\delta, \\ \|1 - u_n\| \leq \|1 - v_{n-1}\|, \\ \varphi(e_{n-1} - e_n) < (\tfrac{1}{2})^n \varepsilon. \end{gathered}$$

Since $u_n^* v_{n-1} \in U(\mathscr{M})$, we find $v_n \in U(\mathscr{M})$ by Lemma 4.16 such that

$$\begin{gathered} v_n e_n = u_n^* v_{n-1} e_n = u_n^* u_{n-1}^* \cdots u_1^* u e_n, \\ \|1 - v_n\| \leq 7\|(1 - u_n^* v_{n-1})e_n\| < (\tfrac{1}{2})^{n-1}\delta. \end{gathered}$$

Thus, by induction, we find sequences $\{u_n\}$, $\{v_n\}$, and $\{e_n\}$ so that conditions (12)–(15) hold.

Putting $e_0 = \text{s-lim}\, e_n$, we have

$$u e_0 = u_1 u_2 \cdots u_n v_n e_0.$$

By the inequality

$$\|1 - u_n\| \leq \|1 - v_{n-1}\| \leq (\tfrac{1}{2})^{n-2}\delta,$$

we have $\sum_{n=1}^{\infty} \|1 - u_n\| < +\infty$. Therefore, $\{u_1 u_2 \cdots u_n\}$ converges uniformly to $u_0 \in U(A)$. Then we have

$$\begin{aligned} \|ue_0 - u_0 e_0\| &\leq \|(u_1 u_2 \cdots u_n) v_n e_0 - (u_1 \cdots u_n) e_0\| + \|(u_1 \cdots u_n) e_0 - u_0 e_0\| \\ &\leq \|1 - v_n\| + \|(u_1 u_2 \cdots u_n) - u_0\| < (\tfrac{1}{2})^{n-1}\delta + \|(u_1 u_2 \cdots u_n) - u_0\|; \end{aligned}$$

hence $ue_0 = u_0 e_0$. Furthermore,

$$\begin{aligned} \|1 - (u_1 \cdots u_n)\| &\leq \|1 - u_1 + u_1 - u_1 u_2 + \cdots - u_1 u_2 \cdots u_n\| \\ &\leq \|1 - u_1\| + \|1 - u_2\| + \cdots + \|1 - u_n\| \\ &\leq \|1 - u\| + 2\delta; \end{aligned}$$

hence $\|1 - u_0\| \leq \|1 - u\| + 2\delta$. Q.E.D.

As an immediate consequence of Theorem 4.15, we obtain the following important result:

Theorem 4.18 (Transitivity Theorem). *Let A be a C^*-algebra of operators on a Hilbert space $\mathfrak{H}$. If A is irreducible on $\mathfrak{H}$, that is, if the identity representation* *of A on $\mathfrak{H}$ is irreducible, then for every projection e of $\mathfrak{H}$ of finite rank,*

$$Ae = \mathscr{L}(\mathfrak{H})e. \tag{16}$$

Furthermore, (16) *is more precisely described as follows:*

(i) *For every $b \in \mathscr{L}(\mathfrak{H})$ and $\varepsilon > 0$, there exists $a \in A$ such that*

$$ae = be \quad \text{and} \quad \|a\| \leq (1 + \varepsilon)\|be\|.$$

(ii) *For every self-adjoint $b \in \mathscr{L}(\mathfrak{H})$ and $\varepsilon > 0$, there exists $a \in A_h$ such that*

$$ae = be \quad \text{and} \quad \|a\| \leq \min\{2(1 + \varepsilon)\|be\|, \|b\| + \varepsilon\}.$$

(iii) *If A_1 is the C^*-algebra obtained by adjunction of the identity 1 to A, then for every unitary $u \in \mathscr{L}(\mathfrak{H})$ and $\varepsilon > 0$, there exists $v \in U(A_1)$ such that*

$$ue = ve \quad \text{and} \quad \|1 - v\| \leq \|1 - u\| + \varepsilon.$$

Therefore, the action of A on $\mathfrak{H}$ is algebraically irreducible. In other words, every irreducible representation of a C^*-algebra is algebraically irreducible.

PROOF. Let $\varphi(x) = \mathrm{Tr}(xe)$, $x \in \mathscr{L}(\mathfrak{H})$. Then apply Theorem 4.15 to φ, $b \in \mathscr{L}(\mathfrak{H})$ and A. Our assertion follows immediately from the fact that if a projection $e_1 \leq e$ satisfies the inequality $\varphi(e - e_1) < \frac{1}{2}$ then $e_1 = e$. Q.E.D.

So far, we have obtained a number of approximation theorems. But we have not discussed the topological nature of the order structure in the approximation problems. Now, we are going to discuss monotone approximations.

For a set A of self-adjoint operators on a Hilbert space $\mathfrak{H}$, we denote by A_σ and A_δ, respectively, the sets of strong limits of *monotone increasing* and *decreasing sequences* in A. We denote by A^m and A_m, respectively, the sets of strong limits of *monotone increasing* and *decreasing nets* in A. Of course, we have, in general, $A \subset A_\sigma \subset A^m$ and $A \subset A_\delta \subset A_m$.

To proceed to the monotone approximation theorems, we need the following preparation.

Definition 4.19. A monotone increasing function f on an interval $[\alpha,\beta]$ is said to be *operator-monotone* if $f(a) \leq f(b)$ for every pair a,b of self-adjoint operators on a Hilbert space such that $a \leq b$ and $\mathrm{Sp}(a) \cup \mathrm{Sp}(b) \subset [\alpha,\beta]$.

Then, Proposition I.6.3 can be rephrased by saying that the function $f(t) = t^\alpha$ on the positive real half-line $[0,\infty)$ is operator-monotone for $0 \leq \alpha \leq 1$.

Lemma 4.20. *For $\alpha > 0$, the function $f_\alpha(t) = (\alpha + (1 - \alpha)t)^{-1}t$ on the interval $[0,1]$ is operator-monotone. Furthermore, if $0 < \alpha \leq 1$, then f_α is operator-monotone as a function on the half-line $[0,\infty)$.*

PROOF. Let $0 < \alpha \leq 1$. Suppose that a and b are positive operators on a Hilbert space $\mathfrak{H}$ such that $a \leq b$. For any $\varepsilon > 0$, we have $(\varepsilon + a)^{-1} \geq (\varepsilon + b)^{-1}$, so that

$$(\varepsilon + a)^{-1}a = 1 - \varepsilon(\varepsilon + a)^{-1} \leq 1 - \varepsilon(\varepsilon + b)^{-1} = (\varepsilon + b)^{-1}b.$$

Therefore, if $0 < \alpha < 1$, we have

$$\begin{aligned} f_\alpha(a) &= (1-\alpha)^{-1}\left(\frac{\alpha}{1-\alpha} + a\right)^{-1} a \\ &\leq (1-\alpha)^{-1}\left(\frac{\alpha}{1-\alpha} + b\right)^{-1} b = f_\alpha(b). \end{aligned}$$

If $\alpha = 1$, then $f_\alpha(a) = a \leq b = f_\alpha(b)$.

Suppose that $0 \leq a \leq b \leq 1$ and $\alpha > 1$. Then we get $\alpha + (1-\alpha)a \geq \alpha + (1-\alpha)b$, so that $[\alpha + (1-\alpha)a]^{-1} \leq [\alpha + (1-\alpha)b]^{-1}$. Therefore, we get

$$\begin{aligned} (\alpha - 1)f_\alpha(a) &= \alpha(\alpha + (1-\alpha)a)^{-1} - 1 \\ &\leq \alpha(\alpha + (1-\alpha)b)^{-1} - 1 = (\alpha - 1)f_\alpha(b). \end{aligned}$$

Hence $f_\alpha(a) \leq f_\alpha(b)$. Q.E.D.

Lemma 4.21. *Let A be a C^*-algebra of operators on a Hilbert space $\mathfrak{H}$. Let S denote the unit ball of $\mathscr{L}(\mathfrak{H})$. For any projection p in A'' and for any sequence $\{\xi_n\}$ of vectors in $\mathfrak{H}$ there exists a projection q in $((A_+ \cap S)_\sigma)_\delta$ such that $q(1-p)\xi_n = 0$ and $(1-q)p\xi_n = 0$, $n = 1,2,\ldots$.*

PROOF. We may assume that $\|\xi_n\| = 1$, $n = 1,2,\ldots$. By Theorem 4.8, p is strongly* approximated by a net $\{x_i\}$ in $A \cap S$. Then $\{x_i^* x_i\}$ converges strongly to p. Hence p is strongly approximated by elements in $A_+ \cap S$. Working with vectors of the form $p\xi_k$ and $(1-p)\xi_k$, we can find a sequence $\{x_n\}$ in $A_+ \cap S$ such that

$$\|p\xi_k - x_n p\xi_k\| < \frac{1}{n},$$

$$\|x_n(1-p)\xi_k\| < \frac{1}{2^n n}, \qquad k = 1,2,\ldots,n.$$

For $n < m$, we define

$$y_{n,m} = \left(1 + \sum_{k=n}^{m} kx_k\right)^{-1} \sum_{k=n}^{m} kx_k.$$

Then it follows that $y_{n,m} \in A_+ \cap S$ and $y_{n,m} \leq \sum_{k=n}^{m} kx_k$. Hence, for $i \leq n$,

$$(y_{nm}(1-p)\xi_i | (1-p)\xi_i) \leq \sum_{k=n}^{m} \frac{1}{2^k} < 2^{-n+1}. \tag{17}$$

Since $\sum_{k=n}^{m} kx_k \geq mx_m$, we have

$$y_{nm} \geq (1 + mx_m)^{-1}mx_m,$$

so that we have, noticing $0 \leq x_m \leq 1$,

$$\begin{aligned} 1 - y_{nm} &\leq 1 - (1 + mx_m)^{-1}mx_m = (1 + mx_m)^{-1} \\ &\leq (1 + m)^{-1}(1 + m(1 - x_m)). \end{aligned}$$

Therefore, we get

$$\begin{aligned} ((1 - y_{nm})p\xi_k | p\xi_k) &\leq \frac{1}{1+m}([1 + m(1 - x_m)]p\xi_k | p\xi_k) \\ &\leq \frac{1}{1+m}(\|p\xi_k\|^2 + m\|(1 - x_n)p\xi_k\|) \\ &\leq \frac{2}{1+m} \quad \text{for} \quad k \leq m. \end{aligned} \tag{18}$$

For fixed n, the sequence $\{y_{n,m}: m = 1,2,\ldots\}$ is monotone increasing, hence strongly converging to an element y_n in $(A_+ \cap S)_\sigma$. Since $y_{n+1,m} \leq y_{n,m}$, we have $y_{n+1} \leq y_n$; hence the sequence $\{y_n\}$ is monotone decreasing, and so converging strongly to an element y in $(A_+ \cap S)_{\sigma\delta}$. From (17) and (18), we get

$$\begin{aligned} (y_n(1 - p)\xi_k | (1 - p)\xi_k) &\leq 2^{-n+1}, \\ ((1 - y_n)p\xi_k | p\xi_k) &\leq 0. \end{aligned}$$

Hence we have, since $0 \leq y \leq 1$,

$$y(1 - p)\xi_k = 0 \quad \text{and} \quad (1 - y)p\xi_k = 0$$

for $k = 1,2,\ldots$.

Let $\{f_n\}$ be the sequence of operator-monotone functions on $[0,1]$ defined by $f_n(t) = (n + (1 - n)t)^{-1}t$. The sequence $\{f_n\}$ is decreasing and convergent pointwise on $[0,1]$ to the characteristic function f_∞ of the point 1. We put $z_{nm} = f_n(y_{nm})$, $z_n = f_n(y_n)$, and $q = f_\infty(y)$. The projection q is the spectral projection of y corresponding to the eigenvalue 1. Each z_{nm} is in $A_+ \cap S$, so z_n is in $(A_+ \cap S)_\sigma$ by Theorem 4.7. Furthermore, $\{z_n\}$ is decreasing since $f_n(y_n) \geq f_{n+1}(y_n) \geq f_{n+1}(y_{n+1})$. Thus $\{z_n\}$ converges strongly to an element $z \in (A_+ \cap S)_{\sigma\delta}$. We have

$$q = f_\infty(y) \leq f_n(y) \leq f_n(y_n) = z_n,$$

so that $q \leq z$. Conversely, we have, again by Theorem 4.7,

$$f_n(y) = \lim_{k\to\infty} f_n(y_k) \geq \lim_{k\to\infty} f_{n+k}(y_{n+k}) = z,$$

so that $q \geq z$. Thus $q = z$. It is now clear that q satisfies the requirements.

Q.E.D.

Theorem 4.22 (Up–Down Theorem). *Let A be a nondegenerate C^*-algebra of operators on a Hilbert space $\mathfrak{H}$ and S be the unit ball of $\mathscr{L}(\mathfrak{H})$. If the von Neumann algebra A'' generated by A is σ-finite, then*

$$A''_+ \cap S = (A_+ \cap S)_{\sigma\delta}. \tag{18}$$

Hence $A''_h = (A_h)_{\sigma\delta}$.

PROOF. We will first verify that $A''_+ \cap S = (A_+ \cap S)_{\sigma\delta}$. Let $\{\xi_k\}$ be a separating sequence of vectors for A'' whose existence is assured by Proposition 3.19. Lemma 4.21 says that every projection in A'' belongs to $(A_+ \cap S)_{\sigma\delta}$. But $1 \in A''$ is the largest element in $A''_+ \cap S$, so that $1 \in (A_+ \cap S)_\sigma$.

Let x be an arbitrary element in $A''_+ \cap S$. Let $x = \int_0^1 \lambda\, de(\lambda)$. Let $\lambda = \sum_{k=1}^\infty \lambda_k/2^k$, $\lambda_k = 0,1$, be the dyadic expansion of $\lambda \in [0,1]$. Let p_k be the spectral projection of x corresponding to the set $\{\lambda \in [0,1]: \lambda_k = 1\}$. Then we have $x = \sum_{k=1}^\infty (1/2^k)p_k$, where the sum converges uniformly. Let $\{z_{k,n}\}$ be a decreasing sequence in $(A_+ \cap S)_\sigma$ which converges to p_k. Define

$$x_n = \sum_{k=1}^n 2^{-k} z_{k,n} + 2^{-n}.$$

Since $(A_+ \cap S)_\sigma$ is convex, x_n is in $(A_+ \cap S)_\sigma$. Furthermore, we have

$$x_n - x_{n+1} = \sum_{k=1}^n 2^{-k}(z_{k,n} - z_{k,n+1}) + 2^{-n} - (2^{-n-1}z_{n+1,n+1} + 2^{-n-1}) \geq 0,$$

so that $\{x_n\}$ is decreasing. Since

$$x_n - x \leq \sum_{k=1}^m 2^{-k}(z_{kn} - p_k) + 2^{-m}$$

for $n > m$, we have $\lim_{n\to\infty} (x_n - x) \leq 2^{-m}$ for $m = 1,2,3,\ldots$; thus we get $x = \text{s-lim}\, x_n$ because $x_n \geq x$ by the definition of x_n. Therefore, x is in $(A_+ \cap S)_{\sigma\delta}$.

For a general element y in A''_h, there exist positive numbers α, β and an element x in $A''_+ \cap S$ such that $y = \alpha x - \beta$. Then αx is in $\alpha(A_+ \cap S)_{\sigma\delta} \subset (A_h)_{\sigma\delta}$. As we have seen above, 1 is in $(A_h)_\sigma$, so $-\beta$ is in $-(A_h)_\sigma = (A_h)_\delta \subset (A_h)_{\sigma\delta}$. The set $(A_h)_{\sigma\delta}$ is closed under addition, so that we conclude that $y = \alpha x - \beta$ is in $(A_h)_{\sigma\delta}$. Q.E.D.

Lemma 4.23. *Let A be a C^*-algebra of operators on a Hilbert space $\mathfrak{H}$ and let $A_{\mathrm{I}} = A + \mathbf{C}1$. Let S denote the unit ball of $\mathscr{L}(\mathfrak{H})$. Then we conclude:*

(i) *For any $\varepsilon > 0$ and $x \in ((A_{\mathrm{I}})_+ \cap S)_\sigma$, $(1+\varepsilon)^{-1}(x+\varepsilon)$ belongs to $(A_+ \cap S)^m$.*
(ii) $(((A_{\mathrm{I}})_+ \cap S)_\sigma)_m \subset ((A_+ \cap S)^m)_m$.

PROOF. Let $\{x_n\}$ be a sequence in $(A_{\mathrm{I}})_+ \cap S$ which increases to x. Each element $x_n - x_{n-1}$ is of the form $y_n + \alpha_n$ where $y_n \in A_n$ and $\alpha_n \in \mathbf{R}$, where

$x_0 = 0$. Since α_n is the canonical image of the positive element $x_n - x_{n-1}$ in the quotient algebra $\mathbf{C} = A_{\mathrm{I}}/A$, we have $\alpha_n \geq 0$. Let $\{u_i\}_{i \in I}$ be an approximate identity of A. Since $\{u_i\}$ is an increasing net, it converges strongly to an element e in $(A_+ \cap S)^m$. But we have $ex = \text{s-lim}_i\, u_i x = x$ for every $x \in A$, so that $e = 1$ since A is nondegenerate. Thus $\{u_i\}_{i \in I}$ converges strongly to 1. Let

$$z_{n,m,i} = y_n + (\alpha_n + \varepsilon 2^{-n})[(m^{-1}2^{-n} + u_i + |y_n|)^{-1}(u_i + |y_n|)].$$

Then $z_{n,m,i}$ is in A_h. Since the function: $t \mapsto (\beta + t)^{-1}t$, $t \geq 0$, is operator-monotone, we have, for $m > 1/\varepsilon$,

$$\begin{aligned} z_{n,m,i} &\geq y_n + (\alpha_n + \varepsilon 2^{-n})(m^{-1}2^{-n} + |y_n|)^{-1}|y_n| \\ &\geq y_n + (\alpha_n + \varepsilon 2^{-n})(\varepsilon 2^{-n} + |y_n|)^{-1}|y_n| \\ &= (\varepsilon 2^{-n} + |y_n|)^{-1}[\varepsilon 2^{-n}(y_n + |y_n|) + |y_n|(y_n + \alpha_n)] \\ &\geq 0. \end{aligned}$$

Define $v_{m,i} = (1 + \varepsilon)^{-1} \sum_{n=1}^{m} z_{n,m,i}$. Then for $m > 1/\varepsilon$, we have

$$\begin{aligned} 0 \leq v_{m,i} &\leq (1 + \varepsilon)^{-1} \sum_{n=1}^{m} (y_n + \alpha_n + \varepsilon 2^{-n}) \\ &\leq (1 + \varepsilon)^{-1}(x_m + \varepsilon) \leq (1 + \varepsilon)^{-1}(x + \varepsilon) \leq 1. \end{aligned}$$

The net $\{v_{m,i}\}_{i \in I}$ is monotone increasing and contained in $A_+ \cap S$. Each net $\{z_{n,m,i}: m = 1,2, \ldots, i \in I\}$ increases to $y_n + \alpha_n + \varepsilon 2^{-n}$, so that $\{v_{m,i}\}$ converges strongly to $(1 + \varepsilon)^{-1}(x + \varepsilon)$. Hence $(1 + \varepsilon)^{-1}(x + \varepsilon)$ belongs to $(A_+ \cap S)^m$, as required.

Let $\{x_j\}_{j \in J}$ be a decreasing net in $((A_{\mathrm{I}})_+ \cap S)_\sigma$. Then by the first assertion, $[1 + (1/n)]^{-1}[x_j + (1/n)] \in (A_+ \cap S)^m$ for each n and j. Since the net $\{[1 + (1/n)]^{-1}[x_j + (1/n)]\}$ is decreasing in n and j and has the same limit as $\{x_j\}_{j \in J}$, the second assertion follows. Q.E.D.

Theorem 4.24 (Up–Down–Up Theorem). *Let A be a C^*-algebra of operators on a Hilbert space $\mathfrak{H}$. Let S denote the unit ball of $\mathscr{L}(\mathfrak{H})$. Then we conclude that*

$$A''_+ \cap S = (((A_+ \cap S)^m)_m)^m. \tag{18}$$

Hence $A''_h = (((A_h)^m)_m)^m$. If A contains the identity 1, *then*

$$A''_+ \cap S = (((A_+ \cap S)_\sigma)_m)^m; \tag{18'}$$

hence $A''_h = (((A_h)_\sigma)_m)^m$.

Remark 4.25. Comparing this result with Theorem 4.22, the last up process may seem quite odd. However, as we will see in Exercise 2, the last up process cannot be dropped in the general case.

PROOF. By Lemma 4.23, we may assume $1 \in A$. We first show that for any set $\{p_i : i \in I\}$ of projections in $(((A_+ \cap S)_\sigma)_m)^m$, the projection $p = \bigvee_{i \in I} p_i$ belongs to $(((A_+ \cap S)_\sigma)_m)^m$, where $\bigvee_{i \in I} p_i$ means the projection of $\mathfrak{H}$ onto the closed subspace $[\bigcup_{i \in I} p_i \mathfrak{H}]$ spanned by $\bigcup_{i \in I} p_i \mathfrak{H}$. For each $i \in I$, let $\{x_j : j \in J_i\}$ be a net in $((A_+ \cap S)_\sigma)_m$ which increases to p_i. We may assume, without loss of generality, that every net has zero as its first element. Let $\Gamma = \{\gamma \in \prod_{i \in I} J_i : \gamma(i) = 0$ except for finitely many $i\}$. For γ and γ' in Γ, we write $\gamma \leq \gamma'$ if $\gamma(i) \leq \gamma'(i)$ for every $i \in I$. Then Γ is a net. For each $\gamma \in \Gamma$, we denote by $|\gamma|$ the number of nonzero values of γ. For $n = 1,2, \ldots$, we define

$$y_{\gamma,n} = \left(|\gamma|^{-1} n^{-1} + |\gamma|^{-1} \sum_{i \in I} x_{\gamma(i)}\right)^{-1} |\gamma|^{-1} \sum_{i \in I} x_{\gamma(i)}.$$

Since $((A_+ \cap S)_\sigma)_m$ is convex, $|\gamma|^{-1} \sum_{i \in I} x_{\gamma(i)}$ belongs to $((A_+ \cap S)_\sigma)_m$. Noticing that the function $f_\alpha : a \in \mathcal{L}(\mathfrak{H})_+ \mapsto f_\alpha(a) = (\alpha + a)^{-1} a$, $0 < \alpha \leq 1$, preserves the ordering and is strongly continuous, we conclude that $y_{\gamma,n}$ is in $((A_+ \cap S)_\sigma)_m$. Suppose $(\gamma_1, n_1) \leq (\gamma_2, n_2)$. Then $\sum_{i \in I} x_{\gamma_1(i)} \leq \sum_{i \in I} x_{\gamma_2(i)}$, so that we have,

$$\begin{aligned} y_{\gamma_1,n_1} &= \left(|\gamma_1|^{-1} n_1^{-1} + |\gamma_1|^{-1} \sum_{i \in I} x_{\gamma_1(i)}\right)^{-1} |\gamma_1|^{-1} \sum_{i \in I} x_{\gamma_1(i)} \\ &= \left(n_1^{-1} + \sum_{i \in I} x_{\gamma_1(i)}\right)^{-1} \sum_{i \in I} x_{\gamma_1(i)} \\ &\leq \left(n_1^{-1} + \sum_{i \in I} x_{\gamma_2(i)}\right)^{-1} \sum_{i \in I} x_{\gamma_2(i)} \\ &\leq \left(n_2^{-1} + \sum_{i \in I} x_{\gamma_2(i)}\right)^{-1} \sum_{i \in I} x_{\gamma_2(i)} = y_{\gamma_2,n_2}; \end{aligned}$$

thus $\{y_{\gamma,n}\}$ is increasing with γ and n. Therefore, $\{y_{\gamma,n}\}$ converges strongly to y in $(((A_+ \cap S)_\sigma)_m)^m$. For each $i \in I$, we have $y_{\gamma,n} \geq (n^{-1} + x_{\gamma(i)})^{-1} x_{\gamma(i)}$, and the right-hand side converges strongly to p_i as γ and n run through Γ and positive integers. Hence we have $y \geq p_i$, $i \in I$. On the other hand, $px_i = x_i$ for every $i \in I$, so that we have $py_{\gamma,n} = y_{\gamma,n}$ for every $\gamma \in \Gamma$ and n; hence $py = y$. Therefore, $p = y$, that is, $p \in (((A_+ \cap S)_\sigma)_m)^m$.

Now we show that for any family $\{q_i : i \in I\}$ of projections in $((A_+ \cap S)_\sigma)_m$, the projection $q = \bigwedge_{i \in I} q_i$ belongs to $((A_+ \cap S)_\sigma)_m$, where $\bigwedge_{i \in I} q_i$ denotes the projection of $\mathfrak{H}$ onto $\bigcap_{i \in I} q_i \mathfrak{H}$. Since $1 - q = \bigvee_{i \in I} (1 - q_i)$ and since $1 - q_i \in ((A_+ \cap S)_\delta)^m$ for every $i \in I$, we conclude, using $((A_+ \cap S)_\delta)^m$ in the above arguments instead of $(((A_+ \cap S)_\sigma)_m)^m$, that $1 - q \in ((A_+ \cap S)_\delta)^m$; and thus $q \in ((A_+ \cap S)_\sigma)_m$.

Let p be any projection in A''. For each vectors $\xi \in p\mathfrak{H}$ and $\eta \in (1 - p)\mathfrak{H}$, there exists a projection $p_{\xi,\eta}$ in $(A_+ \cap S)_{\sigma\delta}$, by Lemma 4.21, such that $p_{\xi,\eta}\xi = \xi$ and $p_{\xi,\eta}\eta = 0$. Let $p_\xi = \bigwedge\{p_{\xi,\eta} : \eta \in (1 - p)\mathfrak{H}\}$. Then from what we have proved above, it follows that p_ξ belongs to $((A_+ \cap S)_\delta)^m$; and $p_\xi \leq p$. It is obvious that $p = \bigvee\{p_\xi : \xi \in p\mathfrak{H}\}$. Hence p belongs to $(((A_+ \cap S)_\sigma)_m)^m$.

As in the proof of Theorem 4.22, each $x \in A''_+ \cap S$ has the form $x = \sum_{k=1}^{\infty} 2^{-k} p_k$, where $\{p_k\}$ is a sequence of spectral projections of x. For each $k = 1,2,\ldots$, there exists a net $\{y_j : j \in J_k\}$ in $((A_+ \cap S)_\delta)_m$ which increases to p_k. Let $\Gamma = \{\gamma \in \prod_{k=1}^{\infty} J_k : \gamma(k) = 0$ except for finitely many $k\}$. Then as before, Γ is a net. For each $\gamma \in \Gamma$, we define $x_\gamma = \sum_{k=1}^{\infty} 2^{-k} p_{\gamma(k)}$. Then $\{x_\gamma\}$ is an increasing net in $((A_+ \cap S)_\delta)_m$ with strong limit x, so that x belongs to $(((A_+ \cap S)_\sigma)_m)^m$. Q.E.D

Corollary 4.26 *Let A be a nondegenerate C^*-algebra of operators on a Hilbert space $\mathfrak{H}$. Then A is a von Neumann algebra on $\mathfrak{H}$ if and only if the limit of every bounded increasing net in A_h falls in A_h.***

Exercises

1. Let $\mathfrak{H}$ be a Hilbert space with $\dim \mathfrak{H} = \infty$, and $\mathscr{U}$ be the group of all unitaries on $\mathfrak{H}$.
   (a) Show that the strong closure of $\mathscr{U}$ is precisely the set of all isometries on $\mathfrak{H}$.
   (b) Show that the weak closure $\tilde{\mathscr{U}}$ of $\mathscr{U}$ is a semigroup with respect to the multiplication, and stable under the *-operation.
   (c) Show that $\tilde{\mathscr{U}}$ contains all projections and partial isometries.
   (d) Show that if a positive operator h on $\mathfrak{H}$ with $0 \leq h \leq 1$ has an infinite dimensional null space, then there exist projections e and f such that $h = efe$. (*Hint*: Consider the following projection in $\mathfrak{H} \oplus \mathfrak{H}$:
$$\left[\begin{matrix} h, & (h-h^2)^{1/2} \\ (h-h^2)^{1/2}, & 1-h \end{matrix}\right].\Bigg)$$
   (e) Show that $\tilde{\mathscr{U}}$ is precisely the closed unit ball of $\mathscr{L}(\mathfrak{H})$.
2. Let $\mathfrak{H}_d = l^2(0,1)$ (resp. $\mathfrak{H}_c = L^2(0,1)$) be the Hilbert space of all square integrable functions on $[0,1]$ with respect to the discrete counting (resp. the Lebesgue) measure on $[0,1]$. Let π_d (resp. π_c) be the representation of $C[0,1]$ on $\mathfrak{H}_d$ (resp. $\mathfrak{H}_c$) by multiplication. Set $\pi = \pi_d \oplus \pi_c$, $A = \pi(C[0,1])$, and $\mathscr{M} = A''$.
   (a) Show that if $a \in (A_h)^m$, then there exists a unique lower semicontinuous function f on $[0,1]$ such that $a\xi = f\xi$ for each $\xi \in \mathfrak{H}_d$ and $\xi \in \mathfrak{H}_c$.
   (b) Show that the projection z_d of $\mathfrak{H}$ onto $\mathfrak{H}_d$ belongs to $\mathscr{M}$.
   (c) Show that if $a \in (A_h)^m$ majorizes z_d, then a must majorize 1. (*Hint*: Use (a).) Hence z_d does not belong to $((A_h)^m)_m$.

Notes

The results in Section 1 were obtained by R. Schatten and J. von Neumann [317–320] in the study of the tensor product of Banach spaces. A further detailed account of the theory was presented by Schatten [28, 29]. The presentation here is, however, taken from J. Dixmier [80]. The weak topology, the strong topology, and the σ-strong topology were introduced by J. von Neumann [254, 255]. It was J. Dixmier [80] who recognized the

σ-weak topology as the weak* topology in the duality between $\mathscr{L}(\mathfrak{H})$ and $\mathscr{L}_*(\mathfrak{H})$, and thus proved that a von Neumann algebra is the conjugate space of a Banach space, Theorem 2.6. As mentioned already, the importance of Theorem 3.9 cannot be overstated. The whole theory of operator algebras hatched out of this theorem. The continuity of functional calculus presented in Section 4, Theorem 4.7, was due to I. Kaplansky [196]. A further generalization of this result was obtained by R. Kadison [189]. Theorem 4.8 is called the Kaplansky density theorem, and indeed due to him [196]. The rest of the sections are direct consequences of this theorem. Theorem 4.11 is due to J. Glimm and R. Kadison [139]. Theorems 4.13 and 4.15 are due to M. Tomita [372] and K. Saitô [302]. A striking consequence of these approximation theorems is the transitivity theorem, Theorem 4.18, due to R. Kadison [182]. The up–down theorem and the up–down–up theorem are due to G. Pedersen [281], and answer the problem raised by R. Kadison's work [183].

Chapter III
Conjugate Spaces

0. Introduction

We study the conjugate space of a C^*-algebra as a Banach space. In this chapter, we will see that a C^*-algebra behaves, as a Banach space, like the space of continuous functions on a compact space, and its conjugate space looks like the L^1-space on a measure space.

In Section 1, we first investigate abelian von Neumann algebras and their spectrum in detail. The spectrum of an abelian von Neumann algebra is characterized as an extremely disconnected compact space with sufficiently many normal measures, Theorem 1.18. The uniqueness of abelian von Neumann algebras with no minimal projection on a separable Hilbert space is also proved in this section, Theorem 1.22. We shall see in Section 2 that the second conjugate space of a C^*-algebra as a Banach space is naturally identified with a von Neumann algebra which has a special meaning for the original C^*-algebra. With this von Neumann algebra, the universal enveloping von Neumann algebra, a large number of problems for C^*-algebras are reduced to problems for von Neumann algebras. Section 3 is devoted to a characterization of a C^*-algebra which admits a faithful representation as a von Neumann algebra. Namely, such a C^*-algebra is characterized as the conjugate space of a uniquely determined Banach space, the predual. Also, the uniqueness of the predual of a von Neumann algebra is established, demonstrating the σ-weak continuity of an isomorphism between von Neumann algebras. The polar decomposition and the absolute value of functionals will be introduced in Section 4. In Section 5, we shall investigate compact sets in the predual of a von Neumann algebra and characterize the Arens–Mackey topology on the bounded parts as the σ-strong* topology. In the last section, we shall investigate the noncommutative analogue of semicontinuous, Borel, and universally measurable functions.

1. Abelian Operator Algebras

Although our main objects are noncommutative infinite dimensional operator algebras, the knowledge of abelian operator algebras is vitally important in later study. For example, without the spectral decomposition theorem for bounded self-adjoint operators, which is essentially a result for abelian operator algebras, it would have been impossible to obtain results in the previous chapter such as Theorem II.1.8 and Theorem II.3.9 (von Neumann double commutation theorem). Thus, in this section, we will systematically study abelian operator algebras. The results in this section will give us some ideas as to how we should look at the general operator algebras.

As we have seen in Theorem I.4.4, every abelian C^*-algebra A may be regarded as the C^*-algebra $C_\infty(\Omega)$ of all continuous functions on a locally compact space Ω vanishing at infinity. Therefore, to show algebraic properties of such a C^*-algebra A, we can reduce problems to those of $C_\infty(\Omega)$. Following this idea, we shall discuss representations of abelian C^*-algebras in this section, although the complete description, such as unitary invariants, of representations of such a C^*-algebra will be given in a later section.

We identify an abelian C^*-algebra A with the C^*-algebra $C_\infty(\Omega)$ of all continuous functions on the spectrum Ω of A, vanishing at infinity, by means of the Gelfand representation. Then the following is a review of results in integration theory on a locally compact space, especially the Riesz–Markov–Kakutani theorem.

Theorem 1.1. *The conjugate space A^* of an abelian C^*-algebra A is the Banach space $\mathfrak{M}^1(\Omega)$ of all finite Radon measures on the spectrum Ω of A. The duality in $\{A,A^*\}$ is given by*

$$\langle a,\mu\rangle = \int_\Omega a(\omega)\,d\mu(\omega), \qquad a \in A, \qquad \mu \in \mathfrak{M}^1(\Omega). \tag{1}$$

Furthermore, a measure $\mu \in \mathfrak{M}^1(\Omega)$ gives rise to a positive linear functional on A if and only if it is a positive measure.

For the proof, we refer the reader to text books of integration theory. As seen in Theorem I.9.14, a positive measure $\mu \in \mathfrak{M}^1(\Omega)$ gives rise to a cyclic representation $\{\pi_\mu,\mathfrak{H}_\mu,\xi_\mu\}$ of A. Considering $A = C_\infty(\Omega)$ as a dense subspace of $L^2(\Omega,\mu)$, we define a map U of $C_\infty(\Omega)$ onto $\pi_\mu(A)\xi_\mu$ by $Ux = \pi_\mu(x)\xi_\mu$, $x \in A$. Then we have

$$(Ux|Uy) = \mu(y^*x) = \int_\Omega x(\omega)\overline{y(\omega)}\,d\mu(\omega);$$

hence U is extended to an isometry of $L^2(\Omega,\mu)$ onto $\mathfrak{H}_\mu$, which is also denoted by U. Define a representation π of A on $L^2(\Omega,\mu)$ by

$$(\pi(a)\xi)(\omega) = a(\omega)\xi(\omega), \qquad a \in A, \qquad \xi \in L^2(\Omega,\mu).$$

Then it is not difficult to see that

$$U\pi(a)U^* = \pi_\mu(a), \qquad a \in A.$$

Therefore, the cyclic representation π_μ of A is unitarily equivalent to the multiplication representation π of A on $L^2(\Omega,\mu)$. Thus, we have proved the first part of the following:

Theorem 1.2. *Let Ω be a locally compact space and $\mu \in \mathfrak{M}^1(\Omega)$ be a positive linear functional on the abelian C^*-algebra $A = C_\infty(\Omega)$. Then we conclude the following*:

(i) *The cyclic representation $\{\pi_\mu,\mathfrak{H}_\mu,\xi_\mu\}$ of A induced by μ is realized on the Hilbert space $L^2(\Omega,\mu)$ of all square integrable functions on Ω with respect to μ, as the multiplication representation, where the cyclic vector ξ_μ is given by the constant function 1 on Ω.*

(ii) *The von Neumann algebra $\{\mathcal{M}(\pi_\mu),\mathfrak{H}_\mu\}$ generated by $\pi_\mu(A)$ is maximal abelian, that is,*

$$\mathcal{M}(\pi_\mu)' = \mathcal{M}(\pi_\mu).$$

(iii) *The algebra $\mathcal{M}(\pi_\mu)$ consists of all the operators $\pi(f)$, $f \in L^\infty(\Omega,\mu)$, defined by*

$$(\pi(f)\xi)(\omega) = f(\omega)\xi(\omega), \qquad \xi \in \mathfrak{H}_\mu = L^2(\Omega,\mu).$$

Proof. Since $\pi_\mu(A) \subset \pi_\mu(A)' = \mathcal{M}(\pi_\mu)'$, we have only to prove that $\mathcal{M}(\pi_\mu)'$ consists of all the operators $\pi(f)$, $f \in L^\infty(\Omega,\mu)$. It is clear that $\pi(f)$, $f \in L^\infty(\Omega,\mu)$, commutes with $\pi_\mu(A)$.

Take an arbitrary positive operator $a \in \mathcal{M}(\pi_\mu)'$. We define a linear functional μ_a on A by

$$\mu_a(x) = (\pi_\mu(x)a\xi_\mu|\xi_\mu), \qquad x \in A.$$

Let $\pi_\mu(x) = uh$ be the polar decomposition of $\pi_\mu(x)$ for a fixed $x \in A$. Then we have

$$\begin{aligned}
|\mu_a(x)| &= |(uha\xi_\mu|\xi_\mu)| \\
&= |(uha^{1/2}\xi_\mu|a^{1/2}\xi_\mu)| \\
&= |(uh^{1/2}a^{1/2}\xi_\mu|h^{1/2}a^{1/2}\xi_\mu)| \\
&\leq \|uh^{1/2}a^{1/2}\xi_\mu\| \, \|h^{1/2}a^{1/2}\xi_\mu\| \\
&= \|h^{1/2}a^{1/2}\xi_\mu\|^2 \\
&\leq \|a\| \, \|h^{1/2}\xi_\mu\|^2 = \|a\| \int_\Omega |x(\omega)| \, d\mu(\omega),
\end{aligned}$$

where the last step of the above calculation follows from the fact that $h = \pi_\mu(|x|)$. Recalling that $A = C_\infty(\Omega)$ is dense in $L^1(\Omega,\mu)$, we see that μ_a defines a bounded linear functional on $L^1(\Omega,\mu)$, which is also denoted by μ_a. Hence there is a function $f \in L^\infty(\Omega,\mu)$ such that

$$\mu_a(x) = \int_\Omega f(\omega)x(\omega) \, d\mu(\omega), \qquad x \in A,$$

which means that

$$(a\pi_\mu(x)\xi_\mu|\pi_\mu(y)\xi_\mu) = \int_\Omega f(\omega)x(\omega)\overline{y(\omega)} \, d\mu(\omega), \qquad x,y \in A.$$

Therefore, for every $\xi,\eta \in \mathfrak{H}_\mu$, we have

$$(a\xi|\eta) = \int_\Omega f(\omega)\xi(\omega)\overline{\eta(\omega)}\, d\mu(\omega);$$

thus we get $a = \pi(f)$. Q.E.D.

The following results are immediate consequences of the above theorem.

Corollary 1.3. *If an abelian von Neumann algebra on a Hilbert space admits a cyclic vector, then it is maximal abelian.*

Corollary 1.4. *Every normal isomorphism of an abelian von Neumann algebra* $\{\mathcal{M},\mathfrak{H}\}$ *onto another* $\{\mathcal{N},\mathfrak{K}\}$ *is spatial if both of them have cyclic vectors.*

Corollary 1.5. *Let A be an abelian C^*-algebra with spectrum Ω. If μ and ν are two positive finite measures on Ω, then the following two statements are equivalent:*

(i) *μ and ν are equivalent in the sense of absolute continuity.*
(ii) *π_μ and π_ν are unitarily equivalent.*

Now, we turn to study the spectrum of an abelian von Neumann algebra. Since an abelian von Neumann algebra contains a lot of projections, its spectrum is obviously totally disconnected.

Definition 1.6. A Hausdorff space is said to be *extremely disconnected* if the closure of every open subset is open. A compact extremely disconnected space is called *stonean.*

Proposition 1.7. *For a compact space Ω, the following three statements are equivalent:*

(i) *Ω is stonean.*
(ii) *$C_{\mathbf{R}}(\Omega)$, the Banach space of all real valued continuous functions on Ω, is a conditionally complete lattice; that is, every bounded subset of $C_{\mathbf{R}}(\Omega)$ has a least upper bound in $C_{\mathbf{R}}(\Omega)$.*
(iii) *Every bounded real valued lower semicontinuous function coincides with a continuous function except possibly on a subset of the first category.*

PROOF. (iii)$\Rightarrow$(ii): Let $\{f_i\}_{i\in I}$ be a bounded family in $C_{\mathbf{R}}(\Omega)$. Put $g(\omega) = \sup_i f_i(\omega)$, $\omega \in \Omega$. Then g is bounded and lower semicontinuous. Hence by assumption, there is a continuous function f which coincides with g except on a set K of the first category. Since $g - f$ is lower semicontinuous too, the set $G = \{\omega : g(\omega) - f(\omega) > 0\}$ is open and contained in K. Since Ω is a Baire space, being compact, no set of the first category can contain a non-empty open set; hence G must be empty, which means that $f \geq g$. If an $h \in C_{\mathbf{R}}(\Omega)$ majorizes $\{f_i\}_{i\in I}$, then $h(\omega) \geq f(\omega)$ for every $\omega \in K^c$, which implies the inequality $h(\omega) \geq f(\omega)$ for every $\omega \in \Omega$ because K^c is dense in Ω. Thus, f is the least upper bound of $\{f_i\}_{i\in I}$ in $C_{\mathbf{R}}(\Omega)$.

(ii) ⇒ (i): Let G be an open subset of Ω. Let χ_G denote the characteristic function of G. By Urysohn's lemma, there exists a family $\{f_i\}$ of positive continuous functions with $\chi_G(\omega) = \sup_i f_i(\omega)$. Let f be the least upper bound of $\{f_i\}$ in $C_{\mathbf{R}}(\Omega)$. Then we have, for every $\omega \in \Omega$, $1 \geq f(\omega) \geq \chi_G(\omega)$. Therefore, by continuity, f majorizes the characteristic function $\chi_{\bar{G}}$ of the closure $\bar{G}$ of G. Suppose $f(\omega_0) > 0$ at some point $\omega_0 \notin \bar{G}$. Then we can choose a positive nonzero function $g \in C_{\mathbf{R}}(\Omega)$, by Urysohn's lemma again, so that $f(\omega) = g(\omega)$ for every $\omega \in \bar{G}$; $f(\omega_0) > g(\omega_0)$; $f(\omega) \geq g(\omega)$ for every $\omega \in \Omega$. Then $g(\omega) \geq f_i(\omega)$ for every $\omega \in \Omega$ because $g \geq \chi_{\bar{G}}$. This contradicts the fact that f is the least upper bound of $\{f_i\}$ in $C_{\mathbf{R}}(\Omega)$. Hence $f(\omega) = 0$ for $\omega \in \bar{G}$. Thus, we get $f = \chi_{\bar{G}}$. Therefore, by the continuity of f, $\bar{G}$ is open and closed.

(i) ⇒ (iii): Suppose f is a bounded real valued lower semicontinuous function on Ω. By a suitable adjustment, we may assume $0 \leq f(\omega) \leq 1$, $\omega \in \Omega$. For each $\lambda \in \mathbf{R}$, let $F(\lambda) = \{\omega \in \Omega : f(\omega) \leq \lambda\}$. Then $F(\lambda)$ is closed for every $\lambda \in \mathbf{R}$. Let $G(\lambda)$ denote the interior of $F(\lambda)$. Then by the dual assertion of our assumption, $G(\lambda)$ is open and closed. Hence the characteristic function χ_λ of $G(\lambda)$ is continuous. For each $n = 1,2, \ldots$, define a function f by

$$f_n = \sum_{k=1}^{2^n} \frac{k}{2^n} (\chi_{k/2^n} - \chi_{(k-1)/2^n}).$$

Then f_n is in $C_{\mathbf{R}}(\Omega)$, and

$$\|f_n - f_{n+1}\| \leq \frac{1}{2^n}, \qquad n = 1,2, \ldots.$$

Hence the sequence $\{f_n\}$ converges uniformly to a continuous function f_0 as $n \to \infty$. Put

$$K = \bigcup_{n=1}^{\infty} \bigcup_{k=1}^{2^n} (F(k/2^n) - G(k/2^n)).$$

Then K is of the first category, since each set in the above union is rare. If $\omega \in G(k/2^n) - F((k-1)/2^n)$, then we have $(k-1)/2^n < f(\omega) \leq k/2^n$, so that

$$|f(\omega) - f_n(\omega)| < \frac{1}{2^n}, \qquad \omega \notin K;$$

thus $f(\omega) = f_0(\omega)$ for every $\omega \notin K$. Q.E.D.

Corollary 1.8. *If f is a bounded continuous function defined on an open dense subset G of a stonean space Ω, then f can be extended to a continuous function defined on the whole space Ω. Therefore, the Stone–Čech compactification of any open dense subset of Ω is the whole space Ω itself.*

PROOF. Considering the real part and the imaginary part of f separately, we may assume f is real valued. Define a lower semicontinuous function f' on Ω by

$$f'(\omega_0) = \begin{cases} f(\omega_0) & \text{if } \omega_0 \in G. \\ \inf\{f(\omega) : \omega \in G\} & \text{if } \omega_0 \notin G. \end{cases}$$

Then the implication (i) $\Rightarrow$ (iii) in the previous proposition yields our assertion. Q.E.D.

Proposition 1.9. *If f is a bounded real valued lower semicontinuous function on a stonean space Ω, then the continuous function f' which coincides with f except on a set of the first category is given by*

$$f'(\omega_0) = \limsup_{\omega \to \omega_0} f(\omega), \qquad \omega_0 \in \Omega.$$

In otherwords f' is the upper semicontinuous regularization of f.

PROOF. By arguments used in the proof of the assertion (i) $\Rightarrow$ (iii) in Proposition 1.7, we have $f \leq f'$. Hence we have

$$\limsup_{\omega \to \omega_0} f(\omega) \leq \limsup_{\omega \to \omega_0} f'(\omega) = f'(\omega_0).$$

Put $A = \{\omega \in \Omega : f'(\omega) - f(\omega) > 0\}$. Then A is of the first category. For $\varepsilon > 0$, there exists a neighborhood U of ω_0 such that $f'(\omega) > f'(\omega_0) - \varepsilon$ for every $\omega \in U$; since $U \cap A^c \neq \emptyset$, there exists an $\omega \in U$ with $f(\omega) > f'(\omega_0) - \varepsilon$. Hence $\limsup_{\omega \to \omega_0} f(\omega) \geq f'(\omega_0) - \varepsilon$, so that

$$\limsup_{\omega \to \omega_0} f(\omega) \geq f'(\omega_0). \qquad \text{Q.E.D.}$$

Definition 1.10. A positive Radon measure μ on a stonean space Ω is called *normal* if for any increasing bounded net $\{f_i\}$ in $C_{\mathbf{R}}(\Omega)$ with $f = \sup f_i$, where $\sup f_i$ means the least upper bound of $\{f_i\}$ in $C_{\mathbf{R}}(\Omega)$, $\mu(f) = \sup \mu(f_i)$. A linear combination of positive normal measures is also called *normal.*

Proposition 1.11. *For a positive Radon measure μ on a stonean space Ω to be normal it is necessary and sufficient that every rare set in Ω be μ-null.*

PROOF. Suppose μ is normal. Let E be a rare set in Ω. Considering the closure of E, we may assume that E is closed. Then the complement E^c is open and dense in Ω. Let $\{G_i\}$ be an increasing net of open and closed sets with $E^c = \bigcup G_i$. Then we have

$$1 - \chi_E(\omega) = \chi_{E^c}(\omega) = \sup \chi_{G_i}(\omega), \qquad \omega \in \Omega,$$

where χ_A denotes the characteristic function of A for each set A. It is clear that the least upper bound of $\{\chi_{G_i}\}$ in $C_{\mathbf{R}}(\Omega)$ is the identity function 1. Hence we have

$$\begin{aligned} \mu(\Omega) &= \mu(1) = \sup \mu(\chi_{G_i}) \\ &= \sup \mu(G_i) = \mu(E^c), \end{aligned}$$

so that $\mu(E) = 0$.

Suppose every rare set is μ-null. Let $\{f_i\}$ be an increasing bounded net in $C_{\mathbf{R}}(\Omega)$ with $f = \sup f_i$. Put $f'(\omega) = \sup f_i(\omega)$. Then $f'(\omega)$ is lower semicontinuous and $\mu(f') = \sup \mu(f_i)$. Clearly, $f \geq f'$. For each $n = 1, 2, \ldots,$ the set $A_n = \{\omega \in \Omega : f(\omega) - f'(\omega) \geq 1/n\}$ is closed and has, by Urysohn's lemma, no interior point, that is, A_n is rare. Hence $A = \bigcup_{n=1}^{\infty} A_n = \{\omega : f(\omega) \neq f'(\omega)\}$ is of the first category. By assumption, $\mu(A) = 0$, which implies that

$$\mu(f) = \mu(f') = \sup \mu(f_i). \qquad \text{Q.E.D.}$$

Proposition 1.12. *Let μ be a positive normal measure on a stonean space Ω and f a bounded μ-measurable real valued function on Ω. Let f_1 (resp. f_2) be the lower (resp. upper) semicontinuous regularization of f, that is,*

$$f_1(\omega_0) = \liminf_{\omega \to \omega_0} f(\omega),$$

$$f_2(\omega_0) = \limsup_{\omega \to \omega_0} f(\omega),$$

and let f'_1 (resp. f'_2) be the upper (resp. lower) semicontinuous regularization of f_1 (resp. f_2). Then f'_1 and f'_2 are both continuous; f, f_1, f_2, f'_1, and f'_2 coincide except on a μ-null set.

PROOF. By Proposition 1.9, f'_1 and f'_2 are both continuous, and f'_1 (resp. f'_2) coincides with f_1 (resp. f_2) except on a set of the first category, which is μ-null. By Lusin's theorem for measurable functions, there exists a disjoint sequence $\{K_n\}$ of compact sets such that the restriction $f|_{K_n}$ of f is continuous and the complement of $\bigcup_{n=1}^{\infty} K_n$ is μ-null. Let K'_n be the interior of K_n, which is open and closed. Since $K_n - K'_n$ is rare, the complement of $\bigcup_{n=1}^{\infty} K'_n$ is μ-null. It is clear that $f = f_1 = f_2$ on K'_n for each n; hence $f = f_1 = f_2$ on $\bigcup_{n=1}^{\infty} K'_n$. Q.E.D.

Corollary 1.13. *If μ is a positive normal measure on a stonean space Ω, then every μ-measurable set A in Ω coincides, within a μ-null difference, with its closure $\bar{A}$, with its interior A°, with the interior of $\bar{A}$, and with the closure of A°.*

Therefore, the support of μ is open and closed.

PROOF. Let f be the characteristic function χ_A of A. Then the lower (resp. upper) semicontinuous regularization f_1 (resp. f_2) is nothing but the characteristic function χ_{A° (resp. $\chi_{\bar{A}}$) of A° (resp. $\bar{A}$). The upper (resp. lower) semicontinuous regularization f'_1 (resp. f'_2) of f_1 (resp. f_2) coincides with the characteristic function of the closure (resp. the interior) of A° (resp. $\bar{A}$). Therefore, the previous proposition yields our assertions. Q.E.D.

Definition 1.14. A stonean space Ω is said to be *hyperstonean* if it admits sufficiently many positive normal measures; that is, if for any nonzero positive $f \in C_{\mathbf{R}}(\Omega)$ there exists a positive normal measure μ with $\mu(f) \neq 0$.

We remark here that a family $\{\mu_i\}$ of positive normal measures on Ω contains sufficiently many measures if and only if the union of the supports of the μ_i is dense in Ω.

Proposition 1.15. *Let Ω be hyperstonean and $\{\mu_i\}$ a family of sufficiently many positive normal measures on Ω. Then a subset A of Ω is rare if and only if $\mu_i(A) = 0$ for every μ_i.*

PROOF. By Proposition 1.11, every set of the first category is μ_i-null. Conversely, if a subset A of Ω is μ_i-null for every i, then so is the closure $\bar{A}$ of A; hence the interior of $\bar{A}$ is μ_i-null too for every i, so that $\bar{A}$ has no interior point by the assumption on $\{\mu_i\}$, which means that A is rare. Q.E.D.

Thus in a hyperstonean space, every set of the first category is rare.

Corollary 1.16. *Let Ω be a hyperstonean space and $\{\mu_i\}$ a family of sufficiently many positive normal measures on Ω. Then we conclude the following:*

(i) *For a bounded real valued measurable function f, with respect to every μ_i, let f_1 (resp. f_2) be the lower (resp. upper) semicontinuous regularization of f and f'_1 (resp. f'_2) be the upper (resp. lower) semicontinuous regularization of f_1 (resp. f_2). Then f'_1 and f'_2 are continuous, and $f'_1 = f'_2$. Furthermore, f, f_1, f_2, f'_1, and f'_2 all coincide except on a rare set.*
(ii) *If A is measurable for every μ_i, then the interior A' of the closure $\bar{A}$ of A coincides with the closure of the interior A° of A. Furthermore, A, A', $\bar{A}$, and A° coincide within a rare difference.*

PROOF. Our assertions follow immediately from Propositions 1.11 and 1.15 and Corollary 1.13. Q.E.D.

Theorem 1.17. *A stonean space Ω admits a unique partition $\{\Omega_1,\Omega_2,\Omega_3\}$ by open and closed subsets Ω_1, Ω_2 and Ω_3 with the following properties:*

(i) *Ω_1 is hyperstonean.*
(ii) *Ω_2 contains a dense subset of the first category; hence there is no nontrivial normal measure on Ω_2.*
(iii) *In Ω_3, every set of the first category is rare and the support of every measure is rare; hence there is no nontrivial normal measure on Ω_3.*

PROOF. Let $\{\mu_i\}_{i\in I}$ be the family of all positive normal measure on Ω. For each $i \in I$, let F_i be the support of μ_i. Then F_i is open and closed by Corollary 1.13. Let $F = \bigcup_{i\in I} F_i$. Then F is open. Let Ω_1 be the closure of F, which is open and closed. Since the union of the support of positive measures is dense in Ω_1, Ω_1 is hyperstonean.

Let $\mathfrak{G}$ be the family of open sets in Ω which contain a dense subset of the first category. Let $\{G_j\}_{j\in J}$ be a maximal disjoint subfamily of $\mathfrak{G}$. For each $j \in J$, let $\{A_{j,n}: n = 1,2,\dots\}$ be a sequence of rare sets whose union is dense in G_j. Put $A_n = \bigcup_{j\in J} A_{j,n}$. Since $\{G_j\}$ is a disjoint family of open sets, A_n is rare. The union $A = \bigcup_{n=1}^{\infty} A_n$ is dense in $G = \bigcup_{j\in J} G_j$. Hence the closure Ω_2 of G, which is open and closed, contains a dense subset A of the first category. Since $F_i \cap \Omega_2 = \varnothing$ for every $i \in I$, Ω_1 and Ω_2 are disjoint.

Put $\Omega_3 = \Omega_1^c \cap \Omega_2^c$. Suppose $\Omega_3 \neq \varnothing$. Then every subset of Ω_3 of the first category is rare. By construction, Ω_3 does not admit nontrivial positive normal measure. Let μ be a positive Radon measure on Ω_3. Put

$$\alpha = \sup\{\mu(B): B \text{ is rare in } \Omega_3\}.$$

Let $\{B_n\}$ be a sequence of rare subsets of Ω_3 with $\alpha = \lim \mu(B_n)$. The union $B = \bigcup B_n$ is of the first category; hence it is rare. Therefore $\mu(\bar{B}) \leq \alpha$. On the other hand, $\mu(B_n) \leq \mu(\bar{B})$, $n = 1,2,\ldots$, so that $\alpha \leq \mu(\bar{B})$. Hence $\alpha = \mu(\bar{B})$. Let G be an open and closed subset of Ω_3 with $G \cap \bar{B} = \varnothing$. For arbitrary rare subset C of G, $\mu(\bar{B} \cup C) = \mu(\bar{B}) + \mu(C) \leq \alpha$. Hence $\mu(C) = 0$. Therefore, μ vanishes on every rare subset of G, so that the restriction $\mu|_G$ of μ onto G is normal by Proposition 1.11. By the construction of Ω_3, $\mu(G) = 0$. Since the open and closed subsets of Ω_3 form a basis of open set in Ω_3, μ is supported by the set $\bar{B}$ of the first category.

Suppose $\{\Omega'_1,\Omega'_2,\Omega'_3\}$ is another partition satisfying (i), (ii), and (iii). By the construction of Ω_1, we have $\Omega'_1 \subset \Omega_1$. If $\Omega'_2 \cap \Omega_2^c \neq \varnothing$, then it is open and closed, and contains a dense subset of the first category, which contradicts the maximality of $\{G_j\}_{j\in J}$. Hence $\Omega'_2 \subset \Omega_2$. By the constructions of Ω_1 and Ω_2, it is almost obvious that $(\Omega_1 \cup \Omega_2) \cap \Omega'_3 = \varnothing$. Therefore, $\Omega_i = \Omega'_i$, $i = 1,2,3$. Q.E.D.

Now we can determine when an abelian C^*-algebra admits a faithful representation whose range is a von Neumann algebra on the representation Hilbert space.

Theorem 1.18. *For an abelian C^*-algebra A with spectrum Ω, the following statements are equivalent*:

(i) *Ω is hyperstonean.*
(ii) *A admits a faithful representation $\{\pi,\mathfrak{H}\}$ such that $\pi(A)$ is an abelian von Neumann algebra on $\mathfrak{H}$.*
(iii) *There exists a locally compact space Γ with a positive Radon measure μ such that A is isomorphic to the algebra $L^\infty(\Gamma,\mu)$ of all essentially bounded μ-measurable functions over Γ.*

PROOF. (ii) $\Rightarrow$ (i): Identifying A and $\pi(A)$, we may assume that A is an abelian von Neumann algebra on $\mathfrak{H}$. Let $\{h_i\}_{i\in I}$ be a bounded increasing net in A_h. Since $(h_i\xi|\xi)$ is bounded and increasing for every $\xi \in \mathfrak{H}$, $(h_i\xi|\eta)$ converges for every pair ξ, η in $\mathfrak{H}$. Put $B(\xi,\eta) = \lim_i (h_i\xi|\eta)$. Then $B(\xi,\eta)$ is a bounded hermitian sesquilinear form on $\mathfrak{H}$, so that there exists a self-adjoint $h \in \mathscr{L}(\mathfrak{H})$ such that $(h\xi|\eta) = B(\xi,\eta)$, that is, $(h\xi|\eta) = \lim_i (h_i\xi|\eta)$. Hence $\{h_i\}$ converges weakly to h; so h is in A. It is clear that h is the least upper bound of $\{h_i\}$ in A. Therefore, the self-adjoint part A_h of A is a conditionally complete lattice, which means, by Proposition 1.7, that the spectrum Ω of A is stonean.

Let μ be the positive Radon measure on Ω induced by a positive linear functional ω_ξ, $\xi \in \mathfrak{H}$, on A, where ω_ξ is defined by $\omega_\xi(x) = (x\xi|\xi)$, $x \in A$. Then we have $\mu(\sup h_i) = \sup \mu(h_i)$ for every bounded increasing net $\{h_i\}$ in A_h. Hence μ is normal. Of course, Ω has sufficiently many such measures μ, so that it is hyperstonean.

(i) $\Rightarrow$ (iii): First of all, let us assume that there is a positive normal measure μ with support Ω. We claim that $C(\Omega) \cong L^\infty(\Omega,\mu)$ as a C^*-algebra. By Proposition 1.12, to each $f \in L^\infty(\Omega,\mu)$ there corresponds a unique $f' \in C(\Omega)$ which coincides with f almost everywhere with respect to μ. Put $\pi(f) = f'$. For a pair f, g in $L^\infty(\Omega,\mu)$, put $\pi(f) = f'$ and $\pi(g) = g'$. Then $\alpha f' + \beta g'$ and $\bar{f}'g'$ coincide respectively with $\alpha f + \beta g$ and $\bar{f}g$ μ-almost everywhere. Hence $\pi(\alpha f + \beta g) = \alpha\pi(f) + \beta\pi(g)$ and $\pi(fg) = \overline{\pi(f)}\pi(g)$. Since $C(\Omega)$ is contained in $L^\infty(\Omega,\mu)$ and $\pi(f) = f$ for every $f \in C(\Omega)$, π is surjective; thus π is an isomorphism of $L^\infty(\Omega,\mu)$ onto $C(\Omega)$.

For the general case, let $\{\mu_i\}_{i\in I}$ be a maximal family of positive normal measures on Ω with disjoint supports. Let Γ_i be the support of each μ_i. By Corollary 1.13, Γ_i is open and closed. Put $\Gamma = \bigcup_{i\in I} \Gamma_i$. Then Γ is open; so the closure $\bar{\Gamma}$ of Γ is open and closed. If $\bar{\Gamma} \neq \Omega$, then there is a positive normal measure μ on Ω with $\mu(\chi_{\Omega-\Gamma}) \neq 0$. Hence the measure μ' defined by $\mu'(f) = \mu(f\chi_{\Omega-\Gamma})$ is normal and has support contained in $\Omega - \bar{\Gamma}$, which contradicts the maximality of $\{\mu_i\}_{i\in I}$. Hence Γ is dense in Ω. Since Γ is open in the compact space Ω, it is a locally compact space. Let $\mathscr{K}(\Gamma)$ denote the space of all continuous functions on Γ with compact support. Let K be a compact subset of Γ. Then $\{\Gamma_i\}_{i\in I}$ is an open covering of K; hence K is covered by a finite family $\{\Gamma_{i_1}, \dots, \Gamma_{i_n}\}$. Let μ be a linear functional on $\mathscr{K}(\Gamma)$ defined by

$$\mu(f) = \sum_{i\in I} \mu_i(f), \qquad f \in \mathscr{K}(\Gamma).$$

Note that the support of f, being compact, meets only with finitely many Γ_i, so that the summation is actually taken for only finitely many nonzero terms. It is then clear that μ is a positive Radon measure on Γ and the restriction $\mu|_{\Gamma_i}$ of μ onto Γ_i is μ_i. Now, by Corollary 1.16, to each $f \in L^\infty(\Gamma,\mu)$ there corresponds a unique $f' \in C(\Omega)$ which coincides with f except on a rare subset of Ω because f can be regarded as a bounded μ_i-measurable function on Ω for every μ_i by putting $f(\Gamma^c) = 0$. The same arguments as before show that the correspondence $\pi: f \in L^\infty(\Gamma,\mu) \mapsto \pi(f) = f' \in C(\Omega)$ is an isomorphism.

(iii) $\Rightarrow$ (ii): We may identify A with $L^\infty(\Gamma,\mu)$. Let $\mathfrak{H}$ denote the Hilbert space $L^2(\Gamma,\mu)$ of all square integrable functions with respect to μ. Define a representation π of A on $\mathfrak{H}$ by

$$\pi(f)\xi(\gamma) = f(\gamma)\xi(\gamma), \qquad f \in A, \qquad \xi \in \mathfrak{H}.$$

Of course, π is a faithful representation of A, and $\pi(1) = 1$. Since $L^\infty(\Gamma,\mu)$ is the conjugate space of $L^1(\Gamma,\mu)$ under the canonical identification, we may consider the $\sigma(L^\infty(\Gamma,\mu),L^1(\Gamma,\mu))$-topology in A. For any sequences $\{\xi_n\}$ and

$\{\eta_n\}$ in $\mathfrak{H}$ with $\sum_{n=1}^\infty \|\xi_n\|_2^2 < +\infty$ and $\sum_{n=1}^\infty \|\eta_n\|_2^2 < +\infty$, we have

$$\left\langle \pi(f), \sum_{n=1}^\infty \omega_{\xi_n,\eta_n} \right\rangle = \sum_{n=1}^\infty (\pi(f)\xi_n|\eta_n)$$
$$= \sum_{n=1}^\infty \int_\Gamma f(\gamma)\xi_n(\gamma)\overline{\eta_n(\gamma)}\, d\mu(\gamma).$$

But

$$\|\xi_n\bar{\eta}_n\|_1 = \int_\Gamma |\xi_n(\gamma)\overline{\eta_n(\gamma)}|\, d\mu(\gamma)$$
$$\leq \left(\int_\Gamma |\xi_n(\gamma)|^2\, d\mu(\gamma)\right)^{1/2} \left(\int_\Gamma |\eta_n(\gamma)|^2\, d\mu(\gamma)\right)^{1/2}$$
$$= \|\xi_n\|_2\|\eta_n\|_2,$$

so that we have

$$\sum_{n=1}^\infty \|\xi_n\bar{\eta}_n\|_1 \leq \sum_{n=1}^\infty \|\xi_n\|_2\|\eta_n\|_2$$
$$\leq \left(\sum_{n=1}^\infty \|\xi_n\|_2^2\right)^{1/2} \left(\sum_{n=1}^\infty \|\eta_n\|_2^2\right)^{1/2} < +\infty.$$

Hence $\sum_{n=1}^\infty \xi_n\bar{\eta}_n$ converges in $L^1(\Gamma,\mu)$, and we have

$$\left\langle \pi(f), \sum_{n=1}^\infty \omega_{\xi_n,\eta_n} \right\rangle = \left\langle f, \sum_{n=1}^\infty \xi_n\bar{\eta}_n \right\rangle.$$

Therefore, we have

$${}^t\pi(\mathscr{L}_*(\mathfrak{H})) \subset L^1(\Gamma,\mu),$$

which says that π is continuous with respect to the $\sigma(L^\infty(\Gamma,\mu),L^1(\Gamma,\mu))$-topology and the σ-weak topology. Therefore, the image $\pi(S)$ of the closed unit ball S of A is σ-weakly compact because S is $\sigma(L^\infty(\Gamma,\mu),L^1(\Gamma,\mu))$-compact. Hence $\pi(A)$ is a von Neumann algebra by Theorem II.4.8. Q.E.D.

Therefore, there are nontrivial hyperstonean spaces, namely, the spectrum of any infinite dimensional abelian von Neumann algebra.

Proposition 1.19. *A hyperstonean space Ω admits a unique partition $\{\Omega_d,\Omega_c\}$ such that*

(i) *Ω_d contains a dense open discrete subset,*
(ii) *Ω_c does not contain any isolated point.*

Proof. Let Γ be the set of all isolated points in Ω. Being the union of open points, Γ is open. Let Ω_d be the closure of Γ. Then Ω_d is open and closed. Let Ω_c be the complement of Ω_d. It is easy to check the required properties of Ω_d and Ω_c. Q.E.D.

Obviously, in the proposition, Ω_d is the Stone–Čech compactification of the discrete space Γ. Hence $C(\Omega_d)$ is isomorphic to the algebra $l^\infty(\Gamma)$ of all bounded functions on Γ. Naturally, Ω_d is called the *discrete* part or *atomic* part of Ω and Ω_c is called the *continuous* part of Ω. To see the structure of a hyperstonean space Ω, it is sufficient to consider only the continuous part Ω_c.

Lemma 1.20. *Let Γ be a compact space. If Γ admits a countable separating family $\{E_n\}$ of open and closed subsets, then the abelian C^*-algebra $C(\Gamma)$ of all continuous functions on Γ is generated by a single real valued continuous function as a C^*-algebra. Here "a family $\{E_n\}$ of subsets of Γ is separating" means that for any distinct pair γ_1 and γ_2 in Γ there exists E_n such that either $\gamma_1 \in E_n$ and $\gamma_2 \notin E_n$ or $\gamma_1 \notin E_n$ and $\gamma_2 \in E_n$.*

Proof. Let e_n be the characteristic function of E_n, $n = 1,2,\ldots$. We define a function $f \in C(\Gamma)$ by

$$f(\gamma) = \sum_{n=1}^{\infty} \frac{1}{3^n}(2e_n(\gamma) - 1).$$

Then by assumption, f separates points of Γ. Therefore, by the Stone–Weierstrass theorem, f generates $C(\Gamma)$. Q.E.D.

Proposition 1.21. *An abelian von Neumann algebra on a separable Hilbert space is generated by a single self-adjoint operator.*

Proof. Let $\mathscr{A}$ be an abelian von Neumann algebra on a separable Hilbert space. By Proposition II.2.7, the unit ball S of $\mathscr{A}$ is metrizable and compact for the weak topology, so that it admits a countable dense subset $\{a_n\}$. Considering the real part and the imaginary part, we may assume that the $\{a_n\}$ are all self-adjoint. Let $a_n = \int_{-1}^{1} \lambda \, de_n(\lambda)$ be the spectral decomposition of a_n for $n = 1,2,\ldots$. Let A be the C^*-subalgebra of $\mathscr{A}$ generated by the countable family $\{e_n(\lambda): \lambda$ runs over all rational numbers in $[-1,1]$, $n = 1,2,\ldots\}$ of projections. Then A is σ-weakly dense in $\mathscr{A}$. By Lemma 1.20, A is generated by a single self-adjoint operator as a C^*-algebra. Thus, the original von Neumann algebra $\mathscr{A}$ is generated by a single self-adjoint operator. Q.E.D.

Theorem 1.22. *Let $\mathscr{A}$ be an abelian von Neumann algebra on a separable Hilbert space. If $\mathscr{A}$ contains no nonzero minimal projection, then $\mathscr{A}$ is isomorphic to the algebra $L^\infty(0,1)$ of all essentially bounded functions on the unit interval* (0,1) *with respect to the Lebesgue measure.*

Proof. By Proposition 1.21, $\mathscr{A}$ is generated by a single self-adjoint operator a. Considering $\lambda a + \mu 1$ if necessary, we may assume $0 \le a \le 1$. Let Ω be the spectrum of $\mathscr{A}$. We identify $\mathscr{A}$ with $C(\Omega)$. Let φ be a faithful state in $\mathscr{A}_*$, whose existence is guaranteed by Proposition II.3.19. Let μ be the normal measure on Ω induced by φ. Then the support of μ is Ω itself, so that $C(\Omega)$, hence $\mathscr{A}$, is isomorphic to $L^\infty(\Omega,\mu)$ by the first half of the proof of assertion

(i) $\Rightarrow$ (iii) in Theorem 1.18. Since the function $a(\cdot)$ maps Ω into $[0,1]$ continuously, it transforms the measure μ on Ω into the measure v on $[0,1]$, which is given by $v([0,\alpha]) = \mu(\{\omega \in \Omega : a(\omega) \leq \alpha\})$. Since the C^*-subalgebra A of $\mathscr{A}$ generated by a is isomorphic to $C(a(\Omega))$ under the canonical map π by Proposition I.4.6, and since $v = {}^t\pi(\mu)$, we conclude that π is extended to an isomorphism of $L^\infty(\Omega,\mu)$ onto $L^\infty([0,1],v)$. We define a function f on $[0,1]$ with values in $[0,1]$ by $f(\alpha) = v([0,\alpha])$. Since μ is not atomic, neither is v. Hence f is a continuous increasing function with $f(0) = 0$ and $f(1) = 1$. Let m be the Lebesgue measure on $[0,1]$. Then we have $m([f(\alpha),f(\beta)]) = v([\alpha,\beta)]$ for $0 \leq \alpha \leq \beta \leq 1$. By continuity, the family $[f(\alpha),f(\beta)]$ exhausts all closed subintervals of $[0,1]$ and we have $m([\alpha,\beta]) = v(f^{-1}([\alpha,\beta]))$ because $f^{-1}([\alpha,\beta])$ is a closed interval $[\alpha',\beta']$. Thus we conclude that $f(v) = m$ and the map f induces an isomorphism of $L^\infty(0,1)$ onto $L^\infty([0,1],v)$. Therefore $\mathscr{A}$ is isomorphic to $L^\infty(0,1)$. Q.E.D.

*Now we are going to construct stonean spaces with property (ii) in Theorem 1.17. We leave the construction of a stonean space with property (iii) to an exercise.

Lemma 1.23. *Let Γ be a Hausdorff space. If f is a bounded real valued lower (resp. upper) semicontinuous function on Γ, then the upper (resp. lower) semicontinuous regularization f' coincides with f except on a subset of the first category.*

Proof. Let f be lower semicontinuous. Let $A = \{\gamma \in \Gamma : f'(\gamma) - f(\gamma) > 0\}$ and $A_n = \{\gamma \in \Gamma : f'(\gamma) - f(\gamma) \geq 1/n\}$. Then $A = \bigcup_{n=1}^\infty A_n$ and A_n is closed because $f' - f$ is upper semicontinuous. If A_n contains an interior point γ_0, then for any neighborhood U of γ_0 contained in A_n, we have $\sup_{\gamma \in U} f'(\gamma) \geq \sup_{\gamma \in U} f(\gamma) + (1/n)$, so that

$$f'(\gamma_0) = \limsup_{\gamma \to \gamma_0} f'(\gamma) \geq \limsup_{\gamma \to \gamma_0} f(\gamma) + \frac{1}{n}$$

$$= f'(\gamma_0) + \frac{1}{n}.$$

This is a contradiction. Thus each A_n must be rare. Q.E.D.

Let Γ be a completely regular space, and $\mathscr{B}(\Gamma)$ denote the C^*-algebra of all bounded Borel functions on Γ with the natural algebraic and metric structures. Let $\mathscr{N}(\Gamma)$ denote the subset of $\mathscr{B}(\Gamma)$ consisting of all bounded Borel functions which vanish outside a set of the first category. Obviously, $\mathscr{N}(\Gamma)$ is a closed ideal of $\mathscr{B}(\Gamma)$. Let $B(\Gamma)$ denote the quotient C^*-algebra $\mathscr{B}(\Gamma)/\mathscr{N}(\Gamma)$.

Lemma 1.24. *Every real valued function f in $\mathscr{B}(\Gamma)$ coincides with a bounded real valued lower semicontinuous function except on a set of the first category.*

PROOF. Let $\mathscr{B}'$ be the set of all bounded real valued functions which coincide, except on a set of the first category, with a lower semicontinuous function. By Lemma 1.23, $\mathscr{B}'$ contains all bounded real valued lower or upper semicontinuous functions. Let $\{f_n\}$ be a sequence in $\mathscr{B}'$ converging pointwise to a bounded function f on Γ. If we prove that $f \in \mathscr{B}'$, then $\mathscr{B}'$ contains the real part $\mathscr{B}_{\mathbf{R}}(\Gamma)$ of $\mathscr{B}(\Gamma)$ because $\mathscr{B}_{\mathbf{R}}(\Gamma)$ is the smallest family which contains all bounded real valued lower semicontinuous functions and is closed under pointwise limits. Let $g_n(\gamma) = \sup_{k>n} f_k(\gamma)$ and $g_{n,p}(\gamma) = \sup_{n<k<n+p} f_k(\gamma)$. Then g_n is the limit of the increasing sequence $\{g_{n,p}: p = 1,2,\ldots\}$, and f is the limit of the decreasing sequence $\{g_n\}$. Let $g'_{n,p}$ be the lower semicontinuous function which coincides, except on a set, say $A_{n,p}$, of the first category, with $g_{n,p}$. For a fixed n and $\gamma \notin \bigcup_{p=1}^{\infty} A_{n,p}$, the sequence $g'_{n,p}(\gamma)$ is increasing and converges to $g_n(\gamma)$. Hence if we define $g'_n(\gamma) = \sup\{g'_{n,p}(\gamma): p = 1,2,\ldots\}$, then g'_n is lower semicontinuous and coincides with g_n except on $\bigcup_{p=1}^{\infty} A_{n,p}$, which is of the first category. It follows then by the previous lemma that g'_n coincides with the upper semicontinuous regularization g''_n except on a set of the first category. Therefore, each g_n coincides with the upper semicontinuous function g''_n except on a set of the first category. Then for the same reason as before, g''_n decreases to the upper semicontinuous function f' which coincides with f except on a set of the first category. Thus f is a member of $\mathscr{B}'$. Q.E.D.

Theorem 1.25. *If Ω is a stonean space, then there exists a completely regular space Γ such that Ω is the spectrum of $B(\Gamma)$.*

Conversely, if Γ is a completely regular space, then the spectrum Ω of $B(\Gamma)$ is stonean.

PROOF. Let Ω be stonean. By Proposition 1.7(iii), every bounded real valued semicontinuous function on Ω coincides with a continuous function except on a set of the first category. Therefore, we have $C(\Omega) + \mathscr{N}(\Omega) = \mathscr{B}(\Omega)$; hence $C(\Omega) = \mathscr{B}(\Omega)/\mathscr{N}(\Omega) = B(\Omega)$.

Now, we consider $B(\Gamma)$ on a completely regular space Γ. It is sufficient to show that the real part $B_{\mathbf{R}}(\Gamma)$ is a conditionally complete lattice. In other words, we will show that every bounded increasing net $\{\tilde{f}_i\}_{i\in I}$ in $B_{\mathbf{R}}(\Gamma)$ has least upper bound in $B_{\mathbf{R}}(\Gamma)$, where $\tilde{f}$ means the canonical image in $B_{\mathbf{R}}(\Gamma)$ of $f \in \mathscr{B}_{\mathbf{R}}(\Gamma)$. For each $i \in I$, let f_i denote the bounded lower semicontinuous function which represents the class $\tilde{f}_i$. Let $f(\gamma) = \sup_{i\in I} f_i(\gamma)$. Then f is bounded and lower semicontinuous. We claim that $\tilde{f}$ is the least upper bound of $\{\tilde{f}_i\}_{i\in I}$ in $B_{\mathbf{R}}(\Gamma)$. Obviously, $\tilde{f} \geq \tilde{f}_i$ because $f \geq f_i$. Suppose $\tilde{g}$ is an element of $B_{\mathbf{R}}(\Gamma)$ such that $\tilde{g} \geq \tilde{f}_i$ for every $i \in I$. Let g be the bounded upper semicontinuous representative of $\tilde{g}$. Then for each i, $g \geq f_i$ except on a set of the first category. Let $A = \{\gamma \in \Gamma : f(\gamma) > g(\gamma)\}$. By the respective semicontinuity properties of f and g, A is open. Let $\{G_j\}_{j\in J}$ be a maximal family of disjoint open subsets such that $G_j \cap A$ is of the first category. Let $G = \bigcup_{j\in J} G_j$. We claim that G is dense in Γ. Suppose G is not dense in Γ. Then there exists an open set G_1 disjoint from G. If $G_1 \cap A = \emptyset$, then we may add G_1 to $\{G_j\}_{j\in J}$, so that $\{G_j\}_{j\in J}$ is not maximal. Hence $G_1 \cap A \neq \emptyset$. Let $\gamma_0 \in G_1 \cap A$. Then

$f(\gamma_0) > g(\gamma_0)$, so that $f_i(\gamma_0) > g(\gamma_0)$ for some i. But f_i and g are respectively lower and upper semicontinuous. Hence we can choose a smaller open set $G_2 \subset G_1 \cap A$ such that $f_i(\gamma) > g(\gamma)$ for every $\gamma \in G_2$. Thus G_2 must be of the first category. Hence $\{G_j\}_{j \in J}$ is not maximal. Thus G must be dense. Therefore G^c is rare. Being the union of sets of the first category which are separated by disjoint open sets, $A = \bigcup_{j \in J} (G_j \cap A) \cup (G^c \cap A)$ is of the first category. Therefore, we have $\tilde{g} \geq \tilde{f}$. Q.E.D.

Remark. If Γ is a Baire space, then the set A in the above arguments must be empty. This is because, if A contains some point γ_0, then $f_i(\gamma_0) > g(\gamma_0)$ for some i, so that A contains an open subset on which $f_i > g$. But in a Baire space, an open set cannot be of the first category; hence A is empty.

Proposition 1.26. *If Γ is a completely regular Baire space which contains a dense subset of the first category, then the spectrum Ω of $B(\Gamma)$ contains a dense subset of the first category. Therefore, Ω enjoys property* (ii) *in Theorem 1.17.*

PROOF. Let $\{A_n\}$ be a sequence of closed rare subsets of Γ such that $A = \bigcup_{n=1}^{\infty} A_n$ is dense in Γ. For each fixed n, let $\{A_{n,i}: i \in I_n\}$ be the family of all open subsets of Γ containing A_n. The characteristic function $\chi_{A_{n,i}}$ of $A_{n,i}$ is a projection in $\mathscr{B}(\Gamma)$, so that it corresponds to an open and closed subset $B_{n,i}$ of Ω. Let $B_n = \bigcap_{i \in I_n} B_{n,i}$.

We claim that B_n is rare and closed. It is obvious that B_n is closed, being the intersection of closed subsets $B_{n,i}$. Suppose B_n contains an open subset. Then B_n contains its closure C which is open and closed. To the characteristic function χ_C of C, there corresponds a projection $\{\tilde{f}\}$ of $B(\Gamma)$. Let f be a lower semicontinuous representative of $\{\tilde{f}\}$. Then f and f^2 coincide except on a set of the first category. Let $D = \{\gamma \in \Gamma: f(\gamma) > 0\}$. Then D is open and the characteristic function χ_D and f coincide except on a subset of the first category. For each $i \in I$, we have $\tilde{\chi}_{A_{n,i}} \geq \tilde{f}$, so that $\chi_{A_{n,i}} \geq f$ except on a subset of the first category, Hence $A_{n,i}$ contains D except for a subset of the first category. Since $f \notin \mathscr{N}(\Gamma)$, D is a nonempty open subset. Hence $D \cap A_n^c$ is a nonempty open subset since $D \cap A_N$ is rare, so that $D \cap A_n^c$ contains a point γ. Since Γ is completely regular and A_n is closed, there exist open disjoint subsets V and W such that $\gamma \in V$ and $A_n \subset W$. Hence W is in the family $\{A_{n,i}: i \in I_n\}$, so that W contains D except for a subset of the first category. Since $V \cap D$ is a nonempty open subset, $V \cap D$ is not of the first category, which contradicts the fact that $V \cap W = \varnothing$. Thus B_n is rare.

We claim now that $B = \bigcup_{n=1}^{\infty} B_n$ is dense in Ω. If not, there exists an open and closed subset F of Ω such that $F \cap B_n = \varnothing$, $n = 1,2,\ldots$. Just as above, there corresponds a nonempty open subset G of Γ to F. Since $F^c \supset B_n$, there exists $i_n \in I_n$ with $F^c \supset B_{n,i_n}$, so that G^c contains A_{n,i_n} except for a subset of the first category. Hence G^c contains, except for a set of the first category, $\bigcup_{n=1}^{\infty} A_{n,i_n}$, which is open and dense. Therefore, G is contained in a subset of the first category, which is a contradiction. Q.E.D.

Therefore, if we take the unit interval $[0,1]$, for example, as the above Γ, then the spectrum Ω of $B([0,1])$ is stonean, but not hyperstonean.**

Now, we turn to the study of the conjugate space of an abelian von Neumann algebra. By Theorem 1.18, it is sufficient to consider the Banach space $L^\infty(\Gamma,\mu)$ of all essentially bounded measurable functions on a locally compact space Γ with respect to a positive Radon measure μ on it. Let Σ be the σ-ring of all μ-measurable subsets of Γ. Let $\mathrm{BV}(\Gamma,\mu)$ denote the space of all bounded additive complex valued functions on Σ which vanish on every locally μ-null set, where the boundedness of an additive function v on Σ is understood as $\sup\{|v(S)|: S \in \Sigma\} < +\infty$. When Γ is discrete and μ is the point measure on Γ, then $\mathrm{BV}(\Gamma,\mu)$ is denoted by $\mathrm{BV}(\Gamma)$. If v is in $\mathrm{BV}(\Gamma,\mu)$, then v is actually of bounded variation. In fact, if v is real, then for an arbitrary partition $\{S_1, \ldots, S_n\}$ of Γ in Σ we have

$$\begin{aligned}\sum_{i=1}^{n} |v(S_i)| &= \sum_{v(S_i)>0} v(S_i) - \sum_{v(S_j)<0} v(S_j)\\ &= v\left(\bigcup_{v(S_i)<0} S_i\right) - v\left(\bigcup_{v(S_j)<0} S_j\right)\\ &\leq 2\sup\{|v(S)|: S \in \Sigma\}.\end{aligned}$$

If v is complex, then we can show that its real and imaginary parts are both of bounded variation. Let $v(v)$ denote the total variation of $v \in \mathrm{BV}(\Gamma,\mu)$. Then $v(v)$ is given by

$$v(v)(S) = \sup \sum_{i=1}^{n} |v(S_i)|, \qquad S \in \Sigma,$$

where the supremum is taken over all finite disjoint sequences $\{S_i\}$ in Σ with $S_i \subset S$. Then $v(v)$ is again a bounded positive additive function on Σ and the quantity $\|v\|$ defined by $\|v\| = v(v)(\Gamma)$ is a norm on $\mathrm{BV}(\Gamma,\mu)$, which makes $\mathrm{BV}(\Gamma,\mu)$ a Banach space.

For each real $f \in L^\infty(\Gamma,\mu)$, there exists a locally μ-null set N with

$$\sup\{|f(\gamma)|: \gamma \in \Gamma - N\} = \|f\| < \infty.$$

For each integer $n > 0$, put

$$E_k^n = \left\{\gamma \in \Gamma - N: \frac{k}{2^n}\|f\| \leq f(\gamma) < \frac{k+1}{2^n}\|f\|\right\}$$

for $k = -2^n, -2^n + 1, \ldots, 2^n$. Then $\{E_k^n: -2^n \leq k \leq 2^n\}$ is a partition of $\Gamma - N$ in Σ. Define a function f_n by

$$f_n = \sum_{k=-2^n}^{2^n} \frac{k}{2^n}\|f\|\chi_{E_k^n}.$$

Then we have

$$|f(\gamma) - f_n(\gamma)| \leq \frac{1}{2^n} \quad \text{for} \quad \gamma \in \Gamma - N.$$

For each $\nu \in \mathrm{BV}(\Gamma,\mu)$, the sequence $\{\nu(f_n)\}$ defined by

$$\nu(f_n) = \sum_{k=-2^n}^{2^n} \frac{k}{2^n} \|f\| \nu(E_k^n)$$

is a Cauchy sequence; hence we can define $\nu(f)$ by

$$\nu(f) = \lim_{n\to\infty} \nu(f_n).$$

For a complex $f \in L^\infty(\Gamma,\mu)$, we define $\nu(f)$ by considering its real part and imaginary part. Then $\nu(\cdot)$ is a bounded linear functional on $L^\infty(\Gamma,\mu)$. We denote it, by an analogy with integration, by

$$\nu(f) = \int_\Gamma f(\gamma)\, d\nu(\gamma).$$

Theorem 1.27 *The conjugate space $L^\infty(\Gamma,\mu)^*$ of the Banach space $L^\infty(\Gamma,\mu)$ is identified with the Banach space* $\mathrm{BV}(\Gamma,\mu)$ *under the canonical bilinear form*:

$$\nu(f) = \int_\Gamma f(\gamma)\, d\nu(\gamma), \qquad f \in L^\infty(\Gamma,\mu), \qquad \nu \in \mathrm{BV}(\Gamma,\mu).$$

Proof. Let φ be a bounded linear functional on $L^\infty(\Gamma,\mu)$. Then the function ν on Σ defined by $\nu(E) = \varphi(\chi_E)$, $E \in \Sigma$, is an additive bounded function on Σ; hence it is in $\mathrm{BV}(\Gamma,\mu)$. Since for any $f \in L^\infty(\Gamma,\mu)$ we have $\|f - f_n\| \le 1/2^n$ and $\varphi(f_n) = \nu(f_n)$ for the function f_n defined above, we have $\varphi(f) = \nu(f)$. It follows from the construction of $\nu(f)$ that $|\nu(f)| \le \|\nu\|\|f\|$; hence $\|\varphi\| \le \|\nu\|$. But the converse inequality $\|\nu\| \le \|\varphi\|$ is trivial. Thus the correspondence: $\varphi \leftrightarrow \nu$ is isometric. Q.E.D.

Theorem 1.28. *Let Γ be a discrete space. If $\{\nu_n\}$ is a bounded sequence in* $\mathrm{BV}(\Gamma)$ *such that* $\lim_{n\to\infty} \nu_n(E) = 0$ *for every subset E of Γ, then*

$$\lim_{n\to\infty} \sum_{\gamma\in\Gamma} |\nu_n(\{\gamma\})| = 0.$$

Proof. Suppose $\sum_{\gamma\in\Gamma} |\nu_n(\{\gamma\})|$ does not converge to zero as $n \to \infty$. Considering a subsequence of $\{\nu_n\}$, we may assume $\sum_{\gamma\in\Gamma} |\nu_n(\{\gamma\})| \ge \varepsilon$, $n = 1,2,\ldots,$ for some $\varepsilon > 0$. Then there exists a finite subset E_1 of Γ such that

$$\sum_{\gamma\in E_1} |\nu_1(\{\gamma\})| \ge \sum_{\gamma\in\Gamma} |\nu_1(\{\gamma\})| - \varepsilon/10.$$

Since $\lim_{n\to\infty} \nu_n(\{\gamma\}) = 0$ for every $\gamma \in E_1$, we can find an index n_2 with $\sum_{\gamma\in E_1} |\nu_n(\{\gamma\})| < \varepsilon/10$ for $n \ge n_2$; then there exists a finite subset $E_2 \subset \Gamma - E_1$ such that

$$\sum_{\gamma\in E_2} |\nu_{n_2}(\{\gamma\})| > \sum_{\gamma\in\Gamma} |\nu_{n_2}(\{\gamma\})| - \varepsilon/10.$$

By induction, we find an infinite sequence $\{E_k\}$ of disjoint finite subsets of Γ and a subsequence $\{\nu_{n_k}\}$ such that

$$\sum_{\gamma\in E_k} |\nu_{n_k}(\{\gamma\})| > \sum_{\gamma\in\Gamma} |\nu_{n_k}(\{\gamma\})| - \varepsilon/10.$$

Observing that $v(\nu)(\{\gamma\}) = |\nu(\{\gamma\})|$ for each $\nu \in \mathrm{BV}(\Gamma)$, we see that the last inequality holds for $v(\nu_{n_k})$. Put $F_1 = E_1$ and $\mu_1 = \nu_1$. Let k be an integer $> 10v(\mu_1)(\Gamma)/\varepsilon$. Consider k sequences $\{E_{kj+p}: j = 1,2,\ldots\}$, $1 \leq p \leq k$. If we have

$$v(\mu_1)\left(\bigcup_{j-1}^{\infty} E_{kj+p}\right) > \varepsilon/10, \qquad p = 1,2,\ldots,k,$$

then

$$v(\mu_1)(\Gamma) \geq v(\mu_1)\left(\bigcup_{p=1}^{k} \bigcup_{j=1}^{\infty} E_{kj+p}\right)$$

$$= \sum_{p=1}^{k} v(\mu_1)\left(\bigcup_{j=1}^{\infty} E_{kj+p}\right) \geq k\frac{\varepsilon}{10} > v(\mu_1)(\Gamma),$$

which is impossible. Hence there is some p with

$$v(\mu_1)\left(\bigcup_{j=1}^{\infty} E_{kj+p}\right) < \varepsilon/10.$$

Let $F_2 = E_{k+p}$ and $\mu_2 = \nu_{n_{k+p}}$. By induction, we find subsequences $\{F_n\}$ of $\{E_n\}$ and $\{\mu_n\}$ of $\{\nu_n\}$ such that

$$\sum_{\gamma \in F_n} |\mu_n(\{\gamma\})| > \sum_{\gamma \in \Gamma} |\mu_n(\gamma)| - \varepsilon/10,$$

$$v(\mu_n)\left(\bigcup_{i=n+1}^{\infty} F_i\right) < \varepsilon/10.$$

Define a function $f \in l^\infty(\Gamma)$ by

$$f(\gamma) = \begin{cases} 0 & \text{for } \gamma \notin \bigcup_{n=1}^{\infty} F_n \\ (-1)^n \arg \overline{\mu_n(\{\gamma\})} & \text{for } \gamma \in F_n. \end{cases}$$

Then we have

$$\left|\mu_n(f) - (-1)^n \sum_{\omega \in F_n} |\mu_n(\{\gamma\})|\right| \leq \left|\sum_{k<n} \int_{F_k} f\, d\mu_n\right| + \left|\int_{\bigcup_{k>n} F_k} f\, d\mu_n\right|$$

$$\leq \sum_{k<n} \sum_{\gamma \in F_k} |\mu_n(\{\gamma\})| + v(\mu_n)\left(\bigcup_{k>n} F_k\right)$$

$$\leq \sum_{\gamma \notin F_n} |\mu_n(\{\gamma\})| + v(\mu_n)\left(\bigcup_{k>n} F_k\right) < \frac{2}{10}\varepsilon.$$

Hence we get

$$|\mu_n(f)| \geq \sum_{\gamma \in \Gamma} |\mu_n(\{\gamma\})| - \frac{3}{10}\varepsilon.$$

On the other hand, every function $f \in l^\infty(\Gamma)$ is uniformly well approximated by a finite linear combinations of characteristic functions and $\{\mu_n\}$ is bounded in $\mathrm{BV}(\Gamma) = l^\infty(\Gamma)^*$, so that $\lim_{n\to\infty} \mu_n(f) = 0$. This is a contradiction. Q.E.D.

Corollary 1.29. *Let Γ be a discrete space. In the Banach space $l^1(\Gamma)$ of all summable functions on Γ, every weakly convergent sequence converges in the norm topology.*

PROOF. Let $\{f_n\}$ be a sequence in $l^1(\Gamma)$ weakly convergent to $f \in l^1(\Gamma)$. Considering $\{f_n - f\}$, we may assume $f = 0$. By the uniform boundedness theorem, $\{f_n\}$ is bounded. Define a bounded sequence $\{v_n\}$ in $\mathrm{BV}(\Gamma)$ by

$$v_n(E) = \sum_{\gamma \in E} f_n(\gamma), \qquad E \subset \Gamma.$$

Then by assumption, $\lim_{n\to\infty} v_n(E) = 0$. Therefore, by the previous theorem, we get

$$\begin{aligned} 0 &= \lim_{n\to\infty} \sum_{\gamma\in\Gamma} |v_n(\{\gamma\})| = \lim_{n\to\infty} \sum_{\gamma\in\Gamma} |f_n(\gamma)| \\ &= \lim_{n\to\infty} \|f_n\|_1. \end{aligned}$$

Q.E.D.

EXERCISES

1. Show that every positive Radon measure μ on a stonean space is decomposed uniquely in the form $\mu = \mu_n + \mu_s$ for a positive normal measure μ_n and a positive Radon measure μ_s concentrated on a set of the first category.
2. Show that if a stonean space Ω has no hyperstonean portion, then every positive Radon measure on Ω is concentrated on a set of the first category.
3. Let Γ be a Hausdorff space. Suppose Γ admits a basis $\mathfrak{G}$ for open subsets such that for any decreasing sequence $\{G_n\}$ in $\mathfrak{G}$, $\bigcap_{n=1}^{\infty} G_n$ contains a nonempty member of $\mathfrak{G}$.
   (a) Show that if $\{G_n\}$ is a sequence of open dense subsets of Γ, then the interior of $\bigcap_{n=1}^{\infty} G_n$ is dense again in Γ.
   (b) Show that the spectrum Ω of $B(\Gamma)$ admits a basis $\tilde{\mathfrak{G}}$ for open subsets such that for any decreasing sequence $\{\tilde{G}_n\}$ in $\tilde{\mathfrak{G}}$, the intersection $\bigcap_{n=1}^{\infty} \tilde{G}_n$ contains a nonempty member of $\tilde{\mathfrak{G}}$.
   (c) Show that every subset of Ω of the first category is rare.
4. Keep the hypothesis and the notations of the previous problem. Suppose Γ has no isolated point. Let μ be a positive Radon measure on Ω with support S.
   (a) Show that every open and closed subset of Ω contains a point with μ-measure zero.
   (b) Show that every open and closed subset E of Ω contains a decreasing sequence $\{\tilde{G}_n\}$ in $\tilde{\mathfrak{G}}$ such that $\mu(\tilde{G}_n) < 1/n$.
   (c) Show that every open and closed subset contains an open subset with μ-measure zero.
   (d) Show that S is rare.
5. Let I be an uncountable set. Let Γ be the set of all functions on I with values 0 and 1. For each countable subset J of I and each function $\alpha = \{\alpha_j\}_{j\in J}$ with values $\alpha_j = 0$ or 1, we define a subset $U(J;\alpha)$ of Γ as the set consisting of all $\gamma \in \Gamma$ such that $\gamma_j = \alpha_j$ for every $j \in J$. Show that the family $\mathfrak{G}$ of all $U(J;\alpha)$ makes Γ a complete regular space with no isolated point and satisfies the property in the previous exercise. Therefore, the spectrum Ω of $B(\Gamma)$ satisfies property (iii) in Theorem 1.17.

6. Let Ω be a stonean space, $\mathscr{A} = C(\Omega)$, and $A = l^\infty(\Omega)$, the C^*-algebra of all bounded functions on Ω.
   (a) Show that if $f \in A$ is real valued, then there exists a unique element $f' \in \mathscr{A}$ such that every $g \in \mathscr{A}_h$ with $g \leq f$ is majorized by f'. Set $f' = \varepsilon(f)$. (*Hint*: Set $f' = \text{l.u.b.}\{g \in \mathscr{A}_h : g \leq f\}$.)
   (b) Show that ε is a linear map of A_h onto $\mathscr{A}_h$ such that $\mathscr{F}\ \varepsilon = \varepsilon$ and $\|\varepsilon\| = 1$.
   (c) Show that ε can be extended to a projection of A onto $\mathscr{A}$ of norm one. Denote it by ε also.
   (d) Show that ε enjoys the following properties: (i) $\varepsilon(f) \geq 0$ if $f \geq 0$; (ii) $\varepsilon(fg) = f\varepsilon(g)$ if $f \in \mathscr{A}$, $g \in A$.
   (e) Let E be a complex Banach space and F a closed subspace of E. Let T be a bounded linear map of F into $\mathscr{A}$. Show, by the Hahn–Banach extension theorem for linear functionals, that T can be extended to a linear map T' of E into A with $\|T'\| = \|T\|$.
   (f) Show that there exists an extension $\tilde{T}$ of T with values in $\mathscr{A}$ such that $\|\tilde{T}\| = \|T\|$. (*Hint*: Set $\tilde{T} = \varepsilon \circ T'$.)

2. The Universal Enveloping von Neumann Algebra of a C^*-Algebra

As seen in the previous chapter, the second conjugate space of the C^*-algebra $\mathscr{L}\mathscr{C}(\mathfrak{H})$ of all compact operators on a Hilbert space $\mathfrak{H}$ is isometric to the von Neumann algebra $\mathscr{L}(\mathfrak{H})$ of all bounded operators on $\mathfrak{H}$. In this section, we shall show that the second conjugate space of any C^*-algebra is isometrically isomorphic to a von Neumann algebra as a Banach space, which has special meaning for the original C^*-algebra.

The following result corresponds to the Jordan decomposition of a finite Radon measure on a locally compact space.

Proposition 2.1 (Jordan Decomposition). *Let A be a C^*-algebra. Every $\omega \in A_h^*$ is represented in the form*

$$\omega = \omega_+ - \omega_- \quad \text{and} \quad \|\omega\| = \|\omega_+\| + \|\omega_-\| \tag{1}$$

by some ω_+ and ω_- in A_+^*. Every $\omega \in A^*$ is represented in the form

$$\omega = \omega(\pi;\xi,\eta) \tag{2}$$

by some representation $\{\pi,\mathfrak{H}\}$ of A and $\xi,\eta \in \mathfrak{H}$, where $\omega(\pi;\xi,\eta)$ means the functional on A defined by

$$\langle x,\omega(\pi;\xi,\eta)\rangle = (\pi(x)\xi|\eta), \qquad x \in A. \tag{3}$$

Proof. Considering the C^*-algebra A_{I}, obtained by adjunction of an identity to A, we may assume that A is unital. Let $\mathfrak{S} = \mathfrak{S}(A)$ be the set of all states of A. Then, $\mathfrak{S}$ is $\sigma(A^*,A)$-compact, due to the existence of the identity, and convex. Furthermore, we have, by Theorem I.9.18,

$$\|h\| = \sup\{|\langle h,\omega\rangle| : \omega \in \mathfrak{S}\}, \qquad h \in A_h^* \tag{4}$$

Therefore, the unit ball $S^* \cap A_h^*$ of the real Banach space A_h^* is the $\sigma(A^*,A)$-closure of the convex hull of $\mathfrak{S} \cup (-\mathfrak{S})$, where S^* denotes the unit ball of A^*. But the compactness and the convexity of $\mathfrak{S}$ yield the compactness of the convex hull of $\mathfrak{S} \cup (-\mathfrak{S})$, so that we get

$$S^* \cap A_h^* = \{\lambda\sigma - \mu\rho ; \sigma,\rho \in \mathfrak{S}, \lambda + \mu = 1, \lambda,\mu \geq 0\}.$$

Take an arbitrary nonzero $\omega \in A_h^*$. Then there exists $\omega_1,\omega_2 \in \mathfrak{S}$ and $\lambda,\mu \geq 0$, $\lambda + \mu = 1$, such that $\omega/\|\omega\| = \lambda\omega_1 - \mu\omega_2$. Clearly, we have

$$\left\|\frac{1}{\|\omega\|}\omega\right\| = 1 = \lambda + \mu = \|\lambda\omega_1\| + \|\mu\omega_2\|;$$

hence we get the desired decomposition of ω, putting $\omega_+ = \lambda\|\omega\|\omega_1$ and $\omega_- = \mu\|\omega\|\omega_2$.

Therefore, each $\omega \in A_h^*$ is written in the form $\omega = \omega_1 - \omega_2 + i(\omega_3 - \omega_4)$ by some positive linear functionals $\omega_1,\omega_2,\omega_3$, and ω_4. Put $\omega_0 = \omega_1 + \omega_2 + \omega_3 + \omega_4$, which will be denoted by $[\omega]$ in the sequel. Let $\{\pi,\mathfrak{H},\xi\}$ be the cyclic representation of A induced by ω_0. By the inequality, $\omega_i \leq \omega_0$, $1 \leq i \leq 4$, ω_i induces a bounded linear functional on the linear manifold $\pi(A)\xi$, so that there exists a vector $\eta_i \in \mathfrak{H}$ with $\omega_i(x) = (\pi(x)\xi|\eta_i)$, $x \in A$. Putting $\eta = \eta_1 - \eta_2 + i(\eta_3 - \eta_4)$, we get the desired expression $\omega = \omega(\pi;\xi,\eta)$.

Q.E.D.

The decomposition (1) of $\omega \in A_h^*$ is unique, as will be seen later, Theorem 4.2.

Lemma 2.2. *Let A be a C^*-algebra and $\{\pi,\mathfrak{H}\}$ be a representation of A. Let $\mathcal{M}(\pi)$ denote the von Neumann algebra $\pi(A)''$ generated by $\pi(A)$. Then there is a unique linear map $\tilde{\pi}$ of the second conjugate space A^{**} of A onto $\mathcal{M}(\pi)$ with the following properties*:

(i) *The diagram*

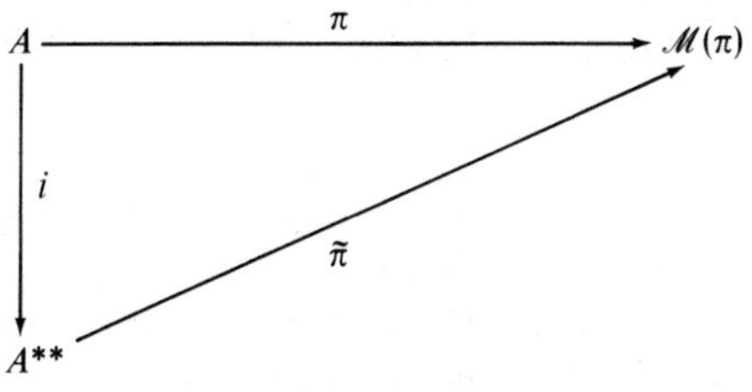

is commutative, where i means the canonical imbedding of A into A^{**}.

(ii) *$\tilde{\pi}$ is continuous with respect to the $\sigma(A^{**},A^*)$-topology and the σ-weak topology of $\mathcal{M}(\pi)$.*

(iii) *$\tilde{\pi}$ maps the unit ball S^{**} of A^{**} onto the unit ball $S_{\mathcal{M}}$ of $\mathcal{M}$.*

PROOF. By Proposition I.5.2, π maps the open unit ball S_0 of A into the open unit ball $S_{\mathscr{M},0}$ of $\mathscr{M}$. Since π is the composed map of the faithful representation π_0 of the quotient C^*-algebra $A/\pi^{-1}(0)$ and the canonical map of A onto $A/\pi^{-1}(0)$, and since any faithful representation of a C^*-algebra is an isometry by Propositions I.5.2 and I.5.3, π maps S_0 onto $S_{\mathscr{M},0} \cap \pi(A)$. Let $\mathscr{M}_*(\pi)$ denote the Banach space of all σ-weakly continuous linear functionals on $\mathscr{M}(\pi)$ and π_* denote the restriction of the transpose ${}^t\pi$ of π onto $\mathscr{M}_*(\pi)$. Since $\mathscr{M}(\pi)$ is the conjugate space of $\mathscr{M}_*(\pi)$ as a Banach space, the transpose ${}^t\pi_*$, say $\tilde{\pi}$, of π_* is a map of A^{**} into $\mathscr{M}(\pi)$. Then we have, for every $a \in A$ and $\omega \in \mathscr{M}_*(\pi)$,

$$\langle \pi(a),\omega \rangle = \langle i(a),\pi_*(\omega) \rangle = \langle \tilde{\pi} \circ i(a),\omega \rangle,$$

so that $\tilde{\pi} \circ i = \pi$. Assertion (ii) follows from the definition of $\tilde{\pi}$. The image $\tilde{\pi}(S^{**})$ of S^{**} is σ-weakly compact by the continuity of $\tilde{\pi}$ and contains $\pi(S)$. On the other hand, the open unit ball $S_{\mathscr{M},0} \cap \pi(A) = \pi(S_0)$ of $\pi(A)$ is σ-weakly dense in $S_{\mathscr{M}}$ by Theorem II.4.8. Hence we have $\tilde{\pi}(S^{**}) = S_{\mathscr{M}}$.
Q.E.D.

Definition 2.3. Let A be a C^*-algebra. For each representation $\{\pi,\mathfrak{H}\}$ of A, we denote by $\mathscr{M}(\pi)$ the von Neumann algebra $\pi(A)''$ generated by $\pi(A)$. A representation $\{\pi,\mathfrak{H}\}$ is called *universal* if for any representation $\{\rho,\mathfrak{K}\}$ there exists a σ-weakly continuous $*$-homomorphism $\tilde{\rho}$ of $\mathscr{M}(\pi)$ onto $\mathscr{M}(\rho)$ such that $\rho(x) = \tilde{\rho} \circ \pi(x)$, $x \in A$. If $\{\pi,\mathfrak{H}\}$ is a universal representation, then the von Neumann algebra $\mathscr{M}(\pi)$ is called the *universal enveloping von Neumann algebra of A*.

The universal enveloping von Neumann algebra is uniquely determined up to isomorphism. In fact, let $\{\pi_1,\mathfrak{H}_1\}$ and $\{\pi_2,\mathfrak{H}_2\}$ be two universal representations of A. Then there is a σ-weakly continuous homomorphism ρ_1 of $\mathscr{M}(\pi_1)$ onto $\mathscr{M}(\pi_2)$ and ρ_2 of $\mathscr{M}(\pi_2)$ onto $\mathscr{M}(\pi_1)$ such that

$$\pi_2(x) = \rho_1 \circ \pi_1(x) \quad \text{and} \quad \pi_1(x) = \rho_2 \circ \pi_2(x), \qquad x \in A.$$

Hence $\rho_2 \circ \rho_1$ and $\rho_1 \circ \rho_2$ are both the identity maps in $\pi_1(A)$ and in $\pi_2(A)$, respectively, so that they are both the identity maps in $\mathscr{M}(\pi_1)$ and in $\mathscr{M}(\pi_2)$, respectively, since $\pi_1(A)$ and $\pi_2(A)$ are respectively σ-weakly dense in $\mathscr{M}(\pi_1)$ and in $\mathscr{M}(\pi_2)$. Therefore, ρ_1 and ρ_2 are both isomorphisms and are inverses of each other.

Theorem 2.4. *A C^*-algebra A admits a universal representation $\{\pi,\mathfrak{H}\}$, hence the universal enveloping algebra $\mathscr{M}(\pi)$. Furthermore, there is a unique isometry $\tilde{\pi}$ of the second conjugate space A^{**} of A onto $\mathscr{M}(\pi)$ which is a homeomorphism with respect to the $\sigma(A^{**},A^*)$-topology and the σ-weak topology in $\mathscr{M}(\pi)$.*

PROOF. For each $\omega \in A_+^*$ let $\{\pi_\omega,\mathfrak{H}_\omega\}$ be the cyclic representation of A induced by ω. Put

$$\{\pi,\mathfrak{H}\} = \sum_{\omega \in A_+^*}^{\oplus} \{\pi_\omega,\mathfrak{H}_\omega\}.$$

By Lemma 1.2, there exists a unique linear map $\tilde{\pi}$ of A^{**} onto $\mathscr{M}(\pi)$. By Proposition 1.1, every $\omega \in A^*$ is of the form $\omega = \omega(\pi;\xi,\eta)$ for some $\xi,\eta \in \mathfrak{H}$. Hence we have $\pi_*(\mathscr{M}_*(\pi)) = A^*$, so that $\tilde{\pi}$ is injective; hence by Lemma 2.2(iii), $\tilde{\pi}$ is an isometry of A^{**} onto $\mathscr{M}(\pi)$ and it is bicontinuous with respect to the $\sigma(A^{**},A^*)$-topology and the σ-weak topology.

Take a representation $\{\rho,\mathfrak{K}\}$ of A. Let $\tilde{\rho}$ be the linear map of A^{**} onto $\mathscr{M}(\rho)$ defined in Lemma 1.2. Put $\rho_0 = \tilde{\rho} \circ \tilde{\pi}^{-1}$. Then ρ_0 is a linear map of $\mathscr{M}(\pi)$ onto $\mathscr{M}(\rho)$ with $\rho_0 \circ \pi(x) = \rho(x)$, $x \in A$, which is σ-weakly continuous. Thus ρ_0 is a σ-weakly continuous homomorphism of $\mathscr{M}(\pi)$ onto $\mathscr{M}(\rho)$. Therefore, π is a universal representation of A. Q.E.D.

Definition 2.5. The representation $\{\pi,\mathfrak{H}\}$ defined in the above theorem is called *the* universal representation of A.

In the sequel, the second conjugate space A^{**} of a C^*-algebra A and the universal enveloping von Neumann algebra of A will be identified.

Let A be a C^*-algebra. For each element a in A and ω in A^*, we define actions of a on ω by

$$\langle x,a\omega\rangle = \langle xa,\omega\rangle \quad \text{and} \quad \langle x,\omega a\rangle = \langle ax,\omega\rangle \tag{5}$$

for every $x \in A$.

Definition 2.6. A subset V of A^* is said to be *left* (resp. *right*) *invariant* if $aV \subset V$ (resp. $Va \subset V$) for every $a \in A$. A two-sided invariant subset is simply called *invariant*.

If $\{\mathscr{M},\mathfrak{H}\}$ is a von Neumann algebra on a Hilbert space, then the Banach space $\mathscr{M}_*$ of all σ-weakly continuous linear functionals is invariant as a subspace of the conjugate space $\mathscr{M}^*$ of $\mathscr{M}$. The second conjugate space A^{**} of a C^*-algebra A is a von Neumann algebra, and the conjugate space A^* is the Banach space of all σ-weakly continuous linear functional on A^{**}, so that the actions of A^{**} on A^* are also considered. This does not create, however, any ambiguity concerning the invariance of closed subspaces of A^*, as seen in the following.

Theorem 2.7. *Let $\{\mathscr{M},\mathfrak{H}\}$ be a von Neumann algebra and $\mathscr{M}_*$ be the Banach space of all σ-weakly continuous linear functionals on $\mathscr{M}$. Let V be a closed subspace of $\mathscr{M}_*$.*

(i) *If there exists a subset A of $\mathscr{M}$, whose linear combinations are σ-weakly dense in $\mathscr{M}$, such that $aV \subset V$ (resp. $Va \subset V$) for every $a \in A$, then V is left (resp. right) invariant.*

(ii) *There exists a bijection $V \leftrightarrow \mathscr{I}$ between the closed left (resp. right) invariant subspaces V of $\mathscr{M}_*$ and the σ-weakly closed right (resp. left)* *ideals $\mathscr{I}$ of $\mathscr{M}$ determined by*

$$V^0 = \mathscr{I} \quad \text{and} \quad \mathscr{I}^0 = V, \tag{6}$$

where V^0 and $\mathscr{I}^0$ mean the polar of V and $\mathscr{I}$ in $\mathscr{M}$ and in $\mathscr{M}_*$, respectively.

(iii) *Every closed left (resp. right) invariant subspace V is of the form*

$$V = \mathscr{M}_* e \quad (\text{resp. } V = e\mathscr{M}_*) \tag{7}$$

for some projection e in $\mathscr{M}$. Furthermore, V is two-sided invariant if and only if e is central.

PROOF. If $\mathscr{I}$ is a σ-weakly closed right ideal of $\mathscr{M}$, then there is a projection $e \in \mathscr{M}$ with $\mathscr{I} = (1 - e)\mathscr{M}$ by Proposition II.3.12. If $\varphi \in \mathscr{M}_*$ vanishes on $\mathscr{I}$, then for every $x \in \mathscr{M}$, we have

$$0 = \langle (1 - e)x, \varphi \rangle = \langle x, \varphi(1 - e) \rangle,$$

so that $\varphi = \varphi e$. Hence $\mathscr{M}_* e \supset \mathscr{I}^0$. Conversely, for every $x \in \mathscr{I}$ and $\varphi \in \mathscr{M}_*$, we have

$$\langle x, \varphi e \rangle = \langle ex, \varphi \rangle = \langle 0, \varphi \rangle = 0,$$

so that $\mathscr{M}_* e \subset \mathscr{I}^0$. Hence $\mathscr{I}^0 = \mathscr{M}_* e$. From the general theory of Banach spaces, it follows that $\mathscr{I}^{00} = \mathscr{I}$ and $V^{00} = V$ for a σ-weakly closed subspace $\mathscr{I}$ of $\mathscr{M}$ and for a closed subspace V of $\mathscr{M}_*$, respectively.

Let V be a closed subspace of $\mathscr{M}_*$ satisfying the condition in (i). Put $\mathscr{I} = V^0$. Then we have

$$\langle \mathscr{I} a, V \rangle = \langle \mathscr{I}, aV \rangle \subseteq \langle \mathscr{I}, V \rangle = \{0\}$$

for every $a \in A$, so that $\mathscr{I} a \subset \mathscr{I}$ for every $a \in A$. Since the subset $\{x \in \mathscr{M} : \mathscr{I} x \subset \mathscr{I}\}$ is a σ-weakly closed subspace containing A by the σ-weak closedness of $\mathscr{I}$, we conclude that $\mathscr{I}$ is a right ideal of $\mathscr{M}$, and then $V = \mathscr{I}^0$ is left invariant. The last assertion in (iii) follows from Proposition II.3.12.
Q.E.D.

Definition 2.8. The projection e in equality (7) is called the *support* projection of a left invariant subspace V of $\mathscr{M}_*$.

Therefore, if A is a C^*-algebra, then equality (6), with respect to the polar considered in the universal enveloping von Neumann algebra $\tilde{A}$ and in A^*, respectively, implements a bijective correspondence between the σ-weakly closed right (resp. left) ideals of $\tilde{A}$ and the closed left (resp. right) invariant subspaces of A^*. Replacing $\mathscr{M}$ and $\mathscr{M}_*$ by $\tilde{A}$ and A^*, respectively, we obtain expression (7) of the closed left (resp. right) invariant subspaces of A^* be means of projections in the universal enveloping von Neumann algebra $\tilde{A}$ of A.

Suppose $\{e_i\}_{i \in I}$ is a family of projections in von Neumann algebra $\{\mathscr{M}, \mathfrak{H}\}$. Then the projection of $\mathfrak{H}$ onto $\bigcap_{i \in I} e_i \mathfrak{H}$ belongs to $\mathscr{M}$, which we denote by $\bigwedge_{i \in I} e_i$. The projection $\bigvee_{i \in I} e_i$ is defined by $\bigvee_{i \in I} e_i = 1 - \bigwedge_{i \in I} (1 - e_i)$.

With these notations, we have the following:

Proposition 2.9. *For a family $\{e_i\}_{i\in I}$ of projections in a von Neumann algebra $\{\mathscr{M},\mathfrak{H}\}$, set $p = \bigwedge_{i\in I} e_i$ and $q = \bigvee_{i\in I} e_i$. Then we have*

(i) $\mathscr{M}_*p = \bigcap_{i\in I} \mathscr{M}_*e_i, \qquad p\mathscr{M}_* = \bigcap_{i\in I} e_i\mathscr{M}_*,$

(ii) $\mathscr{M}_*q = \left[\bigcup_{i\in I} \mathscr{M}_*e_i\right], \qquad q\mathscr{M}_* = \left[\bigcup_{i\in I} e_i\mathscr{M}_*\right].$

PROOF. Since $p \le e_i$, $i \in I$, it is clear that $\mathscr{M}_*p \subset \bigcap_{i\in I} \mathscr{M}_*e_i$. Let $V = \bigcap_{i\in I} \mathscr{M}_*e_i$. Then V is a closed left invariant subspace of $\mathscr{M}_*$. Hence, by Theorem 2.7, there exists a projection p_0 in $\mathscr{M}$ such that $V = \mathscr{M}_*p_0$. Since $V \subset \mathscr{M}_*e_i$, $i \in I$, we have $p_0 \le e_i$, $i \in I$, so that $p_0\mathfrak{H} \subset \bigcap_{i\in I} e_i\mathfrak{H}$; thus $p_0 \le p$. Hence $V \subset \mathscr{M}_*p$, which implies, together with the reversed inclusion shown first, that $V = \mathscr{M}_*p$. The assertion for q is verified similarly. Q.E.D.

Definition 2.10. Let $\{\pi_1,\mathfrak{H}_1\}$ and $\{\pi_2,\mathfrak{H}\}$ be two representations of a C^*-algebra A. If there exists a σ-weakly bicontinuous isomorphism ρ of $\mathscr{M}(\pi_1)$ onto $\mathscr{M}(\pi_2)$ such that $\rho \circ \pi_1(x) = \pi_2(x)$ for every $x \in A$, then $\{\pi_1,\mathfrak{H}_1\}$ and $\{\pi_2,\mathfrak{H}_2\}$ are called *quasi-equivalent* and we write this fact as

$$\{\pi_1,\mathfrak{H}_1\} \sim \{\pi_2,\mathfrak{H}_2\} \quad \text{or merely} \quad \pi_1 \sim \pi_2.$$

Let $\{\pi,\mathfrak{H}\}$ be a representation of a C^*-algebra A. Let $\mathscr{M}_*(\pi)$ denote the Banach space of all σ-weakly continuous linear functional on $\mathscr{M}(\pi)$. Let $V(\pi)$ denote the subspace ${}^t\pi(\mathscr{M}_*(\pi))$ of A^*. Let S_A and $S_{\mathscr{M}}$ denote the unit balls of A and $\mathscr{M}(\pi)$, respectively. Then, by Lemma 1.2, we have, for each $\omega \in \mathscr{M}_*(\pi)$,

$$\begin{aligned}\|\omega\| &= \sup\{|\langle x,\omega\rangle| : x \in S_{\mathscr{M}}\}\\ &= \sup\{|\langle \tilde{\pi}(x),\omega\rangle| : x \in S_{\tilde{A}}\}\\ &= \sup\{|\langle \pi(x),\omega\rangle| : x \in S_A\}\\ &= \|{}^t\pi(\omega)\|,\end{aligned}$$

where $S_{\tilde{A}}$ means the unit ball of the universal enveloping von Neumann algebra $\tilde{A}$ of A. Hence ${}^t\pi$ is an isometry of $\mathscr{M}_*(\pi)$ onto $V(\pi)$, so that $V(\pi)$ is a closed subspace of A^*. For every $a,b,x \in A$ and $\omega \in \mathscr{M}_*(\pi)$, we have

$$\begin{aligned}\langle x,a{}^t\pi(\omega)b\rangle &= \langle bxa,{}^t\pi(\omega)\rangle\\ &= \langle \pi(bxa),\omega\rangle = \langle \pi(b)\pi(x)\pi(a),\omega\rangle\\ &= \langle \pi(x),\pi(a)\omega\pi(b)\rangle\\ &= \langle x,{}^t\pi(\pi(a)\omega\pi(b))\rangle;\end{aligned}$$

hence $a{}^t\pi(\omega)b = {}^t\pi(\pi(a)\omega\pi(b)) \in V(\pi)$, that is, $V(\pi)$ is invariant. Therefore, by Theorem 2.7, there exists a unique central projection $z(\pi)$ in $\tilde{A}$ with $V(\pi) = A^*z(\pi)$.

Definition 2.11. The projection $z(\pi)$ is called the *support projection* of the representation π. The subspace $V(\pi)$ of A^* is said to be *associated with* π.

Conversely, let V be a closed invariant subspace of A^*. Then, by Theorem 1.7, there exists a central projection z in $\tilde{A}_0$ with $V = A^*z$. Let $\{\pi,\mathfrak{H}\}$ be the universal representation of A. Then π is extended to a σ-weakly continuous faithful representation $\tilde{\pi}$ of the universal enveloping von Neumann algebra $\tilde{A}$ of A. Define a representation ρ of A on $\tilde{\pi}(z)\mathfrak{H} = \mathfrak{K}$ by $\rho(x) = \tilde{\pi}(xz)$, $x \in A$. Then the representation $\{\rho,\mathfrak{K}\}$ is nondegenerate, and it is not hard to see that $V = V(\rho)$.

Proposition 2.12. *Let $\{\pi_1,\mathfrak{H}_1\}$ and $\{\pi_2,\mathfrak{H}_2\}$ be two representations of a C^*-algebra A, $V(\pi_1)$ and $V(\pi_2)$ be the closed invariant subspaces of A^* associated with π_1 and π_2, respectively, and, further, let $z(\pi_1)$ and $z(\pi_2)$ be the support projections of π_1 and π_2, respectively. Then the following three statements are equivalent:*

(i) $\pi_1 \sim \pi_2$;
(ii) $V(\pi_1) = V(\pi_2)$;
(iii) $z(\pi_1) = z(\pi_2)$.

PROOF. The equivalence (ii)$\Leftrightarrow$(iii) follows immediately from Theorem 2.7.

(ii)$\Rightarrow$(i): As we have seen already, ${}^t\pi_i$, $i = 1,2$, maps $\mathcal{M}_*(\pi_i)$ onto $V(\pi_i)$ isometrically; hence the map ρ' defined by $\rho' = ({}^t\pi_1)^{-1} \circ {}^t\pi_2$ is an isometry of $\mathcal{M}_*(\pi_2)$ onto $\mathcal{M}_*(\pi_1)$. The transpose $\rho = {}^t\rho'$ of ρ' maps $\mathcal{M}(\pi_1)$ onto $\mathcal{M}(\pi_2)$. For every $x,y \in A$ and $\omega \in \mathcal{M}_*(\pi_2)$, we have

$$\begin{aligned}\langle \rho(\pi_1(x)^*\pi_1(y)),\omega\rangle &= \langle \pi_1(x)^*\pi_1(y),\rho'(\omega)\rangle \\ &= \langle \pi_1(x^*y),\rho'(\omega)\rangle = \langle x^*y,{}^t\pi_2(\omega)\rangle \\ &= \langle \pi_2(x^*y),\omega\rangle = \langle \pi_2(x)^*\pi_2(y),\omega\rangle;\end{aligned}$$

hence $\rho \circ \pi_1 = \pi_2$ and ρ is an isomorphism. The σ-weak bicontinuity of ρ follows from its construction.

(i)$\Rightarrow$(ii): Let ρ be an isomorphism of $\mathcal{M}(\pi_1)$ onto $\mathcal{M}(\pi_2)$ with $\pi_2 = \rho \circ \pi_1$. By the σ-weak bicontinuity of ρ, the transpose ${}^t\rho$ of ρ maps $\mathcal{M}_*(\pi_2)$ isometrically onto $\mathcal{M}_*(\pi_1)$, and ${}^t\pi_1 \circ {}^t\rho = {}^t\pi_2$. Hence we have

$$\begin{aligned}V(\pi_2) &= {}^t\pi_2(\mathcal{M}_*(\pi_2)) = {}^t\pi_1 \circ {}^t\rho(\mathcal{M}_*(\pi_2)) \\ &= {}^t\pi_1(\mathcal{M}_*(\pi_1)) = V(\pi_1).\end{aligned}$$

Q.E.D.

Now, we apply the above arguments to a von Neumann algebra $\{\mathcal{M},\mathfrak{H}\}$. The action of $\mathcal{M}$ on $\mathfrak{H}$ itself is a representation π_0 of the C^*-algebra $\mathcal{M}$, and then the Banach space $\mathcal{M}_*$ of all σ-weakly continuous linear functionals on $\mathcal{M}$ is the invariant subspace of the conjugate space $\mathcal{M}^*$ of $\mathcal{M}$ associated with π_0. Let z_0 be the support projection of π_0 in the universal enveloping von Neumann algebra $\tilde{\mathcal{M}}$ of $\mathcal{M}$. Then we have

$$\mathcal{M}_* = \mathcal{M}^*z_0.$$

Definition 2.13. Each functional in $\mathcal{M}_*$ is called *normal*, and $\mathcal{M}_*$ itself is called the *predual* of $\mathcal{M}$. On the contrary, the functionals in $\mathcal{M}^*(1 - z_0)$ are called *singular* and $\mathcal{M}^*(1 - z_0)$ is denoted by $\mathcal{M}_*^\perp$.

Theorem 2.14. *In the above situation, if V is a closed right (or left) invariant subspace of $\mathcal{M}^*$, then*

$$V = (V \cap \mathcal{M}_*) \oplus_{l^1} (V \cap \mathcal{M}_*^\perp), \tag{8}$$

$$V \cap \mathcal{M}_* = Vz_0, \qquad V \cap \mathcal{M}_*^\perp = V(1 - z_0), \tag{9}$$

where the notation "$\oplus_{l^1}$" means the l^1-direct sum, i.e.,

$$\|\varphi + \psi\| = \|\varphi\| + \|\psi\|, \qquad \varphi \in V \cap \mathcal{M}_*, \qquad \psi \in V \cap \mathcal{M}_*^\perp. \tag{8'}$$

Futhermore, if π is a representation of $\mathcal{M}$ as a C^*-algebra, then there exists a central projection z of the von Neumann algebra $\mathcal{M}(\pi)$ generated by $\pi(\mathcal{M})$ with the properties:

(i) *The map $\pi_n: x \in \mathcal{M} \mapsto \pi(x)z$ is a σ-weakly continuous representation of $\mathcal{M}$ on the Hilbert space $z\mathfrak{H}_\pi$, if $z \neq 0$.*
(ii) *The map $\pi_s: x \in \mathcal{M} \mapsto \pi(x)(1 - z)$ is a representation of $\mathcal{M}$ on the Hilbert space $(1 - z)\mathfrak{H}_\pi$ such that ${}^t\pi_s(\omega_{\xi,\eta})$ is singular for every $\xi,\eta \in (1 - z)\mathfrak{H}_\pi$, if $z \neq 1$.*

Proof. The first half of our assertion follows directly from Theorem 2.7. By Lemma 2.2 and Theorem 2.4, π is extended to a σ-weakly continuous representation $\tilde{\pi}$ of the universal enveloping von Neumann algebra $\tilde{\mathcal{M}}$ of $\mathcal{M}$. Putting $z = \tilde{\pi}(z_0)$, we get a central projection z in $\mathcal{M}(\pi)$ because z_0 is central in $\tilde{\mathcal{M}}$ and $\tilde{\pi}$ is a homomorphism of $\tilde{\mathcal{M}}$ onto $\mathcal{M}(\pi)$. Noticing that

$$a\,{}^t\pi(\omega)b = {}^t\pi(\tilde{\pi}(a)\omega\tilde{\pi}(b)), \qquad a,b \in \tilde{\mathcal{M}}, \qquad \omega \in \mathcal{M}_*(\pi),$$

we have

$$\begin{aligned} V(\pi) \cap \mathcal{M}_* &= z_0\,{}^t\pi(\mathcal{M}_*(\pi)) = {}^t\pi(\tilde{\pi}(z_0)\mathcal{M}_*(\pi)) \\ &= {}^t\pi(z\mathcal{M}_*(\pi)) = {}^t\pi_n(\mathcal{M}_*(\pi_n)), \\ V(\pi) \cap \mathcal{M}_*^\perp &= (1 - z_0)\,{}^t\pi(\mathcal{M}_*(\pi)) \\ &= {}^t\pi_s(\mathcal{M}_*(\pi_s)). \end{aligned}$$

Therefore, z is the desired central projection in $\mathcal{M}(\pi)$. Q.E.D.

Thus, every $\omega \in \mathcal{M}^*$ is uniquely decomposed into the sum

$$\omega = \omega_n + \omega_s, \qquad \omega_n \in \mathcal{M}_*, \qquad \omega_s \in \mathcal{M}_*^\perp. \tag{10}$$

The functions ω_n and ω_s are called respectively, the *normal part* and the *singular part* of ω. Every representation π of $\mathcal{M}$ is also decomposed into the direct sum

$$\pi = \pi_n \oplus \pi_s \tag{11}$$

in such a way that ${}^t\pi_n(\mathcal{M}_*(\pi_n)) \subset \mathcal{M}_*$ and ${}^t\pi_s(\mathcal{M}_*(\pi_s)) \subset \mathcal{M}_*^\perp$. The representations π_n and π_s are also called the *normal* part and the *singular* part of π, respectively.

Definition 2.15. A uniformly continuous linear map π of a von Neumann algebra $\mathcal{M}$ into another von Neumann algebra $\mathcal{N}$ is called *normal* (resp. *singular*) if π is σ-weakly continuous (resp. ${}^t\pi(\mathcal{N}_*) \subset \mathcal{M}_*^\perp$).

Note that decomposition (8) of V preserves the positivity of a functional in V and that $\mathcal{M}_*^\perp$ is algebraically spanned by the positive elements in $\mathcal{M}_*^\perp$ because so is $\mathcal{M}^*$.

Let π be a uniformly continuous linear mapping of a von Neumann algebra $\{\mathcal{M},\mathfrak{H}\}$ into another von Neumann algebra $\{\mathcal{N},\mathfrak{K}\}$; then the transpose $\tilde{\pi}$ of the restriction π_* of ${}^t\pi$ onto $\mathcal{N}_*$ is an extension of π to the universal enveloping von Neumann algebra $\tilde{\mathcal{M}}$ of $\mathcal{M}$. If we put

$$\pi_n(x) = \tilde{\pi}(xz_0), \qquad \pi_s(x) = \tilde{\pi}(x(1 - z_0)), \qquad x \in \mathcal{M},$$

then we have

$${}^t\pi_n(\mathcal{N}_*) \subset {}_*, \qquad {}^t\pi_s(\mathcal{N}_*) \subset \mathcal{M}_*^\perp,$$
$$\pi = \pi_n + \pi_s.$$

Hence π_n is normal and π_s is singular.

We conclude this section with the following applications of the universal enveloping algebra of a C^*-algebra.

Proposition 2.16. *If ω is a pure state of a C^*-algebra A with left kernel N_ω, then the quotient space A/N_ω, as a Banach space, is a Hilbert space whose inner product coincides with the one induced by ω.*

Proof. Let $\{\pi,\mathfrak{H},\xi\}$ be the cyclic representation of A induced by ω. Then π is extended to a σ-weakly continuous representation $\tilde{\pi}$ of the universal enveloping von Neumann algebra $\tilde{A}$ of A. Considering ω as a normal state of $\tilde{A}$, let $\tilde{N}_\omega$ be the left kernel of ω in $\tilde{A}$. Because we have, for every $x \in \tilde{A}$, $\omega(x) = (\tilde{\pi}(x)\xi|\xi)$, $\tilde{\pi}$ is actually a cyclic representation with a cyclic vector ξ corresponding to ω. Furthermore, we have $\tilde{\pi}(\tilde{A}) = \mathcal{M}(\pi) = \mathcal{L}(\mathfrak{H})$. Since $\tilde{N}_\omega$ is a σ-weakly closed left ideal of $\tilde{A}$, there exists a projection e in $\tilde{A}$ with $\tilde{N}_\omega = \tilde{A}(1 - e)$. Hence the quotient space $\tilde{A}/\tilde{N}_\omega$ is isometric to $\tilde{A}e$ under the canonical map. An $x \in \tilde{A}$ is in $\tilde{N}_\omega$ if and only if $\tilde{\pi}(x)\xi = 0$; that is, $x = x(1 - e)$ if and only if $\tilde{\pi}(x)\xi = 0$. Recalling again that $\tilde{\pi}(\tilde{A}) = \mathcal{L}(\mathfrak{H})$, we conclude that $\tilde{\pi}(e)$ is the projection of $\mathfrak{H}$ onto the one-dimensional subspace $\mathbf{C}\xi$ spanned by ξ. Hence we have $e\tilde{A}e = \mathbf{C}e$, and $exe = \omega(x)e$ for every $x \in \tilde{A}$ since $\omega(exe) = \omega(x)$. For each $x \in \tilde{A}$, we have

$$\|xe\|^2 = \|ex^*xe\| = \omega(x^*x)\|e\|$$
$$= \|\tilde{\pi}(x)\xi\|^2,$$

so that the map: $x \in \tilde{A}e \mapsto \tilde{\pi}(x)\xi$ is an isometry of $\tilde{A}e$ onto $\mathfrak{H} = \mathcal{L}(\mathfrak{H})\xi$. Therefore, $\tilde{A}/\tilde{N}_\omega$ is a Hilbert space.

Put $V_\omega = \tilde{N}_\omega^0$ in A^*. Then, by the σ-weak closedness of $\tilde{N}_\omega$, we have $\tilde{N}_\omega = V_\omega^0$. Hence $V_\omega^* = \tilde{A}/\tilde{N}_\omega$ is a Hilbert space. Therefore, V_ω itself is a Hilbert space, so that it is $\sigma(A^*,A)$-closed, being a reflexive subspace of A^* as a Banach space. Thus we get

$$V_\omega = (V_\omega^0 \cap A)^0 = N_\omega^0.$$

Hence the conjugate space $(A/N_\omega)^* = V_\omega$ of the quotient space of A is a Hilbert space, which means that A/N_ω itself is a Hilbert space. Furthermore, the Hilbert space A/N_ω is isometrically isomorphic to the second conjugate space $\tilde{A}/\tilde{N}_\omega$. Hence, the norm in A/N_ω coincides with the one induced by ω.

Q.E.D.

Proposition 2.17. *Let $\mathscr{I}$ be a closed ideal of a C^*-algebra A. If $\{\pi,\mathfrak{H}\}$ is a representation of $\mathscr{I}$, then π is extended to a representation of A, denoted also by π, on the same space $\mathfrak{H}$ in such a way that $\pi(A)'' = \pi(\mathscr{I})''$.*

Proof. Let $V = \mathscr{I}^0$ in A^*. Then V is a closed invariant subspace of A^*. Hence V is of the form $A^*(1 - z)$ with z a central projection of the universal enveloping von Neumann algebra $\tilde{A}$ of A, and $\mathscr{I}^*$ is identified with A^*z. Therefore, $\tilde{A}$ is decomposed as $\tilde{A} = \tilde{A}z \oplus \tilde{A}(1 - z)$, and $\tilde{A}z = V^0 = \mathscr{I}^{00} = \tilde{\mathscr{I}}$, where $\tilde{\mathscr{I}}$ is the $\sigma(\tilde{A},A^*)$-closure of $\mathscr{I}$ which is regarded as the second conjugate space of $\mathscr{I}$, the universal enveloping von Neumann algebra of $\mathscr{I}$. Therefore, π is extended to a normal representation $\tilde{\pi}$ of $\tilde{\mathscr{I}}$ with $\tilde{\pi}(\tilde{\mathscr{I}}) = \pi(\mathscr{I})''$. We define a representation π of A by $\pi(a) = \tilde{\pi}(az)$, $a \in A$. Then since $xz = x$ for any $x \in \mathscr{I}$, the definition of π is coherent. Furthermore, we have

$$\pi(\mathscr{I}) \subset \pi(A) \subset \tilde{\pi}(\tilde{\mathscr{I}}) = \pi(\mathscr{I})''.$$

Hence we get $\pi(A)'' = \pi(\mathscr{I})''$. Q.E.D.

Exercises

1. Show that the unitary group $\mathscr{U}_{\mathscr{M}}$ of a von Neumann algebra $\mathscr{M}$ is compact if and only if $\mathscr{M}$ is isomorphic to the direct sum of $n \times n$-matrix algebras. In particular, the unitary group of an atomic abelian von Neumann algebra is compact.

2. Let $\mathfrak{m}$ be a nonzero minimal closed left ideal of a C^*-algebra A.
   (a) Show that $\dim(\mathfrak{m} \cap \mathfrak{m}^*) = 1$. (*Hint*: If $\dim(\mathfrak{m} \cap \mathfrak{m}^*) \geq 2$, then $\mathfrak{m} \cap \mathfrak{m}^*$ contains two nonzero positive elements h and k with $hk = 0$. Consider $\overline{Ah} \subset \mathfrak{m}$ and show $k \notin \overline{Ah}$.)
   (b) Show that there exists a minimal projection $e \in A$ such that $\mathfrak{m} = Ae$.
   (c) Show that $\mathfrak{m}$ is a Hilbert space, and that the multiplication of A on $\mathfrak{m}$ from the left is indeed an irreducible representation of A.

3. Let φ and ψ be two pure states on a C^*-algebra A. Show that if $\|\varphi - \psi\| < 2$, then $\pi_\varphi \simeq \pi_\psi$.

4. Let ω be a state on a C^*-algebra A. Show that the quotient Banach space A/N_ω of A by the left kernel N_ω is reflexive if and only if ω is a finite convex combination of pure states.

5. Show that every state on a von Neumann algebra $\mathscr{M}$ is normal if and only if $\mathscr{M}$ is finite dimensional.

6. Let G be a group of unitaries on $\mathfrak{H}$ and $\mathscr{M} = G''$. Let e_0 be the projection of $\mathfrak{H}$ onto the fixed points $\mathfrak{H}_0 = \{\xi \in \mathfrak{H}: u\xi = \xi \text{ for every } u \in G\}$.
   (a) Show that e_0 is a central projection of $\mathscr{M}$ and minimal; hence $\mathscr{M}_{e_0} \cong \mathbf{C}$ provided $e_0 \neq 0$.
   (b) Show that if $\mathfrak{L}$ is a closed convex subset of $\mathfrak{H}$ which is invariant under G, then $\mathfrak{L} \cap \mathfrak{H}_0 \neq \emptyset$. (*Hint*: There exists a unique vector $\xi_0 \in \mathfrak{L}$ such that $\|\xi_0\| = \inf\{\|\xi\|: \xi \in \mathfrak{L}\}$ which must be fixed by G due to the uniqueness.)
   (c) Let $\mathscr{K}$ denote the convex weak (resp. strong) closure of G. Show that if π is a normal representation of $\mathscr{M}$, then for every $\xi \in \mathfrak{H}_\pi$, $\pi(e_0)\xi$ lies in $\pi(\mathscr{K})\xi$. (*Hint*: $\mathfrak{L} = \pi(\mathscr{K})\xi$ is a closed convex subset of $\mathfrak{H}_\pi$ invariant under $\pi(G)$. Use (b).)
   (d) Show that e_0 belongs to $\mathscr{K}$. (*Hint*: For any $\xi_1, \ldots, \xi_n \in \mathfrak{H}$, consider $\mathscr{K}_{\xi_1,\ldots,\xi_n} = \{x \in \mathscr{K}: x\xi_i = e_0\xi_i, i = 1,2,\ldots,n\}$. Let π_n be the natural representation of $\mathscr{M}$ on $\mathfrak{H} \oplus \cdots \oplus \mathfrak{H}$, the n-fold direct sum of $\mathfrak{H}$, and $\xi = \xi_1 \oplus \cdots \oplus \xi_n$. By (c), $\pi_n(e_0)\xi \in \pi_n(\mathscr{K})\xi$, i.e., $\mathscr{K}_{\xi_1,\ldots,\xi_n} \neq \emptyset$. Then conclude that $\{e_0\} = \cap \mathscr{K}_{\xi_1,\ldots,\xi_n} \neq \emptyset$ from the compactness.)

3. W^*-Algebras

The set of postulates for C^*-algebras is a characterization of a uniformly closed self-adjoint operator algebra on a Hilbert space as a special class of involutive Banach algebras, which is free from the underlying Hilbert space, as seen in Section I.9. In this section, we consider the same kind of question for von Neumann algebras. Namely, we give criteria for a C^*-algebra to be isomorphic to a von Neumann algebra on a Hilbert space. Together with this, we will show that the σ-weak topology, the σ-strong topology, and the σ-strong* topology are all determined by the algebraic structure of a von Neumann algebra.

Definition 3.1. A C^*-algebra A is called a *W^*-algebra* if it admits a faithful representation $\{\pi,\mathfrak{H}\}$ such that $\pi(A)$ is a von Neumann algebra on the Hilbert space $\mathfrak{H}$.

Therefore, this section is devoted to dealing with these problems: *under which conditions is a C^*-algebra really a W^*-algebra, and how does the topological structure of a W^*-algebra depend on its algebraic structure.* To do this, we need some preparations.

Lemma 3.2. *Let A be a C^*-algebra and $\omega \in A^*$. If there exists $a \in A_+$ with $\|a\| \leq 1$ such that $\omega(a) = \|\omega\|$, then ω is positive. Hence any Hahn–Banach extension of a positive linear functional on a C^*-subalgebra of A is positive.*

PROOF First, we assume that A is unital. Since $0 \leq a \leq 1$, we have $\|a + e^{i\theta}(1-a)\| \leq 1$ for every $\theta \in \mathbf{R}$. Choosing θ such as $e^{i\theta}\omega(1-a) \geq 0$, we get

$$\|\omega\| \leq \omega(a) + e^{i\theta}\omega(1-a) = \omega(a + e^{i\theta}(1-a)) \leq \|\omega\|;$$

hence $\omega(1-a) = 0$. Therefore, we get $\omega(1) = \omega(a) = \|\omega\|$. Replacing ω by $\omega/\|\omega\|$, we may assume $\|\omega\| = 1$. We claim that $\omega(A_h) \subseteq \mathbf{R}$. Suppose $\omega(h) = \alpha + i\beta$, $\alpha,\beta \in \mathbf{R}$, $\beta \neq 0$, for some $h \in A_h$. Replacing h by $-h$ if necessary, we may assume that $\beta > 0$. Furthermore, by the equality $\omega(h - \alpha 1) = i\beta$, we may assume that $\alpha = 0$. Thus we have $\omega(h) = i\beta$, $\beta > 0$, for some $h \in A_h$. Then we get, for each $\lambda \geq 0$,

$$|\omega(h + i\lambda 1)| = \beta + \lambda \leq \|h + i\lambda 1\| \leq (\|h\|^2 + \lambda^2)^{1/2},$$

which is impossible for sufficiently large λ. Therefore, β must be zero. Hence $\omega(A_h) \subseteq \mathbf{R}$, so that $\omega(x^*) = \overline{\omega(x)}$, $x \in A$.

Now, for an arbitrary $h \in A_+$, we have $\| \|h\| \cdot 1 - h\| \leq \|h\|$, so that

$$\|h\| - \omega(h) = \omega(\|h\| \cdot 1 - \|h\|) \leq \|h\|.$$

Therefore, $\omega(h) \geq 0$, that is, ω is positive.

If A is not unital, then we consider the C^*-algebra A_{I}, obtained by adjunction of an identity to A, and the Hahn–Banach extension $\tilde{\omega}$ of ω to A_{I}. Then $\tilde{\omega}$ satisfies the condition in the lemma for A_{I}; hence $\tilde{\omega}$ is positive and so is ω. Q.E.D.

Definition 3.3. Let A be a C^*-algebra and B a C^*-subalgebra of A. A *projection of A onto B of norm one* is a linear map ε of A onto B such that $\varepsilon(x) = x$ for every $x \in B$ and $\|\varepsilon(x)\| \leq \|x\|$ for every $x \in A$.

Theorem 3.4. *Let A be a C^*-algebra and B a C^*-subalgebra of A. A projection ε of A onto B of norm one enjoys the following properties*:

(i) $\varepsilon(x^*x) \geq 0$, $x \in A$;
(ii) $\varepsilon(axb) = a\varepsilon(x)b$, $a,b \in B$, *and* $x \in A$;
(iii) $\varepsilon(x)^*\varepsilon(x) \leq \varepsilon(x^*x)$, $x \in A$.

PROOF. First we assume that A and B have identities 1_A and 1_B, respectively, and that $\varepsilon(1_A) = 1_B$. Let A_+^* and B_+^* denote the set of all positive linear functionals on A and B, respectively. For every $\omega \in B_+^*$, we have $\|{}^t\varepsilon(\omega)\| \leq \|\omega\|$ and

$$\langle 1_A, {}^t\varepsilon(\omega)\rangle = \langle \varepsilon(1_A), \omega\rangle = \langle 1_B, \omega\rangle = \|\omega\|,$$

so that ${}^t\varepsilon(\omega) \in A_+^*$ by Lemma 3.2. Hence ε maps A_+ onto B_+ and preserves the *-operation.

Considering the universal enveloping von Neumann algebras $\tilde{A}$ and $\tilde{B}$ of A and B, respectively, and the bitranspose ${}^{tt}\varepsilon = \tilde{\varepsilon}$ of ε, we may assume that A is a von Neumann algebra containing B as a weakly closed *-subalgebra.

To show equality (ii), we have only to prove, by the spectral decomposition theorem, that

$$\varepsilon(ex) = e\varepsilon(x) \quad \text{and} \quad \varepsilon(xe) = \varepsilon(x)e, \qquad x \in A, \tag{$*$}$$

for every projection $e \in B$. From the fact that the inequality $0 \le a \le 1_A$, $a \in A$, implies $0 \le eae \le e$, it follows that

$$\varepsilon(eae) = e\varepsilon(eae)e, \qquad a \in A.$$

Put $x' = \varepsilon(ex(1_A - e))$ for $x \in A$ with $\|x\| \le 1$. Then, for any $\lambda > 0$, we have

$$\begin{aligned}\|x' + \lambda e\|^2 &\le \|ex(1_A - e) + \lambda e\|^2 \\ &= \|\{ex(1_A - e) + \lambda e\}\{(1_A - e)x^*e + \lambda e\}\| \\ &= \|ex(1_A - e)x^*e + \lambda^2 e\| \le 1 + \lambda^2.\end{aligned}$$

Put $h = (1/2)(x' + x'^*)$ and $k = (1/(2i))(x' - x'^*)$. Suppose $ehe \neq 0$. Considering $-x$ instead of x if necessary, we may assume that $\mathrm{Sp}_B(ehe)$ contains $\alpha > 0$. Then we have

$$\begin{aligned}\|x' + \lambda e\| &= \|ex'e + \lambda e + (x' - ex'e)\| \\ &\ge \|e(x' + \lambda 1_B)e\| \\ &\ge \|ehe + \lambda e\| \ge \lambda + \alpha.\end{aligned}$$

For a sufficiently large λ, we have

$$\|x' + \lambda e\| \ge \lambda + \alpha > (1 + \lambda^2)^{1/2} \ge \|x' + \lambda e\|,$$

which is a contradiction. Therefore, $ehe = 0$. Similarly, $eke = 0$. Hence we get

$$e\varepsilon(ex(1_A - e))e = 0.$$

Similarly,

$$(1_B - e)\varepsilon(ex(1_B - e))(1_B - e) = 0.$$

Next, suppose $(1_B - e)x'e \neq 0$. For a sufficiently large $\lambda > 0$, we have

$$\begin{aligned}\|x' + \lambda(1_B - e)x'e\| &= \|ex'(1_B - e) + (\lambda + 1)(1_B - e)x'e\| \\ &= \max\{\|ex'(1_B - e)\|, (\lambda + 1)\|(1_B - e)x'e\|\} \\ &= (\lambda + 1)\|(1_B - e)x'e\|.\end{aligned}$$

On the other hand, we have also, for a sufficiently large λ,

$$\begin{aligned}\|x' + \lambda(1_B - e)x'e\| &\le \|ex(1_A - e) + \lambda(1_B - e)x'e\| \\ &= \lambda\|(1_B - e)x'e\|,\end{aligned}$$

which contradicts the above inequality. Hence $(1_B - e)x'e = 0$. Finally, we get

$$x' = ex'(1_B - e).$$

Noticing that

$$x = exe + (1_A - e)xe + ex(1_A - e) + (1_A - e)x(1_A - e),$$

we obtain

$$\begin{aligned} e\varepsilon(x)(1_B - e) &= e\varepsilon(ex(1_A - e))(1_B - e) \\ &= \varepsilon(ex(1_A - e)), \\ e\varepsilon(x)e &= e\varepsilon(exe)e, \end{aligned}$$

which implies equality $(*)$.

Inequality (iii) follows from the simple observation that

$$\begin{aligned} 0 &\le \varepsilon((x - \varepsilon(x))^*(x - \varepsilon(x))) \\ &= \varepsilon(x^*x - \varepsilon(x)^*x - x^*\varepsilon(x) + \varepsilon(x)^*\varepsilon(x)) \\ &= \varepsilon(x^*x) - \varepsilon(x)^*\varepsilon(x). \end{aligned}$$

Let us drop the assumption concerning identities. Considering the universal enveloping von Neumann algebras of A and B, respectively, we may assume that A and B have identities. Let e be the identity of B. Then e is a projection of A. We claim that $\varepsilon(1_A) = e$. Put $a = \varepsilon(1_A - e) = \varepsilon(1_A) - e$, and let $h = (1/2)(a + a^*)$ and $k = (1/2i)(a - a^*)$. For any $\lambda \in \mathbf{R}$, we have, for $|\lambda| \ge 1$,

$$\begin{aligned} \|a + \lambda e\| &\le \|(1 - e) + \lambda e\| = \max\{1,|\lambda|\} = |\lambda|, \\ \|a + \lambda e\| &\ge \|ehe + \lambda e\| \ge |\alpha + \lambda|, \end{aligned}$$

if α is in $\mathrm{Sp}_B(ehe)$. Hence we have $|\lambda| \ge |\lambda + \alpha|$ for each $\alpha \in \mathrm{Sp}_B(ehe)$ and $|\lambda| \ge 1$, so that $\mathrm{Sp}_B(ehe) = \{0\}$. Hence $h = ehe = 0$. Similarly, $k = 0$. Therefore we have $a = 0$, that is, $\varepsilon(1_A) = e$. Thus ε maps the identity of A into the identity of B. Q.E.D.

Theorem 3.5. *Let A be a C^*-algebra. Then the following two statements are equivalent:*

(i) *A is a W^*-algebra, that is, A has a faithful representation $\{\pi,\mathfrak{H}\}$ such that $\pi(A)$ is a von Neumann algebra on the Hilbert space $\mathfrak{H}$.*
(ii) *A is the conjugate space F^* of a certain Banach space F. In this case, the representation $\{\pi,\mathfrak{H}\}$ is chosen so that π is bicontinuous with respect to the $\sigma(A,F)$-topology and the σ-weak topology in $\pi(A)$.*

Proof. If $\{\mathscr{M},\mathfrak{H}\}$ is a von Neumann algebra, then $\mathscr{M}$ is the conjugate space of the Banach space $\mathscr{M}_*$ of all σ-weakly continuous linear functionals on $\mathscr{M}$, by Theorem II.2.6.

Suppose $A = F^*$ for some Banach space F. Let i be the canonical injection of F into $A^* = F^{**}$. Then the transpose $\varepsilon = {}^t i$ of i is a projection of the universal enveloping von Neumann algebra $\tilde{A} = A^{**}$ onto A of norm one. Hence ε enjoys properties (i), (ii), and (iii) in Theorem 3.4. Let $\mathscr{I} = \varepsilon^{-1}(0)$. Then $\mathscr{I}$ is nothing but the polar F^0 of F in $\tilde{A}$, so that $\mathscr{I}$ is $\sigma(\tilde{A},A^*)$-closed. For every $x,y \in A$ and $a \in \mathscr{I}$, we have $\varepsilon(xay) = x\varepsilon(a)y = 0$, so that $x\mathscr{I}y \subset \mathscr{I}$, $x,y \in A$. Since A is $\sigma(\tilde{A},A^*)$-dense in $\tilde{A}$, $\mathscr{I}$ is a σ-weakly closed ideal of $\tilde{A}$. By Proposition II.3.12, there exists a central projection $z \in \tilde{A}$ with $\mathscr{I} = \tilde{A}(1 - z)$.

Noticing that $x - \varepsilon(x) \in \mathscr{I}$ for every $x \in \tilde{A}$, we have $\varepsilon((x - \varepsilon(x))y) = 0$, $x,y \in \tilde{A}$, so that we have

$$\varepsilon(xy) = \varepsilon(x)\varepsilon(y), \qquad x,y \in \tilde{A}.$$

Hence ε is a homomorphism of $\tilde{A}$ onto A with the kernel $\mathscr{I}$, so that ε gives rise to an isomorphism of $\tilde{A}z$ onto A. Clearly, $\tilde{A}z$ is a von Neumann algebra on a Hilbert space $z\mathfrak{H}$, where $\mathfrak{H}$ is the representation space of the universal representation of A on which $\tilde{A}$ acts. Hence A admits a faithful representation π of A on $z\mathfrak{H}$ such that $\pi(A) = \tilde{A}_z$. Since ε transforms the $\sigma(\tilde{A},A^*)$-topology of $\tilde{A}z$ into the $\sigma(A,F)$-topology bicontinuously, π is bicontinuous with respect to the $\sigma(A,F)$-topology and the σ-weak topology. Q.E.D.

For the moment, we shall call the Banach space F in the theorem *a predual* of a W^*-algebra A, although we will soon show that the Banach space F is uniquely determined by A. Therefore, F will be called *the* predual of A after the uniqueness of F is verified (Corollary 3.9). Using a predual F of a W^*-algebra A, we define the *normality* and the *singularity* of linear functionals on A.

Lemma 3.6. *Let A be a W^*-algebra with a predual F. If ω is a nonzero normal positive linear functional, then there exists a unique nonzero projection e such that ω is faithful on eAe and $\omega = e\omega = e\omega e$.*

Proof. Let W denote the closed subspace $[\omega A]$ of F spanned by ωa, $a \in A$. Then W is the smallest right invariant closed subspace containing ω. By Theorem 2.7, there exists a projection $e \in A$ with $W = eF$. As ω is in W, we have $\omega = e\omega$ and $\omega = \omega^* = (e\omega)^* = \omega e$; hence $\omega = e\omega = \omega e = e\omega e$. Suppose $\omega(x^*x) = 0$ for some $x \in A$. The Cauchy–Schwarz inequality implies that $\langle x,\omega a\rangle = 0$ for every $a \in A$; hence $\langle x,W\rangle = 0$, that is, $0 = \langle x,eF\rangle = \langle xe,F\rangle$, so that $xe = 0$. Hence ω is faithful on eAe. Let f be another projection in A with the same properties as e. Then we have $\langle 1 - f,\omega\rangle = 0$, so that $(1 - f)e = 0$; hence $e \leq f$. As $\langle f - e,\omega\rangle = 0$, we have $f = e$. Q.E.D.

Definition 3.7. The projection e in the lemma is called the *support* projection of ω and denoted by $s(\omega)$.

Theorem 3.8. *Let A be a W^*-algebra with a predual F, and ω a positive linear functional on A. Then the following two statements are equivalent:*

(i) *ω is singular.*
(ii) *For every nonzero projection $e \in A$, there exists a nonzero projection $e_0 \leq e$ in A with $\omega(e_0) = 0$.*

Proof. (ii) $\Rightarrow$ (i): Let $\omega = \omega_n + \omega_s$ be the decomposition of ω into the normal part ω_n and the singular part ω_s according to Theorem 2.14. If $\omega_n \neq 0$, then the support projection $e = s(\omega_n)$ of ω_n cannot majorize a nonzero projection e_0 with $\omega(e_0) = 0$ by Lemma 3.6. Hence if ω satisfies condition (ii), then ω is singular.

(i) $\Rightarrow$ (ii): Suppose ω is singular. If $\omega(e) = 0$, then there is nothing to prove. Suppose $\omega(e) > 0$. Take and fix a normal positive linear functional ω_0 on A with $\omega_0(e) > \omega(e)$. Let $\mathscr{F}$ be the set of all the projections p in eAe with $\omega_0(p) \leq \omega(p)$. Then $\mathscr{F}$ is an inductive set under the natural ordering. In fact, if $\{p_i\}$ is a linearly ordered subset of $\mathscr{F}$, then $p = \sup p_i$ satisfies the inequality

$$\omega(p) \geq \sup \omega(p_i) \geq \sup \omega_0(p_i) = \omega_0(p),$$

so that $p \in \mathscr{F}$. By Zorn's lemma, $\mathscr{F}$ has a maximal projection p. Since $e \notin \mathscr{F}$, $e_0 = e - p \neq 0$. By the maximality of p, we have, for every projection $q \leq e_0$, $\omega(q) < \omega_0(q)$. Hence by the spectral decomposition theorem, we have $e_0\omega e_0 \leq e_0\omega_0 e_0$. Let z_0 be the support projection of F in the universal enveloping von Neumann algebra $\tilde{A}$. Then we have, since z_0 is central,

$$\begin{aligned} 0 = (1 - z_0)e_0\omega_0 e_0 &\geq (1 - z_0)e_0\omega e_0 \\ &= e_0(1 - z_0)\omega e_0 = e_0\omega e_0 \geq 0, \end{aligned}$$

so that $e_0\omega e_0 = 0$. Therefore, e_0 is a desired projection. Q.E.D.

Corollary 3.9. *If A is a C^*-algebra, then there exists at most one Banach space F with the conjugate space $F^* = A$. In other words, a predual of a W^*-algebra A is uniquely determined by the C^*-algebra structure of A.*

PROOF. Let F_1 and F_2 be two Banach spaces with the same conjugate space $F_1^* = F_2^* = A$. Corresponding to F_1 and F_2, we say F_1-normal (resp. F_1-singular) or F_2-normal (resp. F_2-singular) for linear functionals on A. Statement (ii) in Theorem 3.8 for a positive linear functional on A does not depend on a predual, so that F_1-singularity of a positive linear functional on A is equivalent to F_2-singularity. Every singular functional is a linear combination of singular positive ones, so that F_1-singularity and F_2-singularity are identical. Let z_1 and z_2 be the support projections of F_1 and F_2, respectively, in the universal enveloping von Neumann algebra $\tilde{A}$ of A. Then we have $(1 - z_1)A^* = (1 - z_2)A^*$, so that $z_1 = z_2$; hence $F_1 = z_1A^* = z_2A^* = F_2$. Thus F_1 and F_2 are identical as spaces of functionals on A. Q.E.D.

Thus, the predual of a W^*-algebra A is unique, and will be denoted by A_*. Now the following result is almost immediate.

Corollary 3.10. *Every isomorphism of a von Neumann algebra onto another is σ-weekly (and hence σ-strongly* and σ-strongly) bicontinuous.*

Based on these results, we will not often specify the Hilbert space on which the von Neumann algebra under consideration acts.

Corollary 3.11. *If ω is a bounded linear functional on a W^*-algebra A, then the following two statements are equivalent:*

(i) *ω is normal.*
(ii) *For every orthogonal family $\{e_i\}_{i\in I}$ of projections in A,*

$$\omega\left(\sum_{i\in I} e_i\right) = \sum_{i\in I} \omega(e_i).$$

PROOF. (i) $\Rightarrow$ (ii): For each finite subset J of I, put $e_J = \sum_{i\in J} e_i$. Then the net $\{e_J\}$ converges strongly to $\sum_{i\in I} e_i$; hence

$$\omega\left(\sum_{i\in I} e_i\right) = \omega\left(\lim_J e_J\right) = \lim_J \omega(e_J)$$

$$= \lim_J \omega\left(\sum_{i\in J} e_i\right) = \sum_{i\in I} \omega(e_i).$$

(ii) $\Rightarrow$ (i): Let $\omega = \omega_n + \omega_s$ be the canonical decomposition of ω into the normal part ω_n and the singular part ω_s. Then $\omega_s = \omega - \omega_n$ also satisfies condition (ii) by the implication (i) $\Rightarrow$ (ii) for ω_n if ω does. Hence we have only to show that a singular linear functional ω satisfying condition (ii) must be zero. By the remark following Definition 2.15, ω is represented by a linear combination of singular positive linear functionals as follows:

$$\omega = \omega_1 - \omega_2 + i(\omega_3 - \omega_4).$$

Put $[\omega] = \omega_1 + \omega_2 + \omega_3 + \omega_4$. Then $[\omega]$ is singular too. Let e be an arbitrary projection in A. Then by Theorem 3.8, there exists an orthogonal family $\{e_i\}_{i\in J}$ of projections in A with $e = \sum_{i\in I} e_i$ such that $[\omega](e_i) = 0$ for every $i \in I$. Since $\omega_k \leq [\omega]$, $1 \leq k \leq 4$, $\omega_k(e_i) = 0$. Therefore, $\omega(e_i) = 0$, $i \in I$. By assumption, $\omega(e) = \sum_{i\in I} \omega(e_i) = 0$. Hence ω vanishes on every projection in A, so that the spectral decomposition theorem yields that $\omega = 0$. Q.E.D.

Remark. The corollary says, in other words, that if the restriction $\omega|_B$ of $\omega \in A^*$ on every abelian W^*-subalgebra B of A is normal, then ω itself is normal on the whole algebra A. This fact will be used later.

For a W^*-algebra, we naturally understand the σ-week topology, the σ-strong topology, and the σ-strong* topology. But the weak topology, the strong topology, and the strong* topology on a W^*-algebra do not make sense unless we specify on which Hilbert space the algebra acts.

Proposition 3.12. *Let A be a W^*-algebra. If $\{\pi,\mathfrak{H}\}$ is a normal representation of A, then the image $\pi(A)$ is a von Neumann algebra on $\mathfrak{H}$ and π is σ-weakly continuous. The cyclic representation induced by a normal positive linear functional is normal.*

PROOF. By definition, the transpose ${}^t\pi$ of π maps $\mathscr{L}(\mathfrak{H})_*$ into A_*, so that π is σ-weakly continuous. Hence the image $\pi(S)$ of the unit ball of A is σ-weakly compact, so that it is strongly closed by Theorem II.2.6. By Theorem II.4.8, the image $\pi(A)$ is strongly closed; hence it is a von Neumann algebra by Theorem II.3.9.

Let ω be a normal positive linear functional on A with the associated cyclic representation $\{\pi_\omega, \mathfrak{H}_\omega, \xi_\omega\}$. Let $\mathscr{M}$ be the von Neumann algebra $\pi_\omega(A)''$ on $\mathfrak{H}_\omega$. For each a, b and x in A, we have

$$\begin{aligned}\langle x, {}^t\pi_\omega(\omega_{\pi_\omega(a)\xi_\omega, \pi_\omega(b)\xi_\omega})\rangle &= (\pi_\omega(x)\pi_\omega(a)\xi_\omega | \pi_\omega(b)\xi_\omega) \\ &= (\pi_\omega(b^*xa)\xi_\omega | \xi_\omega) = \langle x, a\omega b^*\rangle.\end{aligned}$$

Hence ${}^t\pi_\omega(\omega_{\pi_\omega(a)\xi_\omega, \pi_\omega(b)\xi_\omega})$ falls in the predual A_* of A. Since the linear combinations of $\omega_{\pi_\omega(a)\xi_\omega, \pi_\omega(b)\xi_\omega}$, $a,b \in A$, are dense in the predual $\mathscr{M}_*$ of $\mathscr{M}$, as $\pi_\omega(A)\xi_\omega$ is dense in $\mathfrak{H}_\omega$, ${}^t\pi_\omega$ maps $\mathscr{M}_*$ into A_*. Therefore, π_ω is continuous with respect to the $\sigma(A, A_*)$-topology and the $\sigma(\mathscr{M}, \mathscr{M}_*)$-topology. Q.E.D.

Now we discuss characterizations of W^*-algebras which are more directly related to the results in Section 1, especially to Theorem 1.18. The following definition is a natural analogue of the abelian case treated in Section 1.

Definition 3.13. A C^*-algebra A is said to be *monotone closed* if every bounded increasing net in A_h has the least upper bound in A_h. A positive linear functional ω on a monotone closed C^*- algebra A is called *normal* if $\omega(\sup x_i) = \sup \omega(x_i)$ for every bounded increasing net $\{x_i\}$ in A_h, where $\sup x_i$ means the least upper bound of $\{x_i\}$ in A_h.

A W^*-algebra A is, of course, monotone closed and the definitions of normality for a positive linear functional are coherent by Corollary 3.11.

An approximate identity $\{u_i\}$ of a monotone closed C^*-algebra A is a bounded increasing net in A_h; hence it has the least upper bound e. Let $\{\pi, \mathfrak{H}\}$ be a faithful representation of A. Then as we have seen in the proof of Theorem II.4.8, $\{\pi(u_i)\}$ converges strongly to the identity operator 1 on $\mathfrak{H}$. But $\pi(u_i) \leq \pi(e)$ for every i, so that $\pi(e) \geq 1$; hence $\pi(e)$ is invertible in $\mathscr{L}(\mathfrak{H})$. Therefore, by Proposition I.4.8, $\pi(e)$ is invertible in $\pi(A) + \mathbf{C}1$. But $\pi(A)$ is an ideal of $\pi(A) + \mathbf{C}1$; hence $\pi(A) = \pi(A) + \mathbf{C}1$. Hence $\pi(A)$ contains the identity 1; hence A has an identity.

As seen in Section 1, an abelian C^*-algebra is monotone closed if and only if its spectrum is stonean; and a positive linear functional is normal if and only if it gives rise to a positive normal measure on the spectrum.

Lemma 3.14. *Let A be a monotone closed C^*-algebra and $\omega \in A_+^*$. It there exists a sequence $\{\omega_n\}$ of normal positive linear functionals on A with*

$$\lim_{n\to\infty} \|\omega - \omega_n\| = 0,$$

then ω is normal.

PROOF. Let $\{x_i\}_{i\in I}$ be a bounded net in A_h with $x = \sup x_i$. Then obviously $\omega(x) \geq \sup \omega(x_i)$. For any $\varepsilon > 0$, there exists an index n such that $\|\omega - \omega_n\| < \varepsilon$. Then there exists an index $i \in I$ such that $\omega_n(x) - \varepsilon < \omega_n(x_i)$, since ω_n is normal. Then we have

$$\begin{aligned}\omega(x_i) &\geq \omega_n(x_i) - \varepsilon\|x_i\| \geq \omega_n(x) - \varepsilon - \varepsilon\|x_i\| \\ &\geq \omega(x) - \varepsilon\|x\| - \varepsilon - \varepsilon\|x_i\| \\ &\geq \omega(x) - (1 + 2\|x\|)\varepsilon.\end{aligned}$$

Therefore, we have $\omega(x) \leq \sup \omega(x_i)$. Thus ω is normal. Q.E.D.

Proposition 3.15. *If ω is a normal positive linear functional on a monotone closed C^*-algebra A, then the image $\pi_\omega(A)$ of A, under the cyclic representation $\{\pi_\omega,\mathfrak{H}_\omega,\xi_\omega\}$, is a von Neumann algebra on $\mathfrak{H}_\omega$.*

PROOF. By Corollary II.4.25, it is sufficient to show that the strong limit of any bounded monotone net in $\pi_\omega(A)$ is contained in $\pi_\omega(A)$. Let $\{x_i\}_{i\in I}$ be any bounded increasing net in A with $x = \sup x_i$. If we have a decreasing net, then we can dualize the following arguments.

Let u be a unitary element in A. Then we have $uxu^* = \sup_{i\in I} ux_iu^*$. Hence we have

$$\begin{aligned}(\pi_\omega(x)\pi_\omega(u)\xi_\omega|\pi_\omega(u)\xi_\omega) &= (\pi_\omega(u^*xu)\xi_\omega|\xi_\omega) \\ &= \omega(u^*xu) = \sup_i \omega(u^*x_iu) \\ &= \sup_i (\pi_\omega(u^*x_iu)\xi_\omega|\xi_\omega) \\ &= \sup_i (\pi_\omega(x_i)\pi_\omega(u)\xi_\omega|\pi_\omega(u)\xi_\omega).\end{aligned}$$

Hence we have

$$\begin{aligned}\lim_i \|[\pi_\omega(x) - \pi_\omega(x_i)]^{1/2}\pi_\omega(u)\xi_\omega\|^2 &= \lim_i ([\pi_\omega(x) - \pi_\omega(x_i)]\pi_\omega(u)\xi_\omega|\pi_\omega(u)\xi_\omega) \\ &= 0.\end{aligned}$$

Since the linear combinations of the unitary group $U(A)$ of A exhaust A and $\pi_\omega(A)\xi_\omega$ is dense in $\mathfrak{H}_\omega$ and since $[\pi_\omega(x) - \pi_\omega(x_i)]^{1/2}$ is bounded, $[\pi_\omega(x) - \pi_\omega(x_i)]^{1/2}$ converges strongly to zero. Since the product operation is strongly continuous on bounded parts of $\mathscr{L}(\mathfrak{H}_\omega)$, we conclude that $\pi(x_i)$ converges strongly to $\pi_\omega(x)$. Hence the strong limit of $\{\pi(x_i)\}_{i\in I}$ is in $\pi_\omega(A)$. Therefore, $\pi_\omega(A)$ is a von Neumann algebra on $\mathfrak{H}_\omega$. Q.E.D.

Theorem 3.16. *For a C^*-algebra A, the following two statements are equivalent:*

(i) *A is a W^*-algebra.*
(ii) *A is monotone closed and admits sufficiently many normal positive linear functionals.*

PROOF. The implication (i) $\Rightarrow$ (ii) is trivial, so we shall verify the implication (ii) $\Rightarrow$ (i). Let $\{\pi,\mathfrak{H}\}$ be the direct sum representation of all cyclic representations induced by normal positive linear functions on A. By assumption, π is faithful. If $\{x_i\}_{i\in I}$ is a bounded increasing net in A_h with $x = \sup x_i$, then for any normal positive linear functionals ω on A, $\pi_\omega(x)$ is the strong limit of $\pi_\omega(x_i)$ on $\mathfrak{H}_\omega$; hence $\pi(x)$ is the strong limit of $\pi(x_i)$ on $\mathfrak{H}$. Therefore, the monotone closure of $\pi(A)$ in $\mathscr{L}(\mathfrak{H})$ is $\pi(A)$ itself; thus it is a von Neumann algebra on $\mathfrak{H}$. Q.E.D.

Exercises

1. Show that if a von Neumann algebra $\mathscr{M}$ is separable in the uniform topology, then $\mathscr{M}$ must be finite dimensional. (*Hint*: If $\mathscr{M}$ contains an infinite family of orthogonal nonzero projections, then $\mathscr{M}$ contains a subalgebra isomorphic to l^∞.)
2. Let A be a C^*-algebra acting on $\mathfrak{H}$. Suppose that there exists a projection ε of norm one from a von Neumann algebra $\mathscr{M}$ containing A onto A.
   (a) Show that every bounded increasing net $\{h_i\}$ in A_+ has the least upper bound in A, which might be different from the strong limit of $\{h_i\}$.
   (b) Show that if A is separable, then A is finite dimensional.
3. Let A be a separable C^*-algebra contained in a von Neumann algebra $\{\mathscr{M},\mathfrak{H}\}$. Suppose that $\mathscr{M} = A''$ and any normal state ω on $\mathscr{M}$ is faithful if the restriction $\omega|_A$ of ω to A is faithful.
   (a) Show that $\mathscr{M}$ admits a faithful normal state φ.
   (b) Show that every nonzero projection e in $\mathscr{M}$ majorizes a nonzero positive element of A, i.e., $e\mathscr{M}e \cap A \neq \{0\}$. (*Hint*: If not, then the functional $e^\perp\varphi e^\perp$ must be faithful on $\mathscr{M}$, where $e^\perp = 1 - e$.)
   (c) Show that every singular state on $\mathscr{M}$ cannot be faithful on A.
   (d) Show that the normal part ω_n of a state ω is faithful on $\mathscr{M}$ if the restriction $\omega|_A$ of ω to A is faithful.
   (e) Show that A admits a faithful state ω which is a convex l^1-sum of a sequence of pure states, i.e., $\omega = \sum_{n=1}^\infty \lambda_n\omega_n$ with $\sum_{n=1}^\infty \lambda_n = 1$, $\lambda_n \geq 0$. (*Hint*: A is separable.)
   (f) Let φ be the normal part of the Hahn–Banach extension $\tilde{\omega}$ of ω to $\mathscr{M}$. Show that the cyclic representation $\{\pi_\varphi, \mathfrak{H}_\varphi\}$ of $\mathscr{M}$ is faithful, normal and that $\pi_\varphi(\mathscr{M})$ is atomic. (*Hint*: The restriction $\pi_\varphi|_A$ to A is a direct summand of π_ω.)
   (g) Show that $\mathscr{M}$ is generated by minimal projections and A contains all minimal projections of $\mathscr{M}$.

4. The Polar Decomposition and the Absolute Value of Functionals

Let $\mathscr{M}$ be a von Neumann algebra with predual $\mathscr{M}_*$. Then the structures of $\mathscr{M}$ and of $\mathscr{M}_*$ are firmly interdependent. In this section, we shall discuss the polar decomposition in $\mathscr{M}_*$. Throughout this section, $\mathscr{M}$ denotes a von Neumann algebra, $\mathscr{M}_*$ its predual, $\mathscr{M}_*^+$ the positive cone of $\mathscr{M}_*$ and S the unit ball of $\mathscr{M}$.

Lemma 4.1. *Let ω be an element of $\mathscr{M}_*$. If a projection $e \in \mathscr{M}$ satisfies the equality $\|e\omega\| = \|\omega\|$, then we have $e\omega = \omega$.*

PROOF. We may assume $\|\omega\| = 1$, replacing ω by $(1/\|\omega\|)\omega$. Putting $f = 1 - e$, we shall show $f\omega = 0$. If $f\omega \neq 0$, then there exists $b \in S$ with $\langle b, f\omega \rangle = \delta > 0$. Since $\mathscr{M} = (\mathscr{M}_*)^*$, there exists $a \in S$ with $\langle a, e\omega \rangle = 1$. From the observation that

$$\begin{aligned}\|ae + \delta bf\|^2 &= \|(ae + \delta bf)(ae + \delta bf)^*\| \\ &= \|aea^* + \delta^2 bfb^*\| \leq 1 + \delta^2,\end{aligned}$$

it follows that

$$\|ae + \delta bf\| \leq (1 + \delta^2)^{1/2} < 1 + \delta^2.$$

On the other hand,

$$\langle ae + \delta bf, \omega \rangle = \langle a, e\omega \rangle + \delta \langle b, f\omega \rangle = 1 + \delta^2,$$

which contradicts $\|\omega\| = 1$. Hence $f\omega = 0$. Q.E.D.

Take an element ω in $\mathscr{M}_*$ and put

$$V_\omega = [\mathscr{M}\omega] \quad \text{and} \quad W_\omega = [\omega\mathscr{M}].$$

Then V_ω and W_ω are respectively left and right invariant closed subspaces of $\mathscr{M}_*$, so that by Theorem 2.7, there exist projections e and f in $\mathscr{M}$ such that

$$V_\omega = \mathscr{M}_* e \quad \text{and} \quad W_\omega = f\mathscr{M}_*.$$

The projections e and f are, respectively, the smallest projections in $\mathscr{M}$ such that $\omega = \omega e$ and $\omega = f\omega$. Thus the projections e and f are called, respectively, the *right* and the *left support projections* of ω and denoted by $s_r(\omega)$ and $s_l(\omega)$. If ω is hermitian, then $s_r(\omega) = s_l(\omega)$, so that we write $s(\omega)$. Corresponding to the polar decompositions of operators, we obtain the analogous decomposition for normal linear functionals.

Theorem 4.2. (i) *If φ is a normal linear functional on a von Neumann algebra $\mathscr{M}$ with predual $\mathscr{M}_*$, then there exist uniquely a partial isometry $v \in \mathscr{M}$ and $\omega \in \mathscr{M}_*^+$ such that*

$$\varphi = v\omega, \qquad v^*v = s(\omega).$$

Furthermore, we have $s(\omega) = s_r(\varphi)$ and $vv^* = s_l(\varphi)$.

(ii) *If φ is hermitian, then φ is uniquely decomposed as follows:*

$$\varphi = \varphi_+ - \varphi_- \quad \textit{and} \quad \|\varphi\| = \|\varphi_+\| + \|\varphi_-\|,$$

where $\varphi_+, \varphi_- \in \mathscr{M}_*^+$. If φ and ψ are normal positive linear functionals on $\mathscr{M}$, then

$$\|\varphi - \psi\| = \|\varphi\| + \|\psi\|$$

is equivalent to the orthogonality of $s(\varphi)$ and $s(\psi)$.

Proof. Since $\mathcal{M} = (\mathcal{M}_*)^*$, there exists $a \in S$ with $\langle a,\varphi\rangle = \|\varphi\|$. Let $a^* = u|a^*|$ be the polar decomposition of a^*. Then we have

$$\|\varphi\| = \langle a,\varphi\rangle = \langle |a^*|u^*,\varphi\rangle = \langle |a^*|,u^*\varphi\rangle.$$

Put $\omega = u^*\varphi$. Then by Lemma 3.2, ω is positive because $0 \le |a^*| \le 1$. Let $e = uu^*$. Then we have $u\omega = uu^*\varphi = e\varphi$ and

$$\|\varphi\| = \langle a,\varphi\rangle = \langle ae,\varphi\rangle = \langle a,e\varphi\rangle,$$

so that $\|e\varphi\| = \|\varphi\|$. Hence Lemma 4.1 says that $e\varphi = \varphi$, that is, $\varphi = u\omega$. By the equality $u^*u\omega = u^*\varphi = \omega$, we have $u^*u \ge \mathrm{s}(\omega)$; hence $v = u\mathrm{s}(\omega)$ is a partial isometry in $\mathcal{M}$ with $\varphi = v\omega$ and $v^*v = \mathrm{s}(\omega)$. Thus, we have obtained a desired decomposition of φ.

Next, we shall show the uniqueness of ω and v. Suppose φ is represented in another form:

$$\varphi = v_1\omega_1, \qquad \omega_1 \ge 0 \quad \text{and} \quad v_1^*v_1 = \mathrm{s}(\omega_1).$$

Put $p = v^*v = \mathrm{s}(\omega)$, $q = vv^*$, and $p_1 = v_1^*v_1 = \mathrm{s}(\omega_1)$, $q_1 = v_1v_1^*$. For every $x \in \mathcal{M}$, we have

$$\langle x,\varphi\rangle = \langle x,v\omega\rangle = \langle x,v_1\omega_1\rangle,$$

and

$$\begin{aligned}\langle x,\omega\rangle &= \langle xp,\omega\rangle = \langle xv^*v,\omega\rangle \\ &= \langle xv^*,\varphi\rangle = \langle xv^*,v_1\omega_1\rangle \\ &= \langle xv^*v_1,\omega_1\rangle = \langle p_1xv^*v_1,\omega_1\rangle.\end{aligned}$$

Putting $x = 1 - p_1$ in the above equality, we have $\langle 1 - p_1,\omega\rangle = 0$. Hence $p_1 \ge \mathrm{s}(\omega) = p$. By symmetry, $p \ge p_1$ and then $p = p_1$. Let $v_1^*v = h + ik$ with h and k self-adjoint. Then the equality $v_1^*v = pv_1^*vp$ implies that h and k are both in $p\mathcal{M}p$. The equality

$$\begin{aligned}\langle v_1^*v,\omega\rangle &= \langle v_1^*,v\omega\rangle = \langle v_1^*,\varphi\rangle = \langle v_1^*,v_1\omega_1\rangle \\ &= \langle p,\omega_1\rangle = \|\omega_1\| = \|\omega\| \\ &= \langle h,\omega\rangle + i\langle k,\omega\rangle\end{aligned}$$

implies that $\langle h,\omega\rangle = \|\omega\|$ and $\langle k,\omega\rangle = 0$. Since ω is faithful on $p\mathcal{M}p$ and $-p \le h \le p$, we have $h = p$ because $\langle p - h,\omega\rangle = 0$; hence $v_1^*v = p$ because $\|v_1^*v\| \le 1$. Hence $v^*v_1 = (v_1^*v)^* = p$ as well. Thus we get

$$\begin{aligned}v &= vp = vv^*v_1 = qv_1, \\ v_1 &= v_1p = v_1v_1^*v = q_1v,\end{aligned}$$

so that we have

$$\begin{aligned}q &= vv^* = qv_1v_1^*q = qq_1q, \\ q_1 &= v_1v_1^* = q_1vv^*q_1 = q_1qq_1,\end{aligned}$$

which means that $q = q_1$ and $v = v_1$. Therefore,

$$\omega = v^*\varphi = v_1^*\varphi = \omega_1.$$

If $\varphi \in \mathcal{M}_*$ is hermitian, then the a in the first paragraph of the proof is chosen in $\mathcal{M}_h \cap S$, so that the partial isometry v in the above is hermitian; hence v is the difference $e - f$ of two orthogonal projections e and f. As $\omega = v^*\varphi = v\varphi = (v\varphi)^* = \varphi v$, every spectral projection of v commutes with φ, so that $e\varphi$ and $f\varphi$ are both hermitian. Since $\mathrm{s}(e\varphi) \leq e$ and $\mathrm{s}(f\varphi) \leq f$, and since $\omega = e\varphi - f\varphi$, $e\varphi$ and $-f\varphi$ are both positive. Putting $\varphi_+ = e\varphi$ and $\varphi_- = -f\varphi$, we get

$$\varphi = v^2\varphi = (e + f)\varphi = \varphi_+ - \varphi_-, \qquad \varphi_+ \geq 0, \qquad \varphi_- \geq 0,$$

$$\begin{aligned} \|\varphi\| &= \langle 1,\omega\rangle = \langle 1,(e - f)\varphi\rangle = \langle e - f,\varphi\rangle \\ &= \langle e,\varphi_+\rangle + \langle f,\varphi_-\rangle = \|\varphi_+\| + \|\varphi_-\|. \end{aligned}$$

Conversely, suppose that φ_1 and φ_2 are positive normal linear functionals such that $\|\varphi_1 - \varphi_2\| = \|\varphi_1\| + \|\varphi_2\|$. Put $\varphi = \varphi_1 - \varphi_2$. Then we have

$$\begin{aligned} \|\varphi_+\| &= \langle \mathrm{s}(\varphi_+),\varphi\rangle = \langle \mathrm{s}(\varphi_+),\varphi_1 - \varphi_2\rangle \\ &\leq \langle \mathrm{s}(\varphi_+),\varphi_1\rangle \leq \|\varphi_1\|, \\ \|\varphi_-\| &= -\langle \mathrm{s}(\varphi_-),\varphi\rangle = -\langle \mathrm{s}(\varphi_-),\varphi_1 - \varphi_2\rangle \\ &\leq \langle \mathrm{s}(\varphi_-),\varphi_2\rangle \leq \|\varphi_2\|, \end{aligned}$$

so that

$$\begin{aligned} \|\varphi_+\| &= \langle \mathrm{s}(\varphi_+),\varphi_1\rangle = \|\varphi_1\|; \\ \|\varphi_-\| &= \langle \mathrm{s}(\varphi_-),\varphi_2\rangle = \|\varphi_2\|. \end{aligned}$$

Hence $\mathrm{s}(\varphi_1) \leq \mathrm{s}(\varphi_+)$ and $\mathrm{s}(\varphi_2) \leq \mathrm{s}(\varphi_-)$, so that $\mathrm{s}(\varphi_1)$ and $\mathrm{s}(\varphi_2)$ are orthogonal. Therefore, we have $\varphi = \varphi_1 - \varphi_2 = [\mathrm{s}(\varphi_1) - \mathrm{s}(\varphi_2)](\varphi_1 + \varphi_2)$. By the uniqueness of the decomposition of φ in (i), we have $\mathrm{s}(\varphi_1) - \mathrm{s}(\varphi_2) = \mathrm{s}(\varphi_+) - \mathrm{s}(\varphi_-)$ and $\varphi_1 + \varphi_2 = \varphi_+ + \varphi_-$. Hence $\mathrm{s}(\varphi_1) = \mathrm{s}(\varphi_2) = \mathrm{s}(\varphi_2) = \mathrm{s}(\varphi_-)$, which implies that

$$\begin{aligned} \varphi_1 &= \mathrm{s}(\varphi_1)\varphi = \mathrm{s}(\varphi_+)\varphi = \varphi_+, \\ \varphi_2 &= -\mathrm{s}(\varphi_2)\varphi = -\mathrm{s}(\varphi_-)\varphi = \varphi_-. \end{aligned}$$

Q.E.D.

Definition 4.3. The expression of φ in (i) is called the *polar decomposition* of φ and ω is called the *absolute value* of φ and denoted by $|\varphi|$.

The uniqueness of the decomposition in (ii) implies the uniqueness of the Jordan decomposition in Proposition 2.1. Note that $vv^* = \mathrm{s}_l(\varphi)$ and $v^*v = \mathrm{s}_r(\varphi)$.

Corollary 4.4. *Let A be a C^*-algebra. Then every closed left (resp. right) invariant subspace V of A^* has the form*

$$V = [A(V \cap A_+^*)] \qquad (\textit{resp. } V = [(V \cap A_+^*)A]).$$

Therefore, if V is nonzero, then V contains a nonzero positive element and is determined by its positive part $V \cap A_+^*$.

Corollary 4.5. *Every closed left (resp. right) ideal of a C^*-algebra is the intersection of the left (resp. right) kernels of pure states which contain it.*

PROOF. Let A be a C^*-algebra and $\mathfrak{m}$ a closed left ideal of A. Then $V = \mathfrak{m}^0$ is a $\sigma(A^*,A)$-closed right invariant subspace of A^*. By Corollary 4.4, we have $V = [(V \cap A_+^*)A]$. Hence $x \in A$ falls in $\mathfrak{m}$ if and only if $\langle x^*x,\omega\rangle = 0$ for every $\omega \in V \cap A_+^*$. Let S^* denote the unit ball of A^*. Then $V \cap A_+^* \cap S^*$ is $\sigma(A^*,A)$-compact and convex, so that it contains sufficiently many extreme points by the Krein–Milman theorem. If ω is an extreme point in $V \cap A_+^* \cap S^*$ and if $\omega = \frac{1}{2}(\omega_1 + \omega_2)$, $\omega_1,\omega_2 \in A_+^* \cap S^*$, then we have, for every $x \in \mathfrak{m}$,

$$0 = \omega(x^*x) = \tfrac{1}{2}(\omega_1(x^*x) + \omega_2(x^*x));$$

hence $\omega_1(x^*x) = \omega_2(x^*x) = 0$, so that ω_1 and ω_2 are both in V; thus $\omega = \omega_1 = \omega_2$. Hence ω is a pure state of A. Therefore, the condition $\omega(x^*x) = 0$ for every $\omega \in V \cap A_+^*$ is equivalent to saying that $\omega(x^*x) = 0$ for every $\omega \in V \cap P(A)$, where $P(A)$ denotes the set of all pure states of A, which means that $\mathfrak{m} = \bigcap\{N_\omega : \omega \in V \cap P(A)\}$, where N_ω denotes the left kernel of ω. Q.E.D.

The absolute value of a normal linear functional is characterized by the following:

Proposition 4.6. *The absolute value* $|\varphi| = \omega$ of $\varphi \in \mathcal{M}_*$ *is uniquely determined by the properties*

$$\|\omega\| = \|\varphi\| \quad \textit{and} \quad |\varphi(x)|^2 \le \|\varphi\|\omega(xx^*), \qquad x \in \mathcal{M}.$$

PROOF. Let $\varphi = u\omega$ be the polar decomposition of φ. Then we have, by the Cauchy–Schwarz inequality,

$$\begin{aligned}|\langle x,\varphi\rangle|^2 &= |\langle x,u\omega\rangle|^2 = |\langle xu,\omega\rangle|^2 \\ &\le \langle xx^*,\omega\rangle\langle u^*u,\omega\rangle = \|\omega\|\langle x^*x,\omega\rangle.\end{aligned}$$

Conversely, suppose ω_1 is a positive linear functional satisfying the condition. Then

$$\begin{aligned}|\langle x,\omega\rangle|^2 &= |\langle xu^*,\varphi\rangle|^2 \le \|\varphi\|\omega_1(xu^*ux^*) \\ &\le \|\varphi\|\omega_1(xx^*),\end{aligned}$$

so that we have, since $\omega(x^*) = \overline{\omega(x)}$,

$$|\omega(x)| \le \|\varphi\|^{1/2}\omega_1(x^*x)^{1/2}.$$

Let $\{\pi,\mathfrak{H},\xi_0\}$ be the cyclic representation of $\mathcal{M}$ induced by ω_1. The above inequality says that the function: $\pi(x)\xi_0 \in \pi(\mathcal{M})\xi_0 \mapsto \omega(x)$ is bounded, so that there exists a vector $\eta \in \mathfrak{H}$ such that

$$\omega(x) = (\pi(x)\xi_0|\eta), \qquad x \in \mathcal{M}.$$

As $\|\omega_1\| = \|\varphi\| = \|\omega\| = \omega(1)$, we have $(\xi_0|\eta) = \|\xi_0\|^2$. But, the inequality $\|\eta\| \le \|\varphi\|^{1/2} = \|\omega_1\|^{1/2} = \|\xi_0\|$ implies that $\eta = \xi_0$. Hence we get $\omega = \omega_1$. Q.E.D.

Remark. In the proposition, we do not assume normality for ω, which follows from the condition.

Proposition 4.7. *For any pair φ, ψ in $\mathcal{M}_*$, we have*

$$|\langle x,|\varphi+\psi|\rangle|^2 \leq (\|\varphi\|+\|\psi\|)(\langle xx^*,|\varphi|\rangle + \langle xx^*,|\psi|\rangle).$$

PROOF. Let $\varphi = u|\varphi|$, $\psi = v|\psi|$ and $\varphi+\psi = w|\varphi+\psi|$ be the polar decompositions of φ, ψ and $\varphi+\psi$, respectively. Then we have

$$\begin{aligned}
|\langle x,|\varphi+\psi|\rangle|^2 &= |\langle x, w^*(\varphi+\psi)\rangle|^2 \\
&= |\langle xw^*, u|\varphi| + v|\psi|\rangle|^2 \\
&= |\langle xw^*u,|\varphi|\rangle + \langle xw^*v,|\psi|\rangle|^2 \\
&\leq (|\langle xw^*u,|\varphi|\rangle| + |\langle xw^*v,|\psi|\rangle|)^2 \\
&\leq (\langle xx^*,|\varphi|\rangle^{1/2}\langle u^*ww^*u,|\varphi|\rangle^{1/2} + \langle xx^*,|\psi|\rangle^{1/2}\langle v^*ww^*v,|\psi|\rangle^{1/2})^2 \\
&\leq (\|\varphi\|^{1/2}\langle xx^*,|\varphi|\rangle^{1/2} + \|\psi\|^{1/2}\langle xx^*,|\psi|\rangle^{1/2})^2 \\
&\leq (\|\varphi\|+\|\psi\|)(\langle xx^*,|\varphi|\rangle + \langle xx^*,|\psi|\rangle).
\end{aligned}$$

Q.E.D.

Proposition 4.8. *Let A be a C^*-algebra and $\varphi \in A_+^*$. If $a\varphi$, $a \in A$, is hermitian, then*

$$|\langle h, a\varphi\rangle| = |\langle ha,\varphi\rangle| \leq \|a\|\langle h,\varphi\rangle, \qquad h \in A_+.$$

PROOF. Since $a\varphi = (a\varphi)^* = \varphi a^*$, $a^2\varphi = a\varphi a^*$ is positive and $a^{2^n+1}\varphi = a^{2^n}\varphi(a^*)^{2^n}$. For each $h \in A_+$, we have

$$\begin{aligned}
|\langle ha,\varphi\rangle| &= |\langle h^{1/2}h^{1/2}a,\varphi\rangle| \leq \langle h,\varphi\rangle^{1/2}\langle a^*ha,\varphi\rangle^{1/2} \\
&= \langle h,\varphi\rangle^{1/2}\langle ha^2,\varphi\rangle^{1/2} \\
&\leq \langle h,\varphi\rangle^{(1/2)+(1/4)}\langle (a^*)^2ha^2,\varphi\rangle^{1/4} \\
&= \langle h,\varphi\rangle^{(1/2)+(1/4)}\langle ha^4,\varphi\rangle^{1/4} \\
&\leq \cdots \\
&\leq \langle h,\varphi\rangle^{\sum_{k=1}^{n}(1/2)^k}\ ha^{2^n},\varphi\rangle^{2^{-n}} \\
&\leq \langle h,\varphi\rangle^{1-(1/2)^n}\langle ha^{2^n},\varphi\rangle^{2^{-n}} \\
&\leq \langle h,\varphi\rangle^{1-(1/2)^n}(\|h\|\,\|a^{2^n}\|\,\|\varphi\|)^{2^{-n}} \\
&\to \|a\|_{sp}\langle h,\varphi\rangle \text{ as } n \to \infty.
\end{aligned}$$

Thus the inequality follows. Q.E.D.

Proposition 4.9. *In a von Neumann algebra $\mathcal{M}$, we have, for $\varphi \in \mathcal{M}_*^+$ and $a \in \mathcal{M}$,*

$$|a\varphi| \leq \|a\|\varphi.$$

PROOF. Let $a\varphi = u|a\varphi|$ be the polar decomposition of φ. By the equality $|a\varphi| = u^*a\varphi$, Proposition 4.8 implies that for every $h \in \mathcal{M}_+$,

$$\begin{aligned}
\langle h,|a\varphi|\rangle &= \langle h,u^*a\varphi\rangle \leq \|u^*a\|\langle h,\varphi\rangle \\
&\leq \|a\|\langle h,\varphi\rangle.
\end{aligned}$$

Q.E.D.

Proposition 4.10. *If $\{\varphi_n\}$ is a sequence in $\mathscr{M}_*$ converging to φ in norm, then the sequence $\{|\varphi_n|\}$ of absolute values converges to $|\varphi|$ in norm.*

PROOF. Considering a suitable representation, we may assume that $\mathscr{M}$ acts on a Hilbert space $\mathfrak{H}$ with a vector ξ such that $|\varphi| = \omega_\xi$. Let $\varphi_n = u_n|\varphi_n|$ and $\varphi = u|\varphi|$ be the polar decompositions. We have then

$$\begin{aligned} |\|\xi\|^2 - (u_n^* u\xi|\xi)| &= |\|\varphi\| - \varphi(u_n^*)| \\ &\le |\|\varphi\| - \varphi_n(u_n^*)| + |\varphi_n(u_n^*) - \varphi(u_n^*)| \\ &\le |\|\varphi\| - \|\varphi_n\|| + \|\varphi_n - \varphi\| \le 2\|\varphi_n - \varphi\|. \end{aligned}$$

Hence we get $\|\xi\|^2 = \lim_{n\to\infty} (u_n^* u\xi|\xi)$. It follows that

$$\begin{aligned} \|u_n^* u\xi - \xi\|^2 &= \|u_n^* u\xi\|^2 + \|\xi\|^2 - 2\mathrm{Re}(u_n^* u\xi|\xi) \\ &\le 2(\|\xi\|^2 - \mathrm{Re}(u_n^* u\xi|\xi)) \to 0. \end{aligned}$$

For any $x \in \mathscr{M}$, we have

$$\begin{aligned} ||\varphi|(x) - |\varphi_n|(x)| &= ||\varphi|(x) - \varphi_n(xu_n^*)| \\ &\le ||\varphi|(x) - \varphi(xu_n^*)| + |\varphi(xu_n^*) - \varphi_n(xu_n^*)| \\ &\le |(x\xi|\xi) - (xu_n^* u\xi|\xi)| + \|\varphi - \varphi_n\|\|x\| \\ &\le \|x^*\xi\|\|\xi - u_n^* u\xi\| + \|\varphi - \varphi_n\|\|x\| \\ &\le (\|\xi - u_n^* u\xi\| + \|\varphi - \varphi_n\|)\|x\|. \end{aligned}$$

Thus we have $\||\varphi_n| - |\varphi|\| \to 0$ as $n \to \infty$. Q.E.D.

Proposition 4.11. *Let $\{\varphi_i\}$ be a net in the conjugate space A^* of a C^*-algebra A converging to $\varphi \in A$ in the $\sigma(A^*,A)$-topology. If, in addition $\|\varphi_i\|$ converges to $\|\varphi\|$, then the net $|\varphi_i|$ converges to $|\varphi|$ in the $\sigma(A^*,A)$-topology.*

PROOF. Since bounded subsets of A^* are relatively $\sigma(A^*,A)$-compact, it suffices to show that every subset of $\{|\varphi_i|\}$ converges to $|\varphi|$ in the $\sigma(A^*,A)$-topology if it converges. Suppose that $\{|\varphi_j|\}$ is a subnet converging to ψ. Since $|\varphi_j| \ge 0$, we have $\psi \ge 0$. The inequality in Proposition 4.6,

$$|\langle x,\varphi_j\rangle|^2 \le \|\varphi_j\|\langle xx^*,|\varphi_j|\rangle, \qquad x \in A,$$

yields the inequality

$$|\langle x,\varphi\rangle| \le \|\varphi\|\langle xx^*,\psi\rangle.$$

As A is σ-weakly dense in the universal enveloping von Neumann algebra $\tilde{A}$, the last inequality holds for every $x \in \tilde{A}$. Since $\lim\|\varphi_j\| = \|\varphi\|$, we have $\|\psi\| \le \|\varphi\|$. But, the inequality

$$\|\varphi\|^2 = |\langle u^*,\varphi\rangle|^2 \le \|\varphi\|\langle u^*u,\psi\rangle \le \|\varphi\|\|\psi\|,$$

where $\varphi = u|\varphi|$ is the polar decomposition of φ, shows the reverse inequality $\|\varphi\| \le \|\psi\|$; hence $\psi = |\varphi|$ by Proposition 4.6. Q.E.D.

Proposition 4.12. *Let $\mathscr{M}$ be a von Neumann algebra with predual $\mathscr{M}_*$. A closed convex subcone V_+ of $\mathscr{M}_*^+$ is the positive part $V \cap \mathscr{M}_*^+$ of a closed left invariant subspace V of $\mathscr{M}_*$ if and only if V_+ is hereditary in the sense that any element in $\mathscr{M}_*^+$ majorized by some element of V_+ belongs to V_+.*

Proof. Suppose that V is a closed left invariant subspace of $\mathscr{M}_*$. By Theorem 2.7, $V = \mathscr{M}_* e$ for some projection $e \in \mathscr{M}$. Given $\varphi \in \mathscr{M}_*^+$, φ belongs to V if and only if $\varphi = \varphi e$ if and only if $\varphi(1 - e) = 0$ by the Cauchy–Schwarz inequality. Hence if $0 \leq \psi \leq \varphi$ and $\varphi \in V$, then $\psi(1 - e) = 0$; so $\psi \in V$. Hence $V \cap \mathscr{M}_*^+$ is hereditary.

Suppose that V_+ is a closed convex hereditary subcone of $\mathscr{M}_*^+$. Let $V = \mathscr{M} V_+$, the set of $a\varphi$ with $a \in \mathscr{M}$ and $\varphi \in V_+$. We must prove that V is a closed left invariant subspace of $\mathscr{M}_*$ with $V \cap \mathscr{M}_*^+ = V_+$. If $\varphi \in V \cap \mathscr{M}_*^+$, then $\varphi = a\psi$ with $a \in \mathscr{M}$ and $\psi \in V_+$. By Proposition 4.8, we have $\varphi \leq \|a\|\psi$; hence $\varphi \in V_+$ by the hereditariness of V_+. Hence $V \cap \mathscr{M}_*^+ = V_+$. If $\{\varphi_n\}$ is a squence in V converging to φ in norm, then the sequence $|\varphi_n|$ converges in norm to $|\varphi|$ by Proposition 4.10. Since V is, by construction, invariant under the multiplication of $\mathscr{M}$ from the left, each $|\varphi_n|$ belongs to V, hence to V_+. Therefore, $|\varphi|$ belongs to V_+ by the closedness of V_+. Hence φ belongs to $\mathscr{M}|\varphi| \subset \mathscr{M} V_+ = V$. Thus V is closed. For any $\varphi \in V_+$, we have $[\mathscr{M}\varphi] \subset V$, where $[\mathscr{M}\varphi]$ means the closed subspace spanned by $\mathscr{M}\varphi$. But $[\mathscr{M}\varphi] = \mathscr{M}_* \mathrm{s}(\varphi)$ for any $\varphi \geq 0$ with $\mathrm{s}(\varphi)$ the support projection of φ. Hence if $\mathrm{s}(\psi) \leq \mathrm{s}(\varphi)$ for $\varphi \in V_+$ and $\psi \in \mathscr{M}_*^+$, then $\psi \in \mathscr{M}_* \mathrm{s}(\psi) \subset \mathscr{M}_* \mathrm{s}(\varphi) \subset V$. Let φ and ψ be two elements of V with polar decompositions $\varphi = u|\varphi|$ and $\psi = v|\psi|$. It follows then that $|\varphi|$ and $|\psi|$ are both in V_+; so $|\varphi| + |\psi| \in V_+$. Putting $e = \mathrm{s}(|\varphi| + |\psi|)$, we have

$$\begin{aligned}(\varphi + \psi)e &= (u|\varphi| + v|\psi|)e = u|\varphi|e + v|\psi|e \\ &= u|\varphi| + v|\psi| = \varphi + \psi,\end{aligned}$$

which means that $\varphi + \psi \in \mathscr{M}_* e = [\mathscr{M}(|\varphi| + |\psi|)] \subset V$. Thus V is an additive subset of $\mathscr{M}_*$. Therefore V is a closed left invariant subspace of $\mathscr{M}_*$ with $V \cap \mathscr{M}_*^+ = V_+$. Q.E.D.

Proposition 4.13. *Let A be a C^*-algebra. A $\sigma(A^*,A)$-closed convex subcone V_+ of A_+^* is the positive part $V \cap A_+^*$ of a $\sigma(A^*,A)$-closed left invariant subspace V of A^* if and only if V_+ is hereditary.*

Proof. Let $\tilde{A}$ be the universal enveloping von Neumann algebra of A. The "only if" part follows from the previous proposition.

Suppose that V_+ is a $\sigma(A^*,A)$-closed convex hereditary subcone of A_+^*. By the previous proposition, $\tilde{A}V_+ = V$ is a norm closed left invariant subspace of A_+^* with $V \cap A_+^* = V_+$. We must show that V is $\sigma(A^*,A)$-closed. It sufficies to show that the unit ball of V is $\sigma(A^*,A)$-closed. Let $\{\varphi_i\}$ be a net in the unit ball of V converging to φ in the $\sigma(A^*,A)$-topology. Let $\varphi_i = u_i\omega_i$ and $\varphi = u\omega$ be the polar decompositions. Choosing a subnet, we may assume that $\{\omega_i\}$ converges to ρ in the $\sigma(A^*,A)$-topology. Since $\omega_i \in V_+$, ρ belongs to V_+. For any $x \in A$, we have

$$|\langle x,\varphi_i\rangle|^2 \leq \|\varphi_i\|\langle xx^*,\omega_i\rangle \leq \langle xx^*,\omega_i\rangle$$

by Proposition 4.6, hence $|\langle x,\varphi\rangle|^2 \leq \langle xx^*,\rho\rangle$, $x \in A$. By the density of A in $\tilde{A}$, the last inequality holds for every $x \in \tilde{A}$. Thus, we get, for any $x \in A$,

$$\begin{aligned} |\langle x,\omega\rangle|^2 &= |\langle x,u^*\varphi\rangle|^2 = |\langle xu^*,\varphi\rangle| \\ &\leq \langle xu^*ux^*,\rho\rangle \leq \langle xx^*,\rho\rangle. \end{aligned}$$

Therefore, $\langle 1 - \mathrm{s}(\rho),\omega\rangle = 0$, so that $\mathrm{s}(\omega) \leq \mathrm{s}(\rho)$. Hence $\omega \in V_+$ by the previous proposition, which yields that $\varphi = u\omega \in V$. In other words, V is $\sigma(A^*,A)$-closed. Q.E.D.

5. Topological Properties of the Conjugate Space

In this section, we shall study operator algebras as Banach spaces. In particular, we shall establish a strong similarity between $C(K)$-spaces, abelian C^*-algebras, and noncommutative operator algebras, which will allow us to consider some part of the theory of operator algebras as noncommutative integration. The Arens–Mackey topology in a von Neumann algebra will be also characterized on bounded parts.

Suppose $\mathcal{M}$ is a von Neumann algebra with predual $\mathcal{M}_*$. Let $\mathcal{M}^* = \mathcal{M}_* \oplus \mathcal{M}_*^\perp$ be the decomposition of the conjugate space $\mathcal{M}^*$ of $\mathcal{M}$ into the normal part $\mathcal{M}_*$ and the singular part $\mathcal{M}_*^\perp$ according to Theorem 2.14. For each $\varphi \in \mathcal{M}^*$, let $\varphi = \varphi^n + \varphi^s$ denote the decomposition of φ into the normal part φ^n and the singular part φ^s,

Theorem 5.1. *If a sequence $\{\varphi_k\}$ in $\mathcal{M}^*$ converges to $\varphi \in \mathcal{M}^*$ in the $\sigma(\mathcal{M}^*,\mathcal{M})$-topology, then the normal part $\{\varphi_k^n\}$ and the singular part $\{\varphi_k^s\}$ of $\{\varphi_k\}$ converge to the normal part φ^n and the singular part φ^s of φ, respectively, in the $\sigma(\mathcal{M},\mathcal{M})$-topology.*

Before going into the proof, we fix a notation, which we have used once already. Every continuous linear functional ω on a C^*-algebra is uniquely decomposed, by Theorem 4.2, into a linear combination of four positive linear functionals,

$$\omega = \omega_1 - \omega_2 + i(\omega_3 - \omega_4),$$

where ω_1 and ω_2 (resp. ω_3 and ω_4) have orthogonal support projections in the universal enveloping von Neumann algebra. Let $[\omega]$ denote the positive linear functional defined by $[\omega] = \omega_1 + \omega_2 + \omega_3 + \omega_4$. If $[\omega](x^*x) = 0$, then $\omega(x) = \omega(x^*x) = 0$.

PROOF. Considering $\{\varphi_k - \varphi\}$, we may assume that $\{\varphi_k\}$ converges to zero. By the uniform boundedness theorem, $\{\varphi_k\}$ is bounded; hence so are $\{\varphi_k^n\}$ and $\{\varphi_k^s\}$. To prove our assertion, it is sufficient to show that $\{\varphi_k^n\}$ converges to zero in the $\sigma(\mathcal{M}^*,\mathcal{M})$-topology. By the spectral decomposition theorem and by the boundedness of $\{\varphi_k^n\}$, we have only to prove that $\lim_{k\to\infty} \varphi_k^n(p) = 0$ for every projection $p \in \mathcal{M}$.

Define $\omega = \sum_{k=1}^{\infty} (1/2^k)[\varphi_k^s]$, which converges in norm by the boundedness of $\{\varphi_k^s\}$. Then ω is singular. Let p be a fixed nonzero projection in $\mathcal{M}$. Note that if $\omega(q) = 0$ for any projection $q \in \mathcal{M}$, then $\varphi_k^s(q) = 0$ for every $k = 1,2,\ldots$. Let $\{p_i\}_{i\in I}$ be a maximal family of orthogonal projections in $\mathcal{M}$ such that $\omega(p_j) = 0$ and $p_i \le p$, $i \in I$. By Theorem 3.8, we have $p = \sum_{i\in I} p_i$. Define a finitely additive set function Δ_k on the subsets of I by

$$\Delta_k(J) = \varphi_k\left(\sum_{i\in J} p_i\right)$$

for each subset J of I. The sequence $\{\Delta_k\}$ is bounded by the boundedness of $\{\varphi_k\}$ and $\lim_{k\to\infty} \Delta_k(J) = 0$ for each subset J of I. Hence by Theorem 1.28, we have

$$\begin{aligned} 0 = \lim_{k\to\infty} \sum_{i\in I} |\Delta_k(\{i\})| &= \lim_{k\to\infty} \sum_{i\in I} |\varphi_k(p_i)| \\ &= \lim_{k\to\infty} \sum_{i\in I} |\varphi_k^n(p_i)|. \end{aligned}$$

Therefore, we have

$$\begin{aligned} \lim_{k\to\infty} \varphi_k^n(p) &= \lim_{k\to\infty} \varphi_k^n\left(\sum_{i\in I} p_i\right) \\ &= \lim_{k\to\infty} \sum_{i\in I} \varphi_k^n(p_i) = 0. \end{aligned}$$

Q.E.D.

Corollary 5.2. *The predual $\mathcal{M}_*$ of a von Neumann algebra $\mathcal{M}$ is weakly sequentially complete.*

PROOF. Let $\{\varphi_k\}$ be a sequence in $\mathcal{M}_*$ which is weakly Cauchy. Then $\{\varphi_k\}$ converges to some $\varphi \in \mathcal{M}^*$ in the $\sigma(\mathcal{M}^*,\mathcal{M})$-topology. By Theorem 5.1, $\{\varphi_k^s\}$ converges to φ^s. But $\varphi_k^s = 0$ for each $k = 1,2,\ldots$; hence $\varphi^s = 0$, which means that φ is in $\mathcal{M}_*$. Q.E.D.

Proposition 5.3. *If ω is a faithful normal positive linear functional on a von Neumann algebra $\mathcal{M}$, then the σ-strong (resp. σ-strong*) topology in the unit ball S of $\mathcal{M}$ is metrized by*

$$d(x,y) = \omega((x-y)^*(x-y))^{1/2}$$
$$(\text{resp. } d^*(x,y) = \omega((x-y)^*(x-y) + (x-y)(x-y)^*)^{1/2}).$$

The metric d (resp. d^*) on S is complete.

PROOF. Let $\{\pi,\mathfrak{H},\xi_0\}$ be the cyclic representation of $\mathscr{M}$ induced by ω. Then by assumption, π is faithful and $\{\pi(\mathscr{M}),\mathfrak{H}\}$ is a von Neumann algebra on $\mathscr{M}$ by Proposition 3.12. Therefore, we may assume that $\mathscr{M}$ acts on $\mathfrak{H}$, and $[\mathscr{M}\xi_0] = \mathfrak{H}$, and ω is given by

$$\omega(x) = (x\xi_0|\xi_0), \qquad x \in \mathscr{M}.$$

The projection p onto $[\mathscr{M}'\xi_0]$ is in $\mathscr{M}$ and $\omega(1 - p) = 0$, so that $p = 1$, that is, $[\mathscr{M}'\xi_0] = \mathfrak{H}$.

It is clear that if a net $\{x_i\}_{i\in I}$ in $\mathscr{M}$ converges σ-strongly (resp. σ-strongly*) to x, then $d(x_i,x) \to 0$ (resp. $d^*(x_i,x) \to 0$). Suppose a sequence $\{x_n\}$ in S converges to x in the d-metric (resp. d^*-metric) topology. Then we have

$$\lim_{n\to\infty} \|(x_n - x)\xi_0\| = 0$$

$$\left(\textit{resp.} \lim_{n\to\infty} (\|(x_n - x)\xi_0\| + \|(x_n - x)^*\xi_0\|) = 0\right).$$

Hence, for every $a \in \mathscr{M}'$, we have

$$\begin{aligned}\lim_{n\to\infty} \|(x_n - x)a\xi_0\| &= \lim_{n\to\infty} \|a(x_n - x)\xi_0\| \\ &\leq \|a\| \lim_{n\to\infty} \|(x_n - x)\xi_0\| = 0,\end{aligned}$$

which means that

$$\lim_{n\to\infty} \|(x_n - x)\xi\| = 0 \quad \text{for every} \quad \xi \in \mathscr{M}'\xi_0.$$

But $\mathscr{M}'\xi_0$ is dense in $\mathfrak{H}$ and $\{x_n\}$ is bounded. Hence $\{x_n\}$ converges strongly to x. The same arguments assure the σ-strong* convergence of $\{x_n\}$ to x if $d^*(x_n,x) \to 0$.

We now show the completeness of (S,d). The metric space (S,d) is isometric to $S\xi_0$ under the map: $x \in \mathscr{M} \mapsto x\xi_0$. Hence we must show that $S\xi_0$ is complete. But $S\xi_0$ is the image of S under the continuous map with respect to the σ-weak topology in $\mathscr{M}$ and the weak topology in $\mathfrak{H}$, so that $S\xi_0$ is weakly compact. Clearly $S\xi_0$ is convex. Therefore, $S\xi_0$ is closed in the norm topology in $\mathfrak{H}$; thus it is complete, so the completeness of S follows. Similarly, the completeness of (S,d^*) follows. Q.E.D.

Theorem 5.4. *For a subset K of the predual $\mathscr{M}_*$ of a von Neumann algebra $\mathscr{M}$, the following statements are all equivalent:*

(i) *K is relatively $\sigma(\mathscr{M}_*,\mathscr{M})$-compact.*
(ii) *The restriction $K|_{\mathscr{A}}$ of K to each maximal abelian *-subalgebra $\mathscr{A}$ of $\mathscr{M}$ is relatively $\sigma(\mathscr{A}_*,\mathscr{A})$-compact.*
(iii) *K is bounded and for any decreasing sequence $\{p_n\}$ of projections in $\mathscr{M}$ with $\inf p_n = 0$, $\lim_{n\to\infty} \varphi(p_n) = 0$ uniformly for $\varphi \in K$.*

(iv) *K is bounded and there exists an $\omega \in \mathcal{M}_*^+$ with the property that for any $\varepsilon > 0$ there exists $\delta > 0$ such that $|\varphi(a)| < \varepsilon$ for every $\varphi \in K$ if $\omega(a^*a + aa^*) < \delta$ and $\|a\| \leq 1$.*

(v) *K is bounded and for any increasing net $\{p_i\}_{i \in I}$ of projections in $\mathcal{M}$, $\lim_{i \in I} \varphi(p_i)$ exists uniformly for $\varphi \in K$.*

(vi) *K is bounded and if $\{p_i\}_{i \in I}$ is an increasing net of projections in $\mathcal{M}$ with $\sup_{i \in I} p_i = 1$, then $\lim_{i \in I} \|(1 - p_i)\varphi(1 - p_i)\| = 0$ uniformly for $\varphi \in K$.*

PROOF. (i)$\Leftrightarrow$(ii): Suppose K is relatively $\sigma(\mathcal{M}_*, \mathcal{M})$-compact. Take an arbitrary maximal abelian *-subalgebra $\mathcal{A}$ of $\mathcal{M}$. The map: $\varphi \in \mathcal{M}_* \mapsto \varphi|_{\mathcal{A}} \in \mathcal{A}_*$ is weakly continuous, being the transpose map of the imbedding of $\mathcal{A}$ into $\mathcal{M}$; hence the image $K|_{\mathcal{A}}$ of K under this map is relatively $\sigma(\mathcal{A}_*, \mathcal{A})$-compact.

Conversely, suppose K enjoys property (ii). Since every self-adjoint element $h \in \mathcal{M}$ is contained in some maximal abelian *-subalgebra $\mathcal{A}$ and $K|_{\mathcal{A}}$ is bounded by weak relative compactness, $\{\varphi(h) : \varphi \in K\}$ is bounded; hence for every $x \in \mathcal{M}$, $\{\varphi(x) : \varphi \in K\}$ is bounded which is seen by the decomposition $x = (1/2)(x + x^*) + (i/2i)(x - x^*)$. Hence the uniform boundedness theorem says that K is bounded. Now consider K as a subset of $\mathcal{M}^*$. Let $\tilde{K}$ denote the closure of K in the $\sigma(\mathcal{M}^*, \mathcal{M})$-topology. By the boundedness of K, $\tilde{K}$ is $\sigma(\mathcal{M}^*, \mathcal{M})$-compact. If $\tilde{K}$ lies in $\mathcal{M}_*$, then $\tilde{K}$ is nothing but the $\sigma(\mathcal{M}_*, \mathcal{M})$-closure of K; hence it is $\sigma(\mathcal{M}_*, \mathcal{M})$-compact. Therefore, we shall show that every $\varphi \in \tilde{K}$ is normal. For each maximal abelian *-subalgebra $\mathcal{A}$ of $\mathcal{M}$, $\varphi|_{\mathcal{A}}$ is in the $\sigma(\mathcal{A}^*, \mathcal{A})$-closure of the restriction $K|_{\mathcal{A}}$ of K to $\mathcal{A}$; hence by assumption, $\varphi|_{\mathcal{A}}$ is in $\mathcal{A}_*$. Therefore, φ is normal by Corollary 3.11.

(iv)$\Rightarrow$(iii): Let $\{p_n\}$ be a decreasing sequence of projections in $\mathcal{M}$ with $\inf p_n = 0$. Then $\lim_{n \to \infty} \omega(p_n) = 0$. Hence there exists an n_0 with $\omega(p_n) < \frac{1}{2}\delta$ for $n \geq n_0$, so that $|\varphi(p_n)| < \varepsilon$ for $n \geq n_0$ and $\varphi \in K$. This means $\lim_{n \to \infty} \varphi(p_n) = 0$ uniformly for $\varphi \in K$.

(iii)$\Rightarrow$(ii): Let $\mathcal{A}$ be any maximal abelian *-subalgebra of $\mathcal{M}$. Considering the restriction of K to $\mathcal{A}$, we may assume that K lies in $\mathcal{A}_*$ and enjoys property (iii) with respect to $\mathcal{A}$. To prove the relative $\sigma(\mathcal{A}_*, \mathcal{A})$-compactness of K, it is sufficient, by the Eberlein–Šmulian theorem, to show that every sequence $\{\varphi_n\}$ in K has a $\sigma(\mathcal{A}_*, \mathcal{A})$-accumulation point in $\mathcal{A}_*$. Let $\omega = \sum_{n=1}^{\infty} (1/2^n)[\varphi_n]$. By the boundedness of K, the summation converges. Let φ be a $\sigma(\mathcal{A}^*, \mathcal{A})$-accumulation point of $\{\varphi_n\}$ in $\mathcal{A}^*$, which exists due to the boundedness of $\{\varphi_n\}$. We shall show that φ is in $\mathcal{A}_*$. We have only to show that for every orthogonal family $\{p_i\}_{i \in I}$ of projections, $\varphi(\sum_{i \in I} p_i) = \sum_{i \in I} \varphi(p_i)$. If $\omega(p) = 0$ for some projection $p \in \mathcal{A}$, then $\varphi_n(p) = 0, n = 1,2,\ldots$; hence $\varphi(p) = 0$. Since $\omega(\sum_{i \in I} p_i) = \sum_{i \in I} \omega(p_i)$, the set $\{p_i : \omega(p_i) \neq 0\}$ is countable, say $\{p_n\}$, and $\omega(\sum_{i \in I} p_i) = \omega(\sum_{n=1}^{\infty} p_n) = \sum_{n=1}^{\infty} \omega(p_n)$. Therefore, it is sufficient to prove only that $\varphi(\sum_{n=1}^{\infty} p_n) = \sum_{n=1}^{\infty} \varphi(p_n)$. Let $q_n = \sum_{k=n}^{\infty} p_k$, $n = 1,2,\ldots$. Then $\{q_n\}$ is a decreasing sequence of projections with $\inf q_n = 0$. By assumption, $\lim_{n \to \infty} \varphi_k(q_n) = 0$ uniformly for $k = 1,2,\ldots$, that is, for any $\varepsilon > 0$, there exists n_0 such that $|\varphi_k(q_n)| < \varepsilon$ for $n \geq n_0$, that is, $\lim_{n \to \infty} \varphi(q_n) = 0$.

Therefore, by the equality

$$\begin{aligned}\varphi\left(\sum_{k=1}^{\infty} p_k\right) &= \varphi\left(\sum_{k=1}^{n} p_k\right) + \varphi(q_{n+1})\\ &= \sum_{k=1}^{n} \varphi(p_k) + \varphi(q_{n+1}),\end{aligned}$$

we have

$$\varphi\left(\sum_{k=1}^{\infty} p_k\right) = \sum_{k=1}^{\infty} \varphi(p_k)$$

as desired.

(vi) $\Rightarrow$ (v) and (v) $\Rightarrow$ (iii): It is clear.

(iv) $\Rightarrow$ (vi): Choose ω by assumption (iv). For $\varepsilon > 0$, let δ be a positive number in assumption (iv). By the normality of ω, there exists an index $i_0 \in I$ such that $\omega(1 - p_i) < \delta/2$ for $i \geq i_0$. For every $a \in (1 - p_i)\mathcal{M}(1 - p_i)$ with $\|a\| \leq 1$, we have $\omega(a^*a + aa^*) \leq 2\omega(1 - p_i) < \delta$, so that $|\varphi(a)| < \varepsilon$ for every $\varphi \in K$. Therefore, we get

$$\begin{aligned}\|(1 - p_i)\varphi(1 - p_i)\| &= \sup\{|\langle a,(1 - p_i)\varphi(1 - p_i)\rangle| : \|a\| < 1\}\\ &= \sup\{|\varphi((1 - p_i)a(1 - p_i))| : \|a\| \leq 1\}\\ &\leq \varepsilon\end{aligned}$$

for $i \geq i_0$.

To prove the implication (i) $\Rightarrow$ (iv), we need some preparations.

Lemma 5.5. *Let $\{\varphi_n\}$ be a sequence in $\mathcal{M}_*$ converging to $\varphi_0 \in \mathcal{M}_*$ in the $\sigma(\mathcal{M}_*,\mathcal{M})$-topology. If a sequence $\{a_n\}$ in S converges σ-strongly* to zero, then*

$$\lim_{n\to\infty} \varphi_k(a_n) = 0 \quad \textit{uniformly for} \quad k = 1,2,\ldots.$$

Proof. Observing that $\{\varphi_n\}$ is bounded by the uniform boundedness theorem, put $\varphi = \sum_{n=1}^{\infty} (1/2^n)[\varphi_n]$. Let e be the support projection of φ. Then $\varphi_n(a) = \varphi_n(eae)$ for every $a \in \mathcal{M}$ and $n = 1,2,\ldots$. Hence restricting our attention to $e\mathcal{M}e$, we may assume that φ is faithful. Considering the cyclic representation induced by φ, we may assume that $\mathcal{M}$ acts on a Hilbert space $\mathfrak{H}$ with a vector ξ_0 such that $\varphi(x) = (x\xi_0|\xi_0)$, $x \in \mathcal{M}$.

Define a metric d in S by $d(a,b) = \varphi((b - a)^*(b - a))^{1/2}$, $a,b \in S$. Then by Proposition 5.3, the strong topology in S is given by the metric d. Take and fix any $\varepsilon > 0$. Define the set

$$S_n = \{a \in S : |\varphi_k(a) - \varphi_0(a)| \leq \varepsilon \quad \text{for} \quad k \geq n\}.$$

Then S_n is closed in (S,d) and $S = \bigcup_{n=1}^{\infty} S_n$. By the Baire category theorem, some S_{k_0} has an interior point a_0, that is, there exists a $\delta > 0$ such that $d(a,a_0) < \delta$ implies $|\varphi_k(a) - \varphi_0(a)| \leq \varepsilon$ for $k \geq k_0$. Since $\{a_n\}$ converges strongly* to zero, there exists, by Lemma II.4.12, a sequence $\{p_n\}$ of projections converging strongly to 1 such that $\|p_n(a_n^*a_n + a_na_n^*)p_n\| < \varepsilon^2$ for $n = 1,2,\ldots$. Put $\psi_k = \varphi_k - \varphi_0$. We have

$$\begin{aligned}|\psi_k(a_n)| &\leq |\psi_k(p_na_np_n)| + |\psi_k((1-p_n)a_np_n)| \\ &\quad + |\psi_k(p_na_n(1-p_n))| + |\psi_k((1-p_n)a_n(1-p_n))| \\ &\leq 3\varepsilon\|\psi_k\| + |\psi_k((1-p_n)a_n(1-p_n))|.\end{aligned}$$

Put $b_n = p_na_0p_n + (1-p_n)a_n(1-p_n)$. Then b_n is in S, and we have

$$\begin{aligned}d(b_n,a_0) &= \|\{(1-p_n)a_0p_n + p_na_0(1-p_n) + (1-p_n)(a_0-a_n)(1-p_n)\}\xi_0\| \\ &\leq 3\|(1-p_n)\xi_0\| + \|(1-p_n)a_0p_n\xi_0\|.\end{aligned}$$

Since $p_n \to 1$ strongly and the product operation in S is jointly strongly continuous, we can choose an n_0 such that

$$\|(1-p_n)\xi_0\| < \delta/4 \quad \text{and} \quad \|(1-p_n)a_0p_n\xi_0\| < \delta/4$$

for $n \geq n_0$. Hence for $n \geq n_0$, we have $d(b_n,a_0) < \delta$, so that for $k \geq k_0$ and $n \geq n_0$, we have

$$\varepsilon \geq |\psi_k(b_n)| = |\psi_k(p_na_0p_n) + \psi_k((1-p_n)a_n(1-p_n))|.$$

Also, we have, for $n \geq n_0$,

$$\begin{aligned}d(p_na_0p_n,a_0) &= \|\{(1-p_n)a_0p_n + p_na_0(1-p_n) + (1-p_n)a_0(1-p_0)\}\xi_0\| \\ &\leq 2\|(1-p_n)\xi_0\| + \|(1-p_n)a_0p_n\xi_0\| < \delta;\end{aligned}$$

hence

$$|\psi_k(p_na_0p_n)| \leq \varepsilon \quad \text{for} \quad k \geq k_0 \quad \text{and} \quad n \geq n_0.$$

Therefore, we have

$$|\psi_k((1-p_n)a_n(1-p_n))| \leq 2\varepsilon \quad \text{for} \quad k \geq k_0 \quad \text{and} \quad n \geq n_0.$$

Thus we get

$$|\psi_k(a_n)| \leq 3\varepsilon\|\psi_k\| + 2\varepsilon = (3\|\psi_k\| + 2)\varepsilon$$

for $k \geq k_0$ and $n \geq n_0$. Since $\lim_{n\to\infty} \psi_k(a_n) = 0$ for k $= 1,2,\ldots,k_0$, we have $\lim_{n\to\infty} \psi_k(a_n) = 0$ uniformly for $k = 1,2,\ldots$. Q.E.D.

Lemma 5.6. *Let K be a relatively $\sigma(\mathcal{M}_*,\mathcal{M})$-compact subset of $\mathcal{M}_*$. For any $\varepsilon > 0$, there exists a finite subset K_ε of K and $\delta > 0$ such that if a is in S and $[\varphi](a^*a + aa^*) < \delta$ for every $\varphi \in K_\varepsilon$, then $|\varphi(a)| < \varepsilon$ for every $\varphi \in K$.*

Proof. Suppose our assertion is false for some $\varepsilon > 0$. We may assume that K is contained in the unit ball of $\mathcal{M}_*$. Then by induction, we can find sequences $\{\varphi_n\}$ in K and $\{a_n\}$ in S such that

$$|\varphi_{n+1}(a_n)| \geq \varepsilon,$$

$$[\varphi_k](a_n^*a_n + a_na_n^*) < 1/2^n, \qquad k = 1,2,\ldots,n.$$

Let $\omega = \sum_{n=1}^{\infty} (1/2^n)[\varphi_n]$ and $e = s(\omega)$, the support projection of ω. Then we have

$$\begin{aligned}\omega(a_n^*a_n + a_na_n^*) &= \sum_{k=1}^{\infty} \frac{1}{2^k}[\varphi_k](a_n^*a_n + a_na_n^*)\\ &\leq \sum_{k=1}^{n} \frac{1}{2^k}[\varphi_k](a_n^*a_n + a_na_n^*) + \sum_{k=n+1}^{\infty} \|\varphi_k\|/2^{k-1}\\ &\leq \frac{1}{2^n} + \frac{1}{2^{n-1}} \to 0 \quad \text{as} \quad n \to \infty,\end{aligned}$$

where the above estimate is verified by the fact $\|[\psi]\| \leq 2\|\psi\|$ for any $\psi \in \mathscr{M}_*$. Therefore, $\{ea_ne\}$ converges σ-strongly* to zero by Proposition 5.3. On the other hand, by the Eberlein–Šmulian theorem, $\{\varphi_k\}$ has a $\sigma(\mathscr{M}_*,\mathscr{M})$-convergent subsequence $\{\varphi_{k_j}\}$. Then by Lemma 5.5,

$$0 = \lim_{n\to\infty} \varphi_{k_j}(ea_ne) = \lim_{n\to\infty} \varphi_{k_j}(a_n) \quad \text{uniformly for} \quad j = 1,2,\ldots.$$

But by the choice of $\{\varphi_n\}$ and $\{a_n\}$, we have

$$|\varphi_{k_j}(a_{k_j-1})| \geq \varepsilon, \qquad j = 1,2,\ldots,$$

which is a contradiction. Q.E.D.

THE PROOF OF (i) $\Rightarrow$ (iv) IN THEOREM 5.4. We choose $\delta_n > 0$ and a finite subset $K_n = \{\varphi_1^n, \ldots, \varphi_{m_n}^n\}$ for $\varepsilon_n = 1/n$ by Lemma 5.6. Put

(i) $\Rightarrow$ (iv)
$$\omega = \sum_{n=1}^{\infty} \frac{1}{2^n} \sum_{k=1}^{m_n} \frac{1}{2^k}[\varphi_k^n].$$

It is now clear that ω is a desired positive linear functional in $\mathscr{M}_*$. Q.E.D.

As we have seen in Section 3, the predual $\mathscr{M}_*$ of a von Neumann algebra $\mathscr{M}$ is uniquely determined by the algebraic structure, and so is the σ-weak topology, being the $\sigma(\mathscr{M},\mathscr{M}_*)$-topology. There is another locally convex topology on $\mathscr{M}$ which is canonically associated with the duality system $\{\mathscr{M}_*,\mathscr{M}\}$ in such a way that any locally convex topology on $\mathscr{M}$ which gives the conjugate space $\mathscr{M}_*$ is coarser than that locally convex topology. This locally convex topology is given as the uniform convergence topology on $\sigma(\mathscr{M}_*,\mathscr{M})$-compact convex subsets of $\mathscr{M}_*$, which is denoted by $\tau(\mathscr{M},\mathscr{M}_*)$, and called the *Arens–Mackey topology*. Then Theorem 5.4 derives the following characterization of the Arens–Mackey topology.

Theorem 5.7. *In a von Neumann algebra $\mathscr{M}$, the Arens–Mackey topology $\tau(\mathscr{M},\mathscr{M}_*)$ coincides with the σ-strong* topology on bounded parts of $\mathscr{M}$.*

PROOF. Since the conjugate space of $\mathscr{M}$ with respect to the σ-strong* topology is the predual $\mathscr{M}_*$, the $\tau(\mathscr{M},\mathscr{M}_*)$-topology is finer than the σ-strong* topology. Hence we have only to show that if a bounded net $\{a_i\}$ converges σ-strongly to zero, then $\{a_i\}$ converges to zero in the $\tau(\mathscr{M},\mathscr{M}_*)$-topology. Let K be a $\sigma(\mathscr{M}_*,\mathscr{M})$-compact subset of $\mathscr{M}_*$. Let ω be the positive linear functional in $\mathscr{M}_*$ defined in Theorem 5.4(iv). Then $\omega(a_i^*a_i + a_ia_i^*)$ tends to 0, so that $|\varphi(a_i)|$ tends to 0 uniformly for $\varphi \in K$; hence $\{a_i\}$ converges to zero in the $\tau(\mathscr{M},\mathscr{M}_*)$-topology. Q.E.D.

Proposition 5.8. *Let $\mathscr{M}$ be a von Neumann algebra. Then the $\sigma(\mathscr{M}^*,\mathscr{M})$-closure of any countable subset of $\mathscr{M}_*^\perp$ is contained in $\mathscr{M}_*^\perp$ again.*

PROOF. Let $\{\varphi_n\}$ be an arbitrary countable subset of $\mathscr{M}_*^\perp$. Put

$$\omega = \sum_{n=1}^{\infty} \frac{1}{2^n\|\varphi_n\|}[\varphi_n].$$

Then $\omega(p) = 0$ implies $\varphi_n(p) = 0$, $n = 1,2,\ldots$ for any projection $p \in \mathscr{M}$. Since $\mathscr{M}_*^\perp$ is closed in the norm topology, ω is in $\mathscr{M}_*^\perp$. Let φ be a $\sigma(\mathscr{M}^*,\mathscr{M})$-accumulation point of $\{\varphi_n\}$. Let $\varphi = \varphi^n + \varphi^s$ be the decomposition of φ into the normal part and the singular part. We shall show that $\varphi^n(p) = 0$ for every projection $p \in \mathscr{M}$.

Let p be a projection in $\mathscr{M}$. Choose an increasing net $\{p_i\}_{i\in I}$ converging σ-strongly to p with $\omega(p_i) = \varphi^s(p_i) = 0$. The existence of such a net $\{p_i\}$ follows from Theorem 3.8. Then we have

$$\begin{aligned}\varphi^n(p) &= \lim_i \varphi^n(p_i) = \lim_i (\varphi^n(p_i) + \varphi^s(p_i))\\ &= \lim_i \varphi(p_i).\end{aligned}$$

But $\varphi_k(p_i) = 0$ for $k = 1,2,\ldots$ and $i \in I$ since $\omega(p_i) = 0$. Hence $\varphi(p_i) = 0$ for every $i \in I$, so that $\varphi^n(p) = 0$. Therefore $\varphi = \varphi^s$ lies in $\mathscr{M}_*^\perp$. Q.E.D.

We have seen many similarities between the noncommutative operator algebras and the theory of integration on locally compact spaces. Now one can guess that a noncommutative analogue of the (l^1)-space might be the predual $\mathscr{L}_*(\mathfrak{H})$ of the von Neumann algebra $\mathscr{L}(\mathfrak{H})$ of all bounded operators on a Hilbert space $\mathfrak{H}$. Actually, there are a lot of similarities between the duality systems $\{(c),(l^1),(l^\infty)\}$ and $\{\mathscr{LC}(\mathfrak{H}),\mathscr{L}_*(\mathfrak{H}),\mathscr{L}(\mathfrak{H})\}$. As in Corollary 1.29, the (l^1)-space has a very special property as a Banach space. We shall now examine this property of $\mathscr{L}_*(\mathfrak{H})$.

Suppose $\mathfrak{H}$ is a separable Hilbert space of infinite dimension. Let $\{\xi_n\}$ be a complete normalized orthogonal system in $\mathfrak{H}$. Put $\varphi_n(x) = (x\xi_1|\xi_n)$, $x \in \mathscr{L}(\mathfrak{H})$. Since $\|x\xi_1\|^2 = \sum_{n=1}^\infty |(x\xi_1|\xi_n)|^2 = \sum_{n=1}^\infty |\varphi_n(x)|^2$, we have

$$\lim_{n\to\infty} \varphi_n(x) = 0, \qquad x \in \mathscr{L}(\mathfrak{H}),$$

that is, $\{\varphi_n\}$ converges to 0 weakly. Define an operator v_n by $v_n\xi = (\xi|\xi_1)\xi_n$ for $\xi \in \mathfrak{H}$. Then $\|v_n\| = 1$ and $\omega_n(v_n) = 1$; hence $\|\varphi_n\| = 1$. Therefore, $\{\varphi_n\}$ does not converge to 0 in the norm topology. Thus, in $\mathscr{L}_*(\mathfrak{H})$, the weak convergence and the norm convergence for sequences do not coincide, while they are the same in the (l^1)-space.

Furthermore, we define an operator u_n by $u_n\xi = (\xi|\xi_n)\xi_1$ for $\xi \in \mathfrak{H}$. Then $u_n = v_n^*$, and

$$\varphi_n(x) = (x\xi_1|\xi_n) = (xu_n\xi_n|\xi_n);$$

hence the functional $\omega_n(x) = (x\xi_n|\xi_n)$, is the absolute value of φ_n in the sense of the polar decomposition. Define a projection e_n by $e_n\xi = (\xi|\xi_n)\xi_n$, $\xi \in \mathfrak{H}$. Then we have $\omega_n(e_n) = 1$. Hence $\omega_n(e_k)$ does not converge to 0 uniformly for $n = 1,2,\dots$ as $k \to \infty$, so that by Theorem 5.4(iii), $\{\omega_n\}$ is not relatively $\sigma(\mathcal{M}_*,\mathcal{M})$-compact. On the other hand, the absolute value $|\varphi_n^*|$ of φ_n^* is $\varphi_1(=\omega_1)$; hence the set of the absolute values of φ_n^* is reduced to one point, which is, of course, compact in any topology.

With the above example in mind, we generalize Corollary 1.29. To do so, we need the following definition, which is the noncommutative version of the (l^∞)-algebra.

Definition 5.9. A von Neumann algebra $\mathcal{M}$ is called *atomic* if every nonzero projection in $\mathcal{M}$ majorizes a nonzero minimal projection. A projection in a von Neumann algebra $\mathcal{M}$ is said to be of *finite rank* in $\mathcal{M}$ if it is the sum of finitely many orthogonal minimal projections in $\mathcal{M}$.

Proposition 5.10. *Let $\mathcal{M}$ be an atomic von Neumann algebra. Suppose a sequence $\{\varphi_n\}$ in $\mathcal{M}_*$ converges to φ in $\sigma(\mathcal{M}_*,\mathcal{M})$-topology. If the sets $\{|\varphi_n|\}$ and $\{|\varphi_n^*|\}$ are both relatively $\sigma(\mathcal{M}_*,\mathcal{M})$-compact, then we have*

$$\lim_{n\to\infty} \|\varphi - \varphi_n\| = 0.$$

Proof. Set $K = \{|\varphi|,|\varphi^*|,|\varphi_n|,|\varphi_n|: n = 1,2,\dots\}$. Then K is relatively $\sigma(\mathcal{M}_*,\mathcal{M})$-compact. By assumption, we can choose an increasing net $\{e_i\}_{i\in I}$ of projections of finite rank in $\mathcal{M}$ with s-lim $e_i = 1$. Multiplying by a scalar, we may assume $\|\varphi_n\| \le 1$, $n = 1,2,\dots$. By Theorem 5.4(vi), for $\varepsilon > 0$ there exists an index $i_0 \in I$ such that $\|(1 - e_i)\varphi(1 - e_i)\| < \varepsilon$ for every $\varphi \in K$ and $i \ge i_0$. Let $e = e_{i_0}$. Since e is of finite rank, $e\mathcal{M}e$ is a finite dimensional algebra over $\mathbf{C}$; hence $e\mathcal{M}_*e$ admits only one topology compatible with the vector space structure. Hence there exists an integer $n_0 > 0$ such that $\|e(\varphi - \varphi_n)e\| < \varepsilon$ for every $n \ge n_0$. Thus, we have, for each $a \in S$ and $n \ge n_0$,

$$\begin{aligned} |(\varphi - \varphi_n)(a)| &\le |(\varphi - \varphi_n)(eae)| + |(\varphi - \varphi_n)(ea(1 - e))| \\ &\quad + |(\varphi - \varphi_n)((1 - e)ae)| \\ &< \varepsilon + |\varphi^*((1 - e)a^*e)| + |\varphi_n^*((1 - e)a^*e)| \\ &\quad + |\varphi((1 - e)a)| + |\varphi_n((1 - e)a)|. \end{aligned}$$

Now, let $\varphi_n = u_n|\varphi_n|$ be the polar decomposition of φ_n. Then by the Cauchy–Schwarz inequality, we have

$$\begin{aligned} |\varphi_n((1 - e)a)| &= \||\varphi_n|((1 - e)au_n)| \\ &\le |\varphi_n|(1 - e)^{1/2}\{|\varphi_n|(u_n^*a^*au_n)\}^{1/2} \\ &\le \||\varphi_n\||\varphi_n|(1 - e)^{1/2} \le \varepsilon^{1/2}. \end{aligned}$$

Similarly, we have

$$|\varphi((1-e)a)| \le \varepsilon^{1/2}, \qquad |\varphi_n^*((1-e)ae)| \le \varepsilon^{1/2},$$
$$|\varphi^*((1-e)ae)| \le \varepsilon^{1/2}.$$

Combining these estimates, we have

$$|(\varphi - \varphi_n)(a)| < \varepsilon + 4\varepsilon^{1/2}, \qquad a \in S, \qquad n \ge n_0;$$

hence we get

$$\lim_{n\to\infty} \|\varphi - \varphi_n\| = 0. \qquad \text{Q.E.D.}$$

Corollary 5.11. *Under the same assumption as in Proposition 5.10, if a sequence $\{\varphi_n\}$ in the positive part $\mathcal{M}_*^+$ of $\mathcal{M}_*$ converges to φ weakly, then we have* $\lim_{n\to\infty} \|\varphi - \varphi_n\| = 0$.

As seen in the above, the set of the absolute values of all elements in a relatively $\sigma(\mathcal{M}_*,\mathcal{M})$-compact subset K of $\mathcal{M}_*$ is not necessarily relatively $\sigma(\mathcal{M}_*,\mathcal{M})$-compact. But the following result shows a special aspect of the positive part $\mathcal{M}_*^+$ of $\mathcal{M}_*$.

Proposition 5.12. *If K is a relatively $\sigma(\mathcal{M}_*,\mathcal{M})$-compact subset of $\mathcal{M}_*^+$, then $\{a\varphi : \varphi \in K, a \in S\}$ is also relatively $\sigma(\mathcal{M}_*,\mathcal{M})$-compact.*

Proof. By assumption, K is bounded. Let $\{p_n\}$ be a decreasing sequence in $\mathcal{M}$ converging σ-strongly to zero. Then by the Cauchy–Schwarz inequality, we have, for every $\varphi \in S$ and $a \in S$,

$$|a\varphi(e_n)| = |\varphi(e_n a)| \le \varphi(e_n)^{1/2}\varphi(a^*a)^{1/2}$$
$$\le \|\varphi\|\varphi(e_n)^{1/2};$$

hence by the relative $\sigma(\mathcal{M}_*,\mathcal{M})$-compactness of K and by Theorem 5.4(iii),

$$\lim_{n\to\infty} a\varphi(e_n) = 0 \quad \text{uniformly for} \quad a \in S \text{ and } \varphi \in K,$$

so that $\{a\varphi : a \in S, \varphi \in K\}$ is also relatively $\sigma(\mathcal{M}_*,\mathcal{M})$-compact by Theorem 5.4. Q.E.D.

Exercises

1. Let $\mathcal{N}$ be a von Neumann subalgebra of a von Neumann algebra $\mathcal{M}$.
   (a) Show that every $\varphi \in \mathcal{N}_*^+$ admits a normal extension $\tilde{\varphi} \in \mathcal{M}_*^+$. (*Hint*: Represent $\mathcal{M}$ on a Hilbert space $\mathfrak{H}$. Then there exists a sequence $\{\xi_n\}$ in $\mathfrak{H}$ such that $\sum_{n=1}^{\infty} \|\xi_n\|^2 < +\infty$ and $\varphi = \sum_{n=1}^{\infty} \omega_{\xi_n}$. The functional $\tilde{\varphi} = \sum_{n=1}^{\infty} \omega_{\xi_n}$ on $\mathcal{M}$ is indeed a normal extension of φ. Note that $\|\varphi\| = \varphi(1) = \sum_{n=1}^{\infty} \|\xi_n\|^2$.)
   (b) Show that every $\varphi \in \mathcal{N}_*$ admits an extension $\tilde{\varphi} \in \mathcal{M}_*$ such that $\|\varphi\| = \|\tilde{\varphi}\|$. (*Hint*: Use the polar decomposition $\varphi = u|\varphi|$ and then apply (a) to $|\varphi|$.)
2. Let $\mathcal{M}$ be a von Neumann algebra containing an infinite orthogonal sequence $\{e_n\}$ of projections. Show that the Arens–Mackey topology $\tau(\mathcal{M},\mathcal{M}_*)$ is strictly finer than the σ-strong* topology. (*Hint*: Set $A = \{\sqrt{n}e_n : n = 1,2,\ldots\}$. The σ-strong* closure of A contains 0, but the $\tau(\mathcal{M},\mathcal{M}_*)$-closure of A does not.)

3. Let A be a C^*-algebra with universal enveloping von Neumann algebra $\tilde{A}$.
   (a) Show that the left multiplication (resp. the right multiplication) operator on A by each $a \in A$ is weakly compact in the sense that aS (resp. Sa), where S is the unit ball of A, is relatively $\sigma(A,A^*)$-compact if and only if A is an ideal of $\tilde{A}$. Such a C^*-algebra A is called *weakly compact*.
   (b) Show that if A is weakly compact, then every closed left (resp. right) ideal $\mathfrak{m}$ is of the form $\mathfrak{m} = Ae$ (resp. $\mathfrak{m} = eA$) with some projection $e \in \tilde{A}$.
   (c) For each closed left (resp. right) ideal $\mathfrak{m}$ (resp. $\mathfrak{n}$), set $\mathfrak{m}^\perp = \{x \in A : \mathfrak{m}x = \{0\}\}$ (resp. $\mathfrak{n}^\perp = \{y \in A : y\mathfrak{n} = \{0\}\}$. Show that if A is weakly compact, then $\mathfrak{m}^{\perp\perp} = \mathfrak{m}$ and $\mathfrak{n}^{\perp\perp} = \mathfrak{n}$ for every closed left (resp. right) ideal $\mathfrak{m}$ (resp. $\mathfrak{n}$) of A. A C^*-algebra with the latter property is called *dual*.
   (d) Show that in a dual C^*-algebra A a closed left (resp. right) ideal $\mathfrak{m}$ (resp. $\mathfrak{n}$) is maximal if and only if $\mathfrak{m}^\perp$ (resp. $\mathfrak{n}^\perp$) is a minimal right (resp. left) ideal.
   (e) Show that if A is dual, then every nonzero closed left (resp. right) ideal of A contains a nonzero minimal left (resp. right) ideal of A. (*Hint*: Use the fact that every closed right (resp. left) ideal is contained in a maximal closed right (resp. left) ideal.)
   (f) Show that if A is dual, then $\tilde{A}$ is atomic (see Definition 5.9). (*Hint*: Consider a closed left ideal $\mathfrak{m} = \{x \in A : xe = x\}$ of A for each projection $e \in \tilde{A}$ and use (e).)
   (g) Show that if A is dual, then every projection of finite rank in $\tilde{A}$ falls in A.
   (h) Show that if A is dual, then the net $\{e_i\}$ of all projections of finite rank in $\tilde{A}$ is an approximate identity of A. (*Hint*: Consider the closed left ideal $\mathfrak{m} = \{x \in A : \lim_i \|xe_i\| = 0\}$ and show that $\mathfrak{m}^\perp = \{0\}$.)
   (k) Show that if A is dual, then A is an ideal of $\tilde{A}$. Thus A is weakly compact if and only if A is dual.

4. Let $\{A_i : i \in I\}$ be a family of C^*-algebras. Let A be the subset of all those elements $x = \{x_i\}$ in $\prod_{i \in I} A_i$ such that for any $\varepsilon > 0$, $\{i \in I : \|x_i\| \geq \varepsilon\}$ is finite. Consider the coordinatewise algebraic operations in A and the supnorm.
   (a) Show that A is a C^*-algebra. We say that A is the *$C(\infty)$-direct sum* of $\{A_i : i \in I\}$.
   (b) Show that the $C(\infty)$-direct sum of dual C^*-algebras is dual.
   (c) Show that the quotient C^*-algebra of a dual C^*-algebra is dual.
   (d) Show that if $\{\pi_i : i \in I\}$ is a family of disjoint representations of a dual C^*-algebra A, then $(\sum_{i \in I}^{\oplus} \pi_i)(A)$ is isomorphic to the $C(\infty)$-direct sum of $\{\pi_i(A) : i \in I\}$. (*Hint*: $\{(\sum_{i \in I}^{\oplus} \pi_i)(A)\}'' = \sum_{i \in I}^{\oplus} \pi_i(A)'' = \mathscr{M}$ due to the disjointness of $\{\pi_i\}$ and $(\sum_{i \in I}^{\oplus} \pi_i)(A)$ is an ideal of $\mathscr{M}$.)
   (e) Show that every dual C^*-algebra is isomorphic to the $C(\infty)$-direct sum of a family of the C^*-algebras of compact operators on Hilbert spaces.

6. Semicontinuity in the Universal Enveloping von Neumann Algebra*

In general, the universal enveloping von Neumann algebra $\tilde{A}$ of a C^*-algebra A is, unfortunately, too large in practice, although it provides a very convenient frame for theoretical treatment of the algebra A, as we have already seen. In other words, $\tilde{A}$ contains too many elements. For example, let $A = C[0,1]$. Then the algebra $l^\infty[0,1]$ of *all* bounded functions on $[0,1]$

is a direct summand of $\tilde{A}$. But the algebra $l^\infty[0,1]$ says very little about the compact space $[0,1]$ apart from the cardinality of $[0,1]$. Therefore, it is desirable to have certain criteria which enable us to measure how closely related to the original algebra A a given element in $\tilde{A}$ is. For example, the smallest projection $e \in \tilde{A}$ with $ea = a$ for some $a \in A$, the range projection of a, is certainly very closely related to A. We study this question by means of semicontinuity. When the C^*-algebra A in question is not unital, an element $a \in \tilde{A}$ with the property that $aA \subset A$, $Aa \subset A$ is also related to A in some sense. We will characterize such an element.

We begin with elementary general facts about convex sets in a locally convex real vector space. Suppose E is a locally convex real vector space. We of course assume the separation axiom for E. Let E^* be the conjugate space. Suppose K is a convex subset of E. A function f on K is called *affine* if

$$f(\lambda x + (1 - \lambda)y) = \lambda f(x) + (1 - \lambda)f(y), \qquad x,y \in K; 0 < \lambda < 1.$$

We denote by $\mathscr{A}(K)$ the set of all continuous real valued affine functions on K. If L is a convex subset of E containing K, then $\mathscr{A}(K; L)$ denotes the space of all continuous affine functions on K which is the restriction of a function in $\mathscr{A}(L)$. Obviously, $\mathscr{A}(K; L)$ is a subspace of $\mathscr{A}(K)$. It is clear that $\mathscr{A}(K)$ is a real vector space with the obvious structure:

$$(f + g)(x) = f(x) + g(x), \qquad f,g \in \mathscr{A}(K); x \in K,$$
$$(\lambda f)(x) = \lambda f(x), \qquad \lambda \in \mathbf{R}.$$

Since every affine function f on E with $f(0) = 0$ is linear, we have $\mathscr{A}(E) = E^* \oplus \mathbf{R}$. A real valued function f on a convex set K is said to be *convex* if

$$f(\lambda x + (1 - \lambda)x) \leq \lambda f(x) + (1 - \lambda)f(y), \qquad x,y \in K, \qquad 0 \leq \lambda \leq 1.$$

If $-f$ is convex, then f is said to be *concave.*

Lemma 6.1. *If f is a real valued lower semicontinuous convex function on a compact convex set K in E, then we have, for any $x \in K$,*

$$f(x) = \sup\{a(x) : a \in \mathscr{A}(K; E), a \leq f\}.$$

From now on, we assume the compactness of K unless otherwise mentioned.

PROOF. Consider the direct sum $E \oplus \mathbf{R}$. We set

$$M = \{(x,\alpha) \in E \oplus \mathbf{R} : x \in K, f(x) \leq \alpha\}.$$

It follows from assumption that M is a closed convex subset of $E \oplus \mathbf{R}$. Let x be any point in K. Let $\beta < f(x)$. Then $(x,\beta) \notin M$. By the Hahn–Banach separation theorem, there exists a closed hyperplane H in $E \oplus \mathbf{R}$ which separates strictly (x,β) and M. Since H separates $(x,f(x)) \in M$ and (x,β), H is not of the form $H_1 \oplus \mathbf{R}$ with H_1 a hyperplane in E. Hence it is the graph of a continuous affine function g on E, and the open half-spaces associated with

H are $\{(y,\lambda):\lambda < g(y)\}$ and $\{(y,\mu):\mu > g(y)\}$. Hence we have $\beta < g(x) < f(x)$ and $g(y) < f(y)$ for every $y \in K$. Q.E.D.

Lemma 6.2. *If f is a real valued lower semicontinuous affine function on K, then there exists an increasing net $\{f_i\}_{i \in I}$ in $\mathscr{A}(K; E)$ with $f(x) = \lim_i f_i(x)$, $x \in E$.*

PROOF. By Lemma 6.1, it suffices to show that the set $\{g \in \mathscr{A}(K; E): g < f\} = I$ is directed upward, where $g < f$ means that $g(x) < f(x)$ for every $x \in K$. Let g_1 and g_2 be two elements in I. Being lower semicontinuous on a compact set K, f is bounded below. We may assume, by translating by a constant, that f, g_1, and g_2 are positive. Put

$$M = \{(y,\lambda): y \in K, f(y) \le \lambda\} \subset E \oplus \mathbf{R},$$
$$M_i = \{(y,\lambda): y \in K, 0 \le \lambda \le g_i(y)\} \subset E \oplus \mathbf{R},$$

for $i = 1,2$. The assumption $g_1, g_2 < f$ implies that

$$M \cap (M_1 \cup M_2) = \emptyset.$$

The affinity of f implies that

$$M \cap \operatorname{co}(M_1 \cup M_2) = \emptyset,$$

where $\operatorname{co}(M_1 \cup M_2)$ means the convex hull of $M_1 \cup M_2$. Since M_1 and M_2 are both compact and convex, $\operatorname{co}(M_1 \cup M_2)$ is compact. Hence we can find a closed hyperplane in $E \oplus \mathbf{R}$ which separates M from $\operatorname{co}(M_1 \cup M_2)$, and which is the graph of a continuous affine function g on E such that

$$g_1, g_2 < g < f \text{ on } K. \qquad \text{Q.E.D.}$$

Corollary 6.3. *If f is a real valued affine continuous function on K, then there exists an increasing sequence $\{f_n\}$ in $\mathscr{A}(K; E)$ such that $\{f_n\}$ converges uniformly to f.*

PROOF. The assertion follows immediately from Lemma 6.2 and Dini's theorem. Q.E.D.

Lemma 6.4. *If f is a real valued lower semicontinuous affine function on K, then f takes its smallest value at an extreme point of K.*

PROOF. For each $\lambda \in \mathbf{R}$, we set $K_\lambda = \{x \in K: f(x) \le \lambda\}$. Then K_λ is a closed convex subset of K. Let $\lambda_0 = \inf f(x)$. It follows that $K_{\lambda_0} = \bigcap_{\lambda > \lambda_0} K_\lambda$ and $K_\lambda \ne \emptyset$ for $\lambda > \lambda_0$. By compactness, $K_{\lambda_0} \ne \emptyset$. It is easy to see that an extreme point of K_{λ_0} is also extreme in K. Q.E.D.

A convex subset F of K is called a *face* of K if $\lambda x + (1 - \lambda)y \in F$ for any $x,y \in K$ and $0 < \lambda < 1$ implies that x and y are both in F. Two faces F and G of K are called *complementary split faces* if $F \cap G = \emptyset$ and any $x \in K$ has a unique expression $x = \lambda y + (1 - \lambda)z$, with $y \in F$, $z \in G$, and $0 \le \lambda \le 1$. There is then a unique affine function e on K with $e(F) = 1$ and $e(G) = 0$. We assume that e is lower semicontinuous. Hence G is a closed split face.

Lemma 6.5. *In the above situation, let a be a bounded lower semicontinuous affine function on F. For a bounded affine extension $\bar{a}$ of a to K, there exists a net $\{a_i\}$ in $\mathscr{A}(K; E)$ such that $a_i \leq \bar{a}$ and $\lim_i a_i(x) = a(x)$ for every $x \in F$.*

PROOF. Let M be the closure in $E \oplus \mathbf{R}$ of

$$\{(x,\lambda) \in E \oplus \mathbf{R}: x \in K, \bar{a}(x) \leq \lambda\}.$$

Let $\{x_1, \ldots, x_n\}$ be a finite subset of F and $\varepsilon > 0$. Let L be the convex hull of the finite set $\{(x_k, \bar{a}(x_k) - \varepsilon): 1 \leq k \leq n\}$. Obviously, L is compact. Suppose $(x,\lambda) \in M \cap L$. Then $x = \sum_{k=1}^n \alpha_k x_k$, a convex combination, and $\lambda = \sum_{k=1}^n \alpha_k(\bar{a}(x_k) - \varepsilon) = \bar{a}(x) - \varepsilon$. There exists a net $\{(x_i,\lambda_i)\}$ converging to (x,λ) such that $\bar{a}(x_i) \leq \lambda_i$. By assumption, $x_i = \beta_i y_i + (1 - \beta_i)z_i$ with $y_i \in F$, $z_i \in G$ and $0 \leq \beta_i \leq 1$. Noticing that $x \in F$, we have

$$1 = e(x) \leq \liminf e(x_i) = \liminf \beta_i \leq \limsup \beta_i \leq 1,$$

so that $\lim \beta_i = 1$. Hence $\lim y_i = \lim x_i = x$; so we have

$$a(x) \leq \liminf a(y_i) = \liminf \bar{a}(x_i) \leq \liminf \lambda_i = \lambda,$$

which contradicts the equality $\lambda = \bar{a}(x) - \varepsilon = a(x) - \varepsilon$. Therefore, $M \cap L = \emptyset$. By the Hahn–Banach separation theorem, there exists a closed hyperplane H which separates strictly M and L. Since $(x_k, a(x_k)) \in M$ and $(x_k, a(x_k) - \varepsilon) \in L$, H is the graph of a continuous affine function g on E. Since $(x,\lambda) \in M$ implies $(x, \lambda + \alpha) \in M$ for any $\alpha \geq 0$, we have

$$M \subset \{(x,\lambda): g(x) < \lambda\}, \quad \text{and} \quad L \subset \{(x,\lambda): g(x) > \lambda\}.$$

Therefore, we have $g(x) \leq \bar{a}(x)$, $x \in K$, and

$$\bar{a}(x_k) - \varepsilon < g(x_k), \qquad k = 1,2, \ldots, n.$$

Let $a_{\{x_1, \ldots, x_n, \varepsilon\}} = g$. Then the net $\{a_i\}$ with $i = (x_1, \ldots, x_n, \varepsilon)$, $x_1, \ldots, x_n \in F$ and $\varepsilon > 0$, has the required property. Q.E.D.

Corollary 6.6. *If G is a singleton, say $\{x_0\}$, and a is a bounded affine function on K such that $a(x_0) = 0$ and $a|_F$ is lower semicontinuous, then a is the pointwise limit of a net $\{a_i + \alpha_i e\}$ of affine functions such that each a_i is continuous, $a_i(x_0) = 0$, and $a_i + \alpha_i e \leq a$.*

PROOF. By the previous lemma, there is a net $\{b_i\}$ in $\mathscr{A}(K; E)$ with $b_i \leq a$ and $\lim b_i(x) = a(x)$ for any $x \in F$. Put $\alpha_i = b_i(x_0)$ and $a_i = b_i - \alpha_i$. Then $a_i(x_0) = 0$. Since $(a_i + \alpha_i e)|_F = b_i|_F \leq a|_F$, and $(a_i + \alpha_i e)(x_0) = 0 = a(x_0)$, we conclude that $a_i + \alpha_i e \leq a$. Since $a|_{F \cup \{x_0\}}$ is the pointwise limit of $\{(a_i + \alpha_i e)|_{F \cup \{x_0\}}\}$, it follows that a is the pointwise limit of $(a_i + \alpha_i e)$ on the whole K. Q.E.D.

We now return to the study of a C^*-algebra A. We take the self-adjoint part A_h^* of the conjugate space A^* of A equipped with the $\sigma(A_h^*, A_h)$-topology as a locally convex real vector space E in the above discussion. We recall here that A_h^* is precisely the conjugate space of the real Banach space A_{h}, the self-adjoint part of A. Next, we take, as a compact convex set K in E, the set $\mathfrak{Q}$ of all positive linear functional with norm ≤ 1, that is,

$$\mathfrak{Q} = \{\omega \in A_+^* : \|\omega\| \leq 1\}. \tag{1}$$

If A contains an identity, we may take the state space $\mathfrak{S}$ instead of $\mathfrak{Q}$. However, if A is not unital, then $\mathfrak{S}$ is not $\sigma(A_h^*, A_h)$-compact. In general, $\mathfrak{Q}$ is the direct sum of $\mathfrak{S}$ and $\{0\}$ in the sense that every nonzero $\omega \in \mathfrak{Q}$ is uniquely written in the form

$$\omega = \lambda\omega_1, \qquad 0 \leq \lambda \leq 1, \qquad \omega_1 \in \mathfrak{S}.$$

Every $x \in A_h$ is naturally regarded as a continuous function on $\mathfrak{Q}: \omega \in \mathfrak{Q} \mapsto \langle x, \omega \rangle \in \mathbf{R}$. We denote this function by ξ_x in order to avoid possible confusion. Since $\|x\| = \sup\{|\langle x, \omega \rangle| : \omega \in \mathfrak{Q}\}$, the correspondence $\xi : x \in A_h \mapsto \xi_x \in C_{\mathbf{R}}(\mathfrak{Q})$ is an isometry. Obviously, $\xi_x \in \mathscr{A}(\mathfrak{Q}; A_h)$ and $\xi_x(0) = 0$. By Corollary 6.3, the image $\xi(A_h)$ is exactly the Banach space $\mathscr{A}_0(\mathfrak{Q})$ of all real valued continuous affine functions on $\mathfrak{Q}$ vanishing at 0. We call $\mathfrak{Q}$ the *quasi-state space* of A, and denote it sometimes by $\mathfrak{Q}(A)$.

We now consider the universal enveloping von Neumann algebra $\tilde{A}$ of A, which is the second conjugate space of A as a Banach space. Obviously, every $x \in \tilde{A}_h$ gives rise to a bounded affine function on $\mathfrak{Q}: \omega \in \mathfrak{Q} \mapsto \langle x, \omega \rangle \in \mathbf{R}$ vanishing at 0, which will be denoted by ξ_x again. Furthermore, we have

$$\|x\| = \sup\{|\langle x, \omega \rangle| : \omega \in \mathfrak{Q}\} = \|\xi_x\|;$$

hence $\xi : x \in \tilde{A}_h \mapsto \xi_x$ is an isometry of $\tilde{A}_{\mathrm{h}}$ into the Banach space of all bounded real valued affine functions on $\mathfrak{Q}$ vanishing at 0.

Lemma 6.7. *The map ξ is surjective. In other words, every bounded real valued affine function f on $\mathfrak{Q}$ vanishing at 0 is of the form $f = \xi_x$ for some $x \in \tilde{A}_h$.*

PROOF. We first extend f to the positive come A_+^* by $f(\varphi) = \|\varphi\| f(\varphi/\|\varphi\|)$ for nonzero $\varphi \in A_+^*$. Since $f(0) = 0$, this extension is consistent with the original f on $\mathfrak{Q}$. For any $\varphi, \psi \in A_+^*$, we have, since the norm on A_+^* is additive,

$$\begin{aligned}
f(\varphi + \psi) &= \|\varphi + \psi\| f\left(\frac{\varphi + \psi}{\|\varphi + \psi\|}\right) \\
&= \|\varphi + \psi\| f\left(\frac{\|\varphi\|}{\|\varphi + \psi\|} \frac{1}{\|\varphi\|} \varphi + \frac{\|\psi\|}{\|\varphi + \psi\|} \frac{1}{\|\psi\|} \psi\right) \\
&= \|\varphi + \psi\| \frac{\|\varphi\|}{\|\varphi + \psi\|} f\left(\frac{1}{\|\varphi\|} \varphi\right) + \|\varphi + \psi\| \frac{\|\psi\|}{\|\varphi + \psi\|} f\left(\frac{1}{\|\psi\|} \psi\right) \\
&= \|\varphi\| f\left(\frac{1}{\|\varphi\|} \varphi\right) + \|\psi\| f\left(\frac{1}{\|\psi\|} \psi\right) = f(\varphi) + f(\psi).
\end{aligned}$$

Hence f is additive on A_+^*. Making use of the fact that $A_h = A_+^* - A_+^*$, we extend f, in the obvious way, to the whole space A_h^*. It is easy to verify that the extended f is real linear on A_h^*. Since $\mathfrak{Q} - \mathfrak{Q}$ contains the unit ball of A_h^*, f is bounded, so that f is an element of A_h^{**}; therefore f defines an element $x \in \tilde{A}_h$ with $\langle x,\omega\rangle$, $\omega \in A_h^*$. Q.E.D.

Therefore, $\tilde{A}_h$ is also represented as the real Banach space $\tilde{\mathscr{A}}_0(\mathfrak{Q})$ of all bounded real valued affine functions on $\mathfrak{Q}$ vanishing at 0. For a subset M of A_h, M^m (resp. M_m) denotes the set of the σ-strong limits of all bounded increasing (resp. decreasing) nets in M. It follows that $M_m = -(-M)^m$. If M is a linear subspace of $\tilde{A}_h$, then M^m and M_m are both additive and invariant under the multiplication by a positive scalar. We denote by $\bar{M}$ the uniform closure of M. If $1 \notin A$, then the C^*-algebra A_{I}, obtained by adjunction of an identity to A, is regarded as the C^*-subalgebra of $\tilde{A}$ generated by A and the identity 1 of $\tilde{A}$. We should note, however, that $\tilde{A} \neq \tilde{A}_{\mathrm{I}} = \tilde{A} \oplus \mathbf{C}$.

Theorem 6.8. *For an element $x \in \tilde{A}_h$, the following four conditions are equivalent:*

(i) $x \in ((A_h)^m)^-$.
(ii) *ξ_x is lower semicontinuous on Q.*
(iii) *There is an increasing net $\{x_i + \alpha_i 1\}$ in $(A_{\mathrm{I}})_h$ with limit x such that $x_i \in A_h$, $\alpha_i \in \mathbf{R}$, and $\alpha_i \nearrow 0$.*
(iv) *$x + \varepsilon 1 \in (A_h)^m$ for every $\varepsilon > 0$.*

PROOF. If $x \in (A_h)^m$, then there exists an increasing net $\{x_i\}$ in A_h such that $\{x_i\}$ converges σ-strongly to x. Hence $\xi_x(\omega) = \langle x,\omega\rangle = \lim_i \langle x_i,\omega\rangle = \lim_i \xi_{x_i}(\omega)$, and $\{\xi_{x_i}\}$ is increasing; so ξ_x is lower semicontinuous. The uniform limit of lower semicontinuous functions is lower semicontinuous; hence ξ_x is lower semicontinuous for any $x \in ((A_h)^m)^-$.

(ii) $\Rightarrow$ (iii): By Lemma 6.2, there is an increasing net $\{a_i\}$ in $\mathscr{A}(Q)$ with $\xi_x(\omega) = \lim_i a_i(\omega)$, $\omega \in \mathfrak{Q}$. Put $\alpha_i = a_i(0)$. We see immediately that $\alpha_i \nearrow 0 = \xi_x(0)$. Let x_i be the unique element in A_h with $\xi_{x_i} = a_i - \alpha_i$. Then $\{x_i + \alpha_i 1\}$ is a net in $(A_{\mathrm{I}})_h$ such that for any $\omega \in \mathfrak{S}$

$$\omega(x_i + \alpha_i 1) = \omega(x_i) + \alpha_i = a_i(\omega);$$

hence x is the σ-strong limit of $\{x_i + \alpha_i 1\}$.

(iii) $\Rightarrow$ (iv): Choose δ with $0 < \delta < (1/3)\varepsilon$. We may assume that $\alpha_i + \delta > 0$ for all i. Let $\{u_\lambda\}_{\lambda \in \Lambda}$ be the approximate identity consisting of the all positive elements in the open unit ball of A (cf. Theorem I.7.4). For each r with $3\delta < r < \varepsilon$, put

$$y_{ir\lambda} = x_i + (\alpha_i + r)u_\lambda \in A_h.$$

We will show that $y_{ir\lambda}$ is an increasing net converging σ-strongly to $x + \varepsilon 1$. Take $y_{ir\lambda}$ and $y_{js\mu}$. Choose k such that $x_k + \alpha_k 1 \geq x_i + \alpha_i 1$ and $x_k + \alpha_k 1 \geq x_j + \alpha_j 1$. Take $t > 0$ with $\max(r,s) < t < \varepsilon$. Choose u_ν so that $u_\nu \geq u_\lambda$,

$u_\nu \geq u_\mu$ and

$$u_\nu \geq (t - r + |x_k - x_j|)^{-1}|x_k - x_i|,$$
$$u_\nu \geq (t - s + |x_k - x_j|)^{-1}|x_k - x_j|,$$

where $|y|$ denotes the absolute value $(y^*y)^{1/2}$. We have then

$$\begin{aligned} y_{kt\nu} - y_{ir\lambda} &= x_k + (\alpha_k + t)u_\nu - x_i - (\alpha_i + r)u_\lambda \\ &\geq x_k - x_i + (\alpha_k - \alpha_i + t - r)u_\nu \\ &\geq x_k - x_i + (\alpha_k - \alpha_i + t - r)(t - r + |x_k - x_i|)^{-1}|x_k - x_i| \\ &= (t - r + |x_k - x_i|)^{-1}((x_k - x_i)(t - r) \\ &\quad + (x_k - x_i + \alpha_k - \alpha_i + t - r)|x_k - x_i|) \\ &= (t - r + |x_k - x_i|)^{-1}[(x_k - x_i + |x_k - x_i|)(t - r) \\ &\quad + (x_k + \alpha_k - x_i - \alpha_i)|x_k - x_i|] \\ &\geq 0. \end{aligned}$$

Similarly, $y_{kt\nu} - y_{js\mu} \geq 0$. Hence $\{y_{ir\lambda}\}$ is an increasing net. From the construction of $\{y_{ir\lambda}\}$, it follows that $x + \varepsilon 1$ is the σ-strong limit of $\{y_{ir\lambda}\}$. Hence $x + \varepsilon 1 \in (A_h)^m$.

(iv) $\Rightarrow$ (i): $\|x + \varepsilon 1 - x\| = \varepsilon \to 0$ as $\varepsilon \to 0$; hence $x \in ((A_h)^m)^-$. Q.E.D.

Proposition 6.9. *If* $x \in (A_h^m)^- \cap \tilde{A}_+$, *then* $x + \varepsilon 1 \in (A_+)^m$ *for every* $\varepsilon > 0$. *Hence* $(A_h^m)^- \cap \tilde{A}_+ = (A_+^m)^-$.

PROOF. Suppose $x \geq 0$ in the proof of the implication (iii) $\Rightarrow$ (iv) in the previous theorem. Since $\xi_1(\omega) = \|\omega\|$, $\omega \in \mathfrak{Q}$, is lower semicontinuous, $\xi_{x_i + (\alpha_i + \delta)1}$ is lower semicontinuous by the assumption $\alpha_i + \delta > 0$. In $\mathfrak{Q} \times \mathbf{R}$, each set $M_i = \{(\omega,\alpha) : \xi_{x_i + (\alpha_i + \delta)1}(\omega) \leq \alpha \leq 2\|x\|\}$ is compact, and $\bigcap_i M_i = \{(\omega,\alpha) : \xi_{x + \delta 1}(\omega) \leq \alpha \leq 2\|x\|\} \subset \mathfrak{Q} \times [0, 2\|x\|]$. Hence eventually $\xi_{x_i + (\alpha_i + \delta)1}(\omega) \geq -\delta$ for every $\omega \in \mathfrak{S}$. Hence $x_i + (\alpha_i + 2\delta)1 \geq 0$. Consider $y_{ir\lambda} = x_i + (\alpha_i + r)u_\lambda$ for which $u_\lambda \geq (|x_i| + \delta)^{-1}|x_i|$. Then this new net $\{y_{ir\lambda}\}$ is cofinal with the original net; so it is an increasing net with σ-strong limit $x + \varepsilon 1$. We have now, for this $y_{ir\lambda}$,

$$\begin{aligned} y_{ir\lambda} &\geq x_i + (\alpha_i + 3\delta)(\delta + |x_i|)^{-1}|x_i| \\ &= (\delta + |x_i|)^{-1}(\delta x_i + |x_i|x_i + (\alpha_i + 3\delta)|x_i|) \\ &= (\delta + |x_i|)^{-1}((x_i + \alpha_i + 2\delta)|x_i| + \delta(x_i + |x_i|)) \geq 0. \end{aligned}$$

It follows that $x + \varepsilon 1 \in (A_+)^m$. Q.E.D.

Corollary 6.10. *If* $x \in [(A_h)^m]^-$, *then* $(1 + \alpha x)^{-1}x \in [(A_h)^m]^-$ *for small* $a > 0$. *If* $x \in [(A_+)^m]^-$, *then* $(1 + \alpha x)^{-1}x \in [(A_+)^m]^-$ *for every* $\alpha > 0$.

PROOF. For any $\varepsilon > 0$, there is an increasing net $\{x_i\}$ in A_h with σ-strong limit $x + \varepsilon 1$. For a small $\alpha > 0$, we have $x_i \geq -1/\alpha$, and $\{(1 + x_i)^{-1}x_i\}$ is an increasing net in A_h with σ-strong limit $(1 + \alpha\varepsilon + \alpha x)^{-1}(x + \varepsilon 1)$. With $\varepsilon \to 0$, we conclude that $(1 + \alpha x)^{-1}x \in [(A_h)^m]^-$. The latter assertion follows similarly. Q.E.D.

Proposition 6.11. *For any $x \in \tilde{A}_h$, the following three conditions are equivalent:*

(i) $x \in (A_{I,h})^m$.
(ii) $x \in \mathbf{R}1 + ((A_h)^m)^-$.
(iii) *The restriction of ξ_x to $\mathfrak{S}$ admits a bounded lower semicontinuous affine extension a to the whole space $\mathfrak{Q}$.*

PROOF. (i) ⇒ (ii): Let $\{x_i + \alpha 1\}$ be an increasing net in $A_{\mathrm{I,h}}$ with σ-strong limit x. Being a bounded increasing net, $\{\alpha_i\}$ converges to a scalar $\alpha \in \mathbf{R}$, so that $\{x_i + (\alpha_i - \alpha)1\}$ converges increasingly to $x - \alpha 1$. It follows then from Theorem 6.8(iii) that $x - \alpha 1 \in ((A_h)^m)^-$.

(ii) ⇒ (iii): Let $x = \alpha 1 + y$ with $y \in ((A_h)^m)^-$. Put $a(\omega) = \xi_y(\omega) + \alpha$. Then a is a bounded lower semicontinuous affine function with $a|_{\mathfrak{S}} = \xi_x|_{\mathfrak{S}}$.

(iii) ⇒ (i): By Lemma 6.2, there exists an increasing net $\{a_i\}$ in $\mathscr{A}(\mathfrak{Q})$ such that $a(\omega) = \lim a_i(\omega)$, $\omega \in \mathfrak{Q}$. Let x_i be a unique element in A_h with $\xi_{x_i} = a_i - a_i(0)$, and $\alpha_i = a_i(0)$. Then $\{x_i + \alpha_i 1\}$ is a net in $A_{I,h}$ such that for any $\omega \in \mathfrak{S}$,

$$\omega(x_i + \alpha_i 1) = a_i(\omega) \nearrow a(\omega) = \xi_x(\omega).$$

Hence $\{x_i + \alpha_i 1\}$ is increasing and converges σ-strongly to x. Hence $x \in (A_h)^m$.
Q.E.D.

Corollary 6.12. $[(A_h)^m + (A_h)_m]^- = [(A_{I,h})^m + (A_{I,h})_m]^-$.

PROOF. Since $1 \in (A_h)^m$, $(A_h)^m + (A_h)_m = (A_h)^m - (A_h)^m$ is a real vector space containing $\mathbf{R}1$, so that its uniform closure is a uniformly closed real vector space containing $\mathbf{R}1$ and $[(A_h)^m]^-$. By the previous proposition, it contains $(A_{I,h})^m$, which means that

$$[(A_{I,h})^m + (A_{I,h})_m]^- = [(A_h)^m + (A_h)_m]^-.$$

The reverse inclusion is obvious. Q.E.D.

Lemma 6.13. *Let $\{x_i\}$ be a net of self-adjoint operators on a Hilbert space $\mathfrak{H}$ converging weakly to x.*

(i) *If $x_i \leq x$ for all i, then $\{(1 - \alpha x_i)^{-1}\}$ converges strongly to $(1 - \alpha x)^{-1}$ whenever $\alpha x < 1$, $\alpha > 0$.*
(ii) *If $x_i \geq x \geq 0$ for all i, then $\{(1 + \alpha x_i)^{-1} x_i\}$ converges strongly to $(1 + \alpha x)^{-1} x$.*

PROOF. Since the proofs of (i) and (ii) are similar, we present a proof only for (i). Choose $\varepsilon > 0$ with $1 - \alpha x \geq \varepsilon$. Let $y_i = (1 - \alpha x)^{-1} - (1 - \alpha x_i)^{-1}$ for each i. Then $0 \leq y_i \leq 1/\varepsilon$, and we have

$$y_i = (1 - \alpha x_i)^{-1}(\alpha x - \alpha x_i)(1 - \alpha x)^{-1} = \alpha(1 - \alpha x_i)^{-1}(x - x_i)(1 - \alpha x)^{-1}.$$

Hence for any $\xi \in \mathfrak{H}$, we have

$$\|y_i\xi\|^2 \leq \frac{1}{\varepsilon}(y_i\xi|\xi) = \frac{\alpha}{\varepsilon}((1-\alpha x_i)^{-1}(x-x_i)(1-\alpha x)^{-1}\xi|\xi)$$

$$\leq \frac{\alpha}{\varepsilon}\|(x-x_i)^{1/2}(1-\alpha x_i)^{-1}\xi\|\|(x-x_i)^{1/2}(1-\alpha x)^{-1}\xi\|$$

$$= \frac{\alpha}{\varepsilon}((1-\alpha x_i)^{-1}(x-x_i)(1-\alpha x_i)^{-1}\xi|\xi)^{1/2}$$

$$\times((1-\alpha x)^{-1}(x-x_i)(1-\alpha x)^{-1}\xi|\xi)^{1/2}.$$

Since $\{x_i\}$ converges weakly to x, the second inner product tends to zero. On the other hand,

$$\begin{aligned} 0 &\leq \alpha(1-\alpha x_i)^{-1}(x-x_i)(1-\alpha x_i)^{-1} \\ &= (1-\alpha x_i)^{-1}((1-\alpha x_i)-(1-\alpha x))(1-\alpha x_i)^{-1} \\ &\leq (1-\alpha x_i)^{-1} \leq (1-\alpha x)^{-1} \leq \frac{1}{\varepsilon}, \end{aligned}$$

so that the first inner product stays bounded. Hence $\|y_i\xi\|^2$ tends to zero, that is, $(1-\alpha x_i)^{-1}$ converges strongly to $(1-\alpha x)^{-1}$. Q.E.D.

For each $\alpha > 0$, we define a function f_α on the open half-line $(-1/\alpha, +\infty)$ by

$$f_\alpha(t) = \frac{t}{1+\alpha t}, \qquad -\frac{1}{\alpha} < t < +\infty. \tag{2}$$

This family $\{f_\alpha\}$ of functions has the following useful properties:

$$f_\alpha\left(\left]-\frac{1}{\alpha}, +\infty\right[\right) = \left]-\infty, \frac{1}{\alpha}\right[; \tag{3}$$

$$f_\beta(t) \leq f_\alpha(t) \leq t \quad \text{for} \quad 0 < \alpha \leq \beta, \qquad t > -\frac{1}{\beta}; \tag{4}$$

$$\lim_{\alpha \to 0} f_\alpha(t) = t; \tag{5}$$

$$f_{\alpha+\beta}(t) = f_\alpha(f_\beta(t)), \qquad t > -\frac{1}{\alpha+\beta}; \tag{6}$$

$$f_\alpha(x) \leq f_\alpha(y) \quad \text{for any self-adjoint operators} \quad y \geq x > -\frac{1}{\alpha}. \tag{7}$$

$$f_\alpha(x) \in (A_h)^m \quad \text{for} \quad x \in (A_h)^m \quad \text{if} \quad x > -\frac{1}{\alpha}. \tag{8}$$

This follows from (7) and the fact that the map: $x \mapsto f_\alpha(x)$ is σ-strongly continuous by Theorem II.4.7.

Proposition 6.14. $[(A_h)^m + (A_h)_m]^-$ *is a uniformly closed Jordan algebra. Hence in particular, the self-adjoint part of the* C^**-subalgebra of* $\tilde{A}$ *generated by* $x \in [(A_h)^m + (A_h)_m]^-$ *is contained in* $[(A_h)^m + (A_h)_m]^-$.

PROOF. We observe first that for any $x \in \tilde{A}_h$,

$$\lim_{\alpha \to 0} \left\| x^2 - \frac{1}{\alpha}(x - f_\alpha(x)) \right\| = 0.$$

Hence if $x \in (A_h)^m$, then x^2 belongs to $[(A_h)^m - (A_h)^m]^-$. If $x = y - z$ with $y,z \in (A_h)^m$, then

$$x^2 = 2y^2 + 2z^2 - (y + z)^2 \in [(A_h)^m - (A_h)^m]^-.$$

Hence $x \in [(A_h)^m + (A_h)_m]$ implies $x^2 \in [(A_h)^m + (A_h)_m]^-$. Therefore, $[(A_h)^m + (A_h)_m]^-$ is closed under square. For any pair x,y in $[(A_h)^m + (A_h)_m]^-$, we have

$$xy + yx = (x + y)^2 - x^2 - y^2 \in [(A_h)^m + (A_h)_m]^-. \qquad \text{Q.E.D.}$$

We now study the functional representation of A_h on the state space $\mathfrak{S}$ instead of $\mathfrak{Q}$. We should remark here, however, that $\mathfrak{Q}(A)$ is canonically identified with the state space $\mathfrak{S}(A_\mathrm{I})$ of A_I. In fact, for any $\omega \in \mathfrak{Q}(A)$, put $\tilde{\omega}(a + \lambda 1) = \omega(a) + \lambda$. Then $\tilde{\omega}(1) = 1$ and

$$\begin{aligned}\tilde{\omega}((a + \lambda 1)^*(a + \lambda 1)) &= \omega(a^*a) + \bar{\lambda}\omega(a) + \lambda\overline{\omega(a)} + |\lambda|^2 \\ &\geq \omega(a^*a) + \bar{\lambda}\omega(a) + \lambda\overline{\omega(a)} + |\lambda|^2\|\omega\| \\ &= \lim_i \omega(u_i(a + \lambda 1)^*(a + \lambda 1)u_i) \geq 0.\end{aligned}$$

The map: $\omega \in \mathfrak{Q}(A) \mapsto \tilde{\omega} \in \mathfrak{S}(A_\mathrm{I})$ is an affine isomorphism. But the functional representation of $A_{\mathrm{I},h}$ on $\mathfrak{S}(A_\mathrm{I})$ and on $\mathfrak{Q}(A)$ are not coherent, due to the fact that the identity $1 \in A_\mathrm{I}$ corresponds to the constant function 1 on $\mathfrak{S}(A_\mathrm{I})$ and to the lower semicontinuous function ξ_1 on $\mathfrak{Q}(A)$, which is not a constant. This "undesirable" incoherence will be clarified when we consider ideals of A.

We now consider the functional representation of A_h on $\mathfrak{S}$ $(= \mathfrak{S}(A))$. Since $\mathfrak{S}$ and $\{0\}$ are complementary split faces of $\mathfrak{Q}$ and $\xi_1 = e$ is lower semicontinuous, $\mathfrak{S}$ satisfies the requirements of Lemma 6.5 and Corollary 6.6.

Theorem 6.15. *For an element* $x \in \tilde{A}_h$, *the following four conditions are equivalent*:

(i) $x \in [(A_{\mathrm{I},h})^m]^-$;
(ii) ξ_x *is lower semicontinuous on* $\mathfrak{S}$;
(iii) $(1 - \alpha x)^{-1} \in [(A_h)^m]^-$ *when* $\alpha x < 1$ *and* $\alpha > 0$;
(iv) $(1 - \alpha x)^{-1}x \in (A_{\mathrm{I},h})^m$ *when* $\alpha x < 1$ *and* $\alpha > 0$.

PROOF. (i) $\Rightarrow$ (ii): If $x \in A_{I,h}$, then ξ_x is continuous on $\mathfrak{S}$. Hence for any $x \in (A_{I,h})^m$, ξ_x is lower semicontinuous on $\mathfrak{S}$. Since $\xi : x \mapsto \zeta_{x|\mathfrak{S}}$ is an isometry of $\tilde{A}_h$, ξ_x is lower semicontinuous on $\mathfrak{S}$ for any $x \in [(A_{I,h})^m]^-$, being a uniform limit of lower semicontinuous functions on $\mathfrak{S}$.

(ii) $\Rightarrow$ (iii): Applying Corollary 6.6 to $\mathfrak{Q}$ and $\mathfrak{S}$ with $K = \mathfrak{Q}$, $F = \mathfrak{S}$, and $G = \{0\}$, we find a net $\{x_i\}$ in $A_{I,h}$ such that $\xi_x(\omega) = \lim_i \xi_{x_i}(\omega)$, $\omega \in \mathfrak{Q}$, and $x_i \le x$ for all i. By Lemma 6.13(i), $(1 - \alpha x_i)^{-1}$ converges σ-strongly to $(1 - \alpha x)^{-1}$ if $\alpha x < 1$ and $\alpha > 0$. Since $(1 - \alpha x_i)^{-1} \in A_{I,+} \subset A_h + \mathbf{R}_+ 1$, and since ξ_1 is lower semicontinuous on $\mathfrak{Q}$, $\xi_{(1-\alpha x_i)^{-1}}$ is lower semicontinuous. Since $\xi_{(1-\alpha x_i)^{-1}}$ converges pointwise to $\xi_{(1-\alpha x)^{-1}}$ from below, $\xi_{(1-\alpha x)^{-1}}$ is lower semicontinuous on $\mathfrak{Q}$. By Theorem 6.8, it follows that $(1 - \alpha x)^{-1} \in [(A_h)^m]^-$.

(iii) $\Rightarrow$ (iv): Since $(1 - \alpha x)^{-1} x = (1/\alpha)((1 - \alpha x)^{-1} - 1)$, we conclude, by Proposition 6.11, that $(1 - \alpha x)^{-1} x \in (A_{I,h})^m$.

(iv) $\Rightarrow$ (i): Since we have

$$\lim_{\alpha \to 0} \|(1 - \alpha x)^{-1} x - x\| = 0,$$

$x \in [(A_{I,h})^m]^-$. Q.E.D.

Corollary 6.16. $([(A_h)^m)^-]^m = [(A_h)^m]^-$ *and* $(((A_{I,h})^m)^-)^m = [(A_{I,h})^m]^-$.

PROOF. Since the elements in $[(A_h)^m]^-$ (resp. $[(A_{I,h})^m]^-$) are characterized by the lower semicontinuity of the representing functions on $\mathfrak{Q}$ (resp. $\mathfrak{S}$) and since the lower semicontinuity is preserved under the monotone increasing limit, we establish the assertions. Q.E.D.

Proposition 6.17. *For any* $x \in \tilde{A}_+$ *and* $\alpha > 0$, $\alpha 1 + x \in [(A_h)^m]^-$ *if and only if* $(\alpha 1 + x)^{-1} x \in ((A_{I,h})^m)^-$.

PROOF. Suppose $(\alpha 1 + x)^{-1} x \in ((A_{I,h})^m)^-$. By Theorem 6.15, we have

$$1 + \alpha^{-1} x = (1 - (\alpha 1 + x)^{-1} x)^{-1} \in ((A_h)^m)^-.$$

Hence $\alpha 1 + x \in ((A_h)^m)^-$.

Conversely, suppose $\alpha 1 + x \in ((A_h)^m)^-$. By Proposition 6.9, for any $\varepsilon > 0$, there is an increasing net $\{x_i\}$ in A_+ such that $\{x_i\}$ converges σ-strongly to $(\alpha + \varepsilon)1 + x$. Clearly, $1 - \alpha(\varepsilon 1 + x_i) \in A_{I,h}$, and $\{1 - \alpha(\varepsilon 1 + x_i)^{-1}\}$ is increasing to

$$1 - \alpha((\alpha + 2\varepsilon 1) + x)^{-1} = ((2\varepsilon + \alpha)1 + x)^{-1}(2\varepsilon 1 + x).$$

Hence the last expression falls in $(A_{I,h})^m$. With $\varepsilon \to 0$, we conclude that $(\alpha 1 + x)^{-1} x \in ((A_{I,h})^m)^-$. Q.E.D.

Theorem 6.18. *A projection* $p \in \tilde{A}$ *belongs to* $(A_+)^m$ *if and only if* $p \in [(A_{I,h})^m]^-$.

PROOF. Suppose $p \in [(A_{I,h})^m]^-$. For any $\varepsilon > 0$,

$$(p + \varepsilon 1)^{-1} p = (1 + \varepsilon)^{-1} p \in [(A_{I,h})^m]^-;$$

hence $p + \varepsilon 1 \in [(A_h)^m]^-$ by Proposition 6.17. With $\varepsilon \to 0$, we conclude that p belongs 0o $[(A_h)^m]^-$. By Theorem 6.8, ξ_p is lower semicontinuous on $\mathfrak{Q}$, so that $\mathfrak{Q}_p = \{\omega \in \mathfrak{Q}, \omega(p) = 0\}$ is a closed face of $\mathfrak{Q}$. Let $V_+ = \;_{\lambda \geq 0} \lambda \mathfrak{Q}_p$. Then V_+ is a convex subcone of A^*_+, which is the set of all positive φ with $\varphi(p) = 0$. Therefore, V_+ is hereditary in A^*_+. Since the intersection of V_+ with the unit ball of A^* is $\mathfrak{Q}_p$, V_+ is $\sigma(A^*,A)$-closed. By Proposition 4.13, $\tilde{A} V_+ = V$ is a $\sigma(A^*,A)$-closed left invariant subspace of A^*. By Theorem 2.7, the polar $V^0 \cap A$ of V in A is a closed right ideal $\mathscr{I}$ such that $V = \mathscr{I}^0$. Since $\varphi p = 0$, $\varphi \geq 0$, is equivalent to $\varphi(p) = 0$, we have $V = A^*(1 - p)$. Hence $V^0 = \mathscr{I}^{00} = p\tilde{A}$. Hence $p\tilde{A}$ is the σ-weak closure of the right ideal $\mathscr{I}$ of A. Let $\{v_i\}$ be an approximate identity of $\mathscr{I} \cap \mathscr{I}^*$. Then $\{v_i\}$ converges σ-strongly to the identity of $(\mathscr{I} \cap \mathscr{I}^*)^{00} = p\tilde{A}p$, which is p of course. Hence p is approximated by positive elements in $\mathscr{I}$ from below, so that $p \in (A_+)^m$.

Q.E.D.

Definition 6.19. Each projection in $(A_+)^m$ is called *open*. The orthogonal complement $1 - p$ of an open projection p is called *closed*.

From the proof of Theorem 6.18, we obtain the following characterization of open (or closed) projections.

Corollary 6.20. *For a projection p in $\tilde{A}$, the following conditions are equivalent*:

(i) *p is open.*
(ii) *There exists a closed right (resp. left) ideal $\mathscr{I}$ of A such that $p\tilde{A} = \mathscr{I}^{00}$ (resp. $\tilde{A}p = \mathscr{I}^{00}$).*
(iii) *$A^*(1 - p)$ is $\sigma(A^*,A)$-closed. Equivalently, $(1 - p)A^*$ is $\sigma(A^*,A)$-closed.*
(iv) *There exists an increasing net in A converging σ-strongly to p.*

Given a closed left (resp. right) ideal $\mathscr{I}$ of A, an open projection p in $\tilde{A}$ with $\tilde{A}_p = \mathscr{I}^{00}$ (resp. $p\tilde{A} = \mathscr{I}^{00}$) is said to *support* $\mathscr{I}$.

Proposition 6.21. *Let $\mathscr{I}$ be a closed ideal of A. If z is the open central projection in $\tilde{A}$ supporting $\mathscr{I}$, then $[(\mathscr{I}_{\mathrm{I},h})^m]^-$ is canonically identified with $[z(A_{\mathrm{I},h})^m]^-$.*

Proof. The universal enveloping von Neumann algebra $\tilde{\mathscr{I}}$ of $\mathscr{I}$ is identified with $\tilde{A}z$. Then $\mathscr{I}_{\mathrm{I}} = \mathscr{I} + \mathbf{C}z$. If $\{u_i\}_{i \in I}$ is an increasing net in $\mathscr{I}$ with limit z, then for any $x \in A_{\mathrm{I},+}$

$$zx = x^{1/2} z x^{1/2} = \lim x^{1/2} u_i x^{1/2} \in (\mathscr{I}_+)^m.$$

By Proposition 6.11, we have

$$zA_{\mathrm{I},h} = z(A_{\mathrm{I},+} + \mathbf{R}1) \subset (\mathscr{I}_{\mathrm{I},h})^m.$$

Hence $z(A_{\mathrm{I},h})^m \subset [(\mathscr{I}_{\mathrm{I},h})^m]^-$ by Corollary 6.16; so $[z(A_{\mathrm{I},h})^m]^- \subset [(\mathscr{I}_{\mathrm{I},h})^m]^-$.

Conversely, making use of Proposition 6.11 twice, we get

$$(\mathscr{I}_{\mathrm{I},h})^m = [(\mathscr{I}_h)^m]^- + \mathbf{R}z = z([(\mathscr{I}_h)^m]^- + \mathbf{R}1)$$
$$\subset z([(A_h)^m]^- + \mathbf{R}1) = z(A_{\mathrm{I},h})^m.$$

Hence it follows that $[(\mathscr{I}_{\mathrm{I},h})^m]^- = [z(A_{\mathrm{I},h})^m]^-$. Q.E.D.

Definition 6.22. Given a C^*-algebra A, an element x of the universal enveloping von Neumann algebra $\tilde{A}$ of A is said to be a *left* (resp. *right*) *multiplier* of A if $xA \subset A$ (resp. $Ax \subset A$). If x is a left and right multiplier of A, then it is called merely a *multiplier*. We denote by $M(A)$ the set of all multipliers of A.

If A contains an identity, then $M(A)$ reduces trivially to A, so that it is not interesting at all. However, if A has no identity, then $M(A)$ plays a role similar to the Stone–Čech compactification of a locally compact space in the abelian case. Obviously, $M(A)$ is a C^*-subalgebra of $\tilde{A}$ containing A as a closed ideal. We call $M(A)$ the *multiplier algebra* of A.

Definition 6.23. In a C^*-algebra B, an ideal $\mathscr{I}$ is said to be *thick* if there is no nonzero ideal $\mathscr{J}$ of B with $\mathscr{I} \cap \mathscr{J} = \{0\}$.

Since a C^*-algebra A is σ-weakly dense in $\tilde{A}$, A is a thick ideal of $M(A)$ because, if $\mathscr{I}$ is an ideal of $M(A)$ with $\mathscr{I} \cap A = \{0\}$, then $xA \subset \mathscr{I} \cap A = \{0\}$ for any $x \in \mathscr{I}$; so $x\tilde{A} = \{0\}$, and hence $x = 0$.

Theorem 6.24. *Given a C^*-algebra A, the self-adjoint part $M(A)_h$ of the multiplier algebra $M(A)$ of A coincides with $(A_{\mathrm{I},h})^m \cap (A_{\mathrm{I},h})_m$.*

Proof. Suppose x is an element in $(A_{\mathrm{I},h})^m \cap (A_{\mathrm{I},h})_m$. There exist then an increasing net $\{y_i\}_{i\in I}$ in $A_{\mathrm{I},h}$ and a decreasing net $\{y_j\}_{j\in J}$ in $A_{\mathrm{I},h}$ with $x = \lim y_i = \lim z_j$. Hence $y_i - z_j \geq 0$ and $\{y_i - z_j\}$ tends to zero along the directed set $I \times J$. Since A is an ideal of A_{I}, for any $a \in A$, $a^*y_ia - a^*z_ja$ are in A_{h}; hence $\{\xi_{a^*y_ia} - \xi_{a^*z_ja}\}$ are continuous on Q, and converge to zero pointwise from above. It follows then from Dini's theorem that

$$\|a^*y_ia - a^*z_ja\| = \sup_{\omega\in\mathfrak{Q}} |\xi_{a^*y_ia}(\omega) - \xi_{a^*z_ja}(\omega)| \to 0.$$

Hence we have

$$\begin{aligned}\|xa - y_ia\|^2 &= \|a^*(y_i - x)^2a\| \\ &\leq \|y_i - x\|\|a^*(y_i - x)a\| \\ &\leq \|y_i - x\|\|a^*(y_i - z_j)a\| \to 0,\end{aligned}$$

so that $xa \in A$. Therefore x is a left multiplier of A. But the equality $ax = (xa^*)^*$ shows that x is a right multiplier also; hence $x \in M(A)$.

Conversely, suppose that x is a multiplier of A and $x = x^*$. Replacing x by $\alpha x + \beta 1$, we may assume that $0 \leq x \leq 1$. With an approximate identity $\{u_i\}_{i \in I}$ in A, we have increasing nets $\{x^{1/2}u_i x^{1/2}\}$ and $\{(1-x)^{1/2}u_i(1-x)^{1/2}\}$ in A. Since $\{u_i\}$ converges to the identity 1 σ-strongly, we have

$$x = \lim_i x^{1/2}u_i x^{1/2} \in (A_h)^m.$$

$$1 - x = \lim_i (1-x)^{1/2}u_i(1-x)^{1/2} \in (A_h)^m.$$

Hence x belongs to $\mathbf{R}1 + (A_h)_m \subset (A_{\mathrm{I},h})_m$. Therefore x belongs to $(A_{\mathrm{I},h})^m \cap (A_{\mathrm{I},h})_m$. Q.E.D.

Proposition 6.25. *Let A and B be two C^*-algebras. If π is an isomorphism of a thick closed ideal $\mathscr{I}$ of B onto A, then π can be extended to an isomorphism of B into the multiplier algebra $M(A)$ of A.*

PROOF. Taking the universal representation of A, we assume that A acts on a Hilbert space $\mathfrak{H}$ and $\tilde{A}$ is the weak closure of A on $\mathfrak{H}$. Then π is viewed as a faithful representation of $\mathscr{I}$ such that $\pi(\mathscr{I}) = A$. By Proposition 2.16, π is extended to a representation of B, denoted also by π, on the space $\mathfrak{H}$ such that $\pi(B)$ is contained in $\pi(\mathscr{I})'' = \tilde{A}$. Since $\mathscr{I}$ is an ideal of B, $A = \pi(\mathscr{I})$ is an ideal of $\pi(B)$; hence $\pi(B) \subset M(A)$. Let $\mathscr{J}$ denote the kernel $\pi^{-1}(0)$ of π in B. Then $\mathscr{I} \cap \mathscr{J} = \{0\}$, π being faithful on $\mathscr{I}$, so that $\mathscr{J} = \{0\}$ by the thickness of $\mathscr{I}$ in B. Thus π is an isomorphism of B into $M(A)$. Q.E.D.

In order to study the structure of monotone limits in representations, we need the following elementary but important result from the theory of convex sets:

Lemma 6.26. *Let K be a compact convex set in a locally convex vector space E. Let F be a closed split face with complementary face F'. Then every lower (resp. upper) semicontinuous bounded affine function f on F is extended to a lower (resp. upper) semicontinuous bounded affine function $\hat{f}$ on K. If f is positive, then a positive extension $\hat{f}$ may be chosen.*

PROOF. Suppose that f is a lower semicontinuous bounded affine function on F with upper bound α. For each $x = \lambda y + (1-\lambda)z \in K$ with $y \in F$, $z \in F'$, $0 \leq \lambda \leq 1$, we have

$$\hat{f}(x) = \lambda f(y) + (1-\lambda)\alpha.$$

It follows that $\hat{f}$ is a bounded affine function on K. Since the upper graph of f, $\{(x,\mu) \in K \times \mathbf{R} : \hat{f}(x) \leq \mu \leq \alpha\}$, is the convex hull of two compact convex sets, $\{(x,\mu) \in F \times \mathbf{R} : f(x) \leq \mu \leq \alpha\}$ and $F \times \{\alpha\}$, in $E \oplus \mathbf{R}$, it is compact, so that $\hat{f}$ is lower semicontinuous. It is clear that if f is positive, then the above $\hat{f}$ is positive.

If f is upper semicontinuous, then we apply the above arguments to the lower bound β of f in place of α, and to the subgraph of $\hat{f}$. Q.E.D.

Let A be a C^*-algebra as before. Let $\mathscr{I}$ be a closed ideal of A with supporting open central projection z in the universal enveloping von Neumann algebra $\tilde{A}$. Naturally, the conjugate space $\mathscr{I}^*$ and the universal enveloping von Neumann algebra $\tilde{\mathscr{I}}$ $(=\mathscr{I}^{**})$ are identified with A^*z and $\tilde{A}z$, respectively. Let B denote the quotient C^*-algebra $A/\mathscr{I}$. Then the conjugate space B^* and the universal enveloping von Neumann algebra $\tilde{B}$ $(=B^{**})$ of B are also naturally identified with $A^*(1-z)$ and $\tilde{A}(1-z)$, respectively. As we have seen in Corollary 6.22, ξ_z is lower semicontinuous on $\mathfrak{Q}$ $(=\mathfrak{Q}(A))$. Put

$$F'_{\mathscr{I}} = \{\omega \in \mathfrak{Q}: \omega(z) = 1\}.$$

It follows then that $F'_{\mathscr{I}}$ is also a face of $\mathfrak{Q}$. Clearly we have $F_{\mathscr{I}} = (1-z)\mathfrak{Q}$ and $F'_{\mathscr{I}} \subset A^*z$. For any $\omega \in \mathfrak{Q}$, put $\sigma = [1/\omega(z)]z\omega$, and $\rho = (1/[1-\omega(z)])(1-z)\omega$ if $\omega(z) \neq 1$. Then $\omega = \omega(z)\sigma + (1-\omega(z))\rho$, and $\sigma \in F'_{\mathscr{I}}$ and $\rho \in F_{\mathscr{I}}$. Hence $F_{\mathscr{I}}$ is a split face of $\mathfrak{Q}(A)$ with complementary face $F'_{\mathscr{I}}$. Under the identifications of B^* with $A^*(1-z)$ and $\mathscr{I}^*$ with $A^*(1-z)$, $F_{\mathscr{I}}$ is identified with the quasi-state space $\mathfrak{Q}(B)$ of B and $F'_{\mathscr{I}}$ is identified with the state space $\mathfrak{S}(\mathscr{I})$ of $\mathscr{I}$. In the above decomposition of ω, if ω is a state, then ρ is a state of B too. Thus, we conclude the following:

Proposition 6.27. *Let $\mathscr{I}$ be a closed ideal of a C^*-algebra A with quotient algebra $A/\mathscr{I} = B$.*

(i) *The quasi-state space $\mathfrak{Q}(B)$ of B is identified with a closed split face of $\mathfrak{Q}(A)$ whose complementary face is identified with the state space $\mathfrak{S}(\mathscr{I})$ of $\mathscr{I}$.*
(ii) *Under the above identification, the state space $\mathfrak{S}(B)$ of B is a closed split face of $\mathfrak{S}(A)$ with complementary face $\mathfrak{S}(\mathscr{I})$.*

Proposition 6.28. *Let A and B be two C^*-algebras. Let π be a homomorphism of A onto B. Then the canonical extension $\tilde{\pi}$ of π to a normal homomorphism of $\tilde{A}$ onto $\tilde{B}$ enjoys the following properties:*

(i) $\tilde{\pi}([(A_h)^m]^-) = [(B_h)^m]^-$;
(ii) $\tilde{\pi}((A_{I,h})^m) = (B_{I,h})^m$;
(iii) $\tilde{\pi}([A_{I,h})^m]^-) = [(B_{I,h})^m]^-$.

Since $\tilde{\pi}$ is an open continuous map of $\tilde{A}$ onto $\tilde{B}$ in the norm topology, (iii) follows from (ii). But by Proposition 6.11 it follows that $(A_{I,h})^m = [(A_h)^m]^- + \mathbf{R}1$ and $(B_{I,h})^m = [(B_h)^m]^- + \mathbf{R}1$. Hence (ii) also follows from (i).

By the normality of $\tilde{\pi}$, we have $\tilde{\pi}((A_h)^m) \subset (B_h)^m$, so that $\tilde{\pi}([(A_h)^m]^-) = [(B_h)^m]^-$. Identifying B with the quotient algebra $A/\pi^{-1}(0)$, the quasi-state space $\mathfrak{Q}(B)$ turns out to be a closed split face of $\mathfrak{Q}(A)$. Hence every lower semicontinuous bounded affine function on $\mathfrak{Q}(B)$ is extended to such a function on $\mathfrak{Q}(A)$ by Lemma 6.26. Therefore, the reverse inclusion $[(B_h)^m]^- \subset \tilde{\pi}([(A_h)^m]^-)$ follows from Theorem 6.8. Q.E.D.

Before proceeding further with our arguments, let us examine abelian C^*-algebras so that we can see the meaning of the above discussion.

Let A be an abelian C^*-algebra with spectrum Ω. It follows that A is identified with the algebra $C_\infty(\Omega)$ of all continuous functions on Ω vanishing at infinity. The quasi-state space Q of A is the set of all positive Radon measures on Ω with total mass ≤ 1. Each point $\omega \in \Omega$ corresponds to a unique pure state of A by the evaluation of $a \in C_\infty(\Omega)$ at ω, which is identified with ω. Thus Ω is the pure state space $P(A)$ of A. The state space $\mathfrak{S}(A)$ is identified with the set of all probability measures on Ω. Suppose f is a real valued lower semicontinuous bounded function Ω. By Urysohn's lemma, there exists an increasing net $\{b_i\}_{i\in I}$ in A_+ which converges to $f + \|f\|1$ (note that -1 cannot be approximated by A from below unless Ω is compact). Let $a_i = b_i - \|f\|1 \in A_{\mathrm{I},h}$. It follows then that $\{a_i\}_{i\in I}$ is increasing and bounded, so that it converges σ-strongly to an element $a_f \in \tilde{A}_h$. By definition, a_f belongs to $(A_{\mathrm{I},h})^m$. The usual extension procedure of integration from continuous functions to lower semicontinuous functions verifies that for any $\mu \in \mathfrak{Q}(A)$,

$$\begin{aligned}\int_\Omega f(\omega)\,d\mu(\omega) &= \sup_i \int_\Omega a_i(\omega)\,d\mu(\omega)\\ &= \lim_i \mu(a_i) = \mu(a_f).\end{aligned}$$

Therefore, the affine function: $\mu \in \mathfrak{Q}(A) \mapsto \int_\Omega f(\omega)\,d\mu(\omega)$ is given by the element a_f in $(A_{\mathrm{I},h})^m$.

Conversely, let a be an element of $(A_{\mathrm{I},h})^m$.Then there exists an increasing net $\{a_i\}$ in $A_{\mathrm{I},h}$ with limit a. Let $f(\omega) = \lim_i a_i(\omega)$, $\omega \in \Omega$. Since each a_i is a continuous function on Ω converging at infinity, f is a lower semicontinuous function Ω. By the boundedness of $\{a_i\}$, f is bounded. It follows from the same arguments as above that $a = a_f$. Therefore, the correspondence: $f \leftrightarrow a_f$ is bijective. Since the set of all bounded lower semicontinuous functions on Ω is uniformly closed, and since the above correspondence is an isometry, we conclude with the uniform closedness of $(A_{\mathrm{I},h})^m$, that is, $(A_{\mathrm{I},h})^m = [(A_{\mathrm{I},h})^m]^-$. Thus, we obtain the following result:

Proposition 6.29. *If A is an abelian C^*-algebra with spectrum Ω, then $(A_{\mathrm{I},h})^m$ (resp. $(A_{\mathrm{I},h})_m$) is uniformly closed, and every element a in $(A_{\mathrm{I},h})^m$ (resp. $(A_{\mathrm{I},h})_m$) corresponds to a unique bounded lower (resp. upper) semicontinuous function f on Ω via the formula*

$$\langle a,\mu\rangle = \int_{\Omega\cdot} f(\omega)\,d\mu(\omega), \qquad \mu \in \mathfrak{Q}(A).$$

Hence an open (resp. closed) projection in $\tilde{A}$ corresponds to the characteristic function of an open (resp. closed) subset of Ω.

Theorem 6.30. *The multiplier algebra $M(A)$ of an algebra C^*-algebra A with spectrum Ω is isomorphic to the algebra $C_b(\Omega)$ of all bounded continuous* *functions Ω; hence its spectrum is homeomorphic to the Stone–Čech compactification of Ω.*

PROOF. Let π_0 be the canonical imbedding of A into $l^\infty(\Omega)$, the algebra of all bounded functions on Ω. Since $l^\infty(\Omega)$ is a von Neumann algebra, π_0 is canonically extended to a normal homomorphism $\tilde{\pi}_0$ of $\tilde{A}$ onto $l^\infty(\Omega)$. Since A is thick in $M(A)$ and π_0 is faithful, $\tilde{\pi}_0$ is faithful on $M(A)$; hence we may regard $M(A)$ as a C^*-subalgebra of $l^\infty(\Omega)$. Namely, each element $a \in M(A)$ is a bounded function on Ω. For any relatively compact open set U in Ω, there exists a function $b \in A$ with $b(\omega) = 1$ for every $\omega \in U$. Hence $a(\omega) = a(\omega)b(\omega) = (ab)(\omega)$ for any $\omega \in U$. But $ab \in A$, so that ab is continuous; hence a is continuous on U. Therefore, every function a in $M(A)$ is continuous on every relatively compact open set in Ω, which means that a is continuous, Ω being locally compact. Therefore, we have $M(A) \subset C_b(\Omega)$. Conversely, $C_\infty(\Omega) = A$ is a thick ideal of $C_b(\Omega)$, so that $C_b(\Omega) \subset M(A)$ by Proposition 6.25. Thus $M(A) = C_b(\Omega)$. Q.E.D.

We now turn to the study of a noncommutative analogue of measurable functions on a locally compact space.

Definition 6.31. Given a C^*-algebra A with universal enveloping von Neumann algebra $\tilde{A}$, we say that an element $a \in \tilde{A}_h$ is *universally measurable* if for any $\varepsilon > 0$ and $\omega \in \mathfrak{S}(A)$ there are elements $x \in (A_h)^m$ and $y \in (A_h)_m$ such that

$$y \leq a \leq x \quad \text{and} \quad \omega(x - y) < \varepsilon.$$

The set of all universally measurable elements in $\tilde{A}_h$ is denoted by M_u or $M_u(A)$.

Proposition 6.32. (i) *The set M_u is a uniformly closed real vector subspace of $\tilde{A}_h$ containing $[(A_h)^m + (A_h)_m]^-$.*

(ii) *An element $a \in \tilde{A}_h$ is universally measurable if and only if for any $\varepsilon > 0$ and $\omega \in \mathfrak{S}$ there exist elements x and y in M_u such that*

$$y \leq a \leq x \quad \textit{and} \quad \omega(x - y) < \varepsilon.$$

PROOF. Let a be an element in $\tilde{A}_h$ with the property that for any $\varepsilon > 0$ and $\omega \in \mathfrak{S}$, there are x and y in M_u such that $y \leq a \leq x$ and $\omega(x - y) < \varepsilon/3$. By definition, there exist two elements $x' \in (A_h)^m$ and $y' \in (A_h)_m$ such that $x \leq x'$, $y' \leq y$ and $\omega(x' - x) < \varepsilon/3$, $\omega(y - y') < \varepsilon/3$. We have then $y' \leq y \leq a \leq x \leq x'$, and $\omega(x' - y') < \varepsilon$. Hence a is in M_u. Since the converse statement in (ii) is trivial, assertion (ii) holds.

Suppose that a is an element in $(A_h)^m$. By definition, there exists an increasing net $\{a_i\}$ in A_h with $a = \lim a_i$. Given $\varepsilon > 0$ and $\omega \in \mathfrak{S}$, there is an index i with $\omega(a - a_i) < \varepsilon$. Put $x = a$ and $y = a_i$. It follows that x and y satisfy the required inequality for the universal measurability of a. Hence M_u contains $(A_h)^m$; in particular, $1 \in M_u$. Since $M_u = -M_u$, M_u contains $(A_h)_m$ as well.

It is clear that $\lambda M_u = M_u$ for any $\lambda > 0$, hence any real λ. Since $(A_h)^m$ and $(A_h)_m$ are both convex cones, it follows that $M_u + M_u = M_u$. Hence M_u is a real vector subspace of $\tilde{A}_h$. Let $\{a_n\}$ be a sequence in M_u with uniform limit a. For any $\varepsilon > 0$, there is an n such that $a_n - \varepsilon 1 \leq a \leq a_n + \varepsilon 1$. Let $x = a_n + \varepsilon 1$ and $y = a_n - \varepsilon 1$. We have then that $x, y \in M_u(A)$ and $\omega(x - y) < 2\varepsilon$ for any $\omega \in \mathfrak{S}$. Hence a is universally measurable from (ii), which was proven above. Hence M_u is uniformly closed. Q.E.D.

We assume from now on that the universal enveloping von Neumann algebra $\tilde{A}$ of the C^*-algebra A in question acts on a Hilbert space $\mathfrak{H}$.

Proposition 6.33. *An element $a \in \tilde{A}_h$ is universally measurable if and only if for any $\varepsilon > 0$ and $\xi \in \mathfrak{H}$ there exist $x \in (A_{I,h})^m$ and $y \in (A_{I,h})_m$ such that*

$$y \leq a \leq x \quad \textit{and} \quad \|(x - a)\xi\| + \|(a - y)\xi\| < \varepsilon.$$

Proof. By symmetry, we have only to prove the strong approximation of $a \in M_u$ by $(A_{I,h})^m$ from above. Since $(A_{I,h})^m$ and M_u are both invariant under the affine transformations: $x \mapsto \alpha x + \beta 1$, $\alpha > 0$ and $\beta \in \mathbf{R}$, we may assume that $0 \leq a \leq 1$. Let $\{f_\alpha\}$ be the family of functions defined by (2). By definition, there exists a net $\{x_i\}$ in $(A_h)^m$ converging weakly to a from above. By Lemma 6.13(ii), $\{f_\alpha(x_i)\}$ converges strongly to $f_\alpha(a)$ from above for any $\alpha > 0$. Hence $\{(1 + \alpha)f_\alpha(x_i)\}$ converges strongly to $(1 + \alpha)f_\alpha(a)$ from above. But for $0 \leq t \leq 1$, we have

$$0 \leq (1 + \alpha)f_\alpha(t) - t = \frac{\alpha(1 - t)t}{1 + \alpha} \leq \alpha,$$

so that $(1 + \alpha)f_\alpha(a) \geq a$ and $\lim_{\alpha \to 0} \|(1 + \alpha)f_\alpha(a) - a\| = 0$. Since f_α preserves the monotone convergence by the operator-monotone property of f_α and it leaves A_h invariant, $f_\alpha(x_i)$ belongs to $(A_h)^m$. Thus a is strongly approximated by the net $\{(1 + \alpha)f_\alpha(x_i) : \alpha > 0, i\}$ in $(A_h)^m$ from above, as desired. Q.E.D.

Theorem 6.34. *The space M_u is sequentially strongly closed.*

Proof. Let $\{a_n\}$ be a strongly convergent sequence in M_u with limit a. By the uniform boundedness theorem, $\{a_n\}$ is bounded; so we may assume that $\|a_n\| \leq 1/2$ for $n = 1, 2, \ldots$. Given $\varepsilon > 0$ and $\varphi \in \mathfrak{S}$, we may further assume, by choosing a subsequence, that

$$\varphi((a - a_n)^2) < \varepsilon^2/4^{n+1}.$$

Putting $b_n = a_{n+1} - a_n$, we have

$$b_n^2 \leq 2[(a_{n+1} - a)^2 + (a_n - a)^2],$$

so that

$$\varphi(|b_n|) \leq \varphi(b_n^2)^{1/2} < \varepsilon/2^n.$$

Since $b_n \in M_u$, there exists, by the previous proposition, a net $\{x_i\}$ in $\{A_{I,h}\}^m$ converging strongly to b_n from above. By Theorem II.4.7, the map: $x \mapsto |x|$ is strongly continuous in $\tilde{A}_h$, so that we can find an $x_n \in (A_{I,h})^m$ such that $b_n \le x_n$ and $\varphi(|x_n|) < \varepsilon/2^n$.

Observing that $t - \alpha/(1-\alpha) \le f_\alpha(t)$ for $|t| \le 1$ and $\|a_{n+1} - a_1\| \le 1$, we have

$$a_{n+1} - a_1 - \frac{\alpha}{1-\alpha} \le f_\alpha(a_{n+1} - a_1) = f_\alpha\left(\sum_{k=1}^{n} b_k\right)$$
$$\le f_\alpha\left(\sum_{k=1}^{n} x_k\right) \le f_\alpha\left(\sum_{k=1}^{n} |x_k|\right).$$

By Proposition 6.14 and Corollary 6.12, all $|x_k|$ belong to the uniformly closed Jordan algebra $[(A_h)^m + (A_h)_m]^-$, so that

$$x_\alpha = \lim_{n\to\infty} f_\alpha\left(\sum_{k=1}^{n} |x_k|\right)$$

belongs to the monotone sequential strong closure of $[(A_h)^m + (A_h)_m]^-$. Therefore, if M_u is monotone sequentially closed, then x_α belongs to M_u. Assume that x_α belongs to M_u. Since $f_\alpha(t) \le t$, we have

$$\varphi(x_\alpha) \le \sum_{k=1}^{\infty} \varphi(|x_k|) < \varepsilon.$$

Put $x'_\alpha = a_1 + x_\alpha + \alpha/(1-\alpha) \in M_u$. It follows that $x'_\alpha \ge a$ and

$$\varphi(x'_\alpha - a) < \varphi(a_1) + \alpha/(1-\alpha) + \varepsilon < 2\varepsilon$$

for a small α. Therefore, a is approximated weakly by M_u from above. The approximation of a from below is by making use of the sequence $\{-a_n\}$ and $-a$. Therefore, a is squeezed between elements in M_u; hence it falls in M_u by Proposition 6.32(ii).

We must show therefore that M_u is monotone sequentially strongly closed. Let $\{a_n\}$ be an increasing sequence in M_u with limit a. We may assume, by an affine transformation: $x \mapsto \alpha x + \beta 1$ with $\alpha > 0$ and $\beta \in \mathbf{R}$, that $0 \le a \le 1$ and $0 \le a_n \le 1$, $n = 1, 2, \ldots$. For any $\varepsilon > 0$ and $\varphi \in \mathfrak{S}$, we have $\omega(a - a_n) < \varepsilon$ for large n, and there is a $y \in (A_h)_m$ such that $y \le a_n$ and $\varphi(a_n - y) < \varepsilon$; hence $\varphi(a - y) < 2\varepsilon$. We have therefore only to prove the approximation of a from above. Put $b_n = a_n - a_{n-1} \ge 0$, and $a_0 = 0$. Then $b_n \in M_u$ and $a = \sum_{n=1}^{\infty} b_n$. For each n, we choose $x_n \in (A_h)^m$ such that $x_n \ge b_n$ and $\varphi(x_n - b_n) < \varepsilon/2^n$. We put

$$x_\alpha = \lim_{n\to\infty} (1+\alpha) f_\alpha\left(\sum_{k=1}^{n} x_k\right).$$

By Proposition 6.9, each x_k belongs to $[(A_+)^m]^-$. Since f_α leaves $[(A_+)^m]^-$ invariant, $f_\alpha(\sum_{k=1}^{n} x_k)$ belongs to $[(A_+)^m]^-$. By Corollary 6.16, x_α belongs to

$[(A_+)^m]^-$. We have now

$$a \leq (1+\alpha)f_\alpha(a) = \lim_{n\to\infty} (1+\alpha)f_\alpha\left(\sum_{k=1}^{n} b_k\right) \leq x_\alpha,$$

$$\begin{aligned}\varphi(x_\alpha - a) &= \lim_{n\to\infty} \varphi\left((1+\alpha)f_\alpha\left(\sum_{k=1}^{n} x_k\right) - a\right)\\ &\leq \lim_{n\to\infty} \varphi\left((1+\alpha)\sum_{k=1}^{n} x_k - a\right)\\ &\leq \lim_{n\to\infty} (1+\alpha)\left(\sum_{k=1}^{n} \varphi(b_k) + \varepsilon/2^k\right) - \varphi(a)\\ &= \alpha\varphi(a) + (1+\alpha)\varepsilon.\end{aligned}$$

Hence we get $\varphi(x_\alpha - a) < 2\varepsilon$ for small $\alpha > 0$. Hence a is approximated weakly by $[(A_+)^m]^-$ from above. Therefore, a is universally measurable. Q.E.D.

Definition 6.35. The *universal atomic representation* $\{\pi_0, \mathfrak{H}_0\}$ of a C^*-algebra A is defined by

$$\{\pi_0, \mathfrak{H}_0\} = \sum_{\omega \in P(A)}^{\oplus} \{\pi_\omega, \mathfrak{H}_\omega\}.$$

A representation π of A is said to be *atomic* if $\pi(A)''$ is an atomic von Neumann algebra in the sense of Definition 5.9.

Proposition 6.36. *The universal atomic representation is atomic. A representation is atomic if and only if it is quasi-equivalent to a direct summand of the universal atomic representation.*

PROOF. Let π_0 be the universal atomic representation of A and z_0 be the support of π_0 in $\tilde{A}$. Then $\pi_0(A)''$ is isomorphic to $\tilde{A}z_0$ by definition. Let ω be a pure state of A and e_ω be the support projection of ω in $\tilde{A}$. As in the proof of Proposition 2.16, $\omega(x)e_\omega = e_\omega x e_\omega$ for any $x \in \tilde{A}$, so that $e_\omega \tilde{A} e_\omega = \mathbf{C}e_\omega$; hence e_ω is a minimal projection in $\tilde{A}$. Conversely, if e is a minimal projection of $\tilde{A}$, then $e\tilde{A}e = \mathbf{C}e$, so that there is a normal state ω of $\tilde{A}$ such that $exe = \omega(x)e$ for every $x \in \tilde{A}$. Since $\omega(e) = 1$ and e has no proper subprojection, e must be the support projection of ω. If $0 \leq \varphi \leq \omega$, then $\varphi = \varphi(e) = \lambda$ and $\varphi(x) = \varphi(exe) = \varphi(\omega(x)e) = \omega(x)\varphi(e) = \lambda\omega(x)$; therefore ω is pure.

If ω is a pure state, then the support projection e_ω of ω in $\tilde{A}$ is majorized by the support projection z_ω of the cyclic irreducible representation π_ω of A induced by ω, so that $e_\omega \leq z_0$. Hence z_0 majorizes all minimal projections in $\tilde{A}$. Let e be a nonzero projection in $\tilde{A}$ with $e \leq z_0$. It follows that $\tilde{\pi}_\omega(e) \neq 0$ for some pure state ω of A, where $\tilde{\pi}_\omega$ denotes the cyclic representation of $\tilde{A}$ induced by ω which coincides with the canonical extension of π_ω to $\tilde{A}$ by means of Theorem 2.4. Since $\tilde{\pi}_\omega(\tilde{A}) = \mathscr{L}(\mathfrak{H}_\omega)$, there is a nonzero projection

f in $\tilde{A}$ such that $\tilde{\pi}_\omega(f) \leq \tilde{\pi}_\omega(e)$ and $\tilde{\pi}_\omega(f)$ is of rank 1. Let $e_0 = fz_\omega$ with z_ω the support of $\tilde{\pi}_\omega$. Then e_0 is a minimal projection of $\tilde{A}$ with $e_0 \leq e$ because $\tilde{\pi}_\omega$ is an isomorphism of $\tilde{A}z_\omega$ onto $\mathscr{L}(\mathfrak{H}_\omega)$. Therefore, π_0 is atomic.

Since any direct summand of $\tilde{A}z_0$ is atomic, any representation which is quasi-equivalent to a direct summand of π_0 is atomic. Let π be an atomic representation of A with support projection z in $\tilde{A}$. Let $\{e_i\}_{i\in I}$ be a maximal orthogonal family of minimal projections in $\tilde{A}z$. It follows then that $z = \sum_{i\in I} e_i$, because otherwise $z - \sum_{i\in I} e_i$ must majorize a nonzero minimal projection by definition, contradicting the maximality of $\{e_i\}_{i\in I}$. Hence $z \leq z_0$. Thus π is quasi-equivalent to a direct summand of π_0. Q.E.D.

Theorem 6.37. *The universal atomic representation π_0 of A is isometric on the class M_u of all universally measurable elements in $\tilde{A}_h$.*

Proof. Suppose $b \in M_u$ and $\varphi(b) < 0$ for some state $\varphi \in \mathfrak{S}$. It follows then that there is an $x \in (A_h)^m$ with $b < x$ such that $\varphi(x) < 0$. By Theorem 6.8, ξ_x is a lower semicontinuous affine function on $\mathfrak{Q}$, so that $\xi_x(\omega) = \omega(x) < 0$ for some pure state ω by Lemma 6.4. Thus $\pi_0(b)$ is not positive. In other words, $\pi_0(b) \geq 0$ for any $b \in M_u$ if and only if $b \geq 0$. For any $a \in M_u$, $\|\pi_0(a)\|1 \in (A_h)^m \subset M_u$, and we have $-\|\pi_0(a)\|1 \leq \pi_0(a) \leq \|\pi_0(a)\|^1$; hence

$$-\|\pi_0(a)\|1 \leq a \leq \|\pi_0(a)\|1,$$

so that $\|a\| \leq \|\pi_0(a)\|$. Thus $\|a\| = \|\pi_0(a)\|$. Q.E.D.

Corollary 6.38. *If $a \in M_u$ majorizes z_0, the support of the universal atomic representation, then a majorizes 1. Hence if $z_0 \neq 1$, then z_0 is not universally measurable.*

Theorem 6.39. *If $\{\pi, \mathfrak{H}\}$ is a representation of A with $\pi(A)''$ σ-finite, then we have*

$$\pi(M_u) = \pi(A)''_h.$$

Proof. Considering the universal representation of A, we assume that A acts on a Hilbert space $\tilde{\mathfrak{H}}$ in such a way that every state of A is of the form ω_ξ, $\xi \in \tilde{\mathfrak{H}}$. Let z be the support of π in $\tilde{A}$. Then $\{\pi, \mathfrak{H}\}$ is quasi-equivalent to the representation: $x \in A \mapsto zx \in \mathscr{L}(z\tilde{\mathfrak{H}})$. By assumption, z is a σ-finite projection in $\tilde{A}$. Therefore, there exists a normal state φ on $\tilde{A}$, by Proposition II.3.19, such that the support $s(\varphi)$ of φ is z. Let ξ be a vector in $z\tilde{\mathfrak{H}}$ such that $\varphi = \omega_\xi$. Let p be an arbitrary projection in $\tilde{A}z$. It follows from Lemma II.4.21 that there exists a projection $q \in M_u$ such that $q(1-p)\xi = 0$ and $(1-q)p\xi = 0$. Since ω_ξ is faithful on $\tilde{A}z$, ξ is a separating vector for $\tilde{A}z$; hence we have $zq(1-p) = 0$; so $zq = zp = p$. Therefore, any projection p in $\tilde{A}z$ is of the form zq for some projection $q \in M_u$.

Let x be an element in $\tilde{A}z$ with $0 \leq x \leq z$. Let $x = \int_0^1 \lambda\, de(\lambda)$ be the spectral decomposition of x on the Hilbert space $z\tilde{\mathfrak{H}}$. Let p_k be the spectral projection of x corresponding to the subset $\{\lambda \in [0,1] : \lambda_k = 1\}$ of $[0,1]$ with dyadic expansion $\lambda = \sum_{k=1}^\infty \lambda_k/2^k$, $\lambda_k = 0$ or 1. We have then $x = \sum_{k=1}^\infty 2^{-k} p_k$. Let

q_k be a projection in M_u with $q_k z = p_k$ for each $k = 1,2,\ldots$. Let $y = \sum_{k=1}^{\infty} 2^{-k} p_k \in M_u$. It follows that $yz = x$. Hence zM_u contains the positive part $z\tilde{A}_+$ of $z\tilde{A}$; thus zM_u contains the self-adjoint part $\tilde{A}_h z$. Therefore, our assertion follows. Q.E.D.

EXERCISES

1. Let A be a C^*-algebra with universal enveloping von Neumann algebra $\tilde{A}$.
   (a) Show that for any $a \in \tilde{A}$, the set $M_a = \{x \in A : ax \in A\}$ is a closed right ideal of A.
   (b) Show that for any $x \in A$, there exist $y,z \in A$ such that $x = yz$ and $\|y\| = \|x\|^{1/2}$, $\|z\| = \|x\|^{1/2}$. (*Hint*: Let $x = uh$ be the polar decomposition of x in $\tilde{A}$, apply (a) to u, and set $y = uh^{1/2}$, $z = h^{1/2}$.)

2. Let A be a C^*-algebra and $\mathscr{I}$ a closed ideal of A, and π be the canonical homomorphism of A onto $A/\mathscr{I}$.
   (a) Show that if $\dot{h} \in A/\mathscr{I}$ is self-adjoint, then there exists a self-adjoint $h \in A$ such that $\dot{h} = \pi(h)$ and $\|\dot{h}\| = \|h\|$. (*Hint*: Take any $x \in A$ with $\dot{h} = \pi(x)$, and set $k = (x + x^*)/2$; then consider the abelian C^*-algebra generated by k and the restriction of π to B, and use Exercise I.8.5.)
   (b) Show that if $\dot{x} \in A/\mathscr{I}$, then there exists $x \in A$ such that $\pi(x) = \dot{x}$ and $\|x\| = \|\dot{x}\|$. (*Hint*: Choose any $y \in A$ with $\pi(y) = \dot{x}$. Let $y = uk$ be the polar decomposition of y in the universal enveloping von Neumann algebra $\tilde{A}$. Considering the hereditary C^*-subalgebra $\overline{kAk} = B$, choose an $h \in B$ such that $\pi(h) = \pi(k)$ and $\|h\| = \|\pi(k)\|$. Note that $\{(k + \varepsilon)^{-1}k : \varepsilon > 0\}$ is an approximate identity for h, set $u_\varepsilon = u(k + \varepsilon)^{-1}k = y(k + \varepsilon)^{-1} \in A$, and show that $u_\varepsilon h$ converges in norm to an element $x \in A$ with the desired property.)

3. Let A be a C^*-algebra. Show that for each $a \in A$, the C^*-subalgebras $\overline{a^*Aa}$ and $\overline{aAa^*}$ of A are isomorphic, where the bar means closure. (*Hint*: Consider the polar decomposition $a = uh = ku$, $h = (a^*a)^{1/2}$ and $k = (aa^*)^{1/2}$ in $\tilde{A}$.)

4. A C^*-algebra A is called *primitive* if A admits a faithful irreducible representation. Let A be primitive.
   (a) Show that $xAy \neq \{0\}$ for every nonzero $x,y \in A$.
   (b) Show that every hereditary C^*-subalgebra B of A is primitive.
   (c) Show that there exists an irreducible faithful representation $\{\pi,\mathfrak{H}\}$ such that $\pi(A) \supset \mathscr{LC}(\mathfrak{H})$ if and only if A admits an element a such that $\dim aAa^* = 1$.
   (d) Let A be a C^*-algebra acting on a Hilbert space $\mathfrak{H}$ such that $A \supset \mathscr{LC}(\mathfrak{H})$. Show that every irreducible representation π such that $\pi(\mathscr{LC}(\mathfrak{H})) \neq \{0\}$ is unitarily equivalent to the original action of A on $\mathfrak{H}$.

5. Let $\mathscr{I}$ be a closed ideal of a C^*-algebra A and π be the canonical homomorphism of A onto $A/\mathscr{I}$.
   (a) Show that if $x,y \in A_+$ and $xy \in \mathscr{I}$, then, with $x_1 = (x - y)_+$ and $y_1 = (x - y)_-$, $\pi(x) = \pi(x_1)$ and $\pi(y) = \pi(y_1)$.
   (b) Show that for an $x \in A$, $\{y \in A : xy \in \mathscr{I}\}$ (resp. $\{y \in A : yx \in \mathscr{I}\}$) is a closed right (resp. left) ideal of A.
   (c) Show that if $xy \in \mathscr{I}$ for a pair x,y in A, then there exist $a,b \in \mathscr{I}$ such that $(x - a)(y - b) = 0$. (*Hint*: by (b), $|x|^{1/2}|y^*|^{1/2} \in \mathscr{I}$, and use (a) [287].)

†6. Let $\varphi_1, \dots, \varphi_n$ be pure states on a separable C^*-algebra A. If $\|\varphi_i - \varphi_j\| = 2$ for $i \neq j$, then there exists a maximal abelian C^*-subalgebra B of A such that (i) $\varphi_i|_B$ is pure for $i = 1,2,\dots,n$, and (ii) $\varphi_i|_B$ has a unique pure state extension to A [43].

†7. Let A be a C^*-algebra and $\mathfrak{H}$ be an infinite dimensional Hilbert space such that every cyclic representation of A has a dimension $\leq \dim \mathfrak{H}$. Let $\mathrm{Rep}(A; \mathfrak{H})$ be the collection of all representations $\{\pi, \mathfrak{H}_\pi\}$ of A such that $\mathfrak{H}_\pi \subset \mathfrak{H}$. Consider the direct sum $\pi_1 \oplus \pi_2$ in $\mathrm{Rep}(A; \mathfrak{H})$ of $\pi_1, \pi_2 \in \mathrm{Rep}(A; \mathfrak{H})$ if $\mathfrak{H}_{\pi_1} \perp \mathfrak{H}_{\pi_2}$. For each $\pi_1, \pi_2 \in \mathrm{Rep}(A; \mathfrak{H})$, let $\mathscr{I}(\pi_1, \pi_2)$ denote the set of all bounded linear operators a from $\mathfrak{H}_{\pi_2}$ into $\mathfrak{H}_{\pi_1}$ such that $\pi_1(x)a = a\pi_2(x)$, $x \in A$. Consider the direct sum $\mathscr{L} = \sum^\oplus \{\mathscr{L}(\mathfrak{H}_\pi) : \pi \in \mathrm{Rep}(A; \mathfrak{H})\}$ and $x = \sum^\oplus x(\pi) \in \mathscr{L}$. We say that x is *admissible* if (i) $x(\pi_1 \oplus \pi_2) = x(\pi_1) \oplus x(\pi_2)$ whenever $\pi_1 \oplus \pi_2$ is defined, and (ii) $x(\pi_1)a = ax(\pi_2)$ for every $a \in \mathscr{I}(\pi_1, \pi_2)$.

(a) Let $\tilde{A}$ denote the universal enveloping von Neumann algebra of A. Then every $\pi \in \mathrm{Rep}(A; \mathfrak{H})$ can be viewed as a normal representation of $\tilde{A}$ such that $\pi(\tilde{A})$ is the weak closure of $\pi(A)$. Show that for each $x \in \tilde{A}$, the map: $\pi \in \mathrm{Rep}(A; \mathfrak{H}) \mapsto \pi(x) \in \mathscr{L}(\mathfrak{H}_\pi)$ gives rise to an admissible element of $\mathscr{L}$ and every admissible element of $\mathscr{L}$ is of this form.

(b) Show that the weak topology and the σ-strong* topology in $\mathscr{L}(\mathfrak{H})$ give rise to the same simple convergence topology in $\mathrm{Rep}(A; \mathfrak{H})$.

(c) Show that every $x \in A$ gives rise to a continuous admissible element: $\pi \in \mathrm{Rep}(A; \mathfrak{H}) \mapsto \pi(x) \in \mathscr{L}(\mathfrak{H}_\pi)$ in $\mathscr{L}$, and that every continuous admissible element of $\mathscr{L}$ is of this form [56], [358].

Notes

The concept of a stonean space was introduced by M. Stone [335], as the spectrum of a complete Boolean algebra, i.e., the spectrum of the abelian C^*-algebra generated by a complete Boolean algebra. The presentation here in Section 1 concerning stonean spaces follows a treatise due to J. Dixmier [81] Theorem 1.22 is due to P. Halmos and J. von Neumann [159]. Theorem 1.28 is known as the Phillips lemma [288]. Theorem 2.4 was announced by S. Sherman, [329] and the proof was presented later by Z. Takeda [345]. For this reason, it is often referred to as the Sherman–Takeda theorem. The decomposition of $\mathscr{M}^*$ into the normal part and the singular part was given by M. Takesaki [352]. The characterization of a W^*-algebra received considerable attention during the 1950s. I. Kaplansky [197], introduced AW^*-algebras to separate the algebraic feature from von Neumann algebras. Pushing Z. Takeda's idea, which proposed the Banach space duality approach to the study of operator algebras [346, 347], S. Sakai succeeded in obtaining a characterization of a W^*-algebra, Theorem 3.5 [304]. The proof presented here is due to J. Tomiyama [376] based on his result, Theorem 3.4. Another characterization of a W^*-algebra as a monotone closed algebra, Theorem 3.16, is due to R. Kadison [181], which is naturally tied up with the up–down–up theorem of G. Pedersen. Theorem 3.8 is due to M. Takesaki [354]. The uniqueness of the predual of a von Neumann algebra, Corollary 3.10, due to J. Dixmier [85], answered completely the question concerning to what extent the algebraic structure of a von Neumann algebra determines its topological structure. Most of Section 4 is essentially due to S. Sakai [308] and M. Tomita [372]. It was S. Sakai [312] who initiated the study of the Arens–Mackey topology of a von Neumann algebra. He showed that the Arens–Mackey topology on the bounded parts of a finite von Neumann algebra (see Chapter V) agrees with the σ-strong topology. C. Akemann [41] gave the complete characterization of the Arens–Mackey topology on the bounded parts, Theorem 5.7. Theorem 5.4 is a combination of results due to several mathematicians: A. Grothendieck [148], S. Sakai [312], M. Takesaki [352], H. Umegaki [395], and, finally, C. Akemann [41].

Section 6 followed the presentation of G. Pedersen [282] and C. Akemann and G. Pedersen [47]. The concept of the multiplier algebra $M(A)$ of a nonunital C^*-algebra was first introduced by R. Busby [63], following a general theory of centralizers of Banach algebra of B. Johnson [169]. One should note here that it is, in principle, possible to develop an integration theory on a locally compact space following the lines set forth in this section. The natural and interesting question in this approach is then whether one can develop a noncommutative counterpart of the theory of Borel spaces. In applications, C^*-algebras appear more frequently than von Neumann algebras as directly associated algebras; von Neumann algebras appear through representations of the C^*-algebras in question. To relate the properties of the von Neumann algebra to the original C^*-algebra, the better understanding of the approximation process from a given C^*-algebra is indispensable. The study of the above question will bring about a new insight into the situation.

Chapter IV
Tensor Products of Operator Algebras and Direct Integrals

0. Introduction

Tensor products of C^*-algebras and von Neumann algebras will be introduced in this chapter, and direct integrals, a continuous analogue of direct sums, will also be discussed. Unlike the finite dimensional case, tensor products of infinite dimensional vector spaces or algebras are quite nontrivial. It is still considered as one of the tricky parts of the theory. The norms in the tensor product of two Banach spaces are highly nonunique. Fortunately however, the norm in the tensor product of Hilbert spaces which makes it a pre-Hilbert space is unique. This uniqueness makes interesting the tensor products of C^*-algebras. However, the norms on the tensor products of C^*-algebras which make it a pre-C^*-algebra are not unique. In this chapter, the problem of defining a suitable norm on the tensor product of C^*-algebras will be attacked. Sections 1 and 2 are just preliminary. In Section 3, completely positive maps are introduced to provide a powerful tool in the succeeding sections. Section 4 is devoted to the problems mentioned above. There we shall show that the tensor product of C^*-algebras admits two natural norms: one is the smallest norm among all possible C^*-norms and the other is the largest. In Section 5, the tensor product of W^*-algebras is discussed, and it is proved that the algebraic type of the tensor product of von Neumann algebras is uniquely determined by those of the component algebras without regard to the underlying Hilbert spaces, Theorem 5.2. The structure of a normal homomorphism of a von Neumann algebra on a Hilbert space onto another is described in terms of tensor products and projections in the commutant, Theorem 5.5, which will allow us later to relate Problems (A) and (B), in the introduction to Chapter I, for von Neumann algebras. In Section 4, Choquet's theory of boundary integrals on compact convex sets is briefly described and applied to integral representations of states. The relation between an integral representation of a state and a direct integral decomposition, (disintegration), of a representation will be discussed in Section 8. The tensor products of a Banach space (or von Neumann algebra) with function spaces on a measure space are studied in Section 7. Section 8 is devoted to direct integrals of operator algebras and Hilbert spaces.

1. Tensor Product of Hilbert Spaces and Operators

Let $\mathfrak{H}_1$ and $\mathfrak{H}_2$ be two Hilbert spaces. Let $\mathfrak{H}_0$ denote the algebraic tensor product of $\mathfrak{H}_1$ and $\mathfrak{H}_2$. A general element ξ of $\mathfrak{H}_0$ is of the form

$$\xi = \sum_{i=1}^{n} \xi_{1,i} \otimes \xi_{2,i}, \qquad \xi_{1,i} \in \mathfrak{H}_1, \qquad \xi_{2,i} \in \mathfrak{H}_2, \qquad 1 \leq i \leq n. \tag{1}$$

In $\mathfrak{H}_0$, we define a sesquilinear form $(\cdot|\cdot)$ by

$$(\xi|\eta) = \sum_{i=1}^{n} \sum_{j=1}^{m} (\xi_{1,i}|\eta_{1,j})(\xi_{2,i}|\eta_{2,j}) \tag{2}$$

for $\xi = \sum_{i=1}^{n} \xi_{1,i} \otimes \xi_{2,i} \in \mathfrak{H}_0$ and $\eta = \sum_{j=1}^{m} \eta_{1,j} \otimes \eta_{2,j} \in \mathfrak{H}_0$.

Lemma 1.1. *The sesquilinear form defined by* (2) *is an inner product in* $\mathfrak{H}_0$.

PROOF. Let $\xi = \sum_{i=1}^{n} \xi_{1,i} \otimes \xi_{2,i}$ be an arbitrary vector in $\mathfrak{H}_0$. Applying the Gram–Schmidt orthogonalization to $\{\xi_{2,i}\}$, we may assume that $\{\xi_{2,i}\}$ is a normalized orthogonal system in $\mathfrak{H}_2$. We have then

$$\begin{aligned}(\xi|\xi) &= \sum_{i,j=1}^{n} (\xi_{1,i}|\xi_{1,j})(\xi_{2,i}|\xi_{2,j}) \\ &= \sum_{i=1}^{n} \|\xi_{1,i}\|^2 \geq 0;\end{aligned}$$

hence the sesquilinear form $(\cdot|\cdot)$ in $\mathfrak{H}_0$ is positive definite, that is, an inner product. Q.E.D.

Definition 1.2. The completion $\mathfrak{H}$ of $\mathfrak{H}_0$ is called the *tensor product* of $\mathfrak{H}_1$ and $\mathfrak{H}_2$, and denoted by $\mathfrak{H}_1 \otimes \mathfrak{H}_2$.

Let $\{\varepsilon_i\}_{i \in I}$ be an orthogonal basis in $\mathfrak{H}_2$. We have then, for any $\{\xi_i : i \in I\} \subset \mathfrak{H}_1$ with $\xi_i = 0$ except for finitely many i,

$$\left\| \sum_{i \in I} \xi_i \otimes \varepsilon_i \right\|^2 = \sum_{i,j \in I} (\xi_i|\xi_j)(\varepsilon_i|\varepsilon_j) = \sum_{i \in I} \|\xi_i\|^2.$$

Hence we can define an isometry U of $\sum_{i\in I}^{\oplus} \mathfrak{H}_i$ onto $\mathfrak{H}_1 \otimes \mathfrak{H}_2$ by

$$U \sum_{i\in I}^{\oplus} \xi_i = \sum_{i\in I} \xi_i \otimes \varepsilon_i \in \mathfrak{H}_1 \otimes \mathfrak{H}_2,$$

where $\mathfrak{H}_i$ is the replica of $\mathfrak{H}_1$ for each $i \in I$. Putting

$$U_i\xi = \xi \otimes \varepsilon_i, \qquad \xi \in \mathfrak{H}_1, \qquad i \in I, \tag{3}$$

we obtain an isometry of $\mathfrak{H}_1$ into $\mathfrak{H}$, whose ranges $\mathfrak{H}^i$ are mutually orthogonal, and where $\mathfrak{H} = \sum_{i\in I}^{\oplus} \mathfrak{H}^i$.

Take arbitrary $x_1 \in \mathscr{L}(\mathfrak{H}_1)$ and $x_2 \in \mathscr{L}(\mathfrak{H}_2)$. We then get an operator x_0 on $\mathfrak{H}_0$ defined by

$$x_0(\xi_1 \otimes \xi_2) = (x_1\xi_1) \otimes (x_2\xi_2),$$

which is denoted by $x_1 \otimes x_2$. If $x_2 = 1$, then

$$\left\|x_0 \sum_{i\in I} \xi_i \otimes \varepsilon_i\right\|^2 = \left\|\sum_{i\in I} x_1\xi_i \otimes \varepsilon_i\right\|^2 = \sum_{i\in I} \|x_1\xi_i\|^2$$

$$\leq \|x_1\|^2 \sum_{i\in I} \|\xi_i\|^2 = \|x_1\|^2 \left\|\sum_{i\in I} \xi_i \otimes \varepsilon_i\right\|^2,$$

so that $x_1 \otimes 1$ is bounded on $\mathfrak{H}_0$; similarly, $1 \otimes x_2$ is bounded, and then so is $x_0 = (x_1 \otimes 1)(1 \otimes x_2)$. Hence $x_1 \otimes x_2$ is extended to a bounded operator on $\mathfrak{H}_1 \otimes \mathfrak{H}_2$, which is also denoted by $x_1 \otimes x_2$ and called the *tensor product* of x_1 and x_2. The following properties are easily verified:

$$\left.\begin{aligned}(\lambda x_1 + \mu y_1) \otimes x_2 &= \lambda(x_1 \otimes x_2) + \mu(y_1 \otimes x_2),\\ x_1 \otimes (\lambda x_2 + \mu y_2) &= \lambda(x_1 \otimes x_2) + \mu(x_1 \otimes y_2);\end{aligned}\right. \tag{4}$$

$$(x_1 \otimes x_2)(y_1 \otimes y_2) = x_1y_1 \otimes x_2y_2; \tag{5}$$

$$(x_1 \otimes x_2)^* = x_1^* \otimes x_2^*; \tag{6}$$

$$\|x_1 \otimes x_2\| = \|x_1\|\|x_2\|. \tag{7}$$

Definition 1.3. Let $\{\mathscr{M}_1,\mathfrak{H}_1\}$ and $\{\mathscr{M}_2,\mathfrak{H}_2\}$ be von Neumann algebras. The von Neumann algebra on $\mathfrak{H}_1 \otimes \mathfrak{H}_2$ generated by $x_1 \otimes x_2$, $x_1 \in \mathscr{M}_1$ and $x_2 \in \mathscr{M}_2$, is called the *tensor product* of $\{\mathscr{M}_1,\mathfrak{H}_1\}$ and $\{\mathscr{M}_2,\mathfrak{H}_2\}$ and denoted by $\{\mathscr{M}_1,\mathfrak{H}_1\} \otimes \{\mathscr{M}_2,\mathfrak{H}_2\}$ or simply by $\mathscr{M}_1 \overline{\otimes} \mathscr{M}_2$. If $\mathscr{M}_2$ (resp. $\mathscr{M}_1$) is reduced to the scalar multiplication algebras $\mathbf{C}1$, then $\mathscr{M}_1 \overline{\otimes} \mathscr{M}_2$ is denoted by $\mathscr{M}_1 \otimes \mathbf{C}$ (resp. $\mathbf{C} \otimes \mathscr{M}_2$).

Now, take an arbitrary $x \in \mathscr{L}(\mathfrak{H})$. Putting $x_{i,j} = U_i^*xU_j$, $i,j \in I$, we get a matrix $(x_{i,j})$ of bounded operators on $\mathfrak{H}_1$. Clearly, the map: $x \mapsto (x_{i,j})$ is injective, so we may write $x = (x_{i,j})$. For each $x = (x_{i,j})$ and $y = (y_{i,j})$,

$$\left.\begin{aligned}\alpha x + \beta y &= (\alpha x_{i,j} + \beta y_{i,j}),\\ xy = (z_{i,j}), &\qquad z_{i,j} = \sum_{k\in I} x_{i,k}y_{k,j},\\ x^* &= (x_{j,i}^*),\end{aligned}\right\} \tag{8}$$

where the summation in the second equality is taken in the strong operator topology.

Proposition 1.4. *If $x = x_1 \otimes 1$, $x_1 \in \mathscr{L}(\mathfrak{H}_1)$, then x commutes with all $U_iU_j^*$, and $x_{i,j} = \delta_{i,j}x_1$, where $\delta_{i,j}$ means the Kronecker symbol: $\delta_{i,j} = 0$, $i \neq j$, $\delta_{i,i} = 1$. Conversely, if $x \in \mathscr{L}(\mathfrak{H})$ commutes with all $U_iU_j^*$, then x is of this form.*

PROOF. The first half of our assertion is almost obvious. We have only to prove the last assertion. Suppose an $x \in \mathscr{L}(\mathfrak{H})$ commutes with all $U_iU_j^*$. Then we have $U_i^*xU_j = 0$, $i \neq j$; $U_ixU_i^* = U_jxU_j^*$. Put $x_1 = U_ixU_i^*$. Then x_1 is in $\mathscr{L}(\mathfrak{H}_1)$ and we have $x_{i,j} = \delta_{i,j}x_1$; hence $x = x_1 \otimes 1$. Q.E.D.

The map $\pi: x_1 \in \mathscr{L}(\mathfrak{H}_1) \mapsto x_1 \otimes 1 \in \mathscr{L}(\mathfrak{H})$ is a faithful representation of the C^*-algebra $\mathscr{L}(\mathfrak{H}_1)$ on $\mathfrak{H}$. As an immediate consequence of the proposition, we get the following:

Corollary 1.5. *For any subset $\mathscr{S}$ of $\mathscr{L}(\mathfrak{H}_1)$,*

$$\pi(\mathscr{S})'' = \pi(\mathscr{S}'').$$

In particular, for any von Neumann algebra $\mathscr{M}_1$ on $\mathfrak{H}_1$, π is an isomorphism of $\mathscr{M}_1$ onto $\mathscr{M}_1 \otimes \mathbf{C}$.

PROOF. Since $U_iU_j^*$ is contained in $\pi(\mathscr{S})'$, $\pi(\mathscr{S})'' \subset \pi(\mathscr{L}(\mathfrak{H}_1))$ by Proposition 1.4. But π is an isomorphism of $\mathscr{L}(\mathfrak{H}_1)$ onto $\pi(\mathscr{L}(\mathfrak{H}_1))$; hence $\pi(\mathscr{S})'' = \pi(\mathscr{S}'')$. Q.E.D.

The above isomorphism π is called the *amplification.*

Proposition 1.6. *If $\{\mathscr{M}_1, \mathfrak{H}_1\}$ is a von Neumann algebra, then*

(i) $(\mathscr{M}_1 \otimes \mathbf{C})' = \mathscr{M}_1' \mathbin{\overline{\otimes}} \mathscr{L}(\mathfrak{H}_2)$,
(ii) $(\mathscr{M}_1 \mathbin{\overline{\otimes}} \mathscr{L}(\mathfrak{H}_2))' = \mathscr{M}_1' \otimes \mathbf{C}$,
(iii) $\mathscr{M}_1 \mathbin{\overline{\otimes}} \mathscr{L}(\mathfrak{H}_2)$ *consists of all the operators* $x = (x_{i,j}) \in \mathscr{L}(\mathfrak{H})$ *with* $x_{i,j} \in \mathscr{M}_1$.

PROOF. For an orthogonal basis $\{\varepsilon_i\}_{i \in I}$ of $\mathfrak{H}_2$, let $u_{i,j} = (\xi|\varepsilon_j)\varepsilon_i$, $\xi \in \mathfrak{H}_2$, $i,j \in I$. Then $1 \otimes u_{i,j}$ is nothing but $U_iU_j^*$ in the previous arguments. If $x = (x_{i,j}) \in \mathscr{L}(\mathfrak{H})$, then

$$x = \sum_{i,j \in I} x_{i,j} \otimes u_{i,j}, \tag{9}$$

where the summation is considered under the strong operator topology. Therefore, if the matrix coefficients of x belong to $\mathscr{M}_1$, then x is in $\mathscr{M}_1 \mathbin{\overline{\otimes}} \mathscr{L}(\mathfrak{H})$.

Let $\rho_{i,j}(x) = x \otimes u_{i,j}$, $x \in \mathscr{L}(\mathfrak{H}_1)$, $i,j \in I$. Then $\rho_{i,j}$ is a σ-weakly bicontinuous isometry of $\mathscr{L}(\mathfrak{H}_1)$ into $\mathscr{L}(\mathfrak{H})$, and $\rho_{i,j}(\mathscr{L}(\mathfrak{H}_1)) = (1 \otimes u_{i,i})\mathscr{L}(\mathfrak{H}) \times$

$(1 \otimes u_{j,j})$. Hence the set

$$\{x \in \mathscr{L}(\mathfrak{H}): \rho_{i,j}^{-1}((1 \otimes u_{i,i})x(1 \otimes u_{j,j})) \in \mathscr{M}_1\} = \{x \in \mathscr{L}(\mathfrak{H}): x_{i,j} \in \mathscr{M}_1\}$$

is σ-weakly closed for each fixed $i,j \in I$, and contains $x \otimes u_{i,j}$ for every $x \in \mathscr{M}_1$. Since $\{u_{i,j}: i,j \in I\}$ generates $\mathscr{L}(\mathfrak{H}_2)$, the above set contains $\mathscr{M}_1 \overline{\otimes} \mathscr{L}(\mathfrak{H}_2)$. Therefore, if $x = (x_{i,j}) \in \mathscr{M}_1 \overline{\otimes} \mathscr{L}(\mathfrak{H}_2)$, then each $x_{i,j}$ belongs to $\mathscr{M}_1$.

Assertions (i) and (ii) follow directly from Proposition 1.4 and Corollary 1.5. Q.E.D.

As a special case of the proposition, we have

$$\mathscr{L}(\mathfrak{H}_1 \otimes \mathfrak{H}_2) = \mathscr{L}(\mathfrak{H}_1) \overline{\otimes} \mathscr{L}(\mathfrak{H}_2). \tag{10}$$

Definition 1.7. A system $\{w_{i,j}: i,j \in I\}$ of elements in a von Neumann algebra $\{\mathscr{M}, \mathfrak{H}\}$ is called a *matrix unit* if

(i) $w_{i,j}^* = w_{j,i}$,

(ii) $w_{i,j} w_{k,l} = \delta_{j,k} w_{i,l}$,

(iii) $\sum_{i \in I} w_{i,i} = 1$,

where the last summation is taken in the strong topology, taking into account the fact that the $\{w_{i,i}: i \in I\}$ are orthogonal projections by (i) and (ii).

Proposition 1.8. *Let $\{\mathscr{M}, \mathfrak{H}\}$ be a von Neumann algebra. Suppose $\{w_{i,j}: i,j \in I\}$ is a matrix unit in $\mathscr{M}$. For a fixed $i_0 \in I$, let $e = w_{i_0,i_0}$, $\mathfrak{H}_1 = e\mathfrak{H}$ and $\mathfrak{H}_2 = l^2(I)$. Then we have*

$$\{\mathscr{M}, \mathfrak{H}\} \cong \{\mathscr{M}_1, \mathfrak{H}_1\} \otimes \{\mathscr{L}(\mathfrak{H}_2), \mathfrak{H}_2\},$$

where $\mathscr{M}_1 = \mathscr{M}_e$.

Proof. Let $\{\varepsilon_i\}_{i \in I}$ be an orthogonal basis in $\mathfrak{H}_2$. Let $w_i = w_{i,i_0}$ for each $i \in I$. Then $w_{i,j} = w_i w_j^*$ and $w_i^* w_i = e$. Put

$$U\left(\sum_{i \in I} \xi_i \otimes \varepsilon_i\right) = \sum_{i \in I} w_i \xi_i$$

for each $\sum \xi_i \otimes \varepsilon_i \in \mathfrak{H}_1 \otimes \mathfrak{H}_2$. We then have

$$\begin{aligned}
\left(U\left(\sum \xi_i \otimes \varepsilon_i\right) \Big| U\left(\sum \eta_j \otimes \varepsilon_j\right)\right) &= \left(\sum w_i \xi_i \Big| \sum w_j \eta_j\right) \\
&= \sum_{i,j} (w_j^* w_i \xi_i | \eta_j) = \sum_i (\xi_i | \eta_i) \\
&= \left(\sum \xi_i \otimes \varepsilon_i \Big| \sum \eta_j \otimes \varepsilon_j\right).
\end{aligned}$$

Hence U is a well-defined isometry of $\mathfrak{H}_1 \otimes \mathfrak{H}_2$ into $\mathfrak{H}$. The range of U contains all $w_i \mathfrak{H}_1 = w_{i,i} \mathfrak{H}$, so that U is surjective since $\sum_i w_{i,i} = 1$. For each

$x = (x_{i,j}) \in \mathscr{M}_1 \otimes \mathscr{L}(\mathfrak{H}_2)$, we have

$$UxU^*\xi = Ux\left(\sum_{j\in I} w_j^*\xi \otimes \varepsilon_j\right) = U\left(\sum_{i,j\in I} x_{i,j}w_j^*\xi \otimes \varepsilon_i\right)$$
$$= \sum_{i,j\in I} w_i x_{i,j} w_j^* \xi;$$

so $UxU^* = \sum_{i,j\in I} w_i x_{i,j} w_j^*$ belongs to $\mathscr{M}$ because $x_{i,j} \in e\mathscr{M}e = \mathscr{M}_1$. Conversely, if x is an operator in $\mathscr{M}$, we have

$$U^*xU\left(\sum_j \xi_j \otimes \varepsilon_j\right) = U^*x \sum_j w_j\xi_j = U^* \sum_j xw_j\xi_j$$
$$= \sum_{i,j} (w_i^* xw_j\xi_j \otimes \varepsilon_i);$$

hence $(U^*xU)_{i,j} = w_i^*xw_j \in e\mathscr{M}e = \mathscr{M}_1$, so that U^*xU belongs to $\mathscr{M}_1 \overline{\otimes} \mathscr{L}(\mathfrak{H}_2)$ by Proposition 1.6(iii). Q.E.D.

Proposition 1.9. *Let $\{\mathscr{M}_1,\mathfrak{H}_1\}$ and $\{\mathscr{M}_2,\mathfrak{H}_2\}$ be von Neumann algebras. Put $\mathfrak{H} = \mathfrak{H}_1 \otimes \mathfrak{H}_2$. If e_1 and e_2 are projections of $\mathscr{M}_1$ and $\mathscr{M}_2$, respectively, then*

$$\{(\mathscr{M}_1 \overline{\otimes} \mathscr{M}_2)_{(e_1\otimes e_2)},(e_1 \otimes e_2)\mathfrak{H}\} = \{\mathscr{M}_{1,e_1},e_1\mathfrak{H}_1\} \otimes \{\mathscr{M}_{2,e_2},e_2\mathfrak{H}_2\},$$
$$\{(\mathscr{M}'_1 \overline{\otimes} \mathscr{M}'_2)_{(e_1\otimes e_2)},(e_1 \otimes e_2)\mathfrak{H}\} = \{\mathscr{M}'_{1,e_1},e_1\mathfrak{H}_1\} \otimes \{\mathscr{M}'_{2,e_2},e_2\mathfrak{H}_2\}$$

Proof. For any $\xi = \sum_{i=1}^n \xi_{1,i} \otimes \xi_{2,i}$, we have $(e_1 \otimes e_2)\xi = \sum_{i=1}^n e_1\xi_{1,i} \otimes e_2\xi_{2,i} \in e_1\mathfrak{H}_1 \otimes e_2\mathfrak{H}_2$. Since such ξ are dense in $\mathfrak{H}$, we have $(e_1 \otimes e_2)\mathfrak{H} \subset e_1\mathfrak{H}_1 \otimes e_2\mathfrak{H}_2$. But the reverse inclusion is trivial; so we get $(e_1 \otimes e_2)\mathfrak{H} = e_1\mathfrak{H}_1 \otimes e_2\mathfrak{H}_2$. By definition, $\mathscr{M}_1 \overline{\otimes} \mathscr{M}_2$ is generated by $x_1 \otimes x_2$, $x_1 \in \mathscr{M}_1$, $x_2 \in \mathscr{M}_2$, so that $(e_1 \otimes e_2)(\mathscr{M}_1 \overline{\otimes} \mathscr{M}_2)(e_1 \otimes e_2)$ is the σ-weak closure of the set of elements of the form

$$(e_1 \otimes e_2)\left(\sum_{i=1}^n x_{1,i} \otimes x_{2,i}\right)(e_1 \otimes e_2)$$
$$= \sum_{i=1}^n (e_1x_{1,i}e_1) \otimes (e_2x_{2,i}e_2) \in e_1\mathscr{M}_1e_1 \otimes e_2\mathscr{M}_2e_2.$$

Hence we have $(e_1 \otimes e_2)(\mathscr{M}_1 \overline{\otimes} \mathscr{M}_2)(e_1 \otimes e_2) \subset (e_1\mathscr{M}_1e_1) \overline{\otimes} (e_2\mathscr{M}_2e_2)$. Since the reverse inclusion is trivial, we have

$$(e_1 \otimes e_2)(\mathscr{M}_1 \overline{\otimes} \mathscr{M}_2)(e_1 \otimes e_2) = (e_1\mathscr{M}_1e_1) \overline{\otimes} (e_2\mathscr{M}_2e_2).$$

Hence $(\mathscr{M}_1 \overline{\otimes} \mathscr{M}_2)_{(e_1\otimes e_2)} = \mathscr{M}_{1,e_1} \overline{\otimes} \mathscr{M}_{2,e_2}$.

For any $y_1 \in \mathscr{M}'_1$ and $y_2 \in \mathscr{M}'_2$, we have $(y_1 \otimes y_2)_{(e_1\otimes e_2)} = y_{1,e_1} \otimes y_{2,e_2}$, so that $\mathscr{M}'_{1,e_1} \otimes \mathscr{M}'_{2,e_2}$ is contained in $(\mathscr{M}'_1 \overline{\otimes} \mathscr{M}'_2)_{(e_1\otimes e_2)}$. The above arguments show that if x is an algebraic element in $\mathscr{M}'_1 \otimes \mathscr{M}'_2$, then $x_{(e_1\otimes e_2)}$ belongs to $\mathscr{M}'_{1,e_1} \otimes \mathscr{M}'_{2,e_2}$. Since the induction map: $x \in \mathscr{M}'_1 \overline{\otimes} \mathscr{M}'_2 \mapsto x_{e_1\otimes e_2} \in \mathscr{L}((e_1 \otimes e_2)\mathfrak{H})$ is a normal homomorphism, the image $(\mathscr{M}'_1 \overline{\otimes} \mathscr{M}'_2)_{(e_1\otimes e_2)}$ of $\mathscr{M}'_1 \overline{\otimes} \mathscr{M}'_2$ is contained in the σ-weak closure $\mathscr{M}'_{1,e_1} \otimes \mathscr{M}'_{2,e_2}$ of $\mathscr{M}'_{1,e_1} \otimes \mathscr{M}'_{2,e_2}$. Hence we get

$$(\mathscr{M}'_1 \overline{\otimes} \mathscr{M}'_2)_{(e_1\otimes e_2)} = \mathscr{M}'_{1,e_1} \otimes \mathscr{M}'_{2,e_2}.$$

Q.E.D.

Exercises

1. Let π be a linear map of a von Neumann algebra $\mathcal{M}$ into another von Neumann algebra $\mathcal{N}$ such that $\pi(x^*) = \pi(x)^*$, $x \in \mathcal{M}$, and $\pi(h^2) = \pi(h)^2$ for each $h \in \mathcal{M}_h$. We say that π is a *Jordan homomorphism*.
   (a) The Jordan product $\{x,y\}$ of $x,y \in \mathcal{M}$ is defined by
   $$\{x,y\} = xy + yx.$$
   Show that π preserves the Jordan product, i.e.,
   $$\pi(\{x,y\}) = \{\pi(x),\pi(y)\}.$$
   (*Hint*: $\{h,k\} = (h+k)^2 - h^2 - k^2$ for $h,k \in \mathcal{M}_h$ and $\{x,y\} = (x+y)^2 - x^2 - y^2$ for $x,y \in \mathcal{M}$ and $x^2 = h^2 + i(hk + kh) - k^2$ if $x = h + ik$ with $h,k \in \mathcal{M}_h$.)
   (b) Show that if x and y commute, then $\pi(xy) = \pi(x)\pi(y) = \pi(y)\pi(x)$.
   (c) Let $\{u_{i,j}: 1 \le i,j \le n\}$ be a system of matrix units in $\mathcal{M}$ and $n \ge 2$. Let $\mathcal{L} = \mathcal{M} \cap \{u_{i,j}: 1 \le i,j \le n\}'$. It follows that every $x \in \mathcal{M}$ is written uniquely as
   $$x = \sum_{i,j=1}^{n} x_{i,j}u_{i,j}, \quad \text{with} \quad x_{i,j} \in \mathcal{L}.$$
   Show that
   $$\pi(x) = \sum_{i,j=1}^{n} \pi(x_{i,j})\pi(u_{i,j}).$$
   (d) Let $v_{i,j} = \pi(u_{i,i})\pi(u_{i,j})\pi(u_{j,j})$ for $i \ne j$, and $w_{i,j} = \pi(u_{i,i})\pi(u_{j,i})\pi(u_{j,j})$. Show that
   $$\pi(u_{i,j}) = v_{i,j} + w_{j,i}, \qquad i \ne j.$$
   (*Hint*: Show that $\pi(axb + bxa) = \pi(a)\pi(x)\pi(b) + \pi(b)\pi(x)\pi(a)$ and use the fact that $u_{i,j} = u_{i,i}u_{i,j}u_{j,j} + u_{j,j}u_{i,j}u_{i,i}$.)
   (e) Show that $\{\pi(u_{i,i}): i = 1,2,\dots,n\}$ are orthogonal projections in $\mathcal{N}$.
   (f) Show that
   $$v_{i,j} = \pi(u_{i,i})\pi(u_{i,j}) = \pi(u_{i,j})\pi(u_{j,j}),$$
   $$w_{i,j} = \pi(u_{i,i})\pi(u_{j,i}) = \pi(u_{j,i})\pi(u_{j,j}).$$
   (g) Show that if $i \ne j$, $j \ne k$ and $i \ne k$ then
   $$v_{i,j}v_{j,k} = v_{i,k},$$
   $$w_{i,j}w_{j,k} = w_{i,k}.$$
   (h) Show that $v_{i,i} = v_{i,j}v_{j,i}$ and $w_{i,i} = w_{i,j}w_{j,i}$ both are independent of the choice of $j \ne i$, and projections in $\mathcal{N}$.
   (i) Put $e = \sum_{i=1}^{n} v_{i,i}$ and $f = \sum_{i=1}^{n} w_{i,i}$. Show that $\{v_{i,j}: 1 \le i,j \le n\}$ (resp. $\{w_{i,j}: 1 \le i,j \le n\}$) is a system of matrix units in $\mathcal{N}_e$ (resp. $\mathcal{N}_f$) and $e + f = \pi(1)$.
   (j) Show that e and f are central projections in $\pi(\mathcal{M})$.
   (k) Show that the map $\pi_1: x \in \mathcal{M} \mapsto \pi(x)e \in \mathcal{N}_e$ is a homomorphism and that the map $\pi_2: x \in \mathcal{M} \mapsto \pi(x)f \in \mathcal{N}_f$ is an antihomomorphism.
   (l) Show that every Jordan homomorphism π of a von Neumann algebra $\mathcal{M}$ into another von Neumann algebra $\mathcal{N}$ is decomposed into the sum of a homomorphism and of an antihomomorphism by a central projection of $\pi(\mathcal{M})$. (*Hint*: Decompose $\mathcal{M}$ into a direct sum of the abelian part and matrix algebras over other von Neumann algebras.) cf. [168].

2. Let π be an isometry of a unital C^*-algebra A onto another C^*-algebra B such that $\pi(1) = 1$.
   (a) Show that π preserves the *-operation and $\pi(h^2) = \pi(h)^2$ for every $h \in A_h$. (*Hint*: Consider the second transpose $\tilde{\pi} = {}^{tt}\pi : \tilde{A} = A^{**} \mapsto \tilde{B} = B^{**}$. Show that $\tilde{\pi}$ maps orthogonal projections of $\tilde{A}$ into those of $\tilde{B}$ by making use of the fact that projections in $\tilde{A}$ are characterized by the extreme points of $\tilde{A}_+ \cap \tilde{S}$, where $\tilde{S}$ is the unit ball of $\tilde{A}$. Then use the spectral decomposition of $h \in \tilde{A}_h$.)
   (b) Show that there exists a central projection $z \in \tilde{B}$ such that $\pi_1(x) = \pi(x)z$, $x \in A$ (resp. $\pi_2(x) = \pi(x)(1 - z)$, $x \in A$) is a homomorphism (resp. an antihomomorphism) of A into $\tilde{B}$. (*Hint*: Use (1).)
   (c) A mapping π is called a *C^*-homomorphism* if the conditions in (a) are satisfied. Show that a positivity preserving isometry of a C^*-algebra A onto another C^*-algebra B is a C^*-isomorphism. (*Hint*: Consider the second transpose $\tilde{\pi} = {}^{tt}\pi$ of π.) cf. [172].

2. Tensor Products of Banach Spaces

For further investigation of tensor products of operator algebras, we review here some general properties of tensor products of Banach spaces.

Suppose E_1 and E_2 are two complex Banach spaces. Let $E_1 \otimes E_2$ be the algebraic tensor product of E_1 and E_2 over the complex number field. If a norm $\|\cdot\|$ on $E_1 \otimes E_2$ satisfies the condition

$$\|x_1 \otimes x_2\| = \|x_1\|\|x_2\|, \qquad x_1 \in E_1, \qquad x_2 \in E_2, \tag{1}$$

then it is called a *cross-norm* of $E_1 \otimes E_2$. In general, *a cross-norm of $E_1 \otimes E_2$ is not a priori determined. We have to specify which cross-norm is considered on* $E_1 \otimes E_2$. The completion of $E_1 \otimes E_2$ under a cross-norm β is denoted by $E_1 \otimes_\beta E_2$.

It is sometimes very useful to identity $E_1 \otimes E_2$ with a subspace of $\mathscr{L}(E_1^*,E_2)$ or $\mathscr{L}(E_2,E_1^*)$. Namely, given an $x = \sum_{i=1}^n x_{1,i} \otimes x_{2,i} \in E_1 \otimes E_2$, we define an operator $\Phi(x)$ of E_1^* into E_2 by

$$\Phi(x)f = \sum_{i=1}^{n} \langle x_{1,i},f \rangle x_{2,i} \in E_2, \qquad f \in E_1^*. \tag{2}$$

If $x \neq 0$, then we may assume that $\{x_{2,1}, \ldots ,x_{2,n}\}$ are linearly independent. If $\langle x_{1,i},f \rangle \neq 0$ for some $i = 1, \ldots ,n$, then $\Phi(x)f = \sum_{i=1}^n \langle x_{1,i},f \rangle x_{2,i} \neq 0$. Hence $\Phi(x) \neq 0$. Therefore, Φ is injective. The adjoint map $\Phi(x)^*$ is given by

$$\Phi(x)^*g = \sum_{i=1}^{n} \langle x_{2,i},g \rangle x_{1,i}, \qquad g \in E_2^*. \tag{2*}$$

We now define two important cross-norms of $E_1 \otimes E_2$. For each $x \in E_1 \otimes E_2$,

$$\|x\|_\lambda = \sup\left\{\left|\sum_{i=1}^{n} f(x_{1,i})g(x_{2,i})\right| : f \in E_1^*, \|f\| \leq 1; g \in E_2^*, \|g\| \leq 1\right\}, \tag{3}$$

where $x = \sum_{i=1}^{n} x_{1,i} \otimes x_{2,i}$, and

$$\|x\|_\gamma = \inf\left\{\sum_{i=1}^{n} \|x_{1,i}\|\|x_{2,i}\| : x = \sum_{i=1}^{n} x_{1,i} \otimes x_{2,i}\right\}. \tag{4}$$

It is clear that γ majorizes all other cross-norms; so it is called the *greatest* (or *projective*) cross-norm. The completion $E_1 \otimes_\gamma E_2$ is called the *projective tensor product* of E_1 and E_2. On the other hand, λ is called the *injective* cross-norm, and $E_1 \otimes_\lambda E_2$ is called the *injective* tensor product. The next result assures that λ is a norm, that is, $\|x\|_\lambda \neq 0$ for any nonzero $x \in E_1 \otimes E_2$, hence so is γ.

Proposition 2.1. *The map $\Phi : x \in E_1 \otimes E_2 \mapsto \Phi(x) \in \mathscr{L}(E_1^*, E_2)$ is an isometry with respect to the injective cross-norm λ in $E_1 \otimes E_2$. Hence Φ is extended to an isometry of $E_1 \otimes_\lambda E_2$ into $\mathscr{L}(E_1^*, E_2)$.*

PROOF. For each $x = \sum_{i=1}^{n} x_{1,i} \otimes x_{2,i}$, we have

$$\begin{aligned}\|x\|_\lambda &= \sup\left\{\left|\sum_{i=1}^{n} \langle x_{1,i}, f_1\rangle\langle x_{2,i}, f_2\rangle\right| : \|f_1\| \le 1, \|f_2\| \le 1\right\}\\ &= \sup\{|\langle \Phi(x) f_1, f_2\rangle| : \|f_1\| \le 1, \|f_2\| \le 1\}\\ &= \|\Phi(x)\|.\end{aligned}$$

Q.E.D.

Now let β be a cross-norm of $E_1 \otimes E_2$. Then β induces naturally a function β^* on $E_1^* \otimes E_2^*$ by

$$\left.\begin{aligned}\|f\|_{\beta^*} &= \sup\{|\langle x, f\rangle| : x \in E_1 \otimes E_2, \|x\|_\beta \le 1\},\\ f &= \sum_{i=1}^{n} f_{1,i} \otimes f_{2,i} \in E_1^* \otimes E_2^*,\end{aligned}\right\} \tag{5}$$

where $\langle x, f\rangle$ means, of course, the value

$$\langle x, f\rangle = \sum_{j=1}^{m} \sum_{i=1}^{n} \langle x_{1,j}, f_{1,i}\rangle\langle x_{2,j}, f_{2,i}\rangle$$

for each $x = \sum_{j=1}^{m} x_{1,j} \otimes x_{2,j} \in E_1 \otimes E_2$.

Proposition 2.2. *The function β^* on $E_1^* \otimes E_2^*$ is a cross-norm if and only if $\lambda \le \beta \le \gamma$.*

PROOF. For each $x \in E_1 \otimes E_2$, we have

$$\begin{aligned}\|x\|_\lambda &= \sup\{|\langle x, f_1 \otimes f_2\rangle| : \|f_1\| \le 1, \|f_2\| \le 1\}\\ &\le \sup\{\|x\|_\beta \|f_1 \otimes f_2\|_{\beta^*} : \|f_1\| \le 1, \|f_2\| \le 1\}.\end{aligned}$$

Hence if β^* is a cross-norm, then $\|f_1 \otimes f_2\|_{\beta^*} = \|f_1\|\|f_2\|$ by definition; hence $\|x\|_\lambda \le \|x\|_\beta$. Conversely, suppose $\lambda \le \beta \le \gamma$. For each $f_1 \in E_1^*$ and $f_2 \in E_2^*$, we have

$$\|f_1\|\|f_2\| = \sup\{|\langle x_1 \otimes x_2, f_1 \otimes f_2\rangle| : \|x_1\| \le 1, \|x_2\| \le 1\}$$
$$\le \|f_1 \otimes f_2\|_{\beta^*} \le \|f_1 \otimes f_2\|_{\lambda^*}.$$

For any $x \in E_1 \otimes E_2$, we have $|\langle x, f_1 \otimes f_2\rangle| \le \|x\|_\lambda \|f_1\|\|f_2\|$, so that $\|f_1 \otimes f_2\|_{\lambda^*} \le \|f_1\|\|f_2\|$; hence $\|f_1 \otimes f_2\|_{\beta^*} = \|f_1\|\|f_2\|$, that is, β^* is a cross-norm of $E_1^* \otimes E_2^*$. Q.E.D.

When $\lambda \le \beta \le \gamma$, the cross-norm β^* on $E_1^* \otimes E_2^*$ is said to be the *adjoint* cross-norm of β.

Let β be a cross-norm of $E_1 \otimes E_2$ with $\beta \ge \lambda$. Let f be an element of $(E_1 \otimes_\beta E_2)^*$. We have then

$$|\langle x_1 \otimes x_2, f\rangle| \le \|x_1 \otimes x_2\|_\beta \|f\|_{\beta^*} = \|x_1\|\|x_2\|\|f\|_{\beta^*},$$

so that there exists a bounded linear operator $\Phi_1'(f)$ of E_1 into E_2^* such that

$$\left.\begin{array}{c}\langle x_2, \Phi_1'(f)x_1\rangle = \langle x_1 \otimes x_2, f\rangle, \qquad x_1 \in E_1, \qquad x_2 \in E_2, \\ \|\Phi_1'(f)\| \le \|f\|_{\beta^*}\end{array}\right\} \tag{6}$$

We may, of course, associate a bounded operator $\Phi_2'(f)$ of E_2 into E_1^* by the following:

$$\left.\begin{array}{c}\langle x_1, \Phi_2'(f)x_2\rangle = \langle x_1 \otimes x_2, f\rangle; \\ \|\Phi_2'(f)\| \le \|f\|_{\beta^*}\end{array}\right\} \tag{6*}$$

It is then clear that $\Phi_1'(f)$ and $\Phi_2'(f)$ are the restrictions of each other's adjoint operator.

Theorem 2.3. *The map Φ_1' is an isometry of $(E_1 \otimes_\gamma E_2)^*$ onto $\mathscr{L}(E_1, E_2^*)$ and Φ_2' is an isometry of $(E_1 \otimes_\gamma E_2)^*$ onto $\mathscr{L}(E_2, E_1^*)$.*

PROOF. For each $x = \sum_{i=1}^n x_{1,i} \otimes x_{2,i} \in E_1 \otimes E_2$ and $f \in (E_1 \otimes_\gamma E_2)^*$, we have

$$|\langle x, f\rangle| = \left|\sum_{i=1}^n \langle x_{1,i} \otimes x_{2,i}, f\rangle\right|$$
$$= \left|\sum_{i=1}^n \langle x_{2,i}, \Phi_1'(f)x_{1,i}\rangle\right| \le \sum_{i=1}^n |\langle x_{2,i}, \Phi_1'(f)x_{1,i}\rangle|$$
$$\le \sum_{i=1}^n \|x_{1,i}\|\|x_{2,i}\|\|\Phi_1'(f)\|;$$

hence $|\langle x, f\rangle| \le \|x\|_\gamma \|\Phi_1'(f)\|$; so $\|f\|_{\gamma^*} \le \|\Phi_1'(f)\|$. Therefore, Φ_1' is an isometry of $(E_1 \otimes_\gamma E_2)^*$ into $\mathscr{L}(E_1, E_2^*)$. Let T be any element of $\mathscr{L}(E_1, E_2^*)$. Define a linear functional f on $E_1 \otimes E_2$ by

$$\left\langle \sum_{i=1}^n x_{1,i} \otimes x_{2,i}, f\right\rangle = \sum_{i=1}^n \langle x_{2,i}, Tx_{1,i}\rangle.$$

Since $|\langle\sum_{i=1}^{n} x_{1,i}\otimes x_{2,i},f\rangle| \le \sum_{i=1}^{n} \|x_{1,i}\|\|x_{2,i}\|\|T\|$, we have $\|f\|_{\gamma^*} \le \|T\|$; hence f is bound with respect to the γ-norm. Therefore, f is considered as an element of $(E_1 \otimes_\gamma E_2)^*$. It is now clear that $T = \Phi'_1(f)$. Q.E.D.

If f is in $E_1^* \otimes E_2^*$, then we have the operator $\Phi(f) \in \mathscr{L}(E_1^{**},E_2^*)$ defined by (2). It is easily seen that $\Phi'_1(f) = \Phi(f)|_{E_1}$ and $\Phi'_2(f) = \Phi(f)^*|_{E_2}$. We have, furthermore,

$$\begin{aligned}\|f\|_\lambda &= \|\Phi(f)\| = \sup\{\|\Phi(f)\tilde{x}\| : \tilde{x} \in E_1^{**}, \|\tilde{x}\| \le 1\}\\ &= \sup\{|\langle y,\Phi(f)\tilde{x}\rangle| : \tilde{x} \in E_1^{**}, \|\tilde{x}\| \le 1, y \in E_2, \|y\| \le 1\}.\end{aligned}$$

Since the function: $\tilde{x} \in E_1^{**} \mapsto \langle y,\Phi(f)\tilde{x}\rangle = \sum_{i=1}^{n} \langle \tilde{x},f_{1,i}\rangle\langle y,f_{2,i}\rangle$, where $f = \sum_{i=1}^{n} f_{1,i} \otimes f_{2,i}$, is $\sigma(E_1^{**},E_1^*)$-continuous, and since the unit ball of E_1 is $\sigma(E_1^{**},E_1^*)$-dense in the unit ball of E_1^{**}, the last supremum can be replaced by

$$\sup\{|\langle y,\Phi(f)x\rangle| : x \in E_1, \|x\| \le 1, y \in E_2, \|y\| \le 1\} = \|\Phi'_1(f)\| = \|f\|_{\gamma^*}.$$

Thus we reach the following conclusion:

Corollary 2.4. *The adjoint cross-norm γ^* of γ on $E_1^* \otimes E_2^*$ is the injective cross-norm λ.*

The converse statement $\lambda^* = \gamma$ is *not true* in general.

We now consider Hilbert spaces $\mathfrak{H}_1$ and $\mathfrak{H}_2$ for E_1 and E_2. The conjugate spaces of $\mathfrak{H}_1$ and $\mathfrak{H}_2$, which we denote by $\bar{\mathfrak{H}}_1$ and $\bar{\mathfrak{H}}_2$ respectively, are Banach spaces. By the Riesz representation theorem, we have a surjective conjugate linear isometry: $\xi \in \mathfrak{H}_i \mapsto \bar{\xi} \in \bar{\mathfrak{H}}_i$, $i = 1,2$, given by $(\eta|\xi) = \langle\eta,\bar{\xi}\rangle$, $\eta \in \mathfrak{H}_i$. Making use of the results in Section II.1, we obtain the following:

Theorem 2.5. (i) $(\mathfrak{H}_1 \otimes_\lambda \mathfrak{H}_2)^* = \bar{\mathfrak{H}}_1 \otimes_\gamma \bar{\mathfrak{H}}_2$.

(ii) *The map Φ given by* (2) *is an isometry of $\mathfrak{H}_1 \otimes_\lambda \mathfrak{H}_2$ onto $\mathscr{L}\mathscr{C}(\bar{\mathfrak{H}}_1,\mathfrak{H}_2)$.*

(iii) *The map Φ'_1 defined by* (6) *is an isometry of $(\mathfrak{H}_1 \otimes_\gamma \mathfrak{H}_2)^*$ onto $\mathscr{L}(\mathfrak{H}_1,\bar{\mathfrak{H}}_2)$.*

We leave the proof to the reader.

The Hilbert space norm in $\mathfrak{H}_1 \otimes \mathfrak{H}_2$ defined by the inner product (2) in Section 1 is certainly a cross-norm, which is sometimes indicated as the σ-norm, and the completion of $\mathfrak{H}_1 \otimes \mathfrak{H}_2$ (in the previous section, we used $\mathfrak{H}_1 \otimes \mathfrak{H}_2$ for the completion of the algebraic tensor product) is denoted by $\mathfrak{H}_1 \otimes_\sigma \mathfrak{H}_2$. However, except for the present discussion, we use $\mathfrak{H}_1 \otimes \mathfrak{H}_2$ for the Hilbert space completion since there is no danger of confusion. The operators in $\Phi(\mathfrak{H}_1 \otimes_\sigma \mathfrak{H}_2)$ are called of the *Hilbert–Schmidt class* and $\Phi(\mathfrak{H}_1 \otimes_\sigma \mathfrak{H}_2)$ is denoted by $\mathscr{L}\mathscr{S}(\bar{\mathfrak{H}}_1,\mathfrak{H}_2)$.

EXERCISES

1. Let E_1 and E_2 be Banach spaces with $E = E_1 \otimes_\gamma E_2$.
   (a) Show that if $\Phi: E_1 \times E_2 \mapsto F$ is a bounded bilinear map of $E_1 \times E_2$ into another Banach space F, then there exists a unique bounded linear map Φ_0 of E into F such that $\Phi(x_1,x_2) = \Phi_0(x_1 \otimes x_2)$ and $\|\Phi\| = \|\Phi_0\|$.
   (b) Let F be a Banach space. Show that if $i: E_1 \times E_2 \mapsto F$ is a bilinear map such that (i) $\|i(x_1,x_2)\| = \|x_1\|\|x_2\|$, (ii) for any Banach space G and a bounded bilinear map $\Phi: E_1 \times E_2 \mapsto G$ there exists a bounded linear map $\Phi_0: F \mapsto G$ such that $\Phi_0 \circ i = \Phi$ and $\|\Phi_0\| = \|\Phi\|$, and (iii) $i(E_1 \times E_2)$ is total in F, then there exists a unique isometry Ψ of $E_1 \otimes_\gamma E_2$ onto F such that $\Psi(x_1 \otimes x_2) = i(x_1,x_2)$.
2. Let $\mathfrak{H}_1$ and $\mathfrak{H}_2$ be Hilbert spaces. Show that $T \in \mathscr{L}(\mathfrak{H}_1,\mathfrak{H}_2)$ is of the Hilbert–Schmidt class $\mathscr{L}\mathscr{S}(\mathfrak{H}_1,\mathfrak{H}_2)$ if and only if $\mathrm{Tr}_{\mathfrak{H}_1}(T^*T) < +\infty$; equivalently, $\mathrm{Tr}_{\mathfrak{H}_2}(TT^*) < +\infty$.

3. Completely Positive Maps

Let A be a C^*-algebra. We denote by $M_n(A)$ the set of all $n \times n$-matrices $a = [a_{ij}]$ with entries a_{ij} in A. With the obvious matrix multiplication, and the *-operation, $M_n(A)$ is an involutive algebra, i.e.,

$$\left.\begin{aligned} (\lambda a + \mu b)_{i,j} &= \lambda a_{i,j} + \mu b_{i,j}, \\ (ab)_{i,j} &= \sum_{k=1}^{n} a_{i,k} b_{k,j}, \\ (a^*)_{i,j} &= a_{j,i}^*, \end{aligned}\right\} \tag{1}$$

for $a = [a_{i,j}]$, $b = [b_{i,j}] \in A$, and $\lambda,\mu \in \mathbf{C}$. Let $\{\pi,\mathfrak{H}\}$ be a faithful representation of A, and $\mathfrak{H}_n$ an n-dimensional Hilbert space with an orthogonal basis $\{\varepsilon_1,\varepsilon_2,\dots,\varepsilon_n\}$. We define a representation $\{\tilde{\pi},\tilde{\mathfrak{H}}\}$ of $M_n(A)$ as follows:

$$\left.\begin{aligned} &\tilde{\mathfrak{H}} = \mathfrak{H} \otimes \mathfrak{H}_n; \\ &\tilde{\pi}(a)(\xi \otimes \varepsilon_j) = \sum_{i=1}^{n} \pi(a_{i,j})\xi \otimes \varepsilon_i, \qquad a = [a_{i,j}] \in M_n(A). \end{aligned}\right\} \tag{2}$$

It is easy to see that $\tilde{\pi}$ is a faithful representation and $\tilde{\pi}(M_n(A)) = \pi(A) \otimes \mathscr{L}(\mathfrak{H}_n)$ on $\tilde{\mathfrak{H}} = \mathfrak{H} \otimes \mathfrak{H}_n$. Making use of U_i defined by (3) in Section 1, we have $\pi(a_{i,j}) = U_i^* \tilde{\pi}(a) U_j$ for each $a = [a_{i,j}] \in M_n(A)$; hence we get

$$\max_{1 \le i,j \le n} \|\pi(a_{i,j})\| \le \|\tilde{\pi}(a)\| \le \sum_{i,j=1}^{n} \|\pi(a_{i,j})\|;$$

therefore, $\tilde{\pi}(M_n(A))$ is uniformly closed in $\mathscr{L}(\tilde{\mathfrak{H}})$, which means that $M_n(A)$ is a C^*-algebra under the norm defined by $\|a\| = \|\tilde{\pi}(a)\|$, $a \in M_n(A)$. Since every faithful representation of a C^*-algebra is an isometry, the norm in $M_n(A)$ defined above does not depend on the choice of a faithful representation π of A.

Lemma 3.1. *An element of $M_n(A)$ is positive if and only if it is a sum of matrices of the form $[a_i^* a_j]$ with $a_1, \dots, a_n \in A$.*

PROOF. If $c = [a_i^* a_j]$, then $c = a^* a$ for $a = [a_{i,j}]$ with $a_{1,j} = a_j$, $1 \le i \le n$, and $a_{i,j} = 0$, $2 \le i \le n$, $1 \le j \le n$; so $c \ge 0$. If $a = [a_{i,j}] \in M_n(A)$ is positive, then there is an element $b = [b_{i,j}] \in M_n(A)$ such that $a = b^* b$. Hence we have

$$a_{i,j} = \sum_{k=1}^{n} b_{k,i}^* b_{k,j}.$$

Putting $c_k = [b_{k,i}^* b_{k,j}] \in M_n(A)$, which is a matrix of the indicated form, we have $a = \sum_{k=1}^n c_k$. Q.E.D.

The following criterion for positivity is also useful.

Lemma 3.2. *A matrix $a = [a_{i,j}] \in M_n(A)$ is positive if and only if $\sum_{i,j=1}^n x_i^* a_{i,j} x_j \ge 0$ for every $x_1, \dots, x_n \in A$.*

PROOF. Suppose $a \ge 0$. By Lemma 3.1, we may assume $a_{i,j} = a_i^* a_j$ for some $a_1, \dots, a_n \in A$. Then we have

$$\sum_{i,j=1}^{n} x_i^* a_{i,j} x_j = \sum_{i,j=1}^{n} x_i^* a_i^* a_j x_j = \left(\sum_{i=1}^{n} a_i x_i \right)^* \left(\sum_{j=1}^{n} a_j x_j \right) \ge 0.$$

Conversely, suppose $\sum_{i,j=1}^n x_i^* a_{i,j} x_j \ge 0$ for every $x_1, \dots, x_n \in A$. Let $\{\pi, \mathfrak{H}, \xi_0\}$ be an arbitrary cyclic representation of A. For each vector $\xi = \sum_{j=1}^n \xi_j \otimes \varepsilon_j \in \tilde{\mathfrak{H}} = \mathfrak{H} \otimes \mathfrak{H}_n$, we have

$$\begin{aligned}
(\tilde{\pi}(a)\xi|\xi) &= \sum_{i,j=1}^{n} (\pi(a)(\xi_j \otimes \varepsilon_j)|\xi_i \otimes \varepsilon_i) \\
&= \sum_{i,j,k=1}^{n} (\pi(a_{k,j})\xi_j \otimes \varepsilon_k|\xi_i \otimes \varepsilon_i) \\
&= \sum_{i,j=1}^{n} (\pi(a_{i,j})\xi_j|\xi_i).
\end{aligned}$$

Choose sequences $\{x_i^m : m = 1,2, \dots\}$ in A with $\xi_i = \lim_{m \to \infty} \pi(x_i^m)\xi_0$. We have then

$$\begin{aligned}
(\tilde{\pi}(a)\xi|\xi) &= \lim_{m \to \infty} \sum_{i,j=1}^{n} (\pi(a_{i,j})\pi(x_j^m)\xi_0|\pi(x_i^m)\xi_0) \\
&= \lim_{m \to \infty} \left(\pi \left(\sum_{i,j=1}^{n} (x_i^m)^* a_{i,j} x_j^m \right) \xi_0 | \xi_0 \right) \ge 0.
\end{aligned}$$

Hence $\tilde{\pi}(a)$ is positive for any cyclic representation π. Since $(\sum_\alpha^{\otimes} \pi_\alpha)^{\sim} = \sum_\alpha^{\otimes} \tilde{\pi}_\alpha$, $\tilde{\pi}(a)$ is positive for any representation π by Proposition I.9.17. Hence $a \ge 0$. Q.E.D.

Definition 3.3. Let A and B be C^*-algebras. For each linear map $\varphi: A \mapsto B$, we define a linear map $\varphi_n: M_n(A) \mapsto M_n(B)$ by

$$\varphi_n[a_{i,j}] = [\varphi(a_{ij})].$$

If φ_n is positive, then φ is said to be *n-positive.* If φ is n-positive for all n, then φ is said to be *completely positive.*

Corollary 3.4. *Let A and B be C^*-algebras. A linear map $\varphi: A \mapsto B$ is n-positive if and only if*

$$\sum_{i,j=1}^{n} y_i^* \varphi(x_i^* x_j) y_j \geq 0 \tag{3}$$

for every $x_1, \ldots, x_n \in A$ and $y_1, \ldots, y_n \in B$.

The assertion here is just a combination of Lemmas 3.1 and 3.2.

Corollary 3.5. *Let A and B be C^*-algebras. If B is abelian, then any positive linear map $\varphi: A \mapsto B$ is completely positive. In particular, a positive linear functional on a C^*-algebra is always completely positive.*

PROOF. By Theorem I.4.4, the C^*-algebra B is identified with the C^*-algebra $C_\infty(\Omega)$ of all continuous functions on a locally compact space Ω vanishing at infinity. For each $x_1, \ldots, x_n \in A$ and $y_1, \ldots, y_n \in B = C_\infty(\Omega)$, we have

$$\begin{aligned}\left(\sum_{i,j=1}^{n} y_i^* \varphi(x_i^* x_j) y_j\right)(\omega) &= \sum_{i,j=1}^{n} \overline{y_i(\omega)} \varphi(x_i^* x_j)(\omega) y_j(\omega) \\ &= \varphi\left(\left(\sum_{i=1}^{n} \overline{y_i(\omega)} x_i^*\right)\left(\sum_{j=1}^{n} y_j(\omega) x_j\right)\right)(\omega) \\ &= \varphi\left(\left(\sum_{i=1}^{n} y_i(\omega) x_i\right)^*\left(\sum_{j=1}^{n} y_j(\omega) x_j\right)\right)(\omega) \geq 0;\end{aligned}$$

hence $\sum_{i,j=1}^{n} y_i^* \varphi(x_i^* x_j) y_j \geq 0$. Hence φ is completely positive. Q.E.D.

We observe that any positive map φ of a C^*-algebra A into another C^*-algebra B is continuous. In fact, if $\{x_n\}$ converges to zero in A and if $\lim_{n\to\infty} \|\varphi(x_n) - y\| = 0$ for some $y \in B$, then for any $\omega \in B_+^*$ $\lim_{n\to\infty} \omega \circ \varphi(x_n) = 0$ since $\omega \circ \varphi$ is continuous by Proposition I.9.12, being a positive linear functional; therefore $\omega(y) = \lim_{n\to\infty} \omega(\varphi(x_n)) = 0$ for every $\omega \in B_+^*$; so $y = 0$. By the closed graph theorem, φ is continuous. In particular, a completely positive map is continuous.

Theorem 3.6. *Let A be a C^*-algebra and $\mathfrak{H}$ a Hilbert space.*

(i) *If $\{\pi, \mathfrak{K}\}$ is a representation of A, and V is a bounded linear operator of $\mathfrak{H}$ into $\mathfrak{K}$, then the map $\varphi: a \in A \mapsto V^*\pi(a)V \in \mathscr{L}(\mathfrak{H})$ is completely positive.*

(ii) *If φ is a completely positive map of A into $\mathscr{L}(\mathfrak{H})$, then there exist a representation $\{\pi,\mathfrak{K}\}$ of A, a normal representation ρ of the von Neumann algebra $\varphi(A)'$ on the same space $\mathfrak{K}$ and a bounded linear operator V of $\mathfrak{H}$ into $\mathfrak{K}$ such that*

$$\left.\begin{aligned} \varphi(a) = V^*\pi(a)V, \qquad a \in A,\\ \rho(x)V = Vx, \qquad x \in \varphi(A)' \quad \text{and} \quad \rho(\varphi(A)') \subset \pi(A)',\\ \mathfrak{K} = [\pi(A)V\mathfrak{H}]. \end{aligned}\right\} \tag{4}$$

(iii) *Given a completely positive map φ of A into $\mathscr{L}(\mathfrak{H})$, the triple of a representation $\{\pi,\mathfrak{K}\}$ of A, a normal faithful representation ρ of $\varphi(A)'$ on $\mathfrak{K}$, and a bounded operator V of $\mathfrak{H}$ into $\mathfrak{K}$ is unique under the minimality condition, $[\pi(A)V\mathfrak{H}] = \mathfrak{K}$. In other words, if $\{\pi',\mathfrak{K}'\}$, ρ' and V' are another triple satisfying* (4), *then there exists a unitary U of $\mathfrak{K}$ onto $\mathfrak{K}'$ such that*

$$\left.\begin{aligned} \pi'(a) = U\pi(a)U^*, \qquad a \in A,\\ \rho'(x) = U\rho(x)U^*, \qquad x \in \varphi(A)',\\ V' = UV. \end{aligned}\right\} \tag{5}$$

PROOF. (i) For each $x_1,\dots,x_n \in A$ and $y_1,\dots,y_n \in \mathscr{L}(\mathfrak{H})$, we have

$$\begin{aligned} \sum_{i,j=1}^{n} y_i^*\varphi(x_i^*x_j)y_j &= \sum_{i,j=1}^{n} y_i^*V^*\pi(x_i^*x_j)Vy_j \\ &= \left(\sum_{i=1}^{n} \pi(x_i)Vy_i\right)^*\left(\sum_{j=1}^{n} \pi(x_j)Vy_j\right) \geq 0. \end{aligned}$$

(ii) Consider the algebraic tensor product $A \otimes \mathfrak{H}$ of A and $\mathfrak{H}$. We define a sesquilinear form on $A \otimes \mathfrak{H}$ as follows:

$$(\xi|\eta) = \sum_{i=1}^{n}\sum_{j=1}^{m} (\varphi(b_j^*a_i)\xi_i|\eta_j) \tag{6}$$

for $\xi = \sum_{i=1}^{n} a_i \otimes \xi_i$, $\eta = \sum_{j=1}^{m} b_j \otimes \eta_j \in A \otimes \mathfrak{H}$. We have then

$$(\xi|\xi) = \sum_{i,j=1}^{n} (\varphi(a_i^*a_j)\xi_j|\xi_i) \geq 0$$

by the complete positivity of φ. We define next actions π_0 of A and ρ_0 of $\varphi(A)'$ on $A \otimes \mathfrak{H}$ as follows:

$$\begin{aligned} \pi_0(a)\xi &= \sum_{i=1}^{n} ax_i \otimes \xi_i, \\ \rho_0(b)\xi &= \sum_{i=1}^{n} x_i \otimes b\xi_i, \end{aligned} \tag{7}$$

for $a \in A$, $b \in \varphi(A)'$ and $\xi = \sum_{i=1}^{n} x_i \otimes \xi_i \in A \otimes \mathfrak{H}$. We claim

$$\left.\begin{aligned} (\pi_0(a)\xi|\pi_0(a)\xi) \leq \|a\|^2(\xi|\xi), \qquad a \in A,\\ (\rho_0(b)\xi|\rho_0(b)\xi) \leq \|b\|^2(\xi|\xi), \qquad b \in \varphi(A)', \qquad \xi \in A \otimes \mathfrak{H}. \end{aligned}\right\} \tag{8}$$

Let $\tilde{a}$ and $\tilde{x}$ be the elements of $M_n(A)$ defined by

$$\tilde{a}_{i,j} = \delta_{i,j}a \quad \text{and} \quad \tilde{x}_{i,j} = \delta_{1,i}x_j, \qquad 1 \leq i,j \leq n.$$

We have then

$$\begin{aligned}
(\pi_0(a)\xi|\pi_0(a)\xi) &= \sum_{i,j=1}^{n} (\varphi(x_j^*a^*ax_i)\xi_i|\xi_j) \\
&= \sum_{i,j=1}^{n} (\varphi_n(\tilde{x}^*\tilde{a}^*\tilde{a}\tilde{x})_{j,i}\xi_i|\xi_j) \\
&= \left(\varphi_n(\tilde{x}^*\tilde{a}^*\tilde{a}\tilde{x})\left(\sum_{i=1}^{n} \xi_i \otimes \varepsilon_i\right)\middle|\left(\sum_{j=1}^{m} \xi_j \otimes \varepsilon_j\right)\right) \\
&\leq \|\tilde{a}\|^2 \left(\tilde{\varphi}_n(\tilde{x}^*\tilde{x})\left(\sum_{i=1}^{n} \xi_i \otimes \varepsilon_i\right)\middle|\left(\sum_{j=1}^{m} \xi_j \otimes \varepsilon_j\right)\right) \\
&= \|a\|^2(\xi|\xi),
\end{aligned}$$

where the last step of the above calculation is justified by the n-positivity of φ and the inequality $\varphi_n(\tilde{x}^*\tilde{a}^*\tilde{a}\tilde{x}) \leq \|\tilde{a}\|^2\varphi_n(\tilde{x}^*\tilde{x})$. Let $\tilde{b}$ be the element in $M_n(\varphi(A)')$ given by $\tilde{b}_{i,j} = \delta_{i,j}b$. Then we have

$$\begin{aligned}
(\rho_0(b)\xi|\rho_0(b)\xi) &= \sum_{i,j=1}^{n} (b^*\varphi(x_j^*x_i)b\xi_i|\xi_j) \\
&= \left(\tilde{b}^*\varphi_n(\tilde{x}^*\tilde{x})\tilde{b}\left(\sum_{i=1}^{n} \xi_i \otimes \varepsilon_i\right)\middle|\left(\sum_{j=1}^{n} \xi_j \otimes \varepsilon_j\right)\right) \\
&\leq \|\tilde{b}\|^2 \left(\varphi_n(\tilde{x}^*\tilde{x})\left(\sum_{i=1}^{n} \xi_i \otimes \varepsilon_i\right)\middle|\left(\sum_{j=1}^{m} \xi_j \otimes \varepsilon_j\right)\right) \\
&= \|b\|^2(\xi|\xi),
\end{aligned}$$

since $\varphi_n(\tilde{x}^*\tilde{x})$ and $\tilde{b}^*\tilde{b}$ commute.

We next claim that

$$\left.\begin{aligned}
(\pi_0(a)\xi|\eta) &= (\xi|\pi_0(a^*)\eta), \qquad a \in A, \qquad \xi,\eta \in A \otimes \mathfrak{H}, \\
(\rho_0(b)\xi|\eta) &= (\xi|\rho_0(b^*)\eta), \qquad b \in \varphi(A)'.
\end{aligned}\right\} \tag{9}$$

But this is verified by a straightforward calculation; so we omit the details. It is then clear that

$$\left.\begin{aligned}
\pi_0(a_1a_2) &= \pi_0(a_1)\pi_0(a_2), \qquad a_1,a_2 \in A, \\
\rho_0(b_1b_2) &= \rho_0(b_1)\rho_0(b_2), \qquad b_1,b_2 \in \varphi(A)'.
\end{aligned}\right\} \tag{10}$$

Let N be the set of all $\xi \in A \otimes \mathfrak{H}$ with $(\xi|\xi) = 0$. The quotient space $\mathfrak{K}_0 = (A \otimes \mathfrak{H})/N$ turns out to be a pre-Hilbert space in the usual fashion. Let $\mathfrak{K}$ denote the completion of $\mathfrak{K}_0$ and P denote the canonical map: $\xi \in A \otimes \mathfrak{H} \mapsto \xi + N \in \mathfrak{K}$. By (8), (9), and (10), there exist representations π of A

and ρ of $\varphi(A)'$ such that

$$\begin{cases} \pi(a)P\xi = P\pi_0(a)\xi, & a \in A, \quad \xi \in A \otimes \mathfrak{H}, \\ \rho(b)P\xi = P\rho_0(b)\xi, & b \in \varphi(A)'. \end{cases}$$

By construction, $\pi_0(A)$ and $\rho_0(\varphi(A)')$ commute, so that $\pi(A)$ and $\rho(\varphi(A)')$ commute. For each $\xi = \sum_{i=1}^n x_i \otimes \xi_i \in A \otimes \mathfrak{H}$, and $b \in \varphi(A)'$, we have

$$\begin{aligned} (\rho(b)P\xi|P\xi) &= \sum_{i,j=1}^{n} (\varphi(x_i^* x_j)b\xi_j|\xi_i) \\ &= \sum_{i,j=1}^{n} (b\varphi(x_i^* x_j)\xi_j|\xi_i); \end{aligned}$$

hence $\omega(\rho;P\xi)$ is a normal linear functional on $\varphi(A)'$. Since $P(A \otimes \mathfrak{H}) = \mathfrak{K}_0$ is dense in $\mathfrak{K}$, $\omega(\rho;\xi)$ is normal for any $\xi \in \mathfrak{K}$; therefore ρ is a normal representation of $\varphi(A)'$. Let $\{u_i\}$ be an approximate identity of A. Then $\{\varphi(u_i)\}_{i\in I}$ is an increasing net in $\mathscr{L}(\mathfrak{H})_+$. Since $\|\varphi(u_i)\| \le \|\varphi\|$, $\{\varphi(u_i)\}$ converges strongly to some positive operator, say a_0. For each $i \in I$, we define $V_i\xi = P(u_i \otimes \xi)$, $\xi \in \mathfrak{H}$. We have then

$$\|V_i\xi\|^2 = (\varphi(u_i^2)\xi|\xi) \le \|\varphi\|\,\|\xi\|^2, \quad \text{so} \quad \|V_i\| \le \|\varphi\|^{1/2};$$

if $i < j$,

$$\begin{aligned} \|(V_i - V_j)\xi\|^2 &= (\varphi((u_j - u_i)^2)\xi|\xi) \le (\varphi(u_j - u_i)\xi|\xi) \\ &\le \|[\varphi(u_j) - \varphi(u_i)]\xi\|\,\|\xi\|; \end{aligned}$$

hence V_i converges strongly to a bounded operator V of $\mathfrak{H}$ into $\mathfrak{K}$. For each $i \in I$, we have

$$\begin{aligned} \left(V_i^*P\left(\sum_{k=1}^n x_k \otimes \xi_k\right)\Big|\eta\right) &= \left(P\left(\sum_{k=1}^n x_k \otimes \xi_k\right)\Big|V_i\eta\right) \\ &= \left(P\left(\sum_{k=1}^n x_k \otimes \xi_k\right)\Big|P(u_i \otimes \eta)\right) \\ &= \sum_{k=1}^n (\varphi(u_i x_k)\xi_k|\eta) = \left(\sum_{k=1}^n \varphi(u_i x_k)\xi_k\Big|\eta\right), \end{aligned}$$

so that

$$V_i^*P\left(\sum_{k=1}^n x_k \otimes \xi_k\right) = \sum_{k=1}^n \varphi(u_i x_k)\xi_k.$$

Therefore, we have

$$V^*P\left(\sum_{k=1}^n x_k \otimes \xi_k\right) = \sum_{k=1}^n \varphi(x_k)\xi_k.$$

Therefore we have, for each $a \in A$ and $\xi \in \mathfrak{H}$,

$$\begin{aligned} V^*\pi(a)V_i\xi &= V^*\pi(a)P(u_i \otimes \xi) \\ &= V^*P(au_i \otimes \xi) = \varphi(au_i)\xi. \end{aligned}$$

Hence we get

$$V^*\pi(a)V\xi = \lim_i V^*\pi(a)V_i\xi$$
$$= \lim_i \varphi(au_i)\xi = \varphi(a)\xi;$$

hence $V^*\pi(a)V = \varphi(a)$. Furthermore, we have, for each $x,y \in A$ and $\xi,\eta \in \mathfrak{H}$,

$$\begin{aligned}(\pi(x)V\xi|P(y \otimes \eta)) &= \lim_i (\pi(x)V_i\xi|P(y \otimes \eta)) \\ &= \lim_i (\pi(x)P(u_i \otimes \xi)|P(y \otimes \eta)) \\ &= \lim_i (P(xu_i \otimes \xi)|P(y \otimes \eta)) \\ &= \lim_i (\varphi(y^*xu_i)\xi|\eta) = (\varphi(y^*x)\xi|\eta) \\ &= (P(x \otimes \xi)|P(y \otimes \eta)).\end{aligned}$$

Since $\{P(y \otimes \eta): y \in A, \eta \in \mathfrak{H}\}$ spans $\mathfrak{K}$, we have

$$\pi(x)V\xi = P(x \otimes \xi).$$

Therefore, we have $[\pi(A)V\mathfrak{H}] = \mathfrak{K}$.

For any $x \in A$, $\xi \in \mathfrak{H}$ and $b \in \varphi(A)'$, we have

$$\begin{aligned}\pi(x)\rho(b)V\xi &= \rho(b)\pi(x)V\xi = \rho(b)P(x \otimes \xi) \\ &= P(x \otimes b\xi) = \pi(x)Vb\xi.\end{aligned}$$

Since $\pi(A)$ is nondegenerate, we have $\rho(b)V = Vb$, $b \in \varphi(A)'$.

(iii) Suppose that $\{\pi',\rho',V'\}$ is another triplet satisfying (4). Let U_0 denote the map defined by

$$U_0\left(\sum_{i=1}^n \pi(x_i)V\xi_i\right) = \sum_{i=1}^n \pi'(x_i)V'\xi_i.$$

We then have

$$\begin{aligned}&\left(U_0 \sum_{i=1}^n \pi(x_i)V\xi_i \middle| U_0 \sum_{j=1}^m \pi(y_j)V\eta_j\right) \\ &= \left(\sum_{i=1}^n \pi'(x_i)V'\xi_i \middle| \sum_{j=1}^m \pi'(y_j)V'\eta_j\right) = \sum_{i,j} (V'^*\pi'(y_j^*x_i)V'\xi_i|\eta_j) \\ &= \sum_{i,j} (\varphi(y_j^*x_i)\xi_i|\eta_j) = \sum_{i,j} (V^*\pi(y_j^*x_i)V\xi_i|\eta_j) \\ &= \left(\sum_{i=1}^n \pi(x_i)V\xi_i \middle| \sum_{j=1}^m \pi(y_j)V\eta_j\right);\end{aligned}$$

therefore, U_0 is extended to an isometry U of $\mathfrak{K}$ into $\mathfrak{K}'$. Since the range of U_0, $\pi'(A)V'\mathfrak{H}$, is dense in $\mathfrak{K}'$, U is indeed a unitary operator of $\mathfrak{K}$ onto $\mathfrak{K}'$.

It is now easy to see that $U\pi(a)U^* = \pi'(a)$, $a \in A$. For any $b \in \varphi(A)'$, we have

$$\begin{aligned}\rho'(b)U\left(\sum_{i=1}^{n} \pi(x_i)V\xi_i\right) &= \rho'(b) \sum_{i=1}^{n} \pi'(x_i)V'\xi_i \\ &= \sum_{i=1}^{n} \pi'(x_i)\rho'(b)V'\xi_i = \sum_{i=1}^{n} \pi'(x_i)V'b\xi_i \\ &= U\left(\sum_{i=1}^{n} \pi(x_i)Vb\xi_i\right) = U\left(\sum_{i=1}^{n} \pi(x_i)\rho(b)V\xi_i\right) \\ &= U\rho(b) \sum_{i=1}^{n} \pi(x_i)V\xi_i.\end{aligned}$$

Hence $U\rho(b)U^* = \rho'(b)$, $b \in \varphi(A)'$. By definition, we have $UV = V'$. Q.E.D.

Remark 3.7. If the C^*-algebra A in question is unital, then the construction of V in the above arguments is easier. Namely, V is defined by $V\xi = P(1 \otimes \xi)$. If $\varphi(1) = 1$, or more generally $a_0 = 1$, then one can show that V is an isometry; hence the representation ρ of $\varphi(A)'$ is faithful because we have

$$V^*V = \lim_i V^*\pi(u_i)V = \lim_i \varphi(u_i) = a_0 = 1.$$

Corollary 3.8. If φ is a completely positive map of a C^*-algebra A into another C^*-algebra B, then we have

$$\varphi(a)^*\varphi(a) \leq \|\varphi\|\varphi(a^*a), \qquad a \in A. \tag{11}$$

Proof. We may assume that B is a C^*-subalgebra of $\mathscr{L}(\mathfrak{H})$ for some Hilbert space $\mathfrak{H}$. The operator V given in Theorem 3.6 has norm $\|V\| \leq \|\varphi\|^{1/2}$, as seen in the second part of the proof of the theorem. Hence we have

$$\begin{aligned}\varphi(a)^*\varphi(a) &= V^*\pi(a)^*VV^*\pi(a)V \\ &\leq \|V\|^2 V^*\pi(a^*a)V \\ &= \|\varphi\|\varphi(a^*a).\end{aligned}$$

Q.E.D.

Proposition 3.9. *If A is an abelian C^*-algebra, then any positive linear map φ of A into another C^*-algebra B is completely positive.*

Proof. The C^*-algebra A is identified with the algebra $C_\infty(\Omega)$ of all continuous functions on a locally compact space Ω vanishing at infinity. We also consider B as a C^*-algebra of operators on a Hilbert space $\mathfrak{H}$. We shall show that $\sum_{i,j} (\varphi(x_i^* x_j)\xi_j|\xi_i) \geq 0$ for each $x_1, \ldots, x_n \in A$ and $\xi_1, \ldots, \xi_n \in \mathfrak{H}$. Let $\mu_{i,j}(x) = (\varphi(x)\xi_i|\xi_j)$, $1 \leq i,j \leq n$. Then each $\mu_{i,j}$ is given by a complex finite Radon measures on Ω. Let $\mu = \sum_{i,j=1}^{n} |\mu_{i,j}|$, where $|\cdot|$ means the absolute value of the measure $\mu_{i,j}$ in the sense of Section III.4. Then there exists $f_{i,j} \in L^1(\Omega,\mu)$ for each i, j such that

$$\mu_{i,j}(x) = \int_\Omega x(\omega)f_{i,j}(\omega)\, d\mu(\omega).$$

For any $\lambda_1, \dots, \lambda_n \in \mathbf{C}$, we have

$$\sum_{i,j=1}^{n} \mu_{i,j}(x^*x)\lambda_i\bar{\lambda}_j = \sum_{i,j} (\varphi(x^*x)\xi_i|\xi_j)\lambda_i\bar{\lambda}_j$$
$$= \left(\varphi(x^*x) \sum_{i=1}^{n} \lambda_i\xi_i \middle| \sum_{j=1}^{n} \lambda_j\xi_j\right) \geq 0;$$

hence $\sum_{i,j=1}^{n} \mu_{i,j}\lambda_i\bar{\lambda}_j$ is a positive measure; so $\sum_{i,j=1}^{n} f_{i,j}(\omega)\lambda_i\bar{\lambda}_j \geq 0$ for μ-almost every ω. Hence we have now

$$\sum_{i,j=1}^{n} (\varphi(x_j^*x_i)\xi_i|\xi_j) = \sum_{i,j=1}^{n} \int_\Omega (x_j^*x_i)(\omega)f_{i,j}(\omega)\, d\mu(\omega)$$
$$= \int_\Omega \sum_{i,j} (f_{i,j}(\omega)x_i(\omega)\overline{x_j(\omega)})\, d\mu(\omega) \geq 0. \qquad \text{Q.E.D.}$$

We are now going to dualize the notion of complete positivity. Let A be a C^*-algebra. We define $M_n(A^*)$ to be the vector space of all $n \times n$-matrices $f = [f_{i,j}]$ with entries $f_{i,j} \in A^*$. We identify $M_n(A^*)$ and $M_n(A)^*$ by putting

$$f(a) = \sum_{i,j=1}^{n} f_{i,j}(a_{i,j}) \quad \text{for} \quad a = [a_{i,j}] \in M_n(A). \tag{12}$$

By Lemma 3.1, f is positive if and only if

$$\sum_{i,j=1}^{n} f_{i,j}(a_i^*a_j) \geq 0 \tag{13}$$

for any $a_1, \dots, a_n \in A$.

For a subspace E of either a C^*-algebra A or its conjugate space A^*, we denote by $M_n(E)$ the space of all matrices $e = [e_{i,j}]$ with $e_{i,j} \in E$. We equip $M_n(E)$ with the relative order given by either $M_n(A)$ or $M_n(A^*)$ as the case may be. If F is another such space, and if φ is a linear map of E into F, we define $\varphi_n : M_n(E) \mapsto M_n(F)$ as before, and φ is said to be *n-positive* if φ_n is positive. If φ is n-positive for all n, then φ is said to be *completely positive.*

It is easy to see that the composition of completely positive maps is again completely positive. Furthermore, for C^*-algebras A, B, and a bounded linear map $\varphi : A \mapsto B$, φ is completely positive if and only if its transpose ${}^t\varphi : B^* \mapsto A^*$ is completely positive. In fact, for each $x = [x_{i,j}] \in M_n(A)$ and $f = [f_{i,j}] \in M_n(B^*)$, we have

$$\langle x, {}^t(\varphi_n)(f)\rangle = \langle \varphi_n(x), f\rangle = \sum_{i,j=1}^{n} \langle \varphi_n(x_{i,j}), f_{i,j}\rangle$$
$$= \sum_{i,j=1}^{n} \langle x_{i,j}, {}^t\varphi(f_{i,j})\rangle = \langle x, ({}^t\varphi)_n(f)\rangle;$$

that is, ${}^t(\varphi_n) = ({}^t\varphi)_n$. Hence the assertion follows from the general fact that a bounded linear map $\varphi: A \mapsto B$ is positive if and only if its adjoint ${}^t\varphi$ is positive.

There is an important example of completely positive map of a C^*-algebra into the conjugate space of another C^*-algebra. Let A be a C^*-algebra and ω a state of A. Let C_ω denote the subspace of A^* spanned linearly by the hereditary convex subcone C_ω^+ of A_+^*:

$$C_\omega^+ = \{\rho \in A_+^* : \rho \leq \alpha\omega \quad \text{for some} \quad \alpha \geq 0\}.$$

Let $\{\pi_\omega, \mathfrak{H}_\omega, \xi_\omega\}$ be the cyclic representation of A induced by ω. Define a map $\theta_\omega : \pi_\omega(A)' \mapsto A^*$ by

$$\langle a, \theta_\omega(x)\rangle = (\pi_\omega(a)x\xi_\omega | \xi_\omega), \qquad a \in A, \qquad x \in \pi_\omega(A)'. \tag{14}$$

Proposition 3.10. *The map θ_ω is a completely positive bijection of $\pi_\omega(A)'$ to C_ω and its inverse is completely positive.*

Proof. Since $\theta_\omega(x^*x) = \omega(\pi; x\xi_\omega) \geq 0$, θ_ω is positive. If h is a positive element of $\pi_\omega(A)'$, we have

$$\begin{aligned}\langle a^*a, \theta_\omega(h)\rangle &= (\pi_\omega(a)^*\pi_\omega(a)h\xi_\omega | \xi_\omega) \\ &= \|h^{1/2}\pi_\omega(a)\xi_\omega\|^2 \leq \|h^{1/2}\|^2\|\pi_\omega(a)\xi_\omega\|^2 \\ &= \|h\|\omega(a^*a);\end{aligned}$$

hence we have $0 \leq \theta_\omega(h) \leq \|h\|\omega$; so $\theta_\omega(\pi_\omega(A)'_+) \subset C_\omega^+$. Therefore, $\theta_\omega(\pi(A)') \subset C_\omega$ since $\pi_\omega(A)'$ is spanned linearly by its positive part $\pi_\omega(A)'_+$. If $\rho \in C_\omega^+$, then the positive sesquilinear form

$$B_\rho(\pi_\omega(a)\xi_\omega, \pi_\omega(b)\xi_\omega) = \rho(b^*a), \qquad a,b \in A,$$

on $\pi_\omega(A)\xi_\omega$ is bounded; hence there exists a bounded positive operator h on $\mathfrak{H}_\omega$ such that

$$\rho(b^*a) = (h\pi_\omega(a)\xi_\omega | \pi_\omega(b)\xi_\omega), \qquad a,b \in A.$$

We have then, for each $a,b,c \in A$,

$$\begin{aligned}(h\pi_\omega(a)\pi_\omega(b)\xi_\omega | \pi_\omega(c)\xi_\omega) &= (h\pi_\omega(ab)\xi_\omega | \pi_\omega(c)\xi_\omega) \\ &= \rho(c^*ab) = \rho((a^*c)^*b) = (h\pi_\omega(b)\xi_\omega | \pi_\omega(a^*c)\xi_\omega) \\ &= (\pi_\omega(a)h\pi_\omega(b)\xi_\omega | \pi_\omega(c)\xi_\omega);\end{aligned}$$

hence $h\pi_\omega(a) = \pi_\omega(a)h$, i.e., $h \in \pi_\omega(A)'$. Furthermore, we have

$$\begin{aligned}\rho(a) &= \lim_i \rho(u_i a) = \lim (h\pi_\omega(a)\xi_\omega | \pi_\omega(u_i)\xi_\omega) \\ &= (h\pi_\omega(a)\xi_\omega | \xi_\omega) = \langle a, \theta_\omega(h)\rangle,\end{aligned}$$

where $\{u_i\}$ is an approximate identity of A. Hence $\rho = \theta_\omega(h)$ is in $\theta_\omega(\pi_\omega(A)'_+)$. Hence θ_ω is surjective. If $\theta_\omega(x) = 0$, then for each $a,b \in A$

$$0 = \langle b^*a, \theta_\omega(x) \rangle = (x\pi_\omega(a)\xi_\omega | \pi_\omega(b)\xi_\omega);$$

hence $x = 0$. Thus θ_ω is injective. We have seen that θ_ω^{-1} is positive. For any $x_1, \ldots, x_n \in \pi_\omega(A)'$ and $a_1, \ldots, a_n \in A$, we have

$$\begin{aligned}
\langle [a_i^* a_j], (\theta_\omega)_n [x_i^* x_j] \rangle &= \langle [a_i^* a_j], [\theta_\omega(x_i^* x_j)] \rangle \\
&= \sum_{i,j=1}^{n} \langle a_i^* a_j, \theta_\omega(x_i^* x_j) \rangle = \sum_{i,j=1}^{n} (x_i^* x_j \pi_\omega(a_i^* a_j)\xi_\omega | \xi_\omega) \\
&= \left\| \sum_{j=1}^{n} x_j \pi_\omega(a_j)\xi_\omega \right\|^2 \geq 0;
\end{aligned}$$

hence $(\theta_\omega)_n$ is positive by Lemma 3.1. Let $f = [f_{i,j}]$ be a positive element of $M_n(C_\omega)$. Let $\tilde{\mathfrak{H}}_\omega = \mathfrak{H}_\omega \otimes \mathfrak{H}_n$, with $\mathfrak{H}_n$ an n-dimensional Hilbert space in which we fix an orthogonal basis $\varepsilon_1, \ldots, \varepsilon_n$. The algebra $M_n(\pi_\omega(A)')$ acts on $\tilde{\mathfrak{H}}_\omega$ as indicated by (2). Choose a vector $\xi \in \tilde{\mathfrak{H}}_\omega$ of the form $\xi = \sum_{i=1}^n \pi_\omega(a_i)\xi_\omega \otimes \varepsilon_i$ for some $a_1, \ldots, a_n \in A$. We have then

$$\begin{aligned}
((\theta_\omega^{-1})_n(f)\xi|\xi) &= \sum_{i,j=1}^{n} (\theta_\omega^{-1}(f_{i,j})\pi_\omega(a_j)\xi_\omega | \pi_\omega(a_i)\xi_\omega) \\
&= \sum_{i,j=1}^{n} (\theta_\omega^{-1}(f_{i,j})\pi_\omega(a_i^* a_j)\xi_\omega | \xi_\omega) \\
&= \sum_{i,j=1}^{n} f_{i,j}(a_i^* a_j) = f([a_i^* a_j]) \geq 0.
\end{aligned}$$

Since such ξ are dense in $\tilde{\mathfrak{H}}_\omega$, ξ_ω being cyclic, it follows that $(\theta_\omega^{-1})_n(f) \geq 0$. Hence $(\theta_\omega^{-1})_n$ is positive. Thus θ_ω^{-1} is completely positive. Q.E.D.

Remark 3.11.(i) In general, θ_ω is not continuous. But, if φ is a completely positive map of a unital C^*-algebra B into A^* such that $\varphi(1) = \omega$, then $\varphi(B) \subset C_\omega$ and $\theta_\omega^{-1} \circ \varphi : B \mapsto \pi_\omega(A)'$ is a complete positive map hence continuous in norm.

(ii) If $A = M_n(\mathbf{C})$, then the map $a = [a_{i,j}] \in M_n(\mathbf{C}) \mapsto f = [a_{i,j}] \in M_n(\mathbf{C})^*$ is not completely positive. However, the map: $a = [a_{i,j}] \in M_n(\mathbf{C}) \mapsto {}^t f = [a_{j,i}] \in M_n(\mathbf{C})^* = A^*$ is completely positive.

Exercises

1. Let π be a C^*-isomorphism of a C^*-algebra A onto another C^*-algebra B. Let M_n, $n \geq 2$, be an $n \times n$-matrix algebra over $\mathbf{C}$. Show that if $\pi \otimes \text{id}$ is a C^*-isomorphism of $A \otimes M_n$ onto $B \otimes M_n$, then π must be an isomorphism, where id means the identity map of M_n onto M_n itself.

2. Let G be a locally compact group and $\mathfrak{H}$ a Hilbert space. An $\mathscr{L}(\mathfrak{H})$-valued strongly continuous function $x(\cdot)$ on G is called *completely positive definite* if for any $s_1,s_2,\ldots,s_n \in G$ the operator matrix $[x(s_i^{-1}s_j)]$ is positive. Show that there exist a strongly continuous unitary representation $\{U,\mathfrak{K}\}$ of G and $T \in \mathscr{L}(\mathfrak{K},\mathfrak{H})$ such that $x(s) = T^*U(s)T$, $s \in G$.

4. Tensor Products of C^*-Algebras

Let A_1 and A_2 be C^*-algebras. The algebraic tensor product $A_1 \otimes A_2$ of A_1 and A_2 turns out to be an involutive algebra over the complex number field $\mathbf{C}$ in the natural fashion:

$$\left.\begin{aligned} (x_1 \otimes x_2)(y_1 \otimes y_2) &= x_1y_1 \otimes x_2y_2; \\ (x_1 \otimes x_2)^* = x_1^* \otimes x_2^*, \qquad & x_1,y_1 \in A_1, \qquad x_2,y_2 \in A_2. \end{aligned}\right\} \tag{1}$$

If a norm β on $A_1 \otimes A_2$ satisfies the C^*-condition:

$$\left.\begin{aligned} \|xy\|_\beta &\le \|x\|_\beta\|y\|_\beta, \\ \|x^*x\|_\beta = \|x\|^2, \qquad & x,y \in A_1 \otimes A_2, \end{aligned}\right\} \tag{2}$$

then it is called a C^*-*norm* of $A_1 \otimes A_2$. The completion $A_1 \otimes_\beta A_2$ of $A_1 \otimes A_2$ under any C^*-norm β is a C^*-algebra. However, it is not *a priori* clear that a C^*-norm is a cross-norm. A representation of $A_1 \otimes A_2$ means here always a *-representation unless otherwise specified. For the moment, we shall denote by A_0 the algebraic tensor product $A_1 \otimes A_2$.

Lemma 4.1. *If $\{\pi,\mathfrak{H}\}$ is a representation of A_0, then there exist unique representations π_1 of A_1 and π_2 of A_2 such that*

$$\pi(x_1 \otimes x_2) = \pi_1(x_1)\pi_2(x_2) = \pi_2(x_2)\pi_1(x_1) \tag{3}$$

for each $x_1 \in A_1$ and $x_2 \in A_2$. Moreover, for any approximate identities $\{u_i\}$ of A_1 and $\{v_j\}$ of A_2, we have

$$\left.\begin{aligned} \pi_1(x_1) &= \text{s-lim}\, \pi(x_1 \otimes v_j), \qquad x_1 \in A_1, \\ \pi_2(x_2) &= \text{s-lim}\, \pi(u_i \otimes x_2), \qquad x_2 \in A_2. \end{aligned}\right\} \tag{4}$$

Therefore, we have

$$\|\pi(x_1 \otimes x_2)\| \le \|\pi_1(x_1)\|\,\|\pi_2(x_2)\|. \tag{5}$$

PROOF. Let $(A_i)_\mathrm{I}$, $i = 1,2$, be the C^*-algebra obtained by adjunction of an identity to A_i. For a vector $\xi = \sum_{k=1}^m \sum_{i=1}^{n_k} \pi(y_{1,i}^k \otimes y_{2,i}^k)\xi_k \in \pi(A_0)\mathfrak{H}$, we define a linear functional φ on $(A_1)_\mathrm{I}$ by

$$\varphi(a) = \sum_{k,l=1}^{m} \sum_{i,j=1}^{n_k} (\pi(ay_{1,i}^k \otimes y_{2,i}^k)\xi_k | \pi(y_{1,y}^l \otimes y_{2,j}^l)\xi_l)$$

for each $a \in (A_1)_{\mathrm{I}}$. Here we should note that A_1 is an ideal of $(A_1)_{\mathrm{I}}$, so that φ is well defined. We have then

$$\varphi(a^*a) = \sum_{k,l=1}^{m} \sum_{i,j=1}^{n_k} (\pi(a^*ay_{1,i}^k \otimes y_{2,i}^k)\xi_k | \pi(y_{1,j}^l \otimes y_{2,j}^l)\xi_l)$$

$$= \sum_{k,l=1}^{m} \sum_{i,j=1}^{n_k} (\pi((y_{1,j}^l \otimes y_{2,j}^l)^*(a^*ay_{1,i}^k \otimes y_{2,i}^k)\xi_k | \xi_l)$$

$$= \sum_{k,l=1}^{m} \sum_{i,j=1}^{n_k} (\pi(ay_{1,i}^k \otimes y_{2,i}^k)\xi_k | \pi(ay_{1,j}^l \otimes y_{2,j}^l)\xi_l)$$

$$= \left\| \sum_{k=1}^{m} \sum_{i=1}^{n_k} \pi(ay_{1,i}^k \otimes y_{2,i}^k)\xi_k \right\|^2 \geq 0;$$

hence φ is positive; so $\varphi(a^*a) \leq \|a\|^2\varphi(1) = \|a\|^2\|\xi\|^2$. Therefore, we get

$$\left\| \sum_{k=1}^{m} \sum_{i=1}^{n_k} \pi(ay_{1,i}^k \otimes y_{2,i}^k)\xi_k \right\|^2 \leq \|a\|^2\|\xi\|^2.$$

Hence there exists a bounded operator, say $\pi_1(a)$, on $\mathfrak{H}$ such that

$$\pi_1(a)\xi = \sum_{k=1}^{m} \sum_{i=1}^{n_k} \pi(ay_{1,i}^k \otimes y_{2,i}^k)\xi_k, \qquad a \in (A_1)_{\mathrm{I}}.$$

It is easily seen that π_1 is a representation of $(A_1)_{\mathrm{I}}$ on $\mathfrak{H}$. Similarly, we can define a representation π_2 of $(A_2)_{\mathrm{I}}$ on $\mathfrak{H}$ by

$$\pi_2(b)\xi = \sum_{k=1}^{m} \sum_{i=1}^{n_k} \pi(y_{1,i}^k \otimes by_{2,i}^k)\xi_k, \qquad b \in (A_2)_{\mathrm{I}}.$$

It is now clear that

$$\pi_1(x_1)\pi_2(x_2)\xi = \pi_2(x_2)\pi_1(x_1)\xi = \pi(x_1 \otimes x_2)\xi$$

for each $x_1 \in A_1$, $x_2 \in A_2$ and $\xi \in \pi(A_0)\mathfrak{H}$. Hence (3) follows because $\pi(A_0)\mathfrak{H}$ is dense in $\mathfrak{H}$. If $\pi_1(x_1)\xi = 0$, $x_1 \in A_1$, for some $\xi \in \mathfrak{H}$, then $\pi(x_1 \otimes x_2)\xi = 0$ for every $x_1 \in A_1$ and $x_2 \in A_2$; so $\pi(x)\xi = 0$, $x \in A_0$; hence $\xi = 0$ since π is nondegenerate. Hence π_1 is nondegenerate. Therefore, the strong limit of $\pi_1(u_i)$ is the identity operator 1 for any approximate identity $\{u_i\}$ of A_1; so (4) follows from (3). Q.E.D.

Definition 4.2. The representation π_i of A_i, $i = 1,2$, obtained in the previous lemma is called the *restriction* of π to A_i.

Corollary 4.3. *Let A_1 and A_2 be C^*-algebras, and $(A_1)_{\mathrm{I}}$ and $(A_2)_{\mathrm{I}}$ be the C^*-algebras obtained by adjunction of identities to A_1 and A_2, respectively. Then any C^*-norm β of $A_1 \otimes A_2$ is extended to a C^*-norm of $(A_1)_{\mathrm{I}} \otimes (A_2)_{\mathrm{I}}$.*

PROOF. Let $A_\beta = A_1 \otimes_\beta A_2$. If $\{\pi, \mathfrak{H}\}$ is a faithful representation of A_β, then $\|x\|_\beta = \|\pi(x)\|$ for any $x \in A_1 \otimes A_2$. Let π_1 and π_2 be the restrictions of π

to A_1 and A_2, respectively. By $\pi_i^0(x + \lambda 1) = \pi_i(x) + \lambda 1$, $x \in A_i$ and $\lambda \in \mathbf{C}$, we can extend π_i to a representation π_i^0 of $(A_i)_{\mathrm{I}}$, $i = 1,2$. Since the ranges of π_1^0 and π_2^0 commute, we can define a representation π_0 of $(A_1)_{\mathrm{I}} \otimes (A_2)_{\mathrm{I}}$ by

$$\pi_0(x_1 \otimes x_2) = \pi_1^0(x_1)\pi_2^0(x_2), \qquad x_1 \in (A_1)_{\mathrm{I}}, \qquad x_2 \in (A_2)_{\mathrm{I}}.$$

It is clear that π_0 extends π. Define a C^*-norm β_0 of $(A_1)_{\mathrm{I}} \otimes (A_2)_{\mathrm{I}}$ by $\|x\|_{\beta_0} = \|\pi_0(x)\|$ for each $x \in (A_1)_{\mathrm{I}} \otimes (A_2)_{\mathrm{I}}$. Then the new norm β_0 of $(A_1)_{\mathrm{I}} \otimes (A_2)_{\mathrm{I}}$ extends the original norm β of $A_1 \otimes A_2$. Q.E.D.

Thus, so far as the norm problem in tensor products of C^*-algebras is concerned, we may assume, without loss of generality, the existence of identities in the C^*-algebras in question. Therefore, we assume, through the rest of this section, that the C^*-algebras under consideration are unital unless otherwise indicated. Hence identifying x_1 and $x_1 \otimes 1$ (resp. x_2 and $1 \otimes x_2$), A_1 and A_2 are considered as subalgebras of $A_1 \otimes A_2$.

Given C^*-algebras A_1 and A_2, we denote by $(A_1 \otimes A_2)_h$ the set of all self-adjoint elements of $A_1 \otimes A_2$, and by $(A_1 \otimes A_2)_+$ the convex cone generated by elements of the form x^*x, $x \in A_1 \otimes A_2$. We consider the order structure in $(A_1 \otimes A_2)_h$ given by $(A_1 \otimes A_2)_+$.

Lemma 4.4. (i) *Identifying the real algebraic tensor product $A_{1,h} \otimes A_{2,h}$ of the self-adjoint parts $A_{1,h}$ of A_1 and $A_{2,h}$ of A_2 with a real subspace of $A_1 \otimes A_2$, we have $A_{1,h} \otimes A_{2,h} = (A_1 \otimes A_2)_h$.*

(ii) *For each $x \in (A_1 \otimes A_2)_h$, there exists an $\alpha > 0$ with $x \leq \alpha 1$.*

Proof. (i) Clearly, $A_{1,h} \otimes A_{2,h} \subset (A_1 \otimes A_2)_h$. If $x = \sum_{j=1}^n x_{1,j} \otimes x_{2,j}$ and $x = x^*$, then

$$\begin{aligned} x &= \tfrac{1}{2}(x + x^*) \\ &= \frac{1}{4}\sum_{i=1}^{n} \{(x_{1,j} + x_{1,j}^*) \otimes (x_{2,j} + x_{2,j}^*) - i(x_{1,j} - x_{2,j}^*) \otimes i(x_{1,j} - x_{2,j}^*)\}; \end{aligned}$$

hence $x \in A_{1,h} \otimes A_{2,h}$.

(ii) Suppose that $x_1 \in A_{1,+}$ and $x_2 \in A_{2,+}$. Then $x_1 \otimes x_2 \leq \|x_1\|(1 \otimes x_2) \leq \|x_1\|\|x_2\|1$. Next, let $x_1 \in A_{1,h}$ and $x_2 \in A_{2,h}$. Then $x_i = x_i^+ - x_i^-$ with $x_i^+, x_i^- \in A_{i,+}$ for $i = 1,2$, and $\|x_i^+\| \leq \|x_i\|$, $\|x_i^-\| \leq \|x\|$. Then

$$\begin{aligned} x_1 \otimes x_2 &= (x_1^+ - x_1^-) \otimes (x_2^+ - x_2^-) \\ &\leq x_1^+ \otimes x_2^+ + x_1^- \otimes x_2^- \leq 2\|x_1\|\|x_1\|1. \end{aligned}$$

If $x \in (A_1 \otimes A_2)_h$, then by (i) we can write $x = \sum_{i=1}^n x_{1,i} \otimes x_{2,i}$ with $x_{1,i} \in A_{1,h}$ and $x_{2,i} \in A_{2,h}$. Then we have $x \leq 2\sum_{i=1}^n \|x_{1,i}\|\|x_{2,i}\|1$. Q.E.D.

Given C^*-algebras A_1 and A_2, we consider the projective cross-norm γ in $A_1 \otimes A_2$. It is easily seen that

$$\left.\begin{aligned} \|xy\|_\gamma &\leq \|x\|_\gamma \|y\|_\gamma, \\ \|x^*\|_\gamma &= \|x\|_\gamma, \qquad x,y \in A_1 \otimes A_2. \end{aligned}\right\} \tag{6}$$

Hence the projective tensor product $A_1 \otimes_\gamma A_2$ turns out naturally to be an involutive Banach algebra. If a linear functional ω on $A_1 \otimes A_2$ is positive in the sense that $\omega(x^*x) \geq 0$, then we have $|\omega(x_1 \otimes x_2)| \leq 4\omega(1)\|x_1\|\|x_2\|$ for $x_1 \in A_1$ and $x_2 \in A_2$ by the proof of Lemma 4.4. Hence ω is continuous with respect to the γ-norm. Therefore, it is extended to a positive linear functional on $A_1 \otimes_\gamma A_2$ which is also denoted by ω. By Theorem I.9.14, there exists a cyclic representation $\{\pi_\omega, \mathfrak{H}_\omega, \xi_\omega\}$ of $A_1 \otimes_\gamma A_2$ such that

$$\omega(x) = (\pi_\omega(x)\xi_\omega|\xi_\omega), \qquad x \in A_1 \otimes_\gamma A_2.$$

We define the *state space* $S(A_1 \otimes A_2)$ of $A_1 \otimes A_2$ as the set of all positive linear functionals ω on $A_1 \otimes A_2$ with $\omega(1) = 1$.

Definition 4.5. The *projective C^*-cross-norm* $\|\cdot\|_{\max}$ on $A_1 \otimes A_2$ is given by

$$\|x\|_{\max} = \sup\{\|\pi(x)\| : \pi \text{ runs through all representations of } A_1 \otimes A_2\}. \tag{7}$$

The completion $A_1 \otimes_{\max} A_2$ of $A_1 \otimes A_2$ under $\|\cdot\|_{\max}$ is called the *projective C^*-tensor product* of A_1 and A_2.

It is not *a priori* obvious that the projective C^*-cross-norm is indeed a cross-norm. What one can say about the projective cross-norm at this moment is that $\|x\|_{\max} \leq \|x\|_\gamma$ by Proposition I.5.2 since any representation of $A_1 \otimes A_2$ is extended to that of $A_1 \otimes_\gamma A_2$. At this point it is not even clear that $\|x\|_{\max} \neq 0$ for a nonzero $x \in A_1 \otimes_\gamma A_2$.

It is easy to see that $\mathfrak{S}(A \otimes B)$ is naturally identified with the state space $\mathfrak{S}(A \otimes_{\max} B)$ of $A \otimes_{\max} B$. The next result explains an important feature of completely positive maps in the theory of tensor products of C^*-algebras.

Proposition 4.6. *An element $\omega \in (A_1 \otimes_\gamma A_2)^*$ is a state if and only if $\Phi'_1(\omega): A_1 \mapsto A_2^*$ is completely positive and $\Phi'_1(\omega)(1) \in \mathfrak{S}(A_2)$.*

Proof. Suppose $\omega \in \mathfrak{S}(A_1 \otimes A_2)$. For each $x_1, \ldots, x_n \in A_1$ and $y_1, \ldots, y_n \in A_2$, we have

$$\begin{aligned}\sum_{i,j=1}^n \langle y_i^* y_j, \Phi'_1(\omega)(x_i^* x_j)\rangle &= \sum_{i,j} \omega(x_i^* x_j \otimes y_i^* y_j)\\ &= \omega\left(\left(\sum_{i=1}^n x_i \otimes y_i\right)^* \left(\sum_{j=1}^n x_j \otimes y_j\right)\right) \geq 0.\end{aligned}$$

Hence $\Phi'_1(\omega)$ is completely positive. In particular, $\Phi'_1(\omega)(1)$ is positive. Furthermore, we have

$$\langle 1, \Phi'_1(\omega)(1)\rangle = \langle 1 \otimes 1, \omega\rangle = \omega(1) = 1,$$

hence $\Phi'_1(\omega)(1)$ is a state of A_2.

Conversely, the above observation shows that, if $\Phi_1'(\omega)$ is completely positive, then ω is positive, and that $\omega(1) = \langle 1 \otimes 1, \omega \rangle = \langle 1, \Phi_1'(\omega)(1) \rangle = 1$ if $\Phi_1'(\omega)(1)$ is a state of A_2. Q.E.D.

The projective C^*-tensor product $A_1 \otimes_{\max} A_2$ has the following universal property:

Proposition 4.7. *Given C^*-algebras A_1, A_2, and B, if $\pi_1 : A_1 \mapsto B$ and $\pi_2 : A_2 \mapsto B$ are homomorphisms with commuting ranges, then there exists a unique homomorphism π of the projective C^*-tensor product $A_1 \otimes_{\max} A_2$ into B such that*

$$\pi(x_1 \otimes x_2) = \pi_1(x_1)\pi_2(x_2), \qquad x_1 \in A_2, \qquad x_2 \in A_2;$$

and the image $\pi(A_1 \otimes_{\max} A_2)$ is the C^*-subalgebra of B generated by $\pi_1(A_1)$ and $\pi_2(A_2)$.

We leave the proof to the reader.

If $\{\pi_1, \mathfrak{H}_1\}$ and $\{\pi_2, \mathfrak{H}_2\}$ are representations of A_1 and A_3, respectively, then there is a unique representation π on the tensor product Hilbert space $\mathfrak{H}_1 \otimes \mathfrak{H}_2$ defined by

$$\pi(x) = \sum_{i=1}^{n} \pi_1(x_{1,i}) \otimes \pi_2(x_{2,i}), \qquad x = \sum_{i=1}^{n} x_{1,i} \otimes x_{2,i} \in A_1 \otimes A_2. \tag{8}$$

We denote this representation by $\pi_1 \otimes \pi_2$.

Definition 4.8. Given C^*-algebras A_1 and A_2, the *injective C^*-cross-norm* $\|\cdot\|_{\min}$ of A_1 and A_2 is defined by

$$\|x\|_{\min} = \sup \|(\pi_1 \otimes \pi_2)(x)\|, \qquad x \in A_1 \otimes A_2, \tag{9}$$

where π_1 and π_2 run over all representations of A_1 and A_2, respectively. The completion $A_1 \otimes_{\min} A_2$ is called the *injective C^*-tensor product* of A_1 and A_2.

It is clear that

$$\|x\|_{\min} \leq \|x\|_{\max}, \qquad x \in A_1 \otimes A_2. \tag{10}$$

In general, the projective C^*-cross-norm and the injective C^*-cross-norm are different; see [357].

As $\|\pi_1(x_1) \otimes \pi_2(x_2)\| = \|\pi_1(x_1)\| \|\pi_2(x_2)\|$ for every representation π_1 of A_1 and π_2 of A_2, the injective C^*-cross-norm satisfies the cross-norm condition (1) in Section 2. Hence inequality (10), together with the inequality $\|x\|_{\max} \leq \|x\|_{\gamma}$, implies that the projective C^*-cross-norm also satisfies the cross-norm condition (1) in Section 2. Let $f_1 \in A_1^*$ and $f_2 \in A_2^*$. By Proposition III.2.1, f_1 and f_2 are of the form

$$f_1 = \omega(\pi_1; \xi_1, \eta_1) \quad \text{and} \quad f_2 = \omega(\pi_2; \xi_2, \eta_2). \tag{11}$$

By the polar decomposition, Theorem III.4.2, we can choose ξ_1, η_1, ξ_2 and η_2 so that $\|f_1\| = \|\xi_1\|\|\eta_1\|$ and $\|f_2\| = \|\xi_2\|\|\eta_2\|$. Hence we have, for $x = \sum_{i=1}^{n} x_{1,i} \otimes x_{2,i} \in A_1 \otimes A_2$,

$$\begin{aligned}|\langle x, f_1 \otimes f_2\rangle| &= \left|\sum_{i=1}^{n} f_1(x_{1,i})f_2(x_{2,i})\right| \\ &= \left|\sum_{i=1}^{n} (\pi_1(x_{1,i})\xi_1|\eta_1)(\pi_2(x_{2,i})\xi_2|\eta_2)\right| \\ &= |((\pi_1 \otimes \pi_2)(x)(\xi_1 \otimes \xi_2)|\eta_1 \otimes \eta_2)| \\ &\leq \|(\pi_1 \otimes \pi_2)(x)\|\|\xi_1 \otimes \xi_2\|\|\eta_1 \otimes \eta_2\| \\ &\leq \|x\|_{\min}\|f_1\|\|f_2\|,\end{aligned}$$

so that we have

$$\|x\|_\lambda \leq \|x\|_{\min}, \qquad x \in A_1 \otimes A_2. \tag{12}$$

Therefore, the injective C^*-cross-norm of $A_1 \otimes A_2$ is indeed a cross-norm; hence so is the projective C^*-cross-norm. Furthermore, inequality (12) entails that the adjoint cross norm $\|\cdot\|_{\min}^*$ in $A_1^* \otimes A_2^*$ is a cross-norm.

By (10) or Proposition 4.7, the identity map: $x \in A_1 \otimes A_2 \mapsto x \in A_1 \otimes A_2 \subset A_1 \otimes_{\min} A_2$ is extended to a homomorphism σ of $A_1 \otimes_{\max} A_2$ onto $A_1 \otimes_{\min} A_2$. In general, for any C^*-norm β of $A_1 \otimes A_2$, there is a unique homomorphism σ_β of $A_1 \otimes_{\max} A_2$ onto $A_1 \otimes_\beta A_2$ which extends the identity map of $A_1 \otimes A_2$. For representations $\{\pi_1, \mathfrak{H}_1\}$ of A_1 and $\{\pi_2, \mathfrak{H}_2\}$ of A_2, the representation $\{\pi_1 \otimes \pi_2, \mathfrak{H}_1 \otimes \mathfrak{H}_2\}$ of the algebraic tensor product $A_1 \otimes A_2$ is then uniquely extended to a representation of $A_1 \otimes_{\min} A_2$, which is denoted again by $\pi_1 \otimes \pi_2$. We denote also by $\pi_1 \otimes \pi_2$ the representation $(\pi_1 \otimes \pi_2) \circ \sigma$ of $A_1 \otimes_{\max} A_2$.

Theorem 4.9. *Given C^*-algebras A_1 and A_2. let $A_{\max}$ and $A_{\min}$ denote the projective C^*-tensor product and the injective C^*-tensor product of A_1 and A_2, respectively.*

(i) *If ω_1 and ω_2 are positive linear functions of A_1 and A_2, respectively, then $\omega = \omega_1 \otimes \omega_2$ is positive on $A_{\max}$, and it induces the tensor product representation $\{\pi_\omega, \mathfrak{H}_\omega, \xi_\omega\} = \{\pi_{\omega_1}, \mathfrak{H}_{\omega_1}, \xi_{\omega_2}\} \otimes \{\pi_{\omega_2}, \mathfrak{H}_{\omega_2}, \xi_{\omega_2}\}$ of the cyclic representations $\{\pi_{\omega_1}, \mathfrak{H}_{\omega_1}, \xi_{\omega_1}\}$ and $\{\pi_{\omega_2}, \mathfrak{H}_{\omega_2}, \xi_{\omega_2}\}$ of A_1 and A_2 induced by ω_1 and ω_2, respectively.*

(ii) *The norm of each $a \in A_{\min}$ is given by*

$$\|a\|_{\min} = \sup\left\{\frac{\omega_1 \otimes \omega_2(x^*a^*ax)^{1/2}}{\omega_1 \otimes \omega_2(x^*x)^{1/2}} : x \in A_{\min}, \omega_1 \in A_{1,+}^*, \omega_2 \in A_{2,+}^*\right\}. \tag{13}$$

(iii) *If $\{\pi_1, \mathfrak{H}_1\}$ and $\{\pi_2, \mathfrak{H}_2\}$ are faithful representations of A_1 and A_2, respectively, then the tensor product representation $\{\pi, \mathfrak{H}\} = \{\pi_1, \mathfrak{H}_1\} \otimes \{\pi_2, \mathfrak{H}_2\}$ of $A_{\min}$ is faithful.*

Proof. (i) Let $\{\pi,\mathfrak{H}\} = \{\pi_{\omega_1},\mathfrak{H}_{\omega_1}\} \otimes \{\pi_{\omega_2},\mathfrak{H}_{\omega_2}\}$, and $\xi_\omega = \xi_{\omega_1} \otimes \xi_{\omega_2}$. We have then, for any $x_1 \in A_1$, $x_2 \in A_2$,

$$\begin{aligned}(\pi(x_1 \otimes x_2)\xi_\omega|\xi_\omega) &= (\pi_{\omega_1}(x_1)\xi_{\omega 1}|\xi_{\omega_1})(\pi_{\omega_2}(x_2)\xi_{\omega_2}|\xi_{\omega_2})\\ &= \omega_1(x_1)\omega_2(x_2) = (\omega_1 \otimes \omega_2)(x_1 \otimes x_2)\\ &= \omega(x_1 \otimes x_2).\end{aligned}$$

Hence we have $(\pi(x)\xi_\omega|\xi_\omega) = \omega(x)$ for any $x \in A_1 \otimes A_2$. Therefore, ω is positive on $A_1 \otimes A_2$, hence on $A_{\max}$. Since $\pi(A_{\max})\xi_\omega$ contains the algebraic tensor product $\pi_{\omega_1}(A_1)\xi_{\omega_1} \otimes \pi_{\omega_2}(A_2)\xi_{\omega_2}$ and the latter is dense in $\mathfrak{H}$, ξ_ω is cyclic under π. Thus assertion (i) follows.

(ii) Take arbitrary representations π_1 of A_1 and π_2 of A_2, respectively. Then π_1 and π_2 are, by Proposition I.9.17, the direct sums of cyclic representations $\{\pi_{1,i}\}$ and $\{\pi_{2,j}\}$, so that we have

$$\|(\pi_1 \otimes \pi_2)(x)\| = \sup_{i,j} \|(\pi_{1,i} \otimes \pi_{2,j})(x)\|, \qquad x \in A_{\min}.$$

Hence we have

$$\|x\|_{\min} = \sup\|(\pi_1 \otimes \pi_2)(x)\|,$$

where π_1 and π_2 run over the cyclic representations of A_1 and A_2, respectively. If $\pi_1 = \pi_{\omega_1}$, $\omega_1 \in A_{1,+}^*$, and $\pi_2 = \pi_{\omega_2}$, $\omega_2 \in A_{2,+}^*$, then we have, for each $a \in A_{\min}$,

$$\|(\pi_1 \otimes \pi_2)(a)\| = \sup\left\{\frac{(\omega_1 \otimes \omega_2)(x^*a^*ax)^{1/2}}{(\omega_1 \otimes \omega_2)(x^*x)^{1/2}} : x \in A_1 \otimes A_2\right\},$$

which implies (13).

(iii) Suppose $\{\pi_1,\mathfrak{H}_1\}$ and $\{\pi_2,\mathfrak{H}_2\}$ are faithful representations of A_1 and A_2, respectively. Let $V_1 = V(\pi_1) \subset A_1^*$ and $V_2 = V(\pi_2) \subset A_2^*$ be the invariant subspaces of A_1^* and A_2^* associated, in the sense of Definition III.2.11, with π_1 and π_2, respectively. Since π_i, $i = 1,2$, is an isomorphism of A_i onto $\pi_i(A_i)$, $x \in A_i$ is positive if and only if $\pi_i(x) \geq 0$ if and only if $\omega(x) \geq 0$ for every $\omega \in V^+$; hence $A_{i,+}^* = \{\omega \in A_i^* : \omega(x) \geq 0$ for every $x \in A_{i,+}\}$ is the $\sigma(A_i^*,A_i)$-closure of V_i^+ by the Hahn–Banach separation theorem. Hence for every $\omega_1 \in A_{1,+}^*$ and $\omega_2 \in A_{2,+}^*$, we can find nets $\{\omega_{1,i}\} \subset V_1^+$ and $\{\omega_{2,j}\} \subset V_2^+$ such that

$$\lim \omega_{1,i} \otimes \omega_{2,j}(x) = \omega_1 \otimes \omega_2(x), \qquad x \in A_1 \otimes A_2;$$

therefore (13) implies that $\|(\pi_1 \otimes \pi_2)(x)\| = \|x\|_{\min}$ for $x \in A_1 \otimes A_2$. Since $A_1 \otimes A_2$ is dense in $A_{\min}$, $\pi_1 \otimes \pi_2$ is faithful, being an isometry. Q.E.D.

Proposition 4.10. *In the algebraic tensor product $A_1^* \otimes A_2^*$ of the conjugate spaces of C^*-algebras A_1 and A_2, the adjoint cross-norm $\|\cdot\|_{\min}^*$ of the injective C^*-cross-norm and the adjoint cross-norm $\|\cdot\|_{\max}^*$ of the projective C^*-cross-norm agree.*

PROOF. Let σ be the canonical homomorphism of $A_{\max} = A_1 \otimes_{\max} A_2$ onto $A_{\min} = A_1 \otimes_{\min} A_2$. By Corollary I.8.2, σ induces canonically an isomorphism σ of the quotient C^*-algebra $A_{\max}/\sigma^{-1}(0)$ onto $A_{\min}$. Hence the transpose ${}^t\sigma$ is an isometry of $A^*_{\min}$ into $A^*_{\max}$. Hence, we get, for each $\omega \in A_1^* \otimes A_2^*$,

$$\begin{aligned}\|\omega\|^*_{\max} &= \sup\{|\omega(x)| : x \in A_1 \otimes A_2, \|x\|_{\max} \le 1\} \\ &= \sup\{|\omega(\sigma(x))| : x \in A_1 \otimes A_2, \|x\|_{\max} \le 1\} \\ &= \sup\{|\langle x, {}^t\sigma(\omega)\rangle| : x \in A_1 \otimes A_2, \|x\|_{\max} \le 1\} \\ &= \|{}^t\sigma(\omega)\| = \|\omega\|^*_{\min}.\end{aligned}$$

Q.E.D.

Hence we denote simply by $\|\omega\|$ the adjoint norm $\|\omega\|^*_{\min} = \|\omega\|^*_{\max}$ in $A_1^* \otimes A_2^*$, and by $A_1^* \overline{\otimes} A_2^*$ the completion of $A_1^* \otimes A_2^*$ under this norm. By the proposition, $A_1^* \overline{\otimes} A_2^*$ is considered as an invariant subspace of $(A_1 \otimes_{\max} A_2)^*$ as well as that of $(A_1 \otimes_{\min} A_2)^*$.

We are now going to study the injective C^*-cross-norm in more detail.

Lemma 4.11. *If A is a C^*-subalgebra of a C^*-algebra B and if the restriction ω_A of a state ω of B to A is a pure state, then we have*

$$\omega(xy) = \omega(x)\omega(y), \qquad x \in A, \qquad y \in A' \cap B, \tag{14}$$

where $A' \cap B$ means the C^*-subalgebra of B consisting of all elements of B which commute with A.

PROOF. Since $A' \cap B$ is linearly spanned by its positive part of norm ≤ 1, we may assume $0 \le y \le 1$, $y \in A' \cap B$. If $\omega(y) = 0$, then the Cauchy–Schwarz inequality

$$\begin{aligned}|\omega(xy)|^2 &\le \omega((xy^{1/2})(xy^{1/2})^*)\omega(y^{1/2}y^{1/2}) \\ &= \omega(xyx^*)\omega(y) = 0\end{aligned}$$

shows equality (14) in this case. If $\omega(y) = 1$, then we apply the above arguments to $1 - y$; so we get (14) too. Suppose $0 < \omega(y) < 1$. We then have

$$\omega_A(x) = \omega(y)\frac{1}{\omega(y)}\omega(xy) + (1 - \omega(y))\frac{1}{1 - \omega(y)}\omega(x(1 - y)), \qquad x \in A.$$

By the commutativity of y and A, the functionals

$$\omega_1(x) = \frac{1}{\omega(y)}\omega(xy), \qquad \omega_2(x) = \frac{1}{1 - \omega(y)}\omega(x(1 - y))$$

are both states of A. Hence by assumption, we have $\omega_A(x) = \omega_1(x) = \omega_2(x)$, so that $\omega(xy) = \omega(x)\omega(y)$. Q.E.D.

Remark 4.12. This lemma provides an alternative proof of the Gelfand representation of an abelian C^*-algebra, Theorem I.4.4.

Proposition 4.13. *Let A_1 and A_2 be C^*-algebras, and let $\{\pi_1,\mathfrak{H}_1\}$ and $\{\pi_2,\mathfrak{H}_2\}$ be their respective representations. We have then*

$$\{(\pi_1 \otimes \pi_2)(A_1 \otimes_{\min} A_2)\}'' = \pi_1(A_1)'' \overline{\otimes} \pi_2(A_2)'',$$

where the latter means the tensor product of the von Neumann algebras $\pi_1(A_1)''$ and $\pi_2(A_2)''$ in Definition 1.3. In particular, if π_1 and π_2 are both irreducible, then so is $\pi_1 \otimes \pi_2$.

Proof. Since $(\pi_1 \otimes \pi_2)(A_1 \otimes_{\min} A_2)$ is the uniform closure of

$$(\pi_1 \otimes \pi_2)(A_1 \otimes A_2) = \pi_1(A_1) \otimes \pi_2(A_2),$$

it follows that $(\pi_1 \otimes \pi_2)(A_1 \otimes_{\min} A_2) \subset \pi_1(A_1)'' \overline{\otimes} \pi_2(A_2)''$. Let x_1 be an operator in $\pi_1(A_1)''$. Then there exists a net $\{x_{1,i}\}$ in A_1 such that $\{\pi_1(x_{1,i})\}$ converges σ-strongly to x_1. Since the map: $y \in \mathscr{L}(\mathfrak{H}_1) \mapsto y \otimes 1 \in \mathscr{L}(\mathfrak{H}_1 \otimes \mathfrak{H}_2)$ is σ-strongly continuous, $\{(\pi_1 \otimes \pi_2)(x_{1,i} \otimes 1)\}$ converges σ-strongly to $x_1 \otimes 1$. Hence $(\pi_1 \otimes \pi_2)(A_1 \otimes_{\min} A_2)''$ contains $x_1 \otimes 1$. By symmetry $1 \otimes x_2$, $x_2 \in \pi_2(A_2)''$, belongs to $(\pi_1 \otimes \pi_2)(A_1 \otimes_{\min} A_2)''$. Hence $\pi_1(A_1)'' \overline{\otimes} \pi_2(A_2)''$ is contained in $(\pi_1 \otimes \pi_2)(A_1 \otimes_{\min} A_2)''$. The last assertion follows from (10) in Section 1. Q.E.D.

Theorem 4.14. *For two C^*-algebras A_1 and A_2, the following three statements are equivalent:*

(i) *Either A_1 or A_2 is abelian.*
(ii) *Every pure state ω of $A_1 \otimes_{\min} A_2$ is of the form $\omega = \omega_1 \otimes \omega_2$ for some pure states ω_1 of A_1 and ω_2 of A_2.*
(iii) *The injective C^*-cross-norm $\|\cdot\|_{\min}$ of $A_1 \otimes A_2$ agrees with the injective cross-norm λ.*

Proof. Let A denote the injective C^*-tensor product $A_1 \otimes_{\min} A_2$.

(i) $\Rightarrow$ (ii): Suppose A_2 is abelian. Then A_2 is contained in the center Z of A. Here we identify $x_2 \in A_2$ and $1 \otimes x_2 \in A$. Every pure state ω of A, by Lemma 4.11, has the property $\omega(ab) = \omega(a)\omega(b)$, $a \in Z$, $b \in A$. In particular, we have $\omega(a_1 \otimes a_2) = \omega(a_1)\omega(a_2)$, $a_1 \in A_1$, $a_2 \in A_2$. Putting $\omega_1(a_1) = \omega(a_1)$, $a_1 \in A_1$, and $\omega_2(a_2) = \omega(a_2)$, $a_2 \in A_2$, we have $\omega = \omega_1 \otimes \omega_2$. Clearly, ω_1 and ω_2 are both pure.

(ii) $\Rightarrow$ (i): Suppose that A_1 and A_2 are both noncommutative. Let $\{\pi_1,\mathfrak{H}_1\}$ and $\{\pi_2,\mathfrak{H}_2\}$ be irreducible representations of A_1 and A_2 with $\dim \mathfrak{H}_1 \geq 2$ and $\dim \mathfrak{H}_2 \geq 2$, respectively. Let $\{\pi,\mathfrak{H}\} = \{\pi_1,\mathfrak{H}_1\} \otimes \{\pi_2,\mathfrak{H}_2\}$. Take orthogonal unit vectors $\xi_{1,1},\xi_{1,2}$ in $\mathfrak{H}_1$ and $\xi_{2,1},\xi_{2,2}$ in $\mathfrak{H}_2$. Putting

$$\xi = \frac{1}{\sqrt{2}}(\xi_{1,1} \otimes \xi_{2,1} + \xi_{1,2} \otimes \xi_{2,2}) \in \mathfrak{H},$$

$$\omega = \omega(\pi; \xi),$$

we get a state ω of A. By Proposition 4.13, ω is pure. If ω is of the form $\omega = \omega_1 \otimes \omega_2$, then $\omega(x_1 \otimes x_2) = \omega(x_1)\omega(x_2)$ for any $x_1 \in A_1$ and $x_2 \in A_2$. Hence

$$(\pi_1(x_1) \otimes \pi_2(x_2)\xi|\xi) = ((\pi_1(x_1) \otimes 1)\xi|\xi)((1 \otimes \pi(x_2))\xi|\xi), \quad x_1 \in A_1, \quad x_2 \in A_2.$$

Therefore, for any $a_1 \in \mathscr{L}(\mathfrak{H}_1)$ and $a_2 \in \mathscr{L}(\mathfrak{H}_2)$,

$$((a_1 \otimes a_2)\xi|\xi) = ((a_1 \otimes 1)\xi|\xi)((1 \otimes a_2)\xi|\xi).$$

Take the projection of $\mathfrak{H}_1$ onto $[\xi_{1,1}]$ as a_1 and the projection of $\mathfrak{H}_2$ onto $[\xi_{2,2}]$ as a_2. Then $a_1 \otimes a_2$ is the projection of $\mathfrak{H}$ onto $[\xi_{1,1} \otimes \xi_{2,2}]$, so that

$$(a_1 \otimes a_2)\xi = (\xi|\xi_{1,1} \otimes \xi_{2,2})(\xi_{1,1} \otimes \xi_{2,2}) = 0.$$

On the other hand, we have

$$((a_1 \otimes 1)\xi|\xi) = \tfrac{1}{2} \quad \text{and} \quad ((1 \otimes a_2)\xi|\xi) = \tfrac{1}{2},$$

which is a contradiction. Hence ω is not of the form $\omega_1 \otimes \omega_2$.

(i) $\Rightarrow$ (iii): Suppose A_2 is abelian. Then A_2 is identified with the C^*-algebra $C(\Omega)$ of all continuous functions on a compact space Ω. To each element $\sum_{i=1}^n x_i \otimes y_i \in A_1 \otimes A_2$, we associate a unique A_1-valued function $\sum_{i=1}^n x_i y_i(\cdot)$ on Ω. Let $C_{A_1}(\Omega)$ denote the space of all A_1-valued continuous functions on Ω. In $C_{A_1}(\Omega)$, we define the following structure:

$$\left.\begin{aligned} \|a\| &= \sup\{\|a(\omega)\| : \omega \in \Omega\}; \\ (\lambda a + \mu b)(\omega) &= \lambda a(\omega) + \mu b(\omega); \\ ab(\omega) &= a(\omega)b(\omega); \\ a^*(\omega) &= a(\omega)^*. \end{aligned}\right\} \tag{15}$$

It is clear that $C_{A_1}(\Omega)$ is a C^*-algebra with this structure. For each $\sum_{i=1}^n x_i \otimes y_i \in A_1 \otimes A_2$, we have

$$\begin{aligned}
\left\|\sum_{i=1}^n x_i y_i(\cdot)\right\| &= \sup\left\{\left\|\sum_{i=1}^n x_i y_i(\omega)\right\| : \omega \in \Omega\right\} \\
&= \sup\left\{\left|\varphi\left(\sum_{i=1}^n x_i y_i(\omega)\right)\right| : \omega \in \Omega, \varphi \in A_1^*, \|\varphi\| \le 1\right\} \\
&= \sup\left\{\left|\sum_{i=1}^n \varphi(x_i) y_i(\omega)\right| : \omega \in \Omega, \varphi \in A_1^*, \|\varphi\| \le 1\right\} \\
&= \sup\left\{\left\|\sum_{i=1}^n \varphi(x_i) y_i(\cdot)\right\| : \varphi \in A_1^*, \|\varphi\| \le 1\right\} \\
&= \sup\left\{\left|\psi\left(\sum_{i=1}^n \varphi(x_i) y_i(\cdot)\right)\right| : \varphi \in A_1^*, \|\varphi\| \le 1, \psi \in A_2^*, \|\psi\| \le 1\right\} \\
&= \sup\left\{\left|\varphi \otimes \psi\left(\sum_{i=1}^n x_i \otimes y_i\right)\right| : \varphi \in A_1^*, \|\varphi\| \le 1, \psi \in A_2^*, \|\psi\| \le 1\right\} \\
&= \left\|\sum_{i=1}^n x_i \otimes y_i\right\|_\lambda;
\end{aligned}$$

hence the natural correspondence between $A_1 \otimes A_2$ and a dense subalgebra of $C_{A_1}(\Omega)$ is an isometry with respect to the λ-norm and the norm in $C_{A_1}(\Omega)$. Therefore, $A_1 \otimes_\lambda A_2$ is identified with the C^*-algebra $C_{A_1}(\Omega)$. Regarding each $\omega \in \Omega$ as a one-dimensional representation of A_2, we get, for a faithful representation π_1 of A_1,

$$\begin{aligned}\left\|\sum_{i=1}^n x_i \otimes y_i\right\|_\lambda &= \left\|\sum_{i=1}^n x_i y_i(\cdot)\right\| \\ &= \sup\left\{\left\|\sum_{i=1}^n x_i y_i(\omega)\right\| : \omega \in \Omega\right\} \\ &= \sup\left\{\left\|\pi_1\left(\sum_{i=1}^n x_i y_i(\omega)\right)\right\| : \omega \in \Omega\right\} \\ &= \sup\left\{\left\|(\pi_1 \otimes \omega)\left(\sum_{i=1}^n x_i \otimes y_i\right)\right\| : \omega \in \Omega\right\} \\ &= \left\|\left(\pi_1 \otimes \sum_{\omega \in \Omega}^{\oplus} \omega\right)\left(\sum_{i=1}^n x_i \otimes y_i\right)\right\| = \left\|\sum_{i=1}^n x_i \otimes y_i\right\|_{\min}\end{aligned}$$

by Theorem 4.9(iii). Hence $\lambda = \|\cdot\|_{\min}$.

(iii) $\Rightarrow$ (ii): Suppose $\lambda = \|\cdot\|_{\min}$. Let S_1^* and S_2^* denote the closed unit balls of A_1^* and A_2^*, respectively. Putting $S_1^* \otimes S_2^* = \{\varphi \otimes \psi : \varphi \in S_1^*, \psi \in S_2^*\}$, we get, by assumption,

$$\|x\| = \sup\{|f(x)| : f \in S_1^* \otimes S_2^*\}.$$

Hence A is isometrically imbedded, as a Banach space, in the Banach space $C(S_1^* \otimes S_2^*)$ of all continuous functions on the compact space $S_1^* \otimes S_2^*$, where $S_1^* \otimes S_2^*$ is equipped with the $\sigma(A^*,A)$-topology. Therefore, the extreme points of the unit ball S^* of A^* are contained in $S_1^* \otimes S_2^*$, so every pure state ω of A is of the form $\omega_1 \otimes \omega_2$; hence (ii) holds since if either ω_1 or ω_2 is not pure then $\omega_1 \otimes \omega_2$ is not pure. Q.E.D.

Lemma 4.15. *Let A be a C^*-algebra and $P(A)$ the set of all pure states of A.*

(i) *If $\mathscr{I}$ is a closed ideal of A, then $K_{\mathscr{I}} = \mathscr{I}^0 \cap P(A)$ is a $\sigma(A^*,A)$-closed subset of $P(A)$ such that $uK_{\mathscr{I}}u^* = K_{\mathscr{I}}$ for any unitary element $u \in A$.*

(ii) *If K is a $\sigma(A^*,A)$-closed subset of $P(A)$ with $uKu^* = K$ for every unitary element $u \in A$, then $K^\perp = \mathscr{I}_K$ is a closed ideal of A, where $K^\perp = \{x \in A : \omega(x) = 0$ for every $\omega \in K\}$.*

(iii) *The correspondences: $\mathscr{I} \mapsto K_{\mathscr{I}}$ and $K \mapsto \mathscr{I}_K$ are each other's inverse.*

Proof. (i) Since $\mathscr{I}^0$ is $\sigma(A^*,A)$-closed, $K_{\mathscr{I}}$ is $\sigma(A^*,A)$-closed in $P(A)$. The invariance of $K_{\mathscr{I}}$ follows from that of $\mathscr{I}$.

(ii) Let $\omega \in P(A)$ and $\{\pi_\omega, \mathfrak{H}_\omega, \xi_\omega\}$ be the cyclic representation of A induced by ω. By Theorem II.4.18, for any unit vector $\eta \in \mathfrak{H}_\omega$, there exists a unitary element $u \in A$ such that $\eta = \pi_\omega(u)\xi_\omega$. Now, put $\{\pi, \mathfrak{H}\} = \sum_{\omega \in K}^{\oplus} \{\pi_\omega, \mathfrak{H}_\omega\}$. If $\pi(x) = 0$, $x \in A$, then $\pi_\omega(x) = 0$ for every $\omega \in K$; hence $\omega(x) = 0$, $\omega \in K$; so $x \in \mathscr{I}_K$. Conversely, if $x \in \mathscr{I}_K$, then we have, for every unitary element

$u \in A$,

$$0 = \langle x, u\omega u^* \rangle = \langle u^*xu, \omega \rangle = (\pi_\omega(u^*xu)\xi_\omega | \xi_\omega)$$
$$= (\pi_\omega(x)u\xi_\omega | u\xi_\omega), \qquad \omega \in K.$$

Hence $\pi_\omega(x) = 0$ for every $\omega \in K$; so $\pi(x) = 0$. Therefore, we have $\mathscr{I}_K = \pi^{-1}(0)$; hence $\mathscr{I}_K$ is a closed ideal.

For each $x \in A_h$, we have

$$\|\pi(x)\| = \sup_{\omega \in K} \|\pi_\omega(x)\|$$
$$= \sup\{|(\pi_\omega(x)\eta|\eta)| : \omega \in K, \eta \in \mathfrak{H}_\omega, \|\eta\| = 1\}$$
$$= \sup\{|(\pi_\omega(x)u\xi_\omega|u\xi_\omega)| : \omega \in K, u \in A \text{ unitary}\}$$
$$= \sup\{|\langle x, u^*\omega u\rangle| : \omega \in K, u \in A \text{ unitary}\}$$
$$= \sup\{|\langle x, \omega\rangle| : \omega \in K\}.$$

Therefore, the self-adjoint part of $A/\mathscr{I}_K \cong \pi(A)$ is isometrically imbedded in the Banach space $C_{\mathbf{R}}(\bar{K})$ of all real valued continuous functions on the $\sigma(A^*,A)$-closure $\bar{K}$ of K. Hence any pure state ω of $A/\mathscr{I}_K$ is given by a point in $\bar{K}$. Since ${}^t\pi(P(A/\mathscr{I}_K)) = K_{\mathscr{I}_K}$, $K_{\mathscr{I}_K}$ is contained in $\bar{K} \cap P(A) = K$. Hence $K \supset K_{\mathscr{I}_K}$; therefore $K = K_{\mathscr{I}_K}$.

(iii) If $\mathscr{I}$ is a closed ideal, then $\mathscr{I}$ is, by Corollary III.4.5, the intersection of all left kernels N_ω of pure states which contain $\mathscr{I}$. Since $\omega \in K_{\mathscr{I}}$ if and only if $N_\omega \supset \mathscr{I}$, we have $\mathscr{I} = \bigcap \{N_\omega : \omega \in K_{\mathscr{I}}\}$; hence $\mathscr{I} = \mathscr{I}_K$. Q.E.D.

We are now going to prove the minimality of the injective C^*-cross-norm among all possible C^*-norms. Let A_1 and A_2 be C^*-algebras. Suppose β is a C^*-norm on $A_1 \otimes A_2$. We denote by A_β the completion $A_1 \otimes_\beta A_2$ of $A_1 \otimes A_2$ under the β-norm. We define a subset S_β of $P(A_1) \times P(A_2)$ as follows:

$$S_\beta = \{(\omega_1,\omega_2) \in P(A_1) \times P(A_2) : \omega_1 \otimes \omega_2 \text{ is continuous in } \beta\text{-norms}\}.$$

Here we remark that if $\omega_1 \otimes \omega_2$ is continuous under β, then $\|\omega_1 \otimes \omega_2\|_\beta^* = 1$ because $\|\omega_1 \otimes \omega_2\|_\beta^* = (\omega_1 \otimes \omega_2)(1)$.

Lemma 4.16. *If $(\omega_1,\omega_2) \in S_\beta$, then $(u^*\omega_1 u, v^*\omega_2 v) \in S_\beta$ for any unitary elements $u \in A_1$ and $v \in A_2$.*

Proof. For each $x \in A_1 \otimes A_2$, we have

$$|\langle x, u^*\omega_1 u \otimes v^*\omega_2 v\rangle| = |\langle (u \otimes v)x(u^* \otimes v^*), \omega_1 \otimes \omega_2\rangle|$$
$$\le \|(u \otimes v)x(u^* \otimes v^*)\|_\beta \|\omega_1 \otimes \omega_2\|_\beta$$
$$= \|(u \otimes v)x(u^* \otimes v^*)\|_\beta = \|x\|_\beta.$$

Hence $(u^*\omega_1 u, v^*\omega_2 v) \in S_\beta$. Q.E.D.

Lemma 4.17. *S_β is closed in $P(A_1) \times P(A_2)$, where we consider the $\sigma(A_i^*,A_i)$-topology in A_i^*, $i = 1,2$.*

PROOF. Let $\{(\omega_{1,\alpha},\omega_{2,\alpha})\}$ be a net in S_β converging to $(\omega_1,\omega_2) \in P(A_1) \times P(A_2)$. For each $x = \sum_{i=1}^n x_{1,i} \otimes x_{2,i} \in A_1 \otimes A_2$, we have

$$\lim_\alpha (\omega_{1,\alpha} \otimes \omega_{2,\alpha})(x) = \lim_\alpha \sum_{i=1}^n \omega_{1,\alpha}(x_{1,i})\omega_{2,\alpha}(x_{2,i})$$
$$= \sum_{i=1}^n \omega_1(x_{1,i})\omega_2(x_{2,j}) = (\omega_1 \otimes \omega_2)(x).$$

Since $|\omega_{1,\alpha} \otimes \omega_{2,\alpha}(x)| \leq \|x\|_\beta$, we get $|(\omega_1 \otimes \omega_2)(x)| \leq \|x\|_\beta$, so that $(\omega_1,\omega_2) \in S_\beta$. Q.E.D.

Lemma 4.18. *If either A_1 or A_2 is abelian, then $\beta = \lambda$.*

PROOF. Suppose A_1 is abelian. By Lemma 4.11, every $\omega \in P(A_\beta)$ is of the form $\omega = \omega_1 \otimes \omega_2$. Hence, by Theorem 4.9, the cyclic representation π_ω of A_β induced by ω is the tensor product representation $\pi_{\omega_1} \otimes \pi_{\omega_2}$ of the cyclic representations π_{ω_1} of A_1 and π_{ω_2} of A_2 induced by ω_1 and ω_2, respectively. Therefore, we get, for each $x \in A_1 \otimes A_2$,

$$\begin{aligned}\|x\|_\beta &= \sup\{\|\pi_\omega(x)\| : \omega \in P(x)\}\\ &\leq \sup\{\|(\pi_1 \otimes \pi_2)(x)\| : \pi_1 \text{ and } \pi_2 \text{ run through all representations}\}\\ &= \|x\|_{\min} = \|x\|_\lambda \text{ by Theorem 4.14.}\end{aligned}$$

Hence we get $\beta \leq \lambda$.

Suppose $S_\beta \neq P(A_1) \times P(A_2)$. By the closedness of S_β, there exist open sets $U_1 \subset P(A_1)$ and $U_2 \subset P(A_2)$ such that $U_1 \times U_2 \cap S_\beta = \emptyset$. Replacing U_1 and U_2 by $\bigcup\{u^*U_1u : u$ runs through all unitary elements of $A_1\}$ and $\bigcup\{v^*U_2v : v$ runs through all unitary elements of $A_2\}$, we may assume that U_1 and U_2 are both unitarily invariant in the sense that $u^*U_1u = U_1$ and $v^*U_2v = U_2$ for any unitary elements $u \in A_1$ and $v \in A_2$. Putting $K_1 = U_1^c$ in $P(A_1)$ and $K_2 = U_2^c$ in $P(A_2)$, we get closed subsets K_1 of $P(A_1)$ and K_2 of $P(A_2)$ satisfying the assumption of Lemma 4.15. Since $U_1 \times U_2 \cap S_\beta = \emptyset$, $S_\beta \subset \{K_1 \times P(A_2)\} \cup \{P(A_1) \times K_2\}$. Let $\mathscr{I}_1 = K_1^\perp$ and $\mathscr{I}_2 = K_2^\perp$. Since $K_1 \neq \emptyset$ and $K_2 \neq \emptyset$, we have $\mathscr{I}_1 \neq \{0\}$ and $\mathscr{I}_2 \neq \{0\}$. Choose nonzero $a_1 \in \mathscr{I}_1$ and $a_2 \in \mathscr{I}_2$. Then $(\omega_1 \otimes \omega_2)(a_1 \otimes a_2) = 0$ for every $(\omega_1,\omega_2) \in S_\beta$. But every $\omega \in P(A_\beta)$ is of the form $\omega_1 \otimes \omega_2$ for some $(\omega_1,\omega_2) \in S_\beta$, as seen above; so $\omega(a_1 \otimes a_2) = 0$ for every $\omega \in P(A_\beta)$, which is a contradiction. Hence $S_\beta = P(A_1) \times P(A_2)$. Therefore, $\omega_1 \otimes \omega_2$ is continuous under the β-norm for any pure states ω_1 and ω_2; hence $\pi_1 \otimes \pi_2$ is continuous under the β-norm for any irreducible representations π_1 of A_1 and π_2 of A_2. Hence we have, for each $x \in A_1 \otimes A_2$,

$$\begin{aligned}\|x\|_\lambda &= \|x\|_{\min}\\ &= \sup\{\|(\pi_1 \otimes \pi_2)(x)\| : \pi_1 \text{ and } \pi_2 \text{ run through all irreducible representations}\}\\ &\leq \|x\|_\beta;\end{aligned}$$

thus $\|x\|_\lambda = \|x\|_\beta$. Q.E.D.

Theorem 4.19. *Let A_1 and A_2 be C^*-algebras. The injective C^*-cross-norm of $A_1 \otimes A_2$ is smallest among all possible C^*-norms on $A_1 \otimes A_2$.*

PROOF. Let β be a C^*-norm on $A_1 \otimes A_2$. We keep the above notations. If $S_\beta = P(A_1) \times P(A_2)$, then we have, for each $a \in A_1 \otimes A_2$,

$$\|a\|_{\min} = \sup\left\{\frac{\omega_1 \otimes \omega_2(x^*a^*ax)^{1/2}}{\omega_1 \otimes \omega_2(x^*x)^{1/2}} : x \in A_1 \otimes A_2, \omega_1 \in P(A_1), \omega_2 \in P(A_2)\right\}$$
$$\leq \|a\|_\beta.$$

Hence it suffices to show that $S_\beta = P(A_1) \times P(A_2)$. Suppose $S_\beta \neq P(A_1) \times P(A_2)$. As in the proof of Lemma 4.18, there exist nonzero positive elements $a_1 \in A_1$ and $a_2 \in A_2$ such that $\omega_1 \otimes \omega_2(a_1 \otimes a_2) = 0$ for every $(\omega_1,\omega_2) \in S_\beta$. Let B be the abelian C^*-subalgebra of A_1 generated by a_2 and 1. By Lemma 4.18, the restriction of the β-norm to $B \otimes A_2$ agrees with the injective C^*-cross-norm $\|\cdot\|_{\min}$ of $B \otimes A_2$. Hence the injective C^*-tensor product $B \otimes_{\min} A_2$ is naturally imbedded in A_β. Choose $\rho \in P(B)$ and $\omega_2 \in P(A_2)$ with $\rho(a_1) \neq 0$ and $\omega_2(a_2) \neq 0$. Then $\rho \otimes \omega_2$ is a pure state of $B \otimes_{\min} A_2$. Let ω be a pure state extension of $\rho \otimes \omega_2$ to A_β. Then the restriction of ω to A_2 is ω_2; hence it is pure. By Lemma 4.11, ω is of the form $\omega = \omega_1 \otimes \omega_2$. Of course, ω_1 is pure; so $(\omega_1,\omega_2) \in S_\beta$ and $(\omega_1 \otimes \omega_2)(a_1 \otimes a_2) = \rho(a_1)\omega_2(a_2) \neq 0$, which is a contradiction. Hence $S_\beta = P(A_1) \times P(A_2)$. Q.E.D.

We can now conclude that any C^*-norm β on $A_1 \otimes A_2$ is a cross-norm because $\|x_1\|\|x_2\| = \|x_1 \otimes x_2\|_{\min} \leq \|x_1 \otimes x_2\|_\beta \leq \|x_1 \otimes x_2\|_{\max} \leq \|x_1 \otimes x_2\|_\gamma = \|x_1\|\|x_2\|$ for any $x_1 \in A_1$ and $x_2 \in A_2$.

Proposition 4.20. *If $\{\mathcal{M},\mathfrak{H}\}$ is a factor, then the map $\pi : \sum_{i=1}^n x_i \otimes x_i' \in \mathcal{M} \otimes \mathcal{M}' \mapsto \sum_{i=1}^n x_i x_i' \in \mathcal{L}(\mathfrak{H})$ is an isomorphism of $\mathcal{M} \otimes \mathcal{M}'$ onto the algebra generated algebraically by $\mathcal{M}$ and $\mathcal{M}'$.*

PROOF. It is clear that π is a homomorphism. Suppose $\sum_{i=1}^n x_i x_i' = 0$. Let $\mathfrak{H}_n$ be an n-dimensional Hilbert space with an orthogonal basis $\{\varepsilon_1, \ldots, \varepsilon_n\}$ and $\tilde{\mathfrak{H}} = \mathfrak{H} \otimes \mathfrak{H}_n$. Let $\mathfrak{M}$ be the closed subspace spanned by $\{\sum_{i=1}^n (ax_i'\xi \otimes \varepsilon_i) : a \in \mathcal{M}', \xi \in \mathfrak{H}\}$. We have then

$$\left(\sum_{i=1}^n ax_i'\xi \otimes \varepsilon_i \,\middle|\, \sum_{j=1}^n x_j^*\eta \otimes \varepsilon_j\right) = \sum_{i=1}^n (ax_i'\xi | x_i^*\eta)$$
$$= \left(\sum_{i=1}^n x_i x_i'\xi \,\middle|\, a^*\eta\right) = 0.$$

Let e be the projection of $\tilde{\mathfrak{H}}$ onto $\mathfrak{M}$. Then e is given by a matrix $e = [e_{ij}]$ with $e_{ij} \in \mathcal{L}(\mathfrak{H})$. We have, for any $a \in \mathcal{M}'$,

$$\sum_{j=1}^n ax_j'\xi \otimes \varepsilon_j = e \sum_{j=1}^n ax_j'\xi \otimes \varepsilon_j$$
$$= \sum_{i,j} (e_{i,j} ax_j'\xi \otimes \varepsilon_i);$$

hence

$$ax_i'\xi = \sum_{j=1}^n e_{i,j}ax_j'\xi, \qquad \xi \in \mathfrak{H} \quad \text{and} \quad a \in \mathscr{M}'.$$

That is,

$$ax_i' = \sum_{j=1}^n e_{i,j}ax_j', \qquad a \in \mathscr{M}'.$$

Next, we have

$$0 = e\sum_{j=1}^n x_j^*\xi \otimes \varepsilon_j = \sum_{i,j=1}^n (e_{i,j}x_j^*\xi \otimes \varepsilon_i), \qquad \xi \in \mathfrak{H},$$

so $\sum_{j=1}^n e_{i,j}x_j^* = 0$; hence we have

$$\sum_{j=1}^n x_j e_{j,i} = \sum_{j=1}^n e_{i,j}x_j^* = 0.$$

Since we have $(a \otimes 1)\mathfrak{M} \subset \mathfrak{M}$ for any $a \in \mathscr{M}'$ and for any $b \in \mathscr{M}$,

$$\begin{aligned}(b \otimes 1)\sum_{i=1}^n ax_i'\xi \otimes \varepsilon_i &= \sum_{i=1}^n bax_i'\xi \otimes \varepsilon_i \\ &= \sum_{i=1}^n ax_i'b\xi \otimes \varepsilon_i \in \mathfrak{M}.\end{aligned}$$

Hence $\mathfrak{M}$ is invariant under $\mathscr{M} \otimes \mathbf{C}$ and $\mathscr{M}' \otimes \mathbf{C}$; hence so is $\mathfrak{M}$ under $(\mathscr{M} \cup \mathscr{M}')'' \otimes \mathbf{C}$. But $\mathscr{M}$ being a factor, $(\mathscr{M} \cup \mathscr{M}')'' = \mathscr{L}(\mathfrak{H})$. Hence $\mathfrak{M}$ is invariant under $\mathscr{L}(\mathfrak{H}) \otimes \mathbf{C}$, so that e belongs to $\mathbf{C} \otimes \mathscr{L}(\mathfrak{H}_n)$. Therefore, each e_{ij} must be a scalar multiple of the identity. Thus the equalities

$$x_i' = \sum_{j=1}^n e_{i,j}x_j', \qquad \sum_{i=1}^n e_{i,j}x_i = 0$$

imply that

$$\begin{aligned}\sum_{i=1}^n x_i \otimes x_i' &= \sum_{i=1}^n x_i \otimes \left(\sum_{j=1}^n e_{i,j}x_j'\right) = \sum_{i,j=1}^n (e_{i,j}x_i \otimes x_j') \\ &= \sum_{j=1}^n \left[\left(\sum_{i=1}^n e_{i,j}x_i\right) \otimes x_j'\right] = 0.\end{aligned}$$

Q.E.D.

Corollary 4.21. *If A_1 and A_2 are simple C^*-algebras (not necessarily unital), then the injective C^*-tensor product $A_1 \otimes_{\min} A_2$ is simple.*

PROOF. Put $A = A_1 \otimes_{\min} A_2$. Let $\{\pi, \mathfrak{H}\}$ be any irreducible representation of A. Let $\|x\|_\beta = \|\pi(x)\|$ for each $x \in A_1 \otimes A_2$. Let π_1 and π_2 denote the restrictions of π to A_1 and A_2, respectively. Since $\mathscr{L}(\mathfrak{H}) = \pi(A)'' = \{\pi_1(A_1) \cup \pi_2(A_2)\}''$, and since $\pi_1(A_1)$ and $\pi_2(A_2)$ commute, $\mathscr{M} = \pi_1(A_1)''$ is a factor. For $x = \sum_{i=1}^n x_{1,i} \otimes x_{2,i} \in A_1 \otimes A_2$, if $\|x\|_\beta = 0$, then $0 = \pi(x) = \sum_{i=1}^n \pi_1(x_{1,i})\pi_2(x_{2,i})$. By Proposition 4.20, we have

$$0 = \sum_{i=1}^n \pi_1(x_{1,i}) \otimes \pi_2(x_{2,i}) = (\pi_1 \otimes \pi_2)(x).$$

Since π_1 and π_2 are faithful, $\pi_1 \otimes \pi_2$ is faithful; so $x = 0$. Thus β is a C^*-norm on $A_1 \otimes A_2$. By Theorem 4.19, we have $\|x\|_{\min} \le \|x\|_\beta \le \|x\|_{\min}$ for each $x \in A_1 \otimes A_2$. Hence π is an isometry. Since every irreducible representation is faithful, A must be simple. Q.E.D.

Proposition 4.22. *Let A_1, A_2, B_1, and B_2 be C^*-algebras. If $\pi_1 : B_1 \mapsto A_1$ and $\pi_2 : B_2 \mapsto A_2$ are homomorphisms, then $\pi_1 \otimes \pi_2 : B_1 \otimes B_2 \mapsto A_1 \otimes A_2$ can be extended to a homomorphism π of $B_1 \otimes_{\min} B_2$ into $A_1 \otimes_{\min} A_2$. Furthermore, if π_1 and π_2 are both injective, then so is π.*

PROOF. Let $\{\rho_1, \mathfrak{H}_1\}$ and $\{\rho_2, \mathfrak{H}_2\}$ be faithful representations of A_1 and A_2, respectively. Then $\rho_1 \otimes \rho_2$ is an isometry of $A_1 \otimes_{\min} A_2$. Hence, for each $x \in B_1 \otimes B_2$, we have

$$\begin{aligned}\|(\pi_1 \otimes \pi_2)(x)\|_{\min} &= \|(\rho_1 \otimes \rho_2) \circ (\pi_1 \otimes \pi_2)(x)\| \\ &= \|(\rho_1 \circ \pi_1) \otimes (\rho_2 \circ \pi_2)(x)\| \le \|x\|_{\min}.\end{aligned}$$

Therefore, $\pi_1 \otimes \pi_2$ is continuous under the injective C^*-cross-norm, so that it is extended to $B_1 \otimes_{\min} B_2$. If π_1 and π_2 are injective, then $\rho_1 \circ \pi_1$ and $\rho_2 \circ \pi_2$ are both faithful; hence the last inequality is replaced by equality. Q.E.D.

Proposition 4.23. *Let A_1 and A_2 be C^*-algebras.*

(i) *If θ_i, $i = 1,2$, are completely positive maps of A_i into another C^*-algebra B_i, then $\theta_1 \otimes \theta_2 : A_1 \otimes A_2 \mapsto B_1 \otimes B_2$ can be extended to a completely positive map of $A_1 \otimes_{\min} A_2$ into $B_1 \otimes_{\min} B_2$.*
(ii) *If θ_1 and θ_2 are completely positive maps of A_1 and A_2 into a C^*-algebra B such that $\theta_1(A_1)$ and $\theta_2(A_2)$ commute, then the map θ_0 of $A_1 \otimes A_2$ into B defined by $\theta_0(x_1 \otimes x_2) = \theta_1(x_1)\theta_2(x_2)$ can be extended to a completely positive map of $A_1 \otimes_{\max} A_2$ into B.*

PROOF. (i) We assume that B_1 and B_2 are acting on Hilbert spaces $\mathfrak{H}_1$ and $\mathfrak{H}_2$, respectively. By Theorem 3.6, there exist representations $\{\pi_1, \mathfrak{K}\}$ and $\{\pi_2, \mathfrak{K}_2\}$ of A_1 and A_2, and bounded operators V_1 and V_2 of $\mathfrak{H}_1$ and $\mathfrak{H}_2$ into $\mathfrak{K}_1$ and $\mathfrak{K}_2$, respectively, such that $\theta_1(x_1) = V_1^*\pi_1(x_1)V_1$, $x_1 \in A_1$ and $\theta_2(x_2) = V_2^*\pi_2(x_2)V_2$. We then have

$$\theta_1(x_1) \otimes \theta_2(x_2) = (V_1 \otimes V_2)^*(\pi_1 \otimes \pi_2)(x_1 \otimes x_2)(V_1 \otimes V_2),$$
$$x_1 \in A_1 \quad \text{and} \quad x_2 \in A.$$

For each $x \in A_1 \otimes_{\min} A_2$, we put $\theta(x) = (V_1 \otimes V_2)^*(\pi_1 \otimes \pi_2)(x)(V_1 \otimes V_2)$. It follows then that θ is completely positive and extends $\theta_1 \otimes \theta_2$. It is also straightforward to check that $\theta(A_1 \otimes_{\min} A_2)$ is contained in $B_1 \otimes_{\min} B_2$.

(ii) Let B_1 be a C^*-subalgebra of B containing $\theta_1(A_1)$ such that B_1 and $\theta_2(A_2)$ commute. Let $x = \sum_{i=1}^n x_{1,i} \otimes x_{2,i} \in A_1 \otimes A_2$. By the complete positivity of θ_1, there exist $\{z_{i,j} : 1 \le i,j \le n\}$ in B_1 such that $\theta_1(x_{1,i}^* x_{1,j}) =$

$\sum_{k=1}^n z_{k,i}^* z_{k,j}$; hence we have

$$\begin{aligned}\theta_0(x^*x) &= \sum_{i,j=1}^n \theta_1(x_{1,i}^* x_{1,j})\theta_2(x_{2,i}^* x_{2,j})\\ &= \sum_{i,j,k=1}^n z_{k,i}^*\theta_2(x_{2,i}^* x_{2,j})z_{k,j}\\ &= \sum_{k=1}^n \sum_{i,j=1}^n z_{k,i}^*\theta_2(x_{2,i}^* x_{2,j})z_{k,j} \geq 0.\end{aligned}$$

Therefore, if $\omega \in B_+^*$, then $\omega \circ \theta_0$ is a positive linear functional on $A_1 \otimes A_2$, so that it is extended uniquely to an element of $(A_1 \otimes_{\max} A_2)_+^*$. Since B^* is linearly spanned by B_+^*, $\omega \circ \theta_0$ is extended to a bounded linear functional on $A_1 \otimes_{\max} A_2$. Hence, the map $\theta' : \omega \in B^* \mapsto \omega \circ \theta_0 \in (A_1 \otimes_{\max} A_2)^*$ is defined. Clearly, θ' is closed. Thus the closed graph theorem yields the boundedness of θ', which means, of course, the boundedness of θ_0. Thus, θ_0 is extended to a bounded map θ of $A_1 \otimes_{\max} A_2$ into B.

The complete positivity of θ follows easily from the next lemma. Q.E.D.

Lemma 4.24 *Let A be a C^*-algebra.*

(i) *If $x = [x_{i,j}]$ and $y = [y_{i,j}]$ are positive $n \times n$-matrices over A such that $\{x_{i,j}\}$ and $\{y_{i,j}\}$ commute, then $z = [x_{i,j}y_{i,j}]$ is positive in $M_n(A)$.*

(ii) *If $x = [x_{i,k;j,l}]$, $1 \leq i,j \leq m$, $1 \leq k,l \leq n$, is a positive $mn \times mn$-matrix on A, then $\bar{x} = [\sum_{k,l=1}^n x_{i,k;j,l}]$ is a positive $m \times m$-matrix over A.*

PROOF. (i) By Lemma 3.1, we may assume that $[x_{i,j}] = [a_i^* a_j]$, $[y_{i,j}] = [b_i^* b_j]$, and $\{a_j\}$ and $\{b_j\}$ commute. We have then

$$x_{i,j}y_{i,j} = a_i^* a_j b_i^* b_j = (a_i b_i)^* a_j b_j.$$

Thus z is positive.

(ii) Let ω be the functional on $M_n(\mathbf{C})$ given by

$$\omega([\lambda_{k,l}]) = \sum_{k,l=1}^n \lambda_{k,l}.$$

Clearly, ω is a positive linear functional on $M_n(\mathbf{C})$. If we write $M_{mn}(A) = M_m(A) \otimes M_n(\mathbf{C})$, then we have $\varphi \otimes \omega(x) = \varphi(\bar{x})$ for any $x \in M_{mn}(A)$ and $\varphi \in M_m(A)^*$. Hence if $x \geq 0$, then $\varphi(\bar{x}) \geq 0$ for every $\varphi \in M_m(A)_+^*$, which means that $\bar{x} \geq 0$. Q.E.D.

Corollary 4.25. *Let A_1 and A_2 be C^*-algebras. If ω_1 is a positive linear functional on A_1, then there exists a completely positive map θ from $A_1 \otimes_{\min} A_2$ to A_2 such that*

$$\theta(x_1 \otimes x_2) = \omega_1(x_1)x_2, \qquad x_1 \in A_1, \qquad x_2 \in A_2.$$

Furthermore, θ enjoys the property

$$\theta(axb) = a\theta(x)b, \qquad a,b \in A_2, \qquad x \in A_1 \otimes_{\min} A_2,$$

where axb means, of course, $(1 \otimes a)x(1 \otimes b)$ which makes sense even if A_1 and A_2 are not unital.

PROOF. In Proposition 4.23, we simply set $B_1 = \mathbf{C}$, $B_2 = A_2$, $\theta_1 = \omega_1$, and $\theta_2 =$ the identity map. Q.E.D.

EXERCISES

1. (a) Let $A_1 = A_2 = M_2(\mathbf{C})$ and θ be the transpose map in A_2. Show that $i \otimes \theta$ on $A_1 \otimes_{\min} A_2 \cong M_4(\mathbf{C})$ is neither an isometry nor of norm one, where i means the identity map on A_1.
   (b) Let $A_1 = A_2 = \mathscr{LC}(\mathfrak{H})$, and θ be the transpose map in A_2 with respect to a fixed normalized orthogonal basis $\{\xi_j\}$ of $\mathfrak{H}$, i.e., $(\theta(x)\xi_i|\xi_j) = (x\xi_j|\xi_i)$, $x \in A_2$. Show that if $\dim \mathfrak{H} = \infty$, then $i \otimes \theta$ on $A_1 \otimes A_2$ is unbounded with respect to the injective C^*-cross-norm.

2. Let A be a unital C^*-algebra generated by commuting unital C^*-subalgebras B and C. Show that if $xy = 0$ implies either $x = 0$ or $y = 0$ for any pair $x \in B$ and $y \in C$ and if either B or C is abelian, then the homomorphism: $\sum_{i=1}^n x_i \otimes y_i \in B \otimes C \mapsto \sum_{i=1}^n x_i y_i \in A$ can be extended to an isomorphism of $B \otimes_{\min} C$ onto A.

3. Let A_1 and A_2 be two unital C^*-algebras and $A = A_1 \otimes_{\min} A_2$. Suppose that $\mathscr{I}$ is a nonzero closed ideal of A.

   (a) Show that there exists a pair $(\varphi_1,\varphi_2) \in P(A_1) \times P(A_2)$ such that $\varphi_1 \otimes \varphi_2(\mathscr{I}) \neq \{0\}$, where $P(A_i)$, $i = 1,2$, means, of course, the set of pure states on A_i.
   (b) Show that there exist open subsets $\mathscr{U}_1$ and $\mathscr{U}_2$ of $P(A_1)$ and $P(A_2)$, respectively, such that $\varphi_1 \otimes \varphi_2(\mathscr{I}) \neq \{0\}$ for every $(\varphi_1,\varphi_2) \in \mathscr{U}_1 \times \mathscr{U}_2$ and $u\mathscr{U}_i u^* = \mathscr{U}_i$ for every unitary element $u \in A_i$, $i = 1,2$. Let $K_1 = \mathscr{U}_1^c \cap P(A_1)$ and $K_2 = \mathscr{U}_2^c \cap P(A_2)$.
   (c) Let $\mathscr{I}_i$, $i = 1,2$, be the closed ideal of A_i corresponding to K_i (Lemma 4.15). Show that $\mathscr{I}_1 \otimes \mathscr{I}_2 \subset \mathscr{I}$.

†4. Let G be the free group on two generators and $\mathfrak{H}$ denote the Hilbert space of all square summable functions on G. Let L and R denote the C^*-algebras generated by the left and right regular representations of G, respectively. Then L and R are both simple [290]. The C^*-algebra A generated by L and R contains the algebra $\mathscr{LC}(\mathfrak{H})$ of all compact operators on $\mathfrak{H}$ as its only closed ideal [46]. The quotient algebra $A/\mathscr{LC}(\mathfrak{H})$ is isomorphic to the injective tensor product $L \otimes_{\min} R$ under the correspondence: $\sum_{i=1}^n x_i y_i \to \sum_{i=1}^n x_i \otimes y_i$, $x_i \in L$ and $y_i \in R$.

5. Tensor Products of W^*-Algebras

Let A_1 and A_2 be W^*-algebras with preduals $(A_1)_*$ and $(A_2)_*$, respectively. Let A_0 be the injective C^*-tensor product $A_1 \otimes_{\min} A_2$. The tensor product $A_1^* \overline{\otimes} A_2^*$ of the conjugate spaces A_1^* and A_2^* with respect to the adjoint cross-norm is considered as a closed subspace of the conjugate space A_0^* of A_0. The algebraic tensor product $(A_1)_* \otimes (A_2)_*$ of the preduals $(A_1)_*$ and $(A_2)_*$ is naturally imbedded in $A_1^* \otimes A_2^*$. The closure $(A_1)_* \overline{\otimes} (A_2)_*$ of $(A_1)_* \otimes (A_2)_*$ in A_0^* is invariant because $(A_1)_* \otimes (A_2)_*$ is invariant under the algebraic tensor product $A_1 \otimes A_2$. Therefore, there exists a unique central projection z in the universal enveloping von Neumann algebra $\tilde{A}_0$ of A_0, by Theorem III.2.7, such that $(A_1)_* \overline{\otimes} (A_2)_* = A_0^* z$. By Theorem 4.9(iii), A_0 is isometrically imbedded into the conjugate space $\{(A_1)_* \otimes (A_2)_*\}^*$, which is isomorphic to the W^*-algebra $\tilde{A}_0 z$. Hence the Banach space $(A_1)_* \overline{\otimes} (A_2)_*$ is the predual of the W^*-algebra $\tilde{A}_0 z$, and A_0 is identified with a dense C^*-subalgebra of $\tilde{A}_0 z$ under the isomorphism: $x \in A_0 \mapsto xz \in \tilde{A}_0 z$.

Definition 5.1. The W^*-algebra $\tilde{A}_0 z$ is called the *W^*-tensor product* of A_1 and A_2, and denoted by $A_1 \overline{\otimes} A_2$.

By definition, we have

$$(A_1 \overline{\otimes} A_2)_* = (A_1)_* \overline{\otimes} (A_2)_*. \tag{1}$$

Theorem 5.2. *Let A_1 and A_2 be two W^*-algebras. If $\{\pi_1, \mathfrak{H}_1\}$ and $\{\pi_2, \mathfrak{H}_2\}$ are faithful normal representations of A_1 and A_2, respectively, then the product representation $\pi_1 \otimes \pi_2$ of the injective C^*-tensor product $A_1 \otimes_{\min} A_2$ is uniquely extended to a faithful normal representation π of the W^*-tensor product $A = A_1 \overline{\otimes} A_2$ whose range $\pi(A)$ is the tensor product $\pi_1(A_1) \overline{\otimes} \pi_2(A_2)$ of $\pi_1(A_1)$ and $\pi_2(A_2)$ as von Neumann algebras.*

Proof. Let $\mathcal{M}_1 = \pi_1(A_1)$ and $\mathcal{M}_2 = \pi_2(A_2)$. Then $\{\mathcal{M}_1, \mathfrak{H}_1\}$ and $\{\mathcal{M}_2, \mathfrak{H}_2\}$ are von Neumann algebras. Let $\{\mathcal{M}, \mathfrak{H}\} = \{\mathcal{M}_1, \mathfrak{H}_1\} \otimes \{\mathcal{M}_2, \mathfrak{H}_2\}$. Let $A_0 = A_1 \otimes_{\min} A_2$ and π_0 denote the product representation $\pi_1 \otimes \pi_2$ of A_0. The image $\pi_0(A_0)$ is the C^*-algebra $\mathcal{M}_0$ generated by $\mathcal{M}_1 \otimes \mathbf{C}$ and $\mathbf{C} \otimes \mathcal{M}_2$. By Theorem 4.9(iii), π_0 is an isometry of A_0 onto $\mathcal{M}_0$, so that the transpose ${}^t\pi_0$ of π_0 is also an isometry of $\mathcal{M}_0^*$ onto A_0^*. Since we have

$$\omega(\pi_0; \xi_1 \otimes \xi_2, \eta_1 \otimes \eta_2) = \omega(\pi_1; \xi_1, \eta_1) \otimes \omega(\pi_2; \xi_2, \eta_2)$$

for every $\xi_1, \eta_1 \in \mathfrak{H}_1$ and $\xi_2, \eta_2 \in \mathfrak{H}_2$, and since we have, in general,

$${}^t\pi(\omega_{\xi,\eta}) = \omega(\pi; \xi, \eta)$$

for any representation $\{\pi, \mathfrak{H}\}$ and $\xi, \eta \in \mathfrak{H}$, ${}^t\pi_0$ maps isometrically the closed subspace of $\mathcal{M}_0^*$ spanned by the $\omega_{\xi_1 \otimes \xi_2, \eta_1 \otimes \eta_2}$ into $(A_1)_* \overline{\otimes} (A_2)_*$. Since the predual $\mathcal{M}_*$ is regarded as a closed subspace of $\mathcal{M}_0^*$ and spanned by the $\omega_{\xi,\eta}$, $\xi, \eta \in \mathfrak{H}$, and since every vector in $\mathfrak{H}$ is well approximated by linear combinations of the $\xi_1 \otimes \xi_2$, $\xi_1 \in \mathfrak{H}_1$ and $\xi_2 \in \mathfrak{H}_2$, we can conclude that ${}^t\pi_0(\mathcal{M}_*) = (A_1)_* \overline{\otimes} (A_2)_*$. Therefore, π_0 is continuous with respect to the σ-weak topologies of A and $\mathcal{M}$, so that π_0 is uniquely extended to a normal isomorphism π of A onto $\mathcal{M}$. Q.E.D.

The representation π in the theorem is denoted also by $\pi_1 \otimes \pi_2$ and called the *tensor product representation* of π_1 and π_2.

As an immediate consequence, we have the following:

Corollary 5.3. *Suppose $\{\mathscr{M}_1,\mathfrak{H}_1\}$, $\{\mathscr{M}_2,\mathfrak{H}_2\}$, $\{\mathscr{N}_1,\mathfrak{K}_1\}$, and $\{\mathscr{N}_2,\mathfrak{H}_2\}$ are von Neumann algebras. If π_1 is an isomorphism of $\mathscr{M}_1$ onto $\mathscr{N}_1$ and π_2 is an isomorphism of $\mathscr{M}_2$ onto $\mathscr{N}_2$, then there exists a unique isomorphism π of $\mathscr{M}_1 \otimes \mathscr{M}_2$ onto $\mathscr{N}_1 \overline{\otimes} \mathscr{N}_2$ such that*

$$\pi(x_1 \otimes x_2) = \pi_1(x_1) \otimes \pi_2(x_2), \qquad x_1 \in \mathscr{M}_1 \quad \text{and} \quad x_2 \in \mathscr{M}_2.$$

Lemma 5.4. *Let $\{\mathscr{M},\mathfrak{H}\}$ be a von Neumann algebra, and $\mathfrak{K}$ a separable infinite dimensional Hilbert space. For any $\omega \in \mathscr{M}_*$, there exist vectors $\xi,\eta \in \mathfrak{H} \otimes \mathfrak{K}$ such that $\omega(x) = ((x \otimes 1)\xi|\eta)$, $x \in \mathscr{M}$.*

Proof. Let $\{\varepsilon_n\}$ be an orthogonal basis of $\mathfrak{K}$. By Theorem II.2.6(ii), ω has the form $\omega = \sum_{n=1}^{\infty} \omega_{\xi_n,\eta_n}$, where $\sum_{n=1}^{\infty} \|\xi_n\|^2 < +\infty$ and $\sum_{n=1}^{\infty} \|\eta_n\|^2 < +\infty$. Put $\xi = \sum_{n=1}^{\infty} \xi_n \otimes \varepsilon_n$ and $\eta = \sum_{n=1}^{\infty} \eta_n \otimes \varepsilon_n$. Then ξ and η are in $\mathfrak{H} \otimes \mathfrak{K}$, and we have

$$\begin{aligned}\omega(x) &= \sum_{n=1}^{\infty} (x\xi_n|\eta_n) = \sum_{n,m=1}^{\infty} ((x \otimes 1)(\xi_n \otimes \varepsilon_n)|(\eta_m \otimes \varepsilon_m)) \\ &= ((x \otimes 1)\xi|\eta), \qquad x \in \mathscr{M}.\end{aligned}$$

Q.E.D.

In the study of the action of a von Neumann algebra on a Hilbert space, the following result provides a powerful tool.

Theorem 5.5. *Let $\{\mathscr{M}_1,\mathfrak{H}_1\}$ and $\{\mathscr{M}_2,\mathfrak{H}_2\}$ be von Neumann algebras. If π is a normal homomorphism of $\mathscr{M}_1$ onto $\mathscr{M}_2$, then there exist a Hilbert space $\mathfrak{K}$, a projection $e' \in \mathscr{M}'_1 \overline{\otimes} \mathscr{L}(\mathfrak{K})$, and an isometry U of $e'(\mathfrak{H}_1 \otimes \mathfrak{K})$ onto $\mathfrak{H}_2$ such that*

$$\pi(x) = U(x \otimes 1_{\mathfrak{K}})_{e'}U^*, \qquad x \in \mathscr{M}_1. \tag{2}$$

In other words, every normal homomorphism of $\mathscr{M}_1$ onto $\mathscr{M}_2$ is decomposed into the composition of an amplification, an induction, and a spatial isomorphism.

Proof. Let $\{\xi_{2,i}\}_{i\in I}$ be a maximal family of nonzero vectors in $\mathfrak{H}_2$ such that the $[\mathscr{M}_1\xi_{2,i}] = \mathfrak{H}_{2,i}$ are mutually orthogonal. From the maximality of $\{\xi_{2,i}\}_{i\in I}$, it follows that $\mathfrak{H}_2 = \sum_{i\in I}^{\otimes} \mathfrak{H}_{2,i}$. Let $\mathfrak{K}$ be the direct sum of replicas $\{\mathfrak{K}_1\}_{i\in I}$ of a separable infinite dimensional Hilbert space $\mathfrak{K}_0$. Let π_1 denote the amplification of $\mathscr{M}_1$ onto $\mathscr{M}_1 \otimes I_{\mathfrak{K}}$. Let $\{\varepsilon_n\}$ be an orthogonal basis of $\mathfrak{K}_0$. By Lemma 5.4, for each $i \in I$, we can find a vector $\xi_{1,i}$ in $\mathfrak{H}_1 \otimes \mathfrak{K}_i$ such that ${}^t\pi_1(\omega_{\xi_{1,i}}) = {}^t\pi(\omega_{\xi_{2,i}})$, that is,

$$(\pi(x)\xi_{2,i}|\xi_{2,i}) = (\pi_1(x)\xi_{1,i}|\xi_{1,i}), \qquad x \in \mathscr{M}_1. \tag{*}$$

Since $[(\mathscr{M}_1 \otimes \mathbf{C})\xi_{1,i}]$ is continued in $\mathfrak{H} \otimes \mathfrak{K}_i$, $i \in I$, the $[(\mathscr{M}_1 \otimes \mathbf{C})\xi_{1,i}]$ are orthogonal. Let e' be the projection of $\mathfrak{H} \otimes \mathfrak{K}$ onto $\sum_{i\in I}^{\oplus} [(\mathscr{M}_1 \otimes \mathbf{C})\xi_{1,i}]$.

Then e' belongs to $(\mathcal{M}_1 \otimes \mathbf{C})' = \mathcal{M}'_1 \overline{\otimes} \mathcal{L}(\mathfrak{K})$. For each $i \in I$, put

$$U_i\pi_1(x)\xi_{1,i} = \pi(x)\xi_{2,i}, \qquad x \in \mathcal{M}_1.$$

By equality (*), U_i is extended to an isometry of $[\pi_1(\mathcal{M}_1)\xi_{1,i}]$ onto $\mathfrak{H}_{2,i}$, which is also denoted by U_i. Let U denote the direct sum $\sum_{i\in I}^{\oplus} U_i$. Putting $\pi_2(x) = x_{e'}$, $x \in \mathcal{M}_1 \otimes \mathbf{C}$, and $\pi_3(x) = UxU^*$, $x \in (\mathcal{M}_1 \otimes \mathbf{C})_{e'}$, we get the desired decomposition $\pi = \pi_3 \circ \pi_2 \circ \pi_1$. Q.E.D.

For a projection e of a von Neumann algebra $\mathcal{M}$, the smallest central projection z of $\mathcal{M}$ majorizing e is called the *central support* of e and denoted by $z(e)$.

Corollary 5.6. *If $\{\mathcal{M}_1,\mathfrak{H}_1\}$ and $\{\mathcal{M}_2,\mathfrak{H}_2\}$ are two isomorphic von Neumann algebras, then there exists a von Neumann algebra $\{\mathcal{M},\mathfrak{H}\}$ and projections $e'_1,e'_2 \in \mathcal{M}'$ with central support* 1 *such that*

$$\{\mathcal{M}_1,\mathfrak{H}_1\} \cong \{\mathcal{M}_{e'_1},e'_1\mathfrak{H}\},$$
$$\{\mathcal{M}_2,\mathfrak{H}_2\} \cong \{\mathcal{M}_{e'_2},e'_2\mathfrak{H}\}.$$

The isomorphism of $\mathcal{M}_1$ onto $\mathcal{M}_2$ is given by $x_{e'_1} \mapsto x_{e'_2}$, $x \in \mathcal{M}$, identifying $\mathcal{M}_1$ with $\mathcal{M}_{e'_1}$ and $\mathcal{M}_2$ with $\mathcal{M}_{e'_2}$, respectively.

We are now going to study the commutant of the tensor product of von Neumann algebras. But we need some preparation for this purpose.

Let $\mathfrak{H}$ be a Hilbert space. If a and b are commuting self-adjoint operators on $\mathfrak{H}$, then $(a\xi|b\xi) = (ab\xi|\xi)$ is real for every $\xi \in \mathfrak{H}$, which means that $\mathrm{Re}(a\xi|ib\xi) = 0$. Hence $a\xi$ and $ib\xi$ are orthogonal with respect to the real inner product $(\cdot|\cdot)_{\mathbf{R}} = \mathrm{Re}(\cdot|\cdot)$. Conversely, if $a\xi$ and $ib\xi$ are real orthogonal for every $\xi \in \mathfrak{H}$, then a and b commute. From this observation, we see that the real Hilbert space structure $\mathfrak{H}_{\mathbf{R}}$ induced from the complex Hilbert space $\mathfrak{H}$ is convenient for studying the commutativity of self-adjoint operators. Thus we introduce the following notations and terminologies:

$$(\xi|\eta)_{\mathbf{R}} = \mathrm{Re}(\xi|\eta), \qquad \xi,\eta \in \mathfrak{H},$$
$$\xi \perp_{\mathbf{R}} \eta \quad \text{if} \quad \mathrm{Re}(\xi|\eta) = 0;$$

and we say that ξ and η are *real orthogonal.* For a subset H of $\mathfrak{H}$, we say that H is a *real* subspace if $H + H \subset H$ and $\mathbf{R}H \subset H$, and we denote by $H_{\mathbf{R}}^{\perp}$ the real orthogonal complement of H which consists of all vectors real orthogonal to every vector in H. We denote here that ξ and η are orthogonal in the usual sense if and only if $\xi \perp_{\mathbf{R}} \eta$ and $i\xi \perp_{\mathbf{R}} \eta$. Under these notations, if $\mathcal{M}$ is a von Neumann algebra on $\mathfrak{H}$, then for any $\xi \in \mathfrak{H}$, $\mathcal{M}_h\xi \perp_{\mathbf{R}} i\mathcal{M}'_h\xi$.

Lemma 5.7. *Let $\{\mathcal{M},\mathfrak{H}\}$ be a von Neumann algebra with a cyclic vector ξ_0. If $A \subset \mathcal{M}$ and $B \subset \mathcal{M}'$ are both nondegenerate *-subalgebras, then the following* *conditions are equivalent*:

(i) $A'' = \mathcal{M}$ *and* $B'' = \mathcal{M}'$;

(ii) $A_h\xi_0 + iB_h\xi_0$ *is dense in* $\mathfrak{H}$;

(iii) $(A_h\xi_0)_{\mathbf{R}}^{\perp} = i\overline{B_h\xi_0}$, *where the bar means the closure.*

PROOF. As mentioned above, $A_h\xi_0$ and $iB_h\xi_0$ are real orthogonal; hence $\overline{A_h\xi_0} + i\overline{B_h\xi_0}$ is a closed real subspace of $\mathfrak{H}$. Hence assertions (ii) and (iii) are equivalent. Suppose condition (ii) or (iii) holds. It follows then that $i(A')_h\xi_0 \subset (A_h\xi_0)_{\mathbf{R}}^{\perp} = i\overline{B_h\xi_0}$; hence $(A')_h\xi_0 \subset \overline{B_h\xi_0}$. If $b \in (A')_h$, then there exists a sequence $\{b_n\}$ in B_h such that $b\xi_0 = \lim b_n\xi_0$. For any $a \in B'$, $x, y \in A$, we have

$$\begin{aligned}(abx\xi_0|y\xi_0) &= (axb\xi_0|y\xi_0) = \lim(axb_n\xi_0|y\xi_0)\\ &= \lim(ax\xi_0|yb_n\xi_0) = (ax\xi_0|yb\xi_0)\\ &= (bax\xi_0|y\xi_0).\end{aligned}$$

On the other hand, we have

$$\begin{aligned}\mathcal{M}_h\xi_0 &\subset (i\mathcal{M}'_h\xi_0)_{\mathbf{R}}^{\perp} \subset (iB_h\xi_0)_{\mathbf{R}}^{\perp} = (A_h\xi_0)_{\mathbf{R}}^{\perp\perp}\\ &= \overline{A_h\xi_0};\end{aligned}$$

hence $\mathcal{M}\xi_0 = \mathcal{M}_h\xi_0 + i\mathcal{M}_h\xi_0 \subset \overline{A_h\xi_0} + i\overline{A_h\xi_0} \subset \overline{A\xi_0}$; thus ξ_0 is cyclic for A. Therefore, the above equality means $ab = ba$ for $a \in B'$ and $b \in (A')_h$. Thus we get

$$B' \subset A'' \subset \mathcal{M} \subset B' \quad \text{and} \quad A' \subset B'' \subset \mathcal{M}' \subset A';$$

thus $A'' = \mathcal{M}$ and $B'' = \mathcal{M}'$.

(i) $\Rightarrow$ (ii): Assume condition (i). Since A_h and B_h are strongly dense in $\mathcal{M}_h$ and $\mathcal{M}'_h$, respectively, we have only to show that $\mathcal{M}_h\xi_0 + i\mathcal{M}'_h\xi_0$ is dense in $\mathfrak{H}$. Suppose an $\eta_0 \in \mathfrak{H}$ is real orthogonal to $\mathcal{M}_h\xi_0 + i\mathcal{M}'_h\xi_0$. We shall show that $\eta_0 = 0$. Let $\tilde{\mathfrak{H}} = \mathfrak{H} \otimes \mathbf{C}^2$ be the tensor product of $\mathfrak{H}$ and a two-dimensional Hilbert space $\mathbf{C}^2$ with normalized orthogonal basis ε_1 and ε_2. Let π be the representation of $\mathcal{M}$ on $\tilde{\mathfrak{H}}$ defined by $\pi(x) = x \otimes 1$, $x \in \mathcal{M}$. By Proposition 1.6, the commutant of $\pi(\mathcal{M}) = \mathcal{M} \otimes \mathbf{C}$ is the algebra of 2×2-matrices with entries from $\mathcal{M}'$. Let $\zeta_0 = \xi_0 \otimes \varepsilon_1 + \eta_0 \otimes \varepsilon_2$ and P be the projection of $\tilde{\mathfrak{H}}$ onto $[\pi(\mathcal{M})\zeta_0]$, which belongs to $\pi(\mathcal{M})'$; hence it is of the form

$$P = \begin{bmatrix} p, & r \\ r^*, & q \end{bmatrix}, \qquad p, q, r \in \mathcal{M}',$$

with $0 \leq p \leq 1$ and $0 \leq q \leq 1$. Since $\zeta_0 \in P\tilde{\mathfrak{H}}$, $P\zeta_0 = \zeta_0$, which yields that

$$p\xi_0 + r\eta_0 = \xi_0. \tag{α}$$

We now use the fact that $\eta_0 \in (\mathcal{M}_h\xi_0)_{\mathbf{R}}^{\perp} \cap (i\mathcal{M}'_h\xi_0)_{\mathbf{R}}^{\perp}$. From the real orthogonality of η_0 to $\mathcal{M}_h\xi_0$, it follows that for each $a \in \mathcal{M}_h$, $\operatorname{Re}(a\xi_0|\eta_0) = 0$, so

$$(a\xi_0|\eta_0) = -(\eta_0|a\xi_0) = -(a\eta_0|\xi_0).$$

But this relation is complex linear in a; it holds for every $a \in \mathcal{M}$, which is equivalent to the fact that $\pi(a)\xi_0$ and $\eta_0 \otimes \varepsilon_1 + \xi_0 \otimes \varepsilon_2$ are orthogonal for every $a \in \mathcal{M}$. Hence

$$P(\eta_0 \otimes \varepsilon_1 + \xi_0 \otimes \varepsilon_2) = 0;$$

so

$$p\eta_0 + r\xi_0 = 0. \tag{β}$$

From the real orthogonality of η_0 to $i\mathcal{M}'_h\xi_0$, it follows that for every $b \in \mathcal{M}'_h\xi_0$, $(b\xi_0|\eta_0)$ is real, so

$$(b\xi_0|\eta_0) = (\eta_0|b\xi_0) = (b\eta_0|\xi_0).$$

Again by the linearity of the above relation in b, we have

$$(b\xi_0|\eta_0) = (b\eta_0|\xi_0) \tag{γ}$$

for every $b \in \mathcal{M}'$; hence in particular this holds for p, q, and r. Thus, we get, from (α), (β), and (γ),

$$\begin{aligned} 0 \le (p\eta_0|\eta_0) &= -(r\xi_0|\eta_0) = -(r\eta_0|\xi_0) \\ &= -((1-p)\xi_0|\xi_0) \le 0. \end{aligned}$$

Hence $p\eta_0 = 0$ and $(1-p)\xi_0 = 0$. Since ξ_0 is separating for $\mathcal{M}'$, being cyclic for $\mathcal{M}$, $p = 1$; thus $\eta_0 = 0$. Thus assertion (ii) follows. Q.E.D.

Let $\mathfrak{H}$ and $\mathfrak{K}$ be Hilbert spaces. For real linear subspaces $H \subset \mathfrak{H}$ and $K \subset \mathfrak{K}$, we denote by $H \odot K$ the real linear span of all the vectors $\xi \otimes \eta$ in $\mathfrak{H} \otimes \mathfrak{K}$ with $\xi \in H$ and $\eta \in K$.

Lemma 5.8. *If H and K are real linear subspaces of Hilbert spaces $\mathfrak{H}$ and $\mathfrak{K}$, respectively, such that $H + iH$ is dense in $\mathfrak{H}$ and $K + iK$ is dense in $\mathfrak{K}$, then $H \odot K + i(H_{\mathbf{R}}^{\perp} \odot K_{\mathbf{R}}^{\perp})$ is dense in $\mathfrak{H} \otimes \mathfrak{K}$.*

Proof. Let ζ be a vector of $\mathfrak{H} \otimes \mathfrak{K}$ which is real orthogonal to both $H \odot K$ and $i(H_{\mathbf{R}}^{\perp} \odot K_{\mathbf{R}}^{\perp})$. We shall show $\zeta = 0$. The real bilinear form on $\mathfrak{H} \times \mathfrak{K}$ given by $(\xi,\eta) \in \mathfrak{H} \times \mathfrak{K} \mapsto (\zeta|\xi \otimes \eta)_{\mathbf{R}} \in \mathbf{R}$ gives rise to a real linear operator t from $\mathfrak{H}$ into $\mathfrak{K}$ by the Riesz representation theorem for real Hilbert spaces such that $(t\xi|\eta)_{\mathbf{R}} = (\zeta|\xi \otimes \eta)_{\mathbf{R}}$. Clearly, t is a bounded operator. Furthermore, we have

$$\begin{aligned} (ti\xi|\eta) &= (\zeta|i\xi \otimes \eta)_{\mathbf{R}} = (\zeta|\xi \otimes i\eta)_{\mathbf{R}} = (t\xi|i\eta)_{\mathbf{R}} \\ &= -(it\xi|\eta)_{\mathbf{R}}, \end{aligned}$$

so that $ti\xi = -it\xi$. That is, t is a conjugately complex linear operator from $\mathfrak{H}$ into $\mathfrak{K}$. Now, we use the orthogonality of ζ to $H \odot K$ and $i(H_{\mathbf{R}}^{\perp} \odot K_{\mathbf{R}}^{\perp})$. For any $\xi \in H$ and $\eta \in K$, we have

$$(t\xi|\eta)_{\mathbf{R}} = (\zeta|\xi \otimes \eta)_{\mathbf{R}} = 0;$$

hence $tH \subset K_{\mathbf{R}}^{\perp}$ and $t'K \subset H_{\mathbf{R}}^{\perp}$, where t' means of course the real transpose of t. If $\xi \in H_{\mathbf{R}}^{\perp}$ and $\eta \in K_{\mathbf{R}}^{\perp}$, then we have

$$0 = (\zeta | i(\xi \otimes \eta))_{\mathbf{R}} = (ti\xi | \eta)_{\mathbf{R}} = (t\xi | i\eta)_{\mathbf{R}};$$

hence $tH_{\mathbf{R}}^{\perp} \subset (i(K_{\mathbf{R}}^{\perp}))_{\mathbf{R}}^{\perp} = iK$ and $t'(iK_{\mathbf{R}}^{\perp}) \subset H$. Thus, we get $t't(H_{\mathbf{R}}^{\perp}) \subset iH_{\mathbf{R}}^{\perp}$. Being the product of two complex conjugately linear operators, $t't$ is complex linear. Furthermore, we have

$$\begin{aligned}(t't\xi | \xi) &= (t't\xi | \xi)_{\mathbf{R}} + i(t't\xi | i\xi)_{\mathbf{R}} \\ &= \|t\xi\|^2 + i(t\xi | -it\xi)_{\mathbf{R}} \\ &= \|t\xi\|^2 + i\,\mathrm{Re}(i\|t\xi\|^2) = \|t\xi\|^2 \geq 0,\end{aligned}$$

so that $t't$ is positive. Thus, $t't$ is approximated by a real polynomial of $(t't)^2$. But $(t't)^2(H_{\mathbf{R}}^{\perp}) \subset H_{\mathbf{R}}^{\perp}$. Hence we get $t't(H_{\mathbf{R}}^{\perp}) \subset H_{\mathbf{R}}^{\perp}$; so $t't(H_{\mathbf{R}}^{\perp}) \subset H_{\mathbf{R}}^{\perp} \cap iH_{\mathbf{R}}^{\perp} = \{0\}$. Hence $t(H_{\mathbf{R}}^{\perp}) = \{0\}$, which means that $t'(\mathfrak{K}) \subset H$. But $t'(K) \subset H_{\mathbf{R}}^{\perp}$; hence $t'(K) = \{0\}$; so $t(\mathfrak{H}) \subset K_{\mathbf{R}}^{\perp}$. Since $it(\mathfrak{H}) = -it(i\mathfrak{H}) = t(\mathfrak{H}) \subset K_{\mathbf{R}}^{\perp}$, we have $t(\mathfrak{H}) \subset K_{\mathbf{R}}^{\perp} \cap iK_{\mathbf{R}}^{\perp} = \{0\}$ by the density of $K + iK$ in $\mathfrak{K}$. Thus $t = 0$, equivalently $\zeta = 0$. Q.E.D.

We can now prove the following commutation theorem for tensor products.

Theorem 5.9. *If $\{\mathscr{M}, \mathfrak{H}\}$ and $\{\mathscr{N}, \mathfrak{K}\}$ are von Neumann algebras, then*

$$(\mathscr{M} \mathbin{\overline{\otimes}} \mathscr{N})' = \mathscr{M}' \mathbin{\overline{\otimes}} \mathscr{N}'.$$

Proof. We first assume that $\mathscr{M}$ and $\mathscr{N}$ both admit cyclic vectors ξ_0 and η_0, respectively. Let $H = \mathscr{M}_h\xi_0$ and $K = \mathscr{N}_h\eta_0$. By assumption, $H + iH$ and $K + iK$ are dense in $\mathfrak{H}$ and $\mathfrak{K}$, respectively. Therefore, by the previous lemma, $H \odot K + i(H_{\mathbf{R}}^{\perp} \odot K_{\mathbf{R}}^{\perp})$ is dense in $\mathfrak{H} \otimes \mathfrak{K}$. By Lemma 5.7, we have

$$H_{\mathbf{R}}^{\perp} = \overline{i\mathscr{M}'_h\xi_0} \quad \text{and} \quad K_{\mathbf{R}}^{\perp} = \overline{i\mathscr{N}'_h\eta_0}.$$

Thus, $\mathscr{M}_h\xi_0 \odot \mathscr{N}_h\eta_0 + i(\mathscr{M}'_h\xi_0 \odot \mathscr{N}'_h\xi_0)$ is dense in $\mathfrak{H} \otimes \mathfrak{K}$. Clearly, $\zeta_0 = \xi_0 \otimes \eta_0$ is cyclic for $\mathscr{M} \mathbin{\overline{\otimes}} \mathscr{N}$, and

$$\begin{aligned}\mathscr{M}_h\xi_0 \odot \mathscr{N}_h\eta_0 &\subset (\mathscr{M} \mathbin{\overline{\otimes}} \mathscr{N})_h\zeta_0, \\ \mathscr{M}'_h\xi_0 \odot \mathscr{N}'_h\eta_0 &\subset (\mathscr{M}' \mathbin{\overline{\otimes}} \mathscr{N}')_h\zeta_0.\end{aligned}$$

Then $(\mathscr{M} \mathbin{\overline{\otimes}} \mathscr{N})_h\zeta_0 + i(\mathscr{M}' \mathbin{\overline{\otimes}} \mathscr{N}')_h\zeta_0$ is dense in $\mathfrak{H} \otimes \mathfrak{K}$. Hence we have $(\mathscr{M} \mathbin{\overline{\otimes}} \mathscr{N})' = \mathscr{M}' \mathbin{\overline{\otimes}} \mathscr{N}'$ by Lemma 5.7.

We now drop the assumption of the existence of cyclic vectors. It is clear that $\mathscr{M} \mathbin{\overline{\otimes}} \mathscr{N} \subset (\mathscr{M}' \mathbin{\overline{\otimes}} \mathscr{N}')'$ and $\mathscr{M}' \mathbin{\overline{\otimes}} \mathscr{N}' \subset (\mathscr{M} \mathbin{\overline{\otimes}} \mathscr{N})'$. We shall show that $(\mathscr{M}' \mathbin{\overline{\otimes}} \mathscr{N}')'$ and $(\mathscr{M} \mathbin{\overline{\otimes}} \mathscr{N})'$ commute. To do this, it suffices to show that

$$(xy(\xi \otimes \eta) | (\xi \otimes \eta)) = (yx(\xi \otimes \eta) | (\xi \otimes \eta))$$

for every $x \in (\mathscr{M}' \mathbin{\overline{\otimes}} \mathscr{N}')'$ and $y \in (\mathscr{M} \mathbin{\overline{\otimes}} \mathscr{N})'$, $\xi \in \mathfrak{H}$ and $\eta \in \mathfrak{K}$. Let e' and f' be, respectively, the projections of $\mathfrak{H}$ onto $[\mathscr{M}\xi]$ and $[\mathscr{N}\eta]$. It follows that

$e' \in \mathscr{M}'$ and $f' \in \mathscr{N}'$. By the first half of the proof, we have

$$\{\mathscr{M}_{e'} \bar{\otimes} \mathscr{N}_{f'}, (e' \otimes f')(\mathfrak{H} \otimes \mathfrak{K})\}' = \{\mathscr{M}'_{e'} \bar{\otimes} \mathscr{N}'_{f'}, (e' \otimes f')(\mathfrak{H} \otimes \mathfrak{K})\}.$$

Let $g' = e' \otimes f' \in \mathscr{M}' \bar{\otimes} \mathscr{N}'$. We have then

$$\begin{aligned} g'x = xg' \in (\mathscr{M}' \bar{\otimes} \mathscr{N}')'g' &= \{g'(\mathscr{M}' \bar{\otimes} \mathscr{N}')g'\}' \qquad \text{by Proposition II.3.10,} \\ &= (\mathscr{M}'_{e'} \bar{\otimes} \mathscr{N}'_{f'})' \\ &= \mathscr{M}_{e'} \bar{\otimes} \mathscr{N}_{f'} = (\mathscr{M} \bar{\otimes} \mathscr{N})_{e' \otimes f'}, \\ g'yg' \in (\mathscr{M} \bar{\otimes} \mathscr{N})'_{g'} &= \{(\mathscr{M} \bar{\otimes} \mathscr{N})_{g'}\}' = (\mathscr{M}_{e'} \bar{\otimes} \mathscr{N}_{f'})' \\ &= \mathscr{M}'_{e'} \bar{\otimes} \mathscr{N}'_{f'} = (\mathscr{M}' \bar{\otimes} \mathscr{N}')_{h\zeta_0} \end{aligned}$$

Thus $g'x$ and $g'yg'$ commute. Therefore, we have

$$\begin{aligned} (xy(\xi \otimes \eta)|\xi \otimes \eta) &= (g'xg'yg'(\xi \otimes \eta)|\xi \otimes \eta) \\ &= (g'yg'g'xg'(\xi \otimes \eta)|\xi \otimes \eta) \\ &= (yx(\xi \otimes \eta)|\xi \otimes \eta). \end{aligned}$$

Q.E.D.

Corollary 5.10. *For $i = 1,2$, let $\mathscr{M}_i$ and $\mathscr{N}_i$ be von Neumann algebras acting on a Hilbert space $\mathfrak{H}_i$. We then have*

$$(\mathscr{M}_1 \bar{\otimes} \mathscr{M}_2) \cap (\mathscr{N}_1 \bar{\otimes} \mathscr{N}_2) = (\mathscr{M}_1 \cap \mathscr{N}_1) \bar{\otimes} (\mathscr{M}_2 \cap \mathscr{N}_2),$$
$$(\mathscr{M}_1 \bar{\otimes} \mathscr{M}_2) \vee (\mathscr{N}_1 \bar{\otimes} \mathscr{N}_2) = (\mathscr{M}_1 \vee \mathscr{N}_1) \bar{\otimes} (\mathscr{M}_2 \vee \mathscr{N}_2),$$

where "$\vee$" means the von Neumann algebra generated by both sides of $\vee$.

Proof. The second formula is a straightforward conclusion from the definition. The first formula follows from the second by taking the commutant and applying the previous commutation theorem. Q.E.D.

Corollary 5.11. *For $i = 1,2$, if $\mathscr{Z}_i$ is the center of a von Neumann algebra $\mathscr{M}_i$, then the center $\mathscr{Z}$ of $\mathscr{M}_1 \bar{\otimes} \mathscr{M}_2$ is given by*

$$\mathscr{Z} = \mathscr{Z}_1 \bar{\otimes} \mathscr{Z}_2.$$

Proof. Representing $\mathscr{M}_i$, $i = 1,2$, on Hilbert spaces, we get

$$\begin{aligned} \mathscr{Z} &= (\mathscr{M}_1 \bar{\otimes} \mathscr{M}_2) \cap (\mathscr{M}_1 \bar{\otimes} \mathscr{M}_2)' \\ &= (\mathscr{M}_1 \bar{\otimes} \mathscr{M}_2) \cap (\mathscr{M}'_1 \bar{\otimes} \mathscr{M}'_2) \qquad \text{by Theorem 5.9} \\ &= (\mathscr{M}_1 \cap \mathscr{M}'_1) \bar{\otimes} (\mathscr{M}_2 \cap \mathscr{M}'_2) \qquad \text{by Corollary 5.10} \\ &= \mathscr{Z}_1 \bar{\otimes} \mathscr{Z}_2. \end{aligned}$$

Q.E.D.

Corollary 5.12. *Let $\mathscr{M}_1$ and $\mathscr{M}_2$ be von Neumann algebras with $\mathscr{M} = \mathscr{M}_1 \bar{\otimes} \mathscr{M}_2$. If φ_1 and φ_2 are normal positive linear functionals on $\mathscr{M}_1$ and $\mathscr{M}_2$, respectively, then the support $s(\varphi)$ of $\varphi = \varphi_1 \otimes \varphi_2$ is $s(\varphi_1) \otimes s(\varphi_2)$. In particular, φ is faithful if and only if both φ_1 and φ_2 are.*

PROOF. Considering the cyclic representations induced by φ_1 and φ_2, we assume that $\mathcal{M}_1$ and $\mathcal{M}_2$ act on $\mathfrak{H}_1$ and $\mathfrak{H}_2$, respectively, and that

$$\varphi_1(x) = (x\xi_1|\xi_1), \qquad x \in \mathcal{M}_1,$$
$$\varphi_2(y) = (y\xi_2|\xi_2), \qquad y \in \mathcal{M}_2$$

with some vectors $\xi_1 \in \mathfrak{H}_1$ and $\xi_2 \in \mathfrak{H}_2$. Let $\xi_0 = \xi_1 \otimes \xi_2$. We then have

$$\varphi(x) = (x\xi_0|\xi_0), \qquad x \in \mathcal{M}.$$

The support $s(\varphi)$ of φ is the projection of $\mathfrak{H}$ onto $[\mathcal{M}'\xi_0]$. But we have $\mathcal{M}' = \mathcal{M}'_1 \overline{\otimes} \mathcal{M}'_2$. Hence we have

$$\begin{aligned}[\mathcal{M}'\xi_0] &= [(\mathcal{M}'_1 \otimes \mathcal{M}'_2)(\xi_1 \otimes \xi_2)] = [\mathcal{M}'_1\xi_1 \otimes \mathcal{M}'_2\xi_2] \\ &= (e_1 \otimes e_2)(\mathfrak{H}_1 \otimes \mathfrak{H}_2),\end{aligned}$$

where $e_1 = s(\varphi_1)$ and $e_2 = s(\varphi_2)$. Q.E.D.

Proposition 5.13. *Let $\mathcal{M}_1$, $\mathcal{M}_2$, $\mathcal{N}_1$, and $\mathcal{N}_2$ be von Neumann algebras. If θ_1 and θ_2 are σ-weakly continuous completely positive maps of $\mathcal{M}_1$ into $\mathcal{N}_1$ and $\mathcal{M}_2$ into $\mathcal{N}_2$, respectively, then there exists a σ-weakly continuous completely positive map θ of $\mathcal{M}_1 \overline{\otimes} \mathcal{M}_2$ into $\mathcal{N}_1 \overline{\otimes} \mathcal{N}_2$ such that*

$$\theta(x_1 \otimes x_2) = \theta_1(x_1) \otimes \theta_2(x_2), \qquad x_1 \in \mathcal{M}_1, \qquad x_2 \in \mathcal{M}_2.$$

PROOF. By Proposition 4.23, there exists a completely positive map θ_0 of $\mathcal{M}_1 \otimes_{\min} \mathcal{M}_2$ into $\mathcal{N}_1 \otimes_{\min} \mathcal{N}_2$ extending $\theta_1 \otimes \theta_2$. For each $\varphi_1 \in (\mathcal{N}_1)_*$ and $\varphi_2 \in (\mathcal{N}_2)_*$, we have

$${}^t\theta_0(\varphi_1 \otimes \varphi_2) = {}^t\theta_1(\varphi_1) \otimes {}^t\theta_2(\varphi_2) \in (\mathcal{M}_1)_* \otimes (\mathcal{M}_2)_*.$$

Therefore, ${}^t\theta_0$ maps a dense part $(\mathcal{N}_1)_* \otimes (\mathcal{N}_2)_*$ of $(\mathcal{N}_1 \overline{\otimes} \mathcal{N}_2)_*$ into a subspace $(\mathcal{M}_1)_* \otimes (\mathcal{M}_2)_*$ of $(\mathcal{M}_1 \overline{\otimes} \mathcal{M}_2)_*$. This means that θ_0 is σ-weakly continuous. Hence θ_0 is extended to a σ-weakly continuous map θ of $\mathcal{M}_1 \overline{\otimes} \mathcal{M}_2$ into $\mathcal{N}_1 \overline{\otimes} \mathcal{N}_2$ by continuity, and the resulting map θ is completely positive. Q.E.D.

EXERCISES

1. Let $\mathcal{M}_1$ and $\mathcal{M}_2$ be subfactors of a factor $\mathcal{M}$ such that $\mathcal{M}_1$ and $\mathcal{M}_2$ commute. Suppose that $\mathcal{M}_1$ and $\mathcal{M}_2$ generate $\mathcal{M}$.
   (a) Show that the map $\pi_0: \sum_{i=1}^n x_i \otimes y_i \in \mathcal{M}_1 \otimes \mathcal{M}_2 \mapsto \sum_{i=1}^n x_i y_i \in \mathcal{M}$ is an isomorphism of the algebraic tensor product $\mathcal{M}_1 \otimes \mathcal{M}_2$ into $\mathcal{M}$.
   (b) Show that the above map π_0 is extended to an isomorphism π of the W^*-algebra tensor product $\mathcal{M}_1 \overline{\otimes} \mathcal{M}_2$ onto $\mathcal{M}$ if and only if there exists a normal nonzero functional φ on $\mathcal{M}$ such that $\varphi(x_1x_2) = \varphi(x_1)\varphi(x_2)$ for every $x_1 \in \mathcal{M}_1$ and $x_2 \in \mathcal{M}_2$.
   (c) Show that the condition on the existence of φ in (b) is also equivalent to the existence of a nonzero normal linear map $\mathcal{E}$ of $\mathcal{M}$ into $\mathcal{M}_1$ (resp. $\mathcal{M}_2$) such that $\mathcal{E}(axb) = a\mathcal{E}(x)b$ for every $a,b \in \mathcal{M}_1$ (resp. $\mathcal{M}_2$) and $x \in \mathcal{M}_2$.

2. Let $\mathscr{A}_1$ and $\mathscr{A}_2$ be maximal abelian von Neumann subalgebras of von Neumann algebras $\mathscr{M}_1$ and $\mathscr{M}_2$, respectively. Show that the tensor product $\mathscr{A}_1 \overline{\otimes} \mathscr{A}_2$ is also maximal abelian in $\mathscr{M}_1 \overline{\otimes} \mathscr{M}_2$.

3. Let $\{\mathscr{M},\mathfrak{H}\}$ be a von Neumann algebra. An operator T on $\mathfrak{H}$ is said to be *affiliated* with $\mathscr{M}$ if $uT = Tu$ for every unitary element $u \in \mathscr{M}$:
   (a) Show that a closed operator T on $\mathfrak{H}$ is affiliated with $\mathscr{M}$ if and only if the projection $g(T)$ of $\mathfrak{H} \oplus \mathfrak{H} = \mathfrak{H} \otimes \mathbf{C}^2$ onto the graph $\mathfrak{G}(T) = \{(\xi, T\xi) : \xi \in \mathscr{D}(T)\} \subset \mathfrak{H} \otimes \mathbf{C}^2$ of T belongs to $M_2(\mathscr{M}) = \mathscr{M} \otimes M_2(\mathbf{C})$, where $M_2(\mathbf{C})$ means the algebra $\mathscr{L}(\mathbf{C}^2)$ of all 2×2-matrices and $\mathscr{D}(T)$ the domain of T.
   (b) Determine the four entries of $g(T)$.

4. Let $\mathscr{A} = L^\infty(\mathbf{R},m)$, where m means the Lebesgue measure on $\mathbf{R}$. For a fixed irrational number α, let $\mathscr{B} = \{f \in \mathscr{A} : f(s) = f(s + \alpha),\ s \in \mathbf{R}\}$ and $\mathscr{C} = \{f \in \mathscr{A} : f(s) = f(s + 1),\ s \in \mathbf{R}\}$.
   (a) Show that $\mathscr{A}$ is generated by $\mathscr{B}$ and $\mathscr{C}$ as a von Neumann algebra.
   (b) Show that the C^*-algebra generated by $\mathscr{B}$ and $\mathscr{C}$ is isomorphic to the tensor product $\mathscr{B} \otimes_{\min} \mathscr{C}$ under the obvious mapping.
   (c) Show that there is no nontrivial $\varphi \in \mathscr{A}_*$ such that $\varphi(fg) = \varphi(f)\varphi(g)$, $f \in \mathscr{B}$ and $g \in \mathscr{C}$.

Notes

The tensor product of Banach spaces was first studied by R. Schatten and J. von Neumann [317]–[320]. Unlike the finite dimensional case, the tensor product of infinite dimensional Banach spaces behaves mysteriously. Cross-norms in the tensor product are highly nonunique. J. von Neumann and R. Schatten proved Theorem 2.5 [319], which indicates that various cross-norms carry important information. Indeed, R. Schatten used cross-norms to study classes of operators [29]. The general theory of tensor products was further developed by A. Grothendieck [6], in the context of locally convex topological vector spaces, which led him to the discovery of nuclear spaces.

The origin of the tensor product of operator algebras goes back to the pioneering work of F. J. Murray and J. von Neumann [240]. They studied when a factor $\mathscr{M}$ on $\mathfrak{H}$ factorizes $\mathscr{L}(\mathfrak{H})$ in the sense that $\mathfrak{H} = \mathfrak{H}_1 \otimes \mathfrak{H}_2$ and $\mathscr{L}(\mathfrak{H}) \cong \mathscr{M} \overline{\otimes} \mathscr{M}'$. It turns out that this is true if and only if $\mathscr{M}$ is a factor of type I; see Chapter V. They called such a factor *direct*. But there are many factors which are not direct. Nonetheless, it is still true that $\mathscr{L}(\mathfrak{H})$ can be viewed as the "tensor product" of $\mathscr{M}$ and $\mathscr{M}'$ with respect to a more complicated "cross topology." It can be said that ever since the pioneering work of Murray and von Neumann, the operator algebraists have been trying to understand this "cross topology."

The tensor product of C^*-algebras was introduced by T. Turumaru [387]. The injective C^*-cross-norm is often referred to as the Turumaru cross-norm. The projective C^*-cross-norm was considered first by A. Guichardet [155].

For the formal treatise, the projective cross-norm behaves more naturally than the injective one. But, very little is known concerning the projective one, although this is the right norm representing the problem of the "cross topology." The theory of nuclear C^*-algebras will be important in this aspect. We will treat them in the forthcoming volume. The tensor product of W^*-algebras was introduced by Y. Misonou [238]. The point was that one can define the tensor product $\mathcal{M}_1 \overline{\otimes} \mathcal{M}_2$ of W^*-algebras $\mathcal{M}_1$ and $\mathcal{M}_2$ independently of the Hilbert spaces where they act.

Theorem 3.6 is due to W. Stinespring [334]. The recognition of the importance of completely positive maps in the tensor products was due to C. Lance [215] and E. Effros and C. Lance [108]. Theorem 4.9 is due to T. Turumaru [389]. Theorem 4.14, Theorem 4.19, and Corollary 4.21 are due to M. Takesaki [351] and [357]. Theorem 5.2 is due to Y. Misonou [238]. The presentation here follows the Takeda–Sakai approach [347] and [305]. Theorem 5.5 is due to J. Dixmier [88]. The tensor product commutation theorem, Theorem 5.9, was first proved for semifinite von Neumann algebras by Y. Misonou [238], for the definition, see Chapter V, which is based on the representation given by a faithful semifinite normal trace. In this special case, the theorem was proved without particular difficulty in the very early stage of the theory. But the general case remained unsolved for quite a while. A partial solution was given by Sakai [314]. It was M. Tomita [374] who solved the problem with full generality. As usually happened, after Tomita's solution quite a few simplified versions of the proof were offered, for example [73], [27], and [360]. The present proof here follows the approach of M. Rieffel and A. van Daele [296].

6. Integral Representations of States

Let A be a C^*-algebra and $\{\pi,\mathfrak{H}\}$ a representation of A. By Proposition I.9.17, $\{\pi,\mathfrak{H}\}$ is decomposed into the direct sum of cyclic representations, so that the study of $\{\pi,\mathfrak{H}\}$ is reduced to that of cyclic representations. If $\{\pi,\mathfrak{H}\}$ is irreducible, then it is obviously impossible to decompose $\{\pi,\mathfrak{H}\}$ into further elementary components. In this section, we shall study how we can relate the study of cyclic representations to that of irreducible ones. By Theorem I.9.22, we have already seen that the irreducibility of the cyclic representation $\{\pi_\omega,\mathfrak{H}_\omega\}$ induced by a state ω of A is equivalent to the indecomposability of ω into a convex combination of states. Furthermore, in the proof of that theorem, we saw some connection between the decompositions of ω and that of the identity operator into the sum of positive operators in $\pi_\omega(A)'$. In this section, we shall examine this connection in detail.

Concerning the above problem, we note first of all that the addition of an identity to a nonunital C^*-algebra has no effect on the nature of a representation. Therefore, we consider throughout this section only unital C^*-algebras. Thus, the state space is a compact convex set with respect to the weak* topology. Furthermore, the self-adjoint part A_h of a unital C^*-algebra A is isomorphic, as a real Banach space, to the space of all affine real valued continuous functions on the state space, as seen in Section III.6. Therefore, our situation fits in perfectly with the so-called Choquet theory of boundary integrals on compact convex sets. We shall explore briefly the part of this general theory related directly to our subject. But, the reader interested in this topic is advised to study the general theory in more detail because our topic is very intimately related to the theory of compact convex sets.

We now continue the study of a compact convex set K contained in a locally convex topological vector space E, and keep the notations established in the beginning of Section III.6. The space $\mathscr{A}(K)$ of all real valued continuous affine functions on K is a Banach space with respect to the pointwise linear operations and the supremum norm. Each point $x \in K$ gives rise to a continuous linear functional: $a \in \mathscr{A}(K) \mapsto a(x) \in \mathbf{R}$, and this association of a linear functional on $\mathscr{A}(K)$ to each $x \in K$ is trivially affine, injective, and continuous with respect to the weak* topology in $\mathscr{A}(K)^*$. Thus, we consider K as a compact convex subset of $\mathscr{A}(K)^*$; in other words the vector space E is replaced by $\mathscr{A}(K)^*$ equipped with the weak* topology.

We recall that a function f on K is said to be *convex* if $f(\lambda x + (1 - \lambda)y) \leq \lambda f(x) + (1 - \lambda)f(y)$, $x,y \in K$ and $0 \leq \lambda \leq 1$. We denote by $\mathscr{P}(K)$ the *space of all continuous real valued convex functions* on K, and by $\mathscr{Q}(K)$ the *space of all lower semicontinuous convex functions* on K with values in $]-\infty,\infty]$.

Lemma 6.1. *The set* $\{f - g : f,g \in \mathscr{P}(K)\} = \mathscr{P}(K) - \mathscr{P}(K)$ *is a uniformly dense linear subspace of* $C_{\mathbf{R}}(K)$, *where* $C_{\mathbf{R}}(K)$ *means the space of all real valued continuous functions on* K.

PROOF. Since $\mathscr{A}(K) \subset \mathscr{P}(K)$, $\mathscr{P}(K)$ separates the points of K. Since $\mathscr{P}(K)$ is a convex cone in $C_{\mathbf{R}}(K)$ and closed under the supremum operation $f \vee g$, $\mathscr{P}(K) - \mathscr{P}(K)$ is a lattice subspace of $C_{\mathbf{R}}(K)$. In fact, if $f_i = g_i - h_i$ with g_i, $k_i \in \mathscr{P}(K)$, $i = 1,2$, then $f'_i = f_i + (h_1 + h_2)$ belongs to $\mathscr{P}(K)$, and so

$$\begin{aligned} f_1 \vee f_2 &= [f'_1 - (h_1 + h_2)] \vee [f'_2 - (h_1 + h_2)] \\ &= f'_1 \vee f'_2 - (h_1 + h_2) \in \mathscr{P}(K) - \mathscr{P}(K). \end{aligned}$$

Thus our assertion follows from Stone's theorem. Q.E.D.

Lemma 6.2. *Each function in* $\mathscr{Q}(K)$ *is a pointwise limit of increasing net in* $\mathscr{P}(K)$.

PROOF. By Lemma III.6.1, if f is a function of $\mathscr{Q}(K)$, then

$$f(x) = \sup\{a(x) : a \in \mathscr{A}(K), a \leq f\}.$$

Hence $f(x)$ is a pointwise limit of the net $\{a_1 \vee a_2 \vee \cdots \vee a_n(x) : a_1, \ldots, a_n \in \mathscr{A}(K)$ and $a_1, \ldots, a_n \leq f\}$, and $a_1 \vee \cdots \vee a_n \in \mathscr{P}(K)$. Q.E.D.

We call the set of all extreme points of K the *extreme boundary* of K and denote it by $\partial_e K$. Given a function f on a subset X of K containing $\partial_e K$ with values in $[\alpha,\infty]$, the *lower envelope* $\check{f}$ of f is defined by

$$\check{f}(x) = \sup\{a(x) | a \in \mathscr{A}(K),\ a|_X \le f\}. \tag{1}$$

Similarly, if $f: X \mapsto [-\infty,\infty[$, then the *upper envelope* of f is given by

$$\hat{f}(x) = \inf\{a(x) | a \in \mathscr{A}(K),\ a|_X \ge f\}. \tag{1'}$$

We note that $\hat{f}$ and $\check{f}$ are both defined on the whole of K, not only on X, and that $\check{f} \in \mathscr{Q}(K)$ and $\hat{f} \in (-\mathscr{Q}(K))$.

To avoid confusion, we make a few remarks on measures on a topological space. In this section, we mean by a *measure* a *real Radon measure* on a compact space X, i.e., a bounded linear functional on $C_{\mathbf{R}}(X)$. A Radon measure μ on X is uniquely extended to a *Baire measure* μ_0 on X, i.e., a σ-additive function μ_0 on the σ-field $\mathscr{B}_0$ generated by G_δ-compact subsets and

$$\mu(f) = \int_X f(x)\, d\mu_0(x), \qquad f \in C_{\mathbf{R}}(X).$$

Every Baire measure μ_0 is *regular* in the sense that

$$\begin{aligned}\mu_0(E) &= \sup\{\mu_0(F): E \supset F,\ F \text{ is closed and } F \in \mathscr{B}_0\}\\ &= \inf\{\mu_0(U): U \supset E,\ U \text{ is open and } U \in \mathscr{B}_0\}.\end{aligned}$$

Furthermore, every Baire measure μ_0 is uniquely extended to a *regular* Borel measure $\tilde{\mu}_0$, i.e., a countably additive function $\tilde{\mu}_0$ on the σ-field $\mathscr{B}$ generated by compact subsets. Regularity means here that

$$\begin{aligned}\tilde{\mu}_0(E) &= \sup\{\tilde{\mu}_0(F): E \supset F,\ F \text{ is closed}\}\\ &= \inf\{\tilde{\mu}_0(U): U \supset E,\ U \text{ is open}\}.\end{aligned}$$

By the uniqueness of $\tilde{\mu}_0$, we denote it by the same symbol μ, and we handle only regular Borel measures unless we explicitly declare otherwise. We denote by $M_{\mathbf{R}}(X)$ the Banach space $C_{\mathbf{R}}(X)^*$, and $M_{\mathbf{R}}^+(X)$ the positive cone in $M_{\mathbf{R}}(X)$. Furthermore, $M_1^+(X)$ denotes the set of positive measures with total mass one. Often a member of $M_1^+(X)$ is called a *probability* measure. Clearly, $M_{\mathbf{R}}^+(X)$ is a compact convex subset of $M_{\mathbf{R}}(X)$ with respect to the weak* topology.

Lemma 6.3. *If μ is a probability measure on a compact convex set K in E, then there exists a unique point $y \in K$ such that*

$$p(y) = \int_K p(x)\, d\mu(x), \qquad p \in E^*. \tag{2}$$

Proof. Let $\varphi(p) = \int_K p(x)\, d\mu(x)$, $p \in E^*$. Clearly, φ defines a linear functional on E^*. Hence it belongs to the algebraic dual $E^{*\prime}$ of E^*. The canonical imbedding of E into $E^{*\prime}$ is continuous with respect to the $\sigma(E^{*\prime},E^*)$-topology and the original topology in E. Hence K is a compact convex subset of $E^{*\prime}$.

Suppose that $\varphi \notin K$. By the Hahn–Banach separation theorem, there exists an element $p \in E^*$ such that

$$\varphi(p) > \sup\{p(x) : x \in K\} = \alpha.$$

Then we have

$$\varphi(p) = \int_K p(x)\, d\mu(x) \leq \int_K \alpha\, d\mu(x) = \alpha < \varphi(p),$$

which is a contradiction. Thus $\varphi \in K$. Q.E.D.

Definition 6.4. The point $y \in K$ given by (2) is called the *barycenter* of μ or the *resultant* of μ and is denoted by $\mathrm{r}(\mu)$. Conversely, the measure μ is called a *representing measure* of the point y. By $M_y^+(K)$, we denote the set of all representing measures of y.

Since $\mathscr{A}(K;E)$ is dense in $\mathscr{A}(K)$ by Corollary III.6.3, we have

$$a(y) = \int_K a(x)\, d\mu(x), \qquad a \in \mathscr{A}(K), \tag{2'}$$

if $y = \mathrm{r}(\mu)$.

Our problem is, given a point $x \in K$, to find a measure $\mu \in M_1^+(K)$ with barycenter x such that μ is concentrated on the extreme boundary $\partial_e K$. To this end, we must compare two representing measures μ and ν of a given point $x \in K$ and determine which of them is concentrated "nearer" to the boundary.

Definition 6.5. Given $\mu, \nu \in M_{\mathbf{R}}(K)$, we write $\nu \prec \mu$ or $\mu \succ \nu$ if $\nu(f) \leq \mu(f)$ for every $f \in \mathscr{P}(K)$.

If $\nu \prec \mu$, then $\nu(a) \leq \mu(a)$ and $\nu(-a) \leq \mu(-a)$ for every $a \in \mathscr{A}(K)$, and so $\mu(a) = \nu(a)$. Hence if ν represents a point $x \in K$, then μ also represents the same point x. If $\nu \prec \mu$ and $\mu \prec \nu$, then $\mu(f) = \nu(f)$ for every $f \in \mathscr{P}(K)$, and so $\mu = \nu$ by Lemma 6.1. Hence the ordering "$\succ$" is a proper ordering in $M_{\mathbf{R}}(K)$. We say that $\mu, \nu \in M_{\mathbf{R}}(K)$ are equivalent, and write $\mu \sim \nu$, if $\mu(a) = \nu(a)$ for every $a \in \mathscr{A}(K)$.

Lemma 6.6. *The ordering "$\prec$" in $M^+(K)$ is inductive; hence every $\nu \in M^+(K)$ is majorized by a maximal $\mu \in M^+(K)$.*

Proof. Let $\{\mu_i\}$ be a linearly ordered subset of $M^+(K)$. Since $1 \in \mathscr{A}(K)$, $\|\mu_i\| = \mu_i(1) = \mu_j(1) = \|\mu_j\|$, so that $\{\mu_i\}$ is bounded. Hence $\{\mu_i(f)\}$ for every $f \in \mathscr{P}(K)$ is increasing and bounded, and so converges. By the boundedness of $\{\mu_i\}$ and Lemma 6.1, $\{\mu_i\}$ converges to $\mu \in M^+(K)$ in the weak* topology, and $\mu_i \prec \mu$. The second assertion is nothing but Zorn's lemma. Q.E.D.

Lemma 6.7. *For any $\mu \in M_1^+(K)$ and $f \in C_{\mathbf{R}}(K)$, there exists $\nu \in M_1^+(K)$ such that $\mu \prec \nu$ and $\nu(f) = \mu(\hat{f})$.*

PROOF. Let $\Phi(g) = \mu(\hat{g})$ for every $g \in C_{\mathbf{R}}(K)$. By the following property of the upper envelop,

$$\left.\begin{array}{lll} (g_1 + g_2)\hat{} \leq \hat{g}_1 + \hat{g}_2, & & g_1, g_2 \in C_{\mathbf{R}}(K), \\ (\alpha g)\hat{} = \alpha \hat{g}, & \alpha \in \mathbf{R}_+, & g \in C_{\mathbf{R}}(K), \end{array}\right\} \tag{3}$$

the function Φ on $C_{\mathbf{R}}(K)$ is sublinear. Over the one-dimensional subspace $\mathbf{R}f$ spanned by f, we define a functional ν_0 by

$$\nu_0(\alpha f) = \alpha\mu(\hat{f}), \qquad \alpha \in \mathbf{R}.$$

For $\alpha > 0$, we have $\nu_0(\alpha f) = \mu(\alpha\hat{f}) = \mu((\alpha f)\hat{}) = \Phi(\alpha f)$. If $\alpha < 0$, then, putting $\beta = -\alpha$, we have

$$0 = \alpha f + \beta f = (\alpha f + \beta f)\hat{} \leq (\alpha f)\hat{} + (\beta f)\hat{},$$

so that $-(\beta f)\hat{} \leq (\alpha f)\hat{}$, and so

$$\nu_0(\alpha f) = \alpha\mu(\hat{f}) = -\mu((\beta f)\hat{}) \leq \mu((\alpha f)\hat{}) = \Phi(\alpha f).$$

Thus ν_0 is majorized by Φ on $\mathbf{R}f$.

By the Hahn–Banach theorem, there exists a linear functional ν on $C_{\mathbf{R}}(K)$ which extends ν_0 and is majorized by Φ. Hence we have

$$\nu(f) = \nu_0(f) = \mu(\hat{f}) \quad \text{and} \quad \nu(g) \leq \mu(\hat{g}), \qquad g \in C_{\mathbf{R}}(K).$$

If $g \in C(K)$, $g \leq 0$, then $\hat{g} \leq 0$, so that $\nu(g) \leq \mu(\hat{g}) \leq 0$. Hence ν is positive. For every $g \in \mathscr{P}(K)$, we have $(-g)\hat{} = -g$, so that $-\nu(g) \leq \mu((-g)\hat{}) = -\mu(g)$; hence $\mu(g) \leq \nu(g)$, and so $\mu \prec \nu$. Q.E.D.

Definition 6.8. To each $f \in C_{\mathbf{R}}(K)$, we associate the *boundary* set B_f of f by

$$\mathrm{B}_f = \{x \in K : \hat{f}(x) = f(x)\}.$$

Since $\hat{f}$ is upper semicontinuous, the boundary set

$$\mathrm{B}_f = \bigcap_{n=1}^{\infty} \left\{x : \hat{f}(x) - f(x) < \frac{1}{n}\right\}, \qquad f \in C_{\mathbf{R}}(K),$$

is a G_δ-set.

Lemma 6.9. $\partial_e K = \bigcap\{B_f : f \in C_{\mathbf{R}}(K)\}$.

PROOF. Suppose that $x \in \partial_e K$ and $f \in C_{\mathbf{R}}(K)$. By Lemma 6.7, there exists $\nu \in M_x^+(K)$ such that $\hat{f}(x) = \nu(f)$. But $M_x^+(K)$ is reduced to the singleton set $\{\delta_x\}$, where δ_x means the point mass measure at x, so that $\nu = \delta_x$. Thus $\hat{f}(x) = f(x)$. Hence $x \in \bigcap \mathrm{B}_f$.

Conversely, suppose that $x \in \bigcap \mathrm{B}_f$, and consider an $f \in C_{\mathbf{R}}(K)$. We then have $f(x) = \hat{f}(x) = \check{f}(x)$ since $-\check{f} = (-f)\hat{}$. For any $\nu \in M_x^+(K)$, we have

$\delta_x \prec \nu$, so that

$$f(x) = \check{f}(x) \leq \nu(\check{f}) \leq \nu(f) \leq \nu(\check{f}) \leq \hat{f}(x) = f(x),$$

where we use the fact that $\mu(g) \leq \nu(g)$ for every $g \in \mathcal{Q}(K)$ if $\mu \prec \nu$, which follows from Lemma 6.2. Hence we have $\nu = \delta_x$, which means that x has no nontrivial convex combination expression. This means, by definition, that x is extreme. Q.E.D.

Definition 6.10. A measure μ on K is said to be a *boundary measure* if $|\mu|(\mathrm{B}_f^c) = 0$ for every $f \in C_{\mathbf{R}}(K)$, where $|\mu|$ means of course the absolute value of μ.

Theorem 6.11. *Let K be a compact convex set in a locally convex vector space E.*

(i) *A measure $\mu \in M^+(K)$ is maximal in the ordering "$\prec$" if and only if μ is a boundary measure.*
(ii) *Every point in K is represented by a boundary measure.*

Proof. (i) Suppose that $\mu \in M_+(K)$ is maximal. By Lemma 6.7, we have $\mu(f) = \mu(\hat{f})$, $f \in C_{\mathbf{R}}(K)$, which means that $\mu(\mathrm{B}_f^c) = 0$. Hence μ is a boundary measure.

Conversely, if μ is a boundary measure, then $\mu(f) = \mu(\hat{f})$ for every $f \in C_{\mathbf{R}}(K)$. By considering $(-f)\hat{} = -f\check{}$, we have $\mu(f) = \mu(\check{f})$. For any $\nu \in M^+(K)$ with $\nu \succ \mu$, and $f \in \mathcal{P}(K)$, we have $f = f$, so that

$$\mu(f) = \mu(\check{f}) \leq \nu(\check{f}) = \nu(f) \leq \nu(\hat{f}) \leq \mu(\hat{f}) = \mu(f).$$

Hence μ and ν coincide on $\mathcal{P}(K)$, and by Lemma 6.1, $\mu = \nu$. Thus μ is maximal.

(ii) This is an immediate consequence of (i) and Lemma 6.6. Q.E.D.

Definition 6.12. A real valued function f on K is said to be *strictly convex* if

$$f(\lambda x + (1 - \lambda)y) < \lambda f(x) + (1 - \lambda)f(y)$$

for any $x \neq y$, $0 < \lambda < 1$.

Lemma 6.13. *If $f \in \mathcal{P}(K)$ is strictly convex then $\partial_e K = B_f$. Hence if there is a strictly convex continuous function, then $\partial_e K$ is a G_δ-set.*

Proof. We know that $\mathrm{B}_f \supset \partial_e K$. If x admits a proper convex combination $x = \lambda y + (1 - \lambda)z$, $y \neq z$, $0 < \lambda < 1$, then

$$\hat{f}(x) \geq \lambda\hat{f}(y) + (1 - \lambda)\hat{f}(z) \geq \lambda f(y) + (1 - \lambda)f(z) > f(x),$$

so that $\hat{f}(x) \neq f(x)$. Hence $x \notin \mathrm{B}_f$. Hence $\mathrm{B}_f \subset \partial_e K$. Q.E.D.

Lemma 6.14. *If K is a metrizable compact convex set, then there exists a strictly convex continuous function.*

PROOF. By assumption, $C_{\mathbf{R}}(K)$ is separable, so that there exists a dense countable sequence $\{f_n\}$ in $\mathscr{P}(K)$. Put

$$f = \sum_{n=1}^{\infty} \frac{1}{2^n \|f_n\|} f_n \in \mathscr{P}(K).$$

We claim that f is strictly convex. If not, there would be a proper convex combination $x = \lambda y + (1 - \lambda)z$ such that

$$f(x) = \lambda f(y) + (1 - \lambda)f(z).$$

Since $f_n(x) \leq \lambda f_n(y) + (1 - \lambda)f_n(z)$, we have $f_n(x) = \lambda f_n(y) + (1 - \lambda)f_n(z)$. By the density of $\{f_n\}$ in $\mathscr{P}(K)$, we have $g(x) = \lambda g(y) + (1 - \lambda)g(z)$. Hence this is true for every $g \in C_{\mathbf{R}}(K)$ by Lemma 6.1. Let $a \in \mathscr{A}(K)$ with $a(y) \neq a(z)$. Then we have

$$a(x)^2 = \lambda a(y)^2 + (1 - \lambda)a(z)^2,$$
$$a(x) = \lambda a(y) + (1 - \lambda)a(z),$$

which is impossible because the function: $t \in \mathbf{R} \mapsto t^2 \in \mathbf{R}$ is strictly convex. Q.E.D.

Theorem 6.15. *If K is a metrizable compact convex set, then every boundary measure is concentrated on the G_δ-set $\partial_e K$. Hence every point in K is represented by a measure concentrated on $\partial_e K$.*

PROOF. By definition, $\mu(B_f^c) = 0$ if μ is a boundary measure. By assumption and the previous lemma, there exists an $f \in \mathscr{P}(K)$ with $\partial_e K = B_f$. Thus $\mu(\partial_e K^C) = 0$ and $\partial_e K$ is a G_δ-set. The last assertion follows from Theorem 6.11(ii). Q.E.D.

Lemma 6.16. *Let K be as before and $\{f_n\}$ be an upper bounded sequence in $\mathscr{Q}(K)$. Let $f(x) = \limsup_{n \to \infty} f_n(x)$. Then we have*

$$\sup\{f(x) : x \in \partial_e K\} = \sup\{f(x) : x \in K\}.$$

PROOF. Let $\alpha = \sup\{f(x) : x \in \partial_e K\}$. Let $x \in K$ be arbitrary. By Lemma III.6.1, there exists a sequence $\{a_n\}$ in $\mathscr{A}(K; E)$ such that

$$a_n \leq f_n \quad \text{and} \quad f_n(x) - \frac{1}{n} \leq a_n(x).$$

Put $\Phi(y) = \{a_n(y)\} \in \mathbf{R}^{\aleph_0}$ for each $y \in E$. In $E' = \mathbf{R}^{\aleph_0}$, we consider the metric d given by

$$d(s,t) = \sum \frac{1}{2^n} \frac{|s_n - t_n|}{1 + |s_n - t_n|}, \qquad s = \{s_n\}, \qquad t = \{t_n\} \in E'.$$

It follows then that E' is a metrizable locally convex space and Φ is a continuous affine map of E into E'. Hence $K' = \Phi(K)$ is a metrizable compact convex subset in E'. Let p_n be the projection of E' to the nth coordinate. Then p_n is a continuous linear functional and $a_n = p_n \circ \Phi$.

If $y' \in \partial_e K'$, then $\Phi^{-1}(y)$ is a closed face of K, so that it contains an extreme point y, which is automatically extreme in K, and so $y' = \Phi(y)$ $y \in \partial_e K$. Therefore

$$\limsup_{n\to\infty} p_n(y') = \limsup_{n\to\infty} a_n(y) \le \alpha, \qquad y' \in \partial_e K'.$$

By Theorem 6.15, there exists a representing measure μ of $x' = \Phi(x)$ in $M^+(K')$ such that $\mu(\partial_e K'^C) = 0$. The sequence $\{p_n\}$ is upper bounded on K' due to the upper boundedness of $\{a_n\}$ on K, so we have, by Fatou's lemma,

$$\begin{aligned} f(x) &= \limsup a_n(x) = \limsup p_n(x') \\ &= \sup_{n\to\infty} \int_{\partial_e K'} p_n(y')\,d\mu(y') \le \int_{\partial_e K'} \limsup_{n\to\infty} p_n'(y)\,d\mu(y') \\ &\le \alpha. \end{aligned}$$

Q.E.D.

Theorem 6.17. *If μ is a boundary measure on a compact convex set K, then $|\mu|(C) = 0$ for every Baire set C disjoint from $\partial_e K$.*

Proof. By the regularity of μ as a Baire measure, it suffices to show that $|\mu|(C) = 0$ for every G_δ-closed set C disjoint from $\partial_e K$. By Urysohn's lemma, there exists a sequence in $C_{\mathbf{R}}(K)$ such that

$$f_n(x) = 1, \qquad x \in C, \qquad 0 \le f_n \le 1,$$

and

$$\lim f_n(x) = 0, \qquad x \notin C.$$

In particular, $|\mu|(C) \le |\mu|(f_n)$. We apply Lemma 6.16 to $\{\check{f}_n\}$. For every $x \in \partial_e K$, we have

$$\limsup \check{f}_n(x) = \limsup f_n(x) = 0.$$

Hence we have $\limsup \check{f}_n(x) \le 0$, $x \in K$; thus we have

$$\begin{aligned} 0 &\le |\mu|(C) \le \limsup |\mu|(f_n) \\ &= \limsup |\mu|(\check{f}_n) \le |\mu|(\limsup \check{f}_n) \text{ by Fatou's lemma,} \\ &\le 0. \end{aligned}$$

Q.E.D.

Instead of discussing further the general account of integral representations in a compact convex set, we shall return to the study of C^*-algebras and will come back to the general theory when necessary.

Now let A be a unital C^*-algebra and $\mathfrak{S}$ be the compact convex set of all states. Applying the above theory to $\mathfrak{S}$, we see that every state φ is represented by a boundary measure μ on $\mathfrak{S}$, i.e.,

$$\varphi(a) = \int_{\mathfrak{S}} \omega(a)\, d\mu(\omega). \tag{4}$$

Furthermore, if A is separable, then this measure μ is concentrated on the extreme boundary $\partial_e\mathfrak{S}$, which in our case is the pure state space $P(A)$, so that (4) is given by

$$\varphi(a) = \int_{P(A)} \omega(a)\, d\mu(\omega). \tag{5}$$

To learn about the nature of integral representations (4) or (5) of a state φ, we must study, of course, the relation between the cyclic representation π_φ and the family $\{\pi_\omega\}$ of cyclic representations and the measure μ.

Proposition 6.18. *Let A be a unital C^*-algebra and $\mu \in M_1^+(\mathfrak{S}(A))$. If φ is the barycenter of μ, i.e.,*

$$\varphi(a) = \int_{\mathfrak{S}} \omega(a)\, d\mu(\omega), \qquad a \in A,$$

then there exists a unique σ-weakly continuous positive map θ_μ of $L^\infty(\mathfrak{S},\mu)$ into $\pi_\varphi(A)'$ such that $\theta_\mu(1) = 1$ and

$$(\theta_\mu(f)\pi_\varphi(a)\xi_\varphi|\xi_\varphi) = \int_{\mathfrak{S}} f(\omega)\omega(a)\, d\mu(\omega), \qquad a \in A, \qquad f \in L^\infty(\mathfrak{S},\mu). \tag{6}$$

PROOF. The uniqueness of θ_μ follows from the cyclicity of ξ_φ for $\pi_\varphi(A)$. Thus we shall show the existence of θ_μ. We recall that the map $\theta_\varphi : x \in \pi_\varphi(A)' \mapsto \theta_\varphi(x) \in A^*$ given by

$$\langle a,\theta_\varphi(x)\rangle = (\pi_\varphi(a)x\xi_\varphi|\xi_\varphi)$$

is a completely positive injection of $\pi_\varphi(A)'$ into A^* with the range C_φ spanned linearly by positive linear functionals dominated by φ, Proposition 3.10. For each $f \in L^\infty(\mathfrak{S},\mu)$, let

$$\langle a,\theta'_\mu(f)\rangle = \int_{\mathfrak{S}} f(\omega)\omega(a)\, d\mu(\omega), \qquad a \in A.$$

It follows that $\theta'_\mu(f)$ is a bounded linear functional on A and $0 \le \theta'_\mu(f) \le \varphi$ if $0 \le f \le 1$. Hence $\theta'_\mu(L^\infty(\mathfrak{S},\mu)) \subset C_\varphi$. Define $\theta_\mu = \theta_\varphi^{-1} \circ \theta'_\mu$. Since θ'_μ is positive, θ_μ is also positive, and $\theta'_\mu(1) = \varphi$ implies that $\theta_\mu(1) = 1$. By construction, (6) follows.

We now show the normality of θ_μ. For any $a,b \in A$, we have

$$\begin{aligned}(\theta_\mu(f)\pi_\varphi(a)\xi_\varphi|\pi_\varphi(b)\xi_\varphi) &= (\theta_\mu(f)\pi_\varphi(b^*a)\xi_\varphi|\xi_\varphi) \\ &= \int_{\mathfrak{S}} f(\omega)\omega(b^*a)\, d\mu(\omega).\end{aligned}$$

Since the function $(b^*a)^\sim : \omega \in \mathfrak{S} \mapsto \omega(b^*a) \in \mathbf{C}$ is integrable, the functional: $f \in L^\infty(\mathfrak{S},\mu) \mapsto (\theta_\mu(f)\pi_\varphi(a)\xi_\varphi|\pi_\varphi(b)\xi_\varphi) \in \mathbf{C}$ is normal. Thus, the normality of θ_μ follows from the density of $\pi_\varphi(A)\xi_\varphi$ in $\mathfrak{H}_\varphi$. Q.E.D.

Theorem 6.19. *Let A be a unital C^*-algebra with state space $\mathfrak{S} = \mathfrak{S}(A)$, and $\mu \in M_1^+(\mathfrak{S})$. Let φ be the barycenter of μ. Then the following conditions are equivalent, where θ_μ is the map defined in Proposition 6.18:*

(i) *θ_μ is multiplicative.*
(ii) *$\theta_\mu(\chi_E)\theta_\mu(1 - \chi_E) = 0$ for every Borel subset E of $\mathfrak{S}$, where χ_E means the characteristic function of E.*
(iii) *The functionals φ_E and φ_{E^c} defined by*

$$\varphi_E(a) = \int_E \omega(a)\, d\mu(\omega), \qquad \varphi_{E^c} = \int_{E^c} \omega(a)\, d\mu(\omega), \qquad a \in A,$$

are orthogonal in the sense that $0 \le \psi \le \varphi_E$ and $0 \le \psi \le \varphi_{E^c}$ imply $\psi = 0$.

If this the case, then θ_μ is an isomorphism of $L^\infty(\mathfrak{S},\mu)$ into $\pi_\varphi(A)'$.

Proof. (i) $\Rightarrow$ (ii): Since χ_E and $1 - \chi_E$ are orthogonal projections in $L^\infty(\mathfrak{S},\mu)$, $\theta_\mu(\chi_E)$ and $\theta_\mu(1 - \chi_E)$ are also orthogonal projections. Hence (ii) follows.

(ii) $\Rightarrow$ (iii): Let θ_φ be the completely positive map of $\pi_\varphi(A)$ into A^* that appeared in the proof of Proposition 6.18. Since $\varphi_E \le \varphi$ and $\varphi_{E^c} = \varphi - \varphi_E$, φ_E and φ_{E^c} are both in the range of θ_φ. If $0 \le \psi \le \varphi_E$ and $0 \le \psi \le \varphi_{E^c}$, then ψ is also in $\theta_\varphi(\pi_\varphi(A)')$, and we have

$$\begin{aligned} 0 \le \theta_\varphi^{-1}(\psi) \le \theta_\varphi^{-1}(\varphi_E) &= \theta_\mu(\chi_E), \\ 0 \le \theta_\varphi^{-1}(\psi) \le \theta_\varphi^{-1}(\varphi_{E^c}) &= \theta_\mu(1 - \chi_E). \end{aligned}$$

Hence the orthogonality of $\theta_\mu(\chi_E)$ and $\theta_\mu(1 - \chi_E)$ imply $\theta_\varphi^{-1}(\psi) = 0$, and so $\psi = 0$.

(iii) $\Rightarrow$ (i): First, we claim that $\theta_\mu(\chi_E)$ is a projection for any Borel subset E of $\mathfrak{S}$. Let $p = \theta_\mu(\chi_E)$ and $q = \theta_\mu(1 - \chi_E)$. By assumption, $0 \le x \le p$ and $0 \le x \le q$, $x \in \pi_\varphi(A)'$, imply $x = 0$, since $\theta_\varphi(x) \le \varphi_E$ and $\theta_\varphi(x) \le \varphi_{E^c}$. Since $p + q = 1$, p and q commute, so that $0 \le pq \le p$ and $0 \le pq \le q$. Hence $pq = 0$, and so

$$p = p(p + q) = p^2 \quad \text{and} \quad q = q(p + q) = q^2.$$

Thus p and q are both projections in $\pi_\varphi(A)'$. If E and F are any Borel subsets of $\mathfrak{S}$, then $\chi_E = \chi_E\chi_F + \chi_E(1 - \chi_F)$ and $\chi_F = \chi_E\chi_F + \chi_F(1 - \chi_E)$, so that

$$\begin{aligned} \theta_\mu(\chi_E)\theta_\mu(\chi_F) &= [\theta_\mu(\chi_{E \cap F}) + \theta_\mu(\chi_{E \cap F^c})][\theta_\mu(\chi_{E \cap F}) + \theta_\mu(\chi_{E^c \cap F})] \\ &= \theta_\mu(\chi_{E \cap F}) = \theta_\mu(\chi_E\chi_F) \end{aligned}$$

because $\theta_\mu(\chi_G)\theta_\mu(\chi_H) = 0$ if $G \cap H = \varnothing$. Therefore, θ_μ is multiplicative on step functions in $L^\infty(\mathfrak{S},\mu)$, and so is on $L^\infty(\mathfrak{S},\mu)$ by continuity.

Suppose that θ_μ is multiplicative. For any $f \in L^\infty(\mathfrak{S},\mu)$, we have

$$\|\theta_\mu(f)\xi_\varphi\|^2 = (\theta_\mu(\bar{f}f)\xi_\varphi|\xi_\varphi) = \int_{\mathfrak{S}} |f(\omega)|^2\, d\mu(\omega).$$

Thus $\|f\|_\infty \ne 0$ implies $\theta_\mu(f)\xi_\varphi \ne 0$, i.e., θ_μ is faithful. Q.E.D.

Definition 6.20. Given a unital C^*-algebra A with $\mathfrak{S} = \mathfrak{S}(A)$, a measure $\mu \in M_1^+(\mathfrak{S})$ is called *orthogonal* if any one of the conditions in Theorem 6.19 is satisfied.

Proposition 6.21. *For a measure $\mu \in M_1^+(\mathfrak{S}(A))$ with barycenter φ, consider the following statements:*

(i) *μ is orthogonal.*
(ii) *$\mathscr{A}_{\mathrm{C}}(\mathfrak{S}) = \{\tilde{a}: a \in A\}$ is dense in $L^1(\mathfrak{S},\mu)$, where $\tilde{a}(\omega) = \omega(a)$, $a \in A$.*
(iii) *μ is extreme in $M_\varphi^+(\mathfrak{S})$.*

Then we have the implications (i) $\Rightarrow$ (ii)$\Leftrightarrow$(iii).

PROOF. (i) $\Rightarrow$ (ii): By Theorem 6.18, θ_μ is injective. If $f \in L^\infty(\mathfrak{S},\mu)$ is orthogonal to $\mathscr{A}_{\mathrm{C}}(\mathfrak{S})$, then

$$(\theta_\mu(f)\pi_\varphi(a)\xi_\varphi | \pi_\varphi(b)\xi_\varphi) = \int_{\mathfrak{S}} f(\omega)\omega(b^*a)\, d\mu(\omega) = 0$$

for every $a,b \in A$. Hence $\theta_\mu(f) = 0$, and so $f = 0$. By the duality $L^1(\mathfrak{S},\mu)^* = L^\infty(\mathfrak{S},\mu)$, $\mathscr{A}_{\mathrm{C}}(\mathfrak{S})$ is dense $L^1(\mathfrak{S},\mu)$.

(ii)$\Leftrightarrow$(iii): Assume that (ii) holds. If $\mu = \frac{1}{2}(\mu_1 + \mu_2)$ for some $\mu_1,\mu_2 \in M_\varphi^+(\mathfrak{S})$, then $0 \le \mu_i \le 2\varphi$, $i = 1,2$, so that there exist $h_1,h_2 \in L^\infty(\mathfrak{S},\mu)$ such that $0 \le h_i \le 2$ and

$$\int_{\mathfrak{S}} f(\omega)\, d\mu_i(\omega) = \int_{\mathfrak{S}} f(\omega)h_i(\omega)\, d\mu(\omega), \qquad f \in L^\infty(\mathfrak{S},\mu).$$

Since $\mu_i \in M_\varphi^+(\mathfrak{S})$, we have, for each $a \in A$,

$$\varphi(a) = \int_{\mathfrak{S}} \tilde{a}(\omega)\, d\mu_i(\omega) = \int_{\mathfrak{S}} \tilde{a}(\omega)h_i(\omega)\, d\mu(\omega).$$

Hence we have $(1 - h_i) \perp \mathscr{A}_{\mathrm{C}}$, which means by assumption that $h_i = 1$ in $L^\infty(\mathfrak{S},\mu)$, $i = 1,2$. Hence $\mu_1 = \mu_2 = \mu$, and so μ is extreme in $M_\varphi^+(\mathfrak{S})$.

Assume that (iii) holds instead. Suppose that an $h \in L^\infty(\mathfrak{S},\mu)$ is orthogonal to $\mathscr{A}_{\mathrm{C}}$. Since $\mathscr{A}_{\mathrm{C}}$ is closed under the conjugation, the real part and the imaginary part of h are both orthogonal to $\mathscr{A}_{\mathrm{C}}$. To show $h = 0$, we may assume that h is real and $\|h\|_\infty \le 1$, by multiplying a scalar. We then have

$$\int_{\mathfrak{S}} \tilde{a}(\omega)h(\omega)\, d\mu(\omega) = 0, \qquad a \in A.$$

Put $\nu(f) = \mu(fh)$, $f \in C(\mathfrak{S})$. Then we have $-\mu \le \nu \le \mu$ and $\mu = \frac{1}{2}\{(\mu + \nu) + (\mu - \nu)\}$. Clearly $\mu \pm \nu$ are both in $M_\varphi^+(\mathfrak{S})$. By the extremality of μ, $\nu = 0$; hence $h = 0$. Thus $\mathscr{A}_{\mathrm{C}}(\mathfrak{S})$ is dense in $L^1(\mathfrak{S},\mu)$. Q.E.D.

Remark 6.22. In the equivalence (ii)$\Leftrightarrow$(iii), we did not use the fact that $\mathfrak{S}$ comes from a C^*-algebra. Indeed, (ii) and (iii) are equivalent for a general compact convex set. A measure satisfying this condition is called *simplicial*. But, there are simplicial but not orthogonal representing measures of a state on a C^*-algebra; see [333].

We now restrict our attention to orthogonal representing measures of a state on a unital C^*-algebra A with state space $\mathfrak{S}$. If $\mu \in M_\varphi^+(\mathfrak{S})$ is orthogonal for a given state φ, then θ_μ is an isomorphism of $L^\infty(\mathfrak{S},\mu)$ into $\pi_\varphi(A)'$ and its range $\mathscr{C}_\mu$ is an abelian von Neumann algebra of $\pi_\varphi(A)'$. We call $\mathscr{C}_\mu$ the *associated abelian von Neumann algebra to the orthogonal integral representation*, $\varphi = \int_\mathfrak{S} \omega \, d\mu(\omega)$, *or to the orthogonal measure* μ. The following result shows that $\mathscr{C}_\mu$ carries all information about μ.

Proposition 6.23. *If φ is a state on A, then for every abelian von Neumann subalgebra $\mathscr{C}$ of $\pi_\varphi(A)'$, there exists a unique orthogonal measure $\mu \in M_\varphi^+(\mathfrak{S})$ such that $\mathscr{C} = \mathscr{C}_\mu$.*

PROOF. We first remark that ξ_φ is separating for $\mathscr{C}$, being cyclic for $\pi_\varphi(A)$. Hence the normal state $\bar{\varphi}$ on $\mathscr{C}$ defined by $\bar{\varphi}(x) = (x\xi_\varphi|\xi_\varphi)$, $x \in \mathscr{C}$, is faithful. Let e be the projection to $[\mathscr{C}\xi_\varphi]$. It follows that $\mathscr{C}_e$ is maximal abelian, since $\mathscr{C}_e$ has a cyclic vector ξ_φ and the map: $x \in \mathscr{C} \mapsto x_e \in \mathscr{C}_e$ is an isomorphism. Since $(\mathscr{C}_e)' = (\mathscr{C}')_e$, we have $e\pi_\varphi(A)e \subset \mathscr{C}_e$. Therefore, for each $a \in A$, there exists a unique element, say $\theta(a)$, of $\mathscr{C}$ such that $\pi_\varphi(a)\xi_\varphi = \theta(a)\xi_\varphi$. It is clear that θ is positive and $\theta(1) = 1$. Let Ω be the spectrum of $\mathscr{C}$. Then the transpose ${}^t\theta$ of θ maps Ω into $\mathfrak{S}$ continuously. Hence its transpose $\bar{\theta} = {}^{tt}\theta$ maps $\mathscr{C}(\mathfrak{S})$ into $\mathscr{C}$, extending the original map θ. Let $\mu = \bar{\varphi} \circ \bar{\theta} \in M_1^+(\mathfrak{S})$. We then have for each $a \in A$,

$$\begin{aligned}\varphi(a) &= (\pi_\varphi(a)\xi_\varphi|\xi_\varphi) = (\theta(a)\xi_\varphi|\xi_\varphi) = \bar{\varphi} \circ \theta(a) \\ &= \bar{\varphi} \circ \bar{\theta}(\tilde{a}) = \int_\mathfrak{S} \tilde{a}(\omega) \, d\mu(\omega) = \int_\mathfrak{S} \omega(a) \, d\mu(\omega).\end{aligned}$$

Therefore, μ is a representing measure of φ. By construction, $\bar{\theta}$ is a homomorphism of $\mathscr{C}(\mathfrak{S})$ into $\mathscr{C}$ which carries $\bar{\varphi}$ into μ. Therefore, $\bar{\theta}$ is extended to a normal isomorphism θ_μ of $L^\infty(\mathfrak{S},\mu)$ into $\mathscr{C}$. We now claim that $\bar{\theta}$ maps $L^\infty(\mu)$ onto $\mathscr{C}$. To see this, let $\mathscr{C}_0$ denote the von Neumann algebra generated by $e\pi_\varphi(A)e$ on $e\mathfrak{H}_\varphi = [\mathscr{C}\xi_\varphi]$. Clearly, $\mathscr{C}_0$ is the von Neumann subalgebra of $\mathscr{C}_e$ generated by $\theta(A)$. But we have $[\mathscr{C}_0\xi_\varphi] \supset [e\pi_\varphi(A)e\xi_\varphi] = [e\pi_\varphi(A)\xi_\varphi] = e\mathfrak{H}_\varphi$, so that ξ_φ is cyclic for $\mathscr{C}_0$ on $e\mathfrak{H}_\varphi$. Hence $\mathscr{C}_0$ is maximal abelian; hence $\mathscr{C}_0 = \mathscr{C}$. Thus $e\pi_\varphi(A)e$ generates $\mathscr{C}_e$, i.e., $\theta(A)$ generates $\mathscr{C}$, which means that $\bar{\theta}(L^\infty(\mu)) = \mathscr{C}$. Next, for each $f \in L^\infty(\mu)$, we have

$$\begin{aligned}(\pi_\varphi(a)\theta(f)\xi_\varphi|\xi_\varphi &= (e\pi_\varphi(a)e\theta(f)\xi_\varphi|\xi_\varphi) \\ &= (\theta(a)\theta(f)\xi_\varphi|\xi_\varphi) = (\bar{\theta}(\tilde{a}f)\xi_\varphi|\xi_\varphi) \\ &= \int_\mathfrak{S} \tilde{a}(\omega)f(\omega) \, d\mu(\omega) = \int_\mathfrak{S} \omega(a)f(\omega) \, d\mu(\omega).\end{aligned}$$

Therefore, we get $\theta = \theta_\mu$.

Suppose that $\mathscr{C}_\mu = \mathscr{C}_\nu$ for orthogonal measures $\mu,\nu \in M_\varphi^+(\mathfrak{S})$. Let e be the projection of $\mathfrak{H}_\varphi$ onto $[\mathscr{C}_\mu\xi_\varphi] = [\mathscr{C}_\nu\xi_\varphi]$. As we have just proved that $\theta_\mu(\tilde{a})\xi_\varphi = e\pi_\varphi(a)\xi_\varphi = \theta_\nu(\tilde{a})\xi_\varphi$ for every $a \in A$, so that $\theta_\mu(\tilde{a})e = e\pi_\varphi(a)e = \theta_\nu(\tilde{a})e$. Thus, we have $\theta_\mu(\tilde{a}) = \theta_\nu(\tilde{a})$, $a \in A$, which means that θ_μ and θ_ν coincide on the C^*-algebra generated by $\{\tilde{a}: a \in A\} = \mathscr{A}_{\mathbf{C}}(\mathfrak{S})$, which is $C(\mathfrak{S})$ by the Stone–Weierstrass theorem. Then we have, for each $f \in C(\mathfrak{S})$,

$$\mu(f) = (\theta_\mu(f)\xi_\varphi|\xi_\varphi) = (\theta_\nu(f)\xi_\varphi|\xi_\varphi) = \nu(f). \qquad \text{Q.E.D.}$$

We are now going to show that $\mu \prec \nu$ is equivalent to $\mathscr{C}_\mu \subset \mathscr{C}_\nu$ for orthogonal measures $\mu, \nu \in M_\varphi^+(\mathfrak{S})$. But we need the following result from general convexity theory.

Lemma 6.24. *Let K be a compact convex set. For positive measures μ and ν on K, the following two conditions are equivalent:*

(i) $\mu \prec \nu$.
(ii) *If* $\mu = \sum_{i=1}^n \mu_i$, *and* $\mu_i \geq 0$, $i = 1, \ldots, n$, *there exist positive measures* $\nu_1, \ldots, \nu_n$ *such that* $\nu_i \sim \mu_i$ *and* $\nu = \sum_{i=1}^n \nu_i$.

If this is the case, then $\{\nu_i\}$ *is chosen so that* $\mu_i \prec \nu_i$, $i = 1,2, \ldots, n$.

PROOF. (i) $\Rightarrow$ (ii): Assume $\mu \prec \nu$ and $\mu = \sum_{i=1}^n \mu_i$ with $\mu_i \geq 0$, $i = 1,2, \ldots, n$. We define a map Φ of $C_{\mathbf{R}}(K)^n$ into $\mathbf{R}$ by

$$\Phi(\tilde{f}) = \sum_{i=1}^n \mu_i(\hat{f}_i), \qquad \tilde{f} = (f_1, \ldots, f_n) \in C_{\mathbf{R}}(K)^n.$$

As in the case of Lemma 6.7, Φ is sublinear. Let F denote the diagonal of $C_{\mathbf{R}}(K)^n$, i.e., $F = \{(f, \ldots, f): f \in C_{\mathbf{R}}(K)\}$, and let Ψ_0 be the linear functional on F given by $\Psi_0(f, \ldots, f) = \nu(f)$. We then have

$$\Psi_0(f, \ldots, f) = \nu(f) \leq \nu(\hat{f}) \leq \mu(\hat{f}) = \sum_{i=1}^n \mu_i(f) = \Phi(f, \ldots, f).$$

Hence Ψ_0 is majorized on F by Φ.

By the Hahn–Banach theorem, there exists a linear functional Ψ on $C_{\mathbf{R}}(K)^n$ which extends Ψ_0 and is majorized by Φ. Defining a norm in $C_{\mathbf{R}}(K)$ by $\|\tilde{f}\| = \max_{1 \leq i \leq n} \|f_i\|$ for $\tilde{f} = (f_1, \ldots, f_n)$, we have

$$\begin{aligned}\Psi(\tilde{f}) \leq \Phi(\tilde{f}) &= \sum_{i=1}^n \mu_i(\hat{f}_i) \leq \sum_{i=1}^n \|\mu_i\| \|\hat{f}_i\|_\infty \\ &\leq \max_{1 \leq i \leq n} \|\hat{f}_i\| \sum_{i=1}^n \|\mu_i\| \leq \max_{1 \leq i \leq n} \|f_i\| \|\mu\| = \|\tilde{f}\| \|\mu\|.\end{aligned}$$

Considering $-\tilde{f}$ in the above calculation, we have $|\Psi(\tilde{f})| \leq \|\tilde{f}\| \|\mu\|$. Thus, Ψ is a bounded linear functional on $C_{\mathbf{R}}(K)^n$, so that there exist $\nu_1, \ldots, \nu_n \in M_{\mathbf{R}}(K) = C_{\mathbf{R}}(K)^*$ such that

$$\Psi(\tilde{f}) = \sum_{i=1}^n \nu_i(f_i), \qquad \tilde{f} = (f_1, \ldots, f_n) \in C_{\mathbf{R}}(K)^n,$$

$$\|\Psi\| = \sum_{i=1}^n \|\nu_i\| \leq \|\mu\|.$$

Putting $\tilde{1} = (1,1,\ldots,1)$, we have

$$\sum_{i=1}^{n} \nu_i(1) = \Psi(\tilde{1}) = \Psi_0(\tilde{1}) = \nu(1)$$
$$= \mu(1) = \|\mu\|,$$

so that $\sum_{i=1}^{n} \|\nu_i\| = \|\mu\|$ and $\nu_i(1) = \|\nu_i\|$, $1 \le i \le n$, which indicates the positivity of each ν_i. Now, we have for each $f \in C_{\mathbf{R}}(K)$ and $i = 1,2,\ldots,n$,

$$\nu_i(f) = \Psi(0,\ldots,0,\overset{i}{\check{f}},0,\ldots,0)$$
$$\le \Phi(0,\ldots,0,f,0,\ldots,0) = \mu_i(\hat{f});$$

hence $\mu_i \prec \nu_i$ since $(-f)\hat{} = -f$ for each $f \in \mathscr{P}(K)$. Thus we get $\mu_i \sim \nu_i$. Since Ψ is an extension of Ψ_0, we have $\sum_{i=1}^{n} \nu_i = \nu$.

(ii) $\Rightarrow$ (i): Suppose condition (ii) is satisfied. We shall prove that $\mu(f) \le \nu(f)$ for every $f \in \mathscr{P}(K)$. Let $\varepsilon > 0$. By the continuity of f and the local convexity of the vector space E containing K, there exists a finite collection $\{G_1,\ldots,G_n\}$ of closed convex subsets of K such that

$$|f(x) - f(y)| < \varepsilon \quad \text{for} \quad x,y \in G_i,$$

$$K = \bigcup_{i=1}^{n} G_i,$$

where we use, of course, the compactness of K. For each $i = 1,\ldots,n$, let $A_i = G_i \bigcup_{j<i} G_j$ and $d\mu_i = \chi_{A_i}\, d\mu$. Then $\mu = \sum_{i=1}^{n} \mu_i$. By assumption, there exists a decomposition $\nu = \sum_{i=1}^{n} \nu_i$ where $\nu_i \ge 0$ and $\mu_i \sim \nu_i$. For each nonzero μ_i, let

$$x_i = \frac{1}{\mu_i(K)} \int_K x\, d\mu_i(x) = \frac{1}{\nu_i(K)} \int_K x\, d\nu_i(x) \in K.$$

It follows then that x_i belongs to the convex closure of A_i, which is contained in G_i; hence $x_i \in G_i$. By the construction of G_i and μ_i, we have

$$\mu_i(f) \le \mu_i(K)(f(x_i) + \varepsilon) \qquad (1 \le i \le n)$$
$$= \nu_i(K)(f(x_i) + \varepsilon).$$

Thus we get

$$\mu(f) = \sum_{i=1}^{n} \mu_i(f) \le \sum_{i=1}^{n} \mu_i(K)(f(x_i) + \varepsilon) = \sum_{i=1}^{n} \nu_i(K)(f(x_i) + \varepsilon).$$

But we have $\delta_{x_i} \prec [1/\nu_i(K)]\nu_i$ for each i with $\nu_i(K) \ne 0$, where δ_{x_i} the point mass measure at x_i. Hence the last term of the above inequality is dominated by

$$\sum_{i=1}^{n} (\nu_i(f) + \varepsilon\nu_i(K)) = \nu(f) + \varepsilon\nu(K).$$

Since $\varepsilon > 0$ is arbitrary, we get $\mu(f) \le \nu(f)$, $f \in \mathscr{P}(K)$. Thus $\mu \prec \nu$. Q.E.D.

Returning to the C^*-algebra case, we can now prove the following:

Theorem 6.25. *Let A be a unital C^*-algebra with state space $\mathfrak{S}$, and $\varphi \in \mathfrak{S}$. For orthogonal representing measures $\mu,\nu \in M_\varphi^+(\mathfrak{S})$, the following conditions are equivalent:*

(i) $\mu \prec \nu$;

(ii) $\int_{\mathfrak{S}} \omega(a)^2 \, d\mu(\omega) \leq \int_{\mathfrak{S}} \omega(a)^2 \, d\nu(\omega), \qquad a \in A_h$;

(iii) $\mathscr{C}_\mu \subset \mathscr{C}_\nu$.

Proof. (i) $\Rightarrow$ (ii): For every $a \in A_h$, $\tilde{a}^2$ is convex and hence belongs to $\mathscr{P}(\mathfrak{S})$. Thus inequality (ii) follows.

(ii) $\Rightarrow$ (iii): Suppose that inequality (ii) holds. Let e_μ (resp. e_ν) be the projection of $\mathfrak{H}_\varphi$ onto $[\mathscr{C}_\mu \xi_\varphi]$ (resp. $[\mathscr{C}_\nu \xi_\varphi]$). We note that $e_\mu \pi_\varphi(a) e_\mu = \theta_\mu(\tilde{a}) e_\mu$ and $e_\nu \pi_\varphi(a) e_\nu = \theta_\nu(\tilde{a}) e_\nu$ for every $a \in A$. With $a = b + ic$, $b,c \in A_h$, we have

$$
\begin{aligned}
(e_\mu \pi_\varphi(a)\xi_\varphi | \pi_\varphi(a)\xi_\varphi) &= (e_\mu \pi_\varphi(a^*) e_\mu \pi_\varphi(a) e_\mu \xi_\varphi | \xi_\varphi) \\
&= (\theta_\mu(\tilde{a}^*)\theta_\mu(\tilde{a}) e_\mu \xi_\varphi | \xi_\varphi) \\
&= \int_{\mathfrak{S}} (\tilde{b}(\omega) + i\tilde{c}(\omega))(\tilde{b}(\omega) - i\tilde{c}(\omega)) \, d\mu(\omega) \\
&= \int_{\mathfrak{S}} (\tilde{b}(\omega)^2 + \tilde{c}(\omega)^2) \, d\mu(\omega) \\
&\leq \int_{\mathfrak{S}} (\tilde{b}(\omega)^2 + \tilde{c}(\omega)^2) \, d\nu(\omega) \quad \text{by (ii)}, \\
&= \cdots = (e_\nu \pi_\varphi(a)\xi_\varphi | \pi_\varphi(a)\xi_\varphi).
\end{aligned}
$$

Therefore we get $e_\mu \leq e_\nu$ since ξ_φ is cyclic for $\pi_\varphi(A)$, which means that $[\mathscr{C}_\mu \xi_\varphi] \subset [\mathscr{C}_\nu \xi_\varphi]$.

Let U_ν be the isometry of $L^2(\mathfrak{S},\nu)$ onto $[\mathscr{C}_\nu \xi_\varphi]$ given by $U_\nu f = \theta_\nu(f)\xi_\varphi$, $f \in L^\infty(\mathfrak{S},\nu)$. Let $x \in \mathscr{C}_\mu$. By the inclusion $[\mathscr{C}_\mu \xi_\varphi] \subset [\mathscr{C}_\nu \xi_\varphi]$, there exists an $f \in L^2(\mathfrak{S},\nu)$ such that $x\xi_\varphi = U_\nu f$. We shall show that f is bounded and $\theta_\nu(f) = x$. For each $n = 1,2,\ldots$, let $E_n = \{\omega \in \mathfrak{S} : |f(\omega)| \leq n\}$ and $e_n = \theta_\nu(\mathscr{X}_{E_n})$. We have then

$$
\begin{aligned}
e_n x \xi_\varphi &= \theta_\nu(\chi_{E_n}) U_\nu f = U_\nu(\chi_{E_n} f) \\
&= \theta_\nu(\chi_{E_n} f)\xi_\varphi.
\end{aligned}
$$

Since ξ_φ is separating for $\pi_\varphi(A)'$, we have $e_n x = \theta_\nu(\chi_{E_n} f) \in \mathscr{C}_\nu$. Since $\{\chi_{E_n}\}$ converges σ-strongly to 1 in $L^\infty(\mathfrak{S},\nu)$, $\{e_n\}$ converges to 1 σ-strongly, so that $e_n x \to x$ σ-strongly, and so $x \in \mathscr{C}_\nu$, and $x\xi_\varphi = U_\nu f$ means that $f \in L^\infty(\mathfrak{S},\nu)$ and $x = \theta_\nu(f)$. Thus, we have proved that $\mathscr{C}_\mu \subset \mathscr{C}_\nu$.

(iii) $\Rightarrow$ (i): Suppose that $\mathscr{C}_\mu \subset \mathscr{C}_\nu$. To conclude (i), we use Lemma 6.24. So let $\mu = \sum_{i=1}^n \mu_i$ be a decomposition of μ with $\mu_i \geq 0$, $i = 1,2,\ldots,n$. Since $0 \leq \mu_i \leq \mu$, for each $i = 1,2,\ldots,n$, there exists an $f_i \in L^\infty(\mathfrak{S},\mu)$ such that $d\mu_i = f_i \, d\mu$, $0 \leq f_i \leq 1$. Clearly, we have $\sum_{i=1}^n f_i = 1$. Since $\theta_\mu(L^\infty(\mu)) \subset \theta_\nu(L^\infty(\nu))$, and since both θ_μ and θ_ν are isomorphisms, there exists a $g_i \in L^\infty(\nu)$, $0 \leq g_i \leq 1$ such that $\theta_\mu(f_i) = \theta_\nu(g_i)$. Put $d\nu_i = g_i \, d\nu$, $i = 1,\ldots,n$. We then have $\sum_{i=1}^{n} \nu_i = \nu$ since $\sum_{i=1}^{n} g_i = 1$, and for each $a \in A$,

$$\begin{aligned}\nu_i(\tilde{a}) &= \int_{\mathfrak{S}} \tilde{a}(\omega) g_i(\omega)\, d\nu(\omega) = (\theta_\nu(g_i)\pi_\varphi(a)\xi_\varphi|\xi_\varphi) \\ &= (\theta_\mu(f_i)\pi_\varphi(a)\xi_\varphi|\xi_\varphi) = \int_{\mathfrak{S}} \tilde{a}(\omega) f_i(\omega)\, d\mu(\omega) \\ &= \mu_i(\tilde{a}).\end{aligned}$$

Hence we have $\nu_i \sim \mu_i$. Thus, we conclude by Lemma 6.24 that $\mu \prec \nu$; hence (i). Q.E.D.

We are now going to study orthogonal representing measures μ of a state φ with the maximal abelian associated von Neumann algebra $\mathscr{C}_\mu$. We begin with the following:

Lemma 6.26. *For a state φ on a unital C^*-algebra A with $\mathfrak{S} = \mathfrak{S}(A)$, the following conditions are equivalent*:

(i) *$\pi_\varphi(A)'$ is abelian.*
(ii) *There exists a unique maximal measure $\mu \in M_\varphi^+(\mathfrak{S})$, namely, the orthogonal measure corresponding to $\pi_\varphi(A)'$.*

PROOF. (i) $\Rightarrow$ (ii): Assume that $\pi_\varphi(A)'$ is abelian. Let $\mu \in M_\varphi^+(\mathfrak{S})$ be the orthogonal measure with $\mathscr{C}_\mu = \pi_\varphi(A)'$. Let $\nu \in M_\varphi^+(\mathfrak{S})$. We then use Lemma 6.24 to conclude that $\nu \prec \mu$. If $\nu = \sum_{i=1}^{n} \nu_i$, $\nu_i \geq 0$, is a decomposition of ν, then there exist $g_1, \ldots, g_n \in L^\infty(\mathfrak{S},\nu)$ such that $0 \leq g_i \leq 1$ and $d\nu_i = g_i\, d\nu$, $i = 1,2, \ldots, n$. Since $\theta_\nu(L^\infty(\nu)) \subset \pi_\varphi(A)' = \mathscr{C}_\mu$, there exist $f_1, \ldots, f_n \in L^\infty(\mathfrak{S},\mu)$ such that $\theta_\mu(f_i) = \theta_\nu(g_i)$, $1 \leq i \leq n$ and $0 \leq f_i \leq 1$, because θ_μ is an isomorphism, and $\sum_{i=1}^{n} f_i = 1$. Put $d\mu_i = f_i\, d\mu$, $1 \leq i \leq n$. We then have $\mu = \sum_{i=1}^{n} \mu_i$ and, for each $a \in A$,

$$\begin{aligned}\mu_i(\tilde{a}) &= \int_{\mathfrak{S}} \tilde{a}(\omega) f_i(\omega)\, d\mu(\omega) = (\theta_\mu(f_i)\pi_\varphi(a)\xi_\varphi|\xi_\varphi) \\ &= (\theta_\nu(g_i)\pi_\varphi(a)\xi_\varphi|\xi_\varphi) = \int_{\mathfrak{S}} \tilde{a}(\omega) g_i(\omega)\, d\nu(\omega) \\ &= \nu_i(\tilde{a}).\end{aligned}$$

Thus we get $\nu_i \sim \nu_i$. Hence by Lemma 6.24, we get $\nu \prec \mu$.

(ii) $\Rightarrow$ (i): Assume that μ is the unique maximal measure in $M_\varphi^+(\mathfrak{S})$. If $\mu = \frac{1}{2}(\mu_1 + \mu_2)$ with $\mu_1,\mu_2 \in M_\varphi^+(\mathfrak{S})$, then $\mu_1,\mu_2 \prec \mu$ by the uniqueness of a maximal measure. Hence for any $f \in \mathscr{P}(\mathfrak{S})$, we have $\mu_1(f) \leq \mu(f)$ and $\mu_2(f) \leq \mu(f)$ while $\mu(f) = \frac{1}{2}(\mu_1(f) + \mu_2(f))$, so that $\mu(f) = \mu_1(f) = \mu_2(f)$. By the density of $\mathscr{P}(\mathfrak{S}) - \mathscr{P}(\mathfrak{S})$, we have $\mu = \mu_1 = \mu_2$. Thus, μ is extreme in $M_\varphi^+(\mathfrak{S})$, i.e., μ is simplicial. By Proposition 6.21, $\mathscr{A}_{\mathbf{C}}(\mathfrak{S}) = \{\tilde{a}: a \in A\}$ is dense in $L^1(\mathfrak{S},\mu)$.

Let $\mathscr{C}$ be an arbitrary abelian von Neumann subalgebra of $\pi_\theta(A)'$, and let ν be the orthogonal representing measure of φ with $\mathscr{C}_\nu = \mathscr{C}$. For any $h \in \mathscr{C}$ with $0 \leq h \leq 1$, let $\nu_1(f) = (\theta_\nu(f)h\xi_\varphi|\xi_\varphi)$ and $\nu_2(f) = (\theta_\nu(f)(1 - h)\xi_\varphi|\xi_\varphi)$

for each $f \in \mathscr{C}(\mathfrak{S})$. We then have $\nu = \nu_1 + \nu_2$ and $\nu_1, \nu_2 \geq 0$. Since $\nu \prec \mu$ by maximality, there exists a decomposition $\mu = \mu_1 + \mu_2$ with $\mu_i \sim \nu_i$, $i = 1,2$, and $\mu_1 \geq 0$, $\mu_2 \geq 0$. Hence there exists $g \in L^\infty(\mathfrak{S},\mu)$ with $0 \leq g \leq 1$ such that $d\mu_1 = g\, d\mu$ and $d\mu_2 = (1 - g)\, d\mu$. We have then for each $a \in A$,

$$\begin{aligned}(\theta_\mu(g)\pi_\varphi(a)\xi_\varphi|\xi_\varphi) &= \int_{\mathfrak{S}} g(\omega)\hat{a}(\omega)\, d\mu(\omega) \\ &= \int_{\mathfrak{S}} \hat{a}(\omega)\, d\mu_1(\omega) = \mu_1(\hat{a}) = \nu_1(\hat{a}) \quad \text{by } \mu_1 \prec \nu_1 \\ &= (\theta_\nu(\hat{a})h\xi_\varphi|\xi_\varphi) = (\theta_\nu(\hat{a})e_\nu h\xi_\varphi|\xi_\varphi) \\ &= (e_\nu\pi_\varphi(a)e_\nu h\xi_\varphi|\xi_\varphi) = (\pi_\varphi(a)h\xi_\varphi|\xi_\varphi) = (h\pi_\varphi(a)\xi_\varphi|\xi_\varphi),\end{aligned}$$

where e_ν is the projection of $\mathfrak{H}_\varphi$ onto $[\mathscr{C}_\nu\xi_\varphi]$. Therefore, we have $\theta_\mu(g) = h$. Hence we get $\theta_\mu(\{f \in L^\infty(\mu): 0 \leq f \leq 1\}) \supset \{h \in \mathscr{C}: 0 \leq h \leq 1\}$. Since $\mathscr{C}$ is arbitrary, the image of the positive part of the unit ball of $L^\infty(\mu)$ under θ_μ contains the positive part of the unit ball of $\pi_\varphi(A)'$. But, θ_μ is a contraction, so that the above two sets must coincide. By the density of $\mathscr{A}_{\mathbf{C}}(\mathfrak{S})$ in $L^1(\mathfrak{S},\mu)$, θ_μ is injective. Hence θ_μ is an order isomorphism of the self-adjoint part of $L^\infty(\mu)$ onto $\pi_\varphi(A)'_h$. By Lemma I.10.1, for each Borel set E in $\mathfrak{S}$, χ_E is an extreme point in the positive part of the unit ball of $L^\infty(\mu)$, so that $\theta_\mu(\chi_E)$ is also an extreme point in that of $\pi_\varphi(A)'$, which means, again by Lemma I.10.1, and the spectral decomposition, that $\theta_\mu(\chi_E)$ is a projection. From this, it follows, as in the proof of Theorem 6.19, that θ_μ is multiplicative. Hence $\pi_\varphi(A)'$ is abelian. Q.E.D.

Definition 6.27. In general, a representation $\{\pi,\mathfrak{H}\}$ of a C^*-algebra A is called *multiplicity-free* if $\pi(A)'$ is abelian.

Theorem 6.28. *Let A be a unital C^*-algebra with $\mathfrak{S} = \mathfrak{S}(A)$, and $\varphi \in \mathfrak{S}$. If μ is an orthogonal representing measure of φ whose associated abelian von Neumann algebra $\mathscr{C}_\mu$ is maximal abelian in $\pi_\varphi(A)'$, then μ is pseudoconcentrated on the pure state space $P(A) = \partial_e\mathfrak{S}$ in the sense that $\mu(E) = 0$ for every Baire subset E of $\mathfrak{S}$ disjoint from $P(A)$. In particular, if A is separable, then μ is concentrated on $P(A)$.*

PROOF. Let $\mathscr{C}_A = C(\mathfrak{S}) \otimes_{\min} A$ be the injective tensor product of A and the abelian C^*-algebra $C(\mathfrak{S})$, which is also the projective tensor product by Lemma 4.18. Since $\theta_\mu(C(\mathfrak{S})) \subset \mathscr{C}_\mu$ and $\pi_\varphi(A)$ commute, there exists, by Proposition 4.7, a representation π of $\mathscr{C}_A$ such that $\pi(f \otimes a) = \theta_\mu(f)\pi(a)$ for each $f \in C(\mathfrak{S})$ and $a \in A$. Let $\tilde{\varphi}$ be the state on $\mathscr{C}_A$ given by $\tilde{\varphi}(x) = (\pi(x)\xi_\varphi|\xi_\varphi)$, $x \in \mathscr{C}_A$. Since ξ_φ is cyclic for $\mathscr{C}_A$, the representation π of $\mathscr{C}_A$ is cyclic with respect to ξ_φ and is unitarily equivalent to $\pi_{\tilde{\varphi}}$. So we identify π and $\pi_{\tilde{\varphi}}$. We now observe that

$$\begin{aligned}\pi_{\tilde{\varphi}}(\mathscr{C}_A)' &= (\pi_\varphi(A) \cup \theta_\mu(\mathscr{C}(\mathfrak{S}))' = \pi_\varphi(A)' \cap \theta_\mu(\mathscr{C}(\mathfrak{S}))' \\ &= \pi_\varphi(A)' \cap \mathscr{C}'_\mu = \mathscr{C}_\mu\end{aligned}$$

by the maximal abelianness of $\mathscr{C}_\mu$ in $\pi_\varphi(A)'$. Thus $\pi_{\tilde{\varphi}}$ is multiplicity-free. Therefore, $\tilde{\varphi}$ admits the unique maximal representing measure $\tilde{\mu}$ by Lemma 6.26, which is pseudoconcentrated on the pure state space $P(\mathscr{C}_A)$ in $\mathfrak{S}(\mathscr{C}_A)$. However, we have $P(\mathscr{C}_A) = \mathfrak{S} \times P(A)$ by Theorem 4.14.

We are now going to determine this unique maximal measure $\tilde{\mu}$ on $\mathfrak{S}(\mathscr{C}_A)$. For each $\omega \in \mathfrak{S}$, put $\Phi(\omega) = \omega \otimes \omega$ on $C(\mathfrak{S}) \otimes_{\min} A$. It follows that Φ is a homeomorphism of $\mathfrak{S}$ into $\mathfrak{S}(\mathscr{C}_A)$. Put $\nu = \Phi(\mu)$. We shall show $\nu = \tilde{\mu}$. Indeed, for each $G \in L^\infty(\mathfrak{S}(\mathscr{C}_A),\nu)$, we have, putting $g = G \circ \Phi$, for each $f \in C(\mathfrak{S})$ and $a \in A$,

$$\begin{aligned}\int_{\mathfrak{S}(\mathscr{C}_A)} G(\tilde{\omega})(f \otimes a)\tilde{}(\tilde{\omega})\, d\nu(\tilde{\omega}) &= \int_{\mathfrak{S}} G(\omega \otimes \omega)(f \otimes a)\tilde{}(\omega \otimes \omega)\, d\mu(\omega)\\ &= \int_{\mathfrak{S}} g(\omega) f(\omega)\tilde{a}(\omega)\, d\mu(\omega) = (\theta_\mu(gf)\pi_\varphi(a)\xi_\varphi|\xi_\varphi)\\ &= (\theta_\mu(g)\pi_{\tilde{\varphi}}(f \otimes a)\xi_\varphi|\xi_\varphi).\end{aligned}$$

By linearizing this, we get

$$\int_{\mathfrak{S}(\mathscr{C}_A)} G(\tilde{\omega})\tilde{x}(\tilde{\omega})\, d\nu(\tilde{\omega}) = (\theta_\mu(g)\pi_{\tilde{\varphi}}(x)\xi_\varphi|\xi_\varphi), \qquad x \in \mathscr{C}_A.$$

Therefore, ν is a representing measure of $\tilde{\varphi}$ with $\theta_\nu = \theta_\mu \circ \Phi^*$, where $\Phi^*(G) = G \circ \Phi$ for each $G \in L^\infty(\mathfrak{S}(\mathscr{C}_A),\nu)$. Since Φ^* is an isomorphism of $L^\infty(\nu)$ onto $L^\infty(\mu)$, θ_ν is an isomorphism of $L^\infty(\nu)$ onto $\mathscr{C}_\mu$. By the uniqueness of $\tilde{\mu}$, we have $\tilde{\mu} = \nu$.

We shall finally show that μ is pseudoconcentrated on $P(A)$. To this end, let $\Psi = \mathrm{i}^*: \mathfrak{S}(\mathscr{C}_A) \mapsto \mathfrak{S}$, where i denotes the imbedding of A into $\mathscr{C}_A$. In other words, Ψ is the restriction map of $\mathfrak{S}(\mathscr{C}_A)$ to $\mathfrak{S}$. It follows that Ψ is continuous and $\Psi \circ \Phi(\omega) = \omega$ for every $\omega \in \mathfrak{S}$. Let E be a G_δ-compact subset of $\mathfrak{S}$ disjoint from $P(A)$. Then $\Psi^{-1}(E)$ is a G_δ-compact subset of $\mathfrak{S}(\mathscr{C}_A)$. As mentioned above, we know $\Psi(P(\mathscr{C}_A)) = \Psi(\mathfrak{S} \times P(A)) = P(A)$, so that $\Psi^{-1}(E) \cap P(\mathscr{C}_A)) = \varnothing$; hence $\tilde{\mu}(\Psi^{-1}(E)) = 0$ because $\tilde{\mu}$ is a boundary measure on $\mathfrak{S}(\mathscr{C}_A)$. Thus we get

$$\mu(E) = \mu((\Psi \circ \Phi)^{-1}(E)) = \mu(\Phi^{-1}(\Psi^{-1}(E)) = \tilde{\mu}(\Psi^{-1}(E)) = 0.$$

Therefore μ is pseudoconcentrated on $P(A)$. Q.E.D.

A disadvantage of disintegrating a state into pure states is the non-uniqueness of measures because there are many maximal abelian subalgebras in $\pi_\varphi(A)'$ which are not conjugate under the inner automorphism group of $\pi_\varphi(A)'$. However, there is a distinguished abelian von Neumann subalgebra of $\pi_\varphi(A)'$, namely, the center $\mathscr{Z}_\varphi$ of $\pi_\varphi(A)''$. We shall study the orthogonal measure corresponding to $\mathscr{Z}_\varphi$.

Definition 6.29 Let A, $\mathfrak{S}$, and φ be as before. The orthogonal measure $\mu \in M_\varphi^+(\mathfrak{S})$ with $\mathscr{C}_\mu =$ the center of $\pi_\varphi(A)''$ is called the *central* (representing) measure of φ.

Definition 6.30. A state φ on a C^*-algebra A is said to be *factorial* or *primary* if $\pi_\varphi(A)''$ is a factor. We shall denote by $\mathfrak{F}$ the set of all factorial states.

Lemma 6.31. *Let A and B be unital C^*-algebras, and put $C = A \otimes_{\max} B$. If φ is a pure state of C, then the restriction ψ of φ to A, identifying A and $A \otimes \mathbf{C} \subset C$, is a factorial state of A.*

PROOF. Let $\{\pi_\varphi, \mathfrak{H}_\varphi, \xi_\varphi\}$ be the cyclic representation of C induced by φ. It follows that $\pi_\varphi(C)$ is generated as a C^*-algebra by $\pi_\varphi(A)$ and $\pi_\varphi(B)$, and is σ-weakly dense in $\mathscr{L}(\mathfrak{H}_\varphi)$. Hence $\pi_\varphi(A)''$ and $\pi_\varphi(B)''$generate $\mathscr{L}(\mathfrak{H}_\varphi)$ as a von Neumann algebra. Since $\pi_\varphi(A)''$ and $\pi_\varphi(B)''$ commute, this means that $\pi_\varphi(A)''$ and $\pi_\varphi(B)''$ are both factors. Let e be the projection of $\mathfrak{H}_\varphi$ to $[\pi_\varphi(A)\xi_\varphi]$. Then the cyclic representation $\{\pi_\psi, \mathfrak{H}_\psi, \xi_\psi\}$ of A induced by ψ is unitarily equivalent to the restriction π_φ^e of $\pi_\varphi|_A$ to the invariant subspace $[\pi_\varphi(A)\xi_\varphi]$. Hence we have $\pi_\psi(A)'' \cong \pi_\varphi(A)_e''$. But $\pi_\varphi(A)_e''$ is a factor, being an induced von Neumann algebra of the factor $\pi_\varphi(A)''$. Thus $\pi_\psi(A)''$ is a factor. Q.E.D.

Theorem 6.32. *Let A, $\mathfrak{S}$, and φ be as before. Then the central representing measure μ of φ is pseudoconcentrated on the factorial state space $\mathfrak{F}$ in the sense that for any Baire subset E of $\mathfrak{S}$ disjoint from $\mathscr{F}$, $\mu(E) = 0$.*

PROOF. Let $C = A \otimes_{\max} \pi_\varphi(A)'$. By Proposition 4.7, there exists a representation π of C on $\mathfrak{H}_\varphi$ such that $\pi(a \otimes b) = \pi_\varphi(a)b$, $a \in A$ and $b \in \pi_\varphi(A)'$. Let ψ be the state on C given by $\psi(x) = (\pi(x)\xi_\varphi|\xi_\varphi)$. It follows that $\{\pi, \mathfrak{H}_\varphi, \xi_\varphi\}$ is unitarily equivalent to the cyclic representation $\{\pi_\psi, \mathfrak{H}_\psi, \xi_\psi\}$ of C induced by ψ. Since $\pi(C)$ is generated, as a C^*-algebra, by $\pi_\varphi(A)$ and $\pi_\varphi(A)'$, we have $\pi(C)' = \pi_\varphi(A)' \cap \pi_\varphi(A)'' = \mathscr{Z}$. Hence the orthogonal representing measure v of ψ on $\mathfrak{S}(C)$ corresponding to $\mathscr{Z}$ is the unique maximal measure of $M_\psi^+(\mathfrak{S}(C))$. Let Φ be the restriction map of $\mathfrak{S}(C)$ onto $\mathfrak{S}$, i.e., $\Phi(\rho)(a) = \rho(a)$ for every $\rho \in \mathfrak{S}(C)$, and let Φ^* be the dual map of $C(\mathfrak{S})$ into $C(\mathfrak{S}(C))$, i.e., $\Phi^* f = f \circ \Phi$ for each $f \in C(\mathfrak{S})$. We have then for each $f \in C(\mathfrak{S})$ and $a \in A$,

$$\begin{aligned}\int_{\mathfrak{S}} f(\omega)\tilde{a}(\omega)\, d\Phi(v)(\omega) &= \int_{\mathfrak{S}(C)} f \circ \Phi(\rho)\tilde{a} \circ \Phi(\rho)\, dv(\rho)\\ &= \int_{\mathfrak{S}(C)} (\Phi^* f)(\rho)\tilde{a}(\rho)\, dv(\rho)\\ &= (\theta_v \Phi^*(f)\pi_\psi(a)\xi_\psi|\xi_\psi) = (\theta_v \Phi^*(f)\pi_\varphi(a)\xi_\varphi|\xi_\varphi).\end{aligned}$$

Hence $\Phi(v)$ is the orthogonal representing measure of φ on $\mathfrak{S}$ with $\theta_{\Phi(v)} = \theta_v \circ \Phi^*$. Therefore, we get $\mathscr{C}_{\Phi(v)} = \mathscr{C}_v = \mathscr{Z} = \mathscr{C}_\mu$, which implies $\Phi(v) = \mu$.

If E is a Baire subset of $\mathfrak{S}$ disjoint from $\mathfrak{F}$, then $\Phi^{-1}(E)$ is a Baire subset $\mathfrak{S}(C)$ and disjoint from $P(C)$ by Lemma 6.30. Hence $\mu(E) = v(\Phi^{-1}(E)) = 0$. Q.E.D.

We are now going to show that the set of all factorial states $\mathfrak{F}$ on a *separable* C^*-algebra A is a Borel subset of $\mathfrak{S}$. To this end, we shall employ the techniques and the notations developed in Section 3. To each $\omega \in \mathfrak{S}$, there corresponds a unique completely positive map θ_ω of $\pi_\omega(A)'$ onto $C_\omega \subset A^*$. Let $I(\omega) = \{\rho \in A^* : 0 \leq \rho \leq \omega\}$. It follows that $I(\omega)$ is the image of the positive part of the unit ball of $\pi_\omega(A)'$ under θ_ω. Let $Z(\omega)$ denote the intersection of $I(\omega)$ and the image of the center $\mathscr{Z}_\omega$ of $\pi_\omega(A)''$ under θ_ω, i.e.,

$$Z(\omega) = I(\omega) \cap \theta_\omega(\pi_\omega(A)' \cap \pi_\omega(A)'').$$

Clearly, ω is factorial if and only if $Z(\omega) = \{\lambda\omega : 0 \leq \lambda \leq 1\}$.

For each $\omega \in A^*$, we associate a linear functional Φ_ω on $A \otimes A$ by

$$\Phi_\omega\left(\sum_{i=1}^n x_i \otimes y_i\right) = \omega\left(\sum_{i=1}^n x_i y_i\right).$$

For each $h \in A^{**}$, $0 \leq h \leq 1$, put

$$\Phi_{\omega,h}\left(\sum_{i=1}^n x_i \otimes y_i\right) = \omega\left(\sum_{i=1}^n x_i h y_i\right),$$

and

$$N = \left\{\sum_{i=1}^n x_i \otimes y_i : \sum_{i=1}^n x_i y_i = 0\right\}.$$

It follows then that a linear functional Φ on $A \otimes A$ is of the form Φ_ρ for some $\rho \in I(\omega)$ if and only if

$$0 \leq \Phi(x^* \otimes x) \leq \omega(x^* x), \qquad x \in A,$$
$$\Phi(N) = \{0\}.$$

Lemma 6.33. *Let N_0 be a dense subset of N with respect to the greatest cross-norm γ in $A \otimes A$ and A_0 be a dense *-subalgebra of A over $\mathbf{Q} + i\mathbf{Q}$. The state ω is factorial if and only if for any $\varepsilon > 0$ and $x \in A_0$ there exist $\delta > 0$ and $u_1, \ldots, u_n \in N_0$ such that*

$$|\Phi_{\omega,h}(u_i)| < \delta, \qquad i = 1,2,\ldots,n, \Rightarrow |\omega(x^* h x) - \omega(h)\omega(x^* x)| < \varepsilon$$

for every $h \in A_0$, $0 \leq h \leq 1$.

Proof. Suppose that the condition does not hold. There exist $\varepsilon > 0$ and $x \in A_0$ such that for any $\delta > 0$ and $u_1, \ldots, u_n \in N_0$ there exists $h = h(\delta; u_1, \ldots, u_n) \in A_0$, $0 \leq h \leq 1$, such that

$$\max_{1<j<n} |\Phi_{\omega,h}(u_j)| < \delta \quad \text{and} \quad |\omega(x^* h x) - \omega(h)\omega(x^* x)| \geq \varepsilon.$$

Since $\{h\}$ is bounded, and the positive part of the unit ball of A^{**} is σ-weakly compact, we can find a subnet $\{h_i\}$ of $\{h\}$ converging to $k \in A^{**}$, $0 \leq k \leq 1$, σ-weakly. Since every element of $A \otimes A$ is an algebraic tensor product element, the map: $k \in A^{**} \mapsto \Phi_{\omega,k}(u) \in \mathbf{C}$ is σ-weakly continuous for every

$u \in A \otimes A$; hence

$$\Phi_{\omega,k}(u) = \lim \Phi_{\omega,h}(u) = 0, \qquad u \in N_0,$$

so by continuity

$$\Phi_{\omega,k}(N) = \{0\},$$

and

$$|\omega(x^*kx) - \omega(k)\omega(x^*x)| = \lim|\omega(x^*hx) - \omega(h)\omega(x^*x)| \geq \varepsilon.$$

For every $y \in A$, we have

$$\Phi_{\omega,k}(y^* \otimes y) = \lim \omega(y^*hy) \leq \omega(y^*y).$$

Thus, $\Phi_{\omega,k}$ is of the form Φ_ρ for some $\rho \in I(\omega)$ by the arguments before the lemma. Hence we get, with $c = \theta_\omega^{-1}(\rho) \in \pi_\omega(A)'$,

$$\begin{aligned}(\tilde{\pi}_\omega(k)\pi_\omega(y)\xi_\omega|\pi_\omega(y)\xi_\omega) &= \omega(y^*ky) = \rho(y^*y)\\ &= (c\pi_\omega(y)\xi_\omega|\pi_\omega(y)\xi_\omega), \qquad y \in A_0,\end{aligned}$$

where $\tilde{\pi}_\omega$ means the canonical extension of π_ω to A^{**}. Thus, $\tilde{\pi}_\omega(k) = c$ falls in $\pi_\omega(A)'' \cap \pi_\omega(A)'$. However, we have

$$|(\tilde{\pi}_\omega(k)\pi_\omega(x)\xi_\omega|\pi_\omega(x)\xi_\omega) - \omega(k)\|\pi_\omega(x)\xi_\omega\|^2| = |\omega(x^*kx) - \omega(k)\omega(x^*x)| \geq \varepsilon,$$

which means that $\tilde{\pi}_\omega(k) \neq \omega(k)1$. Thus, $\tilde{\pi}_\omega(k)$ is not a scalar multiple of the identity. Namely, $\pi_\omega(A)'$ is not a factor.

Suppose that the condition holds. Let k be a central element of A^{**} with $0 \leq k \leq 1$. By Theorem II.4.8, there exists a net $\{a_i\}$ in the unit ball of A_0 converging σ-strongly* to $k^{1/2}$. Let $h_i = a_i^*a_i$. Then $\{h_i\}$ converges σ-strongly to k in A^{**} since the product in the bounded part is σ-strongly* continuous. For any $u = \sum_{i=1}^n x_i \otimes y_i \in N$, we have

$$\begin{aligned}\lim \Phi_{\omega,h_i}(u) &= \lim \omega\left(\sum_{j=1}^n x_j h_i y_j\right) = \omega\left(\sum_{j=1}^n x_j k y_j\right)\\ &= \omega\left(\sum_{j=1}^n x_j y_j k\right) = 0.\end{aligned}$$

By the condition, we have for each $x \in A_0$,

$$\begin{aligned}(\tilde{\pi}_\omega(k)\pi_\omega(x)\xi_\omega|\pi_\omega(x)\xi_\omega) &= \lim(\pi_\omega(h_i)\pi_\omega(x)\xi_\omega|\pi_\omega(x)\xi_\omega)\\ &= \lim \omega(x^*h_ix) = \lim \omega(h_i)\omega(x^*x)\\ &= \omega(k)\|\pi_\omega(x)\xi_\omega\|^2,\end{aligned}$$

which means that $\pi_\omega(k) = \omega(k)1$. Thus the center of $\pi_\omega(A)''$ is trivial, i.e., $\pi_\omega(A)''$ is a factor. Q.E.D.

Theorem 6.34. *The set* $\mathfrak{F}$ *of all factorial states on a seprable unital* C^*-*algebra* A *is a Borel subset of the state space* $\mathfrak{S}$ *of* A.

PROOF. Let A_0 be a countable dense *-subalgebra of A over the "complex" rational field $\mathbf{Q} + i\mathbf{Q}$. In the tensor product $A \otimes A$, we consider the greatest cross-norm $\|\cdot\|_\gamma$, and a countable dense subspace N_0 of N with respect to this norm. For each $m,n = 1,2,\ldots$, $h \in A_0$, $0 \leq h \leq 1$, $u_1,\ldots,u_r \in N_0$, and $x \in A_0$, let

$$\mathfrak{S}(m; n; h; u_1,\ldots,u_r; x) = \left\{\omega \in \mathfrak{S}\colon \max_{1<i<r} |\Phi_{\omega,h}(u_i)| \geq \frac{1}{m} \text{ or } |\omega(x^*hx) - \omega(h)\omega(x^*x)| \leq \frac{1}{n}\right\}.$$

It follows then that $\mathfrak{S}(m; n; h; u_1,\ldots,u_r; x)$ is a closed subset of $\mathfrak{S}$. By the previous lemma, we have

$$\mathfrak{F} = \bigcap_{n=1}^{\infty} \bigcap_{\substack{h\in A_0\\ 0\leq h\leq 1}} \bigcap_{x\in A_0} \bigcup_{m=1}^{\infty} \bigcup_{u_1,\ldots,u_r\in N_0} \mathfrak{S}(m; n; h; u_1,\ldots,u_r; x).$$

Thus $\mathfrak{F}$ is an $F_{\sigma\delta}$-set in $\mathfrak{S}$; hence it is a Borel subset of $\mathfrak{S}$. Q.E.D.

EXERCISES

1. Let A be a unital C^*-algebra equipped with a group G of automorphisms. Let $\mathfrak{S}^G$ denote the space of all G-invariant states on A.

   (a) Show that $\mathfrak{S}^G$ is a compact convex subset of the state space $\mathfrak{S}$ of A.
   (b) Show that if $\varphi \in \mathfrak{S}^G$, then there exists an orthogonal representing measure μ of φ which is pseudoconcentrated on the ergodic state space $\partial_e\mathfrak{S}^G$. (*Hint*: Show that an orthogonal representing measure μ of φ indeed has the desired property if $\mathscr{C}_\mu$ is a maximal abelian subalgebra of $\pi_\varphi(A)' \cap U_\varphi(G)'$, where U_φ is the unitary representation of G associated with the cyclic representation $\{\pi_\varphi,\mathfrak{H}_\varphi,\xi_\varphi\}$ in Exercise I.9.7.)

†2. A compact convex set K in E is called a *simplex* if $\mathscr{A}(K)^*$ is a vector lattice with respect to the order structure dual to the natural order structure in $\mathscr{A}(K)$ [1].

   (a) Show that K is a simplex if and only if every point in K is the barycenter of a unique boundary probability measure [1, Theorem II.3.6].
   (b) Show that the state space $\mathfrak{S}$ of a unital C^*-algebra A is a simplex if and only if A is abelian.

3. Let φ be a state on a unital C^*-algebra A with $\mathfrak{S} = \mathfrak{S}(A)$, and F_φ denote the smallest face of $\mathfrak{S}$ containing φ.

   (a) Show that F_φ is closed in $\mathfrak{S}$ if and only if $\pi(A)'$ is finite dimensional if and only if F_φ is finite dimensional. (*Hint*: Show that $F_\varphi = \{\psi \in \mathfrak{S}\colon \psi(x) = (\pi_\varphi(x)a\xi_\varphi|\xi_\varphi)$, $x \in A$, for some $a \in \pi_\varphi(A)'\}$.)
   (b) Show that the closure $\bar{F}_\varphi$ of F_φ in $\mathfrak{S}$ is a simplex if and only if $\pi_\varphi(A)'$ is abelian.

4. Let $\{\mathscr{M},\mathfrak{H}\}$ be a von Neumann algebra with a cyclic vector ξ_0.
   (a) Show that if $\mathscr{A}$ is an abelian von Neumann subalgebra of $\mathscr{M}'$ and if e is the projection of $\mathfrak{H}$ onto $[\mathscr{A}\xi_0]$, then $e\mathscr{M}e$ is commutative, i.e., $e\mathscr{M}e \subset (e\mathscr{M}e)'$. (*Hint*: Note that an abelian von Neumann algebra with a cyclic vector is maximal abelian. Hence $\mathscr{A}'_e = \mathscr{A}_e \supset e\mathscr{M}e$.)
   (b) Show that if e is a projection in $\mathfrak{H}$ with $e\xi_0 = \xi_0$ such that $e\mathscr{M}e$ is commutative, then $\mathscr{A} = [\mathscr{M} \cup \{e\}]'$ is abelian and e is the projection to $[\mathscr{A}\xi_0]$. (*Hint*: Note that $(e\mathscr{M}e)''_e$ on $e\mathfrak{H}$ is abelian and admits a cyclic vector ξ_0, and that $x \in \mathscr{A} \mapsto x_e \in \mathscr{L}(e\mathfrak{H})$ is an isomorphism.)
   (c) Show that the correspondence between abelian von Neumann subalgebras of $\mathscr{M}'$ and projections with the property in (b) is bijective.

5. Let A be a unital C^*-algebra with $\mathfrak{S} = \mathfrak{S}(A)$ and G a group. Let α be a homomorphism: $s \in G \mapsto \alpha_s \in \mathrm{Aut}(A)$ of G into the automorphism group of A. Let $\mathfrak{S}^G$ be the set of all invariant states on A under the action α, i.e., $\mathfrak{S}^G = \{\omega \in \mathfrak{S} : \omega \circ \alpha_s = \omega$ for all $s \in G\}$. For a fixed $\varphi \in \mathfrak{S}^G$, let $\{\pi_\varphi,\mathfrak{H}_\varphi,\xi_\varphi\}$ be the cyclic representation of A and U_φ be the associated unitary representation of G on $\mathfrak{H}_\varphi$. Let e_0 be the projection of $\mathfrak{H}_\varphi$ onto $\mathfrak{H}_0 = \{\xi \in \mathfrak{H}_\varphi : U_\varphi(s)\xi = \xi, s \in G\}$.
   (a) Show that the equivalence of the following statements: (i) $e_0\pi_\varphi(A)e_0$ is commutative; (ii) for every $x,y \in A$ and $\xi \in \mathfrak{H}_0$ one has
   $$\inf_{s \in G} |(\pi_\varphi(\alpha_s(x)y - y\alpha_s(x))\xi|\xi)| = 0.$$
   (*Hint*: (i) $\Rightarrow$ (ii): Use Exercise 6.4 to conclude that $\mathscr{A} = [\pi_\varphi(A) \cup \{e_0\}]'$ is abelian. Using Exercise III.2.6, conclude that
   $$\begin{aligned} 0 &\le \inf_{s \in G} |(\{\pi_\varphi(\alpha_s(x))\pi_\varphi(y) - \pi_\varphi(y)\pi_\varphi(\alpha_s(x))\}\xi|\xi)| \\ &= \inf_{s \in G} |(\{\pi_\varphi(x)U_\varphi(s)\pi_\varphi(y) - \pi_\varphi(y)U_\varphi(s)\pi_\varphi(x)\}\xi|\xi)| \\ &\le |(\{\pi_\varphi(x)e_0\pi_\varphi(y)e_0 - e_0\pi_\varphi(y)e_0\pi_\varphi(x)\}\xi|\xi)| = 0. \end{aligned}$$
   (ii) $\Rightarrow$ (i): Approximate e_0 strongly by $\sum_{i=1}^n \lambda_i U_\varphi(s_i)$, $\lambda_i \ge 0$ and $\sum \lambda_i = 1$, on the vectors $U_\varphi(s_0)\pi_\varphi(y)\xi$ and $U_\varphi(s_0)\pi_\varphi(x)\xi$, where
   $$|((\pi_\varphi(x)U_\varphi(s_0)y - \pi_\varphi(y)U_\varphi(s_0)\pi_\varphi(x))\xi|\xi)| < \varepsilon,$$
   to conclude that
   $$|(\{\pi_\varphi(x)e_0\pi_\varphi(y)e_0 - e_0\pi_\varphi(y)e_0\pi_\varphi(x)\}\xi|\xi)| < 2\varepsilon.)$$
   (b) Show that either condition (i) or (ii) in (a) implies the commutativity of $U_\varphi(G)' \cap \pi_\varphi(A)'$. (*Hint*: The vector ξ_φ is separating for $U_\varphi(G)' \cap \pi_\varphi(A)' \subset \pi_\varphi(A)'$.)
   (c) Show that the closure F_φ^G of the smallest face F_φ^G of $\mathfrak{S}^G$ containing φ is a simplex if the conditions in (a) hold. Hence φ is the barycenter of a unique orthogonal boundary measure on $\mathfrak{S}^G$ which is pseudoconcentrated on the ergodic states $\partial_e\mathfrak{S}^G$.
   (d) If either condition (i) or (ii) in (a) holds for every $\varphi \in \mathfrak{S}^G$, then the action α on the system $\{A,G,\alpha\}$ is said to be *G-abelian*. Show that $\mathfrak{S}^G$ is a simplex for a G-abelian system $\{A,G,\alpha\}$.

6. Let $A = M(n; \mathbf{C})$ and G be the group of all diagonal unitary matrices. Let $\alpha_s = \mathrm{Ad}(s)$, $s \in G$, and set $\varphi(x) = x_{11}$, $x \in A$. Show that $\pi_\varphi(A)' \cap U_\varphi(G)'$ is abelian (indeed one-dimensional), but that the condition (i) or (ii) in the previous exercise does not hold for $\{A,G,\alpha,\varphi\}$; thus conclude that the simplex property of $\mathfrak{S}^G$ does not yield the G-abelianess of $\{A,G,\alpha\}$ in general.

7. Let $\{A,G,\alpha\}$ be as above. For each $s \in A$, let $K_G(a)$ denote the convex hull of $\{\alpha_s(a): s \in G\}$. We say that $\alpha(G)$ is *large* or simply α is *large* if for every $\varphi \in \mathfrak{S}^G$, $a,b_1,\dots,b_n \in A$ and $x \in A$, one has

$$\inf_{a' \in K_G(a)} |\varphi(x(a'b_i - b_i a')x^*)| = 0, \qquad i = 1,2,\dots,n.$$

(a) Show that α is large if and only if the strong closure $\pi_\varphi(K_G(a))^-$ of $\pi_\varphi(K_G(a))$ meets with the center of $\pi_\varphi(A)''$ for every $\varphi \in \mathfrak{S}^G$. (*Hint*: Notice the weak compactness of $\pi_\varphi(K_G(a))^-$.)

(b) Show that if α is large, then $\{A,G,\alpha\}$ is G-abelian; hence $\mathfrak{S}^G$ is a simplex.

(c) Show that if α is large, then the following statements for $\varphi \in \mathfrak{S}^G$ are equivalent: (i) φ is ergodic, i.e., $\varphi \in \partial_e\mathfrak{S}^G$; (ii) $U_\varphi(G)' \cap \pi_\varphi(A)'' \cap \pi_\varphi(A)' = \mathbf{C}1$; (iii) $\mathfrak{H}_0 = \{\xi \in \mathfrak{H}_\varphi : U_\varphi(s)\xi = \xi, s \in G\} = \mathbf{C}\xi_\varphi$.

7. Representation of $L^2(\Gamma,\mu) \otimes \mathfrak{H}$, $L^1(\Gamma,\mu) \otimes_\gamma \mathcal{M}_*$, and $L^\infty(\Gamma,\mu) \mathbin{\overline{\otimes}} \mathcal{M}$

In this section, we shall represent the tensor products specified above by systems of vector or operator valued functions.

Let Γ be a locally compact space with a fixed positive Radon measure μ.

Definition 7.1. Let E be a Banach space. An E-valued function f on Γ is said to be *measurable* with respect to μ or simply *μ-measurable* if for any compact set K and $\varepsilon > 0$ there exists a compact subset K_0 of K such that $\mu(K - K_0) < \varepsilon$ and f is continuous on K_0.

Sometimes, it is not easy to show the measurability of a given function directly. But the following proposition solves partially this problem:

Proposition 7.2. *An E-valued function f is μ-measurable if and only if the following conditions are satisfied:*

(i) *For each $x^* \in E^*$, the function: $\gamma \in \Gamma \mapsto \langle f(\gamma),x^* \rangle \in \mathbf{C}$ is μ-measurable.*

(ii) *For each compact subset K of Γ, there exists a μ-measurable subset K_∞ of K such that $\mu(K - K_\infty) = 0$ and $f(K_\infty)$ is separable.*

Proof. Suppose that f is μ-measurable. Then the numerical valued function: $\gamma \in \Gamma \mapsto \langle f(\gamma),x^* \rangle \in \mathbf{C}$, $x^* \in E^*$, is μ-measurable by definition. Let K be a compact subset of Γ. For each $n = 1,2,\dots$, let K_n be a compact subset of K such that $\mu(K - K_n) < 1/n$ and f is continuous on K_n. By continuity, $f(K_n)$ is compact in E, hence separable, being metrizable. Put $K_\infty = \bigcup_{n=1}^\infty K_n$. Then $\mu(K - K_\infty) = 0$ and $f(K_\infty) = \bigcup_{n=1}^\infty f(K_n)$ is separable.

Suppose that f satisfies (i) and (ii). Let K be a compact subset of Γ and K_∞ be a subset of K such that $\mu(K - K_\infty) = 0$ and $f(K_\infty)$ is separable. For a given $\varepsilon > 0$, let K'_∞ be a compact subset of K_∞ such that $\mu(K_\infty - K'_\infty) < \varepsilon/2$. Trivially, $f(K'_\infty)$ is separable. Let F be the closed subspace generated by $f(K'_\infty)$. It follows that F is a separable Banach space. By the Hahn–Banach theorem, the numerical valued function: $\gamma \in K'_\infty \mapsto \langle f(\gamma),x^* \rangle$ is μ-measurable for every $x^* \in F^*$. Let $\{x_n^*\}$ be a weakly* dense sequence of the unit ball of F^* whose existence follows from the separability of F. For every $x \in F$, we have

$$\|f(\gamma) - x\| = \sup_n |\langle f(\gamma), x_n^* \rangle - \langle x, x_n^* \rangle|,$$

so that the function $f_x : \gamma \in K'_\infty \mapsto \|f(\gamma) - x\|$ is μ-measurable. Let $\{x_n\}$ be a dense sequence in F. For each n, let K_n be a compact subset of K'_∞ such that that $\mu(K'_\infty - K_n) < \varepsilon/2^{n+1}$ and f_{x_n} is continuous on K_n. Let $K_0 = \bigcap_{n=1}^\infty K_n$. We have then $\mu(K'_\infty - K_0) < \varepsilon/2$, and so $\mu(K - K_0) \leq \mu(K_\infty - K'_\infty) + \mu(K'_\infty - K_0) < (\varepsilon/2) + (\varepsilon/2) = \varepsilon$. On K_0, each f_{x_n} is continuous. If $x \in F$, then there exists a subsequence $\{x_{n_j}\}$ converging to x, so that

$$|f_x(\gamma) - f_{x_{n_j}}(\gamma)| = |\|f(\gamma) - x\| - \|f(\gamma) - x_{n_j}\|| \leq \|x - x_{n_j}\| \to 0.$$

Hence f_x is a uniform limit of a sequence of continuous functions on K_0; thus f_x is continuous. For a fixed $\gamma_0 \in K_0$, the function $\|f(\cdot) - f(\gamma_0)\|$ is continuous on K_0 and vanishes at γ_0, so that for any $\varepsilon > 0$, there exists a neighborhood U of γ_0 such that $\|f(\gamma) - f(\gamma_0)\| < \varepsilon$ for every $\gamma \in U \cap K_0$. Thus f is continuous on K_0. By definition, f is μ-measurable. Q.E.D.

Let $\mathscr{L}$ be a linear space of numerical valued functions on Γ. The algebraic tensor product $\mathscr{L} \otimes E$ of $\mathscr{L}$ and E is identified with a linear space of E-valued functions by the identification: $\sum_{i=1}^n f_i \otimes x_i \leftrightarrow \sum_{i=1}^n f_i(\cdot)x_i$. With this identification, we regard $\mathscr{L} \otimes E$ as a linear space of E-valued functions on Γ. It follows then that $\mathscr{K}(\Gamma) \otimes E$, where $\mathscr{K}(\Gamma)$ is the space of all continuous functions on Γ with compact support, is nothing but the space of all continuous E-valued functions on Γ with compact support and finite dimensional range. Fixing p, $1 \leq p < +\infty$, we define a seminorm $\|\cdot\|_p$ in $\mathscr{K}(\Gamma) \otimes E$ by

$$\|f\|_p = \left\{\int_\Gamma \|f(\gamma)\|^p \, d\mu(\gamma)\right\}^{1/p}, \qquad f \in \mathscr{K}(\Gamma) \otimes E. \tag{1}$$

With $N = \{f \in \mathscr{K}(\Gamma) \otimes E : \|f\|_p = 0\}$, the completion of $(\mathscr{K}(\Gamma) \otimes E)/N$ is denoted by $L_E^p(\Gamma, \mu)$. By definition, $L_E^p(\Gamma, \mu)$ is a Banach space.

Proposition 7.3. *If Γ is compact, then $C(\Gamma) \otimes E$ is dense in the Banach space $C_E(\Gamma)$ of all E-valued continuous functions on Γ with respect to the norm $\|\cdot\|_\infty$ defined by*

$$\|f\|_\infty = \sup \|f(\gamma)\|, \qquad f \in C_E(\Gamma).$$

Furthermore, the norm $\|\cdot\|_\infty$ agrees with the injective cross-norm $\|\cdot\|_\lambda$ on $C(\Gamma) \otimes E$; hence $C_E(\Gamma) = C(\Gamma) \otimes_\lambda E$.

PROOF. Take an $f \in C_E(\Gamma)$ and $\varepsilon > 0$. Since $f(\Gamma)$ is compact in E, being a continuous image of the compact set Γ, there exist $\gamma_1, \ldots, \gamma_n \in \Gamma$ such that $\inf_{1<i<n} \|f(\gamma) - f(\gamma_i)\| < \varepsilon/2$ for every $\gamma \in \Gamma$. Put $U_i = \{\gamma \in \Gamma : \|f(\gamma) - f(\gamma_i)\| \leq \varepsilon/2\}$ and $V_i = \{\gamma \in \Gamma : \|f(\gamma) - f(\gamma_i)\| < \varepsilon\}$ for $i = 1, 2, \ldots, n$. By the choice of $\{\gamma_1, \ldots, \gamma_n\}$, $\bigcup_{i=1}^n U_i = \Gamma$. Since U_i is compact and $U_i \subset V_i$, there exists, by Urysohn's lemma, an $h_i \in C(\Gamma)$ such that $0 \leq h_i(\gamma) \leq 1$, $h_i(\gamma) = 1$ if $\gamma \in U_i$ and $h_i(\gamma) = 0$ if $\gamma \notin V_i$. Since $\{U_i\}$ covers Γ, we have $\sum_{i=1}^n h_i(\gamma) > 0$ for every $\gamma \in \Gamma$. Put $g_i(\gamma) = h_i(\gamma)/\sum_{j=1}^n h_j(\gamma)$, $\gamma \in \Gamma$. We then have that $0 \leq$

$g_i(\gamma) \leq 1$, $\sum_{i=1}^n g_i(\gamma) = 1$ and $\operatorname{supp} g_i \subset V_i$. With $x_i = f(\gamma_i)$, $1 \leq i \leq n$, we have

$$\left\| f(\gamma) - \sum_{i=1}^n g_i(\gamma)x_i \right\| = \left\| \sum_{i=1}^n g_i(\gamma)(f(\gamma) - x_i) \right\|$$
$$\leq \sum_{i=1}^n g_i(\gamma)\|f(\gamma) - x_i\| \leq \sum_{i=1}^n \varepsilon g_i(\gamma) = \varepsilon \qquad (\operatorname{supp} g_i \subset V_i).$$

Hence f is approximated by $\sum_{i=1}^n g_i \otimes x_i$ within ε. Therefore, $C(\Gamma) \otimes E$ is dense in $C_E(\Gamma)$.

Let $M(\Gamma) = C(\Gamma)^*$. To each $\gamma \in \Gamma$, there corresponds a point mass $\delta_\gamma \in M(\Gamma)$ at γ, that is, $\delta_\gamma(f) = f(\gamma)$, $f \in C(\Gamma)$. If $f = \sum_{i=1}^n f_i \otimes x_i \in C(\Gamma) \otimes E$, we have

$$\begin{aligned}
\|f\|_\infty = \sup\|f(\gamma)\| &= \sup \left\| \sum_{i=1}^n f_i(\gamma)x_i \right\| \\
&= \sup \left\{ \left| \left\langle \sum_{i=1}^n f_i(\gamma)x_i, x^* \right\rangle \right| : x^* \in E^*, \|x^*\| \leq 1 \right\} \\
&= \sup \left\{ \left| \left\langle \sum_{i=1}^n f_i \otimes x_i, \delta_\gamma \otimes x^* \right\rangle \right| : x^* \in E^*, \|x^*\| \leq 1, \gamma \in \Gamma \right\} \\
&\leq \sup \left\{ \left| \left\langle \sum_{i=1}^n f_i \otimes x_i, \mu \otimes x^* \right\rangle \right| : \mu \in M(\Gamma), \|\mu\| \leq 1; \right. \\
&\qquad\qquad \left. x^* \in E^*, \|x^*\| \leq 1 \right\} \\
&= \|f\|_\lambda \\
&= \sup \left\{ \left| \int_\Gamma \sum_{i=1}^n f_i(\gamma)\langle x_i, x^*\rangle \, d\mu(\gamma) \right| : \mu \in M(\Gamma), \|\mu\| \leq 1; \right. \\
&\qquad\qquad \left. x^* \in E^*, \|x^*\| \leq 1 \right\} \\
&= \sup \left\{ \left| \int_\Gamma \left\langle \sum_{i=1}^n f_i(\gamma)x_i, x^* \right\rangle d\mu(\gamma) \right| : \mu \in M(\Gamma), \|\mu\| \leq 1; \right. \\
&\qquad\qquad \left. x^* \in E^*, \|x^*\| \leq 1 \right\} \\
&\leq \sup \left\{ \int_\Gamma \left| \left\langle \sum_{i=1}^n f_i(\gamma)x_i, x^* \right\rangle \right| d|\mu|(\gamma) : \mu \in M(\Gamma), \|\mu\| \leq 1; \right. \\
&\qquad\qquad \left. x^* \in E^*, \|x^*\| \leq 1 \right\} \\
&\leq \sup \left\{ \int_\Gamma \|f(\gamma)\| \, d\mu(\gamma) : \mu \in M_+(\Gamma), \|\mu\| \leq 1 \right\} \\
&\leq \|f\|_\infty,
\end{aligned}$$

where $|\mu|$ means the absolute value of the complex measure μ in the sense of the polar decomposition in the conjugate space $C(\Gamma)^*$ of the C^*-algebra $C(\Gamma)$.

Q.E.D.

Proposition 7.4. *The Banach space $L^p_E(\Gamma,\mu)$, $1 \leq p < +\infty$, is identified canonically with the space of E-valued μ-measurable functions f with finite $\|f\|_p$ given by* (1).

PROOF. Take an $f \in L^p_E(\Gamma,\mu)$. By definition, there exists a sequence $\{f_n\}$ in $\mathscr{K}(\Gamma) \otimes E$ converging to f. Replacing $\{f_n\}$ by a subsequence if necessary, we assume that $\sum_{n=0}^{\infty} \|f_{n+1} - f_n\|^p < +\infty$ with $f_0 = 0$. By Fubini's theorem for positive functions, we have

$$\sum_{n=0}^{\infty} \|f_{n+1} - f_n\|_p^p = \sum_{n=0}^{\infty} \int_\Gamma \|f_{n+1}(\gamma) - f_n(\gamma)\|^p \, d\mu(\gamma)$$

$$= \int_\Gamma \sum_{n=0}^{\infty} \|f_{n+1}(\gamma) - f_n(\gamma)\|^p \, d\mu(\gamma),$$

so that $\sum_{n=0}^{\infty} \|f_{n+1}(\gamma) - f_n(\gamma)\|^p < +\infty$ except on a locally μ-null set S. Hence $\{f_n(\gamma): n = 0,1,\dots\}$ converges to some element, say $f'(\gamma)$, of E for every $\gamma \notin S$. Put $f(\gamma) = f'(\gamma)$ if $\gamma \notin S$ and $f(\gamma) = 0$ if $\gamma \in S$. We then have an E-valued function $f(\cdot)$. A routine application of Egoroff's theorem yields that $f(\cdot)$ is indeed μ-measurable. We then get

$$\|f(\gamma) - f_n\|_p^p = \int \|f(\gamma) - f_n(\gamma)\|^p \, d\mu(\gamma)$$

$$\leq \int \sum_{k=n}^{\infty} \|f_{k+1}(\gamma) - f_k(\gamma)\|^p \, d\mu(\gamma)$$

$$= \sum_{k=n}^{\infty} \int \|f_{k+1}(\gamma) - f_k(\gamma)\|^p \, d\mu(\gamma)$$

$$= \sum_{k=n}^{\infty} \|f_{k+1} - f_k\|_p^p \to 0 \quad \text{as} \quad n \to \infty.$$

Thus the function $f(\cdot)$ gives rise to the element f in $L^p_E(\Gamma,\mu)$.

Conversely, suppose f is an E-valued μ-measurable function with $\|f\|_p < +\infty$. It follows then that there exists a disjoint sequence $\{K_n\}$ of compact subsets of Γ such that $f(\gamma) = 0$ locally μ-almost everywhere in $(\bigcup_{n=1}^{\infty} K_n)^C$. So we assume $f(\gamma) = 0$ outside of $\bigcup_{n=1}^{\infty} K_n$. By measurability, we can choose K_n so that f is continuous on K_n. By Proposition 7.3, there exists a sequence $\{f_{n,j}\}$ in $C(K_n) \otimes E$ such that $\lim_{j\to\infty} \sup_{\gamma \in K_n} \|f(\gamma) - f_{n,j}(\gamma)\| = 0$. Hence we have

$$\int_{K_n} \|f(\gamma) - f_{n,j}(\gamma)\|^p \, d\mu(\gamma) \to 0 \quad \text{as} \quad j \to \infty.$$

Choosing a subsequence of $\{f_{n,j}\}$, we may assume that

$$\int_{K_n} \|f(\gamma) - f_{n,j}(\gamma)\|^p \, d\mu(\gamma) < \frac{1}{2^j}, \qquad j = 1,2,\ldots.$$

Put $f_j(\gamma) = f_{n,j}(\gamma)$ for $\gamma \in \bigcup_{n=1}^j K_n$ and $f_j(\gamma) = 0$ for $\gamma \notin \bigcup_{n=1}^j K_n$. Then f_j is continuous on $\bigcup_{n=1}^j K_n$ and $\|f - f_j\|_p \to 0$ as $j \to \infty$. Thus, replacing f by f_j, we must prove that if f is an E-valued continuous function on a compact subset K of Γ and vanishes outside of K, then f is well approximated by $\mathcal{K}(\Gamma) \otimes E$ in $L_E^p(\Gamma)$. Take an $\varepsilon > 0$. By Proposition 7.3, there exists $\sum_{i=1}^n f_i \otimes x_i \in C(K) \otimes E$ such that $\|f(\gamma) - \sum_{i=1}^n f_i(\gamma)x_i\| < \varepsilon$ for every $\gamma \in K$. Let $M = (\max_{1\le j\le n} \|f_j\|_\infty + \varepsilon)^p(\sum_{j=1}^n \|x_j\|)^p$, and choose an open set U such that $U \supset K$ and $\mu(U - K) < \varepsilon/M$. Let g_i, $1 \le i \le n$, be a continuous function such that $g_i(\gamma) = f_i(\gamma)$ for $\gamma \in K$, $g_i(\gamma) = 0$ for $\gamma \notin U$ and $\|g_i\|_\infty \le \|f_i\| + \varepsilon$. We have then

$$\begin{aligned}\left\|f - \sum_{i=1}^n g_i \otimes x_i\right\|_p^p &= \int_K \left\|f(\gamma) - \sum_{i=1}^n f_i(\gamma)x_i\right\|^p d\mu(\gamma) + \int_{U-K} \left\|\sum_{i=1}^n g_i(\gamma)x_i\right\|^p d\mu(\gamma)\\ &\le \varepsilon^p \mu(K)^p + \mu(U - K)M\\ &\le \varepsilon^p \mu(K)^p + \varepsilon.\end{aligned}$$

Since K is a preassigned compact set independent of ε, $\sum_{i=1}^n g_i \otimes x_i \in \mathcal{K}(\Gamma) \otimes E$ approximates f arbitrarily well in $L_E^p(\Gamma)$. Q.E.D.

We now apply the above to Hilbert spaces and operator algebras. Choosing E to be a Hilbert space $\mathfrak{H}$, the tensor product Hilbert space $L^2(\Gamma,\mu) \otimes \mathfrak{H}$ is identified with $L_{\mathfrak{H}}^2(\Gamma,\mu)$ and the inner product there is given by

$$\left.\begin{aligned}(\xi|\eta) &= \int_\Gamma (\xi(\gamma)|\eta(\gamma))\, d\mu(\gamma), \qquad \xi,\eta \in L^2(\Gamma,\mu),\\ \|\xi\| &= \left\{\int_\Gamma \|\xi(\gamma)\|^2 \, d\mu(\gamma)\right\}^{1/2}.\end{aligned}\right\} \tag{2}$$

On $L^2(\Gamma,\mu)$, $L^\infty(\Gamma,\mu)$ acts as a maximal abelian von Neumann algebra by multiplication; hence it acts on $L^2(\Gamma,\mu) \otimes \mathfrak{H}$ by amplification. Let $\pi(f)$ denote the operator on $L^2(\Gamma,\mu) \otimes \mathfrak{H}$ corresponding to $f \in L^\infty(\Gamma,\mu)$. It follows immediately that

$$(\pi(f)\xi)(\gamma) = f(\gamma)\xi(\gamma), \qquad f \in L^\infty(\Gamma,\mu), \qquad \xi \in L_{\mathfrak{H}}^2(\Gamma,\mu). \tag{3}$$

We denote by $\mathcal{A}$ the image $\{\pi(f) : f \in L^\infty(\Gamma,\mu)\}$.

We are now going to study operator valued functions, and start from the following:

Lemma 7.5. *If $x(\cdot)$ is an $\mathcal{L}(\mathfrak{H})$-valued function on Γ such that for each $\xi \in \mathfrak{H}$, the function: $\gamma \in \Gamma \mapsto x(\gamma)\xi \in \mathfrak{H}$ is μ-measurable, then for any $\mathfrak{H}$-valued μ-measurable function $\xi(\cdot)$, the function: $\gamma \in \Gamma \mapsto x(\gamma)\xi(\gamma) \in \mathfrak{H}$ is also μ-measurable.*

PROOF. Let K be a compact subset of Γ such that ξ is continuous on K. By Proposition 7.3, there exists a sequence $\xi_n = \sum_{i=1}^{m_n} f_i^n \otimes x_i^n \in C(K) \otimes \mathfrak{H}$ such that $\xi = \lim \xi_n$ on K. We then have

$$x(\gamma)\xi_n(\gamma) = \sum_{i=1}^{m_n} f_i^n(\gamma)x(\gamma)\xi_i^n \to x(\gamma)\xi(\gamma), \qquad \gamma \in K.$$

By assumption, each $x(\cdot)\xi_i^n$ is μ-measurable, so that $x(\cdot)\xi_n(\cdot)$ is μ-measurable, which implies the μ-measurability of $x(\cdot)\xi(\cdot)$ on K. Since the definition of μ-measurability is concerned only with compact subsets of Γ, the above argument shows that $x(\cdot)\xi(\cdot)$ is measurable on Γ. Q.E.D.

The following example shows that there is a basic difficulty in the study of operator valued functions in nonseparable Hilbert space which forces us to stay in the separable case if we are to advance beyond the formal level.

EXAMPLE 7.6. Let $\mathfrak{H} = l^2[0,1]$ and $\Gamma = [0,1]$ with the Lebesgue measure μ. Let $\{\varepsilon_t : 0 \le t \le 1\}$ be an orthogonal basis of $\mathfrak{H}$. Let $u(t)\xi = (\xi|\varepsilon_t)\varepsilon_0$, $0 \le t \le 1$. Since $u(t)\xi \neq 0$ for only countably many t, $u(t)\xi = 0$ for almost every t; hence $t \in [0,1] \mapsto u(t)\xi \in \mathfrak{H}$ is measurable. But $u(t)^*\xi = (\xi|\varepsilon_0)\varepsilon_t$. Hence $u(t)^*\varepsilon_0 = \varepsilon_t$ and $\|u(t)^*\varepsilon_0 - u(s)^*\varepsilon_0\| = \sqrt{2}$ for every pair s, t with $s \neq t$. Thus, there is no nondiscrete subset of $[0,1]$ on which $t \mapsto u(t)^*\varepsilon_0$ is continuous. Hence $t \in [0,1] \mapsto u(t)^*\varepsilon_0$ is not measurable.

Definition 7.7. An $\mathscr{L}(\mathfrak{H})$-valued function $x(\cdot)$ on Γ is said to be *measurable* (or more precisely *μ-measurable*) if the functions: $\gamma \in \Gamma \mapsto x(\gamma)$ and $\gamma \in \Gamma \mapsto x(\gamma)^*$ both satisfy the condition in Lemma 7.5.

Corollary 7.8. *The set of all $\mathscr{L}(\mathfrak{H})$-valued measurable functions on Γ forms a *-algebra under the natural algebraic operations.*

Let $x(\cdot)$ be an $\mathscr{L}(\mathfrak{H})$-valued measurable bounded function on Γ. If $\xi(\cdot)$ is an element of $L^2_{\mathfrak{H}}(\Gamma,\mu)$, then $\gamma \mapsto x(\gamma)\xi(\gamma)$ is μ-measurable and

$$\int_\Gamma \|x(\gamma)\xi(\gamma)\|^2 \, d\mu(\gamma) \le \int_\Gamma \|x(\gamma)\|^2 \|\xi(\gamma)\|^2 \, d\mu(\gamma)$$
$$\le \left(\sup_{\gamma \in \Gamma} \|x(\gamma)\|\right)^2 \int_\Gamma \|\xi(\gamma)\|^2 \, d\mu(\gamma).$$

Hence the function: $\gamma \in \Gamma \mapsto x(\gamma)\xi(\gamma) \in \mathfrak{H}$ belongs to $L^2_{\mathfrak{H}}(\Gamma,\mu)$ and the map: $\xi \in L^2_{\mathfrak{H}}(\Gamma,\mu) \mapsto x(\cdot)\xi(\cdot) \in L^2_{\mathfrak{H}}(\Gamma,\mu)$ is a bounded operator. We write this operator as

$$x = \int_\Gamma^\oplus x(\gamma) \, d\mu(\gamma) \quad \text{and} \quad (x\xi)(\gamma) = x(\gamma)\xi(\gamma). \tag{4}$$

Definition 7.9. An operator x on $L^2_{\mathfrak{H}}(\Gamma,\mu)$ is called *decomposable* if x is given by (4) with some bounded $\mathscr{L}(\mathfrak{H})$-valued measurable function $x(\cdot)$ on Γ. Each

$\pi(f)$, $f \in L^\infty(\Gamma,\mu)$, given by (3) is, of course, decomposable, but we call it a *diagonal* operator. The algebra $\{\pi(f), f \in L^\infty(\Gamma,\mu)\}$ is called the *diagonal* algebra.

The following formulas are easily seen:

$$\left.\begin{aligned}
\int_\Gamma^\oplus \{x(\gamma) + y(\gamma)\}\, d\mu(\gamma) &= \int_\Gamma^\oplus x(\gamma)\, d\mu(\gamma) + \int_\Gamma^\oplus y(\gamma)\, d\mu(\gamma);\\
\int_\Gamma^\oplus x(\gamma)y(\gamma)\, d\mu(\gamma) &= \left(\int_\Gamma^\oplus x(\gamma)\, d\mu(\gamma)\right)\left(\int_\Gamma^\oplus y(\gamma)\, d\mu(\gamma)\right);\\
\int_\Gamma^\oplus x(\gamma)^*\, d\mu(\gamma) &= \left(\int_\Gamma^\oplus x(\gamma)\, d\mu(\gamma)\right)^*;\\
\left\|\int_\Gamma^\oplus x(\gamma)\, d\mu(\gamma)\right\| &\leq \sup\|x(\gamma)\|.
\end{aligned}\right\} \tag{5}$$

If $\mathfrak{H}$ is separable, then with a countable dense subset $\{\xi_n\}$ of the unit ball of $\mathfrak{H}$, we have $\|x(\gamma)\| = \sup_n \|x(\gamma)\xi_n\|$, so that the function: $\gamma \in \Gamma \mapsto \|x(\gamma)\|$ is measurable for each $\mathcal{L}(\mathfrak{H})$-valued measurable function $x(\cdot)$. Hence we have

$$\left\|\int_\Gamma^\oplus x(\gamma)\, d\mu(\gamma)\right\| = \operatorname{ess\,sup}\|x(\gamma)\| \quad \text{for separable } \mathfrak{H}. \tag{6}$$

Theorem 7.10. *In the above situation, a bounded operator on $L^2_{\mathfrak{H}}(\Gamma,\mu)$ is decomposable if and only if it commutes with the diagonal algebra $\mathcal{A}$.*

PROOF. It is clear that every decomposable operator commutes with $\mathcal{A}$. For the converse, we need to use a lifting of $L^\infty(\Gamma,\mu)$.

A lifting of $L^\infty(\Gamma,\mu)$ is, by definition, an isomorphism ρ of $L^\infty(\Gamma,\mu)$ into the algebra $\mathcal{L}^\infty(\Gamma,\mu)$ of all bounded measurable functions on Γ such that $\rho(f)$, $f \in L^\infty(\Gamma,\mu)$, is a representative of the coset class f. Recall that $L^\infty(\Gamma,\mu)$ is not really an algebra of functions, but the algebra of the classes of functions classified by the locally almost everywhere agreement. For the existence of such ρ we refer the reader to [13].

Suppose that x is an operator of $\mathcal{A}'$. Take arbitrary ξ, η in $\mathfrak{H}$. If f and g are functions in $\mathcal{K}(\Gamma)$, then we have, with $K = \operatorname{supp} f \cup \operatorname{supp} g$,

$$(x(f \otimes \xi)|g \otimes \eta) = (\pi(fg)x(\chi_K \otimes \xi)|\chi_K \otimes \eta),$$

where χ_K means the characteristic function of K. Furthermore, with the polar decomposition $f = u|f|$ and $g = v|g|$, we have

$$\begin{aligned}
|(x(f \otimes \xi)|g \otimes \eta)| &= |(\pi(v^*u)x(|f| \otimes \xi)||g| \otimes \eta)|\\
&= |(\pi(uv^*)x(|f|^{1/2}|f|^{1/2} \otimes \xi)||g|^{1/2}|g|^{1/2} \otimes \eta)|\\
&= |(\pi(uv^*)x(|f|^{1/2}|g|^{1/2} \otimes \xi)||f|^{1/2}|g|^{1/2} \otimes \eta)|\\
&\leq \|\pi(uv^*)x\|\,\||fg|^{1/2}\|_2^2\|\xi\|\,\|\eta\|\\
&= \|x\|\,\|f\bar{g}\|_1\|\xi\|\,\|\eta\|.
\end{aligned}$$

Therefore, there exists a function $F_{\xi,\eta} \in L^{\infty}(\Gamma,\mu)$ such that

$$(x(f \otimes \xi)|g \otimes \eta) = \int_{\Gamma} f(\gamma)\overline{g(\gamma)}F_{\xi,\eta}(\gamma)\, d\mu(\gamma),$$
$$\|F_{\xi,\eta}\|_{\infty} \leq \|x\|\,\|\xi\|\,\|\eta\|.$$

Let $F_{\xi,\eta}(\gamma) = \rho(F_{\xi,\eta})(\gamma)$, $\gamma \in \Gamma$. Notice that $F_{\xi,\eta}(\cdot)$ is defined everywhere on Γ. Since $(\xi,\eta) \mapsto F_{\xi,\eta}$ is sesquilinear and bounded, $F_{\xi,\eta}(\gamma)$ is a bounded sesquilinear form of (ξ,η), so that there exists a bounded operator $x(\gamma)$ on $\mathfrak{H}$ such that $F_{\xi,\eta}(\gamma) = (x(\gamma)\xi|\eta)$, $\gamma \in \Gamma$, and $\|x(\gamma)\| \leq \|x\|$. That is, we have

$$(x(f \otimes \xi)|g \otimes \eta) = \int_{\Gamma} f(\gamma)\overline{g(\gamma)}(x(\gamma)\xi|\eta)\, d\mu(\gamma).$$

Since $F_{\xi,\eta}(\cdot)$ is measurable, the function: $\gamma \in \Gamma \mapsto (x(\gamma)\xi|\eta)$ is measurable for every $\xi,\eta \in \mathfrak{H}$.

Let K be a fixed compact set, and ξ be a fixed element of $\mathfrak{H}$. We then have an element $x(\chi_K \otimes \xi)$ of $L^2_{\mathfrak{H}}(\Gamma,\mu)$ supported by K. Let $\mathfrak{K}$ be a separable subspace of $\mathfrak{H}$ and K_{∞} be a measurable subset of K such that $\mu(K - K_{\infty}) = 0$ and $x(\chi_K \otimes \xi)(\gamma) \in \mathfrak{K}$ for every $\gamma \in K_{\infty}$. Let e be the projection of $\mathfrak{H}$ onto $\mathfrak{K}$. We have then $(1 - e)x(\chi_K \otimes \xi)(\gamma) = 0$ for every $\gamma \in K_{\infty}$. Hence we get $F_{\xi,(1-e)\eta}\chi_K = 0$ in $L^{\infty}(\Gamma,\mu)$, which means that

$$\begin{aligned} 0 = \rho(F_{\xi,(1-e)\eta}\chi_K) &= \rho(\chi_K)(\gamma)(x(\gamma)\xi|(1 - e)\eta) \\ &= \rho(\chi_K)(\gamma)((1 - e)x(\gamma)\xi|\eta). \end{aligned}$$

Since $\rho(\chi_K)$ and χ_K differ only on a locally null set N, we get, for every $\gamma \in K - N$ and $\eta \in \mathfrak{H}$, $((1 - e)x(\gamma)\xi|\eta) = 0$; so $(1 - e)x(\gamma)\xi = 0$, $\gamma \in K - N$. Thus $x(\gamma)\xi$ belongs to the separable subspace $\mathfrak{K}$ for every $\gamma \in K - N$. Therefore, the function $x(\cdot)$ satisfies the condition in Lemma 7.5. By symmetry, $x(\cdot)^*$ also satisfies the condition in Lemma 7.5, where we use the fact that the lifting ρ preserves the conjugation. Thus, $x(\cdot)$ is indeed measurable. It is now clear that

$$x = \int_{\Gamma}^{\oplus} x(\gamma)\, d\mu(\gamma). \qquad \text{Q.E.D.}$$

Remark 7.11. Suppose $x(\cdot)$ and $y(\cdot)$ are two bounded $\mathscr{L}(\mathfrak{H})$-valued measurable functions such that

$$\int_{\Gamma}^{\oplus} x(\gamma)\, d\mu(\gamma) = \int_{\Gamma}^{\oplus} y(\gamma)\, d\mu(\gamma).$$

By definition, it follows that for every $\xi \in \mathfrak{H}$, we have $x(\gamma)\xi = y(\gamma)\xi$ locally μ-almost everywhere. But this does not imply that $x(\gamma) = y(\gamma)$ locally μ-almost everywhere. For example, consider $\mathfrak{H} = l^2[0,1]$, $\Gamma = [0,1]$ and the Lebesgue measure μ on Γ. Let $e(t)$ be the projection to the one-dimensional subspace $\mathbf{C}\varepsilon_t$, where $\varepsilon_t(s) = \delta_{s,t}$, $s,t \in [0,1]$. Since $e(t)\xi = 0$ except for countably many t, $t \in \Gamma \mapsto e(t)$ is measurable and $\int_{\Gamma}^{\oplus} e(t)\, d\mu(t) = 0$. But it is impossible to find a null set N in Γ such that $e(t) = 0$ for every $t \notin N$. This phenomenon indicates that there is a serious difficulty in the general theory of direct integrals for nonseparable Hilbert spaces.

Proposition 7.12. *If $\mathfrak{H}$ is separable, then for bounded measurable $\mathscr{L}(\mathfrak{H})$-valued functions $x(\cdot)$ and $y(\cdot)$*

$$\int_\Gamma^\oplus x(\gamma)\, d\mu(\gamma) = \int_\Gamma^\oplus y(\gamma)\, d\mu(\gamma)$$

implies that $x(\gamma) = y(\gamma)$ locally μ-almost everywhere.

Proof. For each $\xi \in \mathfrak{H}$, there exists a locally null set N_ξ in Γ such that $x(\gamma)\xi = y(\gamma)\xi$ for every $\gamma \notin N_\xi$. Let $\{\xi_n\}$ be a countable dense subset of $\mathfrak{H}$ and put $N = \bigcup_{n=1}^\infty N_{\xi_n}$. We then have $x(\gamma)\xi_n = y(\gamma)\xi_n$ for every $\gamma \notin N$, and N is locally null. Since $x(\gamma)$ and $y(\gamma)$ are both bounded, we get $x(\gamma) = y(\gamma)$ for every $\gamma \notin N$. Q.E.D.

We are now going to show that $L^1_E(\Gamma,\mu)$ is naturally identified with $L^1(\Gamma,\mu) \otimes_\gamma E$ for any Banach space E, hence for the predual $\mathscr{M}_*$ of any von Neumann algebra $\mathscr{M}$, $L^1_{\mathscr{M}_*}(\Gamma,\mu) = L^1(\Gamma,\mu) \otimes_\gamma \mathscr{M}_*$. To do this, we need the following:

Lemma 7.13. *Let E be a Banach space. If φ is a bounded linear functional on the projective tensor product $L^1(\Gamma,\mu) \otimes_\gamma E$, then there exists a bounded E^*-valued function: $\gamma \in \Gamma \mapsto \varphi(\gamma) \in E^*$ such that* (i) *for each fixed element $a \in E$ the function: $\gamma \in \Gamma \mapsto \langle a,\varphi(\gamma)\rangle \in \mathbf{C}$ is μ-measurable,* (ii) *for any $a \in E$ and $f \in L^1(\Gamma,\mu)$ we have*

$$\langle f \otimes a,\varphi\rangle = \int_\Gamma f(\gamma)\langle a,\varphi(\gamma)\rangle\, d\mu(\gamma),$$

$$\|\varphi(\gamma)\| \le \|\varphi\|, \qquad \gamma \in \Gamma.$$

Proof. By Theorem 2.3, the conjugate space $(L^1(\Gamma,\mu) \otimes_\gamma E)^*$ is isometrically isomorphic with the Banach space $\mathscr{L}(E,L^1(\Gamma,\mu)^*) = \mathscr{L}(E,L^\infty(\Gamma,\mu))$. Let Φ denote the isomorphism of $(L^1(\Gamma,\mu) \otimes_\gamma E)^*$ onto $\mathscr{L}(E,L^\infty(\Gamma,\mu))$, and ρ be a lifting of $L^\infty(\Gamma,\mu)$ into $\mathscr{L}^\infty(\Gamma,\mu)$. We then have for each $f \in L^1(\Gamma,\mu)$ and $a \in E$,

$$\langle f \otimes a,\varphi\rangle = \int_\Gamma f(\gamma)\rho(\Phi(\varphi)a)(\gamma)\, d\mu(\gamma).$$

For each $\gamma \in \Gamma$, the map: $a \in E \mapsto \rho(\Phi(\varphi)a)(\gamma)$ is a bounded linear functional on E; so we write this functional as $\varphi(\gamma)$. We then have

$$\langle f \otimes a,\varphi\rangle = \int_\Gamma f(\gamma)\langle a,\varphi(\gamma)\rangle\, d\mu(\gamma).$$

The E^*-valued function: $\gamma \in \Gamma \mapsto \varphi(\gamma) \in E^*$ then enjoys the measurability condition described in (i). The boundedness follows from the inequality

$$\begin{aligned}|\langle a,\varphi(\gamma)\rangle| &= |\rho(\Phi(\varphi)a)(\gamma)| \le \|\Phi(\varphi)a\|\\ &\le \|\varphi\|\|a\|.\end{aligned}$$

Q.E.D.

Proposition 7.14. *For any Banach space E,*

$$L^1(\Gamma,\mu) \otimes_\gamma E = L^1_E(\Gamma,\mu)$$

under the obvious identification.

PROOF. By definition, $L^1(\Gamma,\mu) \otimes E$ is dense in both spaces $L^1(\Gamma,\mu) \otimes_\gamma E$ and $L^1_E(\Gamma,\mu)$. Hence we have only to prove the equality

$$\|x\|_\gamma = \int_\Gamma \|x(\gamma)\| \, d\mu(\gamma), \qquad x \in L^1(\Gamma,\mu) \otimes E.$$

Since the $\|\cdot\|_1$-norm on $L^1(\Gamma,\mu) \otimes E$ is clearly a cross-norm, we have the inequality $\|x\|_1 \leq \|x\|_\gamma$. Let $x \in L^1(\Gamma,\mu) \otimes E$ be a fixed element, and φ be an element of $(L^1(\Gamma,\mu) \otimes_\gamma E)^*$ with $\|\varphi\| = 1$ such that $\langle x,\varphi \rangle = \|x\|_\gamma$. By Lemma 7.13, there exists an E^*-valued function: $\gamma \in \Gamma \mapsto \varphi(\gamma) \in E^*$ such that

$$\langle x,\varphi \rangle = \int_\Gamma \langle x(\gamma),\varphi(\gamma) \rangle \, d\mu(\gamma),$$

$$\|\varphi(\gamma)\| \leq \|\varphi\| = 1.$$

We then have

$$\|x\|_\gamma = \langle x,\varphi \rangle = \int_\Gamma \langle x(\gamma),\varphi(\gamma) \rangle \, d\mu(\gamma)$$

$$\leq \int_\Gamma \|x(\gamma)\| \, d\mu(\gamma) = \|x\|_1.$$

Thus $\|x\|_\gamma \leq \|x\|_1$, as desired. Q.E.D.

Lemma 7.15. *If an E^*-valued function $\varphi: \gamma \in \Gamma \mapsto \varphi(\gamma) \in E^*$ satisfies condition* (i) *in the previous proposition, then for each E-valued measurable function $x(\cdot)$ on Γ, the numerical valued function: $\gamma \in \Gamma \mapsto \langle x(\gamma),\varphi(\gamma) \rangle \in \mathbf{C}$ is measurable.*

PROOF. Let K be a compact subset of Γ and $\varepsilon > 0$. By definition, there exists a compact subset K_0 of K such that $\mu(K - K_0) < \varepsilon$ and x is continuous on K_0. Since the image $x(K_0)$ of K_0 is compact in E, $x(K_0)$ is contained in a separable closed subspace E_0 of E. Let I denote the imbedding map of E_0 into E. We then have $\langle x(\gamma),\varphi(\gamma) \rangle = \langle Ix(\gamma),\varphi(\gamma) \rangle = \langle x(\gamma),{}^tI\varphi(\gamma) \rangle$ for every $\gamma \in K_0$. By Proposition 7.2, the E_0^*-valued function: $\gamma \in \Gamma \mapsto {}^tI\varphi(\gamma) \in E_0^*$ is μ-measurable, so that there exists a compact subset K_1 of K_0 such that $\mu(K_0 - K_1) < \varepsilon$ and the map: $\gamma \in K_1 \mapsto {}^tI\varphi(\gamma) \in E_0^*$ is continuous. Therefore, the numerical valued function: $\gamma \in K_1 \mapsto \langle x(\gamma),\varphi(\gamma) \rangle = \langle x(\gamma),{}^tI\varphi(\gamma) \rangle$ is continuous and $\mu(K - K_1) < 2\varepsilon$. Thus our assertion follows. Q.E.D.

Proposition 7.16. *Let E be a Banach space.*

(i) *If $\varphi(\cdot)$ is a bounded E^*-valued function on Γ such that for any $a \in E$, the function: $\gamma \in \Gamma \mapsto \langle a,\varphi(\gamma) \rangle \in \mathbf{C}$ is μ-measurable, then the integral*

$$\langle x,\varphi \rangle = \int_\Gamma \langle x(\gamma),\varphi(\gamma) \rangle \, d\mu(\gamma), \qquad x(\cdot) \in L^1_E(\Gamma,\mu), \tag{7}$$

gives rise to a bounded linear functional φ on $L^1_E(\Gamma,\mu)$.

(ii) *For any* $\varphi \in L^1_E(\Gamma,\mu)^*$, *there exists a bounded* E^**-valued function* $\varphi(\cdot)$ *satisfying the measurability condition in* (i) *such that* φ *is given by integral* (7) *and*

$$\|\varphi\| = \sup\|\varphi(\gamma)\|. \tag{8}$$

Proof. Assertion (i) is an immediate consequence of Lemma 7.15. We have only to prove (ii). Suppose $\varphi \in L^1_E(\Gamma,\mu)^*$ is given. By Proposition 7.14 and the proof of Lemma 7.13, there exists an isometry Φ of $L^1_E(\Gamma,\mu)^*$ onto $\mathcal{L}(E,L^\infty(\Gamma,\mu))$ such that

$$\langle f \otimes a,\varphi\rangle = \langle f,\Phi(\varphi)a\rangle, \qquad f \in L^1(\Gamma,\mu), \qquad a \in E.$$

Let ρ be a lifting of $L^\infty(\Gamma,\mu)$ onto $\mathcal{L}^\infty(\Gamma,\mu)$, and put

$$\langle a,\varphi(\gamma)\rangle = \rho(\Phi(\varphi)a)(\gamma), \qquad \gamma \in \Gamma, \qquad a \in E.$$

By construction, the E^*-valued function $\varphi(\cdot)$ satisfies the measurability condition in (i). We have

$$\langle f \otimes a,\varphi\rangle = \int f(\gamma)\langle a,\varphi(\gamma)\rangle \, d\mu(\gamma).$$

Since the both sides of integral formula (7) are continuous linear functionals on $L^1_E(\Gamma,\mu)$ and agree on the elements of the form $f \otimes a$, formula (7) follows. Since Φ is an isometry, we have, for any $a \in E$,

$$\begin{aligned} |\langle a,\varphi(\gamma)\rangle| &= |\rho(\Phi(\varphi)a)(\gamma)| \\ &\le \|\rho(\Phi(\varphi)a)\| = \|\Phi(\varphi)a\| \\ &\le \|a\|\,\|\varphi\| \end{aligned}$$

Hence we get $\|\varphi(\gamma)\| \le \|\varphi\|$. On the other hand, integral formula (7) readily yields the reversed inequality, $\|\varphi\| \le \sup\|\varphi(\gamma)\|$. Q.E.D.

Theorem 7.17. *Let* $\mathcal{M}$ *be a von Neumann algebra and* Γ *a locally compact space with a positive Radon measure* μ.

(i) *Considering* $L^\infty(\Gamma,\mu)$ *as an abelian von Neumann algebra, we have*

$$(L^\infty(\Gamma,\mu) \bar{\otimes} \mathcal{M})_* = L^1_{\mathcal{M}_*}(\Gamma,\mu);$$

hence for any $x \in L^\infty(\Gamma,\mu) \bar{\otimes} \mathcal{M}$ *there exists an* $\mathcal{M}$*-valued bounded function* $x(\cdot)$ *on* Γ *such that the numerical valued function*: $\gamma \in \Gamma \mapsto \langle x(\gamma),\varphi\rangle \in \mathbf{C}$ *is measurable for every* $\varphi \in \mathcal{M}_*$ *and*

$$\langle x,\varphi\rangle = \int_\Gamma \langle x(\gamma),\varphi(\gamma)\rangle \, d\mu(\gamma), \qquad \varphi \in L^1_{\mathcal{M}_*}(\Gamma,\mu).$$

(ii) *If* $\mathcal{M}$ *acts on a Hilbert space* $\mathfrak{H}$, *then one can choose the above function* $x(\cdot)$ *so that* $x(\cdot)$ *is measurable, in the sense of Definition 7.7, as an* $\mathcal{L}(\mathfrak{H})$*-valued function.*

Proof. Assertion (i) is an immediate consequence of Proposition 7.16.

To prove (ii), we shall use the notations in the proof of Theorem 7.10. The tensor product $L^\infty(\Gamma,\mu)\,\overline{\otimes}\,\mathcal{M}$ acts on the Hilbert space $L^2(\Gamma,\mu)\otimes\mathfrak{H} = L^2_{\mathfrak{H}}(\Gamma,\mu)$, and any operator of $L^\infty(\Gamma,\mu)\,\overline{\otimes}\,\mathcal{M}$ is decomposable since it commutes with the diagonal algebra $L^\infty(\Gamma,\mu) = \mathcal{A}$. If $\varphi = \omega_{\xi,\eta} \in \mathcal{M}_*$, then we have, by construction, $\Phi(\varphi) = F_{\xi,\eta} \in L^\infty(\Gamma,\mu)$. With ρ a fixed lifting of $L^\infty(\Gamma,\mu)$ into $\mathcal{L}^\infty(\Gamma,\mu)$, we have

$$\begin{aligned}\langle x(\gamma),\varphi\rangle &= \rho(\Phi(\varphi))(\gamma) = \rho(F_{\xi,\eta})(\gamma)\\ &= (x(\gamma)\xi|\eta).\end{aligned}$$

Therefore, the $\mathcal{M}$-valued function chosen in the proof of Proposition 7.16 coincides with the $\mathcal{L}(\mathfrak{H})$-valued function chosen in Theorem 7.10. Thus, $x(\cdot)$ is measurable by Theorem 7.10 and takes values in $\mathcal{M}$. Q.E.D.

8. Direct Integral of Hilbert Spaces, Representations, and von Neumann Algebras

In this section we shall discuss a continuous analogue of the direct sum based on integration. Since measure theory is consistent only with countable operations, the theory of the direct integral has a natural restriction, the countability conditions, on its objects, which will force us to consider only separable Hilbert spaces. We begin with a preparatory discussion of Borel spaces.

A *Borel space* is a set Γ equipped with a set $\mathcal{B}$ of subsets of Γ such that $\emptyset \in \mathcal{B}$, the union of any countably many members of $\mathcal{B}$ is a member of $\mathcal{B}$, and the complement of any member of $\mathcal{B}$ is a member of $\mathcal{B}$. Each subset of Γ belonging to $\mathcal{B}$ is called a *Borel set*. Two Borel spaces Γ_1 and Γ_2 are said to be *isomorphic* if there exists a bijection f of Γ_1 onto Γ_2 such that the image and the inverse image of a Borel set under f are both Borel. A topological space is called *Polish* if it is homeomorphic to a separable complete metric space. Given a topological space Γ, we consider the smallest family $\mathcal{B}$ of subsets of Γ which contains all open subsets of Γ and is closed under the operations of countable union and complement. We then obtain a Borel space $\{\Gamma,\mathcal{B}\}$ which will be called the *Borel space generated by the topology of* Γ. A Borel space is said to be *standard* if it is isomorphic to the Borel space of a Polish space generated by the topology. The fundamental properties of a Polish space and a standard Borel space are cited in the Appendix. A measure μ on a Borel space Γ is said to be *standard* if there exists a μ-null subset N of Γ such that $\Gamma - N$ is a standard Borel space with respect to the relative Borel structure inherited from Γ.

Definition 8.1. Given a separable Banach space E, we equip the space $\mathfrak{W}(E^*)$ of all weakly* closed subspaces of E^* with the smallest Borel structure which makes measurable the function: $F \in \mathfrak{W}(E^*) \mapsto \|x|_F\|$ for every $x \in E$. The Borel structure in $\mathfrak{W}(E^*)$ is called the *Effros Borel structure*.

The next theorem may be regarded as an elaborated form of the usual Hahn–Banach theorem. Let X be a subspace of a real normed space E and f a linear functional on X with norm ≤ 1. To extend f to the subspace $X + \mathbf{R}x$ for any $x \notin X$, we consider the following quantities $L(f)$ and $M(f)$:

$$\left.\begin{aligned} L(f) &= \sup\{-\|x+u\| - f(u): u \in X\} \\ M(f) &= \inf\{\|x+v\| - f(v): v \in X\}. \end{aligned}\right\} \tag{1}$$

The usual Hahn–Banach extension arguments show that $L(f) \leq M(f)$ and any number in the interval $[L(f), M(f)]$ can be assigned to x as $f(x)$, namely, we have

$$L(f) \leq f(x) \leq M(f). \tag{2}$$

In other words,

$$f(x) = tL(f) + (1-t)M(f) \quad \text{for some} \quad 0 \leq t \leq 1. \tag{2'}$$

Furthermore, we have, by construction (1),

$$M(f) = -L(-f). \tag{3}$$

Theorem 8.2. *If E is a separable Banach space, then $\mathfrak{W}(E^*)$ with Effros Borel structure is standard, and admits countably many Borel choice functions $f_n: \mathfrak{W}(E^*) \mapsto E_1^*$ such that for any $F \in \mathfrak{W}(E^*)$, $\{f_n(F): n = 1,2,\ldots\}$ is weakly* dense in F_1, where we denote the unit ball of any normed space X by X_1 and consider the Borel structure in E_1^* generated by the weak* topology.*

Proof. Suppose that E is real. For each $F \in \mathfrak{W}(E^*)$, we identify F with $(E/F^\perp)^*$. For each sequence of real numbers $t = (t_1, t_2, \ldots)$, $0 \leq t_i \leq 1$, we shall construct a function $f_t^F \in (E/F^\perp)_1^* = F_1$. Let $\{x_n\}$ be a fixed dense sequence in E with $x_1 = 0$. Let $x_n(F) = x_n|_F \in E/F^\perp$ and $X_n(F)$ be the linear subspace spanned by $x_1(F), \ldots, x_n(F)$ in $E/F^\perp$. Define $f_{t_1}^F(0) = 0$. Suppose that we have constructed $f_{t_1,\ldots,t_n}^F$ to be an element of $X_n(F)_1^*$. Putting $X_n(F) = X$, $f_{t_1,\ldots,t_n}^F = f$ and $x_{n+1}(F) = x$ in our previous arguments, we define

$$f_{t_1,t_2,\ldots,t_{n+1}}^F(x) = t_{n+1}L(f) + (1-t_{n+1})M(f);$$

so

$$f_{t_1,\ldots,t_{n+1}}^F(\lambda x + w) = \lambda f_{t_1,\ldots,t_{n+1}}^F(x) + f(w), \qquad \lambda \in \mathbf{R}, \qquad w \in X_n(F).$$

Thus we obtain an extension $f_{t_1,\ldots,t_{n+1}}^F$ of f to an element of $X_{n+1}(F)_1^*$. We then define f_t^F on $\bigcup_{n=1}^\infty X_n(F)$ naturally and further extend it to the whole space $E/F^\perp$ by continuity.

Our previous discussion shows that any element in $(E/F^\perp)_1^*$ must be of the form f_t^F for some $t = (t_1, t_2, \ldots)$. We shall show that the countable family of functions f_r^F, $r = (r_1, r_2, \ldots)$ with the r_i rational and all but a finite number equal to 0, is weakly* dense in $(E/F^\perp)_1^* = F_1$. Since $\{x_n(F)\}$ is dense in $E/F^\perp$, it suffices to show that $\{f_r^F(x_n(F))\}_{n=1}^\infty$ approximates $\{f_t^F(x_n(F))\}_{n=1}^\infty$ arbitrarily well termwise for any $t = (t_1, t_2, \ldots)$. This is equivalent to showing that for each $n = 1,2,\ldots$, $\{f_r^F(x_1(F)), \ldots, f_r^F(x_n(F))\}$ approximates $\{f_t^F(x_1(F)), \ldots, f_n^F(x_n(F))\}$ arbitrarily well. Since this is trivial for $n = 1$ because $f_t^F(x_1(F)) = 0$, we assume that this is true for n. Since $\dim X_n(F) \leq n$, $f_r^F|_{X_n(F)}$ can approximate $f_t^F|_{X_n(F)}$ in norm. Since L is a convex function on $X_n(F)^*$, it is continuous, and hence M is also continuous on $X_n(F)^*$. Thus $L(f_{r_1,\ldots,r_n}^F)$ and $M(f_{r_1,\ldots,r_n}^F)$ approximates $L(f)$ and $M(f)$, respectively, where $f = f_t^F$. By a suitable choice of r_{n+1}, $f_{r_1,r_2,\ldots,r_{n+1}}^F|_{X_{n+1}(F)}$ approximates $f|_{X_{n+1}(F)}$. Thus $\{f_r^F\}$ is weakly* dense in F_1.

We now show that the map: $F \in \mathfrak{W}(E^*) \mapsto f_t^F \in E_1^*$ is Borel for any $t = (t_1, t_2, \ldots)$. Since $\{x_n\}$ is dense in E, it suffices to show that each map: $F \in \mathfrak{W}(E^*) \mapsto f_t^F(x_n) \in \mathbf{R}$ is Borel. This is trivial for $n = 1$, being constantly 0. Suppose that it is true for $k \leq n$. We then have

$$\begin{aligned} f_t^F(x_{n+1}) &= f_{t_1,\ldots,t_{n+1}}^F(x_{n+1}(F)) \\ &= t_{n+1}L(f_{t_1,\ldots,t_n}^F) + (1 - t_{n+1})M(f_{t_1,\ldots,t_n}^F), \end{aligned}$$

$$L(f_{t_1,\ldots,t_n}^F) = \sup\{-\|(x_{n+1} + u)|_F\| - f_t^F(u) : u \in X_n(F)\}.$$

By definition, the function: $F \in \mathfrak{W}(E^*) \mapsto \|(x_{n+1} + u)|_F\| \in \mathbf{R}$ is Borel, and hence, by the induction hypothesis, the function:

$$F \in \mathfrak{W}(E^*) \mapsto -\|(x_n + u)|_F\| - f_t^F(u) \in \mathbf{R}$$

is Borel. Since $\dim X_n(F) \leq n$, $L(f_{t_1,\ldots,t_n}^F)$ is the supremum of a countable number of Borel functions of F, so that it is Borel. Thus, $F \mapsto f_t^F(x_{n+1})$ is Borel, as required. Therefore, for each n, $F \mapsto f_t^F(x_n)$ is Borel. Hence the map: $F \mapsto f_t^F \in E_1^*$ is Borel.

Suppose that E is a complex Banach space. Let $E_{\mathfrak{R}}$ be the real Banach space obtained by considering E as a real vector space. For each $f \in E^*$, put

$$Tf(x) = \operatorname{Re} f(x), \qquad x \in E.$$

It follows then that T is a real linear isometry of E^* onto $(E_{\mathbf{R}})^*$ and that the inverse T^{-1} is given by

$$T^{-1}g(x) = g(x) - ig(ix), \qquad x \in E, \qquad g \in (E_{\mathbf{R}})^*.$$

Clearly T is a homomorphism with respect to the weak* topologies. Let $\{g_n\}$ be a countable family of Borel choice functions of $\mathfrak{W}(E_{\mathbf{R}}^*)$ into $(E_{\mathbf{R}}^*)_1$ satisfying the condition in our theorem. Put

$$f_n(F) = T^{-1}g_n(TF), \qquad F \in \mathfrak{W}(E^*).$$

For each $x \in E_{\mathbf{R}}$, we have

$$\begin{aligned} \|x|_{TF}\| &= \sup\{|Tf(x)| : f \in F_1\}, \qquad F \in \mathfrak{W}(E^*) \\ &= \sup\{|\operatorname{Re} f(x)| : f \in F_1\} \\ &= \sup\{|f(x)| : f \in F_1\} = \|x|_F\|. \end{aligned}$$

Hence the function: $F \in \mathfrak{W}(E^*) \mapsto \|x|_{TF}\|$ is Borel, which means that the map: $F \in \mathfrak{W}(E^*) \mapsto TF \in \mathfrak{W}((E_{\mathbf{R}})^*)$ is Borel. Since T^{-1} is weakly* continuous, f_n is a Borel choice function for each $n = 1,2,\ldots$. Since $T(F_1) = (TF)_1$, $\{f_n(F)\}$ is weakly* dense in F_1 because $\{g_n(TF)\}$ is dense in $(TF)_1$.

We now show that $\mathfrak{W}(E^*)$ is standard. Let $\mathscr{N}(E)$ be the space of all closed subspaces of E. It follows that the map: $F \in \mathfrak{W}(E^*) \mapsto F^\perp \in \mathscr{N}(E)$ is a bijection and $\|x|_F\| = \inf\{\|x - y\| : y \in F^\perp\} = d(x,F^\perp)$ for each $x \in E$. Let $\mathscr{C}_0(E)$ be the space of all closed subsets of E equipped with the Borel structure which makes measurable the function: $A \in \mathscr{C}_0(E) \mapsto d(x,A) = \inf\{d(x,y) : y \in A\}$, $x \in E$. Clearly, $\mathscr{N}(E)$ is contained in $\mathscr{C}_0(E)$. Let $\{U_n\}$ be a countable base of open sets in E, and $\{\lambda_n\}$ be a dense sequence in the scalar field (**R** or **C**). For any open set, put $\mathscr{U}(U) = \{A \in \mathscr{C}_0(E) : A \cap U \neq \varnothing\}$. We then have

$$\mathscr{N}(E) = \bigcap_{m,n=1}^{\infty} [(\mathscr{U}(U_m) \cap \mathscr{U}(U_n))^c \cup \mathscr{U}(U_n + U_m)]$$

$$\cap \left\{ \bigcap_{m,n=1}^{\infty} [\mathscr{U}(U_m)^c \cup \mathscr{U}(\lambda_n U_m)] \right\}.$$

Since $\mathscr{U}(U)$ is a Borel set in $\mathscr{C}_0(E)$ (see Corollary A.17, i.e., Corollary 17 in the Appendix), $\mathscr{N}(E)$ is a Borel set of $\mathscr{C}_0(E)$; thus it is a standard Borel space. Q.E.D.

Corollary 8.3. *Let E be a separable Banach space and Γ a Borel space. A map: $\gamma \in \Gamma \mapsto F(\gamma) \in \mathfrak{W}(E^*)$ is Borel if and only if there exists a countable family $\{f_n\}$ of Borel functions of Γ into E_1^* such that for each $\gamma \in \Gamma$, $\{f_n(\gamma)\}$ is weakly* dense in $F(\gamma)_1$.*

Proof. If $\gamma \in \Gamma \mapsto F(\gamma)$ is Borel, then the functions $f_n(\gamma)$ are obtained by composing this map with the choice functions of Theorem 8.2.

Suppose that such functions exist. We then have, for each $x \in E$,

$$\|x|_{F(\gamma)}\| = \sup\{|f_n(\gamma)(x)| : n = 1,2,\ldots\}.$$

Hence $\gamma \in \Gamma \mapsto \|x|_{F(\gamma)}\|$ is Borel for each $x \in E$, which means, by definition, that $\gamma \in \Gamma \mapsto F(\gamma) \in \mathfrak{W}(E^*)$ is Borel. Q.E.D.

We now consider a *separable* Hilbert space $\mathfrak{H}$. Throughout the rest of this section, we assume always separability for Hilbert spaces unless otherwise explicitly indicated. We take $\mathscr{L}(\mathfrak{H})_*$, the Banach space of all σ-weakly continuous linear functionals on $\mathscr{L}(\mathfrak{H})$, as the separable Banach space E in the previous discussion, and consider the space $\mathfrak{W}(\mathscr{L}(\mathfrak{H}))$ with Effros Borel structure. For each $\mathscr{M} \in \mathfrak{W}(\mathscr{L}(\mathfrak{H}))$, we write $\mathscr{M}^*$ for the set of the adjoints of elements in $\mathscr{M}$ (not for the Banach space conjugate space) and, as usual, $\mathscr{M}'$ for the commutant of $\mathscr{M}$.

Theorem 8.4. *If $\mathfrak{H}$ is a separable Hilbert space, then the maps:*

$$\mathscr{M} \in \mathfrak{W}(\mathscr{L}(\mathfrak{H})) \mapsto \mathscr{M}^* \in \mathfrak{W}(\mathscr{L}(\mathfrak{H})) \quad \text{and} \quad \mathscr{M} \in \mathfrak{W}(\mathscr{L}(\mathfrak{H})) \mapsto \mathscr{M}' \in \mathfrak{W}(\mathscr{L}(\mathfrak{H}))$$

are both Borel with respect to the Effros Borel structure.

PROOF. For each $\varphi \in \mathscr{L}(\mathfrak{H})_*$ we have

$$\begin{aligned}\|\varphi|_{\mathscr{M}_*}\| &= \sup\{|\varphi(x)|: x \in \mathscr{M}_1^*\}\\ &= \sup\{|\varphi(x^*)|: x \in \mathscr{M}_1\}\\ &= \sup\{|\varphi^*(x)|: x \in \mathscr{M}_1\} = \|\varphi^*|_{\mathscr{M}}\|.\end{aligned}$$

Hence the function: $\mathscr{M} \in \mathfrak{W}(\mathscr{L}(\mathfrak{H})) \mapsto \|\varphi|_{\mathscr{M}*}\|$, $\varphi \in \mathscr{L}(\mathfrak{H})_*$, is Borel, which means that $\mathscr{M} \mapsto \mathscr{M}^*$ is Borel.

By Theorem 8.2, let $\{a_n(\mathscr{M})\}$ be a countable family of Borel choice functions such that $\{a_n(\mathscr{M})\}$ is σ-weakly dense in $\mathscr{M}_1$. We then have

$$\mathscr{M}' = \{x \in \mathscr{L}(\mathfrak{H}): xa_n(\mathscr{M}) - a_n(\mathscr{M})x = 0,\ n = 1,2,\ldots\}.$$

Put $\mathscr{A} = l^\infty \overline{\otimes} \mathscr{L}(\mathfrak{H})$ and $\mathscr{A}_* = l^1 \otimes_\gamma \mathscr{L}(\mathfrak{H})_*$. It follows then that $\mathscr{A}$ is the algebra of all bounded sequences in $\mathscr{L}(\mathfrak{H})$ and $\mathscr{A}_*$ is the Banach space of all summable sequences in $\mathscr{L}(\mathfrak{H})_*$. The duality between $\mathscr{A}$ and $\mathscr{A}_*$ is given by

$$\langle\{x_n\},\{f_n\}\rangle = \sum_{n=1}^{\infty} f_n(x_n), \qquad \{x_n\} \in \mathscr{A}, \qquad \{f_n\} \in \mathscr{A}_*.$$

For each $\mathscr{M} \in \mathfrak{W}(\mathscr{L}(\mathfrak{H}))$, we define a linear map $T_{\mathscr{M}}: \mathscr{L}(\mathfrak{H}) \mapsto \mathscr{A}$ by the formula

$$T_{\mathscr{M}}(x) = \{xa_n(\mathscr{M}) - a_n(\mathscr{M})x\}.$$

We then have $\mathscr{M}' = T_{\mathscr{M}}^{-1}(0)$. Since

$$T_{\mathscr{M}}(x) = (1 \otimes x)\{a_n(\mathscr{M})\} - \{a_n(\mathscr{M})\}(1 \otimes x),$$

$T_{\mathscr{M}}$ is σ-weakly continuous, and its transpose ${}^tT_{\mathscr{M}}$ on $\mathscr{A}_*$ is given by

$${}^tT_{\mathscr{M}}(\{f_n\}) = \sum_{n=1}^{\infty} \{a_n(\mathscr{M})f_n - f_n a_n(\mathscr{M})\}, \qquad \{f_n\} \in \mathscr{A}_*.$$

Furthermore, we know that $T_{\mathscr{M}}^{-1}(0)^\perp$ is the closure of the range of ${}^tT_{\mathscr{M}}$. Hence if $\{b_n\}$ is a σ-weakly dense sequence in $\mathscr{L}(\mathfrak{H})_1$ and $\{g_j = \{f_n^j\}: j = 1,2,\ldots\}$ is norm dense in $\mathscr{A}_*$, we have, for any $f \in \mathscr{L}(\mathfrak{H})_*$,

$$\begin{aligned}\|f|_{\mathscr{M}'}\| &= \inf\{\|f - g\|: g \in (\mathscr{M}')^\perp\}\\ &= \inf\{\|f - {}^tT_{\mathscr{M}}(g_j)\|: j = 1,2,\ldots\},\end{aligned}$$

$$\begin{aligned}\|f - {}^tT_{\mathscr{M}}(g_j)\| &= \sup\{|f(b_n) - {}^tT_{\mathscr{M}}(g_j)(b_n)|: n = 1,2,\ldots\}\\ &= \sup\left\{\left|f(b_n) - \sum_{k=1}^{\infty} f_k^j(b_n a_k(\mathscr{M}) - a_k(\mathscr{M})b_n)\right|: n = 1,2,\ldots\right\}.\end{aligned}$$

Since $\mathscr{M} \mapsto a_n(\mathscr{M})$ is σ-weakly Borel, $\mathscr{M} \mapsto \|f|_{\mathscr{M}'}\|$ is Borel for every $f \in \mathscr{L}(\mathfrak{H})_*$; thus the map: $\mathscr{M} \mapsto \mathscr{M}'$ is Borel by definition. Q.E.D.

Corollary 8.5. *With the above notation, the set $\mathfrak{A}$ of all von Neumann algebras on a separable Hilbert space $\mathfrak{H}$ is a Borel subset of $\mathfrak{W}(\mathscr{L}(\mathfrak{H}))$, hence a standard Borel space.*

Proof. By Theorem 8.4, $\mathfrak{A}$ is the set of all fixed elements of $\mathfrak{W}(\mathscr{L}(\mathfrak{H}))$ under the two Borel maps: $\mathscr{M} \mapsto \mathscr{M}^*$ and $\mathscr{M} \mapsto \mathscr{M}''$. Let $\{f_n\}$ be a countable norm dense set in $\mathscr{L}(\mathfrak{H})_*$. We then have

$$\mathfrak{A} = \{\mathscr{M} \in \mathfrak{W}(\mathscr{L}(\mathfrak{H})): \|f_n|_{\mathscr{M}}\| = \|f_n|_{\mathscr{M}^*}\|, \text{ and } \|f_n|_{\mathscr{M}}\| = \|f_n|_{\mathscr{M}''}\|, n = 1,2,\dots\}.$$

Hence $\tilde{\mathfrak{A}}$ is a Borel subset of $\mathfrak{W}(\mathscr{L}(\mathfrak{H}))$. Q.E.D.

Corollary 8.6. *The maps of $\mathfrak{A} \times \mathfrak{A}$ defined by $(\mathscr{M},\mathscr{N}) \mapsto \mathscr{M} \cap \mathscr{N}$ and $(\mathscr{M},\mathscr{N}) \mapsto \mathscr{M} \vee \mathscr{N} = (\mathscr{M} \cup \mathscr{N})''$ are both Borel.*

Proof. As $\mathscr{M} \cap \mathscr{N} = (\mathscr{M}' \vee \mathscr{N}')'$, we have only to prove the second assertion. By Theorem 8.2, there exist countably many Borel choice functions $a_i : \mathscr{M} \in \tilde{\mathfrak{A}} \mapsto a_i(\mathscr{M}) \in \mathscr{L}(\mathfrak{H})$ such that $\{a_i(\mathscr{M}) : i = 1,2,\dots\}$ is dense in $\mathscr{M}_1$. Let $\{p_k\}$ be an enumeration of polynomials of noncommuting variables $x_1, x_2, \dots ; x_1^*, x_2^*, \dots ; y_1, \dots, y_n, \dots ;$ and $y_1^*, y_2^*, \dots$ with rational coefficients. For each $(\mathscr{M},\mathscr{N}) \in \mathfrak{A} \times \mathfrak{A}$, we put

$$b_k(\mathscr{M},\mathscr{N}) = p_k(\{a_i(\mathscr{M})\},\{a_i(\mathscr{M})^*\}; \{a_i(\mathscr{N})\},\{a_i(\mathscr{N})^*\})$$

if it is of norm ≤ 1, and $b_k(\mathscr{M},\mathscr{N}) = 0$ if the above expression on the right-hand side is of norm >1. From Theorem II.4.8, $\{b_k(\mathscr{M},\mathscr{N})\}$ is σ-weakly dense in $(\mathscr{M} \vee \mathscr{N})_1$. The map: $(\mathscr{M},\mathscr{N}) \mapsto b_k(\mathscr{M},\mathscr{N})$ is Borel, so that our assertion follows from Theorem 8.2. Q.E.D.

Corollary 8.7. *The set $\mathfrak{F}$ of all factors on a separable Hilbert space $\mathfrak{H}$ is a Borel subset of $\mathfrak{A}$. Hence it is a standard Borel space.*

Proof. Clearly, $\mathfrak{F}$ is the inverse image of **C** under the Borel map: $\mathscr{M} \in \mathfrak{A} \mapsto \mathscr{M} \cap \mathscr{M}' \in \mathfrak{A}$. Hence it is Borel. Q.E.D.

Remark 8.8. The argument in the proof of Corollary 8.6 shows that a map: $\gamma \in \Gamma \mapsto \mathscr{M}(\gamma) \in \mathfrak{A}$ of a Borel space Γ into $\mathfrak{A}$ is Borel if and only if there exists a countable family $\{a_i(\cdot)\}$ of Borel functions of Γ into $\mathscr{L}(\mathfrak{H})$ such that $\{a_i(\gamma)\}$ generates $\mathscr{M}(\gamma)$ for each $\gamma \in \Gamma$.

In order to generalize the notion of a direct sum to a "continuous direct sum," which will be called a direct integral, we must have a proper definition of measurability for functions with values in different spaces. Since we handle spaces indexed by points of a measure space, we have to specify how spaces sitting on distinct points are bound together. This leads us to the following:

Definition 8.9. Let Γ be a Borel space equipped with a σ-finite Borel measure μ. A *measurable field* of Hilbert spaces on $\{\Gamma,\mu\}$ is a family $\{\mathfrak{H}(\gamma) : \gamma \in \Gamma\}$ of Hilbert spaces indexed by Γ together with a subspace $\mathfrak{M}$ of the product vector space $\prod_{\gamma \in \Gamma} \mathfrak{H}(\gamma)$ with the following properties:

(i) For any $\xi \in \mathfrak{M}$, the function: $\gamma \in \Gamma \mapsto \|\xi(\gamma)\|$ is μ-measurable.
(ii) For any $\eta \in \prod \mathfrak{H}(\gamma)$, if the function: $\gamma \in \Gamma \mapsto (\xi(\gamma)|\eta(\gamma)) \in \mathbf{C}$ is μ-measurable for every $\xi \in \mathfrak{M}$, then η belongs to $\mathfrak{M}$.
(iii) There exists a countable subset $\{\xi_1,\xi_2,\ldots,\xi_n,\ldots\}$ of $\mathfrak{M}$ such that for every $\gamma \in \Gamma$, $\{\xi_n(\gamma):n = 1,2,\ldots\}$ is total in $\mathfrak{H}(\gamma)$.

The field of vectors belonging to $\mathfrak{M}$ is called *measurable*. The family in (iii) is called a *fundamental sequence of μ-measurable vector fields.*

It follows from (iii) that each Hilbert space $\mathfrak{H}(\gamma)$ is separable. By the polarization identity and (i), for each $\xi,\eta \in \mathfrak{M}$, the function: $\gamma \in \Gamma \mapsto (\xi(\gamma)|\eta(\gamma))$ is measurable.

Lemma 8.10. *Let $\{\Gamma,\mu\}$ be as above, and $\{\mathfrak{H}(\gamma):\gamma \in \Gamma\}$ be a family of Hilbert spaces. If $\{\xi_n\}$ is a sequence in $\prod_{\gamma\in\Gamma} \mathfrak{H}(\gamma)$ such that* (a) *for every n and m, the function: $\gamma \in \Gamma \mapsto (\xi_n(\gamma)|\xi_m(\gamma)) \in \mathbf{C}$ is μ-measurable, and* (b) *for each $\gamma \in \Gamma$ the sequence $\{\xi_n(\gamma)\}$ of vectors in $\mathfrak{H}(\gamma)$ is total in $\mathfrak{H}(\gamma)$, then the set*

$$\mathfrak{M} = \{\xi \in \prod \mathfrak{H}(\gamma):\gamma \in \Gamma \mapsto (\xi(\gamma)|\xi_n(\gamma)) \quad \text{is } \mu\text{-measurable} \\ \text{for every} \quad n = 1,2,\ldots\}$$

satisfies conditions (i), (ii), *and* (iii) *in Definition* 8.9.

Proof. It is clear that $\mathfrak{M}$ is a linear subspace of $\prod \mathfrak{H}(\gamma)$ containing $\{\xi_n\}$. Hence conditions (ii) and (iii) hold. Considering linear combinations of $\{\xi_n\}$ with complex rational coefficients, we may assume that $\{\xi_n(\gamma)\}$ is dense in $\mathfrak{H}(\gamma)$ for each $\gamma \in \Gamma$. If $\xi \in \mathfrak{M}$, then we have

$$\|\xi(\gamma)\| = \sup_n \frac{|(\xi(\gamma)|\xi_n(\gamma))|}{\|\xi_n(\gamma)\|}, \qquad \gamma \in \Gamma,$$

where we assign value 0 to $|(\xi(\gamma)|\xi_n(\gamma))|/\|\xi_n(\gamma)\|$ if $\|\xi_n(\gamma)\| = 0$. Thus condition (i) holds. Q.E.D.

Corollary 8.11. *If A is a separable C*-algebra, then for any positive Radon measure μ on the quasi-state space $\mathcal{Q}(A)$ of A, the association: $\varphi \in \mathcal{Q}(A) \mapsto \mathfrak{H}_\varphi$ of the cyclic Hilbert space to each $\varphi \in \mathcal{Q}(A)$ turns out to be a measurable field of Hilbert spaces on $\{\mathcal{Q}(A),\mu\}$ by considering $\{\varphi \in \mathfrak{S}(A) \mapsto \eta_\varphi(x_n) \in \mathfrak{H}_\varphi: n = 1,2,\ldots\}$ with $\{x_n\}$ a dense sequence of A.*

Lemma 8.12. *Let $\{\Gamma,\mu\}$ be as before and $\{\mathfrak{H}(\gamma):\gamma \in \Gamma\}$ a measurable field of Hilbert spaces on $\{\Gamma,\mu\}$. Then the function: $\gamma \in \Gamma \mapsto n(\gamma) = \dim \mathfrak{H}(\gamma)$ is measurable and there exists a fundamental sequence $\{\xi_n\}$ of measurable vector fields such that*

(a) *$\{\xi_1(\gamma),\ldots,\xi_{n(\gamma)}(\gamma)\}$ is a normalized orthogonal basis of $\mathfrak{H}(\gamma)$ for every $\gamma \in \Gamma$,*
(b) *$\xi_{n(\gamma)+k}(\gamma) = 0$, $k = 1,2,\ldots$, if $n(\gamma) < +\infty$.*

PROOF. Let $\{\xi_n'\}$ be a fundamental sequence of measurable vector fields with $\xi_0' = 0$, and let $\mathfrak{H}_0(\gamma) = \{0\}$ for each $\gamma \in \Gamma$. Suppose that $\{\xi_1, \dots, \xi_k\}$ are measurable vector fields such that $\{\xi_1(\gamma), \dots, \xi_k(\gamma)\}$ is a normalized orthogonal system if $k \le n(\gamma)$, and $\xi_{n(\gamma)+1}(\gamma) = \cdots = \xi_k(\gamma) = 0$ if $k > n(\gamma)$. Let $\mathfrak{H}_k(\gamma) = [\xi_1(\gamma), \dots, \xi_k(\gamma)] \subset \mathfrak{H}(\gamma)$ for each $\gamma \in \Gamma$, and let $e_k(\gamma)$ denote the projection of $\mathfrak{H}(\gamma)$ onto $\mathfrak{H}_k(\gamma)$. If ξ is a measurable vector field, then the vector field

$$\gamma \in \Gamma \mapsto e_k(\gamma)\xi(\gamma) = \sum_{i=1}^{k} (\xi(\gamma)|\xi_i(\gamma))\xi_i(\gamma)$$

is measurable. Let $E_1 = \{\gamma \in \Gamma : (1 - e_k(\gamma))\xi_1'(\gamma) \neq 0\}$ and

$$E_j = \{\gamma \in \Gamma : (1 - e_k(\gamma))\xi_j'(\gamma) \neq 0,$$
$$(1 - e_k(\gamma))\xi_1'(\gamma) = \cdots = (1 - e_k(\gamma))\xi_{j-1}'(\gamma) = 0\}$$

for $j = 2,3, \dots$. Clearly, $\{E_j\}$ is a disjoint family of measurable subsets of Γ. We put

$$\xi_{k+1}(\gamma) = \begin{cases} 0 & \text{if} \quad \gamma \notin \bigcup_{j=1}^{\infty} E_j; \\ \dfrac{1}{\|(1 - e_k(\gamma))\xi_j'(\gamma)\|}(1 - e_k(\gamma))\xi_j'(\gamma) & \text{if} \quad \gamma \in E_j. \end{cases}$$

It follows that $\xi_{k+1}(\cdot)$ is a measurable vector field and $\{\xi_1, \dots, \xi_{k+1}\}$ satisfies our induction hypothesis. Since $\{\gamma : n(\gamma) \le k\} = (\bigcup_{j=1}^{\infty} E_j)^c$ is measurable, the function: $\gamma \mapsto n(\gamma)$ is measurable. Since our construction of $\{\xi_j(\gamma)\}$ for each $\gamma \in \Gamma$ is nothing else but the Gram–Schmidt orthogonalization of $\{\xi_n'(\gamma)\}$, $\{\xi_j(\gamma)\}$ is a normalized orthogonal basis of $\mathfrak{H}(\gamma)$. Q.E.D.

Theorem 8.13. *Let Γ be a Borel space equipped with a positive Borel measure μ, and $\mathfrak{H}_0$ be a fixed infinite dimensional separable Hilbert space. A field $\{\mathfrak{H}(\gamma) : \gamma \in \Gamma\}$ of Hilbert spaces on Γ is measurable with respect to the measurable vector fields $\mathfrak{M}$ if and only if for each $\gamma \in \Gamma$ there exists an isometry $U(\gamma)$ of $\mathfrak{H}(\gamma)$ into $\mathfrak{H}_0$ such that the map: $\gamma \in \Gamma \mapsto U(\gamma)\mathfrak{H}(\gamma) \in \mathfrak{W}(\mathfrak{H}_0)$ is measurable with respect to the Effros Borel structure in $\mathfrak{W}(\mathfrak{H}_0)$. Furthermore, the family $\mathfrak{M}$ of measurable vector fields is given as the family of vector fields ξ such that $\gamma \in \Gamma \mapsto U(\gamma)\xi(\gamma) \in \mathfrak{H}_0$ is measurable.*

PROOF. From Theorem 8.2, it follows that if a family $\{U(\gamma)\}$ of isometries satisfying our condition is given, then $\mathfrak{M} = \{\xi \in \prod \mathfrak{H}(\gamma) : \gamma \in \Gamma \mapsto U(\gamma)\xi(\gamma) \in \mathfrak{H}_0$ is measurable$\}$ is indeed the family of measurable vector fields satisfying conditions (i), (ii), and (iii) in Definition 8.9.

Conversely, if the family of measurable vector fields $\mathfrak{M}$ is given, we choose a fundamental sequence $\{\xi_n\}$ of measurable vector fields satisfying the conditions in Lemma 8.12. Let $\{\varepsilon_n\}$ be a fixed normalized orthogonal basis of $\mathfrak{H}_0$. For each $\gamma \in \Gamma$, we define

$$U(\gamma)\xi = \sum_{n=1}^{\infty} (\xi|\xi_n(\gamma))\varepsilon_n, \qquad \xi \in \mathfrak{H}(\gamma).$$

Clearly, $U(\gamma)$ is an isometry of $\mathfrak{H}(\gamma)$ into $\mathfrak{H}$, and for each $\xi(\cdot) \in \prod \mathfrak{H}(\gamma)$, ξ is measurable if and only if $\gamma \in \Gamma \mapsto U(\gamma)\xi(\gamma) \in \mathfrak{H}_0$ is also. Q.E.D.

We shall assume from now on that a measurable field $\{\mathfrak{H}(\gamma) : \gamma \in \Gamma\}$ of Hilbert spaces on a Borel space Γ equipped with a positive σ-finite measure μ is given. Unless there is a danger of confusion, we will not specify the family $\mathfrak{M}$ of measurable fields. Let $\mathfrak{H}$ be the collection of measurable vector fields ξ such that

$$\|\xi\| = \left\{ \int_\Gamma \|\xi(\gamma)\|^2 \, d\mu(\gamma) \right\}^{1/2} < +\infty. \tag{4}$$

With respect to the natural pointwise linear operation, $\mathfrak{H}$ is a vector space and the sesquilinear form

$$(\xi|\eta) = \int_\Gamma (\xi(\gamma)|\eta(\gamma)) \, d\mu(\gamma), \qquad \xi,\eta \in \mathfrak{H}, \tag{4'}$$

makes $\mathfrak{H}$ a pre-Hilbert space, where we identify two vector fields $\xi,\eta \in \mathfrak{H}$ if $\xi(\gamma) = \eta(\gamma)$ μ-almost everywhere. The usual Riesz–Fisher type of arguments shows that $\mathfrak{H}$ is indeed complete, hence a Hilbert space.

Definition 8.14. We call the Hilbert space $\mathfrak{H}$ the *direct integral* of measurable fields of Hilbert spaces and denote it by

$$\mathfrak{H} = \int_\Gamma^\oplus \mathfrak{H}(\gamma) \, d\mu(\gamma).$$

Each vector $\xi \in \mathfrak{H}$ is then written as

$$\xi = \int_\Gamma^\oplus \xi(\gamma) \, d\mu(\gamma).$$

When $\mathfrak{H}(\gamma) = \mathfrak{H}_0$ for every $\gamma \in \Gamma$ and $\mathfrak{M}$ is the family of measurable $\mathfrak{H}_0$-valued functions, the field $\{\mathfrak{H}(\gamma) : \gamma \in \Gamma\}$ is said to be a *constant field* of Hilbert spaces. If this happens to be the case, then it is clear that

$$\int_\Gamma^\oplus \mathfrak{H}(\gamma) \, d\mu(\gamma) = L^2_{\mathfrak{H}_0}(\Gamma,\mu).$$

Definition 8.15. Given two measurable fields $\{\mathfrak{H}(\gamma) : \gamma \in \Gamma\}$ and $\{\mathfrak{K}(\gamma) : \gamma \in \Gamma\}$ of Hilbert spaces on $\{\Gamma,\mu\}$, an operator field $x(\cdot) \in \prod_{\gamma \in \Gamma} \mathscr{L}(\mathfrak{H}(\gamma),\mathfrak{K}(\gamma))$ is called *measurable* if for any measurable vector field ξ in $\{\mathfrak{H}(\gamma)\}$, the vector field: $\gamma \in \Gamma \mapsto x(\gamma)\xi(\gamma) \in \mathfrak{K}(\gamma)$ is measurable. If a measurable operator field $x(\cdot)$ is essentially bounded in the sense that $\|x(\cdot)\| \in L^\infty(\Gamma,\mu)$, then for each $\xi = \int_\Gamma^\oplus \xi(\gamma) \, d\mu(\gamma) \in \int_\Gamma^\oplus \mathfrak{H}(\gamma) \, d\mu(\gamma)$ a new vector

$$x\xi = \int_\Gamma^\oplus x(\gamma)\xi(\gamma) \, d\mu(\gamma)$$

is a vector in $\int_\Gamma^\oplus \mathfrak{K}(\gamma) \, d\mu(\gamma)$. We write this operator as

$$x = \int_\Gamma^\oplus x(\gamma) \, d\mu(\gamma)$$

and call it the *direct integral* of the bounded measurable operator field $x(\cdot)$. The operators of this form are said to be *decomposable*. If each $x(\gamma)$ is a scalar, then x is called a *diagonal operator*. The algebra of all diagonal operators is called the *diagonal algebra*.

Let $\{\mathfrak{H}(\gamma):\gamma \in \Gamma\}$ be a measurable field of Hilbert spaces on $\{\Gamma,\mu\}$ as above. Let $\{U(\gamma):\gamma \in \Gamma\}$ be the family of isometries given in Theorem 8.13. With respect to the given measurability in $\{\mathfrak{H}(\gamma):\gamma \in \Gamma\}$ and the constant field $\mathfrak{H}_0$, $\{U(\gamma)\}$ is a measurable field of isometries. Set

$$U = \int_\Gamma^\oplus U(\gamma)\, d\mu(\gamma).$$

It then follows that U is an isometry of $\mathfrak{H} = \int_\Gamma^\oplus \mathfrak{H}(\gamma)\, d\mu(\gamma)$ into $L^2_{\mathfrak{H}_0}(\Gamma,\mu)$ such that UxU^*, $x \in \mathscr{L}(\mathfrak{H})$, is diagonal in $L^2_{\mathfrak{H}_0}(\Gamma,\mu)$ if and only if x is also, and is decomposable if and only if x is also. Thus Theorem 7.10 implies the following:

Corollary 8.16. *Let $\mathfrak{H} = \int_\Gamma^\oplus \mathfrak{H}(\gamma)\, d\mu(\gamma)$. A bounded operator x on $\mathfrak{H}$ is decomposable if and only if it commutes with the algebra of all diagonal operators.*

Definition 8.17. Let $\{\Gamma,\mu\}$ be as before, and $\{\mathfrak{H}(\gamma):\gamma \in \Gamma\}$ be a measurable field of Hilbert spaces. A field $\{\mathscr{M}(\gamma),\mathfrak{H}(\gamma)\}_{\gamma \in \Gamma}$ of von Neumann algebras is said to be *measurable* if there exists a countable family $\{x_n(\gamma):n = 1,2,\ldots\}$ of measurable fields of operators such that $\mathscr{M}(\gamma)$ is generated by $\{x_n(\gamma):n = 1,2,\ldots\}$ for almost every $\gamma \in \Gamma$.

Given a measurable field $\{\mathscr{M}(\gamma),\mathfrak{H}(\gamma):\gamma \in \Gamma\}$ of von Neumann algebras on $\{\Gamma,\mu\}$, let $\Gamma_n = \{\gamma \in \Gamma:\dim \mathfrak{H}(\gamma) = n\}$ for $n = 1,2,\ldots,+\infty$, and let $\mathfrak{H}_n$ be a fixed separable Hilbert space of dimension n. By Theorem 8.13, for each $\gamma \in \Gamma_n$, there exists a unitary $U(\gamma)$ of $\mathfrak{H}(\gamma)$ onto $\mathfrak{H}_n$ such that a vector field $\xi(\cdot) \in \prod_\gamma \mathfrak{H}(\gamma)$ is measurable if and only if $\gamma \in \Gamma_n \mapsto U(\gamma)\xi(\gamma) \in \mathfrak{H}_n$ is measurable. Therefore, by Corollary 8.3 the map: $\gamma \in \Gamma_n \mapsto U(\gamma)\mathscr{M}(\gamma)U(\gamma)^* \in \mathfrak{A}(\mathfrak{H}_n)$ is measurable with respect to the Effros Borel structure in the space $\mathfrak{A}(\mathfrak{H}_n)$ of all von Neumann algebras on $\mathfrak{H}_n$. Therefore, by Theorem 8.4, the field $\{\mathscr{M}(\gamma)':\gamma \in \Gamma\}$ is measurable.

Theorem 8.18. *Under the above situation, let $\mathscr{M}$ denote the set of all decomposable operators*

$$x = \int_\Gamma^\oplus x(\gamma)\, d\mu(\gamma) \quad \text{on} \quad \mathfrak{H} = \int_\Gamma^\oplus \mathfrak{H}(\gamma)\, d\mu(\gamma)$$

such that $x(\gamma) \in \mathscr{M}(\gamma)$ for every $\gamma \in \Gamma$, and write

$$\mathscr{M} = \int_\Gamma^\oplus \mathscr{M}(\gamma)\, d\mu(\gamma).$$

Then $\mathscr{M}$ is a von Neumann algebra on $\mathfrak{H}$ and its commutant $\mathscr{M}'$ is given by

$$\mathscr{M}' = \int_\Gamma^\oplus \mathscr{M}(\gamma)'\, d\mu(\gamma).$$

The diagonal algebra $\mathscr{A}$ is contained in the center $\mathscr{Z}$ of $\mathscr{M}$.

PROOF. It is obvious that $\mathscr{A}$ is contained in the center $\mathscr{Z}$ of $\mathscr{M}$. Let $y \in \mathscr{M}'$. By Collary 8.16, y is decomposable, so that we can write

$$y = \int_\Gamma^\oplus y(\gamma)\, d\mu(\gamma).$$

For each $x = \int_\Gamma^\oplus x(\gamma)\, d\mu(\gamma) \in \mathscr{M}$, $x(\gamma)$ and $y(\gamma)$ commute for almost every $\gamma \in \Gamma$. Let $\{x_n(\gamma): n = 1,2,\ldots\}$ be a sequence of measurable operator fields such that $\{x_n(\gamma): n = 1,2,\ldots\}$ generates $\mathscr{M}(\gamma)$ for every $\gamma \in \Gamma$. Since $\gamma \in \Gamma \mapsto \|x_n(\gamma)\|$ is measurable, we can obtain bounded measurable fields

$$x_{m,n}(\gamma) = \begin{cases} x_n(\gamma) & \text{if } \|x_n(\gamma)\| \le m \\ 0 & \text{otherwise.} \end{cases}$$

Set

$$x_{m,n} = \int_\Gamma^\oplus x_{m,n}(\gamma)\, d\mu(\gamma) \in \mathscr{M},$$

and

$$N_{m,n} = \{\gamma \in \Gamma : x_{m,n}(\gamma)y(\gamma) \ne y(\gamma)x_{m,n}(\gamma)\}.$$

Then $N_{m,n}$ is μ-null, and $\{x_{m,n}(\gamma)\}$ and $y(\gamma)$ commute for every

$$\gamma \notin \bigcup_{m,n=1}^{\infty} N_{m,n} = N.$$

Hence $y(\gamma)$ belongs to $\mathscr{M}(\gamma)'$ if $\gamma \notin N$. Thus we have

$$y = \int_\Gamma^\oplus y(\gamma)\, d\mu(\gamma) \in \int_\Gamma^\oplus \mathscr{M}(\gamma)'\, d\mu(\gamma).$$

By symmetry, we have $\mathscr{M} = \mathscr{M}''$. Since $\mathscr{M}$ is self-adjoint, $\mathscr{M}$ is a von Neumann algebra and

$$\mathscr{M}' = \int_\Gamma^\oplus \mathscr{M}(\gamma)'\, d\mu(\gamma). \qquad \text{Q.E.D.}$$

Definition 8.19. The von Neumann algebra $\mathscr{M}$ in the previous theorem is called the *direct integral* of $\{\mathscr{M}(\gamma): \gamma \in \Gamma\}$.

Corollary 8.20. *Let*

$$\{\mathscr{M},\mathfrak{H}\} = \int_\Gamma^\oplus \{\mathscr{M}(\gamma),\mathfrak{H}(\gamma)\}\, d\mu(\gamma)$$

be a direct integral of von Neumann algebras. Then the center $\mathscr{Z}$ of $\mathscr{M}$ is also expressed as a direct integral

$$\mathscr{Z} = \int_\Gamma^\oplus \mathscr{Z}(\gamma)\, d\mu(\gamma)$$

with $\mathscr{Z}(\gamma) = \mathscr{M}(\gamma) \cap \mathscr{M}(\gamma)'$. *In particular,* $\mathscr{Z}$ *coincides with the diagonal algebra* $\mathscr{A}$ *if and only if* $\mathscr{M}(\gamma)$ *is a factor for almost every* $\gamma \in \Gamma$.

PROOF. First of all, we note that $\{\mathscr{Z}(\gamma):\gamma \in \Gamma\}$ is a measurable field of von Neumann algebras by Corollary 8.6. Let $x = \int_\Gamma^\oplus x(\gamma)\, d\mu(\gamma) \in \mathscr{Z}$. For any

$$a = \int_\Gamma^\oplus a(\gamma)\, d\mu(\gamma) \in \mathscr{M} \quad \text{and} \quad b = \int_\Gamma^\oplus b(\gamma)\, d\mu(\gamma) \in \mathscr{M}',$$

$x(\gamma)$ commutes with $a(\gamma)$ and $b(\gamma)$ for almost every $\gamma \in \Gamma$. Making use of generating sequences of $\{\mathscr{M}(\gamma):\gamma \in \Gamma\}$ and $\{\mathscr{M}(\gamma)':\gamma \in \Gamma\}$, we conclude that $x(\gamma)$ belongs to $\mathscr{Z}(\gamma)$ for almost every $\gamma \in \Gamma$. Conversely, it is clear that if $x(\gamma) \in \mathscr{Z}(\gamma)$ for almost every $\gamma \in \Gamma$ then x commutes with both $\mathscr{M}$ and $\mathscr{M}'$, so that $x \in \mathscr{Z}$. Q.E.D.

Theorem 8.21 (Existence of Disintegration). *If* $\{\mathscr{M},\mathfrak{H}\}$ *is a von Neumann algebra on a separable Hilbert space, then for any von Neumann subalgebra* $\mathscr{A}$ *of the center* $\mathscr{Z} = \mathscr{M} \cap \mathscr{M}'$, *there exists a measurable field* $\{\mathscr{M}(\gamma),\mathfrak{H}(\gamma):\gamma \in \Gamma\}$ *of von Neumann algebras on a standard* σ*-finite measure space* $\{\Gamma,\mu\}$ *such that*

(i) $\{\mathscr{M},\mathfrak{H}\} = \int_\Gamma^\oplus \{\mathscr{M}(\gamma),\mathfrak{H}(\gamma)\}\, d\mu(\gamma)$,
(ii) $\mathscr{A}$ *is the diagonal algebra.*

In particular, any von Neumann algebra on a separable Hilbert space is a direct integral of factors.

We postpone the proof of the theorem, which will be proven later as a special case of disintegration of a representation of a C^*-algebra. To show the uniqueness of the above disintegration, we need the following:

Lemma 8.22. *Let* $\{\Gamma_1,\mu_1\}$ *and* $\{\Gamma_2,\mu_2\}$ *be standard* σ*-finite measure spaces. If* π *is an isomorphism of the von Neumann algebra* $L^\infty(\Gamma_1,\mu_1)$ *onto* $L^\infty(\Gamma_2,\mu_2)$, *then there exist Borel null sets* $N_1 \subset \Gamma_1$, $N_2 \subset \Gamma_2$ *and a Borel isomorphism* Φ *of* $\Gamma_2 - N_2$ *onto* $\Gamma_1 - N_1$ *such that* $\Phi(\mu_2)$ *and* μ_1 *are equivalent in the sense of absolute continuity and, for every* $a \in L^\infty(\Gamma_1,\mu_1)$, $\pi(a)(\gamma) = a(\Phi(\gamma))$ *for almost every* $\gamma \in \Gamma_2 - N_2$.

PROOF. Since a standard Borel space is either countable or isomorphic to $[0,1]$, we may assume that $\Gamma_1 = \Gamma_2 = [0,1]$. Replacing both μ_1 and μ_2 by equivalent probability measures, we assume $\mu_1(\Gamma_1) = \mu_2(\Gamma_2) = 1$. Set $a_1(\gamma) = \gamma$ for $\gamma \in [0,1]$ and $a_2 = \pi(a_1)$. Since $0 \le a_1 \le 1$, we have $0 \le a_2 \le 1$, so that a_2 is regarded as a Borel function on $[0,1] = \Gamma_2$ with $0 \le a_2(\gamma) \le 1$, $\gamma \in \Gamma_2$. Put $\Phi(\gamma) = a_2(\gamma)$, $\gamma \in \Gamma_2$. Setting

$$\omega_2(x) = \int_{\Gamma_2} x(\gamma)\, d\mu_2(\gamma), \qquad x \in L^\infty(\Gamma_2,\mu_2),$$

we obtain a faithful normal state on $L^\infty(\Gamma_2,\mu_2)$. Put $\omega_1 = {}^t\pi(\omega_2)$. It then follows that ω_1 is a faithful normal state on $L(\Gamma_1,\mu_1)$, so that there exists

$f \in L^1(\Gamma_1,\mu_1)$ such that

$$\omega_2(\pi(x)) = \int_{\Gamma_1} x(\gamma)f(\gamma)\, d\mu_1(\gamma), \qquad x \in L^\infty(\Gamma_1,\mu_1).$$

If p is a polynomial with complex coefficients, then we have

$$\pi(p(a_1))(\gamma) = p(a_2(\gamma)) = p(\Phi(\gamma)), \qquad \gamma \in \Gamma_2.$$

Hence if x is a continuous function on $[0,1]$, then we have, by the Weierstrass approximation theorem,

$$\pi(x)(\gamma) = x(\Phi(\gamma)), \qquad \gamma \in \Gamma_2,$$

$$\begin{aligned}\int_{\Gamma_1} x(\gamma)f(\gamma)\, d\mu_1(\gamma) &= \omega_1(x) = \omega_2(\pi(x)) \\ &= \int_{\Gamma_2} \pi(x)(\gamma)\, d\mu_2(\gamma) = \int_{\Gamma_2} x(\Phi(\gamma))\, d\mu_2(\gamma).\end{aligned}$$

Thus we have $f(\gamma)\, d\mu_1(\gamma) = d\Phi(\mu_2)(\gamma)$. Let x be an arbitrary element of $L^\infty(\Gamma_1,\mu_1)$, and $\{x_n\}$ be a bounded sequence in $C[0,1]$ converging σ-strongly to x. We then have

$$\begin{aligned}\int_{\Gamma_1} |x(\gamma) - x_n(\gamma)|^2 f(\gamma)\, d\mu_1(\gamma) &= \omega_1((x - x_n)^*(x - x_n)) \\ &= \int_{\Gamma_2} |\pi(x)(\gamma) - \pi(x_n)(\gamma)|^2\, d\mu_2(\gamma) \to 0 \quad \text{as} \quad n \to \infty.\end{aligned}$$

Hence there exists a subsequence $\{x_{n_j}\}$ such that

$$x(\gamma) = \lim_j x_{n_j}(\gamma) \quad \text{for almost every} \quad \gamma \in \Gamma_1,$$

$$\pi(x)(\gamma) = \lim_j \pi(x_{n_j})(\gamma) \quad \text{for almost every} \quad \gamma \in \Gamma_2,$$

where we use the fact that $f(\gamma) > 0$ for almost every $\gamma \in \Gamma_1$, ω_1 being faithful. We then have

$$\begin{aligned}\int_{\Gamma_2} |x(\Phi(\gamma)) - \pi(x)(\gamma)|^2\, d\mu_2(\gamma) &= \lim_j \int_{\Gamma_2} |x_{n_j}(\Phi(\gamma)) - \pi(x_{n_j}(\gamma))|^2\, d\mu_2(\gamma) \\ &= 0.\end{aligned}$$

Therefore, we get $x(\Phi(\gamma)) = \pi(x(\gamma))$ for almost every $\gamma \in \Gamma_2$.

We now apply the above argument to π^{-1} and obtain a Borel map Ψ from $\Gamma_1 = [0,1]$ into $\Gamma_2 = [0,1]$ such that $y(\Psi(\gamma)) = \pi^{-1}(y)(\gamma)$ for almost every $\gamma \in \Gamma_1$ and $y \in L^\infty(\Gamma_2,\mu_2)$. We then have

$$x(\Phi \circ \Psi(\gamma)) = x(\gamma) \quad \text{for almost every} \quad \gamma \in \Gamma_1,\ x \in L^\infty(\Gamma_1,\mu_1),$$

$$y(\Psi \circ \Phi(\gamma)) = y(\gamma) \quad \text{for almost every} \quad \gamma \in \Gamma_2,\ y \in L^\infty(\Gamma_2,\mu_2).$$

Choosing x to be the function $x(\gamma) = \gamma$, we get $\Phi \circ \Psi(\gamma) = \gamma$ for almost every $\gamma \in \Gamma_1$. Hence there exists a Borel null set $N_1 \subset \Gamma_1$ such that $\Phi \circ \Psi(\gamma) = \gamma$ for every $\gamma \in \Gamma_1 - N_1$. Hence Ψ is injective on $\Gamma_1 - N_1$. Put $N_2 = \Phi^{-1}(N_1) \subset \Gamma_2$. It follows that N_2 is a Borel null set in Γ_2 and Φ is a Borel isomorphism of $\Gamma_2 - N_2$ onto $\Gamma_1 - N_1$. Q.E.D.

Theorem 8.23 (Uniqueness of Disintegration). *Suppose that*

$$\{\mathscr{M}_1,\mathfrak{H}_1\} = \int_{\Gamma_1}^{\oplus} \{\mathscr{M}_1(\gamma),\mathfrak{H}_1(\gamma)\}\, d\mu_1(\gamma),$$

$$\{\mathscr{M}_2,\mathfrak{H}_2\} = \int_{\Gamma_2}^{\oplus} \{\mathscr{M}_2(\gamma),\mathfrak{H}_2(\gamma)\}\, d\mu_2(\gamma)$$

are direct integrals of von Neumann algebras on separable Hilbert spaces with diagonal algebras $\mathscr{A}_1$ and $\mathscr{A}_2$ on standard σ-finite measure spaces $\{\Gamma_1,\mu_1\}$ and $\{\Gamma_2,\mu_2\}$, respectively. If U is a unitary operator of $\mathfrak{H}_1$ onto $\mathfrak{H}_2$ such that

$$U\mathscr{M}_1U^* = \mathscr{M}_2, \qquad U\mathscr{A}_1U^* = \mathscr{A}_2,$$

then there exist null sets $N_1 \subset \Gamma_1, N_2 \subset \Gamma_2$, a Borel isomorphism Φ of $\Gamma_2 - N_2$ onto $\Gamma_1 - N_1$ and a measurable field $\{U(\gamma): \gamma \in \Gamma_1 - N_1\}$ of unitary operators from $\mathfrak{H}_1(\gamma)$ onto $\mathfrak{H}_2(\Phi^{-1}(\gamma))$ such that

(i) $U(\gamma)\mathscr{M}_1(\gamma)U(\gamma)^* = \mathscr{M}_2(\Phi^{-1}(\gamma)),\ \gamma \in \Gamma_1 - N_1,$

(ii) $\Phi(\mu_2)$ *and* μ_1 *are equivalent in the sense of absolute continuity,*

(iii) $U = \displaystyle\int_{\Gamma_1}^{\oplus} U(\gamma)\sqrt{\frac{d\Phi(\mu_2)}{d\mu_1}}(\gamma)\, d\mu_1(\gamma).$

Proof. Let π be the isomorphism of $\mathscr{L}(\mathfrak{H}_1)$ onto $\mathscr{L}(\mathfrak{H}_2)$ implemented by the unitary operator U. It follows from Lemma 8.22 that there exist null sets $N_1 \subset \Gamma_1$, $N_2 \subset \Gamma_2$ and a Borel isomorphism Φ of $\Gamma_2 - N_2$ onto $\Gamma_1 - N_1$ such that $\Phi(\mu_2)$ and μ_1 are equivalent, and $x(\Phi(\gamma)) = \pi(x)(\gamma)$ for every $x \in \mathscr{A}_1$ and almost every $\gamma \in \Gamma_2 - N_2$. Throwing away N_1 and N_2, and replacing μ_1 by $\Phi(\mu_2)$, we may identify the standard σ-finite measure spaces $\{\Gamma_1,\mu_1\}$ and $\{\Gamma_2,\mu_2\}$ under the isomorphism Φ. So our situation is that we have measurable fields $\{\mathscr{M}_1(\gamma),\mathfrak{H}_1(\gamma)\}$ and $\{\mathscr{M}_2(\gamma),\mathfrak{H}_2(\gamma)\}$ of von Neumann algebras on a standard σ-finite measure space $\{\Gamma,\mu\}$ and a unitary operator U from $\mathfrak{H}_1 = \int_\Gamma^\oplus \mathfrak{H}_1(\gamma)\, d\mu(\gamma)$ onto $\mathfrak{H}_2 = \int_\Gamma^\oplus \mathfrak{H}_2(\gamma)\, d\mu(\gamma)$ such that

$$U\mathscr{M}_1U^* = \mathscr{M}_2,$$

$$\mathscr{M}_1 = \int_\Gamma^{\oplus} \mathscr{M}_1(\gamma)\, d\mu(\gamma) \quad\text{and}\quad \mathscr{M}_2 = \int_\Gamma^{\oplus} \mathscr{M}_2(\gamma)\, d\mu(\gamma),$$

and U commutes with the multiplication operators given by functions in $L^\infty(\Gamma,\mu)$. Hence U is decomposable. Let

$$U = \int_\Gamma^{\oplus} U(\gamma)\, d\mu(\gamma)$$

be the decomposition of U. We then have

$$1 = U^*U = \int_\Gamma^{\oplus} U(\gamma)^*U(\gamma)\, d\mu(\gamma),$$

$$1 = UU^* = \int_\Gamma^{\oplus} U(\gamma)U(\gamma)^*\, d\mu(\gamma).$$

Hence $U(\gamma)$ is a unitary operator of $\mathfrak{H}_1(\gamma)$ onto $\mathfrak{H}_2(\gamma)$ for almost every $\gamma \in \Gamma$. Let $\{x_n\}$ be a bounded sequence of bounded measurable operator fields in $\prod \mathcal{M}_1(\gamma)$ such that $\mathcal{M}_1(\gamma) = \{x_n(\gamma)\}''$ for every $\gamma \in \Gamma$. Put

$$y_n(\gamma) = U(\gamma)x_n(\gamma)U(\gamma)^*, \qquad \gamma \in \Gamma, \qquad n = 1,2,\ldots.$$

We then have

$$y_n = \int_\Gamma^\oplus y_n(\gamma)\, d\mu(\gamma) = Ux_nU^* = \pi(x_n) \in \mathcal{M}_2,$$

so that $y_n(\gamma) \in \mathcal{M}_2(\gamma)$ for each n and almost every $\gamma \in \Gamma$. Since $\{x_n\}$ is countable, we have $U(\gamma)\mathcal{M}_1(\gamma)U(\gamma)^* \subseteq \mathcal{M}_2(\gamma)$ for almost every $\gamma \in \Gamma$. Symmetrically, we can prove that $U(\gamma)^*\mathcal{M}_2(\gamma)U(\gamma) \subset \mathcal{M}_1(\gamma)$ for almost every $\gamma \in \Gamma$. Hence we get finally,

$$U(\gamma)\mathcal{M}_1(\gamma)U(\gamma)^* = \mathcal{M}_2(\gamma)$$

for almost every $\gamma \in \Gamma$. This is exactly the assertion in the theorem without identification through Φ. Q.E.D.

Let A be a C^*-algebra and $\{\mathfrak{H}(\gamma):\gamma \in \Gamma\}$ be a measurable field of Hilbert spaces on a Borel space Γ equipped with a σ-finite Borel measure μ. Suppose that to each $\gamma \in \Gamma$, there corresponds a representation π_γ of A on $\mathfrak{H}(\gamma)$ such that for every $x \in A$ the operator field: $\gamma \in \Gamma \mapsto \pi_\gamma(x) \in \mathcal{L}(\mathfrak{H}(\gamma))$ is measurable. The field $\{\pi_\gamma:\gamma \in \Gamma\}$ of representations is said to be *measurable* if this is the case. We then set

$$\pi(x) = \int_\Gamma^\oplus \pi_\gamma(x)\, d\mu(\gamma) \in \mathcal{L}(\mathfrak{H}), \qquad x \in A,$$

$$\mathfrak{H} = \int_\Gamma^\oplus \mathfrak{H}(\gamma)\, d\mu(\gamma).$$

Clearly, π is a representation of A on $\mathfrak{H}$.

Definition 8.24. The representation π is called the *direct integral* of $\{\pi_\gamma\}$ and written as

$$\pi = \int_\Gamma^\oplus \pi_\gamma\, d\mu(\gamma).$$

Theorem 8.25. *Let A be a separable C^*-algebra, and let*

$$\mathfrak{H} = \int_\Gamma^\oplus \mathfrak{H}(\gamma)\, d\mu(\gamma).$$

If π is a representation of A on $\mathfrak{H}$ such that $\pi(A)$ commutes with the diagonal algebra, then there exists a measurable field $\{\pi_\gamma\}$ of representations, unique up to almost everywhere, such that

$$\pi = \int_\Gamma^\oplus \pi_\gamma\, d\mu(\gamma).$$

PROOF. Let B_0 be a countable subring of $\pi(A)$ which is uniformly dense in $\pi(A)$ and closed under the adjoint operation. Each $x \in B_0$ is decomposable, so that there exists a measurable operator field $x(\cdot)$ such that

$$x = \int_\Gamma^\oplus x(\gamma)\, d\mu(\gamma), \quad \text{and} \quad \|x\| = \text{ess sup}\|x(\gamma)\|.$$

Let $N_x = \{\gamma \in \Gamma : \|x(\gamma)\| > \|x\|\}$, $x \in B_0$ and $N'_x = \{\gamma \in \Gamma : x^*(\gamma) \neq x(\gamma)^*\}$. For each $x, y \in B_0$, set

$$N_{x,y} = \{\gamma \in \Gamma : (x + y)(\gamma) \neq x(\gamma) + y(\gamma)\},$$
$$N'_{x,y} = \{\gamma \in \Gamma : (xy)(\gamma) \neq x(\gamma)y(\gamma)\}.$$

By Proposition 7.12 and Theorem 8.13, we see that each N_x, N'_x, $N_{x,y}$, and $N'_{x,y}$ is negligible. Hence

$$N = \Big(\bigcup_{x \in B_0} N_x\Big) \cup \Big(\bigcup_{x \in B_0} N'_x\Big) \cup \Big(\bigcup_{\substack{x \in B_0 \\ y \in B_0}} N_{x,y}\Big) \cup \Big(\bigcup_{\substack{x \in B_0 \\ y \in B_0}} N'_{x,y}\Big)$$

is also negligible. For every $\gamma \notin N$, the map: $x \in B_0 \mapsto x(\gamma) \in \mathscr{L}(\mathfrak{H}(\gamma))$ is a continuous *-homomorphism, so that it is extended to a *-homomorphism: $x \in \pi(A) \mapsto x(\gamma) \in \mathscr{L}(\mathfrak{H}(\gamma))$. We then set

$$\begin{aligned} \pi_\gamma(a) &= \pi(a)(\gamma), & \gamma \notin N \\ &= 0, & \gamma \in N. \end{aligned}$$

It then follows that $\{\pi_\gamma\}$ is a measurable field of representations of A and that

$$\pi = \int_\Gamma^\oplus \pi_\gamma\, d\mu(\gamma).$$

Suppose that $\{\pi'_\gamma\}$ is another measurable field of representations of A such that

$$\pi = \int_\Gamma^\oplus \pi'_\gamma\, d\mu(\gamma).$$

Let A_0 be a countable dense subset of A. For each $x \in A_0$, we have

$$\int_\Gamma^\oplus \pi_\gamma(x)\, d\mu(\gamma) = \pi(x) = \int_\Gamma^\oplus \pi'_\gamma(x)\, d\mu(\gamma).$$

Hence $\pi_\gamma(x) = \pi'_\gamma(x)$ almost everywhere. Set $N_x = \{\gamma : \pi_\gamma(x) \neq \pi'_\gamma(x)\}$, and $N = \bigcup_{x \in A_0} N_x$. It follows that N is negligible, and $\pi_\gamma(x) = \pi'_\gamma(x)$ for every $x \in A_0$ and $\gamma \notin N$. By continuity, we get $\pi_\gamma = \pi'_\gamma$ if $\gamma \notin N$. Thus $\{\pi_\gamma\}$ is unique.

Q.E.D.

Remark 8.26. The separability assumption in Theorem 8.25 for A is indispensable. Even if $\mathfrak{H}$ is separable, we cannot arrive at the result for a nonseparable C^*-algebra A. For example, if $\dim \mathfrak{H}(\gamma) = +\infty$ for all γ, then every representation of $\mathscr{A}'$ on a separable Hilbert space must be normal, where $\mathscr{A}$ denotes the diagonal algebra; see Theorem V.5.1. This means that to have a representation π_γ of $\mathscr{A}'$ on $\mathfrak{H}(\gamma)$, γ must have positive mass, i.e., $\mu(\{\gamma\}) > 0$. In this case, π_γ is a direct summand of π, which is rather uninteresting.

PROOF OF THEOREM 8.21. Let A be a separable C^*-subalgebra of $\mathscr{M}$ such that $A'' = \mathscr{M}$, and π be the identity representation of A on $\mathfrak{H}$. We then decompose π by Theorem 8.25 as

$$\pi = \int_\Gamma^\oplus \pi_\gamma \, d\mu(\gamma).$$

It then follows that $\{\pi_\gamma(A)'' : \gamma \in \Gamma\}$ is a measurable field of von Neumann algebras such that

$$\mathscr{M} = \int_\Gamma^\oplus \pi_\gamma(A)'' \, d\mu(\gamma). \qquad \text{Q.E.D.}$$

Proposition 8.27. *Let A be a separable C^*-algebra and*

$$\{\pi_i, \mathfrak{H}_i\} = \int_{\Gamma_i}^\oplus \{\pi_\gamma^i, \mathfrak{H}_i(\gamma)\} \, d\mu_i(\gamma), \qquad i = 1,2,$$

be direct integrals of representations of A on standard σ-finite measure spaces $\{\Gamma_1, \mu_1\}$ and $\{\Gamma_2, \mu_2\}$ with the diagonal algebras $\mathscr{A}_1$ and $\mathscr{A}_2$, respectively. If U is a unitary operator of $\mathfrak{H}_1$ onto $\mathfrak{H}_2$ such that

$$U\mathscr{A}_1 U^* = \mathscr{A}_2, \qquad U\pi_1(x)U^* = \pi_2(x), \qquad x \in A,$$

then the conclusions of Theorem 8.23 *hold when assertion* (i) *is replaced by*

$$U(\gamma)\pi_\gamma^1(x)U(\gamma)^* = \pi_\gamma^2(x), \qquad x \in A.$$

Since the proof goes exactly similarly, we leave it to the reader.

Conversely, we have the following:

Theorem 8.28. *Let A be a separable C^*-algebra and $\{\Gamma, \mu\}$ a standard σ-finite measure space. For $i = 1,2$, let $\{\pi_\gamma^i, \mathfrak{H}_i(\gamma) : \gamma \in \Gamma\}$ be measurable fields of representations of A (resp. let $\{\mathscr{M}_i(\gamma), \mathfrak{H}_i(\gamma) : \gamma \in \Gamma\}$ be measurable fields of von Neumann algebras) such that*

$$\{\pi^i, \mathfrak{H}\} = \int_\Gamma^\oplus \{\pi_\gamma^i, \mathfrak{H}_i(\gamma)\} \, d\mu(\gamma)$$

$$\left(\text{resp. } \{\mathscr{M}_i, \mathfrak{H}_i\} = \int_\Gamma^\oplus \{\mathscr{M}_i(\gamma), \mathfrak{H}_i(\gamma)\} \, d\mu(\gamma)\right).$$

If $\pi_\gamma^1 \cong \pi_\gamma^2$ (resp. $\{\mathscr{M}_1(\gamma), \mathfrak{H}_1(\gamma)\} \cong \{\mathscr{M}_2(\gamma), \mathfrak{H}_2(\gamma)\}$) for almost every $\gamma \in \Gamma$, then there exists a measurable field $\{U(\gamma) : \gamma \in \Gamma\}$ of unitary operators such that

$$U(\gamma)\pi_\gamma^1(x)U(\gamma)^* = \pi_\gamma^2(x), \qquad x \in A$$
$$(\text{resp. } U(\gamma)\mathscr{M}_1(\gamma)U(\gamma)^* = \mathscr{M}_2(\gamma))$$

for almost every $\gamma \in \Gamma$. Hence the unitary operator

$$U = \int_\Gamma^\oplus U(\gamma) \, d\mu(\gamma)$$

implements the unitary equivalence of π^1 and π^2 (resp. $\mathscr{M}_1$ and $\mathscr{M}_2$).

PROOF. Removing Borel null sets from Γ, we may assume that $\pi_\gamma^1 \cong \pi_\gamma^2$ (resp. $\{\mathscr{M}_1(\gamma),\mathfrak{H}_1(\gamma)\} \cong \{\mathscr{M}_2(\gamma),\mathfrak{H}_2(\gamma)\}$) for every $\gamma \in \Gamma$. Considering each $\Gamma_n = \{\gamma \in \Gamma : \dim \mathfrak{H}_1(\gamma) = \dim \mathfrak{H}_2(\gamma) = n\}$, $n = 1,2,\ldots,\infty$, we may assume, by Theorem 8.13 or Lemma 8.12, that both $\mathfrak{H}_1(\gamma)$ and $\mathfrak{H}_2(\gamma)$ are a constant field $\mathfrak{H}_0$.

We first note that the unitary group $\mathscr{U}$ of $\mathfrak{H}_0$ is complete and separable with respect to the strong* topology, hence is a Polish space. Let $\mathrm{Rep}(A;\mathfrak{H}_0)$ be the set of all representations of A on $\mathfrak{H}_0$ equipped with the simple (pointwise) convergence topology of the strong* topology in $\mathscr{L}(\mathfrak{H}_0)$. Since $\mathrm{Rep}(A;\mathfrak{H}_0)$ is a closed subset of the Polish space $\mathscr{L}(A,\mathscr{L}(\mathfrak{H}_0))$ of all bounded linear maps of A into $\mathscr{L}(\mathfrak{H}_0)$, it is a Polish space. Let A_0 be a countable dense set in A. For each $x \in A_0$, we can choose Borel functions: $\gamma \mapsto \pi_\gamma^1(x)$ and $\pi_\gamma^2(x) \in \mathscr{L}(\mathfrak{H}_0)$, removing Borel null sets from Γ. Hence we can remove Borel null sets from Γ so that the functions: $\gamma \in \Gamma \mapsto \pi_\gamma^1(x)$ and $\pi_\gamma^2(x)$ are Borel for every $x \in A_0$; hence they are Borel for every $x \in A$, being limits of Borel functions. In a similar way, we may assume that $\gamma \in \Gamma \mapsto \mathscr{M}_1(\gamma) \in \mathfrak{A}(\mathfrak{H}_0)$ and $\gamma \in \Gamma \mapsto \mathscr{M}_2(\gamma) \in \mathfrak{A}(\mathfrak{H}_0)$ are both Borel functions, where $\mathfrak{A}(\mathfrak{H}_0)$ denotes the standard Borel space of all von Neumann algebras on $\mathfrak{H}_0$ with the Effros Borel structure.

The Polish group $\mathscr{U}$ acts on both $\mathrm{Rep}(A;\mathfrak{H}_0)$ and $\mathfrak{A}(\mathfrak{H}_0)$ as follows:

$$(u \cdot \pi)(x) = u\pi(x)u^*, \qquad \pi \in \mathrm{Rep}(A;\mathfrak{H}_0), \qquad x \in A;$$
$$u \cdot \mathscr{M} = u\mathscr{M}u^*, \qquad \mathscr{M} \in \mathfrak{A}(\mathfrak{H}_0), \qquad u \in \mathscr{U}.$$

The maps: $(u,\pi) \in \mathscr{U} \times \mathrm{Rep}(A;\mathfrak{H}_0) \mapsto u \cdot \pi \in \mathrm{Rep}(A;\mathfrak{H}_0)$ and $(u,\mathscr{M}) \in \mathscr{U} \times \mathfrak{A}(\mathfrak{H}_0) \mapsto u \cdot \mathscr{M} \in \mathfrak{A}(\mathfrak{H}_0)$ are both Borel. Consider the subset B of $\mathscr{U} \times \Gamma$ defined by

$$B = \{(u,\gamma) : u \cdot \pi_\gamma^1 = \pi_\gamma^2\}$$
$$(\text{resp. } B = \{(u,\gamma) : u \cdot \mathscr{M}_1(\gamma) = \mathscr{M}_2(\gamma)\}.$$

It then follows that B is a Borel subset of the standard Borel space $\mathscr{U} \times \Gamma$, and covers Γ by assumption. Hence there exists a measurable function $U(\gamma)$ on Γ such that $(U(\gamma),\gamma) \in B$ (see Appendix), which means that

$$U(\gamma) \cdot \pi_\gamma^1 = \pi_\gamma^2 \qquad (\text{resp. } U(\gamma)\mathscr{M}_1(\gamma)U(\gamma)^* = \mathscr{M}_2(\gamma)), \qquad \gamma \in \Gamma.$$

Thus we can now form a direct integral

$$U = \int_\Gamma^\oplus U(\gamma)\, d\mu(\gamma),$$

which implements unitary equivalence of π^1 and π^2 (resp. $\mathscr{M}_1$ and $\mathscr{M}_2$).
Q.E.D.

Proposition 8.29. *Suppose that we have two direct integrals*

$$\{\mathscr{M}_1,\mathfrak{H}_1\} = \int_\Gamma^\oplus \{\mathscr{M}_1(\gamma),\mathfrak{H}_1(\gamma)\}\, d\mu(\gamma),$$
$$\{\mathscr{M}_2,\mathfrak{H}_2\} = \int_\Gamma^\oplus \{\mathscr{M}_2(\gamma),\mathfrak{H}_2(\gamma)\}\, d\mu(\gamma)$$

on a standard σ-finite measure space $\{\Gamma,\mu\}$. *If* $\mathscr{M}_1(\gamma) \cong \mathscr{M}_2(\gamma)$ *for almost every* $\gamma \in \Gamma$, *then there exists a measurable field* $\{\pi_\gamma\}$ *of isomorphisms such that* $\pi_\gamma(\mathscr{M}_1(\gamma)) = \mathscr{M}_2(\gamma)$; *hence the direct integral*

$$\pi = \int_\Gamma^\oplus \pi_\gamma \, d\mu(\gamma)$$

is an isomorphism of $\mathscr{M}_1$ *onto* $\mathscr{M}_2$.

PROOF. Let $\mathfrak{K}$ be a fixed separable infinite dimensional Hilbert space. We then consider a new measurable field of von Neumann algebras $\{\mathscr{M}_1(\gamma) \otimes \mathbf{C}, \mathfrak{H}_1(\gamma) \otimes \mathfrak{K}\}$. For almost every $\gamma \in \Gamma$, $\mathscr{M}_2(\gamma)$ is spatially isomorphic to an induced von Neumann algebra $(\mathscr{M}_1(\gamma) \otimes \mathbf{C})_{e(\gamma)}$ with some projection $e(\gamma) \in (\mathscr{M}_1(\gamma) \otimes \mathbf{C})'$. Since $x \in \mathscr{M}_1(\gamma) \mapsto x \otimes 1 \in \mathscr{M}_1(\gamma) \otimes \mathbf{C}$ is a measurable field of isomorphisms, we have only to prove our assertion in the case that $\mathscr{M}_2(\gamma)$ is spatially isomorphic to $[\mathscr{M}_1(\gamma)]_{e(\gamma)}$ for some $e(\gamma) \in \mathscr{M}_1(\gamma)'$ almost everywhere. As in the case of the previous theorem, we may assume that $\mathfrak{H}_1(\gamma) = \mathfrak{H}_0$ and $\mathfrak{H}_2(\gamma) \subset \mathfrak{H}_0$ with some fixed $\mathfrak{H}_0$ for every $\gamma \in \Gamma$. Let $\{x_{i,n}(\cdot)\}$ (resp. $x'_{i,n}(\cdot)$) be a sequence of operator fields such that $\mathscr{M}_i(\gamma) = \{x_{i,n}(\gamma)\}''$ (resp. $\mathscr{M}_i(\gamma)' = \{x'_{i,n}(\gamma)\}''$), $i = 1,2$. Let B be the subset of $\Gamma \times \mathscr{L}(\mathfrak{H}_0) \times \mathscr{L}(\mathfrak{H}_0)$ consisting of all those (γ,p,u) such that

$$\begin{aligned} p &= p^* = p^2, \\ x_{1,n}(\gamma)p &= px_{1,n}(\gamma), \qquad n = 1,2,\ldots, \\ u^*u &= p, \qquad uu^*\mathfrak{H}_0 = \mathfrak{H}_2(\gamma), \\ ux_{1,n}(\gamma)u^*x'_{2,m}(\gamma) &= x'_{2,m}(\gamma)ux_{1,n}(\gamma)u^*, \qquad n,m = 1,2,\ldots, \\ u^*x_{2,n}(\gamma)ux'_{1,m}(\gamma) &= x'_{1,m}(\gamma)u^*x_{2,n}(\gamma)u, \qquad n,m = 1,2,\ldots. \end{aligned}$$

It then follows that B is a Borel set in the standard Borel space $\Gamma \times \mathscr{L}(\mathfrak{H}_0) \times \mathscr{L}(\mathfrak{H}_0)$ and $\mathrm{pr}_\Gamma(B) = \Gamma$, where pr_Γ means the projection onto the Γ-component. Hence, the measurable cross-section theorem (see Theorem A.16) yields that there exist measurable functions: $\gamma \in \Gamma \mapsto e(\gamma) \in \mathscr{L}(\mathfrak{H}_0)$ and $\gamma \in \Gamma \mapsto u(\gamma) \in \mathscr{L}(\mathfrak{H}_0)$ such that $(\gamma,e(\gamma),u(\gamma)) \in B$, $\gamma \in \Gamma$. Hence the isomorphisms: $x \in \mathscr{M}_1(\gamma) \mapsto u(\gamma)xu(\gamma)^* \in \mathscr{M}_2(\gamma)$, $\gamma \in \Gamma$, form a measurable field of isomorphisms $\{\pi_\gamma\}$ such that

$$\pi_\gamma(\mathscr{M}_1(\gamma)) = \mathscr{M}_2(\gamma). \qquad \text{Q.E.D.}$$

Corollary 8.30. *Suppose that*

$$\{\mathscr{M},\mathfrak{H}\} = \int_\Gamma^\oplus \{\mathscr{M}(\gamma),\mathfrak{H}(\gamma)\} \, d\mu(\gamma)$$

is a direct integral on a standard σ-finite measure space $\{\Gamma,\mu\}$. *If each* $\mathscr{M}(\gamma)$ *is isomorphic to a fixed* $\mathscr{M}_0$, *then*

$$\mathscr{M} \cong \mathscr{M}_0 \overline{\otimes} L^\infty(\Gamma,\mu).$$

The next result tells us the relation between disintegrations of a representation and a state of a C^*-algebra.

Theorem 8.31. *Let A be a separable C^*-algebra with state space $\mathfrak{S}$. For a Radon measure μ on $\mathfrak{S}$ with resultant φ, the following two statements are equivalent:*

(i) *μ is orthogonal,*
(ii) *The cyclic representation $\{\pi_\varphi, \mathfrak{H}_\varphi\}$ of A induced by φ is unitarily equivalent to the direct integral*

$$\{\pi, \mathfrak{H}\} = \int_{\mathfrak{S}}^{\oplus} \{\pi_\omega, \mathfrak{H}_\omega\}\, d\mu(\omega)$$

under the obvious correspondence: $\eta_\varphi(x) \leftrightarrow \int_{\mathfrak{S}}^{\oplus} \eta_\omega(x)\, d\mu(\omega)$, $x \in A$.

PROOF. (i) $\Rightarrow$ (ii): It is clear that the map:

$$\eta_\varphi(x) \in \mathfrak{H}_\varphi \mapsto \int_{\mathfrak{S}}^{\oplus} \eta_\omega(x)\, d\mu(\omega) \in \int_{\mathfrak{S}}^{\oplus} \mathfrak{H}_\omega\, d\mu(\omega) = \mathfrak{H}$$

is extended to an isometry of $\mathfrak{H}_\varphi$ into $\mathfrak{H}$, which we denote by U, and that $U\pi_\varphi(x) = \pi(x)U$, $x \in A$. Hence we have only to prove that U is surjective. Set $\mathfrak{K} = U\mathfrak{H}_\varphi$ and $T_f = \int_{\mathfrak{S}}^{\oplus} f(\omega)\, d\mu(\omega)$ for each $f \in L^\infty(\mathfrak{S}, \mu)$. For any $x, y \in A$, we have

$$\begin{aligned}(\theta_\mu(f)\eta_\varphi(x) | \eta_\varphi(y)) &= \int_{\mathfrak{S}} f(\omega)(\eta_\omega(x) | \eta_\omega(y))\, d\mu(\omega)\\ &= (T_f U\eta_\varphi(x) | U\eta_\varphi(y)),\end{aligned}$$

so that $U^*T_fU = \theta_\mu(f)$, $f \in L^\infty(\mathfrak{S}, \mu)$. On the other hand, we have

$$\begin{aligned}\|T_f U\eta_\varphi(x)\|^2 &= (T_{\bar{f}f} U\eta_\varphi(x) | U\eta_\varphi(x))\\ &= \int_{\mathfrak{S}} |f(\omega)|^2 \|\eta_\omega(x)\|^2\, d\mu(\omega) = (\theta_\mu(\bar{f}\cdot f)\eta_\varphi(x) | \eta_\varphi(x))\\ &= \|\theta_\mu(f)\eta_\varphi(x)\|^2 = \|U^*T_fU\eta_\varphi(x)\|^2;\end{aligned}$$

hence we get $UU^*T_fU = T_fU$ since UU^* is the projection of $\mathfrak{H}$ onto $\mathfrak{K}$. Thus T_f leaves $\mathfrak{K}$ invariant. Let $\xi = \int_{\mathfrak{S}}^{\oplus} \xi(\omega)\, d\mu(\omega)$ be orthogonal to $\mathfrak{K}$. For any $x \in A$ and $f \in L^\infty(\mathfrak{S}, \mu)$ we have

$$\begin{aligned}\int_{\mathfrak{S}} f(\omega)(\xi(\omega) | \eta_\omega(x))\, d\mu(\omega) &= \int_{\mathfrak{S}} (\xi(\omega) | \overline{f(\omega)}\eta_\omega(x))\, d\mu(\omega)\\ &= (\xi | T_f^* U\eta_\varphi(x)) = 0,\end{aligned}$$

so that $(\xi(\omega) | \eta_\omega(x)) = 0$ for almost every $\omega \in \mathfrak{S}$; hence $\xi(\omega) = 0$ for almost every $\omega \in \mathfrak{S}$ by the separability of A. Thus $\xi = 0$, which implies that $\mathfrak{K} = \mathfrak{H}$.

(ii) $\Rightarrow$ (i): Suppose that the map U defined above is surjective. We then have $U^*T_fU = \theta_\mu(f)$ as seen above, where we have not yet used the orthogonality of μ. Hence θ_μ is multiplicative, which establishes the orthogonality of μ. Q.E.D.

Theorem 8.32. *Let A be a separable and*

$$\{\pi,\mathfrak{H}\} = \int_\Gamma^\oplus \{\pi_\gamma,\mathfrak{H}(\gamma)\}\, d\mu(\gamma)$$

be a direct integral of representations of A on a standard σ-finite measure space $\{\Gamma,\mu\}$. Then the following statements are equivalent:

(i) *The diagonal algebra $\mathscr{A}$ is maximal abelian in $\pi(A)'$ (resp.*

$$\mathscr{A} = (\mathscr{A}' \cap \pi(A)')' \cap \pi(A)').$$

(ii) *Almost every component π_γ is irreducible (resp. factorial in the sense that $\pi_\gamma(A)''$ is a factor).*

PROOF. Set $\mathscr{M}(\gamma) = \pi_\gamma(A)''$, $\gamma \in \Gamma$. It follows that $\{\mathscr{M}(\gamma)\}$ is a measurable field of von Neumann algebras. Set

$$\mathscr{M} = \int_\Gamma^\oplus \mathscr{M}(\gamma)\, d\mu(\gamma).$$

Then we have

$$\mathscr{M}' = \int_\Gamma^\oplus \mathscr{M}(\gamma)'\, d\mu(\gamma) = \int_\Gamma^\oplus \pi_\gamma(A)'\, d\mu(\gamma).$$

Since $x = \int_\Gamma^\oplus x(\gamma)\, d\mu(\gamma)$ commutes with $\pi(A)$ if and only if $x(\gamma) \in \pi_\gamma(A)'$ for almost every $\gamma \in \Gamma$, we have $\mathscr{M}' = \pi(A)' \cap \mathscr{A}'$, so that $\mathscr{M}$ is generated by $\pi(A)$ and $\mathscr{A}$. Hence $\mathscr{A}$ is maximal abelian in $\pi(A)'$ if and only if $\mathscr{A}' = \mathscr{M}$ if and only if $\mathscr{M}' = \mathscr{A}$ if and only if $\mathscr{M}(\gamma)' = \mathbf{C}$ for almost every $\gamma \in \Gamma$ if and only if π_γ is irreducible for almost every $\gamma \in \Gamma$. By Corollary 8.20, $\mathscr{M}(\gamma)$ is a factor for almost every $\gamma \in \Gamma$ if and only if $\mathscr{A}$ is the center of $\mathscr{M}'$ if and only if $\mathscr{A} = (\mathscr{A}' \cap \pi(A)')' \cap (\pi(A)' \cap \mathscr{A}') = (\mathscr{A}' \cap \pi(A)')' \cap \pi(A)'$, because $\mathscr{M} \subset \mathscr{A}'$.
Q.E.D.

Definition 8.33. Given a measurable field $\{\mathscr{M}(\gamma),\mathfrak{H}(\gamma)\}$ of von Neumann algebras on a Borel space $\{\Gamma,\mu\}$ with a measure, a field; $\gamma \in \Gamma \mapsto \varphi(\gamma) \in \mathscr{M}(\gamma)_*$ is said to be *measurable* if for every measurable operator field $x(\cdot) \in \prod \mathscr{M}(\gamma)$, the function: $\gamma \in \Gamma \mapsto \langle x(\gamma),\varphi(\gamma)\rangle \in \mathbf{C}$ is μ-measurable.

If $\{x_n(\cdot)\}$ is a sequence of measurable operator fields such that $\{x_n(\gamma)\}$ is σ-weakly dense in the unit ball of $\mathscr{M}(\gamma)$, then we have

$$\|\varphi(\gamma)\| = \sup_n |\langle x_n(\gamma),\varphi(\gamma)\rangle|,$$

so that the function $\gamma \in \Gamma \mapsto \|\varphi(\gamma)\|$ is μ-measurable for any measurable field $\varphi(\cdot)$ of normal functionals. If

$$\|\varphi\| = \int_\Gamma \|\varphi(\gamma)\|\, d\mu(\gamma) < +\infty,$$

$\varphi(\cdot)$ is said to be *integrable*.

Proposition 8.34. *If $\{\mathscr{M},\mathfrak{H}\} = \int_\Gamma^\oplus \{\mathscr{M}(\gamma),\mathfrak{H}(\gamma)\}\, d\mu(\gamma)$ is a direct integral of von Neumann algebras on Γ a Borel space Γ with a σ-finite measure μ, then every*

$\varphi \in \mathcal{M}_*$ is of the form

$$\left\langle \int_\Gamma^\oplus x(\gamma)\, d\mu(\gamma), \varphi \right\rangle = \int_\Gamma \langle x(\gamma), \varphi(\gamma)\rangle\, d\mu(\gamma)$$

for a unique integrable field $\varphi(\cdot)$ of normal functionals, and

$$\|\varphi\| = \int_\Gamma \|\varphi(\gamma)\|\, d\mu(\gamma).$$

PROOF. If $\varphi(\cdot)$ is an integrable field of normal functionals, then the functional φ defined by

$$\left\langle \int_\Gamma^\oplus x(\gamma)\, d\mu(\gamma), \varphi \right\rangle = \int_\Gamma \langle x(\gamma), \varphi(\gamma)\rangle\, d\mu(\gamma)$$

makes sense and is an element of $\mathcal{M}_*$, and we have

$$\|\varphi\| \le \int_\Gamma \|\varphi(\gamma)\|\, d\mu(\gamma).$$

If $\varphi \in \mathcal{M}_*$, then there exist sequences $\{\xi_n\}$ and $\{\eta_n\}$ in $\mathfrak{H}$ such that

$$\varphi = \sum_{n=1}^{\infty} \omega_{\xi_n,\eta_n}$$

and $\sum_{n=1}^\infty \|\xi_n\|^2 < +\infty$, $\sum_{n=1}^\infty \|\eta_n\|^2 < +\infty$. Set $\varphi(\gamma) = \sum_{n=1}^\infty \omega_{\xi_n(\gamma),\eta_n(\gamma)}$, $\gamma \in \Gamma$. By Fubini's theorem, we have $\sum_{n=1}^\infty \|\xi_n(\gamma)\|^2 < +\infty$ and $\sum_{n=1}^\infty \|\eta_n(\gamma)\|^2 < +\infty$ for almost every $\gamma \in \Gamma$. Hence $\varphi(\gamma) \in \mathcal{M}(\gamma)_*$ for almost every $\gamma \in \Gamma$ and the function: $\gamma \in \Gamma \mapsto \langle x(\gamma), \varphi(\gamma)\rangle \in \mathbf{C}$ is μ-measurable for every bounded measurable operator field $x(\cdot)$. For a measurable operator field $x(\cdot)$, set $x_n(\gamma) = x(\gamma)$ if $\|x(\gamma)\| \le n$, and $=0$ otherwise. Then $x_n(\cdot)$ is bounded and measurable. Hence $\gamma \in \Gamma \mapsto \langle x(\gamma), \varphi(\gamma)\rangle = \lim_n \langle x_n(\gamma), \varphi(\gamma)\rangle$ is measurable, so that $\varphi(\cdot)$ is measurable. Furthermore, we have

$$\begin{aligned}\int \|\varphi(\gamma)\|\, d\mu(\gamma) &\le \sum_{n=1}^{\infty} \int_\Gamma \|\xi_n(\gamma)\|\,\|\eta_n(\gamma)\|\, d\mu(\gamma)\\ &\le (\textstyle\sum \|\xi_n\|^2)^{1/2} (\sum \|\eta_n\|^2)^{1/2} < +\infty,\end{aligned}$$

so that $\varphi(\cdot)$ is integrable.

Now, we choose an element $a \in \mathcal{M}$ such that $\langle a, \varphi\rangle = \|\varphi\|$ and $\|a\| = 1$. We then have

$$\|\varphi\| = \langle a, \varphi\rangle = \int_\Gamma \langle a(\gamma), \varphi(\gamma)\rangle\, d\mu(\gamma).$$

Since $\|a\| = \text{ess sup}\|a(\gamma)\|$, we have $\|a(\gamma)\| \le 1$ for almost every $\gamma \in \Gamma$, so that $\langle a(\gamma), \varphi(\gamma)\rangle = \|\varphi(\gamma)\|$ for almost every $\gamma \in \Gamma$. Q.E.D.

We shall write

$$\varphi = \int_\Gamma^\oplus \varphi(\gamma)\, d\mu(\gamma),$$

$$\mathcal{M}_* = \int_\Gamma^\oplus \mathcal{M}(\gamma)_*\, d\mu(\gamma).$$

EXERCISES

1. Let $\{\mathfrak{H}(\gamma):\gamma \in \Gamma\}$ be a measurable field of Hilbert spaces on a Borel space Γ equipped with a positive σ-finite measure μ. Show that the direct integral

$$\mathfrak{H} = \int_\Gamma^\oplus \mathfrak{H}(\gamma)\, d\mu(\gamma),$$

the collection of all measurable vector fields $\xi(\cdot)$ with

$$\|\xi\| = \left\{\int_\Gamma \|\xi(\gamma)\|^2\, d\mu(\gamma)\right\}^{1/2} < +\infty,$$

is indeed complete. (*Hint*: Use arguments similar to those in the proof of Proposition 7.4.)

2. For the same situation as in the previous problem, let

$$x_n = \int_\Gamma^\oplus x_n(\gamma)\, d\mu(\gamma) \qquad n = 1,2,\ldots,$$

$$x = \int_\Gamma^\oplus x(\gamma)\, d\mu(\gamma)$$

be a family of decomposable operators. Show that if $\{x_n\}$ converges strongly to x, then there exists a subsequence $\{x_{n_j}\}$ such that $\{x_{n_j}(\gamma)\}$ converges strongly to $x(\gamma)$ for almost every $\gamma \in \Gamma$. (*Hint*: Let $M = \sup_n \|x_n\|$. Then $\|x_n(\gamma)\| \le M$, $n = 1,2,\ldots$ for almost every $\gamma \in \Gamma$. Let $\mathfrak{H}_0$ be a countable dense subset of $\mathfrak{H}$. Show that $\{\xi(\gamma):\xi \in \mathfrak{H}_0\}$ is dense in $\mathfrak{H}(\gamma)$ for almost every $\gamma \in \Gamma$ and that for each $\xi \in \mathfrak{H}_0$, $\{x_n\}$ contains a subsequence $\{x_{n_j}\}$ such that $x_{n_j}(\gamma)\xi(\gamma)$ converges strongly to $x(\gamma)\xi(\gamma)$ for almost every $\gamma \in \Gamma$, and then apply the diagonal process for choosing a subsequence from $\{x_n\}$ which works uniformly for all $\xi \in \mathfrak{H}_0$; then use the fact that $\{x_n(\gamma)\}$ is bounded.)

3. Let $\{\Gamma,\mu\}$ be a standard measure space with $\mu(\Gamma) < +\infty$. Let $\{\mathcal{M},\mathfrak{H}\}$ be a von Neumann algebra on a separable Hilbert space and θ be an isomorphism of $L^\infty(\Gamma,\mu)$ onto a von Neumann subalgebra $\mathcal{A}$ of $\mathcal{M}$. Let

$$\mathfrak{H} = \int_\Gamma^\oplus \mathfrak{H}(\gamma)\, d\mu(\gamma)$$

be the disintegration of $\mathfrak{H}$ with respect to the diagonal algebra $\mathcal{A}$ over $\{\Gamma,\mu\}$. Let A and B be two separable C^*-algebras contained in $\mathcal{M}'$ such that $\mathcal{M} = A' = B'$. Let π^A and π^B denote the identity representations of A and B, respectively, and

$$\pi^A = \int_\Gamma^\oplus \pi_\gamma^A\, d\mu(\gamma), \qquad \pi^B = \int_\Gamma^\oplus \pi_\gamma^B\, d\mu(\gamma)$$

be the disintegrations of π^A and π^B, respectively, relative to the above disintegration of $\mathfrak{H}$. Show that there exists a null set N in Γ such that for each $\gamma_1,\gamma_2 \in \Gamma - N$, $\pi_{\gamma_1}^A \simeq \pi_{\gamma_2}^A$ if and only if $\pi_{\gamma_1}^B \simeq \pi_{\gamma_2}^B$. (*Hint*: Let A_0 and B_0 be countable dense subsets of A and B, respectively. Using the previous problem, show that for each $a \in A_0$, there exists a null set $N_a \subset \Gamma$ such that for every $\gamma \in \Gamma - N_a$, $\pi_\gamma(a)$ is a strong limit of $\pi_\gamma(B_0)$. Choose a similar null set N_b for each $b \in B_0$ and set $N = \bigcup_{a \in A_0} N_a \cup (\bigcup_{b \in B_0} N_b)$. Show that this N has the desired property.)

4. Let $\{\Gamma,\mu\}$ be a standard measure space with $\mu(\Gamma) < +\infty$. Let θ be an isomorphism of $L^\infty(\Gamma,\mu)$ onto a von Neumann subalgebra $\mathscr{A}$ of a von Neumann algebra $\mathscr{M}$ with separable predual. Let $\{\rho_1,\mathfrak{H}_1\}$ and $\{\rho_2,\mathfrak{H}_2\}$ be two faithful normal representations of $\mathscr{M}$. Decompose $\mathfrak{H}_1$ and $\mathfrak{H}_2$ with respect to $\rho_1(\mathscr{A})$ and $\rho_2(\mathscr{A})$ on $\{\Gamma,\mu\}$:

$$\mathfrak{H}_1 = \int_\Gamma^\oplus \mathfrak{H}_1(\gamma)\, d\mu(\gamma); \qquad \mathfrak{H}_2 = \int_\Gamma^\oplus \mathfrak{H}_2(\gamma)\, d\mu(\gamma).$$

Let A and B be separable C^*-algebras on $\mathfrak{H}_1$ and $\mathfrak{H}_2$, respectively, such that $A' = \rho_1(\mathscr{M})$ and $B' = \rho_2(\mathscr{M})$. Disintegrate the identity representations π^A of A and π^B of B relative to the above direct integrals:

$$\pi^A = \int_\Gamma^\oplus \pi_\gamma^A\, d\mu(\gamma); \qquad \pi^B = \int_\Gamma^\oplus \pi_\gamma^B\, d\mu(\gamma).$$

Show that there exists a null set $N \subset \Gamma$ such that for each pair $\gamma_1,\gamma_2 \in \Gamma - N$, $\pi_{\gamma_1}^A \simeq \pi_{\gamma_2}^A$ if and only if $\pi_{\gamma_1}^B \simeq \pi_{\gamma_2}^B$. (*Hint*: If $\rho_1 \simeq \rho_2$, then the answer follows from the previous exercise. It follows also from the previous exercise that one can choose freely a separable σ-weakly dense C^*-algebra from $\rho_1(\mathscr{M})'$ or $\rho_2(\mathscr{M})'$. Theorem 5.5 tells us that one can assume that the isomorphism $\rho_2 \circ \rho_1^{-1}$ has a special form.)

Therefore, the equivalence relation in Γ is determined, up to a null set, by an isomorphism θ of $L^\infty(\Gamma,\mu)$ into $\mathscr{M}$. In other words, one can view this equivalence relation as an algebraic invariant of the pair $\{\mathscr{M},\mathscr{A}\}$.

Notes

After the existence of sufficiently many irreducible representations was established, Theorem I.9.23, a decomposition of a given representation into irreducible components was quickly established by several mathematicians, for example, R. Godement [143], F. Mautner [230], and I. Segal [31]. Their approaches are more or less based on the reduction theory of J. von Neumann [259]. It was M. Tomita [370] who tried first to represent a state as the barycenter of a boundary measure on the state space prior to the development of the theory of boundary integrals, which in turn provides a useful tool for the present subject. The so-called Choquet theory of boundary integrals was first explicitly applied to the present subject by D. Ruelle [299]. The presentation in Section 6 follows C. Skau [333], with some modification. Theorem 6.19 is due to C. Skau [333]. Theorem 6.25 is due to D. Ruelle [299]. Theorem 6.28 in the separable case is due to F. Mautner [230] and I. Segal [31]. The general case is due to E. Bishop and K. de Leeuw [57], who proved the theorem in the context of a general compact convex set. The formulation of the central decomposition of a state was first given by S. Sakai [310], while the central decomposition of a representation was considered earlier by several mathematicians, for example, J. Ernest [110] and J. Dixmier [92]. Theorem 6.32 in the separable case is due to F. Mautner [230, 231] and I. Segal [31]. The general case is due to W. Wils [402]. Theorem 6.34 is due to J. Ernest [110] and J. Feldman [116].

The material presented in Section 7 was adapted from the Ionescu-Tulcea theory of lifting, [13].

The presentation of Section 8 follows E. Effros' treatise [106]. However, besides his modern formulation of the theory, this is an area where von Neumann's original theory remains most unaffected by the subsequent development.

Chapter V
Types of von Neumann Algebras and Traces

0. Introduction

The material presented in this chapter is directly related to the dimension theory of Murray and von Neumann. The projections of a von Neumann algebra form a complete lattice. A partial ordering and an equivalence relation on the projection lattice of a von Neumann algebra are introduced by means of the partial isometries in the algebra in Section 1. According to the structure of the projection lattice with this ordering, the algebras are classified into those of type I, type II_1, type II_∞, and type III. It will be shown that every von Neumann algebra is decomposed into the direct sum of the algebras of these types. We will see that the von Neumann algebras of type I behave most naturally from the classical point of view. The structure of such an algebra will be completely determined in terms of the spectrum of the center and a set of cardinal numbers. At this point, it should be mentioned that the main task of the theory of von Neumann algebras is to establish methods for analyzing von Neumann algebras of type II or type III. Indeed, von Neumann algebras of type I do not introduce anything mysterious into the frame of the classical point of view, while there is no other theory which can handle mathematical phenomena related to von Neumann algebras of nontype I. Section 2 is devoted to the study of traces on semi-finite von Neumann algebras. It will be seen that the relative dimension of projections in a factor of type II takes continuous values. Normal representations of a W^*-algebra are discussed in Section 3 based on a result in Section IV.5. We shall show in Section 4 that any norm closed convex set in a von Neumann algebra invariant under the inner automorphisms intersects with the center. In Section 5, any separable representation of a von Neumann algebra with no finite direct summand of type I is shown to be normal.

Section 6 is devoted to the study of the Borel spaces of von Neumann algebras of various types. In the last section, the so-called group measure space construction of factors will be presented and the existence of factors of type I, type II, and type III will be established.

1. Projections and Types of von Neumann Algebras

The spectral and polar decomposition theorems say that every bounded operator on a Hilbert space is, in principle, decomposed into a combination of partial isometries and projections. In other words, the partial isometries and the projections of a von Neumann algebra form a fundamental structure of the algebra. In this section, we shall examine them. It is not an overstatement to say that the study of the projection lattice of a von Neumann algebra is at the core of the whole theory.

We begin with the following elementary but important fact:

Proposition 1.1. *If $\mathcal{M}$ is a von Neumann algebra, then the set $\mathcal{M}_p$ of all projections of $\mathcal{M}$ is a complete lattice.*

PROOF. Suppose $\mathcal{M}$ acts on a Hilbert space $\mathfrak{H}$. Let $\{e_i\}_{i\in I}$ be a family of projections in $\mathcal{M}$. Let e_0 be the projection of $\mathfrak{H}$ onto the closed subspace $\bigcap_{i\in I} e_i\mathfrak{H}$. Any unitary in $\mathcal{M}'$ leaves each $e_i\mathfrak{H}$ invariant; so it leaves the intersection $\bigcap_{i\in I} e_i\mathfrak{H}$ invariant as well; hence it commutes with e_0, which means that $e_0 \in \mathcal{M}$. It is clear that e_0 is the greatest lower bound of $\{e_i\}_{i\in I}$. We denote it by $\bigwedge_{i\in I} e_i$. The least upper bound of $\{e_i\}_{i\in I}$ is given by $1 - \bigwedge_{i\in I}(1 - e_i)$, which is denoted by $\bigvee_{i\in I} e_i$. Q.E.D.

The set $\mathcal{M}_p$ of all projections in $\mathcal{M}$ is called the *projection lattice of* $\mathcal{M}$.

Definition 1.2. Two projections e and f in a von Neumann algebra $\mathcal{M}$ are said to be *equivalent* if there exists an element $u \in \mathcal{M}$ with $u^*u = e$ and $uu^* = f$. We write this fact as $e \sim f$. The projections e and f are called, respectively, the *initial* projection and the *final* projection of u.

Observe that if u^*u is a projection, then uu^* is automatically a projection too. If a projection $e \in \mathcal{M}$ is equivalent to a projection $f_1 \in \mathcal{M}$ with $f_1 \leq f \in \mathcal{M}$, then we write $e \precsim f$ or $f \succsim e$. If $e \precsim f$ and $e \nsim f$, then we write $e \prec f$ or $f \succ e$. Clearly, the relation $e \sim f$ is an equivalence relation. We shall also use these notations for the subspaces of the underlying Hilbert space $\mathfrak{H}$ of $\mathcal{M}$ when $\mathcal{M}$ is represented on it. In other words, if $\mathfrak{M}$ and $\mathfrak{N}$ are the ranges of projections e and f in $\mathcal{M}$, respectively, $\mathfrak{M} \precsim \mathfrak{N}$ means that $e \precsim f$ in $\mathcal{M}$.

If $\{e_i\}_{i\in I}$ and $\{f_i\}_{i\in I}$ are two families of orthogonal projections such that $e_i \sim f_i$, $i \in I$, then we have

$$e = \sum_{i\in I} e_i \sim \sum_{i\in I} f_i = f.$$

In fact, let u_i be a partial isometry in $\mathcal{M}$ with $u_i^*u_i = e_i$ and $u_iu_i^* = f_i$. The summation $\sum_{i\in I} u_i$ converges in the σ-strong topology in $\mathcal{M}$ and $u = \sum_{i\in I} u_i$ is a partial isometry which sets up the equivalence $e \sim f$.

It is easily seen that if $e \sim f$, then $ze \sim zf$ for any central projection z. In fact, if u is a partial isometry in $\mathcal{M}$ with $u^*u = e$ and $uu^* = f$, then uz sets up the equivalence of ez and fz.

Proposition 1.3. *The relations $e \precsim f$ and $e \succsim f$ imply $e \sim f$.*

PROOF. Choose partial isometries $u, v \in \mathcal{M}$ with

$$\begin{aligned} e &= u^*u, \qquad & uu^* &= f_1 \le f, \\ f &= v^*v, \qquad & vv^* &= e_1 \le e. \end{aligned}$$

By induction, we construct two decreasing sequences $\{e_n\}$ and $\{f_n\}$ of projections as follows:

$$e_{n+1} = vf_nv^*, \qquad f_{n+1} = ue_nu^*, \qquad n = 1,2,\ldots.$$

Putting $e_\infty = \lim e_n$ and $f_\infty = \lim f_n$, we have

$$e = \sum_{n=0}^{\infty} (e_n - e_{n+1}) + e_\infty, \qquad f = \sum_{n=0}^{\infty} (f_n - f_{n+1}) + f_\infty,$$

where $e_0 = e$ and $f_0 = f$. Clearly, we have

$$u(e_n - e_{n+1})u^* = f_{n+1} - f_{n+2}, \qquad v(f_n - f_{n+1})v^* = e_{n+1} - e_{n+2},$$
$$ue_\infty u^* = f_\infty \quad \text{and} \quad vf_\infty v^* = e_\infty.$$

Hence we have

$$e_{2n} - e_{2n+1} \sim f_{2n+1} - f_{2n+2}, \qquad e_{2n+1} - e_{2n+2} \sim f_{2n} \sim f_{2n+1},$$

so that

$$\begin{aligned} e &= \sum_{n=0}^{\infty} (e_{2n} - e_{2n+1}) + \sum_{n=0}^{\infty} (e_{2n+1} - e_{2n+2}) + e_\infty \\ &\sim \sum_{n=0}^{\infty} (f_{2n+1} - f_{2n+2}) + \sum_{n=0}^{\infty} (f_{2n} - f_{2n+1}) + f_\infty = f. \end{aligned}$$

Q.E.D.

Definition 1.4. Given an element $x \in \mathcal{M}$, the smallest projection e in $\mathcal{M}$ with $ex = x$ is called the *left support* of x and denoted by $s_l(x)$. The *right support* $s_r(x)$ is the smallest projection f in $\mathcal{M}$ with $xf = x$.

It follows then that the left (resp. right) support $s_l(x)$ (resp. $s_r(x)$) of $x \in \mathcal{M}$ is the (range) projection onto the closure of the range of x (resp. x^*).

Proposition 1.5. *For each $x \in \mathcal{M}$, we have*

$$s_l(x) \sim s_r(x).$$

PROOF. Let $x = uh$ be the polar decomposition of x. We have then $s_l(x) = uu^*$ and $s_r(x) = u^*u$. Hence $s_l(x) \sim s_r(x)$. Q.E.D.

Proposition 1.6. *For any pair of projections $e, f \in \mathcal{M}$,*

$$(e \vee f) - e \sim f - (e \wedge f).$$

PROOF. We assume that $\mathcal{M}$ acts on a Hilbert space $\mathfrak{H}$. Let $\mathfrak{M}$ and $\mathfrak{N}$ denote the ranges of e and f, respectively. Let $\mathfrak{K}$ denote the null space $\{\xi \in \mathfrak{H}: (1 - e)f\xi = 0\}$ of $(1 - e)f$. It follows then that

$$\mathfrak{K} = (\mathfrak{H} \ominus \mathfrak{N}) \oplus (\mathfrak{M} \cap \mathfrak{N}).$$

Hence the range projection of $f(1 - e) = [(1 - e)f]^*$ is $f - e \wedge f$. Similarly, the projection to the null space of $f(1 - e)$ (the null projection of $f(1 - e)$) is $e + \{(1 - e) \wedge (1 - f)\} = e + \{1 - (e \vee f)\}$. Hence the right support $s_r(f(1 - e))$ is $(e \vee f) - e$. Hence we have $(e \vee f) - e \sim f - (e \wedge f)$ by Proposition 1.5. Q.E.D.

We recall that the central support $z(e)$ of a projection $e \in \mathcal{M}$ means the smallest central projection in $\mathcal{M}$ majorizing e.

Lemma 1.7 *For two projections e and f in $\mathcal{M}$, the following statements are equivalent:*

(i) *$z(e)$ and $z(f)$ are not orthogonal.*
(ii) *$e\mathcal{M}f \neq \{0\}$.*
(iii) *There exist nonzero projections $e_1 \leq e$ and $f_1 \leq f$ in $\mathcal{M}$ such that $e_1 \sim f_1$.*

PROOF. The implications (iii) $\Rightarrow$ (ii) $\Rightarrow$ (i) are almost obvious.

(i) $\Rightarrow$ (ii): Suppose $e\mathcal{M}f = \{0\}$. Put $\mathfrak{m} = \{x \in \mathcal{M}: e\mathcal{M}x = \{0\}\}$. Then $\mathfrak{m}$ is a σ-weakly closed ideal of $\mathcal{M}$, so that there exists a central projection $z \in \mathcal{M}$ with $\mathfrak{m} = \mathcal{M}z$. Since $f \in \mathfrak{m}$, we have $f \leq z$; so $z(f) \leq z$. By definition, $ez = 0$; so $z(f)e = 0$; hence $e \leq 1 - z(f)$. Therefore $z(e) \leq 1 - z(f)$. Thus $z(e)$ and $z(f)$ are orthogonal. Here we remark that $z(e) = 1 - z$.

(ii) $\Rightarrow$ (iii): Take a nonzero $x \in e\mathcal{M}f$. Then $exf = x$, so that $e_1 = s_l(x) \leq e$ and $f_1 = s_r(x) \leq f$ and $e_1 \sim f_1$ by the previous proposition. Q.E.D.

If $z(e)$ and $z(f)$ are orthogonal, then e and f are said to be *centrally orthogonal.*

The following result provides a powerful tool in the study of the projection lattice.

Theorem 1.8 (Comparability Theorem). *For any pair e, f of projections in a von Neumann algebra $\mathcal{M}$, there exists a central projection z such that*

$$ze \precsim zf \quad \text{and} \quad (1 - z)e \succsim (1 - z)f.$$

In particular, if $\mathcal{M}$ is a factor, then one and only one of the following relations holds:

$$e \prec f; \qquad e \sim f; \qquad e \succ f.$$

PROOF. Let $\{e_i\}_{i \in I}$ and $\{f_i\}_{i \in I}$ be a maximal family of orthogonal projections in $\mathcal{M}$ such that

$$e_i \sim f_i, \qquad e_i \leq e, \quad \text{and} \quad f_i \leq f, \qquad i \in I.$$

Put $e_0 = \sum_{i \in I} e_i$ and $f_0 = \sum_{i \in I} f_i$. Then $e_0 \sim f_0$. By maximality, there exist no nonzero projections $e_1 \leq e - e_0$ and $f_1 \leq f - f_0$ with $e_1 \sim f_1$. By Lemma 1.7, $e - e_0$ and $f - f_0$ are centrally orthogonal, that is, there exists a central projection z in $\mathcal{M}$ such that

$$e - e_0 \leq z, \qquad f - f_0 \leq 1 - z.$$

It follows that $(f - f_0)z = 0$ and $(e - e_0)z = e - e_0$; hence

$$fz = f_0 z \sim e_0 z \leq ez,$$
$$e(1 - z) = e_0(1 - z) \sim f_0(1 - z) \leq f(1 - z),$$

where we use the fact that $p \sim q$ implies $zp \sim zq$ for any central projection z and projections p, q in $\mathcal{M}$. Q.E.D.

Definition 1.9. Let $\{\mathcal{M},\mathfrak{H}\}$ be a von Neumann algebra. For each $\xi \in \mathfrak{H}$, we denote by e_ξ (resp. e'_ξ) the projection of $\mathfrak{H}$ onto $[\mathcal{M}'\xi]$ (resp. $[\mathcal{M}\xi]$), and call it the *cyclic* projection of $\mathcal{M}$ (resp. $\mathcal{M}'$) determined by ξ.

The cyclic projection e_ξ is the smallest projection $e \in \mathcal{M}$ with $e\xi = \xi$, and also is the support of ω_ξ.

Theorem 1.10. *If $\{\mathcal{M},\mathfrak{H}\}$ is a von Neumann algebra, then for any vectors $\xi,\eta \in \mathfrak{H}$, the following two conditions are equivalent:*

(i) $e_\xi \succsim e_\eta$;
(ii) $e'_\xi \succsim e'_\eta$.

PROOF. By symmetry, we shall only show the implication (ii) $\Rightarrow$ (i). Suppose that $v \in \mathcal{M}'$ is a partial isometry with $v^*v = e'_\eta$ and $vv^* \leq e'_\xi$. This means that v maps $[\mathcal{M}\eta]$ into $[\mathcal{M}\xi]$ isometrically. Let $\eta_1 = v\eta \in [\mathcal{M}\xi]$. We have then

$$[\mathcal{M}'\eta] = [\mathcal{M}'v^*\eta_1] \subset [\mathcal{M}'\eta_1] = [\mathcal{M}'v\eta] \subset [\mathcal{M}'\eta],$$

so that $[\mathcal{M}'\eta] = [\mathcal{M}'\eta_1]$, equivalently $e_\eta = e_{\eta_1}$. Thus, we shall prove that $e_{\eta_1} \precsim e_\xi$.

We now consider a functional $\varphi(x) = (x\eta_1|\xi)$, $x \in \mathcal{M}$. Let $\varphi = u|\varphi|$ be the polar decomposition, Theorem III.4.2. We then have, for any $x \in \mathcal{M}$,

$$\begin{aligned}(uu^*\eta_1|x\xi) &= (x^*uu^*\eta_1|\xi) = \varphi(x^*uu^*)\\ &= \varphi(x^*) = (x^*\eta_1|\xi) = (\eta_1|x\xi).\end{aligned}$$

Thus $uu^*\eta_1 = \eta_1$ since η_1 and $uu^*\eta_1$ are both in $[\mathcal{M}\xi]$. Let $\eta_2 = u^*\eta_1$. We then have

$$|\varphi|(a) = (a\eta_2|\xi) = (e_\xi a\eta_2|\xi) = |\varphi|(e_\xi a), \qquad a \in \mathcal{M}.$$

Hence $|\varphi| = |\varphi|e_\xi$, so that $|\varphi| = e_\xi|\varphi|$ as well, which means that for each $a \in \mathcal{M}$,

$$\begin{aligned}(e_\xi\eta_2|a\xi) &= (a^*e_\xi\eta_2|\xi) = e_\xi|\varphi|(a^*) = |\varphi|(a^*)\\ &= (a^*\eta_2|\xi) = (\eta_2|a\xi).\end{aligned}$$

Since $e_\xi\eta_2$ and η_2 are both in $[\mathcal{M}\xi]$, we get $\eta_2 = e_\xi\eta_2$, which means that $\eta_2 \in [\mathcal{M}'\xi]$. Therefore, we get

$$\begin{aligned}&[\mathcal{M}'\xi] \supset [\mathcal{M}'\eta_2] = [\mathcal{M}'u^*\eta_1] = u^*[\mathcal{M}'\eta_1] = u^*[\mathcal{M}'\eta],\\ &[\mathcal{M}'\eta] = [\mathcal{M}'\eta_1] = [\mathcal{M}'u\eta_2] = u[\mathcal{M}'\eta_2].\end{aligned}$$

Thus, u^* maps isometrically $[\mathcal{M}'\eta]$ onto $[\mathcal{M}'\eta_2] \subset [\mathcal{M}'\xi]$; hence $e_\eta \precsim e_\xi$. Q.E.D.

Corollary 1.11. *Let A be a C^*-algebra. For a pair φ, ψ of positive linear functionals on A, the following two conditions are equivalent:*

(i) *The cyclic representation π_φ of A induced by φ is unitarily equivalent to a subrepresentation of the cyclic representation π_ψ of A induced by ψ.*
(ii) $s(\varphi) \precsim s(\psi)$ *in $\tilde{A}$, where $s(\varphi)$ and $s(\psi)$ denote the support projections of φ and ψ, respectively, in the universal enveloping von Neumann algebra $\tilde{A}$ of A.*

Proof. Let $\{\pi,\mathfrak{H}\}$ be the universal representation of A. It follows that $\tilde{A} = \pi(A)''$ and that there exist $\xi,\eta \in \mathfrak{H}$ such that

$$\varphi(x) = (\pi(x)\xi|\xi), \qquad \psi(x) = (\pi(x)\eta|\eta), \qquad x \in A.$$

Furthermore, we have that $\{\pi_\varphi,\mathfrak{H}_\varphi\} \cong \{\pi,[\pi(A)\xi]\}$ and $\{\pi_\psi,\mathfrak{H}_\psi\} \cong \{\pi,[\pi(A)\eta]\}$. Let e' and f' be the projections of $\mathfrak{H}$ onto $[\pi(A)\xi]$ and $[\pi(A)\eta]$, respectively, which both belong to $\pi(A)' = \tilde{A}'$. We claim that $e' \precsim f'$ if and only if π_φ is equivalent to a subrepresentation of π_ψ. Suppose $e' \precsim f'$. There exists a partial isometry v in $\tilde{A}'$ which maps $[\pi(A)\xi]$ isometrically onto a subspace $\mathfrak{N}$ of $[\pi(A)\eta]$. For any $\zeta \in [\pi(A)\xi]$, we have

$$\begin{aligned}v\pi_\varphi(x)\zeta &= v\pi(x)\zeta = \pi(x)v\zeta\\ &= \pi_\psi(x)v\zeta,\end{aligned}$$

where we identify π_φ and π_ψ with $\{\pi,[\pi(A)\xi]\}$ and $\{\pi,\pi(A)\eta]\}$, respectively. Thus $v\pi_\varphi(x)v^* = \pi_\psi(x)|_{\mathfrak{N}}$. Hence $\pi_\varphi = \pi_\psi|_{\mathfrak{N}}$. From the above discussion, it is routine to check the converse.

By Theorem 1.10, $e' \precsim f'$ in $\tilde{A}'$ if and only if $e \precsim f$, where e and f are the cyclic projections in $\tilde{A}$ determined by ξ and η, respectively. But e (resp. f) is the smallest projection in $\tilde{A}$ such that $e\xi = \xi$ (resp. $f\eta = \eta$), equivalently $\omega_\xi(1-e) = 0$ (resp. $\omega_\eta(1-f) = 0$). Thus e and f are, respectively, $s(\omega_\xi) = s(\varphi)$ and $s(\omega_\eta) = s(\psi)$. Q.E.D.

Corollary 1.12. *Let $\{\mathcal{M},\mathfrak{H}\}$ be a von Neumann algebra. If there exists a separating vector in $\mathfrak{H}$, then every normal positive linear functional on $\mathcal{M}$ is of the form ω_ξ for some vector $\xi \in \mathfrak{H}$.*

PROOF. Let ξ_0 be a separating vector in $\mathfrak{H}$, and φ a normal positive linear functional on $\mathcal{M}$. Let $\mathfrak{K} = \mathfrak{H} \oplus \mathfrak{H}_\varphi$, where $\mathfrak{H}_\varphi$ denotes, of course, the underlying Hilbert space of the cyclic representation π_φ of $\mathcal{M}$ induced by φ. Let $\pi(x) = x \oplus \pi_\varphi(x)$, $x \in \mathcal{M}$, and $\mathcal{N} = \pi(\mathcal{M})$. Since π is a normal representation, $\mathcal{N}$ is a von Neumann algebra on $\mathfrak{K}$.

Let $\xi_1 = \xi_0 \oplus 0$ and $\xi_2 = 0 \oplus \xi_\varphi$. We have then

$$\omega_{\xi_0}(x) = (\pi(x)\xi_1|\xi_1) \quad \text{and} \quad \varphi(x) = (\pi(x)\xi_2|\xi_2), \qquad x \in \mathcal{M}.$$

Since ξ_0 is separating for $\mathcal{M}$, ω_{ξ_0} is faithful on $\mathcal{M}$, so that ω_{ξ_1} is faithful on $\mathcal{N}$; thus ξ_1 is separating for $\mathcal{N}$; equivalently ξ_1 is cyclic for $\mathcal{N}'$. Thus, we have $[\mathcal{N}'\xi_1] = \mathfrak{K} \supset [\mathcal{N}'\xi_2]$. By Theorem 1.10, $[\mathcal{N}\xi_1] \succsim [\mathcal{N}\xi_2]$ in $\mathcal{N}'$. Thus there exists a partial isometry v in $\mathcal{N}'$ such that $v\xi_2 \in [\mathcal{N}\xi_1] \subset \mathfrak{H}$ and $v^*v\xi_2 = \xi_2$. Put $\xi = v\xi_2$. We have then, for any $x \in \mathcal{M}$,

$$\begin{aligned}\omega_\xi(x) &= (x\xi|\xi) = (\pi(x)v\xi_2|v\xi_2) = (\pi(x)v^*v\xi_2|\xi_2)\\ &= (\pi(x)\xi_2|\xi_2) = \varphi(x).\end{aligned}$$

Q.E.D.

Corollary 1.13. *Let $\{\mathcal{M}_1,\mathfrak{H}_1\}$ and $\{\mathcal{M}_2,\mathfrak{H}_2\}$ be von Neumann algebras. If $\mathcal{M}_1$ and $\mathcal{M}_2$ admit a cyclic and separating vector, then every isomorphism π of $\mathcal{M}_1$ onto $\mathcal{M}_2$ is spatial.*

PROOF. By Corollary IV.5.6, there exists a von Neumann algebra $\{\mathcal{M},\mathfrak{H}\}$ and projections $e_1' \in \mathcal{M}'$ and $e_2' \in \mathcal{M}'$ such that

$$\begin{aligned}&\{\mathcal{M}_1,\mathfrak{H}_1\} \cong \{\mathcal{M}_{e_1'}, e_1'\mathfrak{H}\}, \qquad \{\mathcal{M}_2,\mathfrak{H}_2\} \cong \{\mathcal{M}_{e_2'}, e_2'\mathfrak{H}\},\\ &z(e_1') = z(e_2') = 1 \quad \text{and} \quad \pi(x_{e_1'}) = x_{e_2'}, \qquad x \in \mathcal{M},\end{aligned}$$

under the identification of $\mathcal{M}_1$ with $\mathcal{M}_{e_1'}$ and $\mathcal{M}_2$ with $\mathcal{M}_{e_2'}$. Let ξ_1 (resp. ξ_2) be a cyclic and separating vector for $\mathcal{M}_1$ (resp. $\mathcal{M}_2$). We have then

$$\begin{aligned}\mathfrak{H}_1 &= [\mathcal{M}_1\xi_1] = [\mathcal{M}_1'\xi_1] = [\mathcal{M}\xi_1] = e_1'[\mathcal{M}'\xi_1],\\ \mathfrak{H}_2 &= [\mathcal{M}_2\xi_2] = \cdots = e_2'[\mathcal{M}'\xi_2].\end{aligned}$$

Hence we have $[\mathcal{M}'\xi_1] = [\mathcal{M}'\xi_2] = \mathfrak{H}$ because the inductions: $x \in \mathcal{M} \mapsto x_{e_i'} \in \mathcal{M}_{e_i'}$, $i = 1,2$, are both isomorphisms. By Theorem 1.10, we have $e_{\xi_1}' \sim e_{\xi_2}'$. In other words, there exists a partial isometry $u \in \mathcal{M}'$ with $u^*u = e_{\xi_1}'$ and $uu^* = e_{\xi_2}'$. But $e_{\xi_1}' = e_1'$ and $e_{\xi_2}' = e_2'$; hence we have

$$\pi(x_{e_1'}) = x_{e_2'} = ux_{e_1'}u^*,$$

which establishes that π is special. Q.E.D.

Corollary 1.14. *Let $\{\mathcal{M},\mathfrak{H}\}$ be a von Neumann algebra. If $\mathcal{M}$ admits a cyclic vector and a separating vector, then $\mathcal{M}$ admits a cyclic and separating vector*

PROOF. Let ξ_1 (resp. ξ_2) be a cyclic (resp. separating) vector for $\mathcal{M}$. It follows that $e'_{\xi_1} = 1$ and $e_{\xi_2} = 1$. By Theorem 1.10, we have $e'_{\xi_1} \sim e'_{\xi_2}$, so that there exists a partial isometry $u \in \mathcal{M}'$ isometrically mapping $[\mathcal{M}\xi_2]$ onto $[\mathcal{M}\xi_1] = \mathfrak{H}$. Let $\xi_0 = u\xi_2$. We then have

$$[\mathcal{M}\xi_0] = [\mathcal{M}u\xi_2] = u[\mathcal{M}\xi_2] = \mathfrak{H},$$
$$x\xi_0 = xu\xi_2 = ux\xi_2, \qquad x \in \mathcal{M}.$$

Thus ξ_0 is cyclic for $\mathcal{M}$ and $x\xi_0 = 0$, $x \in \mathcal{M}$, if and only if $x\xi_2 = 0$ if and only if $x = 0$. Therefore ξ_0 is separating as well. Q.E.D.

Definition 1.15. A projection e in a von Neumann algebra $\mathcal{M}$ is said to be *finite* if $e \sim f \leq e$ implies $e = f$. Otherwise, it is said to be *infinite*. A projection e is said to be *purely infinite* if there is no nonzero finite projection $f \leq e$ in $\mathcal{M}$. If ze is infinite for every central projection $z \in \mathcal{M}$ with $ze \neq 0$, then e is called *properly infinite*. If $e\mathcal{M}e$ is abelian, then e is said to be *abelian*.

It is clear that an abelian projection is finite, and a minimal projection is abelian.

Definition 1.16. A von Neumann algebra $\mathcal{M}$ is said to be *finite*, *infinite*, *properly infinite*, or *purely infinite* according to the property of the identity projection 1.

Definition 1.17. A von Neumann algebra $\mathcal{M}$ is said to be *of type I* if every nonzero central projection in $\mathcal{M}$ majorizes a nonzero abelian projection in $\mathcal{M}$. If there is no nonzero finite projection in $\mathcal{M}$, that is, if $\mathcal{M}$ is purely infinite, then it is said to be *of type III*. If $\mathcal{M}$ has no nonzero abelian projection and if every nonzero central projection in $\mathcal{M}$ majorizes a nonzero finite projection of $\mathcal{M}$, then it is said to be *of type II*. If $\mathcal{M}$ is finite and of type II, then it is said to be *of type II_1*. If $\mathcal{M}$ is of type II and has no nonzero central finite projection, then $\mathcal{M}$ is said to be *of type II_∞*.

Lemma 1.18. *The sum of centrally orthogonal abelian (resp. finite) projections $\{e_i\}_{i \in I}$ is abelian (resp. finite).*

PROOF. Let $e = \sum_{i \in I} e_i$ and $z_i = z(e_i)$, $i \in I$. It follows that $e_i = z_i e$, $i \in I$; hence $e\mathcal{M}e = \sum_{i \in I}^{\oplus} e_i\mathcal{M}e_i$. Therefore, if each $e_i\mathcal{M}e_i$ is abelian, then $e\mathcal{M}e$ is abelian. If $e \sim f \leq e$, then $z_i f = f_i \sim z_i e = e_i$ and $f_i \leq e_i$; so $e_i = f_i$ when e_i is finite, so that $e = f$. Hence e is finite. Q.E.D.

Theorem 1.19. *Every von Neumann algebra $\mathcal{M}$ is uniquely decomposable into the direct sum of those of type I, type II_1, type II_∞, and type III. Moreover, every projection e in $\mathcal{M}$ is uniquely written as the sum of centrally orthogonal projections e_1 and e_2 in $\mathcal{M}$ such that e_1 is finite and e_2 is properly infinite.*

PROOF. Let $\{e_i\}_{i \in I}$ be a maximal family of centrally orthogonal abelian projections in $\mathscr{M}$, and let $e = \sum_{i \in I} e_i$. By Lemma 1.18, e is abelian. Put $z_{\mathrm{I}} = z(e)$. If z is a nonzero central projection majorized by z_{I}, then ze is a nonzero abelian projection. Hence $\mathscr{M}z_{\mathrm{I}}$ is of type I. By construction, there is no nonzero abelian projection in $\mathscr{M}(1 - z_{\mathrm{I}})$, so that $\mathscr{M}_{(1-z_{\mathrm{I}})}$ has no nontrivial direct summand of type I. Let $\{f_j\}_{j \in J}$ be a maximal family of centrally orthogonal finite projections in $\mathscr{M}(1 - z_{\mathrm{I}})$, and let $f = \sum_{j \in J} f_j$. By Lemma 1.18, f is finite. Put $z_{\mathrm{II}} = z(f)$. By construction, $\mathscr{M}z_{\mathrm{II}}$ has no nonzero abelian projection, and every nonzero central projection z in $\mathscr{M}z_{\mathrm{II}}$ majorizes a finite projection $zf \neq 0$, so that $\mathscr{M}z_{\mathrm{II}}$ is of type II. By the maximality of $\{f_j\}_{j \in J}$, $1 - z_{\mathrm{II}} = z_{\mathrm{III}}$ does not majorize a nonzero finite projection, so $\mathscr{M}z_{\mathrm{III}}$ is of type III. By construction, $z_{\mathrm{I}} + z_{\mathrm{II}} + z_{\mathrm{III}} = 1$. Let $\{z_k\}_{k \in K}$ be a maximal family of orthogonal central finite projections in $\mathscr{M}z_{\mathrm{II}}$. Put $z_{\mathrm{II}_1} = \sum_{k \in K} z_k$ and $z_{\mathrm{II}_\infty} = z_{\mathrm{II}} - z_{\mathrm{II}_1}$. It follows that $\mathscr{M}z_{\mathrm{II}_1}$ is of type II_1 and $\mathscr{M}_{\mathrm{II}_\infty}$ is of type II_∞. Thus we obtain the direct sum decomposition:

$$\mathscr{M} = \mathscr{M}_{z_{\mathrm{I}}} \oplus \mathscr{M}_{z_{\mathrm{II}_1}} \oplus \mathscr{M}_{z_{\mathrm{II}_\infty}} \oplus \mathscr{M}_{z_{\mathrm{III}}}.$$

Suppose that $1 = z'_{\mathrm{I}} + z'_{\mathrm{II}_1} + z'_{\mathrm{II}_\infty} + z'_{\mathrm{III}}$ is another orthogonal decomposition of 1 with the same property. It follows that $z'_{\mathrm{I}}(1 - z_{\mathrm{I}}) = 0$ because $1 - z_{\mathrm{I}}$ majorizes no nonzero abelian projection while $z'_{\mathrm{I}}(1 - z_{\mathrm{I}})$ is a central projection in $\mathscr{M}z'_{\mathrm{I}}$. Hence $z'_{\mathrm{I}} \leq z_{\mathrm{I}}$. On the other hand, $z_{\mathrm{I}}(1 - z'_{\mathrm{I}}) = 0$ for the same reason, so that $z_{\mathrm{I}} \leq z'_{\mathrm{I}}$; hence $z_{\mathrm{I}} = z'_{\mathrm{I}}$. Similarly, it follows that $z_{\mathrm{II}_1} = z'_{\mathrm{II}_1}$, $z_{\mathrm{II}_\infty} = z'_{\mathrm{II}_\infty}$, and $z_{\mathrm{III}} = z'_{\mathrm{III}}$.

Let e be a nonzero projection in $\mathscr{M}$. Looking at $\mathscr{M}_e$ $(=e\mathscr{M}e)$, we may assume that $e = 1$. Let e_1 be the sum of a maximal orthogonal family of central finite projections. It follows then that e_1 is finite and central, and $1 - e_1$ is properly infinite. The uniqueness of this decomposition follows from the similar arguments for $\{z_{\mathrm{I}}, \ldots, z_{\mathrm{III}}\}$. Q.E.D.

Corollary 1.20. *A factor is either of type I, type II_1, type II_∞, or type III.*

Definition 1.21. In Theorem 1.19, if $z_{\mathrm{III}} = 0$, then $\mathscr{M}$ is said to be *semifinite*.

Proposition 1.22. *Let $\{\mathscr{M},\mathfrak{H}\}$ be a von Neumann algebra. If $\{e_i\}_{i \in I}$ is an orthogonal family of mutually equivalent projections in $\mathscr{M}$ with $\sum_{i \in I} e_i = 1$, then*

$$\{\mathscr{M},\mathfrak{H}\} \cong \{\mathscr{M}_{e_1}, e_1\mathfrak{H}\} \otimes \{\mathscr{L}(\mathfrak{H}_2),\mathfrak{H}_2\},$$

where $\mathfrak{H}_2 = l^2(I)$.

PROOF. Let $\{u_i\}_{i \in I}$ be a family of partial isometries in $\mathscr{M}$ such that $u_i^* u_i = e_1$ and $u_i u_i^* = e_i$, $i \in I$. Put $w_{i,j} = u_i u_j^*$ for each $i,j \in I$. It follows that $\{w_{i,j}: i,j \in I\}$ is a matrix unit in $\mathscr{M}$. Hence Proposition IV.1.8 entails our assertion. Q.E.D.

We are now going to determine the structure of a von Neumann algebra of type I. Since a minimal projection is abelian, the algebra $\mathscr{L}(\mathfrak{H})$ of all bounded operators on a Hilbert space $\mathfrak{H}$ is a factor of type I. More generally, we can easily conclude that every atomic von Neumann algebra is of type I by virtue of Lemma 1.18.

Proposition 1.23. (i) *If $\mathscr{A}$ is an abelian von Neumann algebra, then $\mathscr{A} \mathbin{\overline{\otimes}} \mathscr{L}(\mathfrak{H})$ is of type I for any Hilbert space $\mathfrak{H}$.*

(ii) *If $\mathscr{A}$ is an abelian von Neumann algebra, then $M_n(\mathscr{A})$, the $n \times n$-matrix algebra with entries from $\mathscr{A}$, is a finite von Neumann algebra of type I and the $\mathscr{A}$-valued map Φ:*

$$\Phi(x) = \frac{1}{n} \sum_{i=1}^{n} x_{ii} \quad \textit{for} \quad x = [x_{i,j}] \in M_n(\mathscr{A}) \tag{1}$$

enjoys the properties

$$\left.\begin{aligned} \Phi(a) = a, &\qquad a \in \mathscr{A}, \\ \Phi(axb) = a\Phi(x)b, &\qquad a,b \in \mathscr{A}, \\ \Phi(x^*x) = \Phi(xx^*) \geq 0 \quad \textit{and} \quad \Phi(x^*x) = 0 &\quad \textit{implies} \quad x = 0, \end{aligned}\right\} \tag{2}$$

where each $a \in \mathscr{A}$ is identified with the matrix $[a_{i,j}] = [\delta_{i,j}a]$.

PROOF. (i) Let p be a minimal projection in $\mathscr{L}(\mathfrak{H})$. Let $e = 1 \otimes p$. We have then, by Proposition IV.1.9,

$$e(\mathscr{A} \mathbin{\overline{\otimes}} \mathscr{L}(\mathfrak{H}))e = \mathscr{A} \mathbin{\overline{\otimes}} p\mathscr{L}(\mathfrak{H})p = \mathscr{A} \otimes \mathbf{C}p;$$

hence e is abelian. Since the center of $\mathscr{A} \mathbin{\overline{\otimes}} \mathscr{L}(\mathfrak{H})$ is $\mathscr{A} \otimes \mathbf{C}$ because $(\mathscr{A} \mathbin{\overline{\otimes}} \mathscr{L}(\mathfrak{H}))' = \mathscr{A}' \otimes \mathbf{C}$ by Proposition IV.1.6, the central support $z(e)$ of e is the identity 1. Hence $\mathscr{A} \mathbin{\overline{\otimes}} \mathscr{L}(\mathfrak{H})$ is of type I.

(ii) The first two properties of Φ follow immediately from its definition (1). The last property is also verified immediately by the following:

$$\Phi(x^*x) = \frac{1}{n} \sum_{i,j=1}^{n} x_{i,j}^* x_{i,j} = \frac{1}{n} \sum_{i,j=1}^{n} x_{i,j} x_{i,j}^* = \Phi(xx^*). \qquad \text{Q.E.D.}$$

Definition 1.24. In general, an $\mathscr{A}$-valued map Φ satisfying (2), except for the faithfulness condition, is called an *$\mathscr{A}$-valued trace*. Of particular interest is a *center valued trace*, as we shall study later.

Lemma 1.25. *If e is an abelian projection in a von Neumann algebra $\mathscr{M}$, then for any projection $f \in \mathscr{M}$ with $z(f) \geq e$ we have $e \gtrsim f$.*

PROOF. By Theorem 1.8, we have only to show that $e \gtrsim f$ implies $e \sim f$. In this case, we have $z(e) = z(f)$. Let $\mathscr{Z}$ denote the center of $\mathscr{M}$. By assumption, we have $e\mathscr{M}e = \mathscr{Z}e$; see Proposition II.3.10. Choose a partial isometry $u \in \mathscr{M}$ with $u^*u = f$ and $uu^* = e_1 \leq e$. There exists a central projection z in $\mathscr{M}$ with $e_1 = ez$. Since equivalent projections have the same central support, we have $z(f) = z(e_1) \leq z(e)$; hence we have $z(e_1) = z(e)$. Therefore $z(e) \leq z$, so that $e = e_1$. Thus $f \sim e_1 = e$. Q.E.D.

Lemma 1.26. *Let z be a central projection in $\mathscr{M}$. If $\{e_i\}_{i\in I}$ and $\{f_j\}_{j\in J}$ are two orthogonal families of abelian projections such that $z(e_i) = z(f_j) = z$ for any $i \in I$ and $j \in J$, and if $\sum_{i\in I} e_i = z = \sum_{j\in J} f_j$, then the cardinal numbers of I and J are equal.*

PROOF. Considering $\mathscr{M}_z$, we assume that $z = 1$. By the previous lemma, $\{e_i\}_{i\in I}$ and $\{f_j\}_{j\in J}$ are mutually equivalent and $e_i \sim f_j$. Let α and β be the cardinal numbers of I and J, respectively. By Proposition 1.22, $\mathscr{M} \cong \mathscr{M}_{e_1} \overline{\otimes} \mathscr{L}(\mathfrak{H}_1) \cong \mathscr{M}_{f_1} \overline{\otimes} \mathscr{L}(\mathfrak{H}_2)$, where $\dim \mathfrak{H}_1 = \alpha$ and $\dim \mathfrak{H}_2 = \beta$. Since e_1 and f_1 are both abelian projections with $z(e_1) = z(f_1) = 1$, $\mathscr{M}_{e_1} \cong \mathscr{M}_{f_1} \cong \mathscr{Z}$, where $\mathscr{Z}$ is the center of $\mathscr{M}$. If α is finite, then we may consider the $\mathscr{Z}$-valued trace Φ defined by (1) based on the tensor product decomposition $\mathscr{M} \cong \mathscr{M}_{e_1} \otimes \mathscr{L}(\mathfrak{H}_1) \cong \mathscr{Z} \otimes \mathscr{L}(\mathfrak{H}_1)$. Since $e_i \sim f_j$, we have $\Phi(e_i) = \Phi(f_j)$ for every $(i,j) \in I \times J$. Hence we have

$$\alpha\Phi(e_1) = \sum_{i\in I} \Phi(e_i) = \Phi(1) = 1 = \sum_{j\in J} \Phi(f_j) = \beta\Phi(f_1),$$

so that $\alpha = \beta$.

Suppose α is infinite. By symmetry, β is infinite. Let φ be a normal state of $\mathscr{Z}$. Since the map $\sigma_i : x \in \mathscr{Z} \mapsto xe_i \in e_i\mathscr{M}e_i$ is an isomorphism, we can define a normal state φ_i on $\mathscr{M}$ by $\varphi_i(x) = \varphi \circ \sigma_i^{-1}(e_i x e_i)$, $x \in \mathscr{M}$, $i \in I$. Let z be the support of φ in $\mathscr{Z}$. It follows that the support of φ_i is ze_i for each $i \in I$. For each $i \in I$, let $J_i = \{j \in J : \varphi_i(f_j) \neq 0\}$. Since $\{f_j\}_{j\in J}$ is orthogonal, J_i is countable. If $\varphi_i(f_j) = 0$, then $ze_i f_j z e_i = 0$; hence $ze_i f_j e_i = 0$, and so $zf_j e_i = 0$. Since $\sum_{i\in I} zf_j e_i = zf_j \neq 0$, $\varphi_i(f_j) \neq 0$ for some $i \in I$. Hence $J = \bigcup_{i\in I} J_i$. Therefore we have $\beta \leq \alpha\aleph_0 = \alpha$, so that $\alpha = \beta$ by symmetry. Q.E.D.

Theorem 1.27. *If $\mathscr{M}$ is a von Neumann algebra of type I, then there exists a unique orthogonal family $\{z_\alpha\}$ of central projections in $\mathscr{M}$ indexed by cardinal numbers with $\sum z_\alpha = 1$ such that $\mathscr{M}_{z_\alpha}$ is isomorphic to the tensor product of an abelian von Neumann algebra $\mathscr{A}_\alpha$ and $\mathscr{L}(\mathfrak{H}_\alpha)$ with $\dim \mathfrak{H}_\alpha = \alpha$. Therefore, we have*

$$\mathscr{M} \cong \sum_\alpha{}^{\oplus} \mathscr{A}_\alpha \overline{\otimes} \mathscr{L}(\mathfrak{H}_\alpha).$$

If, in addition, $\mathscr{M}$ is finite, then every α above is finite.

PROOF. For a cardinal number α, if a central projection $z \in \mathscr{M}$ is the sum of α orthogonal abelian projections each of which has the central support z, then z is said to be *α-homogeneous*. If $\{z_i\}_{i\in I}$ is an orthogonal family of α-homogeneous central projections, then $\sum_{i\in I} z_i$ is α-homogeneous. Hence there exists the greatest α-homogeneous central projection z_α. If $z_\alpha \neq 0$, then there exists an orthogonal family $\{e_i\}_{i\in I_\alpha}$ of abelian projections with $z(e_i) = z_\alpha$ such that the cardinal number of I_α is α. By Lemma 1.25, $\{e_i\}_{i\in I}$ are mutually equivalent, so that we have, by Proposition 1.22,

$$\mathscr{M}z_\alpha \cong \mathscr{M}_{e_1} \overline{\otimes} \mathscr{L}(\mathfrak{H}_\alpha),$$

where $\dim \mathfrak{H}_\alpha = \alpha$. By the abelianess of e_1, $\mathcal{M}_{e_1}$ is abelian. Hence we have only to prove the orthogonality of z_α and z_β for distinct α and β. But this follows from Lemma 1.26. If α is infinite, then I_α and $I_\alpha - \{i_0\}$ have the same cardinal numbers; hence $z_\alpha = \sum_{i \in I_\alpha} e_i \sim \sum_{i \in I_\alpha, i \neq i_0} e_i < z_\alpha$. Therefore, z_α is not finite. Thus α must be finite if $\mathcal{M}$ is finite. Q.E.D.

Corollary 1.28. *A factor of type I is isomorphic to $\mathcal{L}(\mathfrak{H})$ for some Hilbert space $\mathfrak{H}$. Therefore, the dimension of $\mathfrak{H}$ determines completely the algebraic type of the factor.*

PROOF. Since the center of a factor $\mathcal{M}$ of type I contains no nontrivial projection, 1 itself is α-homogeneous for some α. In a factor, a projection is abelian if and only if it is minimal. Thus, our assertion follows immediately from the previous theorem. Q.E.D.

Proposition 1.29. *A von Neumann algebra $\mathcal{M}$ is of type I if and only if $\mathcal{M}$ admits a faithful normal representation $\{\pi,\mathfrak{H}\}$ such that $\pi(\mathcal{M})'$ is abelian.*

PROOF. Let $\{\mathcal{M},\mathfrak{H}\}$ be a von Neumann algebra with abelian commutant $\mathcal{M}'$. It follows that $\mathcal{M}'$ coincides with the center $\mathcal{Z}$ of $\mathcal{M}$. For each $\xi \in \mathfrak{H}$, let e_ξ denote the projection of $\mathfrak{H}$ onto $[\mathcal{Z}\xi]$. Since e_ξ commutes with $\mathcal{Z}(=\mathcal{M}')$, e_ξ belongs to $\mathcal{M}$. Since $\{\mathcal{Z}_{e_\xi},[\mathcal{Z}\xi]\}$ admits a cyclic vector ξ, $\mathcal{Z}_{e_\xi}$ is maximal abelian by Corollary III.1.3. Hence $(\mathcal{Z}_{e_\xi})' = \mathcal{M}_{e_\xi}$ is abelian, so that $e_\xi \mathcal{M} e_\xi \cong \mathcal{M}_{e_\xi}$ is abelian. Therefore, e_ξ is abelian for any $\xi \in \mathfrak{H}$. Since we have $p \geq e_\xi$ for any projection $p \in \mathcal{M}$ and $\xi \in p\mathfrak{H}$, any nonzero projection in $\mathcal{M}$ majorizes a non-zero abelian projection. Thus $\mathcal{M}$ is of type I.

Suppose that $\mathcal{M}$ is of type I. By Theorem 1.27,

$$\mathcal{M} \cong \sum_\alpha^\oplus \mathcal{A}_\alpha \overline{\otimes} \mathcal{L}(\mathfrak{H}_\alpha),$$

with abelian von Neumann algebras $\mathcal{A}_\alpha$ and α-dimensional Hilbert space $\mathfrak{H}_\alpha$. We represent $\mathcal{A}_\alpha$ on a Hilbert space $\mathfrak{K}_\alpha$ as a maximal abelian subalgebra of $\mathcal{L}(\mathfrak{K}_\alpha)$, which is possible by the proof of Theorem III.1.18. We represent $\mathcal{M}$ on the Hilbert space $\mathfrak{H} = \sum_\alpha^\oplus \mathfrak{K}_\alpha \otimes \mathfrak{H}_\alpha$ in the obvious way by means of the above isomorphism. Then we have, by Proposition IV.1.6,

$$\mathcal{M}' = \sum_\alpha^\oplus (\mathcal{A}_\alpha' \overline{\otimes} \mathcal{L}(\mathfrak{H}_\alpha))' = \sum_\alpha^\oplus (\mathcal{A}_\alpha' \otimes \mathbf{C}) \cong \sum_\alpha^\oplus \mathcal{A}_\alpha.$$

Thus $\mathcal{M}'$ is abelian. Q.E.D.

Corollary 1.30. *If $\{\mathcal{M},\mathfrak{H}\}$ is a von Neumann algebra of type I, then the commutant $\mathcal{M}'$ is of type I.*

PROOF. Let e be an abelian projection in $\mathcal{M}$ with $z(e) = 1$. The induction: $x \in \mathcal{M}' \mapsto x_e \in \mathcal{M}_e'$ is an isomorphism of $\mathcal{M}$ onto $\mathcal{M}_e'$, and we have $(\mathcal{M}_e')' = \mathcal{M}_e$. Hence $\mathcal{M}'$ is isomorphic to the von Neumann algebra $\{\mathcal{M}_e', e\mathfrak{H}\}$ with abelian commutant $\mathcal{M}_e$. Thus $\mathcal{M}'$ is of type I by Proposition 1.29. Q.E.D.

Suppose that $\{\mathcal{M},\mathfrak{H}\}$ is a von Neumann algebra of type I. By Theorem 1.27, there is an orthogonal central projections $\{z_\alpha\}$ such that $\sum z_\alpha = 1$ and z_α is the greatest α-homogeneous central projection in $\mathcal{M}$. By Corollary 1.30, $\mathcal{M}'$ is of type I as well. Hence $\mathcal{M}'$ is also decomposed by an orthogonal central projection $\{z'_\alpha\}$ such that $\sum z'_\alpha = 1$ and z'_α is the greatest α-homogeneous central projection in $\mathcal{M}'$. But the center $\mathcal{Z}$ of $\mathcal{M}$ coincides with the center of $\mathcal{M}'$, being $\mathcal{M} \cap \mathcal{M}'$. Therefore, $\{z_\alpha\}$ and $\{z'_\alpha\}$ are both in $\mathcal{Z}$. Putting

$$z_{\alpha,\beta} = z_\alpha z'_\beta,$$

we obtain a finer partition $\{z_{\alpha,\beta}\}$ of the identity. Let $\mathfrak{H}_{\alpha,\beta} = z_{\alpha,\beta}\mathfrak{H}$ and $\mathcal{M}_{\alpha,\beta} = \mathcal{M}_{z_{\alpha,\beta}}$. It follows that $\mathcal{M}_{\alpha,\beta} \cong \mathcal{A}_{\alpha,\beta} \overline{\otimes} \mathcal{L}(\mathfrak{H}_\alpha)$ and $\mathcal{M}'_{\alpha,\beta} \cong \mathcal{A}_{\alpha,\beta} \overline{\otimes} \mathcal{L}(\mathfrak{H}_\beta)$ with an abelian von Neumann algebra $\mathcal{A}_{\alpha,\beta}$ isomorphic to the center of $\mathcal{M}_{\alpha,\beta}$ and an α-dimensional (resp. β-dimensional) Hilbert space $\mathfrak{H}_\alpha$ (resp. $\mathfrak{H}_\beta$). By Theorem IV.5.5, representing $\mathcal{A}_{\alpha,\beta}$ on a Hilbert space $\mathfrak{K}_{\alpha,\beta}$ as a maximal abelian von Neumann algebra, we find a Hilbert space $\mathfrak{M}$ and a projection $e' \in (\mathcal{A}_{\alpha,\beta} \overline{\otimes} \mathcal{L}(\mathfrak{H}_\alpha) \otimes \mathbf{C})' = \mathcal{A}_{\alpha,\beta} \otimes \mathbf{C} \overline{\otimes} \mathcal{L}(\mathfrak{M})$ such that

$$\{\mathcal{M}_{\alpha,\beta},\mathfrak{H}_{\alpha,\beta}\} \cong \{(\mathcal{A}_{\alpha,\beta} \overline{\otimes} \mathcal{L}(\mathfrak{H}_\alpha) \otimes \mathbf{C})_{e'}, e'(\mathfrak{K}_{\alpha,\beta} \otimes \mathfrak{H}_\alpha \otimes \mathfrak{M})\},$$

and $z(e') = 1$. Hence we have

$$\{\mathcal{M}'_{\alpha,\beta},\mathfrak{H}_{\alpha,\beta}\} \cong \{(\mathcal{A}_{\alpha,\beta} \overline{\otimes} \mathbf{C} \overline{\otimes} \mathcal{L}(\mathfrak{M}))_{e'}, e'(\mathfrak{K}_{\alpha,\beta} \otimes \mathfrak{H}_\alpha \otimes \mathfrak{M})\}.$$

Thus, there exists an orthogonal family $\{e'_j\}_{j \in J}$ of abelian projections in $\mathcal{A}_{\alpha,\beta} \otimes \mathbf{C} \overline{\otimes} \mathcal{L}(\mathfrak{M})$ such that $\sum_{j \in J} e'_j = e'$ and card $J = \beta$, and $z(e'_j) = 1$. Since $f'_j = 1 \otimes p_j$ is an abelian projection in $\mathcal{A}_{\alpha,\beta} \overline{\otimes} \mathcal{L}(\mathfrak{M})$ for any minimal projection $p_j \in \mathcal{L}(\mathfrak{M})$, we can find, by Lemma 1.26, an orthogonal family $\{f'_j\}_{j \in J}$ of abelian projections of the form $f'_j = 1 \otimes p_j$ for minimal projections $\{p_j\}$ in $\mathcal{L}(\mathfrak{M})$. Put $f' = \sum_{j \in J} f_j$ and $p = \sum_{j \in J} p_j$. Then we have $e' \sim f'$ in $\mathcal{A}_{\alpha,\beta} \otimes \mathbf{C} \otimes \mathcal{L}(\mathfrak{M})$, $f'(\mathfrak{K}_{\alpha,\beta} \otimes \mathfrak{H}_\alpha \otimes \mathfrak{M}) = \mathfrak{K}_{\alpha,\beta} \otimes \mathfrak{H}_\alpha \otimes p\mathfrak{M}$ and $\dim p\mathfrak{M} = \beta$. Hence $p\mathfrak{M} \cong \mathfrak{H}_\beta$. Therefore, we have

$$\begin{aligned}\{\mathcal{M}_{\alpha,\beta},\mathfrak{H}_{\alpha,\beta}\} &\cong \{(\mathcal{A}_{\alpha,\beta} \overline{\otimes} \mathcal{L}(\mathfrak{H}_\alpha) \otimes \mathbf{C})_{e'}, e'(\mathfrak{K}_{\alpha,\beta} \otimes \mathfrak{H}_\alpha \otimes \mathfrak{M})\}\\ &\cong \{(\mathcal{A}_{\alpha,\beta} \overline{\otimes} \mathcal{L}(\mathfrak{H}_\alpha) \otimes \mathbf{C})_{f'}, f'(\mathfrak{K}_{\alpha,\beta} \otimes \mathfrak{H}_\alpha \otimes \mathfrak{M})\}\\ &\cong \{(\mathcal{A}_{\alpha,\beta} \overline{\otimes} \mathcal{L}(\mathfrak{H}_\alpha) \otimes \mathbf{C}), \mathfrak{K}_{\alpha,\beta} \otimes \mathfrak{H}_\alpha \otimes \mathfrak{H}_\beta\}.\end{aligned}$$

Thus, we have

$$\{\mathcal{M}'_{\alpha,\beta},\mathfrak{H}_{\alpha,\beta}\} \cong \{(\mathcal{A}_{\alpha,\beta} \otimes \mathbf{C} \overline{\otimes} \mathcal{L}(\mathfrak{H}_\beta)), \mathfrak{K}_{\alpha,\beta} \otimes \mathfrak{H}_\alpha \otimes \mathfrak{H}_\beta\}.$$

Therefore, we obtain the following description of the spatial type of a von Neumann algebra of type I acting on a Hilbert space.

Theorem 1.31. *A von Neumann algebra $\{\mathcal{M},\mathfrak{H}\}$ of type I has the unique decomposition*

$$\{\mathcal{M},\mathfrak{H}\} \cong \sum_{\alpha,\beta}^{\oplus} (\{\mathcal{A}_{\alpha,\beta},\mathfrak{K}_{\alpha,\beta}\} \overline{\otimes} \{\mathcal{L}(\mathfrak{H}_\alpha),\mathfrak{H}_\alpha\} \overline{\otimes} \{\mathbf{C},\mathfrak{H}_\beta\}),$$

$$\{\mathcal{M}',\mathfrak{H}\} \cong \sum_{\alpha,\beta}^{\oplus} (\{\mathcal{A}_{\alpha,\beta},\mathfrak{K}_{\alpha,\beta}\} \overline{\otimes} \{\mathbf{C},\mathfrak{H}_\alpha\} \overline{\otimes} \{\mathcal{L}(\mathfrak{H}_\beta),\mathfrak{H}_\beta\}),$$

where $\{\mathscr{A}_{\alpha,\beta},\mathfrak{K}_{\alpha,\beta}\}$ *is maximal abelian and* $\mathfrak{H}_\alpha$ *(resp.* $\mathfrak{H}_\beta$*) is an* α*-dimensional (resp.* β*-dimensional) Hilbert space.*

Corollary 1.32. *An abelian von Neumann algebra* $\{\mathscr{A},\mathfrak{H}\}$ *is uniquely decomposed into the direct sum*

$$\{\mathscr{A},\mathfrak{H}\} \cong \sum_\alpha{}^{\oplus} \{\mathscr{A}_\alpha \otimes \mathbf{C},\mathfrak{K}_\alpha \otimes \mathfrak{H}_\alpha\},$$

where $\{\mathscr{A}_\alpha,\mathfrak{K}_\alpha\}$ *is maximal abelian and* $\dim \mathfrak{H}_\alpha = \alpha$.

Definition 1.33. A von Neumann algebra $\mathscr{M}$ of type I is said to be *of type* I_α if 1 is α-homogeneous.

By Corollary 1.28, a factor of type I_n, $n < +\infty$, is nothing but the $n \times n$-matrix algebra over $\mathbf{C}$.

We now turn our attention to the set of finite projections in a general von Neumann algebra $\mathscr{M}$.

Proposition 1.34. *If* $\{e_i\}_{i\in I}$ *is an orthogonal family of equivalent projections in a von Neumann algebra, then there exists a nonzero central projection* z *and a family* $\{f_j\}_{j\in J}$ *of orthogonal equivalent projections in* $\mathscr{M}$ *such that* $I \subset J$; $f_i \sim e_i z$ *for every* $i \in I$, *and*

$$f_0 = z - \sum_{j\in J} f_j \prec f_i.$$

If J *is infinite, then one can choose* $\{f_j\}$ *so that* $f_0 = 0$.

Proof. Let $\{e_j\}_{j\in J}$ be a maximal family of orthogonal equivalent projections in $\mathscr{M}$ containing $\{e_i\}_{i\in I}$. Put $e_0 = 1 - \sum_{j\in J} e_j$. Let z be a central projection in $\mathscr{M}$ such that $e_0 z \precsim e_j z$ and $e_0(1-z) \succsim e_j(1-z)$, $j \in J$. Put $f_0 = e_0 z$ and $f_j = e_j z$, $j \in J$. If $f_j \precsim f_0$, then $e_0 z \sim e_j z_0$, so that $e_0 \succsim e_j$, which contradicts the maximality of $\{e_j\}_{j\in J}$. Hence $f_0 \prec f_j$, $j \in J$. Suppose I is infinite; then so is J. Let j_0 be an arbitrary element of J. We have then

$$z_0 = f_0 + \sum_{j\in J} f_j \leq f_{j_0} + \sum_{\substack{j\in J\\ j\neq j_0}} f_j = \sum_{j\in J} f_j \leq z;$$

hence $z \sim \sum_{j\in J} f_j$ by Proposition 1.3. Replacing each f_j by the corresponding projection in the equivalence $z \sim \sum_{j\in J} f_j$, we have $z = \sum_{j\in J} f_j$. Q.E.D.

Proposition 1.35. *A von Neumann algebra* $\mathscr{M}$ *has no direct summand of type I if and only if every projection in* $\mathscr{M}$ *is the sum of two equivalent orthogonal projections.*

Proof. If e is an abelian projection in $\mathscr{M}$, then any orthogonal projections majorized by e are centrally orthogonal, so that they are not equivalent.

Suppose that $\mathscr{M}$ has no direct summand of type I. Let e be a nonzero projection in $\mathscr{M}$. Since $e\mathscr{M}e$ is not abelian, there exists a nonzero projection $e_1' \le e$ such that $e_1'\mathscr{M}(e - e_1') \ne \{0\}$. Hence there exist, by Lemma 1.7, nonzero projections $e_1 \le e_1'$ and $e_2 \le e - e_1'$ such that $e_1 \sim e_2$. Therefore, every nonzero projection in $\mathscr{M}$ majorizes two orthogonal equivalent nonzero projections. Let $\{e_{1,i}, e_{2,i} : i \in I\}$ be a maximal family of orthogonal projections in $e\mathscr{M}e$ such that $e_{1,i} \sim e_{2,i}$, $i \in I$. Put $\sum_{i\in I} e_{1,i} = e_1$ and $\sum_{i\in I} e_{2,i} = e_2$. If $e > e_1 + e_2$, then there exist orthogonal nonzero projections e_1' and e_2' such that $e_1' \sim e_2'$ and $e_1' + e_2' \le e - (e_1 + e_2)$, which contradicts the maximality of $\{e_{1,i}, e_{2,i} : i \in I\}$. Hence $e = e_1 + e$, and $e_1 \sim e_2$.

Q.E.D.

Proposition 1.36. *Let $\mathscr{M}$ be a von Neumann algebra. If $\mathscr{M}$ is properly infinite, then there exists a projection $e \in \mathscr{M}$ such that $e \sim (1 - e) \sim 1$.*

Proof. Let z be a nonzero central projection in $\mathscr{M}$. By assumption, there exists a partial isometry u in $\mathscr{M}$ with $u^*u = z$ and $uu^* < z$. Put $p_1 = z - uu^*$ and $p_n = u^n p_1 (u^*)^n$. It follows that $\{p_n\}$ are orthogonal and mutually equivalent. Hence by Proposition 1.34, z majorizes a nonzero central projection z_1 which is the sum of infinitely many orthogonal equivalent projections $\{q_i\}_{i\in I}$. Dividing I into the union of two disjoint subsets I_1 and I_2 of the same cardinality, we get $e_1 = \sum_{i\in I_1} q_i \sim (z - e_1) = \sum_{j\in I_2} q_j \sim \sum_{i\in I} q_i = z_1$. Let $\{e_j\}_{j\in J}$ be a maximal family of centrally orthogonal projections in $\mathscr{M}$ such that $e_j \sim (z(e_j) - e_j) \sim z(e_j)$. Put $e = \sum_{j\in I} e_j$. By maximality, we have

$$1 = z(e) = \sum_{j\in J} z(e_j) \sim \sum_{j\in J} e_j \sim \sum_{j\in J} (z(e_j) - e_j) = 1 - e;$$

hence $1 \sim e \sim (1 - e)$. Q.E.D.

Theorem 1.37. *The set of all finite projections in a von Neumann algebra $\mathscr{M}$ is a modular lattice.*

Proof. Let e and f be two finite projections in $\mathscr{M}$. We claim that $e \vee f$ is finite. To show this, we may assume that $e \vee f = 1$. By Proposition 1.6, we have

$$1 - e = (e \vee f) - e \sim f - (e \wedge f) \le f.$$

Hence $1 - e$ is finite. Therefore, we may assume that e and f are orthogonal and $e + f = 1$. Suppose that $\mathscr{M}$ is not finite. It follows from Theorem 1.19 that there exists a nonzero properly infinite central projection z. Looking at ze and zf, we may assume $z = 1$, from which we shall derive a contradiction. By Proposition 1.36, there exists a projection p with $p \sim (1 - p) \sim 1$. By the comparability theorem (Theorem 1.8), we choose a central projection z such that

$$z(e \wedge p) \precsim z(f \wedge (1 - p)),$$
$$(1 - z)(e \wedge p) \succsim (1 - z)(f \wedge (1 - p)).$$

We have then

$$\begin{aligned} zp &= ze \wedge zp + (zp - ze \wedge zp) \\ &\sim ze \wedge zp + (ze \vee zp - ze) \\ &\precsim z(f \wedge (1-p)) + (ze \vee zp - ze) \leq zf. \end{aligned}$$

But zp is properly infinite and zf is finite, so that $zp = 0$; hence $z = 0$ because the central support of p is 1. Thus we have $e \wedge p \succsim f \wedge (1-p)$. But we have

$$\begin{aligned} 1 - p &= f \wedge (1-p) + [1 - p - f \wedge (1-p)] \\ &\sim f \wedge (1-p) + [(1-p) \vee f - f] \\ &\precsim e \wedge p + [(1-p) \vee f - f] \leq e. \end{aligned}$$

Since e is finite, $1 - p$ is finite, which contradicts the equivalence $1 - p \sim 1$. Thus we reach a contradiction. Therefore, the finite projections form a lattice.

We shall prove the modular law. Let e, f, and g be finite projections with $e \leq g$. Put $h = (e \vee f) \wedge g$ and $k = e \vee (f \wedge g)$. We must show $h = k$. It follows that

$$h \geq k, \qquad h \vee f = e \vee f = k \vee f,$$
$$h \wedge f = f \wedge g = k \wedge f.$$

Hence we get

$$h - (f \wedge g) \sim e \vee f - f \sim k - (f \wedge g);$$

thus $h \sim k$. Therefore we get $h = k$ by the finiteness of h. Q.E.D.

Proposition 1.38. *If e are f are finite equivalent projections in $\mathscr{M}$, then $1 - e$ and $1 - f$ are equivalent; hence there exists a unitary element $u \in \mathscr{M}$ with $ueu^* = f$.*

Proof. By Theorem 1.37, $e \vee f$ is finite, so that, considering $\mathscr{M}_{(e \vee f)}$, we may assume the finiteness of $\mathscr{M}$. There exists a central projection z such that $z(1-e) \precsim z(1-f)$ and $(1-z)(1-e) \succsim (1-z)(1-f)$. Hence there are projections h and k with $z(1-e) \sim h \leq z(1-f)$ and $(1-z)(1-f) \sim k \leq (1-z)(1-e)$, so that we have

$$z = ze + z(1-e) \sim zf + h \leq z,$$
$$(1-z) = (1-z)f + (1-z)(1-f) \sim (1-z)e + k \leq (1-z).$$

The finiteness of z and $(1-z)$ imply that $h = z(1-f)$ and $k = (1-z)e$. Thus $(1-e) \sim (1-f)$. Q.E.D.

Proposition 1.39. *Let e be a properly infinite projection in $\mathscr{M}$. If f is a σ-finite projection in $\mathscr{M}$ with $z(f) \leq z(e)$, then we have $f \precsim e$. In particular, every two infinite projections in a σ-finite factor are equivalent.*

Proof. By the comparability theorem, we may assume $e \precsim f$; hence $e \leq f$. We will show $e \sim f$. Looking at $\mathscr{M}_f$, we have only to show that $1 \sim e$ for any properly infinite projection e with $z(e) = 1$ in a σ-finite von Neumann algebra $\mathscr{M}$. By Proposition 1.36, there are orthogonal projections e_1 and e_1' such that $e \sim e_1 \sim e_1'$ and $e = e_1 + e_1'$. Let u be a partial isometry in $\mathscr{M}$ with $u^*u = e$ and $uu^* = e_1'$. Put $e_{n+1} = u^n e_1 (u^n)^*$, $n = 1,2,\ldots$. Then $\{e_n : n = 1,2,\ldots\}$ are orthogonal and $e \sim e_1 \sim e_2 \sim \cdots \sim e_n \sim \cdots$. Let $\{f_j\}_{j \in J}$ be a maximal family of orthogonal projections such that $f_j \precsim e$. By the σ-finiteness of $\mathscr{M}$, J is countable, and $1 - \sum_{j \in J} f_j$ and e are centrally orthogonal by Lemma 1.7; hence $\sum_{j \in J} f_j = 1$ because $z(e) = 1$. Therefore we have

$$e \succsim \sum_{n=1}^{\infty} e_n \succsim \sum_{j \in J} f_j = 1.$$

Thus $e \sim 1$ by Proposition 1.3. Q.E.D.

Proposition 1.40. *If $\mathscr{M}$ is a properly infinite but semifinite von Neumann algebra, then there exist an orthogonal family $\{z_\alpha\}$, indexed by infinite cardinals $\leq$* card $\mathscr{M}$, *of central projections with $\sum z_\alpha = 1$, and a family $\{\mathscr{N}_\alpha\}$ of finite von Neumann algebras such that*

$$\mathscr{M}_{z_\alpha} \cong \mathscr{N}_\alpha \mathbin{\overline{\otimes}} \mathscr{L}(\mathfrak{H}_\alpha),$$

where dim $\mathfrak{H}_\alpha = \alpha$ *and z_α may be zero. This family $\{z_\alpha\}$ is unique while $\mathscr{N}_\alpha$ is not unique. If $\mathscr{M}$ is σ-finite, then for any finite projection f with $z(f) = 1$,*

$$\mathscr{M} \cong \mathscr{M}_f \mathbin{\overline{\otimes}} \mathscr{L}(\mathfrak{H}_0),$$

where dim $\mathfrak{H}_0 = \aleph_0$.

Proof. Let f be a finite projection of $\mathscr{M}$ with $z(f) = 1$. By assumption, every nonzero central projection z of $\mathscr{M}$ is properly infinite. Hence, by Proposition 1.34, there exist a nonzero central projection $z_0 \leq z$ and a family $\{f_i\}_{i \in I}$ of orthogonal equivalent projections such that $f_i \sim f z_0$ and $z_0 = \sum_{i \in I} f_i$, where I must be infinite due to the proper infiniteness of z_0 by Theorem 1.37. Hence we have, by Proposition 1.22, $\mathscr{M}_{z_0} \cong \mathscr{M}_{fz_0} \mathbin{\overline{\otimes}} \mathscr{L}(l^2(I))$. Making use of the usual exhaustion arguments and relabeling the orthogonal family of central projections by the corresponding cardinals, we get the desired decomposition. Suppose $\mathscr{M}$ is σ-finite. It follows then that the index set I in the above arguments is a countably infinite set. Therefore, the usual exhaustion arguments entail that there exists an orthogonal countable family $\{z_n\}$ of central projections and that for each n, there exists a countably infinite family $\{f_{n,j}\}$ of orthogonal equivalent projections such that $f_{n,j} \sim z_n f$ and $z_n = \sum_{j=1}^{\infty} f_{n,j}$. Putting $f_j = \sum_{n=1}^{\infty} f_{n,j}$ for each j, we get a countably infinite family $\{f_j\}$ of orthogonal equivalent projections such that $f \sim f_j$, $j = 1,2,\ldots$, and $1 = \sum_{j=1}^{\infty} f_j$. Hence we have

$$\mathscr{M} \cong \mathscr{M}_f \otimes \mathscr{L}(l^2).$$

The uniqueness of $\{z_\alpha\}$ in the general case will be proven in the next section, after Corollary 2.9. Q.E.D.

*We are now going to analyze the relative position of two projections on a Hilbert space $\mathfrak{H}$. Let e and f be nonzero projections on $\mathfrak{H}$. For short, we shall write $e^\perp = 1 - e$ and $f^\perp = 1 - f$. We have then

$$e = e \wedge f + e \wedge f^\perp + (e - e \wedge f - e \wedge f^\perp),$$
$$f = e \wedge f + e^\perp \wedge f + (f - e \wedge f - e^\perp \wedge f).$$

Put

$$e_0 = e - e \wedge f - e \wedge f^\perp, \qquad e_1 = e \wedge f + e \wedge f^\perp,$$
$$f_0 = f - e \wedge f - e^\perp \wedge f, \qquad f_1 = e \wedge f + e^\perp \wedge f.$$

It is then straightforward to check that e and f_1 (resp. e_1 and f) commute and that

$$1 = e \wedge f + e \wedge f^\perp + e^\perp \wedge f + e^\perp \wedge f^\perp + e_0 \vee f_0,$$
$$e_0 \wedge f_0 = e_0 \wedge f_0^\perp = e_0^\perp \wedge f_0 = 0.$$

Since the relative position of e_1 and f (resp. e and f_1) is so simple that the decompositions $e_1 = e_1 f + (e_1 - e_1 f)$ and $f = e_1 f + (f - e_1 f)$ show a complete description of the pair e_1 and f, the analysis of e_0 and f_0 suffices for understanding the pair e and f. Hence, replacing e and f by e_0 and f_0, and discarding $(e \vee f)^\perp$, we arrive at the following:

$$\left.\begin{aligned} e \wedge f = e^\perp \wedge f = e \wedge f^\perp = 0; \\ e \vee f = 1, \quad \text{hence} \quad e^\perp \wedge f^\perp = 0. \end{aligned}\right\} \tag{$*$}$$

We then have

$$1 = e \vee f = e^\perp \vee f = e \vee f^\perp = e^\perp \vee f^\perp.$$

The situation of e and f can be illustrated by the following figure:

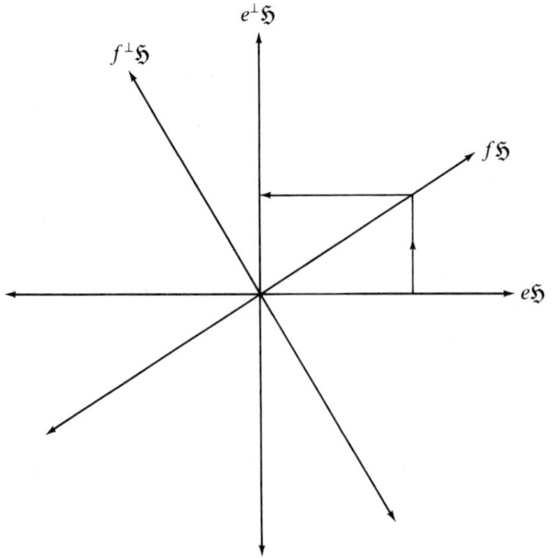

We have then, in the von Neumann algebra $\mathcal{M}$ generated by e and f,

$$\begin{aligned} e &= e - e \wedge f^{\perp} \sim e \vee f^{\perp} - f^{\perp} = f = f - e \wedge f \sim e \vee f - e \\ &= e^{\perp} = e^{\perp} - e^{\perp} \wedge f \sim e^{\perp} \vee f - f = f^{\perp}. \end{aligned}$$

Thus, e, $e^{\perp}$, f, and $f^{\perp}$ are all equivalent in $\mathcal{M}$. Hence there exists a parial isometry $u \in \mathcal{M}$ with $u^*u = e$ and $uu^* = e^{\perp}$. But, we want to have a specific partial isometry which is directly related to this situation. From the special relation (∗) between e and f, it follows that $e^{\perp}fe$ maps $e\mathfrak{H}$ injectively onto a dense subspace of $e^{\perp}\mathfrak{H}$. Let $a = e^{\perp}fe$ and $a = uh$ be the polar decomposition. It follows then that $u^*u = e$ and $uu^* = e^{\perp}$. We then use this u to make a matrix unit $\{e_{11}, e_{12}, e_{21}, e_{22}\}$. Put

$$e_{11} = e, \qquad e_{21} = u, \qquad e_{12} = u^*, \qquad e_{22} = e^{\perp}.$$

With this matrix unit, we have a 2×2-matrix representation of $\mathcal{M}$ over $\mathcal{M}_e$. In other words, every element of $\mathcal{M}$ is represented by a 2×2-matrix with entries from $\mathcal{M}_e$ and $\{e_{ij}\}$ is given by the following:

$$e = e_{11} = \begin{bmatrix} 1, & 0 \\ 0, & 0 \end{bmatrix}; \qquad e_{12} = \begin{bmatrix} 0 & 1 \\ 0, & 0 \end{bmatrix};$$

$$e_{21} = \begin{bmatrix} 0, & 0 \\ 1 & 0 \end{bmatrix}; \qquad e_{22} = \begin{bmatrix} 0, & 0 \\ 0, & 1 \end{bmatrix} = e^{\perp}.$$

Let

$$f = \begin{bmatrix} f_{11}, & f_{12} \\ f_{21}, & f_{22} \end{bmatrix}$$

be the matrix of f. Since $a = e^{\perp}ae$, we have $ehe = h$; thus we get

$$\begin{bmatrix} 0, & 0 \\ f_{21}, & 0 \end{bmatrix} = e^{\perp}fe = uh = \begin{bmatrix} 0, & 0 \\ 1, & 0 \end{bmatrix} \begin{bmatrix} h, & 0 \\ 0, & 0 \end{bmatrix} = \begin{bmatrix} 0, & 0 \\ h, & 0 \end{bmatrix},$$

where

$$h \quad \text{and} \quad \begin{bmatrix} h, & 0 \\ 0, & 0 \end{bmatrix}$$

are identified. Thus we get $f_{21} = h$; by the self-adjointness of f, $h = f_{12}$. Therefore, we get

$$f = \begin{bmatrix} f_{11}, & h \\ h, & f_{22} \end{bmatrix}.$$

From the equality $f = f^2$, we get

$$\left. \begin{aligned} f_{11}^2 + h^2 &= f_{11}, & f_{11}h + hf_{22} &= h, \\ hf_{11} + f_{22}h &= h, & h^2 + f_{22}^2 &= f_{22}. \end{aligned} \right\}$$

Hence f_{11} and f_{22} both commute with h, and so we get $h(f_{11} + f_{22} - 1) = 0$. Since h is injective in $e\mathfrak{H}$, we get $f_{11} + f_{22} = 1$. Since $f_{11} \geq 0$ and $f_{22} \geq 0$, we put $c = f_{11}^{1/2}$ and $s = f_{22}^{1/2}$. Then we get

$$h = (f_{11} - f_{11}^2)^{1/2} = cs.$$

Thus we get the following expression:

$$\left.\begin{aligned} e &= \begin{bmatrix} 1, & 0 \\ 0, & 0 \end{bmatrix}; \\ f &= \begin{bmatrix} c^2, & cs \\ cs, & s^2 \end{bmatrix}; \quad 0 \le c \le 1; \quad 0 \le s \le 1; \quad c^2 + s^2 = 1. \end{aligned}\right\}$$

In the case that dim $\mathfrak{H} = 2$, the above c and s are the cosine and the sine of the angle between $e\mathfrak{H}$ and $f\mathfrak{H}$. Hence c and s are generalizations of the cosine and the sine of the angle between $e\mathfrak{H}$ and $f\mathfrak{H}$. We further observe that

$$|e - f| = \begin{bmatrix} s, & 0 \\ 0, & s \end{bmatrix}, \qquad |e - f^\perp| = \begin{bmatrix} c, & 0 \\ 0, & c \end{bmatrix}.$$

Now, we summarize the above arguments:

Theorem 1.41. *If e and f are two projections on a Hilbert space $\mathfrak{H}$, and if $\mathscr{M}$ is the von Neumann algebra generated by e and f, then*

(i) *$\mathscr{M}$ is of type I,*
(ii) *there exists uniquely a central projections $z \in \mathscr{M}$ such that $\mathscr{M}z$ is of type I_2, and $\mathscr{M}(1 - z)$ is abelian and* dim $\mathscr{M}(1 - z) \le 4$.

Definition 1.42. For two projections e and f, we write $s(e,f) = |e - f|$ and $c(e,f) = |e - f^\perp|$, and call them the *sine* and the *cosine* of e and f, respectively.

EXERCISES

1. Let $\{\mathscr{M},\mathfrak{H}\}$ and $\{\mathscr{N},\mathfrak{K}\}$ be two von Neumann algebras on separable Hilbert spaces. Show that if π is an isomorphism of $\mathscr{M}$ onto $\mathscr{N}$, then $\pi \otimes \mathrm{id}$ is a spatial isomorphism of $\{\mathscr{M} \otimes \mathbf{C}, \mathfrak{H} \otimes \mathfrak{H}_\infty\}$ onto $\{\mathscr{N} \otimes \mathbf{C}, \mathfrak{K} \otimes \mathfrak{H}_\infty\}$, where $\mathfrak{H}_\infty$ means a separable infinite dimensional Hilbert space and id means the identity map of $\mathbf{C}$.

2. Let $\mathscr{M}$ be a von Neumann algebra. Two projections e and f of $\mathscr{M}$ are called *unitarily equivalent* if there exists a unitary element $u \in \mathscr{M}$ such that $ueu^* = f$. Of course, the unitary equivalence of e and f implies $e \sim f$. Conversely, show that if $e \sim f$, then there exist orthogonal decompositions $e = \sum_{i \in I} e_i$ and $f = \sum_{i \in I} f_i$ such that for each i, e_i and f_i are unitarily equivalent.

3. Let $\mathscr{M}$ be a von Neumann algebra, and α an automorphism of $\mathscr{M}$. Show that if there exist a projection $e \in \mathscr{M}$ with $z(e) = 1$ and $u \in \mathscr{M}$ such that $\alpha(x) = uxu^*$ for every $x \in \mathscr{M}_e$, then α is inner.

4. Let $\mathscr{M}$ be a von Neumann algebra of type I. Show that if an automorphism α of $\mathscr{M}$ leaves the center $\mathscr{M} \cap \mathscr{M}'$ elementwise fixed, then α is inner.

5. Let $\mathscr{I}$ be an ideal of a von Neumann algebra $\mathscr{M}$ with projection lattice $\mathscr{P}$.
   (a) Show that for any $e, f \in \mathscr{P}$, if $e \precsim f \in \mathscr{I}$, then $e \in \mathscr{I}$.
   (b) Show that $\mathscr{P} \cap \mathscr{I}$ is a sublattice of $\mathscr{P}$. (*Hint*: $e \vee f - e \sim f - e \wedge f$.)
   (c) Let $\mathscr{P}_0$ be a sublattice of $\mathscr{P}$ such that $e \precsim f \in \mathscr{P}_0$ implies $e \in \mathscr{P}_0$. Show that $\mathscr{I}_0 = \{x \in \mathscr{M} : s_l(x) \in \mathscr{P}_0\}$ is an ideal of $\mathscr{M}$ and $\mathscr{I}_0 \cap \mathscr{P} = \mathscr{P}_0$.
   (d) Given an ideal $\mathscr{I}$ of $\mathscr{M}$, show that $\mathscr{I}_0 = \{x \in \mathscr{M} : s_l(x) \in \mathscr{P} \cap \mathscr{I}\} \subset \mathscr{I} \subset \bar{\mathscr{I}}_0$, where the bar means the uniform closure.
   (e) Show that the collection of uniformly closed ideals of a factor is totally ordered under the inclusion ordering. (*Hint*: If $\mathscr{I}$ and $\mathscr{J}$ are closed ideals of a factor $\mathscr{M}$, then one of the following two cases occurs: (i) for every $e \in \mathscr{I} \cap \mathscr{P}$, there exists an $f \in \mathscr{J} \cap \mathscr{P}$ such that $e \precsim f$; (ii) there exists an $e \in \mathscr{I} \cap \mathscr{P}$ such that $e \succsim f$ for every $f \in \mathscr{J} \cap \mathscr{P}$.)
   (f) Show that a factor is simple if it is either finite or σ-finite and of type III.

6. Let $\mathfrak{m}$ be a left ideal of a von Neumann algebra $\mathscr{M}$.
   (a) Show that $\mathfrak{m} = \mathscr{M}(\mathscr{M}_+ \cap \mathfrak{m})$. (*Hint*: Use the polar decomposition.)
   (b) Let $\mathscr{P}$ denote the projection lattice of $\mathscr{M}$. Show that the linear span of the set $\{xe : x \in \mathscr{M}, e \in \mathscr{P} \cap \mathfrak{m}\}$ is uniformly dense in $\mathfrak{m}$. Hence if $\mathfrak{m}$ is uniformly closed, then $\mathfrak{m} \cap \mathscr{P}$ uniquely determines $\mathfrak{m}$. (*Hint*: If $h = \int_0^{\|h\|} \lambda\, de(\lambda) \in \mathfrak{m} \cap \mathscr{M}_+$ is a spectral decomposition, then $k_\varepsilon h = 1 - e(\varepsilon) \in \mathfrak{m} \cap \mathscr{P}$ with $k_\varepsilon = \int_\varepsilon^{\|h\|} (1/\lambda)\, de(\lambda)$ and $\|h - h(1 - e(\varepsilon))\| \leq \varepsilon$.)

2. Traces on von Neumann Algebras

In this section, we shall discuss the highly important class of positive linear functionals on von Neumann algebras, namely, traces.

Definition 2.1. A *trace* on a von Neumann algebra $\mathscr{M}$ is a function τ on the positive cone $\mathscr{M}_+$ with values in the extended positive reals $[0, +\infty]$ satisfying the following conditions:

(i) $\tau(x + y) = \tau(x) + \tau(y)$, $x, y \in \mathscr{M}_+$;

(ii) $\tau(\lambda x) = \lambda\tau(x)$, $\lambda \geq 0$, $x \in \mathscr{M}_+$;

(iii) $\tau(x^*x) = \tau(xx^*)$, $x \in \mathscr{M}$;

with the usual convention $0(+\infty) = 0$. A trace τ is said to be *faithful* if $\tau(x) > 0$ for any nonzero $x \geq 0$, *semifinite* if every nonzero $x \in \mathscr{M}_+$ majorizes some nonzero $y \geq 0$ with $\tau(y) < +\infty$, *finite* if $\tau(1) < +\infty$, *normal* if $\tau(\sup x_i) = \sup \tau(x_i)$ for every bounded increasing net $\{x_i\}$ in $\mathscr{M}_+$.

We shall first show that the existence of a faithful semifinite normal trace on a given von Neumann algebra $\mathscr{M}$ is equivalent to the semifiniteness of $\mathscr{M}$. We then study the special feature of traces.

First of all we note that, since $\mathscr{M}_+$ spans $\mathscr{M}$ linearly, a finite trace τ is extended uniquely to a positive linear functional on $\mathscr{M}$ which will also be denoted by τ; hence a finite trace is nothing but a positive linear functional enjoying Definition 2.1(iii).

Lemma 2.2. *Let $\mathcal{M}$ be a von Neumann algebra. Let $\{e_n\}$ be an increasing sequence of finite projections. If f is a projection in $\mathcal{M}$ such that $e_n \precsim f$ for $n = 1,2,\ldots$, then $e = \bigvee_{n=1}^{\infty} e_n \precsim f$.*

PROOF. Let $p_n = e_{n+1} - e_n$, $n = 1,2,\ldots$, and $p_0 = e_1$. Then $\{p_n\}$ are orthogonal and $e = \sum_{n=0}^{\infty} p_n$. We shall construct an orthogonal sequence $\{q_n\}$ such that $q_n \sim p_n$, $n = 0,1,\ldots$, and $q_n \leq f$. By assumption, $p_0 = e_1 \precsim f$; hence there exists $q_0 \leq f$ such that $p_0 \sim q_0$. Suppose $\{q_0,\ldots,q_{n-1}\}$ have been defined. It follows that

$$e_n = p_0 + p_1 + \cdots + p_{n-1} \sim q_0 + q_1 + \cdots + q_{n-1} = f_n \leq f.$$

By assumption, there exists $f'_{n+1} \leq f$ such that $f'_{n+1} \sim e_{n+1}$. Since $e_n \leq e_{n+1}$, we get $f_n \precsim f'_{n+1}$; in other words, there exists $f'_n \leq f'_{n+1}$ with $f_n \sim f'_n$. Since e_n is finite, f_n is finite too. By Theorem 1.37, we have $f - f_n \sim f - f'_n$; thus there exists a projection $q_n \leq f - f_n$ such that $q_n \sim f'_{n+1} - f'_n \sim e_{n+1} - e_n = p_n$. By induction, we get $\{q_n\}$. Thus $e = \sum_{n=0}^{\infty} p_n \sim \sum_{n=0}^{\infty} q_n \leq f$. Q.E.D.

Lemma 2.3. *If $\{e_n\}$ is an orthogonal projection in a finite von Neumann algebra $\mathcal{M}$, then any sequence $\{f_n\}$ of projections in $\mathcal{M}$ such that $e_n \sim f_n$, $n = 1,2,\ldots$, converges to zero σ-strongly.*

PROOF. For any p_1, p_2 and q_1, q_2 such that $p_1 \precsim q_1$ and $p_2 \precsim q_2$ and $q_1 q_2 = 0$, we have $p_1 \vee p_2 \precsim q_1 + q_2$ because

$$p_1 \vee p_2 - p_2 \sim p_1 - p_1 \wedge p_2 \precsim q_1,$$
$$p_1 \vee p_2 = (p_1 \vee p_2 - p_2) + p_2 \precsim q_1 + q_2.$$

Hence, by induction, we get, for any $m \leq n$,

$$f_m \vee f_{m+1} \vee \cdots \vee f_n \precsim e_m + e_{m+1} + \cdots + e_n.$$

By Lemma 2.2, we have

$$p_m = (f_m \vee f_{m+1} \vee \cdots) \precsim \sum_{k=m}^{\infty} e_k.$$

By Theorem 1.37, we get

$$1 - p_m \succsim 1 - \sum_{k=m}^{\infty} e_k = e_0 + e_1 + \cdots + e_{m-1},$$

where $e_0 = 1 - \sum_{k=1}^{\infty} e_k$. Putting $p = \bigwedge_{m=1}^{\infty} p_m$, we have

$$1 - p \geq 1 - p_m \succsim e_0 + e_1 + \cdots + e_{m-1};$$

thus by Lemma 2.2, $1 - p \succsim \sum_{j=0}^{\infty} e_j = 1$. By finiteness, $p = 0$. Since $p_m \geq p_{m+1} \geq f_{m+1}$, we have

$$0 = p = \lim p_m \geq \lim f_m.$$

Thus $\{f_m\}$ converges to zero σ-strongly. Q.E.D.

Theorem 2.4. *For a von Neumann algebra $\mathscr{M}$, the following conditions are equivalent*:

(i) *$\mathscr{M}$ is finite*;
(ii) *$\mathscr{M}$ admits sufficiently many finite normal traces.*

PROOF. (ii) $\Rightarrow$ (i): Suppose $\{\tau_i\}$ is a family of sufficiently many finite traces. Let u be an element of $\mathscr{M}$ with $u^*u = 1$. We have then $\tau_i(uu^*) = \tau_i(u^*u) = \tau_i(1)$ for all i; thus $\tau_i(1 - uu^*) = 0$ for every i; hence $1 - uu^* = 0$. Therefore, $e \sim 1$ implies $e = 1$; in other words, $\mathscr{M}$ is finite.

(i) $\Rightarrow$ (ii): Let φ be a normal positive linear functional on $\mathscr{M}$. Let K_φ be the convex norm closure of the set $\mathscr{Q}_\varphi = \{u^*\varphi u : u \in \mathscr{U}_\mathscr{M}\}$, where $\mathscr{U}_\mathscr{M}$ denotes the unitary group of $\mathscr{M}$. We claim that K_φ is $\sigma(\mathscr{M}_*, \mathscr{M})$-compact. By Theorem III.5.4, it suffices to show that for any sequence $\{e_n\}$ of orthogonal projections in $\mathscr{M}$, $\psi(e_n)$ converges to zero uniformly for $\psi \in K_\varphi$. Since the convex hull of $\mathscr{Q}_\varphi$ is norm dense in K_φ, we have only to show that $\lim_{n\to\infty} \varphi(ue_nu^*) = 0$ uniformly for $u \in \mathscr{U}_\mathscr{M}$. Suppose this is not the case. There exists $\delta > 0$, subsequence $\{f_n\}$ of $\{e_n\}$, and a sequence $\{u_n\}$ in $\mathscr{U}_\mathscr{M}$ such that $\varphi(u_nf_nu_n^*) \geq \delta$, $n = 1,2,\ldots$. But trivially, $u_nf_nu_n^* \sim f_n$ and $\{f_n\}$ is orthogonal. Hence, $u_nf_nu_n^*$ converges to zero σ-strongly by Lemma 2.3. Thus $\varphi(u_nf_nu_n^*)$ must converge to zero, contradicting the choice of $\{u_n\}$, $\{f_n\}$, and δ. Thus, K_φ is $\sigma(\mathscr{M}_*, \mathscr{M})$-compact, being norm closed and convex.

For each $u \in \mathscr{U}_\mathscr{M}$, let $T_u\psi = u\psi u^*$, $\psi \in \mathscr{M}_*$. It follows then that $\{T_u : u \in \mathscr{U}_\mathscr{M}\}$ is a group of isometries on $\mathscr{M}_*$ leaving K_φ globally invariant. By the Ryll–Nardzewski fixed point theorem, K_φ contains a fixed point τ_φ under $\{T_u\}$. By definition, any fixed point under $\{T_u\}$ in $\mathscr{M}_*^+$ is a normal finite trace because $\tau(uxu^*) = \tau(x)$ for every $u \in \mathscr{U}_\mathscr{M}$ entails $\tau(ux) = \tau(xu)$, $x \in \mathscr{M}$ and $u \in \mathscr{U}_\mathscr{M}$; so $\tau(xy) = \tau(yx)$ for every $x,y \in \mathscr{M}$ by virtue of Proposition I.4.9.

Let τ be a normal finite trace on $\mathscr{M}$. It follows that the left kernel $N_\tau = \{x \in \mathscr{M} : \tau(x^*x) = 0\}$ is a two-sided ideal, so that the support $s(\tau)$ must be central. Let $\varphi \in \mathscr{M}_*^+$. We then have $\varphi(a) = \varphi(uau^*)$ for every central element $a \in \mathscr{M}$, so that $\varphi(a) = \tau_\varphi(a)$. Thus, $s(\tau_\varphi)$ and the support of the restriction of φ to the center $\mathscr{Z}$ must coincide. Hence we have $\bigvee \{s(\tau_\varphi) : \varphi \in \mathscr{M}_*^+\} = 1$, which means that $\{\tau_\varphi : \varphi \in \mathscr{M}_*^+\}$ is a family of sufficiently many traces.

Q.E.D.

Proposition 2.5. *A finite trace τ on a von Neumann algebra $\mathscr{M}$ is normal if and only if its restriction $\tau|_\mathscr{Z}$ to the center $\mathscr{Z}$ is normal. In particular, a finite trace on a factor is normal.*

PROOF. The "only if" part is trivial. So suppose that the restriction $\tau|_\mathscr{Z}$ is normal. Let $\tau = \tau_n + \tau_s$ be the decomposition of τ into the normal part and the singular part by Theorem III.2.14. Since this decomposition is given by a central projection in the universal enveloping von Neumann algebra $\tilde{\mathscr{M}}$ of $\mathscr{M}$, τ_n and τ_s are both finite traces. The restriction $\tau_s|_\mathscr{Z}$ of τ_s to $\mathscr{Z}$ is $\tau|_\mathscr{Z} - \tau_n|_\mathscr{Z}$, so that $\tau_s|_\mathscr{Z}$ is normal. We shall show that τ_s must be zero. Restricting our attention to τ_s, we assume that $\tau|_{\mathscr{Z}}$ is normal while τ itself is singular, and then drive a contradiction. Let z be the support of $\tau|_{\mathscr{Z}}$ in $\mathscr{Z}$. By Theorem III.3.8, there exists a nonzero projection $e \leq z$ with $\tau(e) = 0$ if z is not zero (if $z = 0$, then $\tau = 0$ automatically). First we claim that z is a finite projection. If z is infinite, then z majorizes a nonzero properly infinite central projection z_0 by Theorem 1.19. By Proposition 1.35, there exists a projection $p \leq z_0$ such that $z_0 \sim p \sim z_0 - p$. We have then $\tau(z_0) = \tau(p) = \tau(z_0 - p)$, so that $\tau(z_0) = 2\tau(z_0)$; so $\tau(z_0) = 0$, which is a contradiction. Therefore, z must be finite. By Proposition 1.34, there exist a central nonzero projection $z_1 \leq z$ and an orthogonal equivalent projection $\{f_i\}$ such that $f_i \sim ez_1$ and $f_0 = z_1 - \sum f_i \prec f_i$. Since z is finite, the family $\{f_i\}$ must be finite, say $\{f_1, \ldots, f_n\}$. We then have that

$$\tau(f_i) = \tau(ez_1) \leq \tau(e) = 0,$$
$$\tau(f_0) \leq \tau(f_i) = 0,$$
$$\tau(z_1) = \sum_{i=1}^{n} \tau(f_i) = 0.$$

This contradicts the choice of $z = s(\tau|_{\mathscr{Z}})$. Q.E.D.

Theorem 2.6. *For a von Neumann algebra $\mathscr{M}$, the following two conditions are equivalent:*

(i) *$\mathscr{M}$ is finite.*
(ii) *There exists a linear map T from $\mathscr{M}$ onto the center $\mathscr{Z}$ with the properties*
   (a) $T(x^*x) = T(xx^*) \geq 0$,
   (b) $T(ax) = aT(x)$, $a \in \mathscr{Z}$, $x \in \mathscr{M}$,
   (c) $T(1) = 1$,
   (d) $T(x^*x) \neq 0$ *for every nonzero* $x \in \mathscr{M}$.
   If this is the case, then T is unique and σ-weakly continuous.

Proof. The implication (ii) $\Rightarrow$ (i) follows from (a) and (d) as in the case of Theorem 2.4.

(i) $\Rightarrow$ (ii): Let $\{\tau_i\}_{i \in I}$ be a maximal family of normal finite traces on $\mathscr{M}$ with orthogonal support $s(\tau_i)$. Since $\mathscr{M}$ is assumed to be finite, we have $\sum_{i \in I} s(\tau_i) = 1$. As seen in the proof of Theorem 2.4, $s(\tau_i)$ is central; hence $\mathscr{M} \cong \sum_{i \in I}^{\oplus} \mathscr{M} s(\tau_i)$. Working on each direct summand, we may assume that $\mathscr{M}$ admits a faithful finite normal trace τ.

For each $x \in \mathscr{M}$, we define an element $\varphi_x \in \mathscr{Z}_*$ by $\varphi_x(a) = \tau(ax)$, $a \in \mathscr{Z}$. Clearly, the map: $x \in \mathscr{M} \mapsto \varphi_x \in \mathscr{Z}_*$ is linear and order preserving, and $\varphi_1 = \tau$. Therefore, the image $\{\varphi_x : x \in \mathscr{M}\}$ is contained in C_τ, where C_τ is the subspace in $\mathscr{Z}_*$ defined in Proposition IV.3.10 based on $\mathscr{Z}$ and τ, that is, C_τ is the subspace of $\mathscr{Z}_*$ spanned linearly by positive functionals on $\mathscr{Z}$ majorized by τ. Since the image of $\mathscr{Z}$ under the cyclic representation of $\mathscr{Z}$ induced by $\tau|_{\mathscr{Z}}$ is maximal abelian by Theorem III.1.2, Proposition IV.3.10

entails that for each $x \in \mathcal{M}$, there exists a unique $T(x) \in \mathcal{Z}$ such that

$$\tau(aT(x)) = \varphi_x(a) = \tau(ax), \qquad a \in \mathcal{Z}.$$

Since $\varphi_x \mapsto T(x)$ is linear and (completely) positive, the map T is linear and positive. Properties (b), (c), and (d) follow from its construction. Property (a) is checked by the following:

$$\begin{aligned}\tau(aT(x^*x)) &= \tau(ax^*x) = \tau(xax^*)\\ &= \tau(axx^*) = \tau(aT(xx^*)).\end{aligned}$$

For any $\psi \in \mathcal{Z}_*^+$, $\psi \circ T$ is a finite trace, and its restriction to $\mathcal{Z}$ is ψ itself, so normal. Thus $\psi \circ T$ is normal by Proposition 2.5. Therefore, T is also normal.

Property (d) follows from properties (a), (b), (c), and the normality of T. Indeed, the set $N = \{x \in \mathcal{M}: T(x^*x) = 0\}$ is a σ-weakly closed (two-sided) ideal of $\mathcal{M}$ by (a) and the normality of T. Hence there exists a central projection $z \in \mathcal{M}$ such that $N = \mathcal{M}z$. But we have $z = zT(1) = T(z) = 0$ by (b) and (c). Hence $N = \{0\}$, establishing (d).

We now prove the uniqueness of T. Let T' be a linear mapping of $\mathcal{M}$ onto $\mathcal{Z}$ enjoying properties (a), (b), and (c). We must prove $T = T'$. To this end, it suffices, by the spectral decomposition theory, that $T(e) = T'(e)$ for every nonzero projection $e \in \mathcal{M}$. By the usual exhaustion arguments, we have only to show that for any nonzero projection e there exists a nonzero projection $f \leq e$ in $\mathcal{M}$ with $T(f) = T'(f)$, where we use the fact that T' is normal by Proposition 2.5. By Theorem 1.19, we may assume that $\mathcal{M}$ is either of type I or type II. Suppose first $\mathcal{M}$ is of type I. By Theorem 1.27, there exists a homogeneous central projection z such that $ze \neq 0$. Let f be a nonzero abelian projection majorized by ze. It follows then that there exist orthogonal abelian projections $f = f_1 \sim f_2 \sim \cdots \sim f_n$ such that $\sum_{i=1}^n f_i = p \in \mathcal{Z}$. We then have

$$\begin{aligned}nT(f) &= T\left(\sum_{i=1}^n f_i\right) = T(p) = p\\ &= T'(p) = T'\left(\sum_{i=1}^n f_i\right) = nT'(f);\end{aligned}$$

thus $T(f) = T'(f)$. Suppose next $\mathcal{M}$ is of type II_1. By Proposition 1.34, there exist a nonzero central projection $z \in \mathcal{M}$ and orthogonal equivalent projections $f_1, \ldots, f_n$ such that $0 \neq ze \sim f_i$, $1 \leq i \leq n$, and $f_0 = z - \sum_{i=1}^n f_i \prec f_i$. Choose a positive integer m with $2^m > n + 1$. Applying Proposition 1.35 successively, we find an orthogonal family $\{p_1, p_2, \ldots, p_{2^m}\}$ of equivalent projections such that $z = \sum_{j=1}^{2^m} p_j$. We claim that $p_j \precsim ze$. By the comparability theorem, Theorem 1.8, we find a central projection h such that $hp_j \succsim hze$ and $(1 - h)p_j \precsim (1 - h)ze$. We then have

$$hz = \sum_{i=0}^{n} hf_i \precsim \sum_{j=1}^{n+1} hp_j < hz;$$

hence $hz = 0$ by finiteness. Thus $(1 - h)p_j = p_j$ and $p_j \precsim ze$. Let f be a projection such that $p_j \sim f \leq ze$. We then have

$$2^m T(f) = 2^m T(p_j) = T\left(\sum_{i=1}^{2^m} p_i\right) = T(z) = z$$
$$= T'(z) = \cdots = 2^m T'(f);$$

thus $T(f) = T'(f)$. Q.E.D.

Definition 2.7. The linear map T in the above theorem is called the *center valued trace* or, more specifically, the *canonical center valued trace*, and is denoted by $T_{\mathscr{M}}$ to indicate in which algebra T is considered.

An immediate consequence of the combination of Theorem 2.6 and the comparability theorem, Theorem 1.8, is:

Corollary 2.8. *In a finite von Neumann algebra $\mathscr{M}$ with center valued trace T, the following two conditions for a pair e, f of projections in $\mathscr{M}$ are equivalent:*

(i) $e \precsim f$;

(ii) $T(e) \leq T(f)$.

Corollary 2.9. *A finite von Neumann algebra is σ-finite if and only if its center is σ-finite. Hence every finite von Neumann algebra is a direct sum of σ-finite ones. In particular, a finite factor is σ-finite.*

Proof. The center of a σ-finite von Neumann algebra is trivially σ-finite. If the center $\mathscr{Z}$ of a finite von Neumann algebra $\mathscr{M}$ is σ-finite then $\varphi \circ T$ is a faithful finite normal trace for any faithful normal state φ on $\mathscr{Z}$. Hence $\mathscr{M}$ is σ-finite. Q.E.D.

End of Proof of Proposition 1.40. It is now easy to complete the proof of the uniqueness of $\{z_\alpha\}$. It suffices to prove that if $\{e_i\}_{i \in I}$ and $\{f_j\}_{j \in J}$ are two infinite families of equivalent orthogonal finite projections such that $\sum_{i \in I} e_i = \sum_{j \in J} f_j$, then $\operatorname{card} I = \operatorname{card} J$. Let $e = \sum_{i \in I} e_i$. Restricting our attention to $\mathscr{M}_e$, we may assume that $e = 1$. Let z be a nonzero central projection of $\mathscr{M}$ which is σ-finite in the center $\mathscr{Z}$ of $\mathscr{M}$. We then have

$$z = \sum_{i \in I} ze_i = \sum_{j \in J} zf_j;$$

and $\{ze_i\}_{i \in I}$ (resp. $\{zf_j\}_{j \in J}$) are equivalent. For each $i \in I$, let τ_i be a faithful finite normal trace on $\mathscr{M}_{ze_i}$. We then have

$$\tau_i(ze_i) = \sum_{j \in J} \tau_i(ze_i f_j ze_i),$$

so that $J_i = \{j \in J : f_j ze_i \neq 0\}$ is countable. Since $f_j z = \sum_{i \in I} f_j ze_i$, we have $J = \bigcup_{i \in I} J_i$. Hence $\operatorname{card} J \leq \aleph_0 \cdot \operatorname{card} I = \operatorname{card} I$. By symmetry, we have $\operatorname{card} I \leq \operatorname{card} J$. Thus we get $\operatorname{card} I = \operatorname{card} J$. Q.E.D.

We are now going to construct a faithful semifinite normal trace on any semifinite von Neumann algebra. But before that, we derive first a few basic facts about general traces.

Proposition 2.10. *If τ is a semifinite normal trace on a von Neumann algebra $\mathcal{M}$, then there exists a unique central projection $z \in \mathcal{M}$ such that $\tau = 0$ on $\mathcal{M}_+(1-z)$ and τ is faithful on $\mathcal{M}z$.*

PROOF. By Proposition 1.6, we have

$$\tau(e) + \tau(f) = \tau(e \vee f) + \tau(e \wedge f)$$

for any projections $e, f \in \mathcal{M}$. Hence $\tau(e \vee f) = 0$ if $\tau(e) = \tau(f) = 0$. Therefore, the set $\{e : \tau(e) = 0\}$ of projections is upward directed. By normality, $p = \bigvee\{e : \tau(e) = 0\}$ is annihilated by τ; that is, p is the greatest projection such that $\tau(p) = 0$. Since $\tau(upu^*) = \tau(p) = 0$ for any unitary element $u \in \mathcal{M}$, we have $upu^* = p$, so that p is central. Let $z = 1 - p$. It follows, from the spectral decomposition theorem, that $\tau = 0$ on $\mathcal{M}_+p$ and τ is faithful on $\mathcal{M}z$. The uniqueness of z follows automatically. Q.E.D.

Definition 2.11. The projection z in the proposition is called the *support* (*projection*) of τ and is denoted by $s(\tau)$.

Lemma 2.12. *If $\{\tau_i\}_{i \in I}$ is a family of semifinite normal traces on a von Neumann algebra $\mathcal{M}$ with orthogonal supports, then the function τ on $\mathcal{M}_+$ given by*

$$\tau(x) = \sum_{i \in I} \tau_i(x), \qquad x \in \mathcal{M}_+$$

is a semifinite normal trace.

PROOF. Being a sum of only nonnegative terms, the summation and the supremum can be interchanged, so that the normality of τ follows that of each τ_i. It is nearly a trivial matter to check the trace property of τ, i.e., Definition 2.1(i)–(iii). The only thing we must check is the semifiniteness of τ. Let $z_i = s(\tau_i)$, $i \in I$. If $x \in \mathcal{M}_+$ is nonzero, then either $\tau(x) = 0$ or $xz_i \neq 0$ for some $i \in I$. Certainly $0 \leq xz_i \leq x$ and $\tau(xz_i) = \tau_i(xz_i)$ by the orthogonality of $\{z_i\}$. If $xz_i \neq 0$, then there exists a nonzero $y \in \mathcal{M}$ such that $y \leq xz_i$ and $\tau_i(y) < +\infty$, by the semifiniteness of τ_i. We then have

$$\tau(y) = \sum_{j \in I} \tau_j(y) = \tau_i(y) < +\infty;$$

thus τ is semifinite. Q.E.D.

Lemma 2.13. *If τ is a normal trace on a von Neumann algebra $\mathcal{M}$, then there exists a unique central projection $z \in \mathcal{M}$ such that τ is semifinite on $\mathcal{M}z$ and $\tau(x) = +\infty$ for every nonzero $x \in \mathcal{M}_+(1-z)$.*

PROOF. Put $\mathfrak{n}_\tau = \{x \in \mathcal{M} : \tau(x^*x) < +\infty\}$. The inequalities

$$(x + y)^*(x + y) \leq 2(x^*x + y^*y),$$
$$(ax)^*ax \leq \|a\|^2 x^*x$$

show that $\mathfrak{n}_\tau$ is a left ideal of $\mathcal{M}$. But 2.1(iii) implies that $\mathfrak{n}_\tau$ is self-adjoint. Hence $\mathfrak{n}_\tau$ is an ideal. The σ-weak closure $\bar{\mathfrak{n}}_\tau$ of $\mathfrak{n}_\tau$ is then of the form $\bar{\mathfrak{n}}_\tau = \mathcal{M}z$ for some unique central projection $z \in \mathcal{M}$ by Proposition II.3.12. By Proposition II.3.13, there exists an increasing net $\{e_i\}$ in $\mathfrak{n}_\tau$ converging σ-strongly to z. For any $x \in \mathcal{M}_+ z$, we have then

$$x = \lim_i x^{1/2} e_i x^{1/2}$$

and $e_i x^{1/2} \in \mathfrak{n}_\tau$; so $\tau(x^{1/2} e_i x^{1/2}) < +\infty$. Thus τ is semifinite on $\mathcal{M}z$. On the other hand, $\mathcal{M}_+(1 - z)$ contains no nonzero element on which τ takes a finite value. The uniqueness of z is trivial. Q.E.D.

Proposition 2.14. *Let $\mathcal{N}$ be a von Neumann algebra and $\mathcal{B}$ a factor of type I. If $\mathcal{N}$ admits a faithful semifinite normal trace, then so does the tensor product $\mathcal{N} \mathbin{\overline{\otimes}} \mathcal{B}$.*

PROOF. Let τ be a faithful semifinite normal trace on $\mathcal{N}$. Fixing a system of matrix units in $\mathcal{B}$, every element in $\mathcal{M} = \mathcal{N} \mathbin{\overline{\otimes}} \mathcal{B}$ is represented by a matrix with entries from $\mathcal{N}$. For each $x \in \mathcal{M}_+$, put

$$\tilde{\tau}(x) = \sum_{i \in I} \tau(x_{i,i}),$$

where $x = [x_{i,j}]_{i,j \in I}$. The additivity and homogeneity follow trivially. For any $x \in \mathcal{M}$, we have

$$(x^*x)_{i,i} = \sum_{j \in I} x_{j,i}^* x_{j,i},$$
$$(xx^*)_{i,i} = \sum_{j \in I} x_{i,j} x_{i,j}^*,$$

where the summation is taken in the σ-strong* topology in $\mathcal{N}$. Hence we have

$$\begin{aligned}
\tilde{\tau}(x^*x) &= \sum_{i \in I} \tau\Big(\sum_{j \in I} x_{j,i}^* x_{j,i}\Big) \\
&= \sum_{i \in I} \sum_{j \in I} \tau(x_{j,i}^* x_{j,i}) \quad \text{by normality} \\
&= \sum_{i \in I} \sum_{j \in I} \tau(x_{j,i} x_{j,i}^*) \\
&= \sum_{j \in I} \sum_{i \in I} \tau(x_{j,i} x_{j,i}^*) \quad \text{by the positivity of all terms} \\
&= \sum_{j \in I} \tau\Big(\sum_{i \in I} x_{j,i} x_{j,i}^*\Big) \quad \text{by normality} \\
&= \sum_{j \in I} \tau((xx^*)_j) = \tilde{\tau}(xx^*).
\end{aligned}$$

Hence $\tilde{\tau}$ is a trace on $\mathscr{M}$. If $\{x_\alpha\}$ is an increasing net in $\mathscr{M}_+$ with $x = \sup_\alpha x_\alpha$, then we have $x_{i,i} = \sup_\alpha x_{\alpha,i,i}$, $i \in I$. Hence we have, by the positivity of the summation terms,

$$\begin{aligned}\tilde{\tau}(x) &= \sum_{i\in I} \tau(x_{i,i}) = \sum_{i\in I} \sup_\alpha \tau(x_{\alpha,i,i})\\ &= \sup_\alpha \sum_{i\in I} \tau(x_{\alpha,i,i}) = \sup_\alpha \tilde{\tau}(x_\alpha).\end{aligned}$$

Thus $\tilde{\tau}$ is normal. If $\tilde{\tau}(x^*x) = 0$, then $\tau(x_{j,i}^*x_{j,i}) = 0$ for every $i,j \in I$, so that $x_{j,i} = 0$ by the faithfulness of τ; hence $x = 0$. Therefore, $\tilde{\tau}$ is faithful.

We now prove the semifiniteness of $\tilde{\tau}$. Recall that the center of $\mathscr{M}$ is $\mathscr{Z} \otimes \mathbf{C}$ with $\mathscr{Z}$ the center of $\mathscr{N}$, by Corollary IV.5.11. If e is any nonzero projection in $\mathscr{Z}$, then there exists a nonzero x such that $0 \le x \le e$ and $\tau(x) < +\infty$. Let y be the element in $\mathscr{M}$ such that $y_{i,j} = 0$ if $i \ne i_0$ or $j \ne i_0$ for some fixed $i_0 \in I$ and $y_{i_0,i_0} = x$. We then have $0 \le y \le e \otimes 1$ and $\tilde{\tau}(y) = \tau(x) < +\infty$. Therefore, the central projection z in Lemma 2.13 must be the identity 1. Hence $\tilde{\tau}$ is semifinite. Q.E.D.

Theorem 2.15. *For a von Neumann algebra $\mathscr{M}$, the following conditions are equivalent*:

(i) *$\mathscr{M}$ is semifinite;*
(ii) *$\mathscr{M}$ admits a faithful semifinite normal trace.*

PROOF. (ii) $\Rightarrow$ (i): Let τ be a faithful semifinite normal trace on $\mathscr{M}$. If e is a projection with $\tau(e) < +\infty$, then the restriction of τ on $e\mathscr{M}e$ is a faithful finite trace; hence e must be finite. If e is any nonzero projection, then there exists a nonzero $x \in \mathscr{M}_+$ such that $x \le e$ and $\tau(x) < +\infty$. By the spectral decomposition theorem, there exist a scalar $\lambda > 0$ and a nonzero projection f such that $\lambda f \le x$. We then have $\tau(f) \le (1/\lambda)\tau(x) < +\infty$ and $f \le e$. Therefore, every nonzero projection majorizes a nonzero finite projection. Hence $\mathscr{M}$ is semifinite.

(i) $\Rightarrow$ (ii): Let $\{\tau_i\}_{i\in I}$ be a maximal family of semifinite normal traces with mutually orthogonal support. Let $\tau = \sum_{i\in I} \tau_i$. By Lemma 2.12, τ is a semifinite normal trace. It is clear that $\mathrm{s}(\tau) = \sum_{i\in I} \mathrm{s}(\tau_i)$. We claim $\mathrm{s}(\tau) = 1$. Let $z = 1 - \mathrm{s}(\tau)$. Suppose $z \ne 0$. It follows that $\mathscr{M}_z$ is semifinite. By Preposition 1.40, there exists a nonzero central projection $z_0 \le z$ such that $\mathscr{M}_{z_0} \cong \mathscr{N} \mathbin{\overline{\otimes}} \mathscr{L}(\mathfrak{H}_0)$ with some finite von Neumann algebra $\mathscr{N}$ and a Hilbert space $\mathfrak{H}_0$. Let τ_0 be a finite normal nontrivial trace on $\mathscr{N}$ whose existence is guaranteed by Theorem 2.4. Let $\tilde{\tau}_0$ be the trace on $\mathscr{M}_{z_0}$ constructed in Proposition 2.14. Then $\tilde{\tau}_0$ is a nontrivial semifinite normal trace on $\mathscr{M}z_0$. This contradicts the maximality of $\{\tau_i\}_{i\in I}$. Thus $\mathrm{s}(\tau) = 1$. Q.E.D.

Therefore, every semifinite von Neumann algebra $\mathscr{M}$ admits a faithful semifinite normal trace τ. Put

$$\begin{aligned}\mathfrak{p}_\tau &= \{x \in \mathscr{M}_+ : \tau(x) < +\infty\},\\ \mathfrak{n}_\tau &= \{x \in \mathscr{M} : \tau(x^*x) < +\infty\}.\end{aligned}$$

We have seen that $\mathfrak{n}_\tau$ is an ideal of $\mathcal{M}$. Hence

$$\mathfrak{m}_\tau = \left\{\sum_{i=1}^{n} x_i y_i : x_i, y_i \in \mathfrak{n}_\tau\right\}$$

is also an ideal of $\mathcal{M}$.

Lemma 2.16. *In the above situation, we have*

$$\mathfrak{p}_\tau = \mathfrak{m}_\tau \cap \mathcal{M}_+,$$

and $\mathfrak{m}_\tau$ is linearly spanned by $\mathfrak{p}_\tau$. The function τ on $\mathfrak{p}_\tau$ can be extended to a linear functional $\dot{\tau}$ on $\mathfrak{m}_\tau$, which enjoys the properties

$$\left.\begin{aligned} &\dot{\tau}(x^*) = \overline{\dot{\tau}(x)}, && x \in \mathfrak{m}_\tau, \\ &\dot{\tau}(ax) = \dot{\tau}(xa), && a \in \mathcal{M}, \quad x \in \mathfrak{m}_\tau, \\ &\dot{\tau}(xy) = \dot{\tau}(yx), && x,y \in \mathfrak{n}_\tau. \end{aligned}\right\}$$

PROOF. Each $x \in \mathfrak{m}_\tau$ is of the form

$$x = \sum_{j=1}^{n} y_j^* z_j, \qquad y_j, z_j \in \mathfrak{n}_\tau.$$

From the polarization identity, it follows that

$$4x = \sum_{j=1}^{n} \sum_{k=0}^{3} i^k (y_j + i^k z_j)^*(y_j + i^k z_j),$$

so that $\mathfrak{m}_\tau$ is linearly spanned by $\mathfrak{m}_\tau \cap \mathcal{M}_+$. If the above x is self-adjoint, then

$$\begin{aligned} 4x &= \sum_{j=1}^{n} (y_j + z_j)^*(y_j + z_j) - \sum_{j=1}^{n} (y_j - z_j)^*(y_j - z_j) \\ &\leq \sum_{j=1}^{n} (y_j + z_j)^*(y_j + z_j) \in \mathfrak{p}_\tau. \end{aligned}$$

Hence if $x \geq 0$, then x is in $\mathfrak{p}_\tau$. Therefore, we get $\mathfrak{m}_\tau \cap \mathcal{M}_+ \subset \mathfrak{p}_\tau$. Conversely, if $x \in \mathfrak{p}_\tau$, then $x^{1/2}$ is in $\mathfrak{n}_\tau$, so that $x = (x^{1/2})^2 \in \mathfrak{m}_\tau \cap \mathcal{M}_+$; thus $\mathfrak{p}_\tau \subset \mathfrak{m}_\tau \cap \mathcal{M}_+$, establishing that $\mathfrak{m}_\tau \cap \mathcal{M}_+ = \mathfrak{p}_\tau$. Therefore, we have also $\mathfrak{m}_\tau \cap \mathcal{M}_h = \mathfrak{p}_\tau - \mathfrak{p}_\tau$ by the above arguments.

We now extend τ. If we have

$$h - k = h' - k' \quad \text{for} \quad h,k,h',k' \in \mathfrak{p}_\tau,$$

then $h + k' = h' + k$, so that $\tau(h) + \tau(k') = \tau(h') + \tau(k)$; hence $\tau(h) - \tau(k) = \tau(h') - \tau(k')$. This allows us to define a real valued function $\dot{\tau}$ on $\mathfrak{m}_\tau \cap \mathcal{M}_h$ by

$$\dot{\tau}(h - k) = \tau(h) - \tau(k), \qquad h,k \in \mathfrak{P}_\tau.$$

This function $\dot{\tau}$ is clearly linear. We then extend further $\dot{\tau}$ to the whole ideal $\mathfrak{m}_\tau = (\mathfrak{m}_\tau \cap \mathcal{M}_h) + i(\mathfrak{m}_\tau \cap \mathcal{M}_h)$ linearly. It is then clear that $\dot{\tau}(x^*) = \overline{\dot{\tau}(x)}$ for any $x \in \mathfrak{m}_\tau$.

For any $x \in \mathfrak{n}_\tau$, we have by definition $\tau(x^*x) = \tau(xx^*)$. By polarizing this, we get

$$\dot{\tau}(y^*x) = \dot{\tau}(xy^*), \qquad x,y \in \mathfrak{n}_\tau.$$

For any $a \in \mathcal{M}$, $x,y \in \mathfrak{n}_\tau$, we have

$$\begin{aligned}\dot{\tau}(a(y^*x)) &= \dot{\tau}((ay^*)x) = \dot{\tau}(x(ay^*)) \\ &= \dot{\tau}((xa)y^*) = \dot{\tau}(y^*(xa)) = \dot{\tau}((y^*x)a).\end{aligned}$$

By linearity, we get

$$\dot{\tau}(ax) = \dot{\tau}(xa), \qquad x \in \mathfrak{m}_\tau, \qquad a \in \mathcal{M}. \qquad \text{Q.E.D.}$$

Remark. Since $\mathfrak{m}_\tau$ and $\mathfrak{n}_\tau$ are both ideal, an $x \in \mathcal{M}$ belongs to $\mathfrak{m}_\tau$ (resp. $\mathfrak{n}_\tau$) if and only if its absolute value $|x|$ belongs to $\mathfrak{m}_\tau$ (resp. $\mathfrak{n}_\tau$). Hence

$$\mathfrak{m}_\tau = \{xy : x,y \in \mathfrak{n}_\tau\}.$$

Definition 2.17. The ideal $\mathfrak{m}_\tau$ is called the *definition ideal* of the trace τ.

In the sequel, we shall not distinguish the extended linear functional τ on $\mathfrak{m}_\tau$ and the original trace τ itself, and we shall omit the dot on top of the τ.

We now fix a faithful semifinite normal trace τ on a von Neumann algebra $\mathcal{M}$. Choose an arbitrary $x \in \mathfrak{m}_\tau$ and $y \in \mathcal{M}$. Let $x = u|x|$ and $y = v|y|$ be their polar decomposition. We then have $|x| = u^*x \in \mathfrak{m}_\tau$ and $|x|^{1/2} \in \mathfrak{n}_\tau$, and

$$\begin{aligned}\tau(yx) &= \tau(v|y|u|x|) = \tau(|x|^{1/2}v|y|u|x|^{1/2}) \\ &= \tau((|x|^{1/2}v|y|^{1/2})(|y|^{1/2}u|x|^{1/2})).\end{aligned}$$

Noticing that $|x|^{1/2}v|y|^{1/2}$ and $|y|^{1/2}u|x|^{1/2}$ are both in $\mathfrak{n}_\tau$, we get, by the Cauchy–Schwarz inequality,

$$\begin{aligned}|\tau(yx)|^2 &\le \tau((|y|^{1/2}v^*|x|^{1/2})^*(|y|^{1/2}v^*|x|^{1/2}))\tau((|y|^{1/2}u|x|^{1/2})^*(|y|^{1/2}u|x|^{1/2})) \\ &= \tau(|x|^{1/2}v|y|v^*|x|^{1/2})\tau(|x|^{1/2}u^*|y|u|x|^{1/2}) \\ &= \tau(v|y|v^*|x|)\tau(|y|u|x|u^*).\end{aligned}$$

Since $|x^*| = u|x|u^*$ and $|y^*| = v|y|v^*$, we have

$$|\tau(yx)|^2 \le \tau(|y^*||x|)\tau(|y||x^*|), \qquad x \in \mathfrak{m}_\tau, \qquad y \in \mathcal{M}. \tag{1}$$

If, in addition, $x \ge 0$ and $y \ge 0$, then

$$\tau(yx) = \tau(x^{1/2}yx^{1/2}) \le \|y\|\tau(x).$$

Hence it follows that

$$\begin{aligned}&\tau(|y^*||x|) \le \||y^*|\|\tau(|x|) = \|y\|\tau(|x|), \\ &\tau(|y||x^*|) \le \||y|\|\tau(|x^*|) = \|y\|\tau(u|x|u^*) = \|y\|\tau(|x|).\end{aligned}$$

Thus (1) entails that

$$|\tau(yx)| \le \|y\|\tau(|x|),$$
$$|\tau(yx)| = |\tau(1(yx))| \le \tau(|yx|).$$

Let $yx = w|yx|$ be the polar decomposition. We then have

$$\tau(|yx|) = \tau(w^*yx) \le \|w^*y\|\tau(|x|)$$
$$\le \|y\|\tau(|x|).$$

Therefore, we obtain the inequality

$$|\tau(yx)| \le \tau(|yx|) \le \|y\|\tau(|x|), \qquad x \in \mathfrak{m}_\tau, \qquad y \in \mathcal{M}. \tag{2}$$

Since $\tau(|x|) = \tau(u^*x)$, we have

$$\tau(|x|) = \sup\{|\tau(yx)|: y \in \mathcal{M}, \|y\| \le 1\}, \qquad x \in \mathfrak{m}_\tau. \tag{3}$$

Put

$$\|x\|_1 = \tau(|x|), \qquad x \in \mathfrak{m}_\tau. \tag{4}$$

Then $(\mathfrak{m}_\tau, \|\cdot\|_1)$ turns out to be a normed space, which is isometric to a subspace of the conjugate space $\mathcal{M}^*$ by the bilinear form

$$(y,x) \in \mathcal{M} \times \mathfrak{m}_\tau \mapsto \tau(yx) \in \mathbf{C}.$$

Let $\omega_x(y) = \tau(yx)$, $y \in \mathcal{M}$, $x \in \mathfrak{m}_\tau$. If $x \ge 0$ and $\{y_i\}$ is an increasing net in $\mathcal{M}_+$ with $y = \sup y_i$, then $x^{1/2}yx^{1/2} = \sup x^{1/2}y_ix^{1/2}$; hence we get

$$\omega_x(y) = \tau(yx) = \tau(x^{1/2}yx^{1/2})$$
$$= \sup \tau(x^{1/2}y_ix^{1/2}) = \sup \tau(y_ix)$$
$$= \sup \omega_x(y_i).$$

Therefore, ω_x is a normal positive linear functional on $\mathcal{M}$; so $\omega_x \in \mathcal{M}_*^+$. Since $\mathfrak{m}_\tau$ is spanned by its positive part $\mathfrak{p}_\tau$, Lemma 2.16, the normed space $(\mathfrak{m}_\tau, \|\cdot\|_1)$ corresponds to a subspace of the predual $\mathcal{M}_*$. Let $L^1(\mathcal{M},\tau)$ denote the completion of $(\mathfrak{m}_\tau, \|\cdot\|_1)$. If $y \ne 0$, $y \in \mathcal{M}$, then there exists an $s \in \mathfrak{m}_\tau \cap \mathcal{M}_+$ with $s|y|s \ne 0$ because $\mathfrak{m}_\tau$ is σ-weakly dense in $\mathcal{M}$. Putting $x = s^2v^*$, we have

$$\omega_x(y) = \tau(yx) = \tau(ys^2v^*) = \tau(v^*ys^2)$$
$$= \tau(s|y|s) \ne 0.$$

Therefore, $\{\omega_x : x \in \mathfrak{m}_\tau\}$ is total in $\mathcal{M}_*$. Since $\mathcal{M} = (\mathcal{M}_*)^*$, the Hahn–Banach theorem entails that $\{\omega_x : x \in \mathfrak{m}_\tau\}$ be norm dense in $\mathcal{M}_*$. Therefore, the Banach space $L^1(\mathcal{M},\tau)$ is isometrically isomorphic to the predual $\mathcal{M}_*$. The function τ on $\mathfrak{m}_\tau$ is bounded in the $\|\cdot\|_1$-norm, so that it is extended to a linear functional on $L^1(\mathcal{M},\tau)$ which will be also denoted by τ. By inequality (2), the bilinear map: $(y,x) \in \mathcal{M} \times \mathfrak{m}_\tau \mapsto yx \in \mathfrak{m}_\tau$ (resp. $xy \in \mathfrak{m}_\tau$) is of norm one; hence it is also extended to an $L^1(\mathcal{M},\tau)$-valued bilinear map of $\mathcal{M} \times L^1(\mathcal{M},\tau)$, whose value at $(y,x) \in \mathcal{M} \times L^1(\mathcal{M},\tau)$ is denoted by yx (resp. xy). Denoting by ω_x the element of $\mathcal{M}_*$ corresponding to an $x \in L^1(\mathcal{M},\tau)$, we have

$$\omega_x(y) = \tau(yx) = \tau(xy), \qquad y \in \mathcal{M}, \qquad x \in L^1(\mathcal{M},\tau). \tag{5}$$

It follows then by the continuity arguments that

$$\omega_{ax} = a\omega_x, \qquad \omega_{xa} = \omega_x a, \qquad a \in \mathcal{M}, \qquad x \in L^1(\mathcal{M},\tau). \tag{6}$$

Thus, we arrive at the following result:

Theorem 2.18. *If τ is a faithful semifinite normal trace on a von Neumann algebra $\mathcal{M}$ with definition ideal $\mathfrak{m}_\tau$, then the bilinear form*

$$(y,x) \in \mathcal{M} \times \mathfrak{m}_\tau \mapsto \tau(yx) \in \mathbf{C}$$

extends to the duality between $\mathcal{M}$ and the completion $L^1(\mathcal{M},\tau)$ of $\mathfrak{m}_\tau$ under the norm $\|x\|_1 = \tau(|x|)$, $x \in \mathfrak{m}_\tau$, under which $\mathcal{M}$ is the conjugate space of $L^1(\mathcal{M},\tau)$.

Thus, the von Neumann algebra $\mathcal{M}$ may be denoted by $L^\infty(\mathcal{M},\tau)$ and the norm $\|a\|$ of $a \in \mathcal{M}$ by $\|a\|_\infty$. The following result allows us to write $\mathfrak{m}_\tau$ as $\mathcal{M} \cap L^1(\mathcal{M},\tau)$ or $L^\infty(\mathcal{M},\tau) \cap L^1(\mathcal{M},\tau)$.

Proposition 2.19. *In the above situation, an $\omega \in \mathcal{M}_*$ is of the form ω_x with $x \in \mathfrak{m}_\tau$ under the notation (5) if and only if*

$$\sup\{|\omega(y)| : y \in \mathfrak{m}_\tau, \|y\|_1 \le 1\} < +\infty.$$

If this is the case, then we have

$$\|x\|_\infty = \sup\{|\omega(y)| : y \in \mathfrak{m}_\tau, \|y\|_1 \le 1\}.$$

PROOF. If $\omega = \omega_x$ for some $x \in \mathfrak{m}_\tau$, then inequality (2) applied to $y \in \mathfrak{m}_\tau$ yields the consequence. Suppose that

$$\|\omega\|_\infty = \sup\{|\omega(y)| : y \in \mathfrak{m}_\tau, \|y\|_1 \le 1\} < +\infty.$$

The linear functional ω on $\mathfrak{m}_\tau$ is bounded in the $\|\cdot\|_1$-norm; so that it is extended to a bounded linear functional on $L^1(\mathcal{M},\tau)$ with norm $\|\omega\|_\infty$. By Theorem 2.18, there exists an $x \in \mathcal{M}$ such that

$$\omega(y) = \tau(xy), \qquad y \in L^1(\mathcal{M},\tau).$$

Let $x = u|x|$ be the polar decomposition. We then have

$$\tau(|x|) = \tau(u^*x) = \tau(xu^*) = \omega(u^*) < +\infty;$$

hence x falls in $\mathfrak{m}_\tau$ and $\omega = \omega_x$. The equality $\|x\|_\infty = \|\omega\|_\infty$ follows from Theorem 2.18. Q.E.D.

To illustrate what we have done above, let us consider an abelian von Neumann algebra $\mathcal{A}$ with a faithful semifinite normal trace τ. As seen in Section III.1, $\mathcal{A}$ is identified with the algebra of all continuous functions on a hyperstonean space Ω. The family of projections in $\mathfrak{m}_\tau$ is upward directed since $\tau(e \vee f) \leq \tau(e) + \tau(f)$ and converges σ-strongly to 1.[1] To each projection $e \in \mathfrak{m}_\tau$, there corresponds an open and closed subset E of Ω. Put $\Gamma = \bigcup\{E : e \in \mathfrak{m}_\tau \text{ projection}\}$. It follows that Γ is an open dense subset of Ω, so that it is a locally compact space. By Corollary III.1.8, $\mathscr{A}$ is identified with the algebra $C_b(\Gamma)$ of all bounded continuous functions on Γ because Ω is the Stone–Čech compactification of Γ. Let $\mathscr{K}(\Gamma)$ be the algebra of all continuous functions on Γ with compact support. It follows that $\mathscr{K}(\Gamma)$ is an ideal of $C_b(\Gamma)$. Since every compact subset of Γ is contained in some E corresponding to an $e \in \mathfrak{m}_\tau$, $\mathscr{K}(\Gamma)$ is contained in $\mathfrak{m}_\tau$. Hence, the positive linear functional: $f \in \mathscr{K}(\Gamma) \mapsto \tau(f)$ gives rise to a Radon measure μ_τ on Γ. By the construction of Γ, any open compact subset of Γ is hyperstonean, and the restriction of μ_τ to such a set is normal in the sense of Definition III.1.10 by the normality of τ. Thus $\mathscr{A}$ is identified with $L^\infty(\Gamma,\mu_\tau)$ as in the proof of Theorem III.1.18. The definition ideal $\mathfrak{m}_\tau$ of τ is then identified with $L^\infty(\Gamma, \mu_\tau) \cap L^1(\Gamma,\mu_\tau)$, and $\mathfrak{n}_\tau$ with $L^\infty(\Gamma,\mu_\tau) \cap L^2(\Gamma,\mu_\tau)$. The norm given by (4) is nothing else but the L^1-norm:

$$\|f\|_1 = \int_\Gamma |f(\gamma)|\, d\mu_\tau(\gamma).$$

Thus $L^1(\mathscr{A},\tau)$ is identified with $L^1(\Gamma,\mu_\tau)$.

From Theorem III.1.2, it follows that $\mathscr{A}$ is represented faithfully as a maximal abelian algebra on the Hilbert space $L^2(\Gamma,\mu)$. This means that on this representation space the commutant $\mathscr{A}'$ of $\mathscr{A}$ is given by $\mathscr{A}$ itself. We are going to generalize this fact to semifinite von Neumann algebras. Of course, we do not expect that $\mathscr{M} = \mathscr{M}'$ for nonabelian $\mathscr{M}$. But we will see something corresponding to the above fact.

Let $\mathscr{M}$ be a semifinite von Neumann algebra equipped with a faithful semifinite normal trace τ. In the ideal $\mathfrak{n}_\tau$, we define a sesquilinear form

$$(x|y)_\tau = \tau(y^*x), \qquad x,y \in \mathfrak{n}_\tau.$$

By the positivity and faithfulness of τ, this sesquilinear form is an inner product in $\mathfrak{n}_\tau$. We denote by $L^2(\mathscr{M},\tau)$ the completion of $\mathfrak{n}_\tau$ and we write the norm of $x \in \mathfrak{n}_\tau$ by

$$\|x\|_2 = \tau(x^*x)^{1/2}, \qquad x \in \mathfrak{n}_\tau. \tag{7}$$

For each $a \in \mathscr{M}$, $x \in \mathfrak{n}_\tau$, we have

$$\begin{aligned}\tau((ax)^*ax) &\leq \|a\|^2\tau(x^*x),\\ \tau((xa)^*xa) &= \tau(a^*x^*xa) = \tau(xaa^*x^*)\\ &\leq \|a\|^2\tau(xx^*) = \|a\|^2\tau(x^*x),\end{aligned}$$

so that we have

$$\left.\begin{aligned}\|ax\|_2 &\leq \|a\|_\infty\|x\|_2,\\ \|xa\|_2 &\leq \|a\|_\infty\|x\|_2, \qquad a \in \mathscr{M}, \qquad x \in \mathfrak{n}_\tau.\end{aligned}\right\} \tag{8}$$

[1] This fact is, of course, independent of the commutativity.

Therefore, $L^2(\mathcal{M},\tau)$ turns out to be a two-sided $\mathcal{M}$-module. The commutativity of the left and right multiplications by $\mathcal{M}$ on $L^2(\mathcal{M},\tau)$ follows from the associativity of the product in $\mathcal{M}$, that is,

$$(ax)b = axb, \qquad a,b \in \mathcal{M}, \qquad x \in \mathfrak{n}_\tau.$$

Let π_τ (resp. π'_τ) be the left (resp. right) multiplication representation of $\mathcal{M}$ on $L^2(\mathcal{M},\tau)$, i.e.,

$$\pi_\tau(a)x = ax, \qquad \pi'_\tau(a)x = xa, \qquad a \in \mathcal{M}, \qquad x \in L^2(\mathcal{M},\tau). \tag{9}$$

The $*$-operation in $\mathfrak{n}_\tau$ is an isometry with respect to the $\|\cdot\|_2$-norm; hence it is extended to a conjugate linear isometry: $x \in L^2(\mathcal{M},\tau) \mapsto x^* \in L^2(\mathcal{M},\tau)$, with period two. When we regard this $*$-operation in $L^2(\mathcal{M},\tau)$ as an isometric conjugate linear operator, we denote it by J, i.e., $Jx = x^*$, $x \in L^2(\mathcal{M},\tau)$.

Lemma 2.20. *For an $x \in \mathcal{M}$, the following conditions are equivalent*:

(i) *x falls in $\mathfrak{n}_\tau$*;
(ii) $\sup\{|\tau(y^*x)| : y \in \mathfrak{m}_\tau, \|y\|_2 \leq 1\} = \|x\|_2 < +\infty$.

PROOF. The implication (i) $\Rightarrow$ (ii) is nothing more than the Cauchy–Schwarz inequality.

(ii) $\Rightarrow$ (i): Suppose that condition (ii) is fulfilled. Let $\{e_i\}$ be an increasing net of projections in $\mathfrak{m}_\tau$ converging σ-strongly to 1. For any $y \in \mathfrak{n}_\tau$, we have that $e_i y \in \mathfrak{m}_\tau$ and

$$\|y - e_i y\|^2 = \tau(y^*(1 - e_i)y) \to 0,$$

so that $\lim e_i y = y$ in $L^2(\mathcal{M},\tau)$. Hence $\mathfrak{m}_\tau$ is dense in $L^2(\mathcal{M},\tau)$. Thus, the linear functional: $y \in \mathfrak{m}_\tau \mapsto \tau(y^*x) \in \mathbf{C}$ is a bounded densely defined conjugate linear functional on $L^2(\mathcal{M},\tau)$. By the Riesz theorem, there exists an $x' \in L^2(\mathcal{M},\tau)$ such that

$$\tau(y^*x) = (x'|y)_\tau, \qquad y \in \mathfrak{m}_\tau.$$

For each e_i, and $y \in \mathfrak{m}_\tau$, we have

$$\begin{aligned}(e_i x'|y)_\tau &= (x'|e_i y)_\tau = \tau(y^* e_i x)\\ &= (e_i x|y)_\tau,\end{aligned}$$

so that $e_i x'$ is in $\mathfrak{m}_\tau$ and $e_i x' = e_i x$. Hence we have

$$\tau(x^*x) = \sup \tau(x^* e_i x) = \sup \|e_i x'\|^2 \leq \|x'\|_2^2 < +\infty,$$

establishing that $x \in \mathfrak{n}_\tau$. Q.E.D.

Lemma 2.21. *For an element $x \in L^2(\mathcal{M},\tau)$, the following three conditions are equivalent*:

(i) *x falls in $\mathfrak{n}_\tau$*;
(ii) $\sup\{|(x|y)_\tau| : y \in \mathfrak{m}_\tau, \|y\|_1 \leq 1\} = \|x\|_\infty < +\infty$;
(iii) $\sup\{\|ax\|_2 : a \in \mathfrak{n}_\tau, \|a\|_2 \leq 1\} = \|x\|_\infty < +\infty$.

PROOF. The implication (ii) ⇒ (i) is exactly inequality (2).

(ii) ⇒ (i): Suppose that the supremum on the left-hand side in (ii) is finite. By Theorem 2.18, there exists an $x' \in \mathcal{M}$ such that

$$(x|y)_\tau = \tau(y^*x'), \qquad y \in \mathfrak{m}_\tau.$$

If $\{e_i\}$ is an increasing net of projections in $\mathfrak{m}_\tau$ converging to 1, then we have, for any $y \in \mathfrak{m}_\tau$,

$$\begin{aligned}(e_ix'|y)_\tau &= \tau(y^*e_ix') = (x|e_iy)_\tau \\ &= (e_ix|y)_\tau,\end{aligned}$$

so that $e_ix' = e_ix$; and then we get

$$\begin{aligned}\tau(x'^*x') &= \sup \tau(x'^*e_ix') = \sup\|e_ix'\|_2^2 \\ &= \sup\|e_ix\|_2^2 \le \|x\|_2^2.\end{aligned}$$

Thus $x' \in \mathfrak{n}_\tau$, and $x = x'$.

(i) ⇒ (iii): If $x \in \mathfrak{n}_\tau$, then $\|ax\|_2 \le \|a\|_2\|x\|_\infty$.

(iii) ⇒ (ii): Suppose that

$$\|x\|_\infty = \sup\{\|ax\|_2 : a \in \mathfrak{n}_\tau, \|a\|_2 \le 1\} < +\infty.$$

Let y be an arbitrary element of $\mathfrak{m}_\tau$, and $y = kv$ be the right polar decomposition, i.e., $k = |y^*|$. We have

$$\begin{aligned}|(x|y)_\tau| &= |(x|k^{1/2}k^{1/2}v)_\tau| = |(k^{1/2}x|k^{1/2}v)_\tau| \\ &\le \|k^{1/2}x\|_2\|k^{1/2}v\|_2 \le \|x\|_\infty\|k^{1/2}\|_2^2 \\ &= \|x\|_\infty\tau(k) = \|x\|_\infty\tau(|y|) = \|x\|_\infty\|y\|_1.\end{aligned}$$

Thus condition (ii) follows. Once we know that $x \in \mathfrak{n}_\tau$, then the above computation shows that

$$\begin{aligned}\|x\|_\infty &= \sup\{|\tau(y^*x)| : y \in \mathfrak{m}_\tau, \|y\|_1 \le 1\} \\ &= \sup\{|\tau(y_2^*y_1x)| : y_1, y_2 \in \mathfrak{n}_\tau, \|y_1\|_2 \le 1, \|y_2\|_2 \le 1\} \\ &= \sup\{\|y_1x\|_2 : y_1 \in \mathfrak{n}_\tau, \|y_1\|_2 \le 1\}.\end{aligned}$$

Q.E.D.

We can now show the following result, which corresponds to the maximal abelianness of $L^\infty(\Gamma,\mu)$ on $L^2(\Gamma,\mu)$.

Theorem 2.22. *If $\mathcal{M}$ is a semifinite von Neumann algebra equipped with a faithful semifinite normal trace τ, then the representation π_τ (resp. π'_τ) of $\mathcal{M}$ on $L^2(\mathcal{M},\tau)$ by the left (resp. right) multiplication is a faithful normal representation (resp. antirepresentation in the sense that it reverses the order of multiplication) such that*

$$\begin{gathered}\pi_\tau(\mathcal{M})' = \pi'_\tau(\mathcal{M}); \qquad \pi'_\tau(\mathcal{M})' = \pi_\tau(\mathcal{M}), \\ J\pi_\tau(a)J = \pi'_\tau(a^*), \qquad a \in \mathcal{M};\end{gathered}$$

hence

$$J\pi_\tau(\mathcal{M})J = \pi'_\tau(\mathcal{M}); \qquad J\pi'_\tau(\mathcal{M})J = \pi_\tau(\mathcal{M}).$$

PROOF. For any $y_1, y_2 \in \mathfrak{n}_\tau$, we have

$$(xy_1|y_2)_2 = \tau(y_2^* x y_1) = \tau(y_1 y_2^* x),$$

so that the linear functional $\omega(\pi_\tau; y_1, y_2) = (\pi_\tau(\cdot)y_1|y_2)_\tau$ of $\mathcal{M}$ is normal since $y_1 y_2^* \in \mathfrak{m}_\tau$. Hence the representation π_τ is normal. The σ-weak density of $\mathfrak{n}_\tau$ in $\mathcal{M}$, together with the faithfulness of τ, imply the faithfulness of π_τ.

The equation $J\pi_\tau(a)J = \pi_\tau(a^*)$ is a trivial matter. Suppose that $b \in \mathcal{L}(L^2(\mathcal{M},\tau))$ commutes with $\pi_\tau(\mathcal{M})$. For any $x, y \in \mathfrak{n}_\tau$, we have

$$\begin{aligned}\|y(bx)\|_2 &= \|\pi_\tau(y)bx\|_2 = \|b\pi_\tau(y)x\|_2 \\ &\leq \|b\|\|yx\|_2 \leq \|b\|\|y\|_2\|x\|_\infty;\end{aligned}$$

hence $bx \in \mathfrak{n}_\tau$, and also we get

$$\begin{aligned}\pi_\tau'(bx)y &= \pi_\tau(y)bx = b\pi_\tau(y)x \\ &= b\pi_\tau'(x)y;\end{aligned}$$

thus $\pi_\tau'(bx) = b\pi_\tau'(x)$, $x \in \mathfrak{n}_\tau$. This means that $b\pi_\tau'(\mathfrak{n}_\tau) \subset \pi_\tau'(\mathfrak{n}_\tau) \subset \pi_\tau'(\mathcal{M})$. If $\{e_i\}$ is an increasing net in $\mathfrak{n}_\tau$ converging σ-strongly to 1, then $\pi_\tau'(e_i) = J\pi_\tau(e_i)J$ converges σ-strongly to 1 on $L^2(\mathcal{M},\tau)$ by the normality of π_τ. Hence we have $b = \lim b\pi_\tau'(e_i) \in \pi_\tau'(\mathcal{M})$, since $\pi_\tau'(\mathcal{M}) = J\pi_\tau(\mathcal{M})J$ is a von Neumann algebra on $L^2(\mathcal{M},\tau)$. Q.E.D.

From this theorem, we can draw a number of consequences for semifinite von Neumann algebras.

Corollary 2.23. *The commutant $\mathcal{M}'$ of a von Neumann algebra $\{\mathcal{M},\mathfrak{H}\}$ is semifinite if and only if $\mathcal{M}$ is.*

PROOF. By symmetry, we have only to prove that if $\mathcal{M}$ is semifinite, then $\mathcal{M}'$ is also. Let τ be a faithful, semifinite, normal trace on $\mathcal{M}$. We then consider the representation π_τ of $\mathcal{M}$ on $L^2(\mathcal{M},\tau)$ constructed in the previous theorem. Since the commutant $\pi_\tau(\mathcal{M})' = \pi_\tau'(\mathcal{M})$ is antiisomorphic to $\mathcal{M}$, $\pi_\tau(\mathcal{M})'$ is semifinite. By Proposition 2.14, $\pi_\tau(\mathcal{M})' \mathbin{\overline{\otimes}} \mathcal{L}(\mathfrak{K})$ is semifinite for any Hilbert space. By Theorem IV.5.5, the commutant $\mathcal{M}'$ is isomorphic to a reduced von Neumann algebra of $\pi_\tau(\mathcal{M})' \mathbin{\overline{\otimes}} \mathcal{L}(\mathfrak{K})$ for some $\mathfrak{K}$ and a projection in $\pi_\tau(\mathcal{M})' \mathbin{\overline{\otimes}} \mathcal{L}(\mathfrak{K})$. But in a semifinite von Neumann algebra, any nonzero projection majorizes a nonzero finite projection. Hence the reduced von Neumann algebra of a semifinite von Neumann algebra by any nonzero projection is semifinite; thus $\mathcal{M}'$ is semifinite. Q.E.D.

Corollary 2.24. *Let $\{\mathcal{M},\mathfrak{H}\}$ be a von Neumann algebra. We then have the following equivalences:*

(i) *$\mathcal{M}$ is of type I* $\Leftrightarrow$ (i') *$\mathcal{M}'$ is of type I;*
(ii) *$\mathcal{M}$ is of type II* $\Leftrightarrow$ (ii') *$\mathcal{M}'$ is of type II;*
(iii) *$\mathcal{M}$ is of type III* $\Leftrightarrow$ (iii') *$\mathcal{M}'$ is of type III.*

PROOF. It is just the dual statement of Corollary 2.23 that $\mathcal{M}$ is of type III if and only if $\mathcal{M}'$ is also of type III. The equivalence (i) $\Leftrightarrow$ (i') was seen in Corollary 1.30. Therefore, the equivalence (ii) $\Leftrightarrow$ (ii') follows from Theorem 1.19. Q.E.D.

Corollary 2.25. *A von Neumann algebra $\mathcal{M}$ is semifinite if and only if $\mathcal{M}$ admits a faithful normal representation $\{\pi,\mathfrak{H}\}$ such that $\pi(\mathcal{M})'$ is finite.*

PROOF. Suppose that $\mathcal{M}$ is semifinite. Let τ be a faithful semifinite normal trace on $\mathcal{M}$. Let $\{\pi_\tau,L^2(\mathcal{M},\tau)\}$ be the representation of $\mathcal{M}$ considered in Theorem 2.22. Let e be a finite projection of $\mathcal{M}$ with $z(e) = 1$. Put $e' = J\pi_\tau(e)J$, i.e., $e' = \pi'_\tau(e)$, $\mathfrak{H} = e'L^2(\mathcal{M},\tau)$, and $\pi(x) = \pi_\tau(x)_{e'}$ for each $x \in \mathcal{M}$. It follows then that π is a faithful normal representation of $\mathcal{M}$ and

$$\pi(\mathcal{M})' = \pi_\tau(\mathcal{M})'_{e'} = \pi'_\tau(\mathcal{M})_{e'} = J\pi_\tau(\mathcal{M}_e)J$$

is finite. The other implication is a special case of Corollary 2.23. Q.E.D.

Proposition 2.26. *Let $\mathcal{M}_1$ and $\mathcal{M}_2$ be von Neumann algebras. The tensor product $\mathcal{M}_1 \overline{\otimes} \mathcal{M}_2$ is finite if and only if $\mathcal{M}_1$ and $\mathcal{M}_2$ are both finite.*

PROOF. Put $\mathcal{M} = \mathcal{M}_1 \overline{\otimes} \mathcal{M}_2$. By definition, a von Neumann subalgebra of a finite von Neumann algebra is finite. Hence if $\mathcal{M}$ is finite, then $\mathcal{M}_1$ and $\mathcal{M}_2$ are both finite.

If τ_1 and τ_2 are finite normal traces on $\mathcal{M}_1$ and $\mathcal{M}_2$, respectively, then the tensor product $\tau_1 \otimes \tau_2$ on $\mathcal{M}$ is clearly a finite normal trace with $s(\tau_1 \otimes \tau_2) = s(\tau_1) \otimes s(\tau_2)$ by Corollary IV.5.12. Therefore, $\mathcal{M}$ admits sufficiently many finite normal traces if $\mathcal{M}_1$ and $\mathcal{M}_2$ both do. Q.E.D.

Lemma 2.27. *Let $\mathcal{M}$ be a von Neumann algebra with a faithful semifinite normal trace τ. If $a \in \mathfrak{n}_\tau$, then the map: $x \in \mathcal{M} \mapsto ax^* \in \mathcal{M}$ is σ-strongly continuous on any bounded part of $\mathcal{M}$.*

PROOF. Let $\{x_i\}$ be a bounded net in $\mathcal{M}$ converging σ-strongly to zero. We then have

$$\lim_i \|ax_i^*\|_2 = \lim_i \|x_i a^*\|_2 = 0.$$

For any $y \in \mathcal{M}$, we have

$$\lim\|ax_i^* y\|_2 \le \|y\|_\infty \lim\|ax_i^*\|_2 = 0;$$

but $\mathfrak{n}_\tau = \mathcal{M} \cap L^2(\mathcal{M},\tau)$ is dense in $L^2(\mathcal{M},\tau)$, so that $\{\pi_\tau(ax_i^*)\}$ converges strongly to zero on $L^2(\mathcal{M},\tau)$ by its boundedness. Hence $\{ax_i^*\}$ converges σ-strongly to zero in $\mathcal{M}$.

Lemma 2.28. *If e is an infinite projection in a von Neumann algebra, then the involution: $x \in e\mathcal{M}e \mapsto x^* \in e\mathcal{M}e$ is σ-strongly discontinuous on the unit ball of $e\mathcal{M}e$.*

PROOF. By assumption, e majorizes a properly infinite projection e_0. By the proof of Proposition 1.36, there exists an infinite orthogonal family $\{e_n\}$ of equivalent projections with $e_0 = \sum_{n=1}^{\infty} e_n$. Hence $\mathscr{M}_{e_0}$ contains, by Proposition 1.22, a factor isomorphic to $\mathscr{L}(\mathfrak{K})$ where $\dim \mathfrak{K} = \infty$. As in Section II.2, the *-operation in $\mathscr{L}(\mathfrak{K})$ is not σ-strongly continuous on the unit ball; hence it is σ-strongly discontinuous on the unit ball of the factor contained in $\mathscr{M}_{e_0}$; therefore the involution in $\mathscr{M}_{e_0}$ is not σ-strongly continuous on the unit ball.
Q.E.D.

Lemma 2.29. *Let $\mathscr{M}$ be a von Neumann algebra and $\mathscr{N}$ a von Neumann subalgebra of $\mathscr{M}$. Suppose that there exists a family $\{\varepsilon_i\}_{i \in I}$ of normal projections of norm one of $\mathscr{M}$ onto $\mathscr{N}$ such that if $x \neq 0$, $x \in \mathscr{M}$, then $\varepsilon_i(x^*x) \neq 0$ for some $i \in I$. If $\mathscr{N}$ is of type III, then $\mathscr{M}$ is of type III.*

PROOF. Suppose $\mathscr{M}$ is not of type III. It follows that there exists a nonzero central projection $z \in \mathscr{M}$ such that $\mathscr{M}z$ is semifinite. Let τ be a faithful semifinite normal trace on $\mathscr{M}_z$. Let b be a nonzero element of $\mathscr{M}_z^+$ such that $\tau(b) < +\infty$. Put $a = b^{1/2} \in \mathfrak{n}_\tau$. By assumption, there exists ε_i, say ε, in the family of projection of norm one of $\mathscr{M}$ onto $\mathscr{N}$ such that $\varepsilon(b) \neq 0$. Choose a scalar $\lambda > 0$ and a nonzero projection $e \in \mathscr{N}$, a spectral projection of $\varepsilon(b)$, such that $e \leq \gamma\varepsilon(b)$. By Theorem III.3.4, we have for any $x \in e\mathscr{N}e$,

$$\begin{aligned} xx^* = xex^* \leq \lambda x\varepsilon(b)x^* &= \lambda\varepsilon(xbx^*) \\ &= \lambda\varepsilon(xa(xa)^*). \end{aligned}$$

If $\{x_i\}$ is a bounded net in $e\mathscr{N}e$ converging σ-strongly to zero, then $\{ax_i^*\}$ converges σ-strongly to zero by Lemma 2.26; hence $\varepsilon((x_ia)(x_ia)^*)$ converges σ-strongly to zero by the σ-strong continuity of ε; thus $\{x_ix_i^*\}$ converges σ-strongly to zero. Hence $\{x_i^*\}$ converges to zero σ-strongly. Therefore, the *-operation: $x \mapsto x^*$ is σ-strongly continuous in a bounded part of $e\mathscr{N}e$. By Lemma 2.27, e must be finite, which contradicts the pure infiniteness of $\mathscr{N}$.
Q.E.D.

Theorem 2.30. *Let $\mathscr{M}_1$ and $\mathscr{M}_2$ be von Neumann algebras and $\mathscr{M} = \mathscr{M}_1 \overline{\otimes} \mathscr{M}_2$. We then conclude the following:*

(i) *$\mathscr{M}$ is of type I if and only if $\mathscr{M}_1$ and $\mathscr{M}_2$ are both of type I;*
(ii) *$\mathscr{M}$ is of type II if and only if both $\mathscr{M}_1$ and $\mathscr{M}_2$ have no direct summand of type III, and either $\mathscr{M}_1$ or $\mathscr{M}_2$ is of type II;*
(iii) *$\mathscr{M}$ is of type III if and only if either $\mathscr{M}_1$ or $\mathscr{M}_2$ is of type III.*

PROOF. Suppose that $\mathscr{M}_1$ and $\mathscr{M}_2$ are both of type I. By Proposition 1.29, we may assume that $\mathscr{M}_1$ and $\mathscr{M}_2$ are acting on Hilbert spaces in such a way that $\mathscr{M}_1'$ and $\mathscr{M}_2'$ are both abelian. We then have that $\mathscr{M}' = \mathscr{M}_1' \overline{\otimes} \mathscr{M}_2'$ is abelian, so that $\mathscr{M}$ is of type I.

Suppose that $\mathscr{M}_1$ and $\mathscr{M}_2$ are both semifinite. By Corollary 2.25, we may assume that $\mathscr{M}_1$ and $\mathscr{M}_2$ act on Hilbert spaces in such a way that $\mathscr{M}_1'$ and

$\mathscr{M}'_2$ are finite. Proposition 2.26 then yields that $\mathscr{M}' = \mathscr{M}'_1 \overline{\otimes} \mathscr{M}'_2$ is finite. Therefore, $\mathscr{M}$ is semifinite by Corollary 2.23.

Identifying $\mathscr{M}_1$ with $\mathscr{M}_1 \otimes \mathbf{C}$ and $\mathscr{M}_2$ with $\mathbf{C} \otimes \mathscr{M}_2$, we regard $\mathscr{M}_1$ and $\mathscr{M}_2$ as von Neumann subalgebras of $\mathscr{M}$. For a normal state φ_2 of $\mathscr{M}_2$, the map $\varepsilon'_{\varphi_2}: \varphi_1 \in (\mathscr{M}_1)_* \mapsto \varphi_1 \otimes \varphi_2 \in \mathscr{M}_*$ is an isometry; hence there exists a map ε_{φ_2} of $\mathscr{M}$ onto $\mathscr{M}_1$ such that

$$\langle \varepsilon_{\varphi_2}(x), \varphi_1 \rangle = \langle x, \varphi_1 \otimes \varphi_2 \rangle, \qquad x \in \mathscr{M}, \qquad \varphi_1 \in (\mathscr{M}_1)_*.$$

Since $\varphi_2(1) = 1$, ε_{φ_2} is a projection of norm one. By construction, it is σ-weakly continuous. If $\varepsilon_{\varphi_2}(x) = 0$ for every normal state φ_2 of $\mathscr{M}_2$, then $\langle x, \varphi_1 \otimes \varphi_2 \rangle = 0$ for every $\varphi_1 \in (\mathscr{M}_1)_*$ and normal state φ_2 of $\mathscr{M}_2$. Since $(\mathscr{M}_2)_*$ is linearly spanned by normal states, we have $\langle x, \varphi_1 \otimes \varphi_2 \rangle = 0$ for every $\varphi_1 \in (\mathscr{M}_1)_*$, and $\varphi_2 \in (\mathscr{M}_2)_*$. But functionals $\varphi_1 \otimes \varphi_2$, $\varphi_1 \in (\mathscr{M}_1)_*$ and $\varphi_2 \in (\mathscr{M}_2)_*$ are total in $\mathscr{M}_*$ by (1) in Section IV.5, so that $x = 0$. Therefore $\{\varepsilon_{\varphi_2}\}$ is a family of sufficiently many normal projections of norm one from $\mathscr{M}$ onto $\mathscr{M}_1$. Thus Lemma 2.29 entails that if $\mathscr{M}_1$ is of type III, then $\mathscr{M}$ must be of type III.

What remains to be proven is now that if $\mathscr{M}$ is of type I, then $\mathscr{M}_1$ and $\mathscr{M}_2$ are both of type I. If either $\mathscr{M}_1$ or $\mathscr{M}_2$ have a direct summand of type III, then $\mathscr{M}$ must have a direct summand of type III by the above arguments. Thus we may assume that $\mathscr{M}_1$ and $\mathscr{M}_2$ are both semifinite. By Corollary 2.25, we may assume that $\mathscr{M}_1$ and $\mathscr{M}_2$ act on Hilbert spaces respectively with finite commutant. Hence $\mathscr{M}' = \mathscr{M}'_1 \overline{\otimes} \mathscr{M}'_2$ is of finite type I. Let T_1 and T_2 be the canonical center valued traces of $\mathscr{M}'_1$ and $\mathscr{M}'_2$, respectively. Since T_1 and T_2 are both completely positive, having abelian ranges, $T_1 \otimes T_2$ is extended by Proposition IV.5.13 to a σ-weakly continuous center valued map of $\mathscr{M}'_1 \overline{\otimes} \mathscr{M}'_2$, which is in turn the canonical center valued trace T of $\mathscr{M}'_1 \overline{\otimes} \mathscr{M}'_2$. Suppose that $\mathscr{M}'_1$ is of type II_1. For any integer $n > 0$, there exists, by Proposition 1.35, a projection $e'_n \in \mathscr{M}'_1$ such that $T_1(e'_n) = 2^{-n}$. Hence $T(e'_n \otimes 1) = 2^{-n}$. If e is an abelian projection in $\mathscr{M}'$, then we have $e \precsim e'_n \otimes 1$ by Lemma 1.25 since $z(e'_n \otimes 1) = 1$. Hence $T(e) \leq 2^{-n}$ for every integer $n > 0$. This means $T(e) = 0$, so that $e = 0$. Therefore, $\mathscr{M}'$ has no nonzero abelian projection, so that it cannot be of type I.

Decomposing $\mathscr{M}_1$ and $\mathscr{M}_2$ into direct sum according to Theorem 1.19, we complete the proof. Q.E.D.

Theorem 2.31. *Let $\mathscr{M}$ be a semifinite von Neumann algebra with a faithful semifinite normal trace τ. If τ' is a semifinite normal trace on $\mathscr{M}$, then*

(i) *$\tau + \tau'$ is a faithful semifinite normal trace on $\mathscr{M}$.*
(ii) *There exists a central element a with $0 \leq a \leq 1$ such that*

$$\tau(x) = (\tau + \tau')(ax),$$
$$\tau'(x) = (\tau + \tau')((1 - a)x), \qquad x \in \mathscr{M}_+,$$

and $\mathrm{s}(a) = 1$ *and* $\mathrm{s}(1 - a) = \mathrm{s}(\tau')$.

PROOF. Let $\tau_0 = \tau + \tau'$. Trivially, τ_0 is faithful and normal. Since $\mathfrak{m}_\tau \cdot \mathfrak{m}_{\tau'} \subset \mathfrak{m}_\tau \cap \mathfrak{m}_{\tau'} = \mathfrak{m}_{\tau_0}$, and $\mathfrak{m}_\tau \cdot \mathfrak{m}_{\tau'}$ is σ-weakly dense in $\mathcal{M}$, τ_0 is semifinite. On $\mathfrak{n}_{\tau_0} = \mathfrak{n}_\tau \cap \mathfrak{n}_{\tau'}$, the sesquilinear form $(\cdot|\cdot)_\tau$ is bounded by the inner product $(\cdot|\cdot)_{\tau_0}$. Hence there exists a positive operator A on $L^2(\mathcal{M},\tau_0)$ such that $0 \le A \le 1$ and

$$\tau(y^*x) = (Ax|y)_{\tau_0}.$$

For any $u \in \mathcal{M}$, we have, for any $x,y \in \mathfrak{n}_{\tau_0}$,

$$\begin{aligned}(\pi_{\tau_0}(u)Ax|y)_{\tau_0} &= (Ax|u^*y)_{\tau_0} = \tau(y^*ux)\\ &= (A\pi_{\tau_0}(u)x|y)_{\tau_0},\\ (\pi'_{\tau_0}(u)Ax|y) &= (Ax|yu^*)_{\tau_0} = \tau(uy^*x)\\ &= \tau(y^*xu) = (A\pi'_{\tau_0}(u)x|y)_{\tau_0}.\end{aligned}$$

Hence we have $A \in \pi_{\tau_0}(\mathcal{M})' \cap \pi'_{\tau_0}(\mathcal{M})' = \pi_{\tau_0}(\mathcal{Z})$, where $\mathcal{Z}$ means, of course, the center of $\mathcal{M}$, so that there exists a unique $a \in \mathcal{Z}$ with $A = \pi_{\tau_0}(a)$. This means that $\tau(y^*x) = \tau_0(ay^*x)$ for every $x,y \in \mathfrak{n}_{\tau_0}$. For any $x \in \mathcal{M}_+$, there exists an increasing net $\{x_i\}$ in $\mathfrak{m}_{\tau_0} \cap \mathcal{M}_+$ such that $x = \sup x_i$. We then have

$$\begin{aligned}\tau(x) &= \sup \tau(x_i) = \sup \tau_0(ax_i)\\ &= \tau_0(ax).\end{aligned}$$

Since $s(\tau) = 1$, we have $s(a) = 1$. The assertion for τ' is now automatic.

Q.E.D.

Corollary 2.32. *On a semifinite factor, any semifinite normal traces are proportional.*

We are now going to generalize the notion of a center valued trace from the finite case to the semifinite case. Let $\mathcal{M}$ be a von Neumann algebra with center $\mathcal{Z}$. Let Ω be the spectrum of $\mathcal{Z}$. As we have seen in Section III.1, and in this section again, there exist an open dense subset Γ of Ω and a Radon measure μ of Γ such that $\mathcal{Z} = L^\infty(\Gamma,\mu)$ and $\mathcal{Z}_* = L^1(\Gamma,\mu)$. Instead of $\int f(\gamma)\,d\mu(\gamma)$, we write $\mu(f)$ for short. We denote by $\hat{\mathcal{Z}}_+$ the space of all $[0,\infty]$-valued continuous functions on Ω. Then each positive element of $L^1(\Gamma,\mu)$ is regarded as an element of $\hat{\mathcal{Z}}_+$ which takes finite values on an open dense subset of Ω. Without difficulties, we can define addition $f+g$ of $f,g \in \hat{\mathcal{Z}}_+$ and the multiplication λf of $f \in \hat{\mathcal{Z}}_+$ and a scalar $\lambda \ge 0$ with the usual convention $0 \cdot (+\infty) = 0$. Recalling that every element of $\hat{\mathcal{Z}}_+$ can be given by the supremum of an increasing net in $\mathcal{Z}_+$, where, the supremum is taken in $\hat{\mathcal{Z}}_+$ instead of the pointwise supremum, we define the product fg of $f \in \mathcal{Z}_+$ and $g \in \hat{\mathcal{Z}}_+$ by $fg = \sup\{fg_i : g_i \in \mathcal{Z}_+, g_i \le g\}$. It follows easily that

$$\left.\begin{aligned}(f_1+f_2)g &= f_1g + f_2g, \qquad f_1,f_2 \in \mathcal{Z}_+, \qquad g \in \hat{\mathcal{Z}}_+,\\ (\sup f_i)g &= \sup f_ig, \qquad g \in \hat{\mathcal{Z}}_+\end{aligned}\right\}$$

for any bounded increasing net $\{f_i\}$ in $\mathscr{Z}_+$. Then the Radon–Nikodym theorem entails that any normal trace ν on $\mathscr{Z}$ (not necessarily semifinite) be given by

$$\nu(f) = \mu(fg), \qquad f \in \mathscr{Z}_+,$$

for some $g \in \hat{\mathscr{Z}}_+$.

We now assume that $\mathscr{M}$ is semifinite, and fix a faithful semifinite normal trace τ on $\mathscr{M}$. Each $x \in \mathscr{M}_+$ gives rise to a normal, not necessarily semifinite, trace ν_x on $\mathscr{Z}$ by $\nu_x(a) = \tau(ax)$, $a \in \mathscr{Z}_+$. Thus, there exists an element $T(x) \in \hat{\mathscr{Z}}_+$ such that

$$\tau(ax) = \mu(aT(x)), \qquad a \in \mathscr{Z}_+. \tag{10}$$

The map $T: x \in \mathscr{M}_+ \mapsto \hat{\mathscr{Z}}_+$ enjoys the following properties:

$$T(x + y) = T(x) + T(y), \qquad x, y \in \mathscr{M}_+; \tag{11}$$

$$T(ax) = aT(x), \qquad a \in \mathscr{Z}_+, \qquad x \in \mathscr{M}_+; \tag{12}$$

$$T(x^*x) = T(xx^*); \tag{13}$$

$$T(x^*x) = 0 \Rightarrow x = 0; \tag{14}$$

$$T(\sup x_i) = \sup T(x_i) \tag{15}$$

for any bounded increasing net $\{x_i\}$ in $\mathscr{M}_+$;

$$\tau(x) = \mu \circ T(x), \qquad x \in \mathscr{M}_+. \tag{16}$$

These properties are routinely derived from the defining equation (10), so we leave the proof to the reader.

Definition 2.33. A map T of $\mathscr{M}_+$ into $\hat{\mathscr{Z}}_+$ is called a *generalized* (or *extended*) *center valued trace* if T satisfies (11)–(13). It is said to be *normal* if it satisfies (15) in addition. If (14) is satisfied, then it is called *faithful*. The set $\{x \in \mathscr{M} : T(x^*x) \in \mathscr{Z}_+\}$ is an ideal. If this ideal is σ-weakly dense in $\mathscr{M}$, then T is said to be *semifinite*.

The extended center valued trace T just constructed above is faithful, semifinite, and normal.

Theorem 2.34. *For a von Neumann algebra $\mathscr{M}$ with center $\mathscr{Z}$, the following statements are equivalent*:

(i) *$\mathscr{M}$ is semifinite*;
(ii) *$\mathscr{M}$ admits a faithful, semifinite, normal extended center valued trace.*

If this is the case, then a faithful, semifinite, normal extended center valued trace is unique up to a multiplication by $\hat{\mathscr{Z}}_+$ in the sense that if T_1 and T_2 are *two such extended center valued traces then there exists* $c \in \hat{\mathscr{Z}}_+$ *such that*

$$T_1(x) = cT_2(x), \qquad x \in \mathscr{M}_+,$$
$$0 < c(\omega) < +\infty \quad \text{on an open dense set.}$$

PROOF. We have proven the implication (i) $\Rightarrow$ (ii).

(ii) $\Rightarrow$ (i): Let T be a faithful, semifinite, normal extended center valued trace of $\mathscr{M}$. For each $\varphi \in \mathscr{Z}_*^+$, let $\tau_\varphi = \varphi \circ T$. It follows that τ_φ is a semifinite normal trace on $\mathscr{M}_+$. If $\tau_\varphi(x) = 0$, $x \in \mathscr{M}_+$, for every $\varphi \in \mathscr{Z}_*^+$, then $T(x) = 0$; hence $x = 0$. Therefore, if z is a central projection of $\mathscr{M}$ such that $\mathscr{M}z$ is of type III, then $z = 0$ because $\tau_\varphi(z) = 0$ for every $\varphi \in \mathscr{Z}_*{}^+$. Thus $\mathscr{M}$ is semifinite.

Let T_1 and T_2 be two faithful, semifinite, normal extended center valued traces. Let μ be a fixed faithful, semifinite, normal trace on $\mathscr{Z}$. Put $\tau_1 = \mu \circ T_1$ and $\tau_2 = \mu \circ T_2$. If z is a central projection with $\mu(z) < +\infty$, then the maps: $x \in \mathscr{M}_+ \mapsto \tau_1(xz)$ and $x \in \mathscr{M}_+ \mapsto \tau_2(xz)$ are both semifinite normal traces. Hence τ_1 and τ_2 are both semifinite, normal traces. By assumption, τ_1 and τ_2 are both faithful. Put $\tau = \tau_1 + \tau_2$. By Theorem 2.31, there exists $a \in \mathscr{Z}$ such that $0 \leq a \leq 1$ $\tau_1(x) = \tau(ax)$ and $\tau_2(x) = \tau((1-a)x)$. By the faithfulness of τ_1 and τ_2, we have $s(a) = 1 = s(1-a)$, which means that $0 < a(\omega) < 1$ on an open dense subset of the spectrum Ω of Z. Put $c = a^{-1}(1-a) \in \hat{\mathscr{Z}}_+$. We then have $0 < c(\omega) < +\infty$ on an open dense set and $\mu \circ T_2(x) = \mu(T_1(x)c)$, $x \in \mathscr{M}_+$. Hence for every $a \in \mathscr{Z}_+$, we have

$$\begin{aligned}\mu(T_2(x)a) &= \mu(T_2(xa)) = \mu(T_1(xa)c)\\ &= \mu(T_1(x)ca).\end{aligned}$$

Since the functions: $\omega \in \Omega \mapsto T_1(x)(\omega)c(\omega)$ and $\omega \in \Omega \mapsto T_2(x)(\omega)$ are continuous on an open dense subset of Ω, we have

$$T_1(x)(\omega)c(\omega) = T_2(x)(\omega)$$

on the open dense subset of Ω, where this open dense subset may depend on $x \in \mathscr{M}_+$. Q.E.D.

Proposition 2.35. *Let* $\mathscr{M}$ *be a semifinite von Neumann with center* $\mathscr{Z}$. *Let* T *be a faithful semifinite normal extended center valued trace of* $\mathscr{M}$. *A projection* $e \in \mathscr{M}$ *is finite if and only if* $T(e)$ *takes finite values on an open dense subset of the spectrum* Ω *of* $\mathscr{Z}$.

PROOF. Suppose that $T(e)(\omega) < +\infty$ for every ω in an open dense subset G of Ω. If $e = u^*u$ and $uu^* = e_1 \leq e$, then $T(e) = T(e_1)$ and $T(e) = T(e_1) + T(e - e_1)$. Since $T(e)$ is finite valued on G, $T(e - e_1) = 0$ on G. By continuity, $T(e - e_1) = 0$; thus $e - e_1 = 0$. Hence e is finite.

Suppose that e is finite. Looking at $\mathscr{M}z$ with $z = z(e)$, we may assume that $z(e) = 1$. Then the center of $\mathscr{M}_e$ is identified with $\mathscr{Z}$ under the correspondence

$x \in \mathscr{Z} \leftrightarrow xe \in \mathscr{M}_e$. Put

$$T_e(x) = T(x)e, \qquad x \in \mathscr{M}_e.$$

It follows that T_e is a faithful, semifinite normal extended center valued trace of the finite von Neumann algebra $\mathscr{M}_e$. Let T_0 be the canonical center valued trace of $\mathscr{M}_e$. By Theorem 2.34, there exists $c \in \hat{\mathscr{Z}}_+$ such that $T_e(x)(\omega) = T_0(x)c(\omega)$ for every ω in an open dense subset of Ω. In particular $T_e(1)(\omega) = c(\omega)$. But the identity in $\mathscr{M}_e$ is the projection e in $\mathscr{M}$. Hence $T(e)(\omega) = c(\omega)$ is finite valued on an open dense subset. Q.E.D.

We close this section with the following noncommutative version of a conditional expectation:

Proposition 2.36. *Let $\mathscr{M}$ be a von Neumann algebra equipped with a faithful semifinite normal trace τ, and $\mathscr{N}$ be a von Neumann subalgebra of $\mathscr{M}$. If the restriction of τ onto $\mathscr{N}$ is semifinite again, then there exists a faithful normal projection E of norm one of $\mathscr{M}$ onto $\mathscr{N}$ such that $\tau = \tau \circ E$.*

PROOF. The imbedding of $\mathfrak{m}_\tau \cap \mathscr{N}$ into $\mathfrak{m}_\tau$ is an isometry with respect to the $\|\cdot\|_1$-norm. Hence $L^1(\mathscr{N},\tau)$ is isometrically imbedded into $L^1(\mathscr{M},\tau)$. Dualizing this imbedding, say E_*, we obtain a projection E of $\mathscr{M} = L^\infty(\mathscr{M},\tau)$ onto $\mathscr{N} = L^\infty(\mathscr{N},\tau)$ which is normal and of norm one by construction. For any $x \in \mathscr{M}$ and $y \in \mathfrak{m}_\tau \cap \mathscr{N}$, we have

$$\tau(E(x)y) = \tau(xE_*(y)) = \tau(xy).$$

Let $\{e_i\}$ be an increasing net in $\mathfrak{m}_\tau \cap \mathscr{N}_+$ with $\sup e_i = 1$. We then have for $x \in \mathscr{M}_+$,

$$\tau(x) = \sup_i \tau(xe_i) = \sup_i \tau(E(x)e_i)$$

$$= \tau \circ E(x). \qquad \text{Q.E.D.}$$

The above projection E is called the *conditional expectation* of $\mathscr{M}$ onto $\mathscr{N}$ with respect to τ.

EXERCISES

1. Show that a semifinite normal trace on a von Neumann algebra is expressed as a sum of normal positive linear functionals; hence it is lower semicontinuous on the positive cone with respect to the σ-weak topology. (*Hint*: The identity 1 is a sum of orthogonal projections in the definition ideal.)

2. Let $\mathscr{M}$ be a von Neumann algebra with a faithful semifinite normal trace τ.
   (a) Show that if e is a projection in $\mathscr{M}$ such that for every $\varepsilon > 0$ there exists a projection $p \in \mathscr{M}$ with $e \wedge p = 0$ and $\tau(1 - p) < \varepsilon$, then $e = 0$. (*Hint*: $e \precsim 1 - p$.)
   (b) Show that if e_1 and e_2 are projections in $\mathscr{M}$ such that for every $\varepsilon > 0$ there exists a projection $p \in \mathscr{M}$ with $e_1 \wedge p = e_2 \wedge p$ and $\tau(1 - p) < \varepsilon$, then $e_1 = e_2$. (*Hint*: Apply (a) to $e_1 - e_1 \wedge e_2$ and $e_2 - e_1 \wedge e_2$.)

3. Let $\{\mathscr{M},\mathfrak{H}\}$ be a von Neumann algebra with a faithful semifinite normal trace τ. Let S and T be closed operators affiliated with $\mathscr{M}$ and with domains $\mathscr{D}(S)$ and $\mathscr{D}(T)$. Show that $S = T$ if there exists a sequence of projections $\{e_n\}$ in $\mathscr{M}$ such that (i) $\tau(1 - e_n) \to 0$ as $n \to \infty$, (ii) $T^{-1}(p_n\mathfrak{H}) \cap p_n\mathfrak{H} \subset \mathscr{D}(S)$, (iii) $S^{-1}(p_n\mathfrak{H}) \cap p_n\mathfrak{H} \subset \mathscr{D}(T)$, (iv) $T\xi = S\xi$ for every $\xi \in (T^{-1}(p_n\mathfrak{H}) \cap (S^{-1}p_n\mathfrak{H})) \cap p_n\mathfrak{H}$. (*Hint*: Consider the 2×2-matrix algebra $M_2(\mathscr{M}) = \mathscr{M} \otimes M_2(\mathbf{C})$ over $\mathscr{M}$ acting on $\mathfrak{H} \oplus \mathfrak{H} = \mathfrak{H} \otimes \mathbf{C}^2$ and the trace $\tilde{\tau} = \tau \otimes \mathrm{Tr}$. Apply Exercise 2.2 to the projections $g(S)$ and $g(T)$ of $\mathfrak{H} \otimes \mathbf{C}^2$ to the graphs $\mathfrak{G}(S)$ and $\mathfrak{G}(T)$ of S and T, and the projections $\tilde{p}_n = p_n \otimes 1$.)

4. Let $\{\mathscr{M},\mathfrak{H}\}$ be a von Neumann algebra with a faithful semifinite normal trace τ. Let $\{e_n\}$ be a sequence of projections in $\mathscr{M}$ with $\lim_{n\to\infty} \tau(1 - e_n) = 0$ and $\mathscr{D}$ the subspace of $\mathfrak{H}$ spanned algebraically by $\bigcup_{n=1}^{\infty} e_n\mathfrak{H}$.

   (a) Show that if T is a closed operator affiliated with $\mathscr{M}$ such that the domain $\mathscr{D}(T)$ contains $\mathscr{D}$, then for every $\xi \in \mathscr{D}(T)$ there exists a sequence $\{\xi_n\}$ in $\mathscr{D}$ such that $\xi = \lim_{n\to\infty} \xi_n$ and $T\xi = \lim_{n\to\infty} T\xi_n$, i.e., T is the closure of $T|_{\mathscr{D}}$. (*Hint*: Consider the closure T_0 of $T|_{\mathscr{D}}$, i.e., the graph $\mathfrak{G}(T_0)$ of T_0 is the closure of the graph of $T|_{\mathscr{D}}$, and then apply Exercise 2.3 to T and T_0.)
   (b) Show that if T is a closed symmetric operator affiliated with $\mathscr{M}$ such that $\mathscr{D}(T) \supset \mathscr{D}$, then T is self-adjoint.
   (c) Show that a closed symmetric operator affiliated with a finite von Neumann algebra is self-adjoint.

5. Prove the equivalence of the following statements for a von Neumann algebra $\mathscr{M}$:

   (a) $\mathscr{M}$ is finite.
   (b) The *-operation in $\mathscr{M}$ is σ-strongly continuous.
   (c) The σ-strong topology on bounded parts of $\mathscr{M}$ agrees with the Arens–Mackey topology $\tau(\mathscr{M},\mathscr{M}_*)$.
   (d) For every relatively $\sigma(\mathscr{M}_*,\mathscr{M})$-compact subset K of $\mathscr{M}_*$, $|K| = \{|\varphi| : \varphi \in K\}$ is also relatively $\sigma(\mathscr{M}_*,\mathscr{M})$-compact. (*Hint*: Use Theorem III.5.4 for the implication (c) $\Rightarrow$ (d). If $|K|$ is not relatively $\sigma(\mathscr{M}_*,\mathscr{M})$-compact, then there exist a sequence $\{|\varphi_n|\}$ in $|K|$, a decreasing sequence $\{e_n\}$ of projections and $\varepsilon > 0$ such that (i) $e_n \to 0$ σ-strongly as $n \to \infty$, (ii) $|\varphi_n|(e_n) \geq \varepsilon$. Set $a_n = u_n e_n$, where $\varphi_n = u_n|\varphi_n|$ is the polar decomposition. Then $a_n \leq 1$ and $a_n \to 0$ σ-strongly. Hence $a_n \to 0$ in the $\tau(\mathscr{M},\mathscr{M}_*)$-topology, so that $\varphi^*(a_n) \to 0$ uniformly for $\varphi \in K$, since $K^* = \{\varphi^* : \varphi \in K\}$ is relatively $\sigma(\mathscr{M}_*,\mathscr{M})$-compact. But $\varphi_n^*(a_n) = |\varphi_n|(e_n) \geq \varepsilon$. For the implication (d) $\Rightarrow$ (a), use the fact that if $\mathscr{M}$ is not finite, then $\mathscr{M}$ contains σ-weakly closed *-subalgebra $\mathscr{B}$ isomorphic to a factor of type I_∞. If $\mathscr{B} = \mathscr{L}(\mathfrak{H})$ with $\dim \mathfrak{H} = \infty$, then $\varphi_n = \omega_{\xi_1,\xi_n} \to 0$ weakly for any normalized orthogonal system $\{\xi_n\}$ in $\mathfrak{H}$, while $|\varphi_n| = \omega_{\xi_n}$ does not converge to zero weakly.)

6. Given a von Neumann algebra $\mathscr{M}$, let $\mathscr{L}(\mathscr{M})$ (resp. $\mathscr{L}_*(\mathscr{M})$) denote the Banach space (algebra) of all bounded (resp. σ-weakly continuous) linear operators on $\mathscr{M}$. The weak* topology in $\mathscr{L}(\mathscr{M})$ means the $\sigma(\mathscr{L}(\mathscr{M}),\mathscr{M} \otimes_\gamma \mathscr{M}_*)$-topology; see Theorem IV.2.3. Show the equivalence of the following conditions for $\mathscr{M}$:

   (a) $\mathscr{M}$ is finite.
   (b) The group $\mathrm{Int}(\mathscr{M})$ of all inner automorphisms of $\mathscr{M}$ is relatively weakly* compact in $\mathscr{L}_*(\mathscr{M})$, i.e., the weak* closure of $\mathrm{Int}(\mathscr{M})$ in $\mathscr{L}(\mathscr{M})$ is contained in $\mathscr{L}_*(\mathscr{M})$. (*Hint*: Modify the proof of Theorem 2.4.)

7. Let $\mathcal{M}$ be a von Neumann algebra. For a pair h,k in $\mathcal{M}_+$, define $h \approx k$ by the existence of a family $\{x_i\}$ in $\mathcal{M}$ such that $h = \sum_{i \in I} x_i^* x_i$ and $k = \sum_{i \in I} x_i x_i^*$.
   (a) Show that the relation "$\approx$" is an equivalence relation in $\mathcal{M}_+$. (*Hint*: Use the asymmetric Riesz decomposition for transitivity.)
   (b) Show that if $h = \sum_{i \in I} h_i$, $k = \sum_{i \in I} k_i$ in $\mathcal{M}_+$ and $h_i \approx k_i$, $i \in I$, then $h \approx k$.
   (c) Show that if $h \approx k$ then $\lambda h \approx \lambda k$ for $\lambda > 0$.
   (d) Show that if $h \approx \sum_{i \in I} k_i$, then there exists a decomposition $h = \sum_{i \in I} h_i$ such that $h_i \approx k_i$, $i \in I$. (*Hint*: Use the asymmetric Riesz decomposition.)
   (e) Show that if $h \approx k$ and $a \in \mathcal{Z}_+$, the center of $\mathcal{M}$, then $ah \approx ak$.
   (f) Show that if $h\mathcal{M}k \neq \{0\}$, then there exists a pair $h_1,k_1 \in \mathcal{M}_+$ such that $0 \neq h_1 \approx k_1$, $h_1 \leq h$ and $k_1 \leq k$.
   (g) Show that for any pair h,k in $\mathcal{M}_+$ there exists a projection $z \in \mathcal{Z}$ such that $zh \approx k_1 \leq zk$ and $(1 - z)k \approx h_1 \leq (1 - z)h$.
   (h) Show that if $\mathcal{M}$ is finite, then $h \approx T(h)$, where T is the center valued trace in $\mathcal{M}$.
   (i) Show that if $\mathcal{M}$ is semifinite, then $h \approx k$ is equivalent to $T(h) = T(k)$ provided that either $T(h)$ or $T(k)$ is finite on an open dense set in the spectrum of $\mathcal{Z}$, where T is a faithful semifinite normal center valued trace.
   (j) Show that if e and f are projections in $\mathcal{M}$ and either e or f is finite, then $e \approx f$ implies $e \sim f$.
   (k) Show that if e and f are projections and either e or f is σ-finite, the $e \approx f$ implies $e \sim f$. (*Hint*: Decomposing $\mathcal{M}$ into a direct sum and applying (e), it suffices, by Proposition 1.39, to show that f is σ-finite if e is also. If $f = \sum_{j \in J} f_j$, then there exists a decomposition $e = \sum_{j \in J} h_j$ with $h_j \approx f_j$ by (d). If φ is a normal state with $s(\varphi) = e$, then $\varphi(e) = \sum_{j \in J} \varphi(h_j)$. Thus J must be countable. Hence f is σ-finite.)
   (l) We say that an $h \in \mathcal{M}_+$ is σ-finite if $h = \sum_{i \in I} h_i$ in $\mathcal{M}_+$ implies the countability of I, where one excludes, of course, the trivial element from $\{h_i\}_{i \in I}$. Show that if h is σ-finite and $h \approx k$, then k is also σ-finite.
   (m) Show that if $h = \sum_{i \in I} h_i \approx k = \sum_{j \in J} k_j$ and if each h_i and k_j are σ-finite, then $\aleph_0 \cdot \operatorname{card} I = \aleph_0 \cdot \operatorname{card} J$. (*Hint*: Use the fact that every $x \in \mathcal{M}$ has a decomposition $x = \sum_\alpha e_\alpha x f_\alpha$ such that $\{e_\alpha\}$ and $\{f_\alpha\}$ are both orthogonal families of σ-finite projections, so that the equivalence $h = \sum_\alpha x_\alpha^* x_\alpha$ and $k = \sum_\alpha x_\alpha x_\alpha^*$ is given by a family $\{x_\alpha\}$ such that $x_\alpha^* x_\alpha$ and $x_\alpha x_\alpha^*$ are both σ-finite.)
   (n) Show that if e and f are projections, then $e \approx f$ implies $e \sim f$.

8. Let $\{\mathcal{M},\mathfrak{H}\}$ be a von Neumann algebra.
   (a) Show that the existence of sufficiently many normal positive mappings $\mathcal{E}$ of $\mathcal{L}(\mathfrak{H})$ onto $\mathcal{M}$ such that $\mathcal{E}(axb) = a\mathcal{E}(x)b$ for every $a,b \in \mathcal{M}$ and $x \in \mathcal{L}(\mathfrak{H})$ is an algebraic invariant of $\mathcal{M}$. (*Hint*: Use Theorem IV.5.5.)
   (b) Show that the property in (a) characterizes atomic von Neumann algebras. (*Hint*: If $\mathcal{M}$ is atomic, then there exists a normal faithful representation such that $\pi(\mathcal{M})' = \mathcal{A}$ is abelian and atomic. Hence the unitary group G of $\mathcal{A}$ is σ-strongly compact. Thus $\mathcal{E}(x) = \int_G uxu^*\, du$, $x \in \mathcal{L}(\mathfrak{H}_\pi)$, gives a faithful normal projection of norm one from $\mathcal{L}(\mathfrak{H}_\pi)$ onto $\pi(\mathcal{M})$, where du is the normalized Haar measure of G. If $\mathcal{M}$ is not atomic, then there exists a nonzero projection $z \in \mathcal{M}$ such that $\mathcal{M}_z$ has no nontrivial minimal projection; so every pure state ω of $\mathcal{M}_z$ must be singular. Set $\varphi = \omega \circ \mathcal{E}$. Then $\varphi(a) = \omega(a\mathcal{E}(z))$ for every $a \in \mathcal{M}_z$, so that the restriction of φ to $\mathcal{M}_z$ is singular, which yields that φ itself is singular. Thus

$\varphi(e) = 0$ for every minimal projection e of $\mathscr{L}(z\mathfrak{H})$, so that $\omega(\mathscr{E}(e)) = 0$. Hence $\omega(\mathscr{E}(e)) = 0$ for every pure state ω on $\mathscr{M}_z$; thus $\mathscr{E}(e) = 0$ for every minimal projection $e \in \mathscr{L}(z\mathfrak{H})$. The normality of $\mathscr{E}$ entails $\mathscr{E} = 0$.)

9. Let $\mathscr{U}$ denote the unitary group of a factor $\mathscr{M}$ of type II_1. The first homotopy group $\pi_1(\mathscr{U})$ of $\mathscr{U}$ is isomorphic to the additive group $\mathbf{R}$, where one considers the uniform topology in $\mathscr{U}$ [50].

Notes

The material presented in Section 1 is today called the Murray–von Neumann dimension theory. When F. J. Murray and J. von Neumann laid the foundation for the theory, they first developed it for factors. It was this part of the theory which inspired I. Kaplansky [197], by developing a theory of AW^*-algebras and, later, Baer* rings [16], to abstract the algebraic content of dimension theory. It is also this part of the theory which distinguishes operator algebras sharply from other algebras. Despite its importance, it has remained unchanged ever since Murray–von Neumann's time.

The theory of trace has been one of the most attractive parts of the theory. After the original proof of Murray and von Neumann for the existence of a trace in a finite von Neumann algebra, there have been several attempts to simplify the proof [79], [179]. It has been called the additivity problem of a trace. It is not difficult to construct a center valued dimension function on the projection lattice in a semifinite von Neumann algebra. For example, if $\mathscr{M}$ is a semifinite properly infinite factor, then the equivalence classes of finite projections in $\mathscr{M}$ form a totally ordered complete additive semigroup isomorphic to a closed subsemigroup of $\mathbf{R}_+$, where the addition $[e] + [f]$ is taken as the class $[e + f]$ by choosing e and f orthogonal. From this one can construct a functional τ on $\mathscr{M}$ by means of the spectral decomposition theorem, which is linear on commuting operators. The problem is how to prove in a simple and natural manner the additivity of τ for a noncommuting pair. It is still an open question whether the dimension function D of an AW^*-factor extends to a trace. More generally, does linearity on the commutative subalgebras imply the linearity of a positive functional? The affirmative answer for a factor of type I was given by Gleason [136]. But the general case is still open. The proof for the existence of a trace presented here is due to Yeadon [414]. Theorem 2.18 is a highlight of the theory of traces. One can construct not only an L^1-space but also L^p-spaces, $1 \leq p \leq +\infty$, from a faithful semifinite normal trace and show, for instance, the duality between L^p-spaces. Traditionally, noncommutative integration meant this part of the theory of operator algebras, which was notably developed by J. Dixmier [85], H. Dye [99], R. Kunze [214], and I. Segal [326]. A beautiful and concise expository treatise was recently given by

E. Nelson [253]. Today, noncommutative integration means, however, more general theory of states, weights, and related topics on operator algebras, which will be handled thoroughly in the subsequent volume. Based on the theory of traces, the tensor product of semifinite von Neumann algebras was proved to be semifinite at the early stage by Y. Misonou [238]. The case involving algebras of type III was settled by S. Sakai [307]. Thus we have now the full result of Theorem 2.30.

3. Multiplicity of a von Neumann Algebra on a Hilbert Space

In this section, we study normal representations of a von Neumann algebra. Let $\mathcal{M}$ be a von Neumann algebra. We recall Theorem IV.5.5, or Corollary IV.5.6, which says that if $\{\pi_1,\mathfrak{H}_1\}$ and $\{\pi_2,\mathfrak{H}_2\}$ are normal representations of $\mathcal{M}$, then there exist a faithful normal representation $\{\pi,\mathfrak{H}\}$ and projections e_1 and e_2 in $\pi(\mathcal{M})'$ such that

$$\{\pi_1,\mathfrak{H}_1\} \cong \{\pi^{e_1},e_1\mathfrak{H}\},$$
$$\{\pi_2,\mathfrak{H}_2\} \cong \{\pi^{e_2},e_2\mathfrak{H}\};$$

and furthermore, $\pi_1 \cong \pi_2$ if and only if $e_1 \sim e_2$ in $\pi(\mathcal{M})'$. Hence Proposition 1.39 yields at once the following fact:

Proposition 3.1. *Let $\mathcal{M}$ be a von Neumann algebra. If two normal faithful representations $\{\pi_1,\mathfrak{H}_1\}$ and $\{\pi_2,\mathfrak{H}_2\}$ of $\mathcal{M}$ have σ-finite properly infinite commutants, then they are unitarily equivalent.*

Corollary 3.2. *If $\mathcal{M}$ is a von Neumann algebra of type III with σ-finite center $\mathcal{Z}$, then it has a unique, up to unitary equivalence, faithful normal representation with σ-finite commutant.*

Proof. By Corollary 2.24, $\pi(\mathcal{M})'$ is of type III for any normal representation π of $\mathcal{M}$, hence is properly infinite. Therefore, $\mathcal{M}$ admits at most only one faithful normal representation with σ-finite commutant by the previous proposition. By Propositions II.3.17 and II.3.19, the commutant of any cyclic representation is σ-finite. Let $\{\varphi_i\}_{i\in I}$ be a maximal family of normal states on $\mathcal{M}$ with centrally orthogonal support. By the σ-finiteness of $\mathcal{Z}$, I is countable. Let $\varphi = \sum_{i=1}^{\infty} (1/2^i)\varphi_i$. It follows that φ is a normal state and the central support of $s(\varphi)$ is 1, so that the cyclic representation π_φ of $\mathcal{M}$ induced by φ is faithful. Q.E.D.

This corollary shows that the representation theory of a von Neumann algebra of type III is essentially trivial. Accordingly, we restrict ourself to the semifinite case.

Definition 3.3. Given a von Neumann algebra $\{\mathscr{M},\mathfrak{H}\}$ with center $\mathscr{Z}$, a unitary involution of $\{\mathscr{M},\mathfrak{H}\}$ is a conjugate linear operator J of $\mathfrak{H}$ onto $\mathfrak{H}$ itself with the following properties:

$$(J\xi|J\eta) = (\eta|\xi), \qquad \xi,\eta \in \mathfrak{H};$$
$$J^2 = 1;$$
$$J\mathscr{M}J = \mathscr{M}', \text{ equivalently } J\mathscr{M}'J = \mathscr{M};$$
$$JaJ = a^*, \qquad a \in \mathscr{Z}.$$

The conjugate linear operator J that appeared in Theorem 2.22 is indeed a prototype of a unitary involution, and is called the *canonical unitary involution* of $L^2(\mathscr{M},\tau)$.

Lemma 3.4. *Let $\{\mathscr{M},\mathfrak{H}\}$ be a von Neumann algebra. If J is a unitary involution of $\{\mathscr{M},\mathfrak{H}\}$, then for any $\xi \in \mathfrak{H}$,*

$$J[\mathscr{M}'\xi] = [\mathscr{M}J\xi] \quad \text{and} \quad e_\xi \sim e_{J\xi}.$$

PROOF. We have $J\mathscr{M}'\xi = J\mathscr{M}'JJ\xi = \mathscr{M}J\xi$, so that $J[\mathscr{M}'\xi] = [\mathscr{M}J\xi]$. Since J commutes with central projections, we may assume, by the comparability theorem, Theorem 1.8, that $e_\xi \precsim e_{J\xi}$. By Theorem 1.10, we have $e'_\xi \precsim e'_{J\xi}$. Since J induces an antiisomorphism of $\mathscr{M}$ onto $\mathscr{M}'$, the relation $e_\xi \precsim e_{J\xi}$ implies $Je_\xi J \precsim Je_{J\xi}J$; hence by the first assertion we have $Je_\xi J = e'_{J\xi} \precsim Je_{J\xi}J = e'_\xi$. Therefore, we get, by Theorem 1.10 again, $e_{J\xi} \precsim e_\xi$, so that $e_\xi \sim e_{J\xi}$. Q.E.D.

Corollary 3.5. *Let $\{\mathscr{M},\mathfrak{H}\}$ be a semifinite von Neumann algebra with a unitary involution J. If normal extended center valued traces T of $\mathscr{M}$ and T' of $\mathscr{M}'$ are related by $T'(x) = T(JxJ)$, $x \in \mathscr{M}'_+$, then we have*

$$T(e_\xi) = T'(e'_\xi).$$

PROOF. By the above lemma, $e_\xi \sim e_{J\xi} = Je'_\xi J$, so that

$$T(e_\xi) = T(e_{J\xi}) = T(Je'_\xi J) = T'(e'_\xi). \qquad \text{Q.E.D.}$$

Proposition 3.6. *Let $\mathscr{M}$ be a semifinite von Neumann algebra equipped with a faithful semifinite normal trace τ. Let $\mathfrak{K}$ be a Hilbert space and τ be the trace on $\tilde{\mathscr{M}} = \mathscr{M} \mathbin{\overline{\otimes}} \mathscr{L}(\mathfrak{K})$ in the proof of Proposition 2.14. Then we have*

$$\{\pi_{\tilde{\tau}}(\tilde{\mathscr{M}}),L^2(\tilde{\mathscr{M}},\tilde{\tau})\} \cong \{\pi_\tau(\mathscr{M}) \mathbin{\overline{\otimes}} \mathscr{L}(\mathfrak{K}) \otimes \mathbf{C}, L^2(\mathscr{M},\tau) \otimes \mathfrak{K} \otimes \overline{\mathfrak{K}}\},$$

where $\overline{\mathfrak{K}}$ means the conjugate Hilbert space of $\mathfrak{K}$.

PROOF. Let $\{\xi_i\}_{i \in I}$ be a complete orthogonal basis of $\mathfrak{K}$. Based on the notations in Section II.1, $1 \otimes t_{\xi_i,\xi_j} = e_{i,j}$ form a system of matrix units in $\tilde{\mathscr{M}}$. If $x = a \otimes t_{\xi,\eta}$ with $a \in \mathscr{M}$ and $\xi,\eta \in \mathfrak{K}$, then its matrix representation $x = [x_{i,j}]$ is given by

$$x_{i,j} = (\xi|\xi_i)(\xi_j|\eta)a, \qquad i,j \in I.$$

If $x = a \otimes t_{\xi,\eta}$ and $y = b \otimes t_{\xi',\eta'}$ with $a,b \in \mathfrak{n}_\tau$, then we have

$$\begin{aligned}\tilde{\tau}(y^*x) &= \tilde{\tau}(b^*a \otimes t_{\eta',\xi'}t_{\xi,\eta}) \\ &= (\xi|\xi')\tilde{\tau}(b^*a \otimes t_{\eta',\eta}) \\ &= (\xi|\xi')\tau(b^*a)\sum_{i\in I}(\eta'|\xi_i)(\xi_i|\eta) \\ &= (\xi|\xi')(\eta'|\eta)\tau(b^*a) \\ &= (\xi \otimes \bar{\eta}|\xi' \otimes \bar{\eta}')\tau(b^*a),\end{aligned}$$

where $\bar{\eta}$ and $\bar{\eta}'$ denote the elements in $\bar{\mathfrak{K}}$ corresponding to η and η', respectively, in $\mathfrak{K}$ under the canonical correspondence. By linearity, we obtain an isometry U from the algebraic tensor product $\mathfrak{n}_\tau \otimes \mathscr{LF}(\mathfrak{K})$ onto the algebraic tensor product $\mathfrak{n}_\tau \otimes \mathfrak{K} \otimes \bar{\mathfrak{K}}$, where $\mathscr{LF}(\mathfrak{K})$ means the ideal of $\mathscr{L}(\mathfrak{K})$ consisting of all operators of finite rank and the isometry refers to the corresponding inner product. If $x = [x_{i,j}] \in \mathfrak{n}_\tau$, then $\sum_{i,j\in I}\tau(x_{i,j}^*x_{i,j}) < +\infty$ by definition, so that $\sum_{i,j} x_{i,j} \otimes t_{\xi_i,\xi_j}$ converges to x in the $\|\cdot\|_2$-norm, which means that $\mathfrak{n}_\tau \otimes \mathscr{LF}(\mathfrak{K})$ is dense in $L^2(\tilde{\mathscr{M}},\tilde{\tau})$. Trivially, $\mathfrak{n}_\tau \otimes \mathfrak{K} \otimes \bar{\mathfrak{K}}$ is dense in the Hilbert space tensor product $L^2(\mathscr{M},\tau) \otimes \mathfrak{K} \otimes \bar{\mathfrak{K}}$. Therefore, U is extended to an isometry of $L^2(\tilde{\mathscr{M}},\tilde{\tau})$ onto $L^2(\mathscr{M},\tau) \otimes \mathfrak{K} \otimes \bar{\mathfrak{K}}$. It is now routine to check that $U\pi_{\tilde{\tau}}(\tilde{\mathscr{M}})U^* = \pi_\tau(\mathscr{M}) \otimes \mathscr{L}(\mathfrak{K}) \otimes \mathbf{C}$, by looking at the generators $\pi_{\tilde{\tau}}(x \otimes 1)$ and $\pi_{\tilde{\tau}}(1 \otimes t_{\xi,\eta})$. Q.E.D.

Proposition 3.7. *Let $\{\mathscr{M},\mathfrak{H}\}$ be a semifinite von Neumann algebra. Then there exist a semifinite von Neumann algebra $\mathscr{N}$ and projections $e,f \in \mathscr{N}$ such that $z(e) = z(f) = 1$ and*

$$\{\mathscr{M},\mathfrak{H}\} \cong \{\pi_\tau(\mathscr{N}_e)_{\pi_\tau(f)}, eL^2(\mathscr{N},\tau)f\},$$

where τ is a faithful, semifinite normal trace on $\mathscr{N}$.

Proof. Let τ_0 be a faithful semifinite normal trace on $\mathscr{M}$. By Theorem IV.5.5, there exists a Hilbert space $\mathfrak{K}$ and a projection $f' \in \pi_{\tau_0}(\mathscr{M})' \bar{\otimes} \mathscr{L}(\bar{\mathfrak{K}})$ such that

$$\{\mathscr{M},\mathfrak{H}\} \cong \{(\pi_{\tau_0}(\mathscr{M}) \otimes \mathbf{C})_{f'}, f'(L^2(\mathscr{M},\tau_0) \otimes \bar{\mathfrak{K}})\}.$$

Let $\mathscr{N} = \mathscr{M} \bar{\otimes} \mathscr{L}(\mathfrak{K})$ and τ be the trace on $\mathscr{N}$ constructed from τ_0 by Proposition 2.14. By the previous proposition, we have

$$\{\pi_\tau(\mathscr{N}), L^2(\mathscr{N},\tau)\} \cong \{\pi_{\tau_0}(\mathscr{M}) \bar{\otimes} \mathscr{L}(\mathfrak{K}) \otimes \mathbf{C}, L^2(\mathscr{M},\tau_0) \otimes \mathfrak{K} \otimes \bar{\mathfrak{K}}\}.$$

Let e_0 be a minimal projection in $\mathscr{L}(\mathfrak{K})$, i.e., a projection of rank one, and put $e = 1 \otimes e_0 \in \mathscr{N}$. We then have

$$\{\pi_{\tau_0}(\mathscr{M}) \otimes \mathbf{C}, L^2(\mathscr{M},\tau) \otimes \bar{\mathfrak{K}}\} = \{\pi_\tau(\mathscr{N}_e), eL^2(\mathscr{N},\tau)\}.$$

Furthermore, there exists a projection $f \in \mathscr{N}$ such that $\pi_\tau'(f)\pi_\tau(e) = f'$ under the identification of $L^2(\mathscr{M},\tau) \otimes \mathfrak{K}$ and $eL^2(\mathscr{N},\tau) = \pi_\tau(e)L^2(\mathscr{N},\tau)$. Thus, we conclude finally that

$$\{\mathscr{M},\mathfrak{H}\} \cong \{\pi_\tau(\mathscr{N}_e)_{\pi_\tau(f)}, eL^2(\mathscr{N},\tau)f\}.$$

Q.E.D.

Theorem 3.8. *Let $\{\mathscr{M},\mathfrak{H}\}$ be a semifinite von Neumann algebra with center $\mathscr{Z}$. If T (resp. T') is a faithful semifinite normal extended center valued trace of $\mathscr{M}$ (resp. $\mathscr{M}'$), then there exists uniquely an element $c \in \hat{\mathscr{Z}}_+$ such that*

$$T(e_\xi) = cT'(e'_\xi), \qquad \xi \in \mathfrak{H},$$
$$0 < c(\omega) < \infty$$

except on a rare subset of the spectrum Ω of $\mathscr{Z}$.

Proof. By Proposition 3.7, there exist a semifinite von Neumann algebra $\{\mathscr{N},\mathfrak{K}\}$ with a unitary involution J and projections $e \in \mathscr{N}$ and $f \in \mathscr{N}'$ with $z(e) = z(f) = 1$ such that $\{\mathscr{M},\mathfrak{H}\} \cong \{\mathscr{N}_{ef}, ef\mathfrak{K}\}$. By Corollary 3.5, $\mathscr{N}$ and $\mathscr{N}'$ admit faithful semifinite normal center valued traces T_1 and T'_1 respectively, such that $T_1(p_\xi) = T'_1(p'_\xi)$, where p_ξ and p'_ξ mean the projection of $\mathfrak{K}$ onto $[\mathscr{N}'\xi]$ and $[\mathscr{N}\xi]$, respectively. By a spatial isomorphism, we identify $\mathscr{M}$ with $\mathscr{N}_{ef}$. Since $z(f) = 1$, $\mathscr{N}_{ef}$ is isomorphic to $\mathscr{N}_e$ under the map $\pi: x \in \mathscr{N}_e \mapsto x_f \in \mathscr{N}_{ef}$. Therefore, we can define a faithful semifinite normal extended center valued trace $\tilde{T}$ of $\mathscr{N}_{ef}$ $(=\mathscr{M})$ by $\tilde{T}(x) = T_1 \circ \pi^{-1}(x)$, $x \in (\mathscr{N}_{ef})_+$, where $\mathscr{N}_e$ is identified with $e\mathscr{N}e \subset \mathscr{N}$. Since $(\mathscr{N}_{ef})' = (\mathscr{N}'_f)_e$, we can define a faithful semifinite normal extended center valued trace $\tilde{T}'$ on $(\mathscr{N}_{ef})'$ $(=\mathscr{M}')$ by the same way. For any $\xi \in ef\mathfrak{K} = \mathfrak{H}$, we have

$$e_\xi\mathfrak{H} = [\mathscr{M}'\xi] = [\mathscr{N}'_{ef}\xi] = f[\mathscr{N}'\xi],$$
$$e'_\xi\mathfrak{H} = [\mathscr{M}\xi] = [\mathscr{N}_{ef}\xi] = e[\mathscr{N}\xi],$$

so that $e_\xi = \pi(p_\xi)$ and $e'_\xi = \pi'(p'_\xi)$, where π' means the isomorphism of $\mathscr{N}'_f$ to $\mathscr{N}'_{ef}$ given by $\pi'(y) = ye$, $y \in \mathscr{N}'_f$. Therefore, we have

$$\tilde{T}(e_\xi) = \tilde{T}'(e'_\xi), \qquad \xi \in \mathfrak{H}.$$

By Theorem 2.34, bearing on the uniqueness of a center valued trace, there exist $d \in \hat{\mathscr{Z}}_+$ and $d' \in \hat{\mathscr{Z}}_+$ such that $\tilde{T}(x) = T(x)d$, $x \in \mathscr{M}_+$ and $\tilde{T}'(x) = T'(x)d'$, $x \in \mathscr{M}'_+$, $0 < d(\omega) < +\infty$ except on a rare subset of Ω. Putting $c = d^{-1}d'$, we get $T(e_\xi) = cT'(e'_\xi)$, $\xi \in \mathfrak{H}$.

We shall show the uniqueness of c. Suppose that there exists another $c_1 \in \hat{\mathscr{Z}}_+$ such that $T(e_\xi) = c_1T'(e'_\xi)$. If $c_1 \neq c$, there exists a nonzero projection $z \in \mathscr{Z}$ such that $c_1z > cz$ (or $c_1z < cz$) and $c_1z \in \mathscr{Z}$. By the semifiniteness of T, there exists $\xi \in \mathfrak{H}$ such that $e_\xi \leq z$ and $T(e_\xi) \in \mathscr{Z}$. We then have

$$T(e_\xi)(\omega) = c(\omega)T'(e_\xi)(\omega) < c_1(\omega)T'(e_\xi)(\omega) = T(e_\xi)(\omega)$$

for every ω with $z(\omega) = 1$, if $T(e_\xi)(\omega) \neq 0$, so that $T(e_\xi) \neq T(e_\xi)$, which is a contradiction. Q.E.D.

Definition 3.9. Let $\{\mathscr{M},\mathfrak{H}\}$ be a *finite* von Neumann algebra with *finite* commutant $\mathscr{M}'$. If T and T' are both the canonical center valued trace of $\mathscr{M}$ and $\mathscr{M}'$, respectively, then the element c of $\hat{\mathscr{Z}}_+$ in the above theorem is called the *coupling function* between $\mathscr{M}$ and $\mathscr{M}'$, and denoted by $c(\mathscr{M},\mathfrak{H})$. If $\mathscr{M}$ is a factor, then $c(\mathscr{M},\mathfrak{H})$ is a positive number, which is called the *coupling constant.*

Proposition 3.10. *If $\{\mathcal{M},\mathfrak{H}\}$ is a finite von Neumann algebra with finite commutant $\mathcal{M}'$, then we have*

(i) $c(\mathcal{M}',\mathfrak{H}) = c(\mathcal{M},\mathfrak{H})^{-1}$,
(ii) $c(\mathcal{M}_{e'},e'\mathfrak{H}) = c(\mathcal{M},\mathfrak{H})T_{\mathcal{M}'}(e')$ *for any projection* $e' \in \mathcal{M}'$ *with* $z(e') = 1$, *where* $T_{\mathcal{M}'}$ *is the canonical center valued trace of* $\mathcal{M}'$.

PROOF. Assertion (i) follows from Theorem 3.8. Let π be the induction: $x \in \mathcal{M} \mapsto x_{e'} \in \mathcal{M}_{e'}$. It follows from $z(e') = 1$ that π is an isomorphism. For each $\xi \in e'\mathfrak{H}$, let f_ξ and f'_ξ denote the projections of $e'\mathfrak{H}$ onto $[\mathcal{M}'_{e'}\xi]$ and $[\mathcal{M}_{e'}\xi]$, respectively. We then have

$$f_\xi\mathfrak{H} = [\mathcal{M}'_e\xi] = e'[\mathcal{M}'\xi],$$
$$f'_\xi\mathfrak{H} = [\mathcal{M}_{e'}\xi] = [\mathcal{M}\xi],$$

so that $f_\xi = \pi(e_\xi)$ and $f'_\xi = e'_\xi$. Identifying $\mathcal{M}'_{e'}$ and $e'\mathcal{M}'e'$, we apply $T_{\mathcal{M}'}$ to $\mathcal{M}'_{e'}$. We then have, recalling $\pi \circ T_{\mathcal{M}} = T_{\mathcal{M}_{e'}} \circ \pi$,

$$\begin{aligned} T_{\mathcal{M}_{e'}}(f_\xi) &= \pi(T_{\mathcal{M}}(e_\xi)) \\ &= \pi(cT_{\mathcal{M}'}(f'_\xi)) = \pi(c)\pi(T_{\mathcal{M}'}(f'_\xi)), \end{aligned}$$

where $c = c(\mathcal{M},\mathfrak{H})$. But $T_{\mathcal{M}'}|_{\mathcal{M}'_{e'}}$ is not the canonical center valued trace while it is a center valued trace of $\mathcal{M}'_{e'}$. Hence there exists an element $d \in \hat{\mathscr{Z}}_+$ such that $\pi(T_{\mathcal{M}'}(x)) = dT_{\mathcal{M}'_{e'}}(x)$, $x \in \mathcal{M}_{e'}$. Putting $x = 1_{e'\mathfrak{H}}$, we get $d = \pi(T_{\mathcal{M}'}(e'))$. Thus we get

$$T_{\mathcal{M}_{e'}}(f_\xi) = \pi(cT_{\mathcal{M}'}(e'))T_{\mathcal{M}'_{e'}}(f'_\xi).$$

However, we identify the center $\mathscr{Z}$ of $\mathcal{M}$ and that of $\mathcal{M}_{e'}$ under the isomorphism π. Thus we conclude assertion (ii). Q.E.D.

Theorem 3.11. *Let $\{\mathcal{M}_1,\mathfrak{H}_1\}$ and $\{\mathcal{M}_2,\mathfrak{H}_2\}$ be von Neumann algebras with finite commutants. An isomorphism π of $\mathcal{M}_1$ onto $\mathcal{M}_2$ is spatial if and only if $\pi(c(\mathcal{M}_1,\mathfrak{H}_1)) = c(\mathcal{M}_2,\mathfrak{H}_2)$, where we consider π as an isomorphism of the extended positive cone $\hat{\mathscr{Z}}_{1,+}$ onto $\hat{\mathscr{Z}}_{2,+}$.*

PROOF. The "only if" part is trivial. Suppose that $\pi(c(\mathcal{M}_1,\mathfrak{H}_1)) = c(\mathcal{M}_2,\mathfrak{H}_2)$. There exist a von Neumann algebra $\{\mathcal{M},\mathfrak{H}\}$ and projections $e'_1,e'_2 \in \mathcal{M}'$ such that $z(e'_1) = z(e'_2) = 1$, $\{\mathcal{M}_1,\mathfrak{H}_1\} \cong \{\mathcal{M}_{e'_1},e'_1\mathfrak{H}\}$ and $\{\mathcal{M}_2,\mathfrak{H}_2\} \cong \{\mathcal{M}_{e'_2},e'_2\mathfrak{H}\}$ and $\pi(x_{e'_1}) = x_{e'_2}$ under the identification of $\mathcal{M}_1$ with $\mathcal{M}_{e'_1}$ and $\mathcal{M}_2$ with $\mathcal{M}_{e'_2}$. By assumption, e'_1 and e'_2 are both finite. Replacing $\mathcal{M}$ by $\mathcal{M}_{(e'_1 \vee e'_2)}$, we may assume, by Theorem 1.37, that $\mathcal{M}'$ is also finite. Denoting by T' the canonical center valued trace $T_{\mathcal{M}'}$ of $\mathcal{M}'$, we have, by Proposition 3.10,

$$c(\mathcal{M}_1,\mathfrak{H}_1) = c(\mathcal{M},\mathfrak{H})T'(e'_1),$$
$$c(\mathcal{M}_2,\mathfrak{H}_2) = c(\mathcal{M},\mathfrak{H})T'(e'_2).$$

Therefore, $T'(e'_1) = T'(e'_2)$, so that $e'_1 \sim e'_2$ in $\mathcal{M}'$. Hence π is spatial. Q.E.D.

Proposition 3.12. *If $\{\mathcal{M},\mathfrak{H}\}$ is a σ-finite von Neumann algebra, then there exists a central projection $z \in \mathcal{M}$ such that $\mathcal{M}_z$ (resp. $\mathcal{M}_{(1-z)}$) admits a cyclic (resp. separating) vector.*

Proof. Let $\{\xi_i\}_{i\in I}$ be a maximal family of nonzero vectors in $\mathfrak{H}$ such that $\{e_{\xi_i}\}_{i\in I}$ and $\{e'_{\xi_i}\}_{i\in I}$ are both orthogonal families of cyclic projections. Put

$$e = \sum_{i\in I} e_{\xi_i}, \qquad e' = \sum_{i\in I} e'_{\xi_i}, \qquad f = 1 - e, \qquad f' = 1 - e'.$$

If $ff' \neq 0$, then $f\mathfrak{H} \cap f'\mathfrak{H}$ contains a nonzero vector η and $[\mathcal{M}'\eta]$ (resp. $[\mathcal{M}\eta]$) is orthogonal to $[\mathcal{M}'\xi_i]$ (resp. $[\mathcal{M}\xi_i]$). But this is impossible by the maximality of $\{\xi_i\}$. Hence $ff' = 0$. Thus $z(f)f' = 0$ and $z(f)z(f') = 0$. This means that $z(f) \leq 1 - f' = e'$, and $1 - z(f) \leq e$. By the σ-finiteness, I is countable, so that we assume $\sum_{i\in I} \|\xi_i\|^2 < +\infty$ by replacing each ξ_i by some scalar multiple of ξ_i. Put $\xi = \sum_{i\in I} \xi_i \in \mathfrak{H}$. Then we have $\xi_i = e_{\xi_i}\xi = e'_{\xi_i}\xi$; hence $[\mathcal{M}\xi]$ and $[\mathcal{M}'\xi]$ both contain each ξ_i, so that $e_\xi \geq \sum_{i\in I} e_{\xi_i} = e$ and $e'_\xi \geq \sum_{i\in I} e'_{\xi_i} = e'$. But $\xi = e\xi = e'\xi$; hence $e_\xi \leq e$ and $e'_\xi \leq e$. Therefore, we get $e = e_\xi$ and $e' = e'_\xi$. We then have for $z = z(f)$,

$$z = ze' = ze'_\xi = e'_{z\xi}.$$

Hence $z\xi$ is cyclic for $\mathcal{M}_z$. Next, we have

$$1 - z = (1 - z)e = (1 - z)e_\xi = e_{(1-z)\xi}.$$

Thus $(1 - z)\xi$ is cyclic for $\mathcal{M}'_{(1-z)}$, so that it is separating for $\mathcal{M}_{(1-z)}$. Q.E.D.

Proposition 3.13. *Let $\{\mathcal{M},\mathfrak{H}\}$ be a σ-finite, finite von Neumann algebra with finite commutant. There exists a cyclic (resp. separating) vector if and only if $c(\mathcal{M},\mathfrak{H}) \leq 1$ (resp. $c(\mathcal{M},\mathfrak{H}) \geq 1$).*

Proof. If there exists a cyclic vector ξ, then $e'_\xi = 1$ and $e_\xi \leq 1$, so that

$$1 \geq T_{\mathcal{M}}(e_\xi) = T_{\mathcal{M}'}(e'_\xi)c(\mathcal{M},\mathfrak{H}) = c(\mathcal{M},\mathfrak{H}).$$

Conversely, suppose $c(\mathcal{M},\mathfrak{H}) \leq 1$. By the previous proposition and by decomposing $\mathcal{M}$ into a direct sum, we may assume that $\mathcal{M}$ admits a separating vector ξ. Then ξ is cyclic for $\mathcal{M}'$, so that $c(\mathcal{M}',\mathfrak{H}) \leq 1$; hence $1 \geq c(\mathcal{M},\mathfrak{H}) = c(\mathcal{M}',\mathfrak{H})^{-1} \geq 1$. Therefore, $c(\mathcal{M},\mathfrak{H}) = 1$. Therefore, we have $T_{\mathcal{M}}(e_\xi) = T_{\mathcal{M}'}(e'_\xi) = 1$, which means that $e'_\xi = 1$. Hence ξ is cyclic for $\mathcal{M}$. Q.E.D.

Proposition 3.14. *Let $\{\mathcal{M},\mathfrak{H}\}$ be a σ-finite von Neumann algebra. If $\mathcal{M}$ admits a unitary involution J, then there exists a cyclic separating vector in $\mathfrak{H}$.*

Proof. By Proposition 3.12, there exist vectors ξ_1 and ξ_2 in $\mathfrak{H}$ such that $z_1 = e_{\xi_1}$ and $z_2 = e'_{\xi_2}$ are central, and $z_1 + z_2 = 1$. Let $\eta_1 = J\xi_1$ and $\eta_2 = J\xi_2$. We have then, by Lemma 3.4,

$$e'_{\eta_1} = Je_{\xi_1}J = z_1 \quad \text{and} \quad e_{\eta_2} = Je'_{\xi_2}J = z_2.$$

Therefore, the von Neumann algebras $\mathscr{M}_{z_1}$ (resp. $\mathscr{M}_{z_2}$) has a cyclic vector and a separating vector. By Corollary 1.14, $\mathscr{M}_{z_1}$ (resp. $\mathscr{M}_{z_2}$) has a cyclic and separating vector. Hence there exist $\zeta_1 \in z_1\mathfrak{H}$ and $\zeta_2 \in z_2\mathfrak{H}$ such that $[\mathscr{M}_{z_1}\zeta_1] = [\mathscr{M}'_{z_1}\zeta_1] = z_1\mathfrak{H}$ and $[\mathscr{M}_{z_2}\zeta_2] = [\mathscr{M}'_{z_2}\zeta_2] = z_2\mathfrak{H}$. Let $\zeta = \zeta_1 + \zeta_2$. Then we have $[\mathscr{M}\zeta] = [\mathscr{M}'\zeta] = \mathfrak{H}$. Q.E.D.

Theorem 3.15. *If $\{\mathscr{M},\mathfrak{H}\}$ is a von Neumann algebra with a unitary involution J, then every $\varphi \in \mathscr{M}_*$ is of the form $\omega_{\xi,\eta}$. If $\varphi \in \mathscr{M}_*$ is positive, then $\varphi = \omega_\xi$ for some $\xi \in \mathfrak{H}$.*

PROOF. Let $\varphi = u|\varphi|$ be the polar decomposition. If $|\varphi| = \omega_\xi$, then $\varphi = u\omega_\xi = \omega_{u\xi,\xi}$. Hence we must prove $\varphi = \omega_\xi$ for $\varphi \geq 0$. Put $e = s(\varphi)$ and $e' = JeJ$. We then have $Jee'J = JeJJe'J = e'e = ee'$. Thus ee' and J commute, so that $J|_{\mathfrak{K}}$ is a conjugate linear unitary element in $\mathfrak{K} = ee'\mathfrak{H}$. We further have

$$\begin{aligned} Je'e\mathscr{M}ee'J &= Je'JJeJJ\mathscr{M}JJeJJe'J \\ &= ee'\mathscr{M}'e'e, \end{aligned}$$

so that $J\mathscr{M}_{ee'}J = \mathscr{M}'_{ee'}$. Clearly, J commutes with central projections of $\mathscr{M}_{ee'}$. Thus J gives rise to a unitary involution of $\{\mathscr{M}_{ee'},\mathfrak{K}\}$. Since e is σ-finite, $\mathscr{M}_{ee'}$ is σ-finite too. Hence it follows from Proposition 3.14 and Corollary 1.12, that every element in $(\mathscr{M}_{ee'})_*^+$ is of the form ω_ξ for some $\xi \in \mathfrak{K}$. Since $\mathscr{M}_e \cong \mathscr{M}_{ee'}$, we have $\varphi(x) = (x\xi|\xi)$ for every $x \in \mathscr{M}_e$. For every $x \in \mathscr{M}$, we get $exe \in \mathscr{M}_e$ and $\varphi(x) = \varphi(exe) = (exe\xi|\xi) = (x\xi|\xi)$. Thus $\varphi = \omega_\xi$ on $\mathscr{M}$. Q.E.D.

Theorem 3.16. *If $\{\mathscr{M}_1,\mathfrak{H}_1\}$ and $\{\mathscr{M}_2,\mathfrak{H}_2\}$ are von Neumann algebras with unitary involutions J_1 and J_2, respectively, then any isomorphism π of $\mathscr{M}_1$ onto $\mathscr{M}_2$ is spatial.*

PROOF. If the algebras in question are σ-finite, then the assertion follows immediately from Corollary 1.13 and the previous proposition.

We claim that every nonzero central projection $z \in \mathscr{M}_1$ majorizes a nonzero central projection $z_1 \in \mathscr{M}_1$ such that the restriction of π to $\mathscr{M}_{1,z_1}$ is spatial. If this is the case, then Zorn's lemma implies that there exists an orthogonal family $\{z_i : i \in I\}$ of central projections in $\mathscr{M}_1$ such that $\sum z_i = 1$ and the restriction of π to each $\mathscr{M}_{1,z_i}$ is spatial. It follows then that π is spatial, being a direct sum of spatial isomorphisms. So we shall prove the above claim. Considering $\mathscr{M}_{1,z}$, we may assume $z = 1$, and have to prove that there exists a nonzero central projection z such that $\pi|_{\mathscr{M}_{1,z}}$ is spatial. The first remark in this proof shows that any nonzero σ-finite central projection satisfies our requirement. Therefore, we have to handle the case where any nonzero central projection is not σ-finite. As usual, we take a von Neumann algebra $\{\mathscr{M},\mathfrak{H}\}$ and projections $e',e' \in \mathscr{M}'$ such that $\{\mathscr{M}_1,\mathfrak{H}_1\}$ and $\{\mathscr{M}_2,\mathfrak{H}_2\}$ are identified with $\{\mathscr{M}_{e'_1},e'_1\mathfrak{H}\}$ and $\{\mathscr{M}_{e'_2},e'_2\mathfrak{H}\}$, respectively, $z(e'_1) = z(e'_2) = 1$, and $\pi(x_{e'_1}) = x_{e'_2}$, $x \in \mathscr{M}$. By the comparability theorem, Theorem 1.8, we may assume that $e'_1 \gtrsim e'_2$. Hence, replacing e'_2 by an equivalent projection, we may assume that $e_2' \leq e_1'$. Thus, considering $\mathscr{M}_{e_1'}$, we have

$$\{\mathscr{M}_1,\mathfrak{H}_1\} = \{\mathscr{M},\mathfrak{H}\},$$
$$\{\mathscr{M}_1,\mathfrak{H}_1\} = \{\mathscr{M}_{e'},e'_1\mathfrak{K}\},$$
$$\pi(x) = x_{e'}, \qquad x \in \mathscr{M}$$

for some projection $e' \in \mathscr{M}'$ with $z(e') = 1$. The absence of a σ-finite central projection in $\mathscr{M}$ means, by Corollary 2.9, that $\mathscr{M}$ is properly infinite, so that $\mathscr{M}'$ is properly infinite also because it is antiisomorphic to $\mathscr{M}$. Let ξ be a nonzero vector in $e'\mathfrak{H}$. We then apply Proposition 1.34 twice to e'_ξ, once in $\mathscr{M}'_{e'}$ and then in $\mathscr{M}'$, and then obtain a family $\{f'_i : i \in I\}$ of orthogonal equivalent projections in $\mathscr{M}'$, a nonzero central projection $z \in \mathscr{M}'$, and a subset J of I such that

$$f'_i \sim e'_\xi z, \qquad \sum_{i \in J} f'_i = e'z, \qquad \sum_{i \in I} f'_i = z.$$

Since $\mathscr{M}'_{e'}$ has no σ-finite central projection, J is uncountable. If $\operatorname{card} J = \operatorname{card} I$, then $e'z \sim z$; so we are through. Thus, we have only to prove that $\operatorname{card} I \leq \operatorname{card} J$. Since $\mathscr{M}'_z$ and $\mathscr{M}'_{e'z}$ are antiisomorphic to $\mathscr{M}_z$ and $\mathscr{M}_{e'z}$, respectively, by the unitary involutions, $\mathscr{M}'_z$ and $\mathscr{M}'_{e'z}$ are isomorphic. Hence our assertion, $\operatorname{card} I \leq \operatorname{card} J$, follows from the next lemma. Q.E.D.

Lemma 3.17. *Let $\mathscr{M}$ be a von Neumann algebra. If $\{e_j : j \in J\}$ is an infinite orthogonal family of σ-finite projections in $\mathscr{M}$ with $\sum_{j \in J} e_j = 1$, then for any family $\{f_i : i \in I\}$ of orthogonal projections in $\mathscr{M}$ we have* $\operatorname{card} I \leq \operatorname{card} J$.

Proof. For each $j \in J$, let φ_j be a normal states of $\mathscr{M}$ with $s(\varphi_j) = e_j$. For each $j \in J$, put $I_j = \{i \in I : \varphi_j(f_i) \neq 0\}$. It follows that I_j, $j \in J$, is countable. Since $\sum_{j \in J} e_j = 1$, $f_i e_j \neq 0$ for some $j \in J$, $\varphi_j(f_i) \neq 0$. Hence $I = \bigcup_{j \in J} I_j$; thus

$$\operatorname{card} I \leq \aleph_0 \cdot \operatorname{card} J = \operatorname{card} J. \qquad \text{Q.E.D.}$$

Definition 3.18. A positive linear functional φ on a C^*-algebra A is said to be *central* or *tracial* if $\varphi(x^*x) = \varphi(xx^*)$ for every $x \in A$.

Proposition 3.19. *If φ is a tracial positive linear functional on a C^*-algebra A, then ξ_φ gives rise to a faithful finite normal trace on $\pi_\varphi(A)''$, where $\{\pi_\varphi,\mathfrak{H}_\varphi,\xi_\varphi\}$ is the cyclic representation of A induced by φ.*

Proof. Let $\mathscr{M} = \pi_\varphi(A)''$ and $\tilde{\varphi}(x) = (x\xi_\varphi|\xi_\varphi)$, $x \in \mathscr{M}$. It follows that $\tilde{\varphi}$ is a finite normal trace on $\mathscr{M}$. By the tracial property of φ, the conjugate linear map: $\pi_\varphi(x)\xi_\varphi \mapsto \pi_\varphi(x^*)\xi_\varphi$, $x \in A$, is well defined and extended to a conjugate linear unitary operator J on $\mathfrak{H}_\varphi$. Trivially, we have $J\pi_\varphi(A)\xi_\varphi = \pi_\varphi(A)\xi_\varphi$. For any $a,b,x \in A$, we have

$$\begin{aligned} J\pi_\varphi(a)J\pi_\varphi(b)\pi_\varphi(x)\xi_\varphi &= J\pi_\varphi(a)J\pi_\varphi(bx)\xi_\varphi \\ &= J\pi_\varphi(a)\pi_\varphi(x^*b^*)\xi_\varphi = \pi_\varphi(bx^*a)\xi_\varphi \\ &= \pi_\varphi(b)J\pi_\varphi(a)J\pi_\varphi(x)\xi_\varphi. \end{aligned}$$

Thus, $J\pi_\varphi(a)J$ and $\pi_\varphi(b)$ commute, so that $J\pi_\varphi(A)J \subset \mathscr{M}'$. Therefore, we get

$$\mathscr{M}'\xi_\varphi \supset J\pi_\varphi(A)J\xi_\varphi = J\pi_\varphi(A)\xi_\varphi = \pi_\varphi(A)\xi_\varphi,$$

so that ξ_φ is cyclic for $\mathscr{M}'$ or, equivalently, separating for $\mathscr{M}$. Thus, the trace $\tilde{\varphi}$ on $\mathscr{M}$ is faithful. Q.E.D.

Exercises

1. Let $\{\mathscr{M},\mathfrak{H}\}$ (resp. $\{\mathscr{N},\mathfrak{K}\}$) be a factor of type II_∞ with finite commutant.
   (a) Show that $\mathscr{M}$ (resp. $\mathscr{N}$) admits a cyclic vector ξ_0 (resp. η_0).
   (b) Show that an ismorphism π of $\mathscr{M}$ onto $\mathscr{N}$ is spatial if and only if $\tau_{\mathscr{M}}(e_{\xi_0}) = \tau_{\mathscr{N}}(e_{\eta_0})$, where $\tau_{\mathscr{M}}$ and $\tau_{\mathscr{N}}$ are respectively faithful semifinite normal traces on $\mathscr{M}$ and $\mathscr{N}$ such that $\tau_{\mathscr{M}} = \tau_{\mathscr{N}} \circ \pi$.

2. Let $\mathscr{M}$ be a von Neumann algebra with σ-finite center $\mathscr{Z}$.
   (a) Show that $\mathscr{M}$ admits a "largest" σ-finite projection e in the sense that if f is a σ-finite projection of $\mathscr{M}$, then $f \precsim e$.
   (b) Show that if $\mathscr{M}$ acts on a Hilbert space $\mathfrak{H}$, then $\mathscr{M}$ admits a "largest" cyclic projection e_ξ in the sense that for any $\eta \in \mathfrak{H}$, $e_\eta \precsim e_\xi$.

3. Let $\{e_i\}$ be a net of projections in a von Neumann algebra $\mathscr{M}$ converging σ-strongly to e. Let f be a projection of $\mathscr{M}$ such that $e_i \precsim f$ for every i.
   (a) Show that if f is finite, then $e \precsim f$. (*Hint*: The central support $z(f)$, say z, of f is a semifinite projection and $z \succsim e_i$ for every i, so that $e \leq z$. For any semifinite normal trace τ on $\mathscr{M}$, $\tau(e_i) \leq \tau(f)$, so that $\tau(e) \leq \liminf \tau(e_i) \leq \tau(f)$; thus $e \precsim f$.)
   (b) Show that if $\{e_i\}$ is a sequence, then $e \precsim f$. (*Hint*: By (a), one may assume that f is properly infinite. Then f is a sum of orthogonal sequence $\{f_i\}$ of projections such that $f_i \sim f$.)
   (c) Let $\mathfrak{H}$ be a nonseparable Hilbert space. Let $\{e_i\}$ be the net of all finite dimensional projections ordered by the natural ordering. If f is a projection in $\mathfrak{H}$ with separable infinite dimensional range, then $e_i \precsim f$ for every i, but $\lim e_i = 1 \not\precsim f$ [187].

4. Let $\{\mathscr{M},\mathfrak{H}\}$ be a von Neumann algebra. Show that if $\{e_{\xi_n}\}$ is a sequence of cyclic projections of $\mathscr{M}$ converging σ-strongly to a projection $e \in \mathscr{M}$, then e is also cyclic [187]. (*Hint*: Let $p = \bigvee_{n=1}^{\infty} e_{\xi_n}$. Then $e \leq p$ and p is σ-finite. Considering $\{\mathscr{M}_p, p\mathfrak{H}\}$, one may assume that $\mathscr{M}$ is σ-finite. By Exercise 3.2, $\mathscr{M}$ admits a "largest" cyclic projection f. Show that $e \precsim f$ by Exercise 3.3.)

4. Ergodic Type Theorem for von Neumann Algebras*

In this section, we shall show first that in a von Neumann algebra $\mathscr{M}$ with center $\mathscr{Z}$, every norm closed unitarily invariant convex set K has nonempty intersection with $\mathscr{Z}$, and then apply this to uniformly closed ideals of $\mathscr{M}$.

Let us fix a von Neumann algebra $\mathcal{M}$ with center $\mathcal{Z}$. For each $h \in \mathcal{M}_h$ and a projection $e \in \mathcal{M}$ commuting with h, we introduce the following notations:

$$\left.\begin{aligned} \mu_e(h) &= \sup\{\lambda \in \mathrm{Sp}_{\mathcal{M}_e}(h_e)\}; \\ \nu_e(h) &= \inf\{\lambda \in \mathrm{Sp}_{\mathcal{M}_e}(h_e)\}; \\ \omega_e(h) &= \mu_e(h) - \nu_e(h) = \text{diameter of } \mathrm{Sp}_{\mathcal{M}_e}(h_e); \\ \lambda_e(h) &= \tfrac{1}{2}\{\mu_e(h) + \nu_e(h)\}. \end{aligned}\right\} \tag{1}$$

If $e = 1$, then we write $\mu(h)$, $\nu(h)$, $\omega(h)$, and $\lambda(h)$. If $\mathcal{F}$ is a family of nonzero projections commuting with h, then we write

$$\omega_{\mathcal{F}}(h) = \sup\{\omega_e(h) : e \in \mathcal{F}\}. \tag{2}$$

Lemma 4.1. *For any $h \in \mathcal{M}_h$, there exists a central projection $z \in \mathcal{M}$ and a $u \in \mathcal{U}_{\mathcal{M}}$ such that*

$$\left.\begin{aligned} \omega_z(\tfrac{1}{2}(h + uhu^*)) &\le \tfrac{3}{4}\omega(h), \\ \omega_{(1-z)}(\tfrac{1}{2}(h + uhu^*)) &\le \tfrac{3}{4}\omega(h). \end{aligned}\right\} \tag{3}$$

Proof. Let $h = \int_{\nu(h)}^{\mu(h)} \lambda \, de(\lambda)$ be the spectral decomposition of h. Put

$$e = \int_{\nu(h)}^{\lambda(h)} de(\lambda).$$

We then have

$$\mu_e(h) \le \lambda(h) \quad \text{and} \quad \nu_{(1-e)}(h) \ge \lambda(h).$$

Let z be a projection in $\mathcal{Z}$ such that $ez \precsim (1-e)z$ and $e(1-z) \succsim (1-e)(1-z)$. Let v and w be partial isometries in $\mathcal{M}$ such that

$$v^*v = ez, \qquad vv^* = f \le (1-e)z,$$
$$w^*w = (1-e)(1-z), \qquad ww^* = g \le e(1-z).$$

Define an element u of $\mathcal{M}$ by

$$u = v + v^* + w + w^* + (1-e)z - f + e(1-z) - g.$$

It follows then that u is a self-adjoint unitary element and

$$\begin{aligned} hz &\ge \nu(h)ez + \lambda(h)(1-e)z \\ &= \nu(h)ez + \lambda(h)f + \lambda(h)((1-e)z - f); \end{aligned}$$

hence

$$(uhu)z \ge \nu(h)f + \lambda(h)ez + \lambda(h)((1-e)z - f).$$

Therefore, we get

$$\begin{aligned} \tfrac{1}{2}(h + uhu)z &\ge \tfrac{1}{2}(\nu(h) + \lambda(h))(ez + f) + \lambda(h)((1-e)z - f) \\ &\ge \tfrac{1}{2}(\nu(h) + \lambda(h))(z + f + (1-e)z - f) \\ &= \tfrac{1}{2}(\nu(h) + \lambda(h))z = \{\mu(h) - \tfrac{3}{4}\omega(h)\}z. \end{aligned}$$

Since it is clear that $\frac{1}{2}(h + uhu)z \le \mu(h)z$, we obtain the first inequality in (3). The second inequality in (3) follows similarly. Q.E.D.

Lemma 4.2. *If $\mathscr{F}$ is a finite family of orthogonal central projections with sum* 1, *then there exist a finite family $\mathscr{F}'$ of orthogonal central projections with sum* 1 *and a $u \in \mathscr{U}_{\mathscr{M}}$ such that*

$$\omega_{\mathscr{F}'}(\tfrac{1}{2}(h + uhu^*)) \leq \tfrac{3}{4}\omega_{\mathscr{F}}(h).$$

PROOF. Considering each direct summand $\mathscr{M}_z$, $z \in \mathscr{F}$, we can conclude the assertion from the previous lemma. Q.E.D.

For each $a \in \mathscr{M}$, let $K(a)$ denote the convex hull of uau^*, $u \in \mathscr{U}_{\mathscr{M}}$, and $\bar{K}(a)$ (resp. $\tilde{K}(a)$) denote the uniform (resp. σ-weak) closure of $\bar{K}(a)$. Of course, $\bar{K}(a) \subset \tilde{K}(a)$ and $\tilde{K}(a)$ is σ-weakly compact. Let $\mathscr{K}$ be the space of all positive functions on $\mathscr{U}_{\mathscr{M}}$ such that $f(u) = 0$ except for finitely many $u \in \mathscr{U}_{\mathscr{M}}$ and $\sum_{u \in \mathscr{U}_{\mathscr{M}}} f(u) = 1$. It follows then that $\mathscr{K}$ is closed under the convolution product on the discrete group $\mathscr{U}_{\mathscr{M}}$. For each $f \in \mathscr{K}$ and $a \in \mathscr{M}$, let

$$T_f(a) = \sum_{u \in \mathscr{U}_{\mathscr{M}}} f(u)uau^*.$$

We then have $K(a) = \{T_f(a) : f \in \mathscr{K}\}$. It is clear that $T_{f*g}(a) = T_f(T_g(a))$ for $f,g \in \mathscr{K}$, where

$$(f * g)(v) = \sum_{u \in \mathscr{U}_{\mathscr{M}}} f(v)g(v^*u).$$

Lemma 4.3. *For any $h \in \mathscr{M}_{\mathrm{h}}$ and $\varepsilon > 0$, there exist an $f \in \mathscr{K}$ and a $k \in \mathscr{Z}_{\mathrm{h}}$ such that*

$$\|T_f(h) - k\| < \varepsilon.$$

PROOF. For any integer $p > 0$, there exist a family $\{z_1, \ldots, z_n\} = \mathscr{F}$ of orthogonal central projections with $\sum_{i=1}^{n} z_i = 1$ and an $f \in \mathscr{K}$ such that

$$\omega_{\mathscr{F}}(T_f(h)) \leq (\tfrac{3}{4})^p \omega(h),$$

where the existence of $\{z_1, \ldots, z_n\}$ and f follow from the inductive use of Lemma 4.2. Hence we have, putting $\lambda_i = \lambda_{z_i}(T_f(h))$,

$$\left\| T_f(h) - \sum_{i=1}^{n} \lambda_i z_i \right\| \leq (\tfrac{3}{4})^p \omega(h).$$

Choosing p large enough, we complete the proof. Q.E.D.

For each $a \in \mathscr{M}$, let

$$d(a,\mathscr{Z}) = \inf\{\|a - x\| : x \in \mathscr{Z}\}. \tag{4}$$

Obviously, $d(a,\mathscr{Z})$ is nothing but the norm of the image $a + \mathscr{Z}$ of a in the quotient Banach space $\mathscr{M}/\mathscr{Z}$. Hence we have

$$d(a + b,\mathscr{Z}) \leq d(a,\mathscr{Z}) + d(b,\mathscr{Z}). \tag{5}$$

For every $f \in \mathscr{K}$, T_f leaves $\mathscr{Z}$ elementwise fixed, so that

$$d(T_f(a),\mathscr{Z}) \leq d(a,\mathscr{Z}), \qquad a \in \mathscr{M}, \qquad f \in \mathscr{K}. \tag{6}$$

Lemma 4.4. *For any* $a_1, \dots, a_n \in \mathscr{M}$ *and* $\varepsilon > 0$, *there exists an* $f \in \mathscr{K}$ *such that*

$$d(T_f(a_i), \mathscr{Z}) \leq \varepsilon, \qquad i = 1, \dots, n.$$

PROOF. Considering the real part and the imaginary part of each a_i, we may assume, by (5), that each a_i is self-adjoint. We construct $f_1, \dots, f_n \in \mathscr{K}$ by induction so that

$$d(T_{f_i}(a_k), \mathscr{Z}) < \varepsilon, \qquad k = 1, \dots, i.$$

By Lemma 4.3, there exists an $f_1 \in \mathscr{K}$ such that

$$d(T_{f_1}(a_1), \mathscr{Z}) < \varepsilon.$$

Suppose $f_1, \dots, f_{i-1}$ are constructed. We apply Lemma 4.3 to $T_{f_{i-1}}(a_i)$ to find a $g \in \mathscr{K}$ such that

$$d(T_g(T_{f_{i-1}}(a_i)), \mathscr{Z}) < \varepsilon.$$

Put $f_i = g * f_{i-1}$. By (6), f_i fulfills the requirement. By induction, we obtain $\{f_1, \dots, f_n\}$. Let $f = f_n$. We then complete the proof. Q.E.D.

Lemma 4.5. *For any* $a_1, \dots, a_n \in \mathscr{M}$, *there exist* $b_1, \dots, b_n \in \mathscr{Z}$ *and a sequence* $\{f_j\}$ *in* $\mathscr{K}$ *such that*

$$\lim_{j \to \infty} \|T_{f_j}(a_i) - b_i\| = 0, \qquad i = 1, 2, \dots, n.$$

PROOF. By Lemma 4.4, there exist $b_{1,1}, \dots, b_{n,1} \in \mathscr{Z}$ and $g_1 \in \mathscr{K}$ such that

$$\|T_{g_1}(a_i) - b_{i,1}\| < 2^{-1}.$$

For a positive integer k, suppose that we found $b_{i,j} \in \mathscr{Z}$, $j = 1, 2, \dots, k$, and $g_1, \dots, g_k \in \mathscr{K}$ such that

$$\|T_{g_k} \cdots T_{g_1}(a_i) - b_{i,k}\| < 2^{-k}. \tag{7}$$

Let $f_k = g_k * \cdots * g_1$. Applying Lemma 4.4 to $\{T_{f_k}(a_i)\}$, we find $g_{k+1} \in \mathscr{K}$ and $b_{i,k+1} \in \mathscr{Z}$, $i = 1, 2, \dots, n$, such that

$$\|T_{g_{k+1}} \cdot T_{f_k}(a_i) - b_{i,k+1}\| < 2^{-(k+1)}.$$

By induction, we find $\{g_k\}$, hence $\{f_k\}$, and $\{b_{i,k}\} \subset \mathscr{Z}$ such that (7) holds for $k = 1, 2, \dots$. We then have

$$\begin{aligned}
\|T_{f_{k+1}}(a_i) - T_{f_k}(a_i)\| &\leq \|T_{g_{k+1}} \cdot T_{f_k}(a_i)) - T_{f_k}(a_i)\| \\
&\leq \|T_{g_{k+1}}(T_{f_k}(a_i)) - b_{i,k} + b_{i,k} - T_{f_k}(a_i)\| \\
&\leq 2\|T_{f_k}(a_i) - b_{i,k}\| < 2^{-k+1},
\end{aligned}$$

so that $\{T_{f_k}(a_i) : k + 1, 2, \dots\}$ is a Cauchy sequence for each $i = 1, 2, \dots, n$, and

$$b_i = \lim_{k \to \infty} T_{f_k}(a_i) = \lim_{k \to \infty} b_{i,k} \in \mathscr{Z}. \qquad \text{Q.E.D.}$$

Theorem 4.6. *For every $a \in \mathcal{M}$, $\bar{K}(a) \cap \mathcal{Z}$ is nonempty. Furthermore, we have the following:*

(i) *If $\mathcal{M}$ is finite, then $\tilde{K}(a) \cap \mathcal{Z}$ is reduced to the singleton $\{T_{\mathcal{M}}(a)\}$.*
(ii) *If $\bar{K}(a) \cap \mathcal{Z}$ is a singleton, then $\mathcal{M}$ is finite.*
(iii) *If $\mathcal{M}$ is σ-finite and of type III, then $\bar{K}(a) \cap \mathcal{Z} \neq \{0\}$ for every nonzero $a \in \mathcal{M}$.*

PROOF. Lemma 4.5 implies immediately that $\bar{K}(a) \cap \mathcal{Z} \neq \varnothing$. If $\mathcal{M}$ is finite, then $T_{\mathcal{M}}(uxu^*) = T_{\mathcal{M}}(a)$ for every $u \in \mathcal{U}_{\mathcal{M}}$, so that $T_{\mathcal{M}}(K(a)) = \{T_{\mathcal{M}}(a)\}$. Since $T_{\mathcal{M}}$ is normal, we have $T_{\mathcal{M}}(\tilde{K}(a)) = \{T_{\mathcal{M}}(a)\}$. But $T_{\mathcal{M}}$ leaves $\tilde{K}(a) \cap \mathcal{Z}$ pointwise invariant. Therefore, $\tilde{K}(a) \cap \mathcal{Z} = \{T_{\mathcal{M}}(a)\}$.

Suppose that $\bar{K}(a) \cap \mathcal{Z}$ is reduced to a singleton for every $a \in \mathcal{M}$. Let $\{T(a)\}$ denote the one-point set $\bar{K}(a) \cap \mathcal{Z}$. We get a map: $a \in \mathcal{M} \mapsto T(a) \in \mathcal{Z}$. This map has the properties

$$\left.\begin{aligned} T(x^*x) &\geq 0, && x \in \mathcal{M}, \\ T(a) &= a, && a \in \mathcal{Z}, \\ T(uxu^*) &= T(x), && x \in \mathcal{M}, \quad u \in \mathcal{U}_{\mathcal{M}}, \\ T(\lambda x) &= \lambda T(x), && x \in \mathcal{M}, \quad \lambda \in \mathbf{C}. \end{aligned}\right\}$$

Let x and y be elements of $\mathcal{M}$. There exists a sequence $\{f_n\}$ in $\mathcal{K}$ such that

$$\lim_{n\to\infty} \|T_{f_n}(x) - T(x)\| = 0, \qquad \lim_{n\to\infty} \|T_{f_n}(y) - T(y)\| = 0;$$

hence we have

$$\lim_{n\to\infty} \|T_{f_n}(x+y) - T(x) - T(y)\| \leq \lim_{n\to\infty} \|T_n(x) - T(x)\| + \lim_{n\to\infty} \|T_n(y) - T(y)\| = 0.$$

Thus, we have

$$T(x + y) = T(x) + T(y).$$

Since the unitary invariance of T yields $T(x^*x) = T(xx^*)$, $x \in \mathcal{M}$, T is a center valued trace of $\mathcal{M}$. Since T leaves $\mathcal{Z}$ elementwise fixed, Proposition 2.5 yields that T is normal and faithful. Thus T is the canonical center valued trace of $\mathcal{M}$, and $\mathcal{M}$ is finite.

Suppose that $\mathcal{M}$ is σ-finite and of type III. Let a be a fixed nonzero element of $\mathcal{M}$. We shall show that $\bar{K}(a) \cap \mathcal{Z} \neq \{0\}$. To this end, we may assume that a is self-adjoint. We recall first that any projections $e, f \in \mathcal{M}$ are equivalent if and only if $z(e) = z(f)$. We may also assume $\|a\| = 1$. Replacing a by $-a$ if necessary, we can find a nonzero spectral projection e of a and an integer $n > 0$ such that $ae \geq (1/n)e$ and $a \geq (1/n)e - (1 - e)$. Considering $\mathcal{M}_{z(e)}$, we assume $z(e) = 1$. If e majorizes any nonzero central projection z, then $a \geq (1/n)e - (1 - e) \geq (1/n)z - (1 - z)$, so that every element in $\bar{K}(a)$ majorizes $(1/n)z - (1 - z)$; hence $\bar{K}(a) \cap \mathcal{Z} \neq \{0\}$. Thus, we assume that e does not majorize any non-zero central projection, that is, $z(1 - e) = 1 = z(e)$. Hence $e \sim (1 - e) \sim 1$. Let $\{e_1, e_2, \ldots, e_{n+1}\}$ be orthogonal equivalent projections such that $e = \sum_{i=1}^{n+1} e_i$. Put $e_0 = 1 - e$ and let v_i, $0 \le i \le n$, be a partial isometry of $\mathscr{M}$ such that $v_i^* v_i = e_i$, $v_i v_i^* = e_{i+1}$, and let $v_{n+1} = v_0^* v_1^* \cdots v_n^*$. Put $u = v_0 + v_1 + \cdots + v_{n+1}$. It follows that u is a unitary element such that $ue_iu^* = e_{i+1}$, $0 \le i \le n$, and $ue_{n+1}u^* = e_0$. Put

$$b = \frac{1}{n+2} \sum_{i=0}^{n+1} u^i a u^{-i}.$$

Observing that $\sum_{i=0}^{n+1} u^i e_j u^{-i} = 1$, $j = 0,1,\ldots,n+1$, we have

$$(n+2)b \ge \sum_{i=0}^{n+1} u^i \left(\frac{1}{n}(e_1 + e_2 + \cdots + e_n) - e_0 \right) u^{-i}$$

$$= \frac{n+1}{n} - 1 = \frac{1}{n},$$

so that $b \ge 1/[n(n+2)]$. Therefore, every $c \in \bar{K}(b)$ majorizes $1/[n(n+2)]$; hence $\bar{K}(b) \cap \mathscr{Z} \ne \{0\}$. But b is in $\bar{K}(a)$; hence $\bar{K}(b) \subset \bar{K}(a)$; so $\bar{K}(a) \cap \mathscr{Z} \ne \{0\}$. Q.E.D.

Corollary 4.7. *If $\mathfrak{m}$ is a closed, in norm, ideal of a finite von Neumann algebra $\mathscr{M}$ with center $\mathscr{Z}$, then*

$$\mathfrak{m} \cap \mathscr{Z} = T_{\mathscr{M}}(\mathfrak{m}).$$

In particular, a finite factor is simple.

PROOF. If $a \in \mathfrak{m}$, then $\bar{K}(a) \subset \mathfrak{m}$, so that $T_{\mathscr{M}}(a) \in \mathfrak{m} \cap \mathscr{Z}$. Hence $T_{\mathscr{M}}(\mathfrak{m}) \subset \mathfrak{m} \cap \mathscr{Z} \subset T_{\mathscr{M}}(\mathfrak{m})$.

Let $\mathscr{M}$ be a finite factor and $\mathfrak{m}$ be a nonzero closed ideal. If $x \in \mathfrak{m}$ and $x \ne 0$, then $T_{\mathscr{M}}(x^*x) = \lambda 1 \in \mathfrak{m} \cap \mathscr{Z}$ and $\lambda \ne 0$. Hence $\mathfrak{m}$ contains $\lambda 1$, so that $\mathfrak{m} = \mathscr{M}$. Thus, every nonzero ideal of $\mathscr{M}$ is uniformly dense in $\mathscr{M}$. But the general linear group $G_{\mathscr{M}}$ of $\mathscr{M}$ is open in the uniform topology. Thus every nonzero ideal contains an invertible element, so that it must be $\mathscr{M}$ itself. Q.E.D.

Theorem 4.8. *If $\mathscr{M}$ is a von Neumann algebra with center $\mathscr{Z}$, then the map: $\mathfrak{m} \mapsto \mathfrak{m} \cap \mathscr{Z} = \mathfrak{n}$ is a bijection of the set of all maximal ideals of $\mathscr{M}$ onto the set of all maximal ideals of $\mathscr{Z}$. Furthermore, the inverse of this map is given by*

$$\mathfrak{n} \mapsto \mathfrak{m} = \{x \in \mathscr{M} : \bar{K}(axb) \cap \mathscr{Z} \subset \mathfrak{n} \text{ for all } a,b \in \mathscr{M}\}$$

for any maximal ideal $\mathfrak{n}$ of $\mathscr{Z}$.

PROOF. Let $\mathfrak{n}$ be a closed, in norm, ideal of $\mathscr{Z}$. Put

$$\mathfrak{m}_{\mathfrak{n}} = \{x \in \mathscr{M} : \bar{K}(axb) \cap \mathscr{Z} \subset \mathfrak{n} \text{ for all } a,b \in \mathscr{M}\}.$$

By definition, we have $\mathscr{M}\mathfrak{m}_{\mathfrak{n}}\mathscr{M} \subset \mathfrak{m}_{\mathfrak{n}}$. Let x and y be elements of $\mathfrak{m}_{\mathfrak{n}}$. We shall show that $x + y \in \mathfrak{m}_{\mathfrak{n}}$. For fixed $a,b \in \mathscr{M}$, and any $s \in \bar{K}(a(x+y)b) \cap \mathscr{Z}$ and $\varepsilon > 0$, there exists an $f \in \mathscr{K}$ such that $\|T_f(a(x+y)b) - s\| < \varepsilon$. Applying

Lemma 4.5 to $T_f(axb)$ and $T_f(ayb)$, we find a $g \in \mathscr{K}$, $s_1 \in \bar{K}(axb) \cap \mathscr{Z}$, and $s_2 \in \bar{K}(ayb) \cap \mathscr{Z}$ such that

$$\|T_gT_f(axb) - s_1\| < \varepsilon \quad \text{and} \quad \|T_gT_f(ayb) - s_2\| < \varepsilon.$$

We then have

$$\begin{aligned}\|s_1 + s_2 - s\| &\le \|s_1 - T_gT_f(axb)\| + \|T_g(T_f(axb + ayb) - s)\| \\ &\quad + \|T_gT_f(ayb) - s_2\| < 3\varepsilon.\end{aligned}$$

Since $s_1,s_2 \in \mathfrak{n}$, we have $s \in \mathfrak{n}$. Hence $\bar{K}(a(x + y)b) \cap \mathscr{Z} \subset \mathfrak{n}$. Thus $x + y \in \mathfrak{m}_\mathfrak{n}$. Therefore, $\mathfrak{m}_\mathfrak{n}$ is an ideal of $\mathscr{M}$.

Let Ω be the spectrum of $\mathscr{Z}$ and $K_\mathfrak{n}$ be the set of all $\omega \in \Omega$ at which every element of $\mathfrak{n}$ vanishes. It follows that $\mathfrak{n}$ is the set of all those elements of $\mathfrak{n}$ that vanish on $K_\mathfrak{n}$, Proposition I.8.3. Together with this, we have, for any $x \in \mathfrak{n}$ and $a,b \in \mathscr{M}$, $\bar{K}(axb) = \bar{K}(xab) = x\bar{K}(ab)$, so that $\mathfrak{n} \subset \mathfrak{m}_\mathfrak{n}$. If $x \in \mathfrak{m}_\mathfrak{n} \cap \mathscr{Z}$, then $K(x) = \{x\}$; so $\bar{K}(x) \cap \mathscr{Z} = \{x\}$ which implies $x \in \mathfrak{n}$. Therefore $\mathfrak{m}_\mathfrak{n} \cap \mathscr{Z} \subset \mathfrak{n}$. Thus, we get $\mathfrak{m}_\mathfrak{n} \cap \mathscr{Z} = \mathfrak{n}$.

Suppose that $\mathfrak{m}$ is a closed ideal of $\mathscr{M}$ with $\mathfrak{m} \cap \mathscr{Z} \subset \mathfrak{n}$. If $x \in \mathfrak{m}$, then $K(axb) \subset \mathfrak{m}$ for any $a,b \in \mathscr{M}$, so that $\bar{K}(axb) \subset \mathfrak{m}$ by the closedness of $\mathfrak{m}$; hence $\bar{K}(axb) \cap \mathscr{Z} \subset \mathfrak{n}$; then $x \in \mathfrak{m}_\mathfrak{n}$. Therefore $\mathfrak{m} \subset \mathfrak{m}_\mathfrak{n}$. Hence $\mathfrak{m}_\mathfrak{n}$ contains every closed ideal $\mathfrak{m}$ of $\mathscr{M}$ with $\mathfrak{m} \cap \mathscr{Z} \subset \mathfrak{n}$, which means that $\mathfrak{m}_\mathfrak{n}$ is a maximal ideal for every maximal ideal $\mathfrak{n}$ of $\mathscr{Z}$ if $\mathfrak{m}_\mathfrak{n}$ is closed. Thus we shall show the closedness of $\mathfrak{m}_\mathfrak{n}$. To this end, it suffices to prove that

$$\bar{\mathfrak{m}} \cap \mathscr{Z} = \overline{\mathfrak{m} \cap \mathscr{Z}}$$

for any ideal $\mathfrak{m}$ of $\mathscr{M}$. Because if this were done, then $\overline{\mathfrak{m}_\mathfrak{n}} \cap \mathscr{Z} = \overline{\mathfrak{m}_\mathfrak{n} \cap \mathscr{Z}} = \mathfrak{n}$ so that $\overline{\mathfrak{m}_\mathfrak{n}} \subset \mathfrak{m}_\mathfrak{n}$. Of course, $\bar{\mathfrak{m}} \cap \mathscr{Z} \supset \overline{\mathfrak{m} \cap \mathscr{Z}}$. Let K be the hull of $\overline{\mathfrak{m} \cap \mathscr{Z}}$, i.e., the set of all those $\omega \in \Omega$ such that $x(\omega) = 0$ for every $x \in \overline{\mathfrak{m} \cap \mathscr{Z}}$. Suppose that there exists an element $a \in \bar{\mathfrak{m}} \cap \mathscr{Z}$ which is not in $\overline{\mathfrak{m} \cap \mathscr{Z}}$. Replacing a by a^*a, we may assume $a \ge 0$. Then there exists a point $\omega_0 \in K$ with $a(\omega_0) > 0$. Multiplying by a scalar, we assume $a(\omega_0) > 1$. Let $E = \{\omega \in \Omega : a(\omega) > 1\}$ and $\bar{E}$ be the closure of E. Since Ω is hyperstonean, $\bar{E}$ is open and closed. Let e be the projection of $\mathscr{Z}$ corresponding to $\bar{E}$. We then have $e \le a$, so that $e \in \bar{\mathfrak{m}} \cap \mathscr{Z}$ and $e \notin \overline{\mathfrak{m} \cap \mathscr{Z}}$ because $e(\omega_0) = 1$. There exists $b \in \mathfrak{m}$ such that $\|b - e\| < 1$. Replacing b by be, we find an element $b \in \mathfrak{m}$ such that $\|b - e\| < 1$ and $b = be$. Hence there exists $c \in \mathscr{M}$ such that $bc = e$ because e is the identity of the algebra $\mathscr{M}e$. Hence $e = bc \in \mathfrak{m}$, which contradicts the fact that $e \notin \overline{\mathfrak{m} \cap \mathscr{Z}}$. Thus, we finally proved $\bar{\mathfrak{m}} \cap \mathscr{Z} = \overline{\mathfrak{m} \cap \mathscr{Z}}$. Q.E.D.

Corollary 4.9. *A factor has only one maximal ideal.*

Corollary 4.10. *Let $\mathscr{M}$ be a finite von Neumann algebra with center $\mathscr{Z}$. Let Ω be the spectrum of $\mathscr{Z}$. The correspondence between maximal ideals of $\mathscr{M}$ and $\mathscr{Z}$ is given by the following:*

$$\mathfrak{m}_\omega = \{x \in \mathscr{M} : T_\mathscr{M}(x^*x)(\omega) = 0\}, \qquad \omega \in \Omega.$$

PROOF. Let $\mathfrak{n}_\omega = \{x \in \mathscr{Z} : x(\omega) = 0\}$ be the maximal ideal of $\mathscr{Z}$ corresponding to $\omega \in \Omega$. We then have

$$\begin{aligned} \mathfrak{m}_{\mathfrak{n}_\omega} &= \{x \in \mathscr{M} : \bar{K}(axb) \cap \mathscr{Z} \subset \mathfrak{n}_\omega \text{ for every } a,b \in \mathscr{M}\} \\ &= \{x \in \mathscr{M} : T_{\mathscr{M}}(axb) \in \mathfrak{n}_\omega \text{ for every } a,b \in \mathscr{M}\} \\ &= \{x \in \mathscr{M} : T_{\mathscr{M}}(axb)(\omega) = 0 \text{ for every } a,b \in \mathscr{M}\} \\ &= \{x \in \mathscr{M} : T_{\mathscr{M}}(x^*x)(\omega) = 0\} = \mathfrak{m}_\omega, \end{aligned}$$

where the last step follows from the Cauchy–Schwarz inequality. Q.E.D.

EXERCISES

1. Let $\mathscr{M}$ be a von Neumann algebra and $\operatorname{Aut}(\mathscr{M})$ the group of all automorphisms of $\mathscr{M}$. For a subgroup G of $\operatorname{Aut}(\mathscr{M})$, let $\mathscr{M}^G$ denote the algebra of all fixed elements under G. For each $x \in \mathscr{M}$, let $K_G(x)$ denote the σ-weakly closed convex closure of the G-orbit $\{g(x) : g \in G\}$ of x.

   (a) Show that if $\varphi \in \mathscr{M}_*^+$ is G-invariant, then the σ-weakly lower semicontinuous function $F_\varphi : y \in K_G(x) \mapsto \|\pi_\varphi(y)\xi_0\| = \varphi(y^*y)^{1/2}$ takes its minimal value on $K_G(x)$ at $y_0 \in K_G(x)$ if only if $\pi_\varphi(y_0)\xi_0 = E_\varphi \pi_\varphi(x)\xi_0$, where E_φ is the projection of $\mathfrak{H}_\varphi$ onto the subspace $\mathfrak{H}_\varphi^G$ of all fixed points under the unitary representation $\{U_\varphi, \mathfrak{H}_\varphi\}$ of G associated with the cyclic representation $\{\pi_\varphi, \mathfrak{H}_\varphi\}$.

   (b) For each G-invariant $\varphi \in \mathscr{M}_*^+$, let $K_G^\varphi(x)$ be the set of all $y \in K_G(x)$ such that $F_\varphi(y) = \varphi(y^*y)^{1/2} = \inf\{F_\varphi(z) : z \in K_G(x)\}$. Show that if $\varphi_1, \ldots, \varphi_n \in \mathscr{M}_*^+$ are G-invariant, then $K_G^\varphi(x) = K_G^{\varphi_1}(x) \cap \cdots \cap K^{\varphi_n}(x)$ with $\varphi = \varphi_1 + \varphi_2 + \cdots + \varphi_n$. (*Hint*: There exist $h_1, h_2, \ldots, h_n \in \pi_\varphi(\mathscr{M})' \cap U_\varphi(G)'$ such that $\varphi_i(y) = (\pi_\varphi(y)h_i\xi_\varphi | h_i\xi_\varphi)$, $i = 1,2,\ldots,n$, and the restrictions of π_φ and U_φ to $[\pi_\varphi(\mathscr{M})h_i\xi_\varphi]$ are unitarily equivalent to $\{\pi_{\varphi_i}, U_{\varphi_i}, \mathfrak{H}_{\varphi_i}\}$. Hence $E_\varphi \pi_\varphi(y)\xi_\varphi = \pi_\varphi(y)\xi_\varphi$ if and only if $E_\varphi \pi_\varphi(y)h_i\xi_\varphi = \pi_\varphi(y)h_i\xi_\varphi$, $i = 1,2,\ldots,n$.)

   (c) Show that if $\mathscr{M}$ admits sufficiently many G-invariant normal states, then $K_G(x)$ intersects with $\mathscr{M}^G$ at exactly one point, say $\mathscr{E}(x)$. In this case, we say that $\mathscr{M}$ is *G-finite* (*Hint:* To show $K_G(x) \cap \mathscr{M}^G \neq \emptyset$, apply the compactness of $K_G(x)$ and (b). To show the uniqueness of the intersection, use the fact that $\pi_\varphi(K_G(x))\xi_\varphi \cap E_\varphi\mathfrak{H}_\varphi = \{E_\varphi \pi_\varphi(x)\xi_\varphi\}$ for every G-invariant $\varphi \in \mathscr{M}_*^+$.)

   (d) Show that $\mathscr{M}$ is G-finite if and only if there exists a faithful G-invariant normal projection $\mathscr{E}$ of norm one from $\mathscr{M}$ onto $\mathscr{M}^G$, and that if this is the case, then $\mathscr{E}$ is unique.

   (e) Show that $\mathscr{M}$ is G-finite if and only if the G-orbit $G(\varphi) = \{\varphi \circ g : g \in G\}$ of each $\varphi \in \mathscr{M}_*^+$ is relatively $\sigma(\mathscr{M}_*, \mathscr{M})$-compact. (*Hint*: If $G(\varphi)$ is relatively $\sigma(\mathscr{M}_*, \mathscr{M})$-compact, then its convex closure $K_G(\varphi)$ is $\sigma(\mathscr{M}_*, \mathscr{M})$-compact; hence the Ryll–Nardzewski theorem applies to yield an invariant functional in $K_G(\varphi)$. If $G(\varphi)$ is not relatively $\sigma(\mathscr{M}_*, \mathscr{M})$-compact, then there exist a decreasing sequence $\{e_n\}$ of projections in $\mathscr{M}$ convergent to zero σ-strongly, and a sequence $\{g_n\}$ in G and $\varepsilon > 0$ such that $\varphi \circ g_n(e_n) \geq \varepsilon$. If $\mathscr{M}$ is G-finite, then set $\omega = \varphi \circ \mathscr{E}$. Then $\omega(g_n(e_n)) = \omega(e_n) \to 0$. Hence $s(\omega)g_n(e_n)s(\omega) \to 0$ σ-strongly; thus $\varphi(s(\omega)g_n(e_n)s(\omega)) \to 0$. But $s(\omega) \geq s(\varphi)$, which is a contradiction.)

   (f) Show that $\mathscr{M}$ is G-finite if and only if G is relatively compact in $\mathscr{L}_*(\mathscr{M})$ with respect to the topology considered in Exercise 2.6.

2. Show that a factor $\mathscr{M}$ is finite if and only if Aut($\mathscr{M}$), the group of all automorphisms of $\mathscr{M}$, is relatively compact in $\mathscr{L}_*(\mathscr{M})$ with respect to the topology considered in Exercise 2.6.

3. Let A be a unital C^*-algebra. Let $\mathfrak{T}$ be the set of all tracial states, i.e.,

$$\mathfrak{T} = \{\varphi \in \mathfrak{S}(A) : \varphi(xy) = \varphi(yx),\ x, y \in A\}.$$

Show that $\mathfrak{T}$ is a simplex. (*Hint*: The group of all inner automorphisms of A is large. Apply Exercise IV.6.7.)

5. Normality of Separable Representations*

Theorem 5.1. *Let $\mathscr{M}$ be a σ-finite von Neumann algebra with no direct summand of finite type I, and $\{\pi, \mathfrak{H}\}$ a representation of $\mathscr{M}$.*

(i) *If $\mathfrak{H}$ is separable, then π is normal.*
(ii) *If $\mathscr{M}$ is properly infinite and if the von Neumann algebra $\pi(\mathscr{M})''$ generated by $\pi(\mathscr{M})$ is σ-finite, then π is normal.*

Proof. Let $\mathscr{N}$ denote the von Neumann algebra $\pi(\mathscr{M})''$. By Theorem III.2.14, we have only to show that the singular part π_s of π is zero. Assuming $\pi_s \neq 0$, we shall derive a contradiction. Looking at π_s, we may assume that π is singular because π_s satisfies the same assumption as π. In either case (i) or (ii), $\mathscr{N}$ is σ-finite, so that it admits a faithful normal state φ. Put $\psi = {}^t\pi(\varphi)$. The kernel $\pi^{-1}(0)$ of π is the left kernel of ψ because for $x \in \mathscr{M}$, $\pi(x) = 0$ if and only if $\pi(x^*x) = 0$ if and only if $\varphi(\pi(x^*x)) = 0 = \psi(x^*x)$. Since ψ is singular, there exists, by Theorem III.3.8, a family $\{e_i\}_{i \in I}$ of orthogonal projections in $\mathscr{M}$ such that $\psi(e_i) = 0$ and $\sum_{i \in I} e_i = 1$. Hence $\pi^{-1}(0)$ contains $\{e_i\}_{i \in I}$, so that $\pi^{-1}(0)$ is σ-weakly dense in $\mathscr{M}$. By the σ-finiteness of $\mathscr{M}$, $\{e_i\}_{i \in I}$ is countable.

Suppose $\mathscr{M}$ is properly infinite. There exists a sequence $\{p_n\}$ of orthogonal projections such that $p_n \sim 1$. For each n, let v_n be a partial isometry with $v_n^* v_n = 1$ and $v_n v_n^* = p_n$. Put $q_n = v_n(\sum_{i=1}^n e_i)v_n^*$. Then $\{q_n\}$ are orthogonal projections. Let $\{j_n\}$ be an increasing sequence of positive integers. Then

$$q_{j_{n+1}} = v_{j_{n+1}}\left(\sum_{k=1}^{j_{n+1}} e_k\right)v_{j_{n+1}}^* \geq v_{j_{n+1}}\left(\sum_{k=j_n+1}^{j_{n+1}} e_k\right)v_{j_{n+1}}^* \sim \sum_{k=j_n+1}^{j_{n+1}} e_k.$$

Hence $\sum_{n=1}^\infty q_{j_n} \gtrsim \sum_{k=1}^\infty e_k = 1$; so $\sum_{n=1}^\infty q_{j_n} \sim 1$. Therefore $\pi(\sum_{n=1}^\infty q_{j_n}) \neq 0$. On the other hand,

$$\pi(q_n) = \pi(v_n)\left(\sum_{k=1}^{n} \pi(e_k)\right)\pi(v_n)^* = 0.$$

Let $\{r_j\}$ be a fixed enumeration of all rational numbers. To each real number s we associate an increasing sequence $\{j_n\}$ such that

$$s = \lim_{n \to \infty} r_{j_n}.$$

Put $\bar{q}_s = \sum_{n=1}^{\infty} q_{j_n}$. Then as we have seen above, $\pi(\bar{q}_s) \neq 0$ for all real number s. But if $s \neq s'$, then

$$\pi(\bar{q}_s) \cdot \pi(\bar{q}_{s'}) = \pi(\bar{q}_s \bar{q}_{s'})$$

$$= \pi\left(\sum_{k \in \{j_n\} \cap \{j'_m\}}\right) q_k = 0$$

because $\{j_n\} \cap \{j'_m\}$ is a finite set. Hence $\{\pi(\bar{q}_s)\}$ is an uncountable family of nonzero orthogonal projections in $\mathcal{N}$, which is a contradiction.

Now suppose $\mathfrak{H}$ is separable. Then the singular representation π annihilates the properly infinite direct summand of $\mathcal{M}$, as we have shown. Hence we may assume that $\mathcal{M}$ is finite. In other words, we assume that $\mathcal{M}$ is of type II_1.

By Corollary 4.7, $T_{\mathcal{M}}(\pi^{-1}(0)) = \pi^{-1}(0) \cap \mathcal{Z}$, where $\mathcal{Z}$ is the center of $\mathcal{M}$. Since $e_i \in \pi^{-1}(0)$ and $\sum_{i=1}^{\infty} e_i = 1, \sum_{i=1}^{n} e_i \in \pi^{-1}(0) \cap \mathcal{Z}$ converges σ-strongly to 1; hence $\pi^{-1}(0) \cap \mathcal{Z}$ is σ-weakly dense in $\mathcal{Z}$. Therefore, the restriction of ψ onto $\mathcal{Z}$ is singular because the left kernel of $\psi|_{\mathcal{Z}}$ is $\pi^{-1}(0) \cap \mathcal{Z}$. Therefore, if we define a trace τ by $\tau(x) = \psi(T_{\mathcal{M}}(x))$, then τ is a singular finite trace of $\mathcal{M}$.

Since the cyclic representation π_ψ of $\mathcal{M}$ induced by ψ is nothing but $\pi_\varphi \circ \pi$, where π_φ is the cyclic representation of $\mathcal{N}$ induced by φ, $\mathfrak{H}_\psi$ has to be separable because any cyclic normal representation of $\mathcal{N}$ is on a separable Hilbert space. We will show that $\mathfrak{H}_\psi$ cannot be separable for any singular state ψ.

Let $\mathcal{A}$ be any fixed maximal abelian *-subalgebra of $\mathcal{M}$, and let μ_ψ and μ_τ be the Radon measures on the spectrum Γ of $\mathcal{A}$ induced by ψ and τ, respectively. Furthermore, let μ be the Radon measure on the spectrum Ω of $\mathcal{Z}$ induced by τ $(=\psi)$. We shall divide the proof into two cases according to the relation between μ_φ and μ_τ.

Case I: μ_ψ is absolutely continuous with respect to μ_τ. By the Radon–Nikodym theorem, there exists a μ_τ-integrable function $f(\gamma)$ on Γ such that

$$\mu_\psi(E) = \int_E f(\gamma)\, d\mu_\tau(\gamma)$$

for each Borel subset E of Γ. Since $\mu_\psi(\Gamma) = \psi(1) > 0$, f is strictly positive on some compact subset K with $\mu_\tau(K) > 0$, so that the restrictions μ_ψ and μ_τ on K are equivalent.

Since μ_τ is a regular measure, there exists a decreasing sequence $\{G_n\}$ of open sets such that

$$G_n \supset \bar{G}_{n+1} \supset K \quad \text{and} \quad \mu_\tau(K) = \lim_n \mu_\tau(G_n).$$

Since Γ is hyperstonean, $\bar{G}_n$ is open and closed. Hence we conclude that there exists a decreasing sequence $\{E'_n\}$ of open and closed subsets such that

$$K \subset E'_n \quad \text{and} \quad \mu_\tau(K) = \lim \mu_\tau(E'_n).$$

Let e_n' be the projection of $\mathscr{A}$ corresponding to E_n'. Since τ is singular on $\mathscr{Z}$, there exists, by Theorem III.3.8, a decreasing sequence $\{z_n\}$ of central projections such that $\tau(z_n) = 1$ and converging σ-strongly to zero.

Put $e_n = e_n' z_n$ and let E_n be the open and closed subset of Γ corresponding to e_n. Then $\{e_n\}$ is a decreasing sequence of projections and converges σ-strongly to zero. We have

$$\lim \mu_\tau(E_n) = \mu_\tau\left(\bigcap_{n=1}^{\infty} E_n\right) = \mu_\tau\left(\bigcap_{n=1}^{\infty} E_n \cap K\right)$$
$$= \mu_\tau(K),$$
$$ee_n = \sum_{k=n}^{\infty} (e_k - e_{k+1}).$$

Hence we have

$$\lim_{n\to\infty} \int g(\gamma)e_n(\gamma)\, d\mu_\tau(\gamma) = \lim_{n\to\infty} \int_{E_n} g(\gamma)\, d\mu_\tau(\gamma)$$
$$= \int_K g(\gamma)\, d\mu_\tau(\gamma)$$

for every μ_τ-integrable function g on Γ.

We construct, by induction, a partition $\{p_{n,j,k}\}_{1\le k\le 2^j}$ of $e_n - e_{n+1}$ consisting of orthogonal equivalent projections in $\mathscr{A}$ such that

$$e_n - e_{n+1} = \sum_{k=1}^{2^j} p_{n,j,k}, \qquad p_{n,j,k} \sim p_{n,j,l},$$
$$p_{n,j,k} = p_{n,j+1,2k-1} + p_{n,j+1,2k}.$$

Put

$$u_{n,j} = \sum_{k=1}^{2^j} (-1)^k p_{n,j,k}$$

for each n and j. Then $u_{n,j_1}u_{n,j_2}$ is expressed, for distinct j_1 and j_2, as the difference of two orthogonal equivalent projections with sum $e_n - e_{n+1}$.

For each real number $0 < s < 1$, we associate an increasing sequence $\{j_{s,n}\}$ of integers such that

$$0 \le j_{s,n} \le 2^n \quad \text{and} \quad \lim_{n\to\infty} \frac{j_{s,n}}{2^n} = s.$$

If $s \neq t$, then there is an integer n_0 with $j_{s,n} \neq j_{t,n}$ for $n \ge n_0$. Let $u_s = \sum_{n=1}^{\infty} u_{n,j_{s,n}}$ for each real number $0 < s < 1$. Then we have

$$u_s^* u_s = \sum_{n=1}^{\infty} (e_n - e_{n+1}) = e_1.$$

If $j_{s,n} \neq j_{t,n}$ for $n \geq n_0$, then

$$e_n u_s^* u_t e_n = p_n - q_n, \qquad n \geq n_0,$$

where p_n and q_n are orthogonal equivalent projections such that $p_n + q_n = e_n$. Hence it follows that

$$\int_K |u_s(\gamma)|^2 \, d\mu_\tau(\gamma) = \lim_{n\to\infty} \tau(e_n u_s^* u_s e_n)$$
$$= \lim_{n\to\infty} \tau(e_n) = \mu_\tau(K) > 0;$$
$$\int_K \overline{u_s(\gamma)} u_t(\gamma) \, d\mu_\tau(\gamma) = \lim \tau(e_n u_s^* u_t e_n)$$
$$= \lim_{n\to\infty} \tau(p_n - q_n) = 0.$$

Therefore, $\{u_s\}$ is an uncountable family of orthogonal nonzero functions in $L^2(K,\mu_\tau)$; hence $L^2(K,\mu_\tau)$ is not separable. Since μ_ψ and μ_τ are equivalent on K, $L^2(K,\mu_\psi)$ is nonseparable. But $L^2(K,\mu_\psi)$ is a subspace of $L^2(\Gamma,\mu_\psi)$ and $L^2(\Gamma,\mu_\psi)$ is a subspace of $\mathfrak{H}_\psi$, which implies the nonseparability of $\mathfrak{H}_\psi$. This contradicts the separability of $\mathfrak{H}_\varphi$.

Case II: μ_ψ is not absolutely continuous with respect to μ_τ. There exists a compact subset K of Γ such that $\mu_\psi(K) > 0$ and $\mu_\tau(K) = 0$. By arguments similar to those above, we can find a decreasing sequence $\{E_n\}$ of open and closed sets in Γ such that

$$E_n \supset K \quad \text{and} \quad \lim \mu_\tau(E_n) = 0.$$

Let e_n be the projection in $\mathscr{A}$ corresponding to E_n. Then $\{T_{\mathscr{M}}(e_n)\}$ is decreasing and

$$\lim_{n\to\infty} \int_\Omega T_{\mathscr{M}}(e_n)(\omega) \, d\mu(\omega) = \lim_{n\to\infty} \tau(e_n)$$
$$= \lim \mu_\tau(E_n) = 0.$$

It follows that the sequence $\{T_{\mathscr{M}}(e_n)(\omega)\}$ of functions on Ω converges to zero μ-almost everywhere. Hence, by Egoroff's theorem, for any $\varepsilon > 0$, $T_{\mathscr{M}}(e_n)(\omega)$ converges uniformly to zero on some compact subset F of Ω with $\mu(F) > 1 - \varepsilon$. Therefore, considering a subsequence of $\{e_n\}$, we may assume $T_{\mathscr{M}}(e_n)(\omega) < 1/4^{n+2}$ for every $\omega \in F$. Put

$$G_n = \{\omega \in \Omega : T_{\mathscr{M}}(e_n)(\omega) < 1/4^{n+2}\}.$$

Then G_n is open and contains F. We have $T_{\mathscr{M}}(e_n)(\omega) \leq 1/4^{n+2}$ on the closure $\bar{G}_n$ of G_n. Consider the projection g_n in $\mathscr{Z}$ corresponding to open and closed set $\bar{G}_1 \cap \cdots \cap \bar{G}_n$, and put $f_n = e_n g_n$. Then we have

$$g_n \geq g_{n+1}, \qquad f_n \geq f_{n+1}, \quad \text{and} \quad T_{\mathscr{M}}(f_n) \leq 1/4^{n+2},$$

so that f_n converges to zero σ-strongly. Let U_n be the open and closed subset of Γ corresponding to g_n and $U = \bigcap_{n=1}^{\infty} U_n$. Then we get

$$\begin{aligned}\mu_\psi(U) &= \lim \mu_\psi(U_n) = \psi(g_n)\\ &= \mu\left(\bigcap_{n=1}^{\infty} \bar{G}_n\right) \geq \mu(F) > 1 - \varepsilon,\end{aligned}$$

which implies that

$$\begin{aligned}\mu_\psi(U \cap K) &= \mu_\psi(U) + \mu_\psi(K) - \mu_\psi(U \cup K)\\ &> 1 - \varepsilon + \mu_\psi(K) - \mu_\psi(K)\\ &> \mu_\psi(K) - \varepsilon > 0\end{aligned}$$

for sufficiently small $\varepsilon > 0$.

Let $\{\pi_\psi, \mathfrak{H}_\psi, \xi_\psi\}$ be the cyclic representation of $\mathscr{M}$ induced by ψ as before. Then we have

$$\begin{aligned}\pi_\psi(f_n) &\geq \pi_\psi(f_{n+1}),\\ \|\pi_\psi(f_n)\xi_\psi\|^2 &= \psi(f_n)\\ &= \mu_\psi(U_n \cap E_n) \geq \mu_\psi(U \cap K) > 0\end{aligned}$$

for every n. It follows that $\pi_\psi(f_n)\xi_\psi$ converges to a nonzero vector $\xi \in \mathfrak{H}_\psi$ belonging to $\bigcap_{n=1}^{\infty} \pi_\psi(f_n)\mathfrak{H}_\psi$.

Put $h_n = f_n - f_{n+1}$, $p_{1,1} = h_1$ and suppose that orthogonal projections $\{p_{k,j}\}$ such as

$$h_k = p_{k,1} \sim p_{k,j}, \qquad j = 1,2,\dots,2^k,$$

are constructed for $k = 1,\dots,n-1$ and $1 \leq j \leq 2^k$, where f_n is orthogonal to $p_{k,j}$. Put

$$p_n = \sum_{k=1}^{n-1} \sum_{j=1}^{2k} p_{k,j} + f_n.$$

Then we have

$$\begin{aligned}T_{\mathscr{M}}(p_n) &= \sum_{k=1}^{n-1} \sum_{j=1}^{2^k} T_{\mathscr{M}}(p_{k,j} + f_n)\\ &\leq \sum_{k=1}^{n-1} \frac{2^k}{4^{k+2}} + \frac{1}{4^{n+2}}\\ &= \frac{1}{16} \sum_{k=1}^{n-1} \frac{1}{2^k} + \frac{1}{4^{n+2}} < \frac{1}{8}\end{aligned}$$

so that $T_{\mathscr{M}}(1 - p_n) \geq 7/8$; hence there exist orthogonal equivalent projections $p_{n,j}$, $1 \leq j \leq 2^n$, such that

$$h_n = p_{n,1} \sim p_{n,j} \leq 1 - p_n, \qquad 1 \leq j \leq 2^n.$$

Therefore, by induction, we get orthogonal projections $\{p_{n,j}\}$ such as above.

Let $u_{n,j}$ be a partial isometry with $u_{n,j}^* u_{n,j} = p_{n,1} = h_n$ and $u_{n,j} u_{n,j}^* = p_{n,j}$. For different j_1 and j_2, we have

$$u_{n,j_1}^* u_{n,j_2} = u_{n,j_1}^* p_{n,j_1} p_{n,j_2} u_{n,j_2} = 0.$$

For each real number $0 < s < 1$, let $\{j_{s,n}\}$ be an increasing sequence of positive integers such as in case I. Let

$$u_s = \sum_{n=1}^{\infty} u_{n,j_{s,n}}$$

for each real number $0 < s < 1$. Then we have

$$u_s^* u_s = \sum_{n=1}^{\infty} h_n = f_1,$$

$$f_{n_0} u_s^* u_t f_{n_0} = 0$$

if $j_{s,n} \neq j_{t,n}$ for $n \geq n_0$. Hence we have

$$\begin{aligned} \|\pi_\psi(u_s)\xi\| &= (\pi_\psi(u_s^* u_s)\xi | \xi) \\ &= (\pi_\psi(f_1)\xi | \xi) \\ &= \|\xi\|^2 > 0, \end{aligned}$$

$$\begin{aligned} (\pi_\psi(u_t)\xi | \pi_\psi(u_s)\xi) &= (\pi_\psi(u_s^* u_t)\xi | \xi) \\ &= \lim_{n\to\infty} (\pi_\psi(f_n)\pi_\psi(u_s^* u_t)\pi(f_n)\xi | \xi) \\ &= 0. \end{aligned}$$

Therefore $\{\pi_\psi(u_s)\xi\}$ is an uncountable orthogonal family of nonzero vectors, so that $\mathfrak{H}_\psi$ is nonseparable; hence $\mathfrak{H}_\varphi$ is nonseparable. This contradicts the separability of the underlying Hilbert space of $\mathscr{N}$. Q.E.D.

Theorem 5.2. *If $\mathscr{M}$ is a finite von Neumann algebra, then the quotient C^*-algebra $\mathscr{M}/\mathfrak{m}$ by a maximal ideal $\mathfrak{m}$ is a finite factor.*

PROOF. Let $\mathscr{Z}$ be the center of $\mathscr{M}$ and Ω its spectrum. Let $\omega \in \Omega$ be the point corresponding to the maximal ideal $\mathfrak{m}$. Since $\mathfrak{m}$ is maximal, $\mathscr{M}/\mathfrak{m}$ is simple. Hence $\mathscr{M}/\mathfrak{m}$ is a factor if it is a W^*-algebra. Let $\tau_\omega(x) = T_{\mathscr{M}}(x)(\omega)$, $x \in \mathscr{M}$. By Corollary 4.10, we have $\mathfrak{m} = \{x \in \mathscr{M} : \tau_\omega(x^*x) = 0\}$. Let $\{\pi, \mathfrak{H}, \xi_0\}$ be the cyclic representation of $\mathscr{M}$ induced by the tracial state τ_ω. As seen in the proof of Corollary 4.10, we have $\mathfrak{m} = \pi^{-1}(0)$. Hence we have only to prove that $\pi(\mathscr{M})'' = \pi(\mathscr{M})$ because the vector ξ_0 gives rise to a faithful trace on $\pi(\mathscr{M})''$ by Proposition 3.19. To this end, it suffices to show that $\pi(S)\xi_0$ is closed in $\mathfrak{H}$ by Theorem II.4.8, where S denotes the unit ball of $\mathscr{M}$, because ξ_0 is separating for $\pi(\mathscr{M})''$ by Proposition 3.19.

Let $\{x_n\}$ be a sequence in S such that $\lim_{n\to\infty} \|\pi(x_n)\xi_0 - \eta\| = 0$ for some $\eta \in \mathfrak{H}$. Considering a subsequence of $\{x_n\}$, we may assume that $\|\pi(x_n)\xi_0 - \pi(x_{n+1})\xi_0\|^2 < 2^{-n}$, $n = 1,2,\dots$. In other words,

$$T_{\mathscr{M}}((x_n - x_{n+1})^*(x_n - x_{n+1}))(\omega) < 2^{-n}, \qquad n = 1,2,\dots.$$

Let $\{U_n\}$ be a decreasing sequence of open neighborhoods of ω such that

$$T_{\mathscr{M}}((x_n - x_{n+1})^*(x_n - x_{n+1}))(\sigma) < 2^{-n}, \qquad \sigma \in U_n.$$

Let z_n be the projection of $\mathscr{Z}$ corresponding to the closure of U_n, $n = 1,2,\dots$. We then have $z_n(\omega) = 1$ for every n. Putting $y_n = z_n x_n + (1 - z_n)$, $n = 1,2,\dots$, we get

$$T_{\mathscr{M}}((y_{n+1} - y_n)^*(y_{n+1} - y_n)) < 2^{-n},$$
$$\pi(y_n)\xi_0 = \pi(x_n)\xi_0, \qquad n = 1,2,\dots.$$

For each normal state φ of $\mathscr{Z}$, put $\tau_\varphi(x) = \varphi \circ T_{\mathscr{M}}(x)$, $x \in \mathscr{M}$. It then follows that τ_φ is a finite normal trace and

$$\tau_\varphi((y_{n+1} - y_n)^*(y_{n+1} - y_n)) < 2^{-n},$$

so that $\{y_n s(\varphi)\}$ is a σ-strongly Cauchy sequence in S by Proposition III.5.3. Hence by Proposition III.5.3, $\{y_n s(\varphi)\}$ converges σ-strongly to $y_\varphi \in \mathscr{M}_{s(\varphi)} \cap S$. Choosing a maximal family $\{\varphi_i\}_{i\in I}$ of normal states of $\mathscr{Z}$ with orthogonal supports, we conclude that $\{y_n\}$ converges σ-strongly to $y = \sum_{i\in I} y_{\varphi_i} \in S$ because $\sum_{i\in I} s(\varphi_i) = 1$. Now we have

$$T_{\mathscr{M}}((y_n - y)^*(y_n - y))^{1/2} \le \sum_{k=n}^{\infty} \tau_{\mathscr{M}}((y_{k+1} - y_k)^*(y_{k+1} - y_k))^{1/2} < 2^{-n+1},$$

so that

$$\|\pi(y_n)\xi_0 - \pi(y)\xi_0\|^2 = T_{\mathscr{M}}((y_n - y)^*(y_n - y))(\omega) < 2^{-n+1};$$

hence

$$\pi(y)\xi_0 = \lim_{n\to\infty} \pi(y_n)\xi_0 = \lim_{n\to\infty} \pi(x_n)\xi_0.$$

Therefore, $\pi(S)\xi_0$ is closed in $\mathfrak{H}$. Q.E.D.

For each integer $n \ge 1$, let $M_n = M_n(\mathbf{C})$ be the factor of all $n \times n$-matrices over $\mathbf{C}$. Let $\mathscr{M} = \sum_{n=1}^{\infty\oplus} M_n$. It follows that $\mathscr{M}$ is a finite von Neumann algebra of type I with center $\mathscr{Z}$ isomorphic to l^∞. The spectrum Ω of $\mathscr{Z}$ is the Stone–Cech compactification of the set $\mathbf{N}$ of positive integers. Let ω be a limit point of Ω, i.e., $\omega \in \Omega\backslash\mathbf{N}$. By induction, we choose a decreasing sequence $\{e_{n,j}\}_{j=1}^{\infty}$ of projections in M_n such that

$$\dim e_{n,j} = \left[\frac{n}{2^j}\right],$$

where $[\cdot]$ means the Gauss symbol. Of course, if $n < 2^j$, then $e_{n,j} = 0$. Put $e_j = \sum_{n=1}^{\infty} e_{n,j} \in \mathscr{M}$. Then $\{e_j\}$ is a decreasing sequence of projections in $\mathscr{M}$ and $T_{\mathscr{M}}(e_j)(\omega) = 1/2^j$. Hence the images $\{e_j(\omega)\}$ of $\{e_j\}$ in the quotient algebra

$\mathcal{M}/\mathcal{M}_\omega$, where $\mathfrak{m}_\omega$ is the maximal ideal of $\mathcal{M}$ corresponding to ω, are strictly decreasing. Hence $\mathcal{M}(\omega) = \mathcal{M}/\mathfrak{m}_\omega$ is a finite factor of infinite dimension, so that it is a factor of type II_1. Therefore, the quotient algebra of an algebra of type I is not necessarily of type I.

EXERCISES

1. Let $\mathfrak{H}$ be an infinite dimensional separable Hilbert space. Show that the quotient C^*-algebra $\mathcal{L}(\mathfrak{H})/\mathcal{LC}(\mathfrak{H})$, called the *Calkin algebra*, has no nontrivial representation on a separable Hilbert space.
2. Let $\mathcal{M}$ be a properly infinite von Neumann algebra with separable predual. Show that if π is a homomorphism of $\mathcal{M}$ *onto* another von Neumann algebra $\mathcal{N}$, then π is normal. (*Hint*: Show, using the arguments in the proof of Theorem 5.1, that if π is not continuous, then $\mathcal{N}$ must contain an orthogonal family $\{e_i\}_{i\in I}$ of projections with $\operatorname{card} I = c$. Then show from the completeness of the projection lattice of $\mathcal{N}$ that $\operatorname{card}\mathcal{N} \geq 2^{\operatorname{card} I} > \operatorname{card}\mathcal{M}$. But this is impossible since π is surjective.)

6. The Borel Spaces of von Neumann Algebras

We are now going to study the Borel space of von Neumann algebras of various types and its applications to direct integrals. Let $\mathfrak{H}$ be a fixed infinite dimensional separable Hilbert space, and $\mathfrak{A}$ (resp. $\mathfrak{F}$) be the space of all von Neumann algebras (resp. factors) on $\mathfrak{H}$ equipped with the Effros Borel structure. Let $\mathfrak{A}_f$ (resp. $\mathfrak{A}_{s,f}$, $\mathfrak{A}_{\mathrm{I}}$, $\mathfrak{A}_{\mathrm{II}}$, and $\mathfrak{A}_{\mathrm{III}}$) denote the collection of all finite (resp. semifinite, type I, type II, and type III) von Neumann algebras on $\mathfrak{H}$, and $\mathfrak{F}_f = \mathfrak{F} \cap \mathfrak{A}_f$, $\mathfrak{F}_{s,f} = \mathfrak{F} \cap \mathfrak{A}_{s,f}$, and so on. We shall determine the type of these Borel spaces, namely, that they are all standard. But in this section, we will not finish the program; instead we shall show that $\mathfrak{A}_f$ and $\mathfrak{A}_{\mathrm{I}}$ are standard, and $\mathfrak{A}_{s,f}$ and $\mathfrak{A}_{\mathrm{III}}^c$ are Souslin spaces. We begin with the following:

Lemma 6.1. *Let $\mathcal{P}$ denote the set of all projections in $\mathcal{L}(\mathfrak{H})$ equipped with the Borel structure generated by the strong operator topology. Then the map: $e \in \mathcal{P} \mapsto e\mathfrak{H} \in \mathfrak{W}(\mathfrak{H})$, the space of all closed subspaces of $\mathfrak{H}$ with Effros Borel structure, is a Borel isomorphism.*

PROOF. Since the map is bijective and both spaces are standard, we have only to prove that the map is Borel. For each $\xi \in \mathfrak{H}$, we have $\|\xi|_{e\mathfrak{K}}\| = \|e\xi\|$, where $\|\xi|_{e\mathfrak{K}}\|$ means the functional norm of the restriction of ξ to $e\mathfrak{H}$. Hence, the function: $e \in \mathcal{P} \mapsto \|\xi|_{e\mathfrak{K}}\|$ is continuous, so that the map: $e \in \mathcal{P} \mapsto e\mathfrak{H} \in \mathfrak{W}(\mathfrak{H})$ is Borel. Q.E.D.

For each $p \in \mathcal{P}$ and $\mathcal{M} \in \mathfrak{A}$, let $e_{\mathcal{M}}(p)$ denote the smallest projection in $\mathcal{M}$ majorizing p, which is given as the projection of $\mathfrak{H}$ onto $[\mathcal{M}'p\mathfrak{H}]$.

Lemma 6.2. *The map*: $(\mathcal{M},p) \in \mathfrak{A} \times \mathcal{P} \mapsto e_{\mathcal{M}}(p)$ *is Borel.*

PROOF. Let $\{x_n(\cdot)\}$ be a sequence of Borel choice functions on $\tilde{\mathfrak{A}}$ such that $\{x_n(\mathcal{M})\}$ is strongly* dense in $\mathcal{M}$. We then have

$$e_{\mathcal{M}}(p)\mathfrak{H} = [x_n(\mathcal{M}')p\mathfrak{H}: n = 1,2,\ldots],$$

and for each $\xi \in \mathfrak{H}$, the map: $(\mathcal{M},p) \in \tilde{\mathfrak{A}} \times \mathcal{P} \mapsto x_n(\mathcal{M}')p\xi \in \mathfrak{H}$ is Borel. Hence $(\mathcal{M},p) \in \tilde{\mathfrak{A}} \times \mathcal{P} \mapsto e_{\mathcal{M}}(p)\mathfrak{H} \in \mathfrak{W}(\mathfrak{H})$ is a Borel map, so that $(\mathcal{M},p) \mapsto e_{\mathcal{M}}(p) \in \mathcal{P}$ is a Borel map by Lemma 6.1. Q.E.D.

Corollary 6.3. *The map*: $(\mathcal{M},\varphi) \in \tilde{\mathfrak{A}} \times \mathcal{L}(\mathfrak{H})_*^+ \mapsto s(\varphi_{\mathcal{M}}) \in \mathcal{P}$ *is a Borel map, where* $s(\varphi_{\mathcal{M}})$ *means the support of the restriction* $\varphi_{\mathcal{M}}$ *of* φ *to* $\mathcal{M}$.

PROOF. Our assertion follows immediately from a combination of Corollary IV.8.11, Theorem IV.8.13 and the fact that $s(\varphi_{\mathcal{M}}) = e_{\mathcal{M}}(s(\varphi))$. Q.E.D.

Lemma 6.4. *Let G be a Polish topological group in the sense that G is a Polish space as a topological space. If H is a closed subgroup of G, then there exists a Borel subset E of G which meets every left (resp. right) H-coset at exactly one point.*

PROOF. We define an equivalence relation $g \sim h$ by $g^{-1}h \in H$. Then each equivalence class is closed and the saturation of any open set is open. Hence the Borel cross-section theorem, Theorem A.15, yields the existence of E. Q.E.D.

For each $\mathcal{M} \in \mathfrak{A}$, we set

$$[\mathcal{M}] = \{\mathcal{N} \in \mathfrak{A}: u\mathcal{M}u^* = \mathcal{N} \text{ for some unitary element } u\},$$
$$[[\mathcal{M}]] = \{\mathcal{N} \in \mathfrak{A}: \mathcal{N} \cong \mathcal{M}\}.$$

Theorem 6.5. *For each* $\mathcal{M} \in \mathfrak{A}$, *the unitary equivalence class* $[\mathcal{M}]$ *(resp. the isomorphism class* $[[\mathcal{M}]]$*) is a Borel subset of* $\mathfrak{A}$.

PROOF. Let $\mathcal{U}$ denote the group of all unitary operators on $\mathfrak{H}$ equipped with the strong* operator topology. Then $\mathcal{U}$ is a Polish topological group. Put $\mathcal{G}(\mathcal{M}) = \{u \in \mathcal{U}: u\mathcal{M}u^* = \mathcal{M}\}$. Then $\mathcal{G}(\mathcal{M})$ is a closed subgroup. Hence there exists a Borel set $\mathcal{E}$ in $\mathcal{U}$ which meets each left $\mathcal{G}(\mathcal{M})$-coset at exactly one point. We then have $[\mathcal{M}] = \{u\mathcal{M}u^*: u \in \mathcal{E}\}$. Hence $[\mathcal{M}]$ is the image of $\mathcal{E}$ under an injective Borel map: $u \in \mathcal{E} \mapsto u\mathcal{M}u^* \in \mathfrak{A}$. Since $\mathcal{E}$ is a standard Borel space and $\mathfrak{A}$ is also, $[\mathcal{M}]$ is a Borel subset of $\mathfrak{A}$.

Let $\mathfrak{H}_0 = \mathfrak{H} \otimes \mathfrak{H}$ and $\Phi(\mathcal{M}) = \mathcal{M} \otimes \mathbf{C}$ for each $\mathcal{M} \in \mathfrak{A}$. It then follows that Φ is an injective Borel map of $\mathfrak{A}$ into $\mathfrak{A}(\mathfrak{H}_0)$, the space of all von Neumann algebras on $\mathfrak{H}_0$. Since $\Phi(\mathcal{M})' = \mathcal{M}' \otimes \mathcal{L}(\mathfrak{H})$ is properly infinite, $\mathcal{M}_1 \cong \mathcal{M}_2$ if and only if $[\Phi(\mathcal{M}_1)] = [\Phi(\mathcal{M}_2)]$. Hence $[[\mathcal{M}]] = \Phi^{-1}([\Phi(\mathcal{M})])$ is a Borel subset of $\mathfrak{A}$. Q.E.D.

Theorem 6.6. $\mathfrak{A}_{\mathrm{I}}$ *and* $\mathfrak{A}_f$ *are Borel sets in* $\mathfrak{A}$; $\mathfrak{A}_{s,f}$ *and* $\mathfrak{A}_{\mathrm{III}}^c$ *are Souslin sets in* $\mathfrak{A}$. *Hence* $\mathfrak{F}_{\mathrm{I}}$ *and* $\mathfrak{F}_{\mathrm{II}_1}$ *are standard Borel spaces, and* $\mathfrak{F}_{\mathrm{II}_\infty}$ *is a Souslin space.*

PROOF. Let $\{x_n(\cdot)\}$ be a sequence of Borel choice functions such that $\{x_n(\mathscr{M})\}$ is σ-strongly* dense in $\mathscr{M}$. Let $\mathfrak{A}_c$ denote the collection of all abelian von Neumann algebras on $\mathfrak{H}$. Then

$$\mathfrak{A}_c = \{\mathscr{A} \in \mathfrak{A} : x_n(\mathscr{A})x_m(\mathscr{A}) = x_m(\mathscr{A})x_n(\mathscr{A}),\ m,n = 1,2,\ldots\}$$

is a Borel subset of $\mathfrak{A}$. By Theorem III.1.22, $\{[[\mathscr{A}]] : \mathscr{A} \in \mathfrak{A}_c\}$ is countable. Let Φ be as above. Then $\bigcup\{[\Phi(\mathscr{A}')] : \mathscr{A} \in \mathfrak{A}_c\}$ exhaust all von Neumann algebras of type I on $\mathfrak{H}_0$ with properly infinite commutant, and is a countable union of unitary equivalence classes, so that it is Borel subset of $\tilde{\mathfrak{A}}(\mathfrak{H}_0)$ by Theorem 6.5. Hence

$$\mathfrak{A}_{\mathrm{I}} = \Phi^{-1}\left(\bigcup_{\mathscr{A} \in \mathfrak{A}_c} [\Phi(\mathscr{A}')]\right)$$

is a Borel set.

Next, we set

$$\begin{aligned}\mathscr{B}_1 = \{(\mathscr{M},\varphi) \in \mathfrak{A} \times \mathscr{L}(\mathfrak{H})_*^+ &: \varphi(x_m(\mathscr{M})x_n(\mathscr{M})) = \varphi(x_n(\mathscr{M})x_m(\mathscr{M})),\\ &n,m = 1,2,\ldots,\ s(\varphi_{\mathscr{M}}) = 1\},\\ \mathscr{B}_2 = \{(\mathscr{M},u) \in \mathfrak{A} \times \mathscr{L}(\mathfrak{H}) &: x_n(\mathscr{M}')u = ux_n(\mathscr{M}'),\ n = 1,2,\ldots,\ u^*u = 1,\\ &uu^* \neq 1\}.\end{aligned}$$

It then follows that $\mathscr{B}_1$ and $\mathscr{B}_2$ are Borel sets in the respective Borel spaces, We have

$$\mathfrak{A}_f = \mathrm{pr}(\mathscr{B}_1) \quad \text{and} \quad \mathfrak{A}_f^c = \mathrm{pr}(\mathscr{B}_2),$$

where $\mathrm{pr}_{\mathfrak{A}}$ denotes the projection to the $\mathfrak{A}$-component. By the Borel separation theorem, Theorem A.3, $\mathfrak{A}_f$ is a Borel set in $\mathfrak{A}$. Let $\mathscr{B}_3$ (resp. $\mathscr{B}_4$) be the set of all those $(\mathscr{M},p,\varphi) \in \mathfrak{A} \times \mathscr{P} \times \mathscr{L}(\mathfrak{H})_*^+$ such that

$$\begin{gathered} px_n(\mathscr{M}') = x_n(\mathscr{M}')p, \qquad n = 1,2,\ldots,\\ \varphi(px_m(\mathscr{M})px_n(\mathscr{M})p) = \varphi(px_n(\mathscr{M})px_m(\mathscr{M})p), \qquad n,m = 1,2,\ldots,\\ e_{\mathscr{M}\cap\mathscr{M}'}(p) = 1, \qquad s(\varphi_{\mathscr{M}}) = p \qquad (\text{resp. } \varphi(p) > 0). \end{gathered}$$

Then we have

$$\mathfrak{A}_{s,f} = \mathrm{pr}_{\mathfrak{A}}(\mathscr{B}_3) \quad \text{and} \quad \mathfrak{A}_{\mathrm{III}}^c = \mathrm{pr}(\mathscr{B}_4).$$

Since $\mathscr{B}_3$ and $\mathscr{B}_4$ are both Borel sets in $\mathfrak{A} \times \mathscr{P} \times \mathscr{L}(\mathfrak{H})_*^+$, $\mathfrak{A}_{s,f}$ and $\mathfrak{A}_{\mathrm{III}}^{\mathrm{C}}$ are Souslin sets in $\tilde{\mathfrak{A}}$. Q.E.D.

Corollary 6.7. *Let*

$$\{\mathscr{M},\mathfrak{H}\} = \int_\Gamma^{\oplus} \{\mathscr{M}(\gamma),\mathfrak{H}(\gamma)\}\, d\mu(\gamma)$$

be a direct integral of factors on a standard σ-finite measure space $\{\Gamma,\mu\}$.

(i) *$\mathscr{M}$ is of type i if and only if almost every $\mathscr{M}(\gamma)$ is of type i, i = I, II, III;*
(ii) *$\mathscr{M}$ is finite (resp. semifinite, properly infinite) if and only if almost every $\mathscr{M}(\gamma)$ is also.*

PROOF. As usual, we may assume that $\mathfrak{H}(\gamma) = \mathfrak{H}$ for some fixed separable Hilbert space $\mathfrak{H}_0$. Replacing μ by a finite equivalent measure, we assume that μ is finite and consider the measure μ on $\mathfrak{A}$ which is the image of μ under the map: $\gamma \mapsto \mathscr{M}(\gamma) \in \mathfrak{A}$. Then each $\mathfrak{F}_{\mathrm{I}}$, $\mathfrak{F}_{\mathrm{II}}$, $\mathfrak{F}_{\mathrm{III}}$, $\mathfrak{F}_f$, and so on are $\tilde{\mu}$-measurable, so that their inverse images Γ_{I}, Γ_{II}, Γ_{III}, Γ_f, and so on, respectively, are μ-measurable. Hence

$$\mathscr{M}_i = \int_{\Gamma_i}^{\oplus} \mathscr{M}(\gamma)\, d\mu(\gamma), \qquad i = \mathrm{I},\mathrm{II}, \ldots, f, s.f, \ldots$$

is a direct summand of $\mathscr{M}$. Thus we have only to prove that if almost every $\mathscr{M}(\gamma)$ is finite, semifinite, ..., or of type III, then $\mathscr{M}$ is also finite, semifinite, ..., or type III, respectively.

Suppose that $\mathscr{M}(\gamma)$ is properly infinite for almost every $\gamma \in \Gamma$. Then $\mathscr{M}(\gamma) \cong \mathscr{M}(\gamma) \mathbin{\overline{\otimes}} \mathscr{L}(\mathfrak{K})$ for an infinite dimensional separable Hilbert space $\mathfrak{K}$. Hence $\mathscr{M} \cong \mathscr{M} \otimes \mathscr{L}(\mathfrak{K})$ by Corollary IV.8.29, so that $\mathscr{M}$ is properly infinite.

Suppose that $\mathscr{M}(\gamma)$ is finite almost everywhere. Then $\tilde{\mu}$ is concentrated on $\mathfrak{F}_f$. In the proof of the previous theorem, there exists a $\tilde{\mu}$-measurable function: $\mathscr{M} \in \mathfrak{F}_f \mapsto \varphi_{\mathscr{N}} \in \mathscr{L}(\mathfrak{H}_0)_*^+$ such that $(\mathscr{N}, \varphi_{\mathscr{N}}) \in \mathscr{B}_1$. Let $\varphi = \int_\Gamma^{\oplus} \varphi_{\mathscr{M}(\gamma)}\, d\mu(\gamma)$. Then φ is a faithful finite normal trace on $\mathscr{M}$, so that $\mathscr{M}$ must be finite.

If $\mathscr{M}(\gamma)$ is semifinite almost everywhere, then we consider a $\tilde{\mu}$-measurable function: $\mathscr{N} \in \mathfrak{F}_{s.f} \mapsto (p_{\mathscr{N}}, \varphi_{\mathscr{N}}) \in \mathscr{P} \times \mathscr{L}(\mathfrak{H}_0)_*^+$ such that $(\mathscr{N}, p_{\mathscr{N}}, \varphi_{\mathscr{N}}) \in \mathscr{B}_3$ for every $\mathscr{N} \in \mathscr{F}_{s.f}$. Putting

$$p = \int_\Gamma^{\oplus} p_{\mathscr{M}(\gamma)}\, d\mu(\gamma) \quad \text{and} \quad \varphi = \int_\Gamma^{\oplus} \varphi_{\mathscr{M}(\gamma)}\, d\mu(\gamma),$$

we get a finite projection $p \in \mathscr{M}$ with central support 1. Hence $\mathscr{M}$ is semifinite.

For the case of type I, we consider the set $\mathscr{B} \subset \mathfrak{F} \times \mathscr{P}$ consisting all those $(\mathscr{F}, p) \in \mathfrak{F} \times \mathscr{P}$ such that $x_k(\mathscr{N}')p = px_k(\mathscr{N}')$, $k = 1,2, \ldots, px_m(\mathscr{N})px_n(\mathscr{N})p = px_n(\mathscr{N})px_m(\mathscr{N})p$, $n,m = 1,2, \ldots, p \neq 0$. Then $\mathscr{B}$ is a Borel set in $\mathfrak{F} \times \mathscr{P}$. We then apply the measurable cross-section theorem to $\mathscr{B}$ and $\mathfrak{F}_{\mathrm{I}}$ when $\mathscr{M}(\gamma)$ is of type I almost everywhere, in order to obtain an abelian projection with central support 1, and conclude that $\mathscr{M}$ is of type I. Q.E.D.

7. Construction of Factors of Type II and Type III

As we have seen already, the factor of all bounded operators on a Hilbert space is indeed a factor of type I, and every factor of type I is of this form. Therefore, we do not have to puzzle at all over the existence of a factor of type I. However, the existence of factors of type II and III is not at all *a priori* trivial. We thus show a construction of factors of type II and type III in this section.

Let $\mathscr{A}$ be an abelian von Neumann algebra. We denote by Aut($\mathscr{A}$) the group of all automorphisms of $\mathscr{A}$.

Definition 7.1. Given an $\alpha \in \text{Aut}(\mathscr{A})$, a projection $e \in \mathscr{A}$ is said to be *absolutely invariant* (or *fixed*) under α if $\alpha(e) = e$ and the restriction of α to $\mathscr{A}_e$ is the identity automorphism. The automorphism α is called *free* if α admits no nonzero absolutely invariant projection.

It follows that any $\alpha \in \text{Aut}(\mathscr{A})$ admits a projection e such that $\alpha(e) = e$, the restriction of α to $\mathscr{A}_e$ is free, and the restriction of α to $\mathscr{A}_{(1-e)}$ is trivial, because the supremum of any number of absolutely invariant projections is again absolutely invariant; hence there exists a greatest absolutely invariant projection, say $(1 - e)$. Let Ω be the spectrum of $\mathscr{A}$. Each $\alpha \in \text{Aut}(\mathscr{A})$ gives rise to a homeomorphism T_α on Ω such that

$$\alpha(a)(\omega) = a(T_\alpha^{-1}\omega), \qquad a \in \mathscr{A}, \qquad \omega \in \Omega.$$

In this setting, the absolute invariance of a projection e is equivalent to saying that the restriction of T_α on the open and closed set E corresponding to e is the trivial homeomorphism. Also, α is free if and only if $N_\alpha = \{\omega \in \Omega : T_\alpha\omega = \omega\}$ is rare.

Lemma 7.2. *An automorphism $\alpha \in \text{Aut}(\mathscr{A})$ is free if and only if the condition*

$$xa = a\alpha(x), \qquad x \in \mathscr{A},$$

implies $a = 0$.

Proof. If e is an absolutely invariant projection of $\mathscr{A}$, then we have $xe = \alpha(xe) = \alpha(x)e = e\alpha(x)$ for every $x \in \mathscr{A}$. Hence if α is not free, then there exists a nonzero a satisfying the condition. Suppose that there exists a nonzero $a \in \mathscr{A}$ such that $xa = \alpha(x)a$ for every $x \in \mathscr{A}$. We then have $x(\omega)a(\omega) = x(T_\alpha^{-1}\omega)a(\omega)$, $\omega \in \Omega$ and $x \in \mathscr{A}$. Hence $x(\omega) = x(T_\alpha^{-1}\omega)$ if $a(\omega) \neq 0$. Let $E = \{\omega : a(\omega) \neq 0\}$. It follows then that $\omega = T_\alpha^{-1}\omega$ for every $\omega \in E$, because $\mathscr{A}$ separates the points of E. Since E is open, E cannot be rare but $E \subset N_\alpha$. Thus α is not free. Q.E.D.

Let G be a countably infinite discrete group. A homomorphism $\alpha : g \in G \mapsto \alpha_g \in \text{Aut}(\mathscr{A})$ is called an *action* of G on $\mathscr{A}$, and the triplet $(\mathscr{A}, G, \alpha)$ is called a *covariant system.*

Definition 7.3. An action α of G on $\mathscr{A}$ is called *free* if for every $g \in G$, $g \neq e$, α_g is free; *ergodic* if there is no invariant projection in $\mathscr{A}$ under $\alpha(G)$ other than 0 and 1.

The ergodicity of α is equivalent to the fact that the fixed point algebra $\mathscr{A}^\alpha = \{x \in \mathscr{A} : \alpha_g(x) = x \text{ for every } g \in G\}$ is reduced to the scalar $\mathbf{C}$. We note that if G is abelian, then the ergodicity of α implies automatically the freeness of α.

Suppose that $\mathscr{A}$ acts on a Hilbert space $\mathfrak{H}$, and an action α of G on $\mathscr{A}$ is given. Let $\mathfrak{K} = \mathfrak{H} \otimes l^2(G)$. We regard $\mathfrak{K}$ as the Hilbert space of all square summable $\mathfrak{H}$-valued functions on G. We consider the representations π of $\mathscr{A}$ and u of G on $\mathfrak{K}$ given by the following:

$$\begin{aligned} \pi(a)\xi(g) &= \alpha_g^{-1}(a)\xi(g), \qquad \xi \in \mathfrak{K}, \qquad a \in \mathscr{A}, \qquad g \in G; \\ u(g)\xi(h) &= \xi(g^{-1}h), \qquad h \in G. \end{aligned} \tag{1}$$

It follows that

$$u(g)\pi(a)u(g)^* = \pi \circ \alpha_g(a), \qquad a \in \mathscr{A}, \qquad g \in G. \tag{2}$$

In general, a pair (π,u) of representations of $\mathscr{A}$ and G on the same Hilbert space is called *covariant* if (2) holds.

Definition 7.4. The von Neumann algebra generated by $\pi(\mathscr{A})$ and $\{u(g) : g \in G\}$ is called the *crossed product* of $\mathscr{A}$ by G with respect to α or the *covariance von Neumann algebra* of $(\mathscr{A},G,\alpha)$, and we denote it by $\mathscr{R}(\mathscr{A},G,\alpha)$, or, simply, $\mathscr{R}(\mathscr{A},G)$ or $\mathscr{R}(\mathscr{A},\alpha)$. When $\mathscr{A} = \mathbf{C}$, we denote it by $\mathscr{R}(G)$, and call it the *(left) group von Neumann algebra* of G.

By the following lemma, the definition of $\mathscr{R}(\mathscr{A},G,\alpha)$ does not depend on the choice of the underlying Hilbert space $\mathfrak{H}$.

Lemma 7.5. *Let $\{\mathscr{A}_1,\mathfrak{H}_1\}$ and $\{\mathscr{A}_2,\mathfrak{H}_2\}$ be abelian von Neumann algebras equipped with actions α^1 and α^2 of G, respectively. If σ is an isomorphism of $\mathscr{A}_1$ onto $\mathscr{A}_2$ such that $\sigma \circ \alpha_g^1 = \alpha_g^2 \circ \sigma$ for every $g \in G$, then there exists an isomorphism $\bar{\sigma}$ of $\mathscr{R}(\mathscr{A}_1,G,\alpha^1)$ onto $\mathscr{R}(\mathscr{A}_2,G,\alpha^2)$ such that $\bar{\sigma} \circ \pi_1(x) = \pi_2 \circ \sigma(x)$ and $\bar{\sigma}u_1(g) = u_2(g)$, $g \in G$, where π_i and u_i, $i = 1,2$, are the representations of $\mathscr{A}_i$ and G on $\mathfrak{H}_i \otimes l^2(G)$ defined by* (1) *based on α^i.*

PROOF. Since our assertion is trivial if σ is spatial, we may assume by Theorem IV.5.5 that there exists an abelian von Neumann algebra $\{\mathscr{A},\mathfrak{H}\}$ and projections e_1' and e_2' in $\mathscr{A}'$ with central support 1 such that $\{\mathscr{A}_1,\mathfrak{H}_1\} = \{\mathscr{A}_{e_1'},e_1'\mathfrak{H}\}$, $\{\mathscr{A}_2,\mathfrak{H}_2\} = \{\mathscr{A}_{e_2'},e_2'\mathfrak{H}\}$ and $\sigma(x_{e_1'}) = x_{e_2'}$, $x \in \mathscr{A}$. Since $z(e_1') = 1$, there exists an action α of G on $\mathscr{A}$ such that $\alpha_g(x)_{e_1'} = \alpha_g^1(x_{e_1'})$, $g \in G$, $x \in \mathscr{A}$. Since σ intertwines α^1 and α^2, we have $\alpha_g(x)_{e_2'} = \alpha_g^2(x_{e_2'})$, $x \in \mathscr{A}$, $g \in G$. Thus, the situation is reduced to the following: Given an abelian von Neumann algebra $\{\mathscr{A},\mathfrak{H}\}$ equipped with an action α of G and a projection $e' \in \mathscr{A}'$ with $z(e') = 1$, we must prove the existence of an isomorphism σ of $\mathscr{R}(\mathscr{A},G,\alpha)$ to $\mathscr{R}(\mathscr{A}_{e'},G,\alpha^{e'})$ with the appropriate property, where $\alpha_g^{e'}(x_{e'}) = \alpha_g(x)_{e'}$, $x \in \mathscr{A}$, $g \in G$. Let $\mathfrak{K} = \mathfrak{H} \otimes l^2(G)$ and $\bar{e}' = e' \otimes 1$. It follows then that $\mathscr{R}(\mathscr{A},G,\alpha)$ acts on $\mathfrak{K}$ and $\mathscr{R}(\mathscr{A}_{e'},G,\alpha^{e'})$ acts on $\bar{e}'\mathfrak{K} = e'\mathfrak{H} \otimes l^2(G)$. Trivially, $\bar{e}'$ belongs to $\mathscr{R}(\mathscr{A},G,\alpha)'$ and

$$\mathscr{R}(\mathscr{A}_{e'},G,\alpha^{e'}) = \mathscr{R}(\mathscr{A},G,\alpha)_{\bar{e}'}.$$

Hence we define σ by $\sigma(x) = x_{\bar{e}'}$, $x \in \mathscr{R}(\mathscr{A},G,\alpha)$. It is clear that σ enjoys the required properties, except possibly the faithfulness. But $\mathscr{A}' \otimes \mathbf{C}$ is contained in $\mathscr{R}(\mathscr{A},G,\alpha)'$ and $[(\mathscr{A}'\otimes \mathbf{C})(\bar{e}'\mathfrak{K})] = [\mathscr{A}'e'\mathfrak{H}]\otimes l^2(G) = \mathfrak{H}\otimes l^2(G)$. Therefore, the central support of $\bar{e}'$ in $\mathscr{R}(\mathscr{A},G,\alpha)'$ must be 1. Hence σ is an isomorphism. Q.E.D.

We now continue the study of $\mathscr{R}(\mathscr{A},G,\alpha)$. For short, we denote $\mathscr{R}(\mathscr{A},G,\alpha)$ by $\mathscr{R}$. Let $\mathscr{R}_0$ denote the set of all finite linear combinations of $\pi(\mathscr{A})$ and $u(G)$. Equation (2) yields that $\mathscr{R}_0$ is a *-algebra, hence σ-weakly dense in $\mathscr{R}$. Each element x of $\mathscr{R}_0$ is of the form

$$x = \sum_{g\in G} \pi(x(g))u(g), \tag{3}$$

with some $\mathscr{A}$-valued function $x(\cdot)$ on G of finite support. With the same notations as before Definition 7.4, we define for each $g \in G$ an operator P_g of $\mathfrak{K}$ onto $\mathfrak{H}$ as follows:

$$P_g\xi = \xi(g^{-1}); \qquad \xi \in \mathfrak{K} = \mathfrak{H}\otimes l^2(G). \tag{4}$$

It follows easily that

$$P_g u(h) = P_{gh}, \qquad g,h\in G, \tag{5}$$

$$\left.\begin{aligned} &P_g\pi(a) = \alpha_g(a)P_g; \qquad P_g\pi(a)P_g^* = \alpha_g(a),\\ &\sum_{g\in G} P_g^*\alpha_g(a)P_g = \pi(a), \qquad a\in\mathscr{A}. \end{aligned}\right\} \tag{6}$$

Furthermore, $\{P_g^*P_g\}$ is an orthogonal family of projections with sum 1 and $P_gP_g^* = 1$, $g \in G$. Hence for each $x \in \mathscr{R}_0$ of the form (3) we get $P_exP_e^* = x(e) \in \mathscr{A}$. Since $\mathscr{R}_0$ is σ-weakly dense in $\mathscr{R}$, we have

$$E(x) = P_exP_e^* \in \mathscr{A}, \qquad x\in\mathscr{R}. \tag{7}$$

If $x \in \mathscr{R}_0$ is of the form (3), then we get

$$\begin{aligned} u(g)xu(g)^* &= \sum_{h\in G} u(g)\pi(x(h))u(h)u(g)^*\\ &= \sum_{h\in G} \pi\circ\alpha_g(x(h))u(ghg^{-1}), \end{aligned}$$

so that $E(u(g)xu(g)^*) = \alpha_g(x(e)) = \alpha_g(E(x))$. By the density of $\mathscr{R}_0$ in $\mathscr{R}$, we get

$$E(u(g)xu(g)^*) = \alpha_g(E(x)), \qquad x\in\mathscr{A}, \qquad g\in G. \tag{8}$$

We now put, for each $x \in \mathscr{R}$,

$$x(g) = E(xu(g)^*) \in \mathscr{A}, \qquad g \in G. \tag{9}$$

We have then

$$\begin{aligned} P_gxP_h^* &= P_eu(g)xu(h)^*P_e^* \qquad \text{by (5)},\\ &= P_eu(g)xu(g^{-1}h)^*u(g)^*P_e\\ &= E(u(g)xu(g^{-1}h)^*u(g)^*)\\ &= \alpha_g(E(xu(g^{-1}h)^*)) = \alpha_g(x(g^{-1}h)). \end{aligned}$$

Thus, we get, with respect to the strong* convergence,

$$\begin{aligned}
x &= \left(\sum_{g\in G} P_g^* P_g\right) x \left(\sum_{h\in G} P_h^* P_h\right)\\
&= \sum_{g,h\in G} P_g^* P_g x P_h^* P_h = \sum_{g,h\in G} P_g^* \alpha_g(x(g^{-1}h))P_h\\
&= \sum_{g,h\in G} P_g^* \alpha_g(x(h))P_{gh} = \sum_{g,h\in G} P_g^* \alpha_g(x(h))P_g u(h) \qquad \text{by (5),}\\
&= \sum_{h\in G}\left(\sum_{g\in G} P_g^* \alpha(x(h))P_g\right) u(h)\\
&= \sum_{h\in G} \pi(x(h))u(h) \qquad \text{by (6).}
\end{aligned}$$

Therefore, every $x \in \mathscr{R}$ is of the form (3) with the $\mathscr{A}$-valued function given by (9), and the summation (3) converges in the strong* topology. Since this convergence does not depend on the Hilbert space $\mathfrak{H}$, it converges in the σ-strong* topology. The next formulas follow by straightforward computation:

$$\left.\begin{aligned}
(xy)(g) &= \sum_{h\in G} x(h)\alpha_h(y(h^{-1}g)), \qquad x,y\in\mathscr{R};\\
x^*(g) &= \alpha_g(x(g^{-1})^*).
\end{aligned}\right\} \tag{10}$$

We now summarize our discussion as follows:

Proposition 7.6. *Every element x of $\mathscr{R}(\mathscr{A},G,\alpha)$ is given uniquely by formulas* (9) *and* (3). *The arithmetic in $\mathscr{R}(\mathscr{A},G,\alpha)$ is governed by* (10).

PROOF. The only thing we have to prove is the uniqueness of expression (3) for each $x \in \mathscr{R} = \mathscr{R}(\mathscr{A},G,\alpha)$. Suppose that an $x \in \mathscr{R}$ is of the form

$$x = \sum_{g\in G} \pi(x'(g))u(g)$$

with respect to the σ-strong* convergence for some $\mathscr{A}$-valued function $x'(\cdot)$ on G. For each $g \in G$, we have

$$\begin{aligned}
x(g) = E(xu(g)^*) &= P_e \sum_{h\in G} \pi(x'(h))u(h)u(g)^* P_e^*\\
&= \sum_{h\in G} P_e \pi(x'(h))u(hg^{-1})P_e^*\\
&= \sum_{h\in G} x'(h)P_e P_{hg^{-1}}^* \qquad \text{by (5) and (6)}\\
&= x'(g).
\end{aligned}$$

Thus, the components of expression (3) must be given by formula (9). Q.E.D.

Corollary 7.7. *The canonical image $\pi(\mathscr{A})$ of $\mathscr{A}$ is maximal abelian in $\mathscr{R}(\mathscr{A},G,\alpha)$ if and only if the action α of G on $\mathscr{A}$ is free.*

PROOF. Suppose that $x = \sum_{g \in G} \pi(x(g))u(g) \in \mathscr{R}$ commutes with $\pi(\mathscr{A})$. For any $a \in \mathscr{A}$, we have

$$\pi(a)x = \sum_{g \in G} \pi(ax(g))u(g),$$
$$x\pi(a) = \sum_{g \in G} \pi(x(g)\alpha_g(a))u(g),$$

hence

$$ax(g) = x(g)\alpha_g(a), \qquad g \in G.$$

By Lemma 7.2, if α is free, $x(g) = 0$ for every $g \neq e$. Hence $x = \pi(x(e)) \in \pi(\mathscr{A})$; thus $\pi(\mathscr{A})$ is maximal abelian in $\mathscr{R}$.

Suppose that α_g is not free for some $g \neq e, g \in G$. Let $p \in \mathscr{A}$ be an absolutely invariant projection under α_g. We then have $ap = \alpha_g(a)p$ for every $a \in \mathscr{A}$. Let $x = \pi(p)u(g)$. It follows then that x commutes with $\pi(\mathscr{A})$, while $x \notin \pi(\mathscr{A})$. Q.E.D.

Corollary 7.8. *If the action α of G is free, then the following two statements are equivalent:*

(i) *The action α is ergodic.*
(ii) *$\mathscr{R}(\mathscr{A},G,\alpha)$ is a factor.*

PROOF. Let $\mathscr{R} = \mathscr{R}(\mathscr{A},G,\alpha)$. By assumption, $\pi(\mathscr{A})$ is maximal abelian in $\mathscr{R}$. Hence the center $\mathscr{Z}$ of $\mathscr{R}$ is contained in $\pi(\mathscr{A})$. Our assertion then follows from the fact that $\pi(a)$, $a \in \mathscr{A}$, commutes with $u(g)$, $g \in G$, if and only if $\alpha_g(a) = a$. Q.E.D.

For each $x \in \mathscr{R}(\mathscr{A},G,\alpha) = \mathscr{R}$, we have by (10)

$$x^*x(e) = \sum_{g \in G} x^*(g)\alpha_g(x(g^{-1})) = \sum_{g \in G} \alpha_g(x(g^{-1})^*x(g^{-1})),$$
$$xx^*(e) = \sum_{g \in G} x(g)\alpha_g(x^*(g^{-1})) = \sum_{g \in G} x(g)x(g)^*.$$

Thus we get, under the σ-strong* convergence,

$$\begin{aligned} E(x^*x) &= \sum_{g \in G} \alpha_g(x(g^{-1})^*x(g^{-1})), \\ E(xx^*) &= \sum_{g \in G} x(g)x(g)^*, \qquad x \in \mathscr{R}. \end{aligned} \tag{11}$$

From this formula, we conclude the following:

Proposition 7.9. *The group von Neumann algebra $\mathscr{R}(G)$ of a countably infinite discrete group G is finite. It is a factor if and only if every conjugacy class*

$$C(g) = \{hgh^{-1} : h \in G\}, \qquad g \neq e,$$

is infinite except for $g = e$. In this case, $\mathscr{R}(G)$ is a factor of type II_1.

PROOF. In this case, we have $\mathscr{A} = \mathbf{C}$, and $\alpha_g =$ the identity automorphism of $\mathscr{A}$. It follows therefore that for each $x \in \mathscr{R}(G)$, the $\mathscr{A}$-valued function $x(\cdot)$ is indeed a numerical valued function on G, and the linear map E of $\mathscr{R}(G)$ onto $\mathscr{A} = \mathbf{C}$ is then a faithful finite normal trace on $\mathscr{R}(G)$ by (11). Thus $\mathscr{R}(G)$ is finite. We note also that formula (11) means that $x(\cdot)$ for each $x \in \mathscr{R}(G)$ is square summable.

Let $x = \sum_{g \in G} x(g)u(g) \in \mathscr{R}(G)$. Since $\mathscr{R}(G)$ is generated by $u(G)$, x is in the center $\mathscr{Z}$ of $\mathscr{R}(G)$ if and only if x commutes with $u(h)$, $h \in G$. We have, however,

$$u(h)xu(h)^* = \sum_{g \in G} x(g)u(hgh^{-1}) = \sum_{g \in G} x(h^{-1}gh)u(g),$$

so that x is in $\mathscr{Z}$ if and only if $x(\cdot)$ is constant on every conjugacy class $C(g)$. By the square summability of $x(\cdot)$, $x(\cdot)$ must vanish on each infinite conjugacy class $C(g)$. Hence if G has infinite conjugacy class except the trivial one, then $\mathscr{R}(G)$ must be a factor. Conversely, if $C(g)$ is finite for some $g \neq e$, then $x = \sum_{h \in C(g)} u(h)$ is a nonscalar central element of $\mathscr{R}(G)$.

If $\mathscr{R}(G)$ is a factor, then $\mathscr{R}(G)$ must be of type II_1 because a finite factor of type I is finite dimensional and $\mathscr{R}(G)$ is clearly infinite dimensional. Q.E.D.

Definition 7.10. An infinite countable discrete group G is said to be of *infinite conjugacy class* if it has only infinite conjugacy classes except the trivial one $C(e) = \{e\}$. We often abbreviate such a group as an ICC-group.

There are abundance of ICC-groups. For example, the following groups are of infinite conjugacy class:

- the group of all finite permutations of an infinite countable set;
- the free group of two or more generators;
- the cartesian product of finite number of ICC-groups;
- the restricted cartesian product of infinitely many ICC-groups.

Before going to continue the study of $\mathscr{R}(\mathscr{A},G,\alpha)$, we prepare some technical results:

Lemma 7.11. *Let $\mathscr{M}$ be a von Neumann algebra equipped with a faithful semifinite normal trace τ, and $\mathscr{A}$ be a maximal abelian *-subalgebra of $\mathscr{M}$. If there exists a normal projection E of norm one from $\mathscr{M}$ onto $\mathscr{A}$, then the restriction of τ to $\mathscr{A}$ is semifinite and $\tau = \tau \circ E$.*

PROOF. For each $x \in \mathscr{M}$, let $K(x)$ denote the convex hull of $\{uxu^* : u \in \mathscr{U}_{\mathscr{A}}\}$, where $\mathscr{U}_{\mathscr{A}}$ means, of course, the unitary group of $\mathscr{A}$, and let $\tilde{K}(x)$ be the σ-weak closure of $K(x)$. It follows that $\tilde{K}(x)$ is a σ-weakly compact convex subset of $\mathscr{M}$ on which $\mathscr{U}_{\mathscr{A}}$ acts as an affine transformation group by the map: $y \in \tilde{K}(x) \mapsto uyu^* \in \tilde{K}(x)$. Since $\mathscr{U}_{\mathscr{A}}$ is abelian, $\tilde{K}(x)$ admits a fixed point x_0 under this transformation group. This means, however, that x_0 belongs to

$\mathscr{A}$ by the maximal abelianess of $\mathscr{A}$. Hence $\tilde{K}(x) \cap \mathscr{A} \neq \emptyset$. On the other hand, for any $u \in \mathscr{U}_{\mathscr{A}}$ we have

$$E(x) = uE(x)u^* = E(uxu^*),$$

so that E is constant on $K(x)$. By normality, E is also constant on $\tilde{K}(x)$. Thus, we have $E(\tilde{K}(x)) = \{E(x)\}$ and

$$\tilde{K}(x) \cap \mathscr{A} = E(K(x) \cap \mathscr{A}) \subset E(\tilde{K}(x)) = \{E(x)\},$$

so that $\tilde{K}(x) \cap \mathscr{A}$ is a singleton set $\{E(x)\}$.

Let $\mathfrak{m}_\tau$ be the definition ideal of τ. If $\{e_i\}$ is an increasing net in $\mathfrak{m}_\tau$ with $\sup e_i = 1$, then we have for any $x \in \mathscr{M}_+$,

$$\tau(x) = \sup_i \tau(xe_i),$$

so that τ is σ-strongly lower semicontinuous on $\mathscr{M}_+$ since each linear function: $x \in \mathscr{M} \mapsto \tau(xe_i)$ is normal. If $x \in \mathfrak{m}_\tau^+$ then we have $\tau(uxu^*) = \tau(x)$ for every $u \in \mathscr{U}_{\mathscr{A}}$, so that τ is constant on $K(x)$. By the lower semicontinuity of τ, we have $\tau(y) \leq \tau(x)$ for every $y \in \tilde{K}(x)$. In particular, $\tau(E(x)) \leq \tau(x)$. Hence $E(x) \in \mathscr{A} \cap \mathfrak{m}_\tau^+$. Since $\mathfrak{m}_\tau$ is spanned linearly by $\mathfrak{m}_\tau^+$, we have $E(\mathfrak{m}_\tau) = \mathscr{A} \cap \mathfrak{m}_\tau$. By normality again, we have $\sup E(e_i) = 1$, which means that $E(\mathfrak{m}_\tau)$ is σ-weakly dense in $\mathscr{A}$. Thus the restriction of τ onto $\mathscr{A}$ is semifinite. Therefore, there exists, by Proposition 2.36, a conditional expectation E_τ of $\mathscr{M}$ onto $\mathscr{A}$ with respect to the trace τ. But the above arguments show that $E_\tau(\tilde{K}(x)) = \tilde{K}(x) \subset \mathscr{A} = \{E(x)\}$. Hence $E = E_\tau$ and $\tau = \tau \circ E$. Q.E.D.

Theorem 7.12. *Let $\mathscr{A}$ be an abelian von Neumann algebra equipped with a free ergodic action of an infinite countable discrete group G, and let $\mathscr{R} = \mathscr{R}(\mathscr{A},G,\alpha)$. We arrive at the following conclusions:*

(i) *$\mathscr{R}$ is of type I if and only if $\mathscr{A}$ contains a minimal projection p such that $\sum_{g\in G} \alpha_g(p) = 1$.*
(ii) *$\mathscr{R}$ is of type II_1 if and only if $\mathscr{A}$ admits a faithful finite normal trace invariant under the action α.*
(iii) *$\mathscr{R}$ is of type II_∞ if and only if $\mathscr{A}$ is not atomic and admits a faithful semifinite, but infinite, normal trace invariant under α.*
(iv) *$\mathscr{R}$ is of type III if and only if $\mathscr{A}$ does not admit a faithful semifinite normal trace invariant under α.*

Before going into the proof, we note here that a semifinite normal trace invariant under α is necessarily faithful and unique up to multiplication by a scalar. In fact, if φ is such a trace on $\mathscr{A}$, then the support $s(\varphi)$ is invariant under α, so that $s(\varphi) = 1$ if $\varphi \neq 0$. If φ and ψ are invariant semifinite normal traces on $\mathscr{A}$, then $\varphi + \psi$ are also an invariant semi-finite normal trace on $\mathscr{A}$. There exists h, $0 \leq h \leq 1$, such that $\varphi(x) = (\varphi + \psi)(xh)$, $x \in \mathscr{A}_+$. Since φ and ψ are both invariant, $\alpha_g(h) = h$ for every $g \in G$. Hence $h = \lambda 1$ for some scalar, which means that φ and ψ are proportional.

PROOF. By the assumption on the action α, $\mathscr{R}$ is a factor of infinite dimension. So the case of finite type I is excluded.

Suppose that $\mathscr{A}$ admits a faithful semifinite normal trace φ invariant under α. Put

$$\tau(x) = \varphi \circ E(x), \qquad x \in \mathscr{R}_+.$$

Since E is normal, τ also is normal. By (11) we have

$$\begin{aligned}\tau(x^*x) &= \varphi\left(\sum_{g \in G} \alpha_g(x(g^{-1})^*x(g^{-1}))\right)\\ &= \sum_{g \in G} \varphi \circ \alpha_g(x(g^{-1})^*x(g^{-1})) = \sum_{g \in G} \varphi(x(g^{-1})^*x(g^{-1}))\\ &= \sum_{g \in G} \varphi(x(g)^*x(g)) = \varphi\left(\sum_{g \in G} x(g)x(g)^*\right)\\ &= \tau(xx^*),\end{aligned}$$

so that τ is a normal trace and faithful. Since $\tau \circ \pi(a) = \varphi(a)$ for every $a \in \mathscr{A}$, τ is finite if and only if φ is finite. Furthermore, the semifiniteness of φ entails that of τ because the definition ideal of φ, which is contained in $\mathfrak{m}_\tau$, contains a net converging to the identity.

Suppose that $\mathscr{R}$ admits a (faithful) semifinite normal trace τ. Identifying $\mathscr{A}$ with its canonical image $\pi(\mathscr{A})$, we consider $\mathscr{A}$ as a maximal abelian *-subalgebra of $\mathscr{R}$. It follows then that E is a normal projection of norm one from $\mathscr{R}$ to $\mathscr{A}$. By Lemma 7.11, the restriction of τ to $\mathscr{A}$, say φ, is a faithful semifinite normal trace on $\mathscr{A}$ and $\tau = \varphi \circ E$. For any $a \in \mathscr{A}_+$, we have

$$\begin{aligned}\varphi \circ \alpha_g(a) &= \tau(u(g)au(g)^*) = \tau(a)\\ &= \varphi(a);\end{aligned}$$

thus φ must be invariant under α. Therefore, we have completed the proof of assertions (ii) and (iv).

Since $\mathscr{A}$ is maximal abelian, a minimal projection of $\mathscr{A}$, if there exists any, is automatically minimal in $\mathscr{R}$. Hence if $\mathscr{A}$ is atomic, then so is $\mathscr{R}$; hence $\mathscr{R}$ is of type I. Suppose that $\mathscr{R}$ is of type I. Then, there exists a Hilbert space $\mathfrak{H}_0$ such that $\mathscr{R} \cong \mathscr{L}(\mathfrak{H}_0)$. The canonical trace Tr on $\mathscr{L}(\mathfrak{H}_0)$, identifying $\mathscr{R}$ with $\mathscr{L}(\mathfrak{H}_0)$, is then semifinite on $\mathscr{A}$. Hence $\mathscr{A} \cap \mathscr{L}\mathscr{C}(\mathfrak{H}) \neq \{0\}$, which means that $\mathscr{A}$ contains a nonzero finite dimensional projection; thus $\mathscr{A}$ has a minimal projection p. Since α is free, $\{\alpha_g(p)\}$ is orthogonal; since α is ergodic, $\sum_{g \in G} \alpha_g(p) = 1$. Therefore, assertion (i) holds. Assertion (iii) is now automatic.

Q.E.D.

Let Γ be a separable locally compact infinite group with right invariant Haar measure μ, and let $\mathscr{A} = L^\infty(\Gamma, \mu)$. Clearly, Γ acts on $\mathscr{A}$ by right translation:

$$\alpha_{\gamma_0}(a)(\gamma) = a(\gamma\gamma_0), \qquad \gamma, \gamma_0 \in \Gamma.$$

Furthermore, map: $\gamma \in \Gamma \mapsto \alpha_\gamma(a)$ is σ-weakly continuous and each α_γ is free on $\mathscr{A}$ except the trivial one. Since the action of Γ on Γ is transitive, the fixed point algebra under $\alpha(\Gamma)$ is the scalar multiples of the identity. Suppose G is a dense countable subgroup of Γ, and consider the restriction of α to G, denoted by α again. It follows then that the action α of G is free and ergodic because the fixed point algebra of $\alpha(G)$ coincides with the fixed point algebra of $\alpha(\Gamma)$ due to the density of G in Γ and the continuity of α. Thus, we obtain $(\mathscr{A},G,\alpha)$ satisfying the basic assumption of Theorem 7.12. If we take a compact group Γ, for example the torus of one dimension, then the measure μ gives rise to an invariant finite normal trace on $\mathscr{A}$. If we take a noncompact group Γ, for example $\Gamma = \mathbf{R}$ and $G = \mathbf{Q}$, then the measure μ gives rise to an invariant semifinite, but infinite, normal trace on $\mathscr{A}$. In these both cases, $\mathscr{A}$ is not atomic, so that $\mathscr{R}(\mathscr{A},G,\alpha)$ is of type II_1 and of type II_∞, respectively.

To construct a factor of type III as $\mathscr{R}(\mathscr{A},G,\alpha)$, we note the fact that if G contains a subgroup G_0 such that the restriction of α to G_0 is already ergodic, then the invariant semifinite normal trace under $\alpha(G_0)$ is unique; thus if G contains a single element g_0 which does not preserve this unique trace, then $\mathscr{A}$ has no invariant semifinite normal trace. With this remark, let $\Gamma = \mathbf{R}$ with the Lebesgue measure μ, and g_0 be the transformation of Γ by a number not of absolute value 1; more precisely, G is the set of transformations: $\gamma \mapsto a^n\gamma + b$, $b \in \mathbf{Q}$, $n \in \mathbf{Z}$, with a fixed rational number a, $|a| \neq 1$. Lifting this transformation group G up to $\mathscr{A} = L^\infty(\Gamma,\mu)$, we obtain a system $(\mathscr{A},G,\alpha)$ which has no invariant semifinite normal trace. Thus, we conclude the following existence theorem of factors in all types:

Theorem 7.13. *There exists a factor of each type I, II_1, II_∞, and III acting on a separable Hilbert space.*

We conclude this section by examining the commutant of $\mathscr{R}(\mathscr{A},G,\alpha)$ when $\mathscr{A}$ acts on a Hilbert space $\mathfrak{H}$ on which G has a unitary representation $v(\cdot)$ such that

$$\alpha_g(a) = v(g)av(g)^*, \qquad g \in G, \qquad a \in \mathscr{A}. \tag{12}$$

Such a unitary representation $v(\cdot)$ of G is called a *unitary implementation* of the action α of G.

Suppose that $\mathscr{A}$ acts on a Hilbert space $\mathfrak{H}$ on which the action α of G on $\mathscr{A}$ is unitarily implemented by $v(\cdot)$. Let $\mathfrak{K} = \mathfrak{H} \otimes l^2(G)$ and $\mathscr{R} = \mathscr{R}(\mathscr{A},G,\alpha)$. We then employ the established notations and distinguish $\mathscr{A}$ and $\pi(\mathscr{A})$ to avoid any possible confusion. First of all we define operators on $\mathfrak{K}$ as follows:

$$\left.\begin{aligned} (\pi'(b)\xi)(h) &= b\xi(h), \qquad h \in \mathscr{A}'; \\ (\tilde{v}(g)\xi)(h) &= v(g)\xi(hg), \qquad g,h \in G, \qquad \xi \in \mathfrak{K}. \end{aligned}\right\} \tag{13}$$

Clearly, π' is a faithful normal representation of $\mathscr{A}'$ on $\mathfrak{K}$ and $\tilde{v}(\cdot)$ is a unitary representation of G and

$$\tilde{v}(g)\pi'(b)\tilde{v}(g)^* = \pi'(v(g)bv(g)^*), \qquad g \in G, \qquad b \in \mathscr{A}'. \tag{14}$$

It is easily seen that $\pi'(\mathscr{A}')$ and $\tilde{v}(G)$ commute with $\pi(\mathscr{A})$ and $u(G)$, hence with $\mathscr{R}$. Furthermore, we have

$$\left.\begin{aligned} P_h\tilde{v}(g) = v(g)P_{g^{-1}h}, \qquad P_g\pi'(b) = bP_g, \qquad b \in \mathscr{A}',\\ \sum_{g\in G} P_g^* b P_g = \pi'(b). \end{aligned}\right\} \tag{15}$$

Suppose that x is an arbitrary fixed element of $\mathscr{R}'$. By (6), we have, for each $a \in \mathscr{A}$,

$$\begin{aligned} aP_e x P_e^* &= P_e\pi(a)xP_e^* = P_e x\pi(a)P_e^* \\ &= P_e x P_e^* a, \end{aligned}$$

so that $E'(x) = P_e x P_e^*$ belongs to $\mathscr{A}'$. Put

$$x(g) = E'(x\tilde{v}(g)^*) \in \mathscr{A}'. \tag{16}$$

Observing that

$$\begin{aligned} P_h x P_g^* &= P_e u(h) x P_g^* = P_e x u(h) P_g^* \\ &= P_e x P_{gh^{-1}}^* \qquad \text{by (5)}, \end{aligned}$$

we have, with respect to the strong convergence,

$$\begin{aligned} x &= \left(\sum_{h\in G} P_h^* P_h\right) x \left(\sum_{g\in G} P_g^* P_g\right) \\ &= \sum_{h,g\in G} P_h^* P_h x P_g^* P_g = \sum_{h,g\in G} P_h^* P_e x P_{gh^{-1}}^* P_g \\ &= \sum_{h,g\in G} P_h^* P_e x P_g^* P_{gh} = \sum_{h,g\in G} P_h^* P_e x\tilde{v}(g) P_e^* v(g)^* P_{gh} \\ &= \sum_{h,g\in G} P_h^* x(g^{-1}) P_h \tilde{v}(g^{-1}) = \sum_{g\in G}\left(\sum_{h\in G} P_h^* x(g) P_h\right)\tilde{v}(g) \\ &= \sum_{g\in G} \pi'(x(g))\tilde{v}(g). \end{aligned}$$

Thus, every $x \in \mathscr{R}'$ has the expression

$$x = \sum_{g\in G} \pi'(x(g))\tilde{v}(g) \tag{17}$$

with some $\mathscr{A}'$-valued function $x(\cdot)$ determined by (16). Thus, we conclude the following:

Proposition 7.14. *If the action α of G on $\{\mathscr{A},\mathfrak{H}\}$ is implemented by a unitary representation $v(\cdot)$ of G, then the commutant $\mathscr{R}'$ of $\mathscr{R} = \mathscr{R}(\mathscr{A},G,\alpha)$ is generated by $\pi'(\mathscr{A})$ and $\tilde{v}(G)$, where π' and $\tilde{v}$ are given by* (13). *Furthermore, each element $x \in \mathscr{R}'$ has the unique expression* (17).

The uniqueness of expression (17) is proven in a similar way to the proof of the uniqueness of (3).

We now define a unitary operator W by

$$(W\xi)(g) = v(g)^*\xi(g^{-1}), \qquad g \in G, \qquad \xi \in \mathfrak{K}. \tag{18}$$

We then have $W^2 = 1$ and

$$W\pi'(b)W^*\xi(h) = \alpha_h^{-1}(b)\xi(h), \qquad b \in \mathscr{A}',$$
$$W\tilde{v}(g)W^*\xi(h) = \xi(g^{-1}h), \qquad g,h \in G, \qquad \xi \in \mathfrak{K}.$$

Therefore, the commutant $\mathscr{R}'$ is isomorphic to the "crossed product" of $\mathscr{A}'$ by the action α' of G on $\mathscr{A}'$ implemented by the unitary representation v of G.

Exercises

1. A maximal abelian von Neumann subalgebra $\mathscr{A}$ of a factor $\mathscr{M}$ is called *regular* if the *normalizer* $N(\mathscr{A}) = \{u \in \mathscr{U}_{\mathscr{M}} : u\mathscr{A}u^* = \mathscr{A}\}$ generates $\mathscr{M}$. It is called *semiregular* if $N(\mathscr{A})$ generates a subfactor, *singular* if $N(\mathscr{A}) \subset \mathscr{A}$.
   (a) Show that if $\mathscr{M} = \mathscr{L}(\mathfrak{H})$ with $\mathfrak{H}$ separable and if $\mathscr{A}$ is either atomic or totally nonatomic, i.e., it contains no minimal projections, then $\mathscr{A}$ is regular.
   (b) Let G be the free group of two generators a and b, and set $\mathscr{M} = \mathscr{R}(G)$. Show that the von Neumann subalgebra $\mathscr{A}$ generated by $u(a)$ is a singular maximal abelian subalgebra.
   (c) Show that if $\mathscr{M} = \mathscr{R}(\mathscr{A},G,\alpha)$ with $(\mathscr{A},G,\alpha)$ an ergodic free convariant system, then $\pi(\mathscr{A})$ is a regular maximal abelian subalgebra of $\mathscr{M}$.

2. Let Γ be a metrizable compact space with a probability measure μ. Let G be a countable discrete group acting freely on Γ as a group of homeomorphisms and leaving μ quasi-invariant. Let α be the action of G on $\mathscr{A} = L^\infty(\Gamma,\mu)$ given naturally by the action of G on Γ. Consider $\mathscr{R} = \mathscr{R}(\mathscr{A},G,\alpha)$. Consider $\{v(g) : g \in G\}$, the unitary representation of G on $L^2(\Gamma,\mu)$, defined by

$$(v(g)\xi)(\gamma) = \sqrt{\frac{d\mu \circ g^{-1}}{d\mu}}(\gamma)\xi(g^{-1}\gamma).$$

Let A be the separable C^*-algebra generated by $\pi'(C(\Gamma))$ and $\{\tilde{v}(g) : g \in G\}$, which is σ-weakly dense in $\mathscr{R}'$ on $\mathfrak{H} = L^2(G \times \Gamma, \delta \times \mu)$, where δ is the counting measure on G. Let φ be the state on $\mathscr{R}'$ given by the vector $\xi_0 \in \mathfrak{H}_0$ such that

$$\xi_0(g,\gamma) = \begin{cases} 1, & g = e \\ 0, & g \neq e. \end{cases}$$

   (a) Show that for each $x = \sum_{g \in G} \pi'(x(g))\tilde{v}(g) \in \mathscr{R}'$,

$$\varphi(x) = \int_\Gamma x(e,\gamma)\, d\mu(\gamma).$$

   (b) Show that for each $\gamma \in \Gamma$,

$$\omega_\gamma(x) = x(e,\gamma), \qquad x \in A,$$

   is a pure state on A and that cyclic representation of A induced by ω_γ is unitarily equivalent to the representation π_γ of A on $l^2(G\gamma)$ given by

$$\left(\pi_\gamma\left(\sum_{g \in G} \pi'(x(g))\tilde{v}(g)\right)\xi\right)(h\gamma) = \sum_{g \in G} x(g; h\gamma)\xi(g^{-1}h\gamma).$$

   (c) Show that the integral $\varphi = \int_\Gamma \omega_\gamma\, d\mu(\gamma)$ is a orthogonal representation of φ with respect to the maximal abelian subalgebra $\pi(\mathscr{A})$ in $\mathscr{R}$.
   (d) Show that $\pi_{\gamma_1} \simeq \pi_{\gamma_2}$ if and only if $G\gamma_1 = G\gamma_2$.

Notes

Multiplicity theory is one of the few areas in the study of operator algebras where the theory is complete. It is this theory which ties up Problems (A) and (B) for von Neumann algebras in the introduction of Chapter I. In other words, the problem of how a given von Neumann algebra acts on a Hilbert space is reduced to the problem of finding the algebraic structure of the algebra plus the determination of the coupling function occurring in the specified Hilbert space.

F. Murray and J. von Neumann developed multiplicity theory for finite factors with finite commutants. In the early 1950s, the theory took its present form. Indeed, much of the effort of the specialists at that time was spent on completing the theory and it worked out successfully in a relatively short period. References for this section are [146,147,195,237,274].

The results in Section 4 are mostly due to J. Dixmier [76]. He developed the theory to show the existence of the center valued trace via this approach. Except for this ergodic type result, we have, at the present, no technique for handling uniformly closed convex sets in a von Neumann algebra. Theorem 4.8 is due to Y. Misonou [235].

Theorem 5.1 is due to J. Feldman and J. M. G. Fell [117] and M. Takesaki [352], [355]. It indicates a special feature of the separability for von Neumann algebras. The author is rather pessimistic for the future of the nonseparable theory of von Neumann algebras. On the other hand, Theorem 5.2, due to F. B. Wright [408] and J. Feldman [111], forces us not to exclude non-separable von Neumann algebras even if we handle only problems related to separable von Neumann algebras.

The Borel space of von Neumann algebras was introduced by E. Effros [105,106]. The presentation of Section 5 follows his treatise. The measurability of A_{III} is due to J. T. Schwartz [323]. To day, it is known that A_{III} is indeed a Borel set; see 0. Nielsen [260].

The construction of factors in Section 7 is nowadays standard. It is called the *group measure space construction* (of Murray–von Neumann). More careful analysis of factors will be given in the subsequent volume.

Appendix
Polish Spaces and Standard Borel Spaces

Let X be a given set. A collection $\mathscr{B}$ of subsets is called a *σ-field* if (a) any union of countably many members of $\mathscr{B}$ is a member of $\mathscr{B}$, (b) the complement of any member of $\mathscr{B}$ belongs to $\mathscr{B}$ and (c) $\phi \in \mathscr{B}$. Given a family $\mathscr{S}$ of subsets of X, there exists the smallest σ-field $\mathscr{B}_{\mathscr{S}}$ of subsets of X containing $\mathscr{S}$, which will be called the σ-field *generated by* $\mathscr{S}$. A *Borel space* is a set X equipped with a specified σ-field $\mathscr{B}$. The members of $\mathscr{B}$ are called *Borel sets* of X. When X is a topological space, the Borel space X generated by the topology means the Borel space X equipped with the σ-field generated by the family of open (hence closed) subsets of X. Given a Borel space X, a collection of Borel subsets is said to be *generating* if the σ-field of Borel subsets is generated by the family; it is said to be *separating* if any distinct pair of points of X are separated by a member of the family. A Borel space is said to be *countably generated* (resp. *separated*) if it admits a countable generating and separating (resp. separating) family of Borel sets. A (positive) measure on a Borel space X means an extended positive real valued countably additive set function μ on the Borel sets in X. Given two Borel spaces X_1 and X_2, a map f of X_1 into X_2 is said to be *Borel* if the inverse image $f^{-1}(S)$ of every Borel set S in X_2 is Borel in X_1. If f is bijective and f^{-1} is Borel also, then f is called a *Borel isomorphism*, and in this case X_1 and X_2 are said to be (Borel) *isomorphic*. If X_1 and X_2 are topological spaces, then any continuous map f of X_1 into X_2 is Borel, i.e., $f^{-1}(S)$ is Borel in X_1 for every Borel set S in X_2. We note, however, that *the image $f(S)$ of a Borel set S in X_1 need not be a Borel set in X_2.*

A topological space is called *Polish* if it is homeomorphic to a separable complete metric space.

Theorem A.1. *In a Polish space, a subspace, with respect to the relative topology, is a Polish space if and only if it is a G_δ-subset of the whole Polish* *space. A topological space is a Polish space if and only if it is homeomorphic to a G_δ-subset of the cube $I^{\mathbf{N}}$, where $I = [0,1]$ and $\mathbf{N}$ denotes the set of all positive integers.*

PROOF. Let X be a Polish space and Y be a subspace. Suppose that Y is a Polish space. Let d be a complete metric on Y compatible with the topology in Y. Let $\bar{Y}$ be the closure of Y in X. For each $n = 1,2,\ldots,$ we denote by Y_n the set of all those points $x \in \bar{Y}$ such that there exists an open neighborhood U of x with diameter $U \cap Y \leq 1/n$. It is clear that Y_n is an open subset of $\bar{Y}$ with respect to the relative topology of $\bar{Y}$, and that $Y_n \supset Y$. Hence we have $\bigcap_{n=1}^\infty Y_n \supset Y$. Suppose that $x \in \bigcap_{n=1}^\infty Y_n$. For each n, there exists an open neighborhood U_n of x such that the diameter of $U_n \cap Y \leq 1/n$. Replacing U_n by $U_1 \cap \cdots \cap U_n$, we may assume that $\{U_n\}$ is decreasing. It then follows that $\{U_n \cap Y\}$ is a sequence of decreasing open subsets of Y whose diameters tend to zero. Hence $\bigcap_{n=1}^\infty \{U_n \cap Y\}$ is a singleton, say $\{y\}$. Since $\{U_n\}$ can be chosen to be a fundamental system of neighborhoods of x, we have $x = y \in Y$. Thus $Y = \bigcap_{n=1}^\infty Y_n$. Since Y_n is open in $\bar{Y}$, there exists an open set V_n in X such that $Y_n = \bar{Y} \cap V_n$. On the other hand, $\bar{Y}$ is a G_δ-set in X, being closed, so that there exists a sequence of open sets $\{W_n\}$ in X such that $\bar{Y} = \bigcap_{n=1}^\infty W_n$. Thus $Y = \bigcap_{n,m} (V_n \cap W_m)$ is a G_δ-set in X.

Suppose that Y is a G_δ-set of X. Let $\{G_n\}$ be a sequence of open sets in X such that $\bigcap G_n = Y$. Let d be a complete metric in X compatible with the topology in X. For each $x \in Y$ and $n = 1,2,\ldots,$ let

$$f_n(x) = 1/d(x,G_n^c) > 0.$$

We set

$$d_0(x,y) = d(x,y) + \sum_{n=1}^\infty \frac{1}{2^n} \frac{|f_n(x) - f_n(y)|}{1 + |f_n(x) - f_n(y)|}$$

for each $(x,y) \in Y \times Y$. Then it is not difficult to prove that d_0 is a complete metric in Y which is compatible with the topology in Y. Being a subspace of a separable metric space, Y is separable. Thus Y is a Polish space.

Suppose that X is a Polish space with a complete metric d. Let $\{a_n\}$ be a dense sequence in X. For a point $x \in X$, we set

$$\varphi(x) = \left\{\frac{d(a_n,x)}{1 + d(a_n,x)}\right\} \in [0,1]^{\mathbf{N}}.$$

It is not too hard to show that φ is a homeomorphism of X into the compact metrizable space $[0,1]^{\mathbf{N}}$. By the first half of our assertion, $\varphi(X)$ is a G_δ-set in $[0,1]^{\mathbf{N}}$. Q.E.D.

Let $\mathbf{N}$ denote the set of all positive integers as usual. Set $\Lambda = \mathbf{N}^{\mathbf{N}}$ and define a metric d in Λ by

$$d(\{m_k\},\{n_k\}) = \sum_{k=1}^\infty \frac{1}{2^k} \frac{|m_k - n_k|}{1 + |m_k - n_k|}.$$

It then follows that Λ is separable and complete with respect to the metric d. Let X be a separable complete metric space with a metric d; hence a Polish space. For each subset $A \subset X$, let $\delta(A)$ be the diameter of A, i.e., $\delta(A) = \sup\{d(x,y): x, y \in A\}$. Let $\{F(n): n \in \mathbf{N}\}$ be a closed covering of X with $\delta(F(n)) \leq 1/2$. Suppose that a closed covering $\{F(n_1, \ldots, n_k): n_1, \ldots, n_k \in \mathbf{N}\}$ of X has been defined for $k = 1, \ldots, m$ with the properties

$$F(n_1, \ldots, n_j) = \bigcup_{n=1}^{\infty} F(n_1, n_2, \ldots, n_j, n),$$

$$\delta(F(n_1, \ldots, n_{j+1})) \leq \tfrac{1}{2}\delta(F(n_1, \ldots, n_j)), \qquad j = 1, 2, \ldots, m-1.$$

Then there exists a closed covering $\{F(n_1, \ldots, n_m, n): n \in \mathbf{N}\}$ of $F(n_1, \ldots, n_m)$ such that

$$\delta(F(n_1, \ldots, n_m, n)) \leq \tfrac{1}{2}\delta(F(n_1, \ldots, n_m)).$$

Now for any sequence $\{n_k\} \in \Lambda$, $\{F(n_1, \ldots, n_k): k \in \mathbf{N}\}$ is a decreasing sequence of closed sets and $\lim_{k \to \infty} \delta(F(n_1, \ldots, n_k)) = 0$. Since X is complete, $\bigcap_{k=1}^{\infty} F(n_1, \ldots, n_k)$ is a singleton which we denote by $\{\varphi(\{n_k\})\}$. Thus we obtain a mapping φ of Λ into X. It is easy to show that φ is continuous and surjective. Thus we obtain the following:

Lemma A.2. *If X is a Polish space, then there exists a continuous map from $\mathbf{N}^{\mathbf{N}} = \Lambda$ onto X.*

A *Souslin space* (or sometimes analytic space) means a metrizable space which is a continuous image of a Polish space. A subset of a topological space is said to be a *Souslin set* if it is a Souslin space as a topological space. By Lemma A.2, any Souslin space is a continuous image of Λ.

Theorem A.3 (Separation of Souslin Sets). *Let X be a metrizable space. If $\{X_n\}$ is a disjoint countable family of Souslin subsets of X, then there exists a disjoint family $\{B_n\}$ of Borel subsets of X such that $X_n \subset B_n$, $n = 1, 2, \ldots$.*

Proof. First, we claim that if $\{A_n\}$ and $\{A'_m\}$ are sequences of subsets of X such that for any n, m there exists a Borel set $B_{n,m}$ with the property $B_{n,m} \supset A_n$ and $B_{n,m} \cap A'_m = \emptyset$, then there exists a Borel set B such that $B \supset \bigcup_{n=1}^{\infty} A_n$ and $B \cap (\bigcup_{m=1}^{\infty} A'_m) = \emptyset$. Indeed, the Borel set $B = \bigcup_{n=1}^{\infty} (\bigcap_{m=1}^{\infty} B_{n,m})$ does the job.

We now prove the theorem for two disjoint Souslin sets Y and Y' in X. Let f and f' be continuous maps of Λ onto Y and Y', respectively. For each

$$(n_1, \ldots, n_k) \in \underbrace{\mathbf{N} \times \cdots \times \mathbf{N},}_{k \text{ times}}$$

we set

$$\Lambda(n_1, \ldots, n_k) = \{\{m_i\} \in \Lambda: m_1 = n_1, \ldots, m_k = n_k\}.$$

We then have

$$\Lambda(n_1, \ldots, n_k) = \bigcup_{n=1}^{\infty} \Lambda(n_1, n_2, \ldots, n_k, n).$$

Set

$$Y(n_1, \ldots, n_k) = f(\Lambda(n_1, \ldots, n_k)),$$
$$Y'(m_1, \ldots, m_k) = f(\Lambda(m_1, \ldots, m_k)).$$

We now assume that there exists no Borel set B such that $B \supset Y$ and $B \cap Y' = \varnothing$. It then follows from the first argument that for some n_1 and m_1, $Y(n_1)$ and $Y'(m_1)$ are not separated by a Borel set. If $Y(n_1, \ldots, n_k)$ and $Y'(m_1, \ldots, m_k)$ are not separated by a Borel set, then for some n_{k+1} and m_{k+1}, $Y(n_1, \ldots, n_k, n_{k+1})$ and $Y'(m_1, \ldots, m_k, m_{k+1})$ are not separated by a Borel set. By induction, we obtain two sequences $\{n_i\}$ and $\{m_j\}$ such that $Y(n_1, \ldots, n_k)$ and $Y'(m_1, \ldots, m_k)$ are not separated by a Borel set. Set $\bar{n} = \{n_i\} \in \Lambda$ and $\bar{m} = \{m_i\} \in \Lambda$ and $y = f(\bar{n}) \in Y$, $y' = f'(\bar{m}) \in Y'$. Since $y \neq y'$, there exist adjoint open sets V and W such that $y \in V$ and $y' \in W$. It follows that $f^{-1}(V)$ and $f'^{-1}(W)$ are open neighborhoods of $\bar{n}$ and $\bar{m}$ in Λ, respectively. Since $\{\Lambda(n_1, \ldots, n_k): k = 1, 2, \ldots\}$ is a fundamental system of neighborhoods of $\bar{n}$, there exists k_0 such that $\Lambda(n_1, \ldots, n_k) \subset f^{-1}(V)$ for $k \geq k_0$. Also, the same is true for $\Lambda(m_1, \ldots, m_k)$ with f' and W in place of f and V, respectively. Therefore, we have

$$Y(n_1, n_2, \ldots, n_k) \subset V; \qquad Y'(m_1, \ldots, m_k) \subset W;$$

thus $Y(n_1, \ldots, n_k)$ and $Y'(m_1, \ldots, m_k)$ are separated by the Borel set V. This contradicts the construction of $Y(n_1, \ldots, n_k)$ and $Y'(m_1, \ldots, m_k)$. Therefore, Y and Y' are separated by a Borel set.

Now, we complete the proof for the general case. By the above argument, for distinct n and m, there exists a Borel set $B_{n,m}$ such that $X_n \subset B_{n,m}$ and $B_{n,m} \cap X_m = \varnothing$. Setting $B_1 = \bigcap_{m=2}^{\infty} B_{1,m}$, we define $\{B_n\}$ by induction as follows:

$$B_n = (B_1 \cup B_2 \cup \cdots \cup B_{n-1})^c \cap \left(\bigcap_{m=n+1}^{\infty} B_{n,m} \right).$$

It then follows that $\{B_n\}$ is a disjoint sequence of Borel sets such that $X_n \subset B_n$, $n = 1, 2, \ldots$. Q.E.D.

Corollary A.4. *In a metrizable space, a Souslin set is a Borel set if its complement is also a Souslin set.*

A metrizable topological space X is called a *Lusin space* if there exists a Polish space P and a bijective continuous map f of P onto X such that:

(∗) *each point of P has a fundamental system of neighborhood of open and closed sets.*

A subset of a topological space is called a *Lusin set* if it is a Lusin space as a topological space. Any *countable intersection of Lusin sets* in a topological space is a Lusin set. Because, if f_n, $n = 1,2,\ldots$, is a bijective continuous map from a Polish space P_n with property $(*)$ into X, then we set $P = \{(y_n) \in \prod_{n=1}^{\infty} P_n : f_1(y_1) = f_2(y_2) = \cdots = f_n(y_n) = \cdots\}$ and $f(\{y_n\}) = f_1(y_1)$. Then P satisfies condition $(*)$, being a closed subspace of $\prod P_n$, and f is a continuous bijection from P onto $\bigcap_{n=1}^{\infty} f_n(P_n)$. Any countable disjoint union of Lusin sets in a metrizable topological space is again a Lusin set. Because, if f_n is a bijective continuous map of a Polish space P_n with property $(*)$ into a metrizable space X such that $f_n(P_n) \cap f_m(P_m) = \emptyset$ for $n \neq m$, then the disjoint sum $P = \bigcup_{n=1}^{\infty} P_n$ is a Polish space with property $(*)$ and the map f of P into X such that $f(x) = f_n(x)$ if $x \in P_n$ is a continuous bijection of P onto $\bigcap_{n=1}^{\infty} f_n(P_n)$. Since any open closed subset of a Lusin space is a Lusin set, *any G_δ-set* in a Lusin space is again a Lusin set. The subset J of all irrational numbers in the unit interval $I = [0,1]$ is a G_δ-set in I, and hence a Polish space. Since $J \cap]r,s[$ with r and s rational numbers forms a fundamental system of neighborhoods of open and closed sets in J, J satisfies condition $(*)$. Therefore, J is a Lusin set in I. Being a disjoint countable union of J and one point sets of rational numbers, I is a Lusin space. Therefore, the cube $I^{\mathbf{N}}$ is a Lusin space. Being homeomorphic to a G_δ-set in $I^{\mathbf{N}}$ by Theorem A.1, *any Polish space* is a Lusin space.

Lemma A.5. *For any Lusin space X (hence for any Polish space), there exists a continuous bijection φ from $\prod_{k=1}^{\infty} \mathbf{N}_k$ onto X, where $\mathbf{N}_k = \mathbf{N}$ or $\{1,2,\ldots,N_k\}$.*

PROOF. We may assume that X is a Polish space with property $(*)$ because we can replace X by P such that P satisfies $(*)$ and $X = f(P)$ for some continuous bijection f. In the argument preceding Lemma A.2, we can choose $F(n_1,n_2,\ldots,n_k,n)$, $1 \le n \le N_{k+1}$, disjoint at each step k. Our assertion then follows automatically. Q.E.D.

Theorem A.6. *A Borel subset of a Lusin space is a Lusin set. Conversely, a Lusin set in a metrizable topological space is a Borel set.*

PROOF. Let X be a Lusin space. Let $\mathscr{F}$ be the collection of all Lusin subsets of X with complement also Lusinian. If $\{A_n\}$ is a sequence in $\mathscr{F}$, then $\bigcap A_n$ is a Lusin set. Put $B_n = A_n \cap (\bigcap_{k<n} A_k^c)$. It follows that $\{B_n\}$ is a disjoint sequence of Lusin sets and $\bigcup A_n = \bigcup B_n$. Thus, $\bigcup A_n$ is a Lusin set. Thus, $\mathscr{F}$ is closed under countable union, intersection and complement, hence is a σ-field. Furthermore, $\mathscr{F}$ contains all open sets. Thus $\mathscr{F}$ contains all Borel sets in X.

Suppose, conversely, that Y is a Lusin set in a metrizable space X. Let φ be a continuous bijection from $N = \prod_{k=1}^{\infty} \mathbf{N}_k$ onto Y, where $\mathbf{N}_k = \mathbf{N}$ or $\{1,2,\ldots,N_k\}$. For each $\bar{n} = \{n_k\} \in N$, set

$$N(n_1,\ldots,n_k) = \{\bar{m} = \{m_j\} \in N : m_1 = n_1,\ldots,m_k = n_k\},$$
$$Y(n_1,\ldots,n_k) = \varphi(N(n_1,\ldots,n_k)).$$

It then follows that $\bigcap_{k=1}^{\infty} Y(n_1,\ldots,n_k) = \{\varphi(\bar{n})\}$ and $\{Y(n_1,\ldots,n_k) : 1 \le n_s \le N_s\}$, is a disjoint family of Lusin sets, hence Souslin sets, in X due to the injectivity of φ. By Theorem A.3, $\{Y(j):1 \le j \le N_1\}$ is separated by a disjoint family $\{B(j):1 \le j \le N_1\}$ of Borel sets, i.e., $Y(j) \subset B(j)$ and $B(i) \cap B(j) = \emptyset$ for $i \neq j$. Replacing $B(j)$ by $B(j) \cap \overline{Y(j)}$, we may assume that $Y(j) \subset B(j) \subset \overline{Y(j)}$, $1 \le j \le N_1$. Set $B_1 = \bigcup B(j)$. Suppose that we could find a disjoint family $\{B(j_1,\ldots,j_k):1 \le j_s \le N_s, 1 \le s \le k\}$ of Borel sets in X for $k = 1,2,\ldots,n$ such that

$$Y(j_1,\ldots,j_s) \subseteq B(j_1,\ldots,j_s) \subseteq \overline{Y(j_1,\ldots,j_s)},$$
$$B(j_1,\ldots,j_s) \subset B(j_1,\ldots,j_{s-1}).$$

Since $\{Y(j_1,\ldots,j_n,j_{n+1})\}$ is a disjo)nt family of Lusin sets in $B(j_1,\ldots,j_n)$, it is separated by a disjoint family $\{B'(j_1,\ldots,j_{n+1})\}$ of Borel subsets of $B(j_1,\ldots,j_n)$ by Theorem A.3. Setting $B(j_1,\ldots,j_{n+1}) = B'(j_1,\ldots,j_{n+1}) \cap \overline{Y(j_1,\ldots,j_{n+1})}$, we obtain a disjoint family $\{B(j_1,\ldots,j_{n+1})\}$ of Borel sets with the above properties. Set

$$B_n = \bigcup B(j_1,\ldots,j_n) \quad \text{and} \quad B = \bigcap_{n=1}^{\infty} B_n.$$

Clearly, B is a Borel set and $B \supset Y$. If $y \in B$, then there exists a unique sequence $\bar{n} = (n_1,n_2,\ldots)$ such that $y \in B(n_1,n_2,\ldots,n_k)$ for $k = 1,2,\ldots$. Hence $y \in \overline{Y(n_1,\ldots,n_k)}$ for $k = 1,2,\ldots$. Since $n_k \le N_k$, $\bar{n}$ is a point of N. Put $x = \varphi(\bar{n})$. Let V be an arbitrary closed neighborhood of x. Since $\{N(n_1,\ldots,n_k):k = 1,2,\ldots\}$ is a fundamental system of neighborhoods of $\bar{n}$, there exists k such that $Y(n_1,\ldots,n_k) = \varphi(N(n_1,\ldots,n_k)) \subset V$, so that $\overline{Y(n_1,\ldots,n_k)} \subset V$, V being closed. Hence $y \in V$. Since V is arbitrary, $x = y$. Thus y belongs to Y. Therefore, we get $B \subset Y$, i.e., $B = Y$; so Y is a Borel set in X. Q.E.D.

Corollary A.7. *The range of an injective continuous map of a Lusin space into a metrizable topological space is a Borel subset, and the map itself is a Borel isomorphism of the domain and its range.*

Corollary A.8. *A Borel subset of a Souslin space is a Souslin set.*

PROOF. Let f be a continuous map from a Polish space Y onto a Souslin space X. If B is a Borel subset of X, then $f^{-1}(B)$ is a Borel subset of Y, hence a Lusin set by Theorem A.6. Being a continuous image of a Lusin set, B is a Souslin set. Q.E.D.

Lemma A.9. *Let X and Y be metrizable topological spaces. Then the graph of a Borel map of X into Y is a Borel subset of $X \times Y$.*

PROOF. If Z is a metric space, then the set A of all fixed points under a Borel map φ of Z into Z is a Borel subset because $A = \{z \in Z:d(z,\varphi(z)) = 0\}$, where d is the metric in Z, and $z \in Z \mapsto d(z,\varphi(z)) \in \mathbf{R}$ is a Borel map. Let f be a Borel map of X into Y and G be the graph of f. Set $\varphi(x,y) = (x,f(x))$, $(x,y) \in X \times Y$. For any open sets $U \subset X$ and $V \subset Y$, we have $\varphi^{-1}(U \times V) = \{U \cap f^{-1}(V)\} \times Y$. Hence φ is a Borel map of $X \times Y$ into $X \times Y$. Thus the set of fixed points of φ is a Borel subset. But the set of fixed points of φ is indeed the graph G of f. Q.E.D.

Corollary A.10. *Let X be a Souslin space and Y a separable metrizable topological space. If f is a Borel map of X into Y, then $f(X)$ is a Souslin set in Y. If f is injective in addition, then f is a Borel isomorphism from X onto $f(X)$.*

PROOF. Consider the product space $X \times \bar{Y}$ with $\bar{Y}$ the completion of Y. It follows that $X \times \bar{Y}$ is a Souslin space. The graph G of f in $X \times \bar{Y}$ is a Borel subset by Lemma A.9; hence a Souslin set by Corollary A.8. The projection of $X \times \bar{Y}$ onto the second space $\bar{Y}$ continuously maps G onto $f(X)$. Hence $f(X)$ is a Souslin set. Therefore, if A is a Souslin set in X, then $f(A)$ is a Souslin set in Y. Suppose that f is injective. If B is a Borel set in X, then $f(B)$ and $f(B^C)$ are both Souslin sets in Y. By Theorem A.3, $f(B)$ is a Borel subset of $f(X)$ (regarding $f(X)$ as a topological space). Thus, f is a Borel isomorphism from X onto $f(X)$. Q.E.D.

Corollary A.11. *A standard Borel space is either countable or isomorphic to the interval $I = [0,1]$.*

PROOF. A standard Borel space is, by definition, isomorphic to the Borel space of a Polish space. Hence it is Borel isomorphic to $N = \prod_{k=1}^{\infty} \mathbf{N}_k$, where $\mathbf{N}_k = \mathbf{N}$ or $\{1,2,\ldots,N_k\}$. Suppose that N is uncountable. Then $N_k \geq 2$ for infinitely many k. Hence there exists a homeomorphism f from $C = \{0,1\}^{\mathbf{N}}$ into N. Furthermore, N is a closed subset of $\Lambda = \mathbf{N}^{\mathbf{N}}$. Let g be an injection of $\mathbf{N}$ into C, which is, of course, a continuous map because $\mathbf{N}$ is discrete. It then follows that the product map $h:\bar{n} = (n_1,n_2,\ldots) \in \Lambda \mapsto h(\bar{n}) = (g(n_1),g(n_2),\ldots) \in C^{\mathbf{N}}$ is an injective continuous map. Since $C^{\mathbf{N}}$ is homeomorphic to C, h can be viewed as an injective continuous map from Λ into C. Thus we obtain two injective continuous maps $f:C \mapsto N \subset \Lambda$ and $h:\Lambda \mapsto C$. By Theorem A.6, $f(C)$ and $h(N)$ are Borel subsets of Λ and C, respectively. The usual Cantor–Bernstein theorem type argument shows that C and N are Borel isomorphic to Λ. Therefore, every uncountable standard Borel space is isomorphic to Λ, and in particular I is Borel isomorphic to Λ. Q.E.D.

A countably separated Borel space is called a *Souslin–Borel space* if it is the range of a Borel map from a standard Borel space. Let X be a countably separated Borel space and $\{B_n\}$ be a countable separating family of Borel sets in X. Let χ_n denote the characteristic function of B_n. Set $\varphi(x) = \{\chi_n(x)\} \in \{0,1\}^{\mathbf{N}} = C$. It then follows that φ is an injective Borel mapping from X into C. Suppose that X is a Souslin–Borel space. Let f be a Borel map from a standard Borel space Y onto X. Since $\varphi \circ f$ is a Borel map from Y into C with range $\varphi(X)$, the image $\varphi(X)$ is a Souslin set in C. Let B be an arbitrary Borel set in X. Set $\psi(x) = \{\chi_B(x),\chi_1(x),\ldots,\chi_n(x),\ldots\} \in C$, where χ_B is the characteristic function of B. It then follows that ψ is an injective Borel map from X into C and that $\psi(X)$ is also a Souslin subset of C. By the definition of the Borel structure in C, $\varphi \circ \psi^{-1}$ is an injective Borel map from $\psi(X)$ onto $\varphi(X)$; hence it is a Borel isomorphism by Corollary A.10. Therefore, the function: $\varphi(x) \in \varphi(X) \mapsto \chi_B(x) \in \{0,1\}$ is a Borel function on $\varphi(X)$, which means that B belongs to the σ-field generated by $\{B_n\}$. Therefore, in a Souslin–Borel space X any separating family of Borel sets generates the σ-field of Borel sets, and the above map φ is a Borel isomorphism of X onto a Souslin set $\varphi(X)$ in C.

Corollary A.12. *Every separating family of Borel sets in a Souslin–Borel space is also generating.*

Theorem A.13. *Let X be a standard Borel space. If μ is a σ-finite measure on X, then every Souslin set is μ-measurable.*

Proof. Since μ is σ-finite, there exists a finite measure on X which is equivalent to μ in the sense of absolute continuity. Hence we may assume that μ is finite. By Corollary A.11, we may assume that X is a compact metrizable space. Let A be a Souslin set in X. Let g be a continuous map from a Polish space P into X with $A = g(P)$. Let B be the graph of g in $P \times X$, which is closed, and thus a Polish space. By Theorem A.1, P is considered as a subspace of $I^{\mathbf{N}}$ with $I = [0,1]$. Let $Y = I^{\mathbf{N}} \times X$. The Polish space B is naturally imbedded in the compact space Y. Let f be the projection of Y onto the second component X. We then have $f(B) = A$. Being a Polish space, B is a G_δ-set in Y, so that there exists a sequence $\{B_n\}$ of open sets in Y with $B = \bigcap_{n=1} B_n$. Since each B_n is open in the compact metrizable space Y, there exists an increasing sequence $\{B_{n,m}\}$ of compact subsets with $B_n = \bigcup_{m=1}^\infty B_{n,m}$. Let μ^* denote the outer measure on X associated with the given finite measure μ. We shall show that for any $\alpha < \mu^*(A)$, there exists a compact subset $C \subset B$ such that $\alpha \leq \mu(f(C))$; thus

$$\mu^*(A) = \sup\{\mu(K): K \subset A \text{ compact}\}.$$

Suppose that we have chosen integers $j_1, \ldots, j_{n-1}$ such that with $C_{n-1} = B \cap B_{1,j_1} \cap \cdots \cap B_{n-1,j_{n-1}}$, $\mu^*(f(C_{n-1})) > \alpha$. As $C_{n-1} \subset B \subset B_n$, we have

$$C_{n-1} = \bigcup_{k=1}^{\infty} (C_{n-1} \cap B_{n,k}), \qquad B_{n,k} \subset B_{n,k+1}.$$

By the monotone continuity of an outer measure, we have

$$\lim_{k\to\infty} \mu^*(f(C_{n-1} \cap B_{n,k})) = \mu^*(f(C_{n-1})) > \alpha.$$

Hence there exists j_n with $\mu^*(f(C_{n-1} \cap B_{n,j_n})) > \alpha$. Set $C_n = C_{n-1} \cap B_{n,j_n}$, and $C = \bigcap_{n=1}^\infty C_n$. We then have

$$\bigcap_{n=1}^{\infty} B_{n,j_n} \subset \bigcap_{n=1}^{\infty} B_n = B,$$

so that $C = \bigcap_{n=1}^{\infty} (B_{n,j_n} \cap B) = \bigcap_{n=1}^{\infty} B_{n,j_n}$. Hence C is compact, and we have

$$\mu(f(C)) = \lim \mu\left(f\left(\bigcap_{k=1}^{n} B_{k,j_k}\right)\right) \geq \lim \mu^*(f(C_n)) \geq \alpha. \qquad \text{Q.E.D.}$$

A measure μ on a Borel space X is said to be *standard* if there exists a null set N in X such that $X - N$ is a standard Borel space.

Corollary A.14. *A σ-finite measure on a Souslin–Borel space is standard.*

PROOF. Let X be a Souslin–Borel space. It is then isomorphic to a Souslin subset of a standard Borel space Y. So we may assume that X is a Souslin set in a standard Borel space Y. Let μ be a σ-finite measure on X. Set $\nu(B) = \mu(B \cap X)$ for every Borel set B in Y. It follows that ν is a σ-finite measure on Y. By Theorem A.13, X is ν-measurable. Hence there exists a Borel set X_0 in Y such that $X_0 \subset X$ and $\nu(X - X_0) = 0$. Let $N = X - X_0$. Then $\mu(N) = 0$ and $X - N = X_0$ is a standard Borel space by Theorem A.6. Q.E.D.

Theorem A.15. *Let R be an equivalence relation in a Polish space X such that each equivalence class under R is closed. If the saturation under R of each open set (resp. closed set) in X is a Borel set, then there exists a Borel set S in X which meets each equivalence class at exactly one point.*

PROOF. Let d be a complete metric on X. Let $\{X(n_1,n_2,\ldots,n_k): n_i \in \mathbf{N}, 1 \leq i \leq k, k = 1,2,\ldots\}$ be a system of open covering of X such that

$$\delta(X(n_1,\ldots,n_k)) \leq 1/2^k,$$

$$X(n_1,\ldots,n_k) = \bigcup_{j=1}^{\infty} X(n_1,\ldots,n_k,j),$$

$$\overline{X(n_1,\ldots,n_k,j)} \subset X(n_1,\ldots,n_k).$$

We consider the lexicographic ordering in each $\mathbf{N}^k$, $k = 2,3,\ldots$. For each subset A of X, we denote by $R(A)$ the saturation of A under R. Put $S(1) = X(1)$ (resp. $\overline{X(1)}$) and

$$S(n) = X(n) - \bigcup_{j=1}^{n-1} R(X(j)) \qquad \left(\text{resp. } \overline{X(n)} - \bigcup_{j=1}^{n-1} R(\overline{X(j)})\right).$$

By assumption, $S(n)$ is a Borel set (but may be empty). For each k, we set $S(1,\ldots,1) = X(1,\ldots,1)$ (resp. $X(1,\ldots,1)$) and

$$S(n_1,\ldots,n_k) = X(n_1,\ldots,n_k) - \bigcup\{R(X(m_1,\ldots,m_k)):(m_1,\ldots,m_k) < (n_1,\ldots,n_k)\}$$

resp.

$$\overline{X(n_1,\ldots,n_k)} - \bigcup\{R(\overline{X(m_1,\ldots,m_k)}):(m_1,\ldots,m_k) < (n_1,\ldots,n_k)\}.$$

It follows that $S(n_1, \ldots, n_k)$ is a Borel set. Set

$$S_k = \bigcup S(n_1, \ldots, n_k) \quad \text{and} \quad S = \bigcap_{k=1}^{\infty} S_k.$$

Let H be an equivalence class under R. For each $k = 1,2, \ldots,$

$$\varnothing \neq H \cap S_k = H \cap S(n_1, \ldots, n_k) = H \cap X(n_1, \ldots, n_k)$$

(resp. $= H \cap \overline{X(n_1, \ldots, n_k)}$) for a unique $(n_1, \ldots, n_k)$. Hence we have that $H \cap S_k$ is relatively open (resp. closed) in H and that $\delta(H \cap S_k) \leq 1/2^k$, $(H \cap S_{k+1})^- \subset H \cap S_k$. Hence $H \cap S = \bigcap_{k=1}^{\infty} H \cap S_k$ is exactly a singleton set since H is complete, being closed. Q.E.D.

Theorem A.16 (Measurable Cross Section). *Let X and Y be Souslin spaces, and f be a Borel map from X onto Y. For any σ-finite measure μ on Y, there exists a μ-measurable map φ from Y into X such that $f \circ \varphi(y) = y$ for every $y \in Y$, where the μ-measurability of φ means that $\varphi^{-1}(B)$ is μ-measurable for every Borel set B in X.*

Proof. Let G be the graph of f in $X \times Y$, which is a Borel set in the Souslin space $X \times Y$ by Lemma A.9. Hence there exists a continuous map g from $\mathbf{N}^{\mathbf{N}} = \Lambda$ onto G. Let h denote the composed map $\mathrm{pr}_Y \circ g: \Lambda \mapsto Y$, which is a surjective continuous map. For each $y \in Y$, $h^{-1}(y)$ is a closed subset of Λ. Consider the lexicographic ordering in Λ. It then follows that Λ is totally ordered and every closed set in Λ admits a least element. Indeed, if A is a closed set in Λ, let n_k be the least number among the kth coordinates of elements of A. Then $(n_1, n_2, \ldots, n_k, \ldots)$ belongs to A by the closedness of A. Let $\psi(y)$ be the least element in the closed set $h^{-1}(y)$ for every $y \in Y$. We then define φ to be $\mathrm{pr}_X \circ g \circ \psi$. It is then clear that $f \circ \varphi(y) = y$ for every $y \in Y$. Thus we must show the measurability of φ. By the continuity of g and pr_X, it suffices to show that ψ is measurable. Let $\Lambda(n_1, \ldots, n_k)$ be the set defined in the proof of Theorem A.3. We then have that

$$\Lambda(n_1, \ldots, n_k) = \{\overline{m} = \{m_i\} \in \Lambda : (n_1, \ldots, n_k, 0, 0, \ldots) \geq \overline{m} < (n_1, \ldots, n_{k-1}, n_k + 1, 0, \ldots, 0)\}.$$

But we have, for any $\overline{n} \in \Lambda$,

$$\psi^{-1}(\{\overline{m} : \overline{m} < \overline{n}\}) = h(\{\overline{m} : \overline{m} < \overline{n}\}).$$

Being the image of an open set under the continuous map h, $\psi^{-1}(\{\overline{m} : \overline{m} < \overline{n}\})$ is a Soulsin set. Hence by Theorem A.13, it is measurable. Therefore, $\psi^{-1}(\Lambda(n_1, \ldots, n_k))$ is measurable, being the intersection of two measurable sets. Since $\{\Lambda(n_1, \ldots, n_k)\}$ is a separating family in Λ, it is a generating family of Borel sets in Λ by Corollary A.12; hence ψ is a measurable map in the sense that $\psi^{-1}(B)$ is μ-measurable for every Borel set B in Λ. Thus φ is also a measurable map in the same sense. Q.E.D.

Theorem A.17. *Let X be a Polish space and $\mathscr{C}_0(X)$ denote the collection of all nonempty closed subsets of X. The Borel structure in $\mathscr{C}_0(X)$ generated by the subsets of $\mathscr{C}_0(X)$ of the form $\{A \in \mathscr{C}_0(X): A \cap U \neq \varnothing\} = \mathscr{U}(U)$ with U open sets in X is standard.*

PROOF. We first show that $\mathscr{C}_0(X)$ is countably separated. Let $\{U_n\}$ be a countable base of open sets in X. If A, B are members of $\mathscr{C}_0(X)$ such that $x \in A$ and $x \notin B$ for some x, then there exists n with $U_n \cap A \neq \varnothing$ and $U_n \cap B = \varnothing$. Hence $\mathscr{U}(U_n)$ separates A and B. Thus $\{\mathscr{U}(U_n): n \in \mathbf{N}\}$ is a separating family of Borel sets.

By Theorem A.1, X is imbedded in a compact metric space as a G_δ-subset. Let $\bar{X}$ denote the closure of X in the compact space, and let ρ be the metric in $\bar{X}$. We consider the space $\mathscr{C}_0(\bar{X})$ of all nonempty closed subsets of $\bar{X}$. For each $K \in \mathscr{C}_0(\bar{X})$, set $f_K(x) = \inf\{\rho(x,y): y \in K\} = \rho(x,K)$, $x \in \bar{X}$. We then have

$$|f_K(x) - f_K(y)| \leq \rho(x,y).$$

Hence $\mathscr{K} = \{f_K: K \in \mathscr{C}_0(\bar{X})\}$ is uniformly equicontinuous, so that it is relatively compact in $\mathscr{C}(\bar{X})$ by the Ascoli–Arzela theorem. Let $\{f_{K_n}\}$ be a sequence in $\mathscr{K}$ converging to f in the sup norm. Set $K = \{x \in \bar{X}: f(x) = 0\}$. Trivially, K is a closed subset of $\bar{X}$. If $x \in K$, then $\rho(x,K_n) = \rho(x,x_n)$ for some $x_n \in K_n$ by the compactness of K_n, so that

$$\lim \rho(x,x_n) = \lim \rho(x,K_n) = \lim f_{K_n}(x) = f(x) = 0.$$

Hence $x = \lim x_n$. Conversely, if $x = \lim x_{n_j}$ for some $\{x_{n_j}\} \in \prod_{j=1}^{\infty} K_{n_j}$, then

$$f(x) = \lim_j \rho(x,K_{n_j}) \leq \lim \rho(x,x_{n_j}) = 0;$$

thus $x \in K$. Now for any $x \in \bar{X}$, we have $\rho(x,K_n) = \rho(x,y_n)$ for some $y_n \in K_n$. Let $\{y_{n_j}\}$ be a convergent subsequence of $\{y_n\}$ and $y = \lim y_{n_j} \in K$. We then have

$$\begin{aligned} f(x) &= \lim \rho(x,K_n) = \lim \rho(x,y_n) = \lim \rho(x,y_{n_j}) \\ &= \rho(x,y) \geq \rho(x,K). \end{aligned}$$

Choosing $z \in K$ with $\rho(x,z) = \rho(x,K)$, we have a sequence $\{z_n\} \in \prod_{n=1}^{\infty} K_n$ such that $z = \lim z_n$, so that

$$\rho(x,K) = \rho(x,z) = \lim \rho(x,z_n) \geq \lim \rho(x,K_n) = f(x).$$

Therefore, $f = f_K \in \mathscr{K}$. Thus $\mathscr{K}$ is closed in $\mathscr{C}(\bar{X})$; so it is compact. With this fact in mind, we define a metric ρ in $\mathscr{C}_0(\bar{X})$ by

$$\rho(K,L) = \|f_K - f_L\| = \sup|f_K(x) - f_L(x)|, \qquad K,L \in \mathscr{C}_0(\bar{X}).$$

We obtain a compact metric space $\{\mathscr{C}_0(\bar{X}),\rho\}$. It is easy to see that

$$\rho(K,L) = \max\left\{\sup_{x \in K} \rho(x,L), \sup_{y \in L} \rho(y,K)\right\}.$$

We now define an injection φ of $\mathscr{C}_0(X)$ into $\mathscr{C}_0(\bar{X})$ by $\varphi(A) = \bar{A}$, $A \in \mathscr{C}_0(X)$. We shall show that $\varphi(\mathscr{C}_0(X))$ is a G_δ-set in the compact space $\mathscr{C}_0(\bar{X})$. Obviously, $\varphi(\mathscr{C}_0(X))$ is the collection of all those $K \in \mathscr{C}_0(\bar{X})$ such that $K \cap X$ is dense in K. Let $\{G_n\}$ be a sequence of open sets in $\bar{X}$ such that $\bigcap G_n = X$. By Baire's theorem, we have

$$\varphi(\mathscr{C}_0(X)) = \bigcap_{n=1}^{\infty} \{K \in \mathscr{C}_0(\bar{X}) : K \cap G_n \text{ is dense in } K\}.$$

For each fixed $x \in \bar{X}$, the function: $K \in \mathscr{C}_0(\bar{X}) \mapsto f_{(K \cap G_n)^-}(x)$ is upper semi-continuous and $f_{(K \cap G)_n)^-}(x) \geq f_K(x)$. Let $\{a_n\}$ be a dense sequence in $\bar{X}$. We then have

$$\{K \in \mathscr{C}_0(\bar{X}) : K \cap G_n \text{ is dense in } K\}$$

$$= \bigcap_{k,j} \left\{K \in \mathscr{C}_0(\bar{X}) : f_{(K \cap G_n)^-}(a_j) < f_K(a_j) + \frac{1}{k}\right\},$$

which is certainly a G_δ-set. Thus $\varphi(\mathscr{C}_0(X))$ is a G_δ-subset of the compact space $\mathscr{C}_0(\bar{X})$.

We now prove that $\varphi(\mathscr{U}(U))$ is a Borel set in $\varphi(\mathscr{C}_0(X))$. But the family of open sets in X is $\{U \cap X : \mathscr{U}$ runs all open sets in $\bar{X}\}$. Hence we must show that $\{K \in \mathscr{C}_0(X) : (K \cap X)^- = K, K \cap X \cap U \neq \varnothing\}$ is Borel for each open set U in $\bar{X}$. But for each $K \in \varphi(\mathscr{C}_0(X))$, $(K \cap X) \cap U \neq \varnothing$ if and only if $K \cap U \neq \varnothing$. In $\mathscr{C}_0(\bar{X})$, $\{K \in \mathscr{C}_0(\bar{X}) : K \cap U \neq \varnothing\}$ is open. Thus, $\varphi(\mathscr{U}(U))$ is a G_δ-set in $\mathscr{C}_0(\bar{X})$. Therefore, $\{\varphi(\mathscr{U}(U_n))\}$ is a countable separating family of Borel sets of the standard Borel space $\varphi(\mathscr{C}_0(X))$. Hence it generates the Borel structure of $\varphi(\mathscr{C}_0(X))$, which means that $\mathscr{C}_0(X)$ is indeed a standard Borel space. Q.E.D.

Corollary A.18. *If $\{X,d\}$ is a separable complete metric space, then the space $\mathscr{C}_0(X)$ of all nonempty closed subsets of X is a standard Borel space with respect to the smallest σ-field which makes measurable the function: $A \in \mathscr{C}_0(X) \mapsto d(x,A) = \inf\{d(x,y) : y \in A\}$ for every $x \in X$.*

Bibliography

The author has made no effort to list all relevant references because it would be impossible to do so within a reasonable space. Thus, many important references have been left out of the list. In particular, the references concerning derivations, automorphism groups, crossed products and modular Hilbert algebras are, with a few exceptions, excluded since these topics will be treated in the succeeding volume.

Monographs

[1] Alfsen, E., *Compact Convex Sets and Boundary Integrals*, Ergebnisse der Mathematik No. 57 Springer-Verlag, Berlin and New York, 1971.

[2] Arveson, W., *An Invitation to C*-algebras*, Graduate Texts in Mathematics, No. 39, Springer-Verlag, Berlin and New York, 1976.

[3] Bonsall, F. F., and J. Duncan, *Complete Normed Algebras*, Ergebnisse der Mathematik No. 80, Springer-Verlag, Berlin and New York, 1973.

[4] Diximer, J., *Les algèbres d'opérateurs dans l'espace Hilbertien*, 2nd ed., Gauthier—Villars, Paris, 1969.

[5] Diximer, J., *Les C*-algèbres et leurs représentations*, 2nd ed., Gauthier–Villars, Paris, 1969.

[6] Grothendieck, A., *Produits tensoriels topologiques et espaces nucléaires*, Mem. Amer. Math. Soc. No. 16, Amer. Math. Soc., Providence, Rhode Island, 1955.

[7] Guichardet, A., *Leçons sur certaines algèbres topologiques: Algèbres de von Neumann; Algèbres topologiques et fonctions homomorphes; Algèbres de Banach commutatives*, Gordon and Breach, New York, 1967.

[8] Guichardet, A., *Special Topics in Topological Algebras*, Gordon and Breach, New York, 1968.

[9] Guichardet, A., *Tensor products of C*-Algebras*, Aarhus Lecture Notes Series, Denmark, 1969.

[10] Halmos, P. R., *A Hilbert Space Problem Book*, Graduate Texts in Mathematics No. 19, Springer-Verlag, Berlin and New York, 1951.

[11] Hewitt, E., and K. A. Ross, *Abstract Harmonic Analysis, I*, Grundlehren der Mathematischen Wissenschaften No. 115, Springer-Verlag, Berlin and New York, 1963.

[12] Hille, E., *Functional Analysis and Semi-Groups*, Amer, Math. Soc. Colloquium Publications No. 31, Amer. Math. Soc., Providence, Rhode Island, 1948.

[13] Ionescu-Tulcea, A., and C. Ionescu-Tulcea, *Topics in the Theory of Lifting*, Ergebnisse der Mathematik No. 48, Springer-Verlag, Berlin and New York, 1970.

[14] Kadison, R. V., *A Representation Theory for Commutative Topological Algebra*, Mem. Amer. Math. Soc. No. 7, Amer. Math. Soc., Providence, Rhode Island, 1951.

[15] Kaplansky, I., *Functional Analysis, Some Aspects of Analysis and Probability*, Wiley, New York, 1958.

[16] Kaplansky, I., *Rings of Operators*, Benjamin, New York, 1968.

[17] Kaplansky, I., *Algebraic and Analytic Aspects of Operator Algebras*, Amer. Math. Soc., Providance, Rhode Island, 1970.

[18] Kato, T., *Perturbation Theory for Linear Operators*, 2nd ed., Grundlehren der Mathematischen Wissenschaften No. 132, Springer-Verlag, Berlin and New York, 1966.

[19] Loomis, L. H., *An Introduction to Abstract Harmonic Analysis*, University Series in Higher Mathematics, Van Nostrand, New York, 1953.

[20] Loomis, L. H., *The Lattice Theoretic background of the Dimension Theory of Operator Algebras*, Mem. Amer. Math. Soc. No. 18, Amer. Math. Soc., Providence, Rhode Island, 1955.

[21] Mackery, G. W., The Theory of Group Representations, Mimeographed Notes, University of Chicago, 1955.

[22] Mackey, G. W., *The Mathematical Foundations of Quantum Mechanics*, Benjamin, New York, 1963.

[23] Naimark, M. A., *Normed Rings* (English translation by L. F. Boron), P. Noordhoff, Groningen, 1964.

[24] Prosser, R. T., *On the Ideal Structure of Operator Algebras*, Mem. Amer. Math. Soc. No. 45, Amer. Math. Soc., Providence, RhodeIsland, 1963.

[25] Rickart, C., *General Theory of Banach Algebras*, University Series in Higher Mathematics, Van Nostrand, New York, 1960.

[26] Ruelle, D., *Statistical Mechanics*, Benjamin, New York, 1969.

[27] Sakai, S., *C*-Algebras* and *W*-Algebras*, Ergebnisse der Mathematik No. 60, Springer-Verlag, Berlin and New York, 1971.

[28] Schatten, R., *A Theory of Cross-Space*, Ann. Math. Studies No. 26, Princeton Univ. Press, Princeton, New Jersey, 1950.

[29] Schatten, R. *Norm ideals of Completely Continuous Operators*, Ergebnisse der Mathematik No. 27, Springer-Verlag, Berlin and New York, 1970.

[30] Schwartz, J. T., *W*-Algebras*, Notes on Mathematics and Its Applications, Gordon and Breach, New York.

[31] Segal, I. E., *Decomposition of Operator algebras, I and II. Multiplicity Theory*, Mem. Amer. Math. Soc. No. 9, pp. 1–67 and 1–66, Amer. Math. Soc., Providence, Rhode Island, 1951.

[32] Sinclair, A., *Automatic Continuity of Linear Operators*, Lecture Notes Series No. 21, London Math. Soc., London, 1977.

[33] Sz.-Nagy, B., *Spektraldarstellung Linearer Transformationen des Hilbertschen Raumes*, Ergebniss der Mathematik No. 5, Springer-Verlag, Berlin and New York, 1942.

[34] Takesaki, M., *The Theory of Operator Algebra*, Lecture Notes, Univ. of California, Los Angles, 1969–1970.

[35] Takesaki, M., *Tomita's Theory of Modular Hilbert Algebras and Its Applications*, Lecture Notes in Mathematics No. 128, Springer-Verlag, Berlin and New York, 1970.

[36] Topping, D. M., *Lectures on von Neumann Algebras*, Van Nostrand, New York, 1971.

[37] Weil, A., *L' intégration dans les groupes topologiques et ses applications*, 2nd ed., Act. Sc. Ind. No. 1145, Hermann, Paris, 1953.

[38] Yosida, K., *Functional Analysis*, 4th ed., Grundlehren der Mathematischen Wissenschaften No. 123, Springer-Verlag, Berlin and New York, 1974.

Papers

[39] Aarnes, J., The Vitali–Hahn–Saks theorem for von Neumann algebras, *Math. Scand.* **18** (1966), 87–92.

[40] Aarnes, J. F., and R. V. Kadison, Pure states and approximate identity, *Proc. Amer. Math. Soc.*, **21** (1969), 749–752.

[41] Akemann, C. A., The dual space of an operator algebra, *Trans. Amer. Math. Soc.* **126** (1967), 268–302.

[42] Akemann, C. A., Sequential convergence in the dual of a W^*-algebra, *Comm. Math. Phys.* **7** (1968), 222–224.

[43] Akemann, C. A., Approximate units and maximal abelian C^*-subalgebras, *Pacific J. Math.* **33** (1970), 543–550.

[44] Akemann, C. A., Left ideal structure of C^*-algebras, *J. Funct. Anal.* **6** (1970), 305–317.

[45] Akemann, C. A., Separable representations of a W^*-algebra, *Proc. Amer. Math. Math. Soc.* **24** (1970), 354–355.

[46] Akemann, C. A., and P. Ostrand, On a tensor product C^*-algebra associated with the free group on two generators, *J. Math. Soc.* Japan, **27** (1975), 589–599.

[47] Akemann, C. A., and G. K. Pedersen, Complications of semicontinuity in C^*-algebra theory, *Duke Math. J.* **40** (1973), 785–795.

[48] Anastasio, S., Maximal abelian subalgebras in hyperfinite factors, *Amer. J. Math.* **87** (1965), 955–971.

[49] Araki, H., A lattice of von Neumann algebras associated with the quantum theory of a free Bose field, *J. Math. Phys.* **4** (1963), 1343–1362.

[50] Araki, H., S. B. Smith, and L. Smith, On the homotopical significance of the type of von Neumann factors, *Comm. Math. Phys.* **22** (1971), 71–88.

[51] Arens, R., On a theorem of Gelfand and Neumark, *Proc. Nat. Acad. U.S.A.* **32** (1946), 237–239.

[52] Arens, R., Representation of *-algebras, *Duke Math. J.* **14** (1947), 269–282.

[53] Berberian, S. K., On the projection geometry of a finite AW^*-algebra, *Trans. Amer. Math. Soc.* **83** (1956), 493–509.

[54] Berberian, S. K., The regular ring of a finite AW^*-algebra, *Ann. Math.* **65** (1957), 224–240.

[55] Berberian, S. K. Trace and the convex hull of the spectrum in a von Neumann algebra of finite class, *Proc. Amer. Math. Soc.* **23** (1969), 211–212.

[56] Bichteler, K., A generalization to the non-separable case of Takesaki's duality theorem for *C^*-algebras*, *Invent. Math.* **9** (1969–1970), 89–98.

[57] Bishop, E., and K. de Leeuw, The representation of linear functionals by measures on sets of extreme points, *Ann. Inst. Fourier* (*Grenoble*) **9** (1959), 305–331.

[58] Bohnenblust, H. F., and S. Karlin, Geometrical properties of the unit sphere of Banach algebras, *Ann. Math.* **62** (1955), 217–239.

[59] Bonsall, F. F., A minimal property of the norm in some Banach algebras, *J. London Math. Soc.* **29** (1954), 156–164.

[60] Broise, M., Sur les isomorphismes de certaines algèbres de von Neumann, *Ann. Sci. Éc. Norm. Sup.* **83** (1966), 91–111.

[61] Brown, A., and C. Pearcy, commutators in factors of type III, *Canad. J. Math.* **18** (1966), 1152–1160.

[62] Bures, D., Tensor products of W^*-algebras, *Pacific J. Math.* **27** (1968), 13–37.

[63] Busby, R. C., Double centralizers and extensions of C^*-algebras, *Trans. Amer. Math. Soc.* **132** (1968), 79–99.

[64] Calkin, J. W., Two-sided ideals and congruences in the ring of bounded operators in Hilbert space, *Ann. Math.* **42** (1941), 839–873.

[65] Combes, F., Étude des représentations tracées d'une C^*-algébre, *C. R. Acad. Sci.* Paris **262** (1966), 114–117.

[66] Combes, F., Étude des poids définis sur une C^*-algèbre, *C. R. Acad. Sci. Paris*, **265** (1967), 340–343.

[67] Combes, F., Sur les états factoriels d'une C^*-algèbre, *C. R. Acad. Sci.* **265** (1967), 736–739.

[68] Combes, F., Eléments semicontinus associés à une C^*-algèbre, *C. R. Acad Sci.* **267** (1968), 986–989.

[69] Combes, F., Poids sur une C^*-algèbre, *J. Math. Pures Appl.* **47** (1968), 57–100.

[70] Combes, F., Sur les faces d'une C^*-algèbre, *Bull. Sci. Math.* (2)**93** (1969), 37–62.

[71] Combes, F., Quelques propriétés des C^*-algèbres, *Bull. Sci. Math.* **94** (1970), 165–192.

[72] Connes, A., Une classification des facteurs de Type III, *Ann. Sci. Éc. Norm. Sup., 4e Ser.*, (2)**6** (1973), 133–252.

[73] Cuculescu, I., A proof of $(A \otimes B)' = A' \otimes B'$ for von Neumann algebras, *Rev. Roumaine Math. Pures Appl.* **16** (1971), 665–670.

[74] Davies, E. B., On the Borel structure of C^*-algebras (with an appendix by R. V. Kadison), *Comm. Math. Phys.* **8** (1968), 147–163.

[75] Davies, E. B., Decomposition of traces on separable C^*-algebras, *Quart. J. Math. Oxford Ser.* (2)**20** (1969), 97–111.

[76] Davies, E. B., The structure of Σ^*-algebras, *Quart. J. Math. Oxford Ser.* (2)**20** (1969), 351–366.

[77] Dell'antonio, G. F., On the limits of sequences of normal states, *Comm. Pure Appl. Math.* **20** (1967), 413–429.

[78] Dixmier, J., Position relative de deux variétés linéaires fermées dans un espace de Hilbert, *Rev. Sci.* **86** (1948), 387–399.

[79] Dixmier, J., Les anneaux d'operateurs de classe finie, *Ann. Éc. Norm. Sup.* **66** (1949), 209–261.

[80] Dixmier, J., Les fonctionnelles linéaires sur l'ensemble des opérateurs bornés d'un espace de Hilbert, *Ann. Math.* **51** (1950), 387–408.

[81] Dixmier, J., Sur certains espaces considetes par M. H. Stone, *Summa Brasil. Math.* (2)**11** (1951), 151–182.

[82] Dixmier, J., Sur la réduction des anneaux d'opérateurs, *Ann. Sci. Éc. Norm. Sup.* **68** (1951), 185–202.

[83] Dixmier, J., Applications $\natural$ dans les anneaux d'opérateurs, *Compos. Math.* **10** (1952), 1–55.

[84] Dixmier, J., Remarques sur les applications $\natural$, *Archiv Math.* **3** (1952), 290–297.

[85] Dixmier, J., Formes linéaires sur un anneau d'operateurs, *Bull. Soc. Math. Fr.* **81** (1953), 9–39.

[86] Dixmier, J., Sur une inégalité de E. Heinz, *Math. Ann.* **126** (1953), 75–78.

[87] Dixmier, J., Sous-anneaux abéliens maximaux dans les facteurs de type fini, *Ann. Math.* **59** (1954), 279–286.

[88] Dixmier, J., Sur les anneaux d'operateurs dans les espaces hilbertiens, *C. R. Acad. Sci.* Paris **238** (1954), 439–441.

[89] Dixmier, J., Sur les C^*-algèbres, *Bull. Soc. Math. Fr.* **88** (1960), 95–112.

[90] Dixmier, J., Sur les structures boréliennes du spectre d'une C^*-algèbre, *Inst. Hautes Études Sci. Publ. Math.* **6** (1960), 5–11.

[91] Dixmier, J., Points séparés dans le spectre d'une C^*-algèbre, *Acta Sci. Math.* **22** (1961), 115–128.

[92] Diximier, J., Dual et quasi-dual d'une algèbre de Banach involutive, *Trans. Amer. Math. Soc.* **104** (1962), 278–283.

[93] Dixmier, J., Traces sur les C^*-algèbres, *Ann. Inst. Fourier* (*Grenoble*) **13** (1963), 219–262.

[94] Dixmier, J., Traces sur les C^*-algèbres, II, *Bull. Sci. Math.* **88** (1964), 39–57.

[95] Dixmier, J., Existence de traces non normales, *C. R. Acad. Sci.* Paris **262** (1966), 1107–1108.

[96] Doplicher, S., and D. Kastler, Ergodic states in a noncommutative ergodic theory, *Comm. Math. Phys.* **7** (1968), 1–20.

[97] Doplicher, S., R. V. Kadison, and D. W. Robinson, Asymptotically abelian systems, *Comm. Math. Phys.* **6** (1967), 101–120.

[98] Dunford, N., and B. J. Pettis, Linear operations on summable functions, *Trans. Amer. Math. Soc.* **47** (1940), 323–392.

[99] Dye, H. A., The Radon–Nikodym theorem for finite rings of operators, *Trans. Amer. Math. Soc.* **72** (1952), 243–280.

[100] Dye, H. A., The unitary structure in finite rings of operators, *Duke Math. J.* **20** (1953), 55–69.

[101] Dye, H. A., On the geometry of projections in certain operator algebras, *Ann. Math.* **61** (1955), 73–89.

[102] Dye, H. A., and B. Russo, A note on unitary operators in C^*-algebras, *Duke Math. J.* **33** (1966), 413–416.

[103] Effros, E. G., A decomposition theory for representations of C^*-algebras, *Trans. Amer. Math. Soc.* **107** (1963), 83–106.

[104] Effros, E. G., Order ideals in a C^*-algebra and its dual, *Duke Math. J.* **30** (1963), 391–412.

[105] Effros, E. G., The Borel space of von Neumann algebras on a separable Hilbert space, *Pacific J. Math.* **15** (1965), 1153–1164.

[106] Effros, E. G., Global structure in von Neumann algebras, *Trans. Amer. Math. Soc.* **121** (1966), 434–454.

[107] Effros, E. G., The canonical measures for a separable C^*-algebra, *Amer. J. Math.* **92** (1970), 56–60.

[108] Effros, E., and C. Lance, Tensor products of operator algebras, *Adv. Math.*, **25** (1977), 1–34.

[109] Elliott, G. A., An extension of some results of Takesaki in the reduction theory of von Neumann algebras, *Pacific J. Math.* **39** (1971), 145–148.

[110] Ernest, J., A decomposition theory for unitary representations of locally compact groups, *Trans. Amer. Math. Soc.* **104** (1962), 252–277.

[111] Feldman, J., Embedding of AW^*-algebras, *Duke Math. J.* **23** (1956), 303–308.

[112] Feldman, J., Isomorphisms of finite type II rings of operators, *Ann. Math.* **63** (1956), 565–571.

[113] Feldman, J., Nonseparability of certain finite factors, *Proc. Amer. Math. Soc.* **7** (1956), 23–26.

[114] Feldman, J., Some connections between topological and algebraical properties in rings of operators, *Duke Math. J.* **23** (1956), 365–370.

[115] Feldman, J., The uniformly closed ideals in an AW^*-algebra, *Bull. Amer. Math. Soc.* **62** (1958), 245–246.

[116] Feldman, J., Borel sets of states and of representations, *Michigan Math. J.* **12** (1965), 363–366.

[117] Feldman, J., and J. M. G. Fell, Separable representations of rings of operators, *Ann. Math.* **65** (1957), 241–249.

[118] Feldman, J., and R. V. Kadison, The closure of the regular operators in a ring of operators, *Proc. Amer. Math. Soc.* **5** (1954), 909–916.

[119] Fell, J. M. G., Representations of weakly closed algebras, *Math. Ann.* **133** (1957), 118–126.

[120] Fell, J. M. G., The dual spaces of C^*-algebras, *Trans. Amer. Math. Soc.* **94** (1960), 365–403.

[121] Fell, J. M. G., The structure of algebras of operator fields, *Acta Math* **106** (1961), 233–280.

[122] Fell, J. M. G., Weak containment and induced representations of groups, *Canad. J. Math.* **14** (1962), 237–268.

[123] Fillmore, P., and D. Topping, A direct integral decomposition for certain operator algebras, *Amer. J. Math.* **91** (1969), 11–17.

[124] Flensted-Jensen, M., A note on disintegration, type and global type of von Neumann algebras, *Math. Scand.* **24** (1969), 232–238.

[125] Foias, C., and I. Kovacs, Une charactérisation nouvelle des algèbres de von Neumann finies, *Acta Univ. Szeged* **23** (1962), 274–278.

[126] Fuglede, B., and R. V. Kadison, On a conjecture of Murray and von Neumann, *Proc. Nat. Acad. Sci. U.S.A.* **37** (1951), 420–425.

[127] Fuglede, B., and R. V. Kadison, On determinants and a property of the trace in finite factors, *Proc. Nat. Acad. Sci. U.S.A.* **37** (1951), 425–431.

[128] Fuglede, B., and R. V. Kadison, Determinant theory in finite factors," *Ann. Math.* **55** (1952), 520–530.

[129] Fukamiya, M., On B^*-algebras, *Proc. Jap. Acad.* **27** (1951), 321–327.

[130] Fukamiya, M., On a theorem of Gelfand and Neumark and the B^*-algebra, *Kumamoto J. Sci.* **1** (1952), 17–22. (See also the review by Schatz, J. A., *Math. Rev.* **14** (1953), 884.)

[131] Fukamiya, M., M. Misonou, and Z. Takeda, On order and commutativity of B^*-algebras, *Tôhoku Math. J.* **6** (1954), 89–93.

[132] Garling, D. J. H., On ideals of operators in Hilbert space, *Proc. London Math. Soc.* **17** (1967), 115–138.

[133] Gelfand, I. M., Normierte Ringe, *Mat. Sbornik, N.S.* (51) **9** (1941), 3–24.

[134] Gelfand, I., and M. Naimark, On the imbedding of normed rings into the ring of operators in Hilbert space, *Mat. Sbornik* **12** (1943), 197–213.

[135] Gelfand, I. M., and D. Raikov, Irreducible unitary representations of locally bicompact groups, *Mat. Sbornik, N.S.* (55)**13** (1943), 301–316.

[136] Gleason, A. M., Measures on the closed subspaces of a Hilbert space, *J. Math. Mech.* **6** (1957), 885–894.

[137] Glimm, J., A Ston–Weierstrass theorem for C^*-algebras, *Ann. Math.* **72** (1960), 216–244.

[138] Glimm, J., Type I C^*-algebras, *Ann. Math.* **73** (1961), 572–612.

[139] Glimm, J., and R. V. Kadison, Unitary operators in C^*-algebras, *Pacific J. Math.* **10** (1960), 547–556.

[140] Godement, R., Théorie générale des sommes continues d'espaces de Banach, *C. R. Acad. Sci.* Paris **228** (1949), 1321–1323.

[141] Godement, R., Sur la théorie des caractères, I. Definition et classification des caractères, *C. R. Acad. Sci.* Paris **229** (1949), 967–969.

[142] Godement, R., Memoire sur la theorie des caracteres dans les groupes localement compacts unimodulaires, *J. Math. Pures Appl.* **30** (1951), 1–110.

[143] Godement, R., Sur la thèorie des représentations untaires, *Ann, Math* **53** (1951), 68–124.

[144] Goldman, M., Structure of AW^*-algebras, *I, Duke Math. J.* **23** (1956), 23–34.

[145] Goldman, M., On subfactors of factors of type II_1, *Michigan Math. J.* **6** (1959), 167–172.

[146] Griffin, E. L., Some contributions to the theory of rings of operators, *Trans. Amer. Math. Soc.* **75** (1953), 471–504.

[147] Griffin, E. L., Some contributions to the theory of rings of operators, II, *Trans. Amer. Math. Soc.* **79** (1955), 389–400.

[148] Grothendieck, A., Sur les applications linéaires faiblement compactes d'espaces du type $C(K)$, *Canad. J. Math.* **5** (1953), 129–173.

[149] Grothendieck, A., Un résultat sur le dual d'une C^*-algèbre, *J. Math. Pures Appl.* **36** (1957), 97–108.

[150] Guichardet, A., Sur un problème posé par G. W. Mackey, *C. R. Acad. Sci.* Paris **250** (1960), 962–963.

[151] Guichardet, A., Sur les structures boréliannes du dual et du quasi-dual d'une C^*-algebre, *C. R. Acad. Sci. Paris* **253** (1961), 2030–2032.

[152] Guichardet, A., Une caractérisation des algèbres de von Neumann discrètes, *Bull. Soc. Math. Fr.* **89** (1961), 77–101.

[153] Guichardet, A., Caractères et représentations des produits tensoriels de C^*-algèbres, *Ann. Sci Éc. Norm. Sup.*, **81** (1964), 189–206.

[154] Guichardet, A., Sur la décomposition des représentations des C^*-algèbres, *C. R. Acad. Sci.* Paris **258** (1964), 768–770.

[155] Guichardet, A., Tensor products of C^*-algebras, *Dokl. Akad. Nauk SSSR* **160** (1965), 986–989.

[156] Guichardet, A., Disintegration of quasi-invariant states on C^*-algebras, *Proc. Funct. Anal. Week, March 3–7, 1969, Aarhus Univ.*

[157] Guichardet, A., and D. Kastler, Désintegration des états quasi-invariants des C^*-algèbres, *J. Math. Pures Appl.* **49** (1970), 349–380.

[158] Haag, R., R. V. Kadison, and D. Kastler, Nets of C^*-algebras and classification of states, *Comm. Math. Phys.* **16** (1970), 81–104.

[159] Halmos, P. R., and J. von Neumann, Operator methods in classical mechanics, II, *Ann. Math.* **43** (1942), 332–350.

[160] Halperin, I., A remark on a preceding paper by J. von Neumann, *Ann. Math.* **41** (1940), 554–555.

[161] Halperin, I., Introduction to von Neumann algebras and continuous geometry, *Canad. Math. Bull.* **3** (1960), 273–288, and **5** (1960), 59.

[162] Halpern, H., An integral representation of a normal functional on a von Neumann algebra, *Trans. Amer. Math. Soc.* **125** (1966), 32–46.

[163] Harris, R. T., A direct integral construction, *Duke Math. J.* **33** (1966), 535–537.

[164] Hasumi, M., The extension property of complex Banach spaces, *Tôhoku Math. J.* 10 (1958), 135–142.

[165] Heinz, E., Beiträge zur Störungstheorie der Spektralzerlegung, *Math. Ann.* **123** (1951), 415–438.

[166] Herman, R., and M. Takesaki, The comparability theorem for cyclic projections, *Bull. London Math. Soc.* **9** (1977), 186–187.

[167] Hugenholtz, N. M., On the factor type of equilibrium states in quantum statistical mechanics, *Comm. Math. Phys.* **6** (1967), 189–193.

[168] Jacobson, N., and C. Rickart, Homorphisms of Jordan rings, *Trans. Amer. Math. Soc.* **69** (1950), 479–502.

[169] Johnson, B., An introduction to the theory of centralizers, *Proc. London Math. Soc.* **14** (1964), 299–320.

[170] Johnson, B. E., Continuity of homomorphisms of algebras of operators, *J. London Math. Soc.* **42** (1967), 537–541.

[171] Johnson, B. E., The uniqueness of the (complete) norm topology, *Bull. Amer. Math. Soc.* **73** (1967), 537–539.

[172] Kadison, R. V., Isometries of operator algebras, *Ann. Math.* **54** (1951), 325–338.

[173] Kadison, R. V., Order properties of bounded self-adjoint operators, *Proc. Amer. Math. Soc.* **2** (1951), 505–510.

[174] Kadison, R. V., A generalized Schwarz inequality and algebraic invariants for operator algebras, *Ann. Math.* **56** (1952), 494–503.

[175] Kadison, R. V., Infinite unitary groups, *Trans. Amer. Math. Soc.* **72** (1952), 386–399.

[176] Kadison, R. V., Infinite general linear groups, *Trans. Amer. Math. Soc.* **76** (1954), 66–91.

[177] Kadison, R. V., Isomorphisms of factors of infinitive type, *Canad. J. Math.* **7** (1955), 322–327.

[178] Kadison, R. V., Multiplicity theory for operator algebras, *Proc. Nat. Acad. Sci. U.S.A.* **41** (1955), 169–173.

[179] Kadison, R. V., On the additivity of the trace in finite factors, *Proc. Nat. Acad. Sci. U.S.A.* **41** (1955), 385–387.

[180] Kadison, R. V., The general linear group of infinite factors, *Duke Math. J.* **22** (1955), 119–122.

[181] Kadison, R. V., Operator algebras with a faithful weakly-closed representation, *Ann. Math.* **64** (1956), 175–181.

[182] Kadison, R. V., Irreducible operator algebras, *Proc. Nat. Acad. Sci. U.S.A.* **43** (1957), 273–276.

[183] Kadison, R. V., Unitary invariants for representations of operator algebras, *Ann. Math.* **66** (1957), 304–379.

[184] Kadison, R. V., Theory of operators; Part II Operator algebras, *Bull. Amer. Math. Soc.* **64** (1958), 61–85.

[185] Kadison, R. V., The trace in finite operator algebras, *Proc. Amer. Math. Soc.* **12** (1961), 973–977.

[186] Kadison, R. V., Normalcy in operator algebras, *Duke Math. J.* **29** (1962), 459–464.

[187] Kadison, R. V., States and represention, *Trans. Amer. Math. Soc.* **103** (1962), 304–319.

[188] Kadison, R. V., Transformations of states in operator theory and dynamics, *Topology* **3** (1965), 177–198.

[189] Kadison, R. V., Strong continuity of operator functions, *Pacific J. Math.* **26** (1968), 121–129.

[190] Kadison, R. V., and G. K. Pedersen, Equivalence in operator algebras, *Math. Scand.* **27** (1970), 205–222.

[191] Kadison, R. V., and I. M. Singer, Extensions of pure states, *Amer. J. Math.* **81** (1959), 383–400.

[192] Kakutani, S., Concrete representation of abstract (*L*)-spaces and the mean ergodic theorem, *Ann. Math.* **42** (1941), 523–537.

[193] Kakutani, S., Concrete representation of abstract (*M*)-spaces (A characterization of the space of continuous functions), *Ann. Math.* **42** (1941), 994–1024.

[192] Kaplansky, I., Normed algebras, *Duke Math. J.* **16** (1949), 399–418.

[195] Kaplansky, I., Quelques résultats sur les anneaux d'opérateurs, *C. R. Acad. Sci.* Paris **231** (1950), 485–486.

[194] Kaplansky, I., A theorem on rings of operators, *Pacific J. Math.* **1** (1951), 227–232.

[197] Kaplansky, I., Projections in Banach algebras, *Ann. Math.* **53** (1951), 235–249.

[198] Kaplansky, I., The structure of certain operator algebras, *Trans. Amer. Math. Soc.* **70** (1951), 219–255.

[199] Kaplansky, I., Algebras of type I, *Ann. Math.* **56** (1952), 460–472.

[200] Kaplansky, I., Any orthocomplemented complete modular lattice is a continuous geometry, *Ann. Math.* **61** (1955), 524–541.

[201] Kastler, D., Topics in the algebraic approach to field theory, *Cargèse Lectures in Theoretical Physics: Application of Mathematics to Problems in Theoretical Physics* (*Cargèse, 1955*), pp. 289–302, Gordon and Breach, New York, 1967.

[202] Kastler, D., M. Mebkhout, G. Loupias, and L. Michel, Central decomposition of invariant states. Applications to the groups of time translations and of Euclidean transformations in algebraic field theory, *Comm. Math. Phys.* **27** (1972), 195–222.

[203] Kastler, D., and W. Robinson, Invariant states in statistical mechanics, *Comm. Math. Phys.* **3** (1966), 151–180.

[204] Kehlet, E. T., On the monotone sequential closure of a C^*-algebra, *Math. Scand.* **25** (1969), 59–70.

[205] Kelley, J. L., Banach spaces with the extension property, *Trans. Amer. Math. Soc.* **72** (1952), 323–326.

[206] Kelley, J. L., and R. L. Vaught, The positive cone in Branch algebras, *Trans. Amer. Math. Soc.* **74** (1953), 44–55.

[207] Kondo, M., Sur la notion de dimension, *Proc. Imp. Acad. Tokyo* **19** (1943), 215–223.

[208] Korányi, A., On a theorem of Löwner and its connections with resolvents of self-adjoint transformations, *Acta Sci. Math. Szeged* **17** (1956), 63–70.

[209] Kovács, I., Un complement à la théorie de l'integration noncommutative, *Acta Sci. Math. Szeged* **21** (1960), 7–11.

[210] Kovács, I., Théorèmes ergodiques non commutatifs, *C. R. Acad. Sci.* Paris **253** (191), 770–771.

[211] Kovács, I., Ergodic theorems for gages, *Acta Sci. Math. Szeged* **24** (1963), 103–118.

[212] Kovács, I., and J. Szücs, Ergodic type theorems in von Neumann algebras, *Acta Sci, Math, Szeged* **27** (1966), 233–246.

[213] Kuiper, N., The homotopy type of the unitary group of Hilbert space, *Topology* **3** (1965), 19–30.

[214] Kunze, R. A., L^p–Fourier transforms on locally compact unimodular groups, *Trans Amer. Math. Soc.* **89** (1958), 519–540.

[215] Lance, C., On nuclear C^*-algebras, *J. Funct. Anal.* **12** (1973), 157–176.

[216] Lanford, O., and D. Ruelle, Integral representations of invariant states on B^*-algebras, *J. Math. Phys.* **8** (1966), 1460–1463.

[217] Leptin, H., Reduktion linearer Funktionale auf Operatorringen, *Abh. Math. Sem. Univ. Hamburg* **22** (1958), 98–113.

[218] Leptin, H., Zur Reduktionstheorie Hilbertscher Räume, *Math. Z.* **69** (1958), 40–58.

[219] Loomis, L. H., Unique direct integral decompositions of convex sets, *Amer. J. Math.* **84** (1962), 509–526.

[220] Löwner, K., Uber monotone Matrixfunctionen, *Math. Z.* **38** (1934), 177–216.

[221] McDuff, D., Uncountably many II_1 factors, *Ann. Math.* **90** (1969), 372–377.

[222] Mackey, G. W., Borel structure in groups and their duals, *Trans, Amer. Math. Soc.* **85** (1957), 134–165.

[223] Mackey, G. W., Point realizations of transformation groups, *Illinois J. Math.* **6** (1962), 327–335.

[224] Maeda, F., Relative dimensionality in operator rings, *J. Sci. Hiroshima Univ.* 11 (1941), 1–6.

[225] Maeda, S., Dimension functions on certain general lattices, *J. Sci. Hiroshima Univ.* **19** (1955), 211–237.

[226] Maeda, S., Lengths of projections in rings of operators, *J. Sci. Hiroshima Univ.* **20** (1956), 5–11.

[227] Maréchal, O., Champs mesurables, d'espaces hilbertiens, *Bull. Sci. Math.* **93** (1969), 113–143.

[228] Maréchal, O., Champs measurables d'espaces hilbertiens, *C. R. Acad. Sci. Paris* **270** (1970), 1316–1319.

[229] Mautner, F. I., The completeness of the irreducible unitary representations of a locally compact group, *Proc. Nat. Acad. Sci. U.S.A.* **34** (1948), 52–54.

[230] Mautner, F. I., Unitary representations of locally compact groups, I, *Ann. Math.* **51** (1950), 1–25.

[231] Mautner, F. I., Unitary representations of locally compact groups, II, *Ann. Math.* **52** (1950), 528–556.

[232] Mazur, S., Sur les anneaux linéaires, *C. R. Acad. Sci. Paris, Ser. A-B* **207** (1938), 1025–1027.

[233] Miles, P. E., B^*-algebra unit ball extremal points, *Pacific J. Math.* **14** (1964), 627–637.

[234] Miles, P. E., Order isomorphisms of B^* algebras, *Trans. Amer. Math. Soc.* **107** (1963), 217–236.

[235] Misonou, Y., On a weakly central operator algebra, *Tôhoku Math. J.* **4** (1952), 194–202.

[236] Misonou, Y., Operator algebras of type I, *Kōdai Math. Sem. Rep.* **5** (1953), 87–90.

[237] Misonou, Y., Unitary equivalence of factors of type III, *Proc. Jap. Acad.* **29** (1953), 482–485.

[238] Misonou, Y., On the direct product of W^*-algebras, *Tôhoku Math. J.* **6** (1954), 189–204.

[239] Misonou, Y., On divisors of factors, *Tôhoku Math. J.* **8** (1956), 63–69.

[240] Murray, F. J., and J. von Neumann, On rings of operators, *Ann. Math.* **37** (1936), 116–229.

[241] Murray, F. J., and J. von Neumann, On rings of operators, II, *Trans. Amer. Math. Soc.* **41** (1937), 208–248.

[242] Murray, F. J., and J. von Neumann, On rings of operators, IV, *Ann. Math.* **44** (1943), 716–808.

[243] Nagumo, M., Einige analytische Unetersuchungen in linearen metrischen Ringen, *Japanese J. Math.* **13** (1936), 61–80.

[244] Naimark, M. A., and S. V. Fomin, Sommes directes continues d'espaces hibertiens et quelques applications, *Usp. Mat. Nauk* **10** (1955), 111–142.

[245] Nakamura, M., On the direct product of finite factors, *Tôhoku Math. J.* **6** (1954), 205–207.

[246] Nakamura, M. A proof of a theorem of Takesaki, *Kōdai Math. Sem. Rep.* **10** (1958), 189–190.

[247] Nakamura, M., and Z. Takeda, Normal states of commutative operator algebras, *Tôhoku Math. J.* **5** (1953), 109–121.

[248] Nakamura, M., M. Takesaki, and H. Umegaki, A remark on the expectation of operator algebras, *Kōdai Math. Sem. Rep.* **12** (1960), 82–90.

[249] Nakamura, M., and T. Turumaru, Simple algebras of completely continuous operators, *Tôhoku Math. J.* **4** (1952), 303–308.

[250] Nakamura, M., and T. Turumaru, Expectations in an operator algebra, *Tôhoku Math. J.* **6** (1954), 182–188.

[251] Nakamura, M., and T. Turumaru, On extensions of pure states of an abelian operator algebra, *Tôhoku Math. J.* **6** (1954), 253–257.

[252] Namioka, I., and E. Asplund, A geometric proof of Ryll-Nardzewski's fixed point theorem, *Bull. Amer. Math. Soc.* **73** (1967), 443–445.

[253] Nelson, E., Non-commutative integration theory, *J. Funct. Anal.* **15** (1974), 103–116.

[254] Von Neumann, J., Zur Algebra der Funktionaloperationen und Theorie der normalen Operatoren, *Math. Ann.* **102** (1929), 370–427.

[255] Von Neumann, J., On a certain topology for rings of operators, *Ann. Math.* **37** (1936), 111–115.

[256] Von Neumann, J., On infinite direct products, *Compos. Math.* **6** (1938), 1–77.

[257] Von Neumann, J., On rings of operators, III, *Ann. Math.* **41** (1940), 94–161.

[258] Von Neumann, J., On some algebraical properties of operator rings, *Ann. Math.* **44** (1943), 709–715.

[259] Von Neumann, J., On rings of operators. Reduction theory, *Ann. Math.* **50** (1949), 401–485.

[260] Nielsen, O., Borel sets of von Neumann algebras, *Amer. J. Math.* **95** (1973), 145–164.

[261] Niiro, F., Sur l'unicité de la décomposition d'une trace, *Sci. Papers College Gen. Ed. Univ. Tokyo* **13** (1963), 159–162.

[262] Nussbaum, A. E., On the integral representation of positive linear functionals, *Trans. Amer. Math. Soc.* **128** (1967), 460–473.

[263] Ogasawara, T., Finite-dimensionality of certain Banach algebras, *J. Sci. Hiroshima Univ.* **17** (1954), 359–364.

[264] Ogasawara, T., A theorem on operator algebras, *J. Sci. Hiroshima Univ.* **18** (1955), 307–309.

[265] Ogasawara, T., Topologies on rings of operators, *J. Sci. Hiroshima Univ.* **19** (1956), 255–272.

[266] Ogasawara, T., and S. Maeda, A generalization of a theorem of Dye, *J. Sci. Hiroshima Univ.* **20** (1957), 1–4.

[267] Ogasawara, T., and K. Yoshinaga, A characterization of dual B^*-algebras, *J. Sci. Hiroshima Univ.* **18** (1954), 179–182.

[268] Ogasawara, T., and K. Yoshinaga, Weakly completely continuous Banach $*$-algebras, *J. Sci. Hiroshima Univ.* **18** (1954), 15–36.

[269] Ogasawara, T., and K. Yoshinaga, A non-commutative theory of integration for operators, *J. Sci. Hiroshima Univ.* **18** (1955), 311–347.

[270] Ogasawara, T., and K. Yoshinaga, Extension of $\natural$-application to unbounded operators, *J. Sci. Hiroshima Univ.* **19** (1956), 273–299.

[271] Okayasu, T., On the tensor product of C^*-algebras, *Tôhoku Maths. J.* **18** (1966), 325–331.

[272] Okayasu, T., Some crossed norms which are not uniformly cross, *Proc. Jap. Acad.* **46** (1970), 54–57.

[273] Orihara, M., Rings of operators and their traces, *Mem. Fac. Sci. Kyushu Univ., Ser. A* **5** (1950), 107–138, and **8** (1953), 89–91.

[272] Pallu de la Barrière, R., Sur les algèbres d'opérateurs dans les espaces hilbertiens, *Bull. Soc. Math. Fr.* **82** (1954), 1–52.

[275] Pearcy, C., and J. R. Ringrose, Trace-preserving isomorphisms in finite operator algebras, *Amer. J. Math.* **90** (1968), 444–455.

[276] Pedersen, G. K., Measure theory for C^*-algebras, *Math. Scand.* **19** (1966), 131–145.

[277] Pedersen, G. K., Measure theory for C^*-algebras, II, *Math. Scand.* **22** (1968), 63–74.

[278] Pedersen, G. K., A decomposition theorem for C^*-algebras, *Math. Scand.* **22** (1968), 266–268.

[279] Pedersen, G. K., On weak and monotone σ-closures and C^*-algebras, *Comm. Math. Phys.* **11** (1968–1969), 221–226.

[280] Pedersen, G. K., Measure theory for C^*-algebras, III, IV, *Math. Scand.* **25** (1969), 71–93, and **25** (1969), 121–127.

[281] Pedersen, G. K., The 'up–down' problem for operator algebras, *Proc. Nat. Acad. Sci. U.S.A.* **68** (1971), 1896–1897.

[282] Pedersen, G. K., Applications of weak*-semi-continuity in C^*-algebra theory, *Duke Math. J.* **39** (1972), 431–450.

[283] Pedersen, G. K., C^*-Integrals, an Approach to Non-Commutative Measure Theory, Ph.D. Thesis, Univ. of Copenhagen, 1972.

[284] Pedersen, G. K., Operator algebras with weakly closed abelian sub-algebras, *Bull. London Math. Soc.* **4** (1972), 171–175.

[285] Pedersen, G. K., Some operator monotone functions, *Proc. Amer. Math. Soc.* (1972), 309–310.

[286] Pedersen, G. K., Borel structure in operator algebras, *Danske Vid. Selsk. Mat.-Fys. Medd.* (5) **39** (1974), 1–13.

[287] Pedersen, G. K., and N. H. Petersen, Ideals in a C^*-algebra, *Math. Scand.* **27** (1970), 193–204.

[288] Phillips, R., On linear transformations, *Trans. Amer. Math. Soc.* **48** (1940), 516–541.

[289] Powers, R. T., Representations of uniformly hyperfinite algebras and their associated von Neumann rings, *Ann. Math.* **86** (1967), 138–171.

[290] Powers, R. T., Simplicity of the C^*-algebra associated with the free group on two generators, *Duke Math. J.* **49** (1975), 151–156.

[291] Pukanszky, L., The theorem of Radon–Nikodym in operator rings, *Acta Sci. Math. Szeged* **15** (1954), 149–156.

[292] Pukanszky, L., Some examples of factors, *Publications Math.* **4** (1956), 135–156.

[293] Pukanszky, L., On maximal abelian subrings of factors of type II_1, *Canad. J. Math.* **12** (1960), 289–296.

[294] Rickart, C., Banach algebras with an adjoint operation, *Ann. Math.* **47** (1946), 528–550.

[295] Rickart, C., The uniqueness of norm problem in Banach algebras, *Ann. Math.* **51** (1950), 615–628.

[296] Rieffel, M. A., and A. van Daele, The commutation theorem for tensor products of von Neumann algebras, *Bull. London Math. Soc.* **7** (1975), 257–260.

[297] Robinson, D. W., and D. Ruelle, Extremal invariant states, *Ann. Inst. H. Poincaré* **6** (1967), 299–310.

[298] Rosenberg, A., The number of irreducible representations of simple rings with no minimal ideals, *Amer. J. Math.* **75** (1953), 523–530.

[299] Ruelle, D., States of physical systems, *Comm. Math. Phys.* **3** (1966), 133–150.

[300] Ruelle, D., Integral representation of states on a C^*-algebra, *J. Funct. Anal.* **6** (1970), 116–151.

[301] Ryll-Nardzewski, C., On fixed points of semi-groups of endomorphisms of linear spaces, *Proc. Fifth Berkeley Sympos. Math. Statist. and Prob., Berkeley*, California, 1965–1966, *Vol. II: Contributions to Probability Theory, Part I*, pp. 55–61, Univ. Calif. Press, Berkeley, California, 1967.

[302] Saitô, K., Non-commutative extension of Lusin's theorem, *Tôhoku Math. J.* (1967), 332–340.

[303] Saitô, K., On the preduals of W^*-algebras, *Tôhoku Math. J.* **19** (1967), 324–331.

[304] Saitô T., and J. Tomiyama, Some results on the direct product of W^*-algebras, *Tôhoku Math. J.* **12** (1960), 455–458.

[305] Sakai, S., A characterization of W^*-algebras, *Pacific J. Math.* **6** (1956), 763–773.

[306] Sakai, S., On the σ-weak topology of W^*-algebras, *Proc. Jap. Acad.* **32** (1956), 329–332.

[307] Sakai, S., On topological properties of W^*-algebras, *Proc. Jap. Acad.* **33** (1957), 439–444.

[308] Sakai, S., On linear functionals of W^*-algebras, *Proc. Jap. Acad.* **34** (1958), 571–574.

[309] Sakai, S., On the reduction theory of von Neumann, *Bull. Amer. Math. Soc.* **70** (1964), 393–398.

[310] Sakai, S., Weakly compact operators on operator algebras, *Pacific J. Math.* **14** (1964), 659–664.

[311] Sakai, S., On the central decomposition for positive functionals on C^*-algebras, *Trans. Amer. Math. Soc.* **118** (1965), 406–419.

[312] Sakai, S., On topologies of finite W^*-algebras, *Illinois J. Math.* **9** (1965), 236–241.

[313] Sakai, S., On pure states of C^*-algebras, *Proc. Amer. Math. Soc.* **17** (1966), 86–87.

[314] Sakai, S., On the tensor product of W^*-algebras, *Amer. J. Math.* **90** (1968), 335–341.

[315] Sankaran, S., Decomposition of von Neumann algebras of type I, *Math. Ann.* **142** (1961), 399–406.

[316] Sasaki, U., Lattices of projections in AW^*-algebras, *J. Sci. Hiroshima Univ.* **19** (1955), 1–30.

[317] Schatten, R., On the direct product of Banach spaces, *Trans. Amer. Math. Soc.* **53** (1943), 195–217.

[318] Schatten, R., The cross-space of linear transformations, *Ann. Math.* **47** (1946), 73–84.

[319] Schatten, R., and J. Von Neumann, The cross-space of linear transformations, II, *Ann. Math.* **47** (1946), 608–630.

[320] Schatten, R., and J. von Neumann, The cross-space of linear transformations, III, *Ann. Math.* **49** (1948), 557–582.

[321] Schwartz, J. T., Non-isomorphism of a pair of factors of type III, *Comm. Pure Appl. Math.* **16** (1963), 111–120.

[322] Schwartz, J. T., Two finite, non-hyperfinite, non-isomorphic factors, *Comm. Pure Appl. Math.* **16** (1963), 19–26.

[323] Schwartz, J. T., Type II factors in a central decomposition, *Comm. Pure Appl. Math.* **16** (1963), 247–252.

[324] Segal, I. E., Irreducible representations of operator algebras, *Bull. Amer. Math. Soc.* **53** (1947), 73–88.

[325] Segal, I. E., Two-sided ideals in operator algebras, *Ann. Math.* **50** (1949), 856–865.

[326] Segal, I. E., A non-commutative extension of abstract integration, Ann. Math., **57** (1953), 401–457; 595–596.

[327] Segal, I. E., Abstract probability spaces and a theorem of Kolmogoroff, *Amer. J. Math.* **76** (1954), 721–732.

[328] Segal, I. E., Algebraic integration theory, *Bull. Amer. Math. Soc.* **71**(1965), 419–489.

[329] Sherman, S., The second adjoint of a C^*-algebra, *Proc. In. Cong. Math. Cambridge* **1** (1950), 470.

[330] Sherman, S., Order in operator algebras, *Amer. J. Math.* **73** (1951), 227–232.

[331] Shields, P. C., The new topology for von Neumann algebras, *Bull. Amer. Math. Soc.* **65** (1959), 267–269.

[332] Singer, I. M., Automorphisms of finite factors, *Amer. J. Math.* **77** (1955), 117–133.

[333] Skau, C., Orthogonal measures on the state space of a C^*-algebra. Algebras in Analysis, *Proc. Instr. Conf. and NATO Advanced Study Inst., Birmingham, 1973*, pp. 272–303, Academic Press, New York, 1975.

[334] Stinespring, W. F., Positive functions on C^*-algebras, *Proc. Amer. Math. Soc.* **6** (1955), 211–216.

[335] Stone, M. H., Applications of the theory of Boolean rings to general topology, *Trans. Amer. Math. Soc.* **41** (1937), 375–481.

[336] Størmer, E., Positive linear maps of operator algebras, *Acta Math.* **110** (1963), 233–278.

[337] Størmer, E., Large groups of automorphisms of C^*-algebras, *Comm. Math. Phys.* **5** (1967), 1–22.

[338] Størmer, E., Two-sided ideals in C^*-algebras, *Bull. Amer. Math. Soc.* **73** (1967), 254–257.

[339] Størmer, E., A characterization of pure states of C^*-algebras, *Proc. Amer. Math. Soc.* **19** (1968), 1100–1102.

[340] Sunouchi, H., A characterization of the maximal ideal in a factor of the case (II_∞), *Kōdai Math. Sem. Rep* **7** (1955), 65–66.

[341] Sunouchi, H., Infinite Lie rings, *Tôhoku Math. J.* **8** (1956), 291–307.

[342] Suzuki, N., On automorphisms of W^*-algebras leaving the center elementwise invariant, *Tôhoku Math. J.* **7** (1955), 186–191.

[343] Suzuki, N., On the invariants of W^*-algebras, *Tôhoku Math. J.* **7** (1955), 177–185.

[344] Takeda, Z., Perfection of measure spaces and W^*-algebras, *Kōdai Math. Sem. Rep.* **5** (1953), 23–26.

[345] Takeda, Z., Conjugate spaces of operator algebras, *Proc. Jap. Acad.* **30** (1954), 90–95.

[346] Takeda, Z., On the representation of operator algebras, *Proc. Jap. Acad.* **30** (1954), 299–304.

[347] Takeda, Z., On the representation of operator algebras II, *Tôhoku Math. J.* **6** (1954), 212–219.

[348] Takeda, Z., and T. Turumaru, On the property "Position p'", *Math. Jap.* **2** (1952), 195–197.

[349] Takemoto, H., W^*-algebra with a nonseparable cyclic representation, *Tôhoku Math. J.* **20** (1968), 567–576.

[350] Takemoto, H., On the homomorphism of von Neumann algebra, *Tôhoku Math. J.* **21** (1969), 152–157, and **22** (1970), 210–211.

[351] Takesaki, M., A note on the cross-norm of the direct product of operator algebras, *Kōdai Math. Sem. Rep.* **10** (1958), 137–140.

[352] Takesaki, M., On the conjugate space of an operator algebra, *Tôhoku Math. J.* **10** (1958), 194–203.

[353] Takesaki, M., On the direct product of W^*-factors, *Tôhoku Math. J.* **10** (1958), 116–119.

[354] Takesaki, M., On the singularity of a positive linear functional on operator algebra, *Proc. Jap. Acad.* **35** (1959), 365–366.

[355] Takesaki, M., On the nonseparability of singular representations of operator algebras, *Kōdai Math. Sem. Rep.* **12** (1960), 102–108.

[356] Takesaki, M., On the unitary equivalence among the components of decompositions of representations of involutive Banach algebras and the associated diagonal algebras, *Tôhoku Math. J.* **15** (1963), 365–393, and **16** (1964), 226–227.

[357] Takesaki, M., On the cross-norm of the direct product of C^*-algebras, *Tôhoku Math. J.* **16** (1964), 111–122.

[358] Takesaki, M., A duality in the representation theory of C^*-algebras, *Ann. Math.* **85** (1967), 370–382.

[359] Takesaki, M., Remarks on the reduction theory of von Neumann algebras, *Proc. Amer. Math. Soc.* **20** (1969), 434–438.

[360] Takesaki, M., A Short Proof for the Commutation Theorem $(\mathcal{M}_1 \overline{\otimes} \mathcal{M}_2)' = \mathcal{M}'_1 \overline{\otimes} \mathcal{M}'_2$, *Lecture Notes in Mathematics*, No. **247**, 1971, Springer-Verlag, Berlin and New York, 780–786.

[361] Takesaki, M., The quotient algebra of a finite von Neumann algebra, *Pacific J. Math.* **36** (1971), 827–831.

[362] Takesaki, M., Duality for crossed products and the structure of von Neumann algebras of type III, *Acta Math.* **131** (1973), 249–310.

[363] Takesaki, M., Faithful states on a C^*-algebra, *Pacific J. Math.* **52** (1974), 605–610.

[364] Tauer, R. J., Maximal abelian subalgebras in finite factors of type II, *Trans. Amer. Math. Soc.* **114** (1965), 281–308.

[365] Tauer, R. J., Semi-regular maximal abelian subalgebras in hyperfinite factors, *Bull. Amer. Math. Soc.* **71** (1965), 606–608.

[366] Taylor, J. L., The Tomita decomposition of rings of operators, *Trans. Amer. Math. Soc.* **113** (1964), 30–39.

[367] Teleman, S., Sur les algèbres de J. von Neumann, *Bull. Sci. Math* **82** (1958), 117–126.

Thoma, E., Zur Reduktionstheorie in separablen Hilbert-Räumen, *Math. Z.* **67** (1957), 1–9.

[369] Thoma, E., Zur Reduktionstheorie in allgemeinen Hilbert-Räumen, *Math. Z.* **68** (1957), 153–188.

[370] Tomita, M., On rings of operators in non-separable Hilbert spaces, *Mem. Fac. Sci. Kyushu Univ.* **7** (1953), 129–168.

[371] Tomita, M., Representations of operator algebras, *Math. J. Okayama Univ.* **3** (1954), 147–173.

[372] Tomita, M., Spectral theory of operator algebras, I, *Math. J. Okayama Univ.* **9** (1959), 63–98.

[373] Tomita, M., Spectral theory of operator algebras, II, *Math. J. Okayama Univ.* **10** (1960), 19–60.

[374] Tomita, M., Quasi-Standard von Neumann Algebras, Mimeographed Notes, Kyushu Univ., 1967.

[375] Tomita, M., The second dual of a C^*-algebra, *Mem. Fac. Sci. Kyushu Univ.* **21** (1967), 185–193.

[376] Tomiyama, J., On the projection of norm one in W^*-algebras, *Proc. Jap. Acad.* **33** (1957), 608–612.

[377] Tomiyama, J., A remark on the invariants of W^*-algebras, *Tôhoku Math. J.* 10 (1958), 37–41.

[378] Tomiyama, J., Generalized dimension function for W^*-algebras of infinite type, *Tôhoku Math. J.* **10** (1958), 121–129.

[379] Tomiyama, J., On the projection of norm one in W^*-algebras, II, *Tôhoku Math. J.* **10** (1958), 204–209.

[380] Tomiyama, J., On the projection of norm one in W^*-algebras, III, *Tôhoku Math. J.* **11** (1959), 125–129.

[381] Tomiyama, J., On the projection of norm one in the direct product of operator algebras, *Tôhoku Math. J.* **11** (1959), 305–313.

[382] Tomiyama, J., A characterization of C^*-algebras whose conjugate spaces are separable, *Tôhoku Math. J.* **15** (1963), 96–102.

[383] Tomiyama, J., Applications of Fubini type theorem to the tensor product of C^*-algebras, *Tôhoku Math. J.* **19** (1967), 213–226.

[384] Tomiyama, J., On the tensor products of von Neumann algebras, *Pacific J. Math.* **30** (1969), 263–270.

[385] Tsuji, K., The J. von Neumann commutation theorem, *Bull. Kyushu Inst. Tech.* **13** (1966), 1–3.

[386] Turumaru, T., On the commutativity of the C^*-algebra, *Kōdai Math. Sem. Rep.* **3** (1951), 51.

[387] Turumaru, T., On the direct-product of operator algebras, I, *Tôhoku Math. J.* **4** (1952), 242–251.

[388] Turumaru, T., On the direct-product of operator algebras, II, *Tôhoku Math. J.* **5** (1953), 1–7.

[389] Turumaru, T., On the direct product of operator algebras, III, *Tôhoku Math. J.* **6** (1954), 208–211.

[390] Turumaru, T., On the direct product of operator algebras, IV, *Tôhoku Math. J.* **8** (1956), 281–285.

[391] Umegaki, H., Operator algebra of finite class II, *Kōdai Math. Sem. Rep.* **5** (1953), 61–63.

[392] Umegaki, H., Conditional expectation in an operator algebra, I, *Tôhoku Math. J.* **6** (1954), 177–181.

[393] Umegaki, H., Ergodic decomposition of stationary linear functional, *Proc. Jap. Acad.* **30** (1954), 358–362.

[394] Umegaki, H., Positive definite function and direct product Hilbert space, *Tôhoku Math. J.* **7** (1955), 206–211.

[395] Umegaki, H., Weak compactness in an operator space, *Kōdai Math. Sem. Rep.* **8** (1956), 145–151.

[396] Varopoulos, N. T., Sur les formes positives d'une algèbre de Banach, *C. R. Acad. Sci.* Paris **258** (1964), 2465–2467.

[397] Vesterstrøm, J., and W. Wills, Direct integral of Hilbert spaces, II, *Math. Scand.* **26** (1970), 89–102.

[398] Vesterstrøm, J., Quotients of finite W^*-algebras, *Bull. Amer. Math. Soc.* **77** (1971), 235–238; *J. Funct. Anal.* **9** (1972), 322–335.

[399] Vidav, I., Quelques propriétés de la norme dans les algèbres de Banach, *Acad. Serbe Sci. Publ. Inst. Math.* **10** (1956), 53–58.

[400] Vowden, B. J., A new proof in the spatial theory of von Neumann algebras, *J. London Math. Soc.* **44** (1969), 429–432.

[401] Widom, H., Embedding in algebras of type I, *Duke Math. J.* **23** (1956), 309–324.

[402] Wils, W., Désintégration centrale des formes positives sur les C^*- algèbres, *C. R. Acad. Sci. Paris* **267** (1968), 810–812.

[403] Wils, W., Central decomposition of C^*-algebras, *Proc. Funct. Anal. Week, March 3–7, 1969, Aarhus Univ.*

[404] Wils, W., Direct integrals of Hilbert spaces, I, *Math. Scand.* **26** (1970), 73–88.

[405] Wogen, W., On generators for von Neumann algebras, *Bull. Amer. Math. Soc.* **75** (1969), 95–99.

[406] Woods, E. J., The classification of factors is not smooth, *Canad. J. Math.* **25** (1973), 96–102.

[407] Woronowicz, S., On a theorem of Mackey, Stone and von Neumann, *Studia Math.* **24** (1964–65), 101–105.

[408] Wright, F. B., A reduction for algebras of finite type, *Ann. Math.* **60** (1954), 560–570.

[409] Wright, F. B., The ideals in a factor, *Ann. Math.* **68** (1958), 475–483.

[410] Wright, J. D. M., An extension theorem and a dual proof of a theorem of Gleason, *J. London Math. Soc.* **43** (1968), 699–702.

[411] Wulfsohn, A., Produit tensoriel de C^*-algèbres, *Bull. Sci. Math.* **87** (1963), 13–21.

[412] Wulfsohn, A., Le produit tensoriel de certaines C^*-algèbres, *C. R. Acad. Sci.* Paris **258** (1964), 6052–6054.

[413] Wulfsohn, A., The primitive spectrum of a tensor product of C^*-algebras, *Proc. Amer. Math. Soc.* **19** (1968), 1094–1096.

[414] Yeadon, F. J., A new proof of the existence of a trace in a finite von Neumann algebra, *Bull. Amer. Math. Soc.* **77** (1971), 257–260.

[415] Yen, T., Trace on finite AW^*-algebras, *Duke Math. J.* **22** (1955), 207–222.

[416] Yen, T., Quotient algebra of a finite AW^*-algebra, *Pacific J. Math.* **6** (1956), 389–395.

[417] Yood, B., On Kadison's condition for extreme points of the unit ball in a C^*-algebra, *Proc. Edinburg Math. Soc.* (2)**16** (1968–1969), 245–250.

Notation Index

A^* 8
A_h^* 25, 121
A_+^* 120
A_h 17
$\mathcal{A}(\mathfrak{H})$ 55, 67
A_I 3
$\mathcal{A}(K)$ 158
$\mathcal{A}(K;L)$ 158
A_m 93
A^m 93
$A_1 \overline{\otimes} A_2$ 221
$A_1 \otimes_{\max} A_2$ 206
$A_1 \otimes_{\min} A_2$ 207
A_+ 24
$(A_+ \cap S)_\sigma$ 95
$(A_+ \cap S)_{\sigma\delta}$ 94, 96
B_f 234
$\mathcal{B}(\Gamma)$ 113
$\mathcal{B}(\Gamma)/\mathcal{N}(\Gamma)$ 113
$\mathcal{B}(\mathfrak{H})$ 59
$\mathrm{BV}(\Gamma,\mu)$ 116
$\mathrm{BV}(\Gamma)$ 116
$C^*(A)$ 42
$C^*(G)$ 45
$C_r^*(G)$ 45
$c(e,f)$ 308
$C_E(\Gamma)$ 254
$C_\infty(\Omega)$ 3
$\mathcal{C}_\mu$ 241
$C_{\mathbf{R}}(\Omega)$ 104
(c_0) 60
$e \sim f$ 290
$e \prec f$ 290
$e \precsim f$ 290
$E_1 \otimes E_2$ 189
$E_1 \otimes_\beta E_2$ 188
$E_1 \otimes_\gamma E_2$ 189
$\exp x$ 12
f^* 36
$f \succsim e$ 290
$f \succ e$ 290
$f \geqslant g$ 36
$\hat{f}$ 232
$\tilde{\hat{f}}$ 232
$\mathrm{Fred}(\mathfrak{H})$ 55
$f(x)$ for $f \in A(\mathrm{Sp}_A(x))$ and $x \in A$ 9
$f(x)$ for $f \in C(\mathrm{Sp}_A(x))$ 19
$|h|$ 19
h_- 19
$H_{\mathbf{R}}^\perp$ 223
$\mathfrak{H}_1 \otimes \mathfrak{H}_2$ 182
$\mathfrak{H}(\pi)$ 36
h_+ 19
$G(A)$ 4
$G_0(A)$ 12
$\mathcal{LC}(\mathfrak{H})$ 61
$\mathcal{L}(\mathfrak{H})$ 3, 59
(l^∞) 60
$\mathcal{L}_G^n$ 80
$\log x$ 12
(l^1) 60
$L^1(G)$ 6
$L^1(\mathcal{M},\tau)$ 320
$L_E^p(\Gamma,\mu)$ 254
$\mathcal{LT}(\mathfrak{H})$ 63
$L_{\mathfrak{H}}^2(\Gamma,\mu)$ 257
$L^2(\mathcal{M},\tau)$ 322
$M(A)$ 169
$\mathcal{M}_*$ 70, 127
$\mathcal{M}_*^\perp$ 127
$\mathcal{M}_e$ 75
$\{\mathcal{M},\mathfrak{H}\}$ 72
$\mathcal{M}_{\mathfrak{R}}$ 75
$(((\cdot)^m)_m)^m$ 97
$M_n(A^*)$ 200
$M_n(\mathbf{C})$ 51
$\mathcal{M}_1 \overline{\otimes} \mathcal{M}_2$ 183
$\mathcal{M}(\pi)$ 121
$M_\varphi^+(\mathfrak{S})$ 240, cf. 233
$M_y^+(K)$ 233
$\mathcal{M}'$ 71
$\mathcal{M}_e'$ 75
$\mathcal{M}_{\mathfrak{R}}'$ 75
M_u 173
$M_u(A)$ 173

$N(\mathfrak{A})$ 373
$\mathfrak{N}(\Gamma)$ 113
N_ω 37
$P(A)$ 43
$\mathfrak{P}(K)$ 231
$\mathfrak{Q}(A)$ 161
$\mathfrak{Q}(K)$ 231
$\mathfrak{R}(\mathfrak{A}, G, \alpha)$ 364
$\mathfrak{R}(G)$ 364
$r(\mu)$ 233
$\mathfrak{S}(A)$ 120
$s(e,f)$ 308
$S_l(\omega)$ 140
$S_l(x)$ 291
$s(\omega)$ 134
$\mathrm{Sp}_A(x)$ 6
$\mathrm{Sp}'_A(x)$ 7
$s_r(\omega)$ 140
$s_r(x)$ 291
$s(\tau)$ 315
$t_{\xi,\eta}$ 60
$t(\omega)$ 61
$\mathrm{Tr}(t(\omega))$ 65
$U(A)$ 17
$\mathfrak{U}(\mathrm{H})$ 84
V^0 124
ωa and $a\omega$ 63, 123
$|x|$ 24
$\|x\|_\gamma$ 189
$\|x\|_\lambda$ 188
$\|x\|_{\max}$ 206
$\|x\|_{\min}$ 207
x^{-1} 4
$\|x\|_1$ 321
$x^{1/2}$ 23
$x_1 \otimes x_2$ 183
$\|x\|_{\mathrm{sp}}$ 7
$\|x\|_2$ 322
$\|x\|_\infty$ 321
$z(e)$ 223
$z(\pi)$ 126

$\nu \prec \mu$ 233
$(\xi|\eta)_{\mathbf{R}}$ 223
$\xi \perp_{\mathbf{R}} \eta$ 223
ξ_ω 40
$\{\pi, \mathfrak{H}\}$ 36
$\pi \cong \pi_2$ 36
$\pi_1 \otimes \pi_2$ 222
$\{\pi_1, \mathfrak{H}_1\} \cong \{\pi_2, \mathfrak{H}_2\}$ 36
$\{\pi_\omega, \mathfrak{H}_\omega\}$ 39
$\tau(\mathfrak{M}, \mathfrak{M}_*)$ 153
$|\varphi|$ 142
φ_+ 140
φ_- 140
$[\omega]$ 121
ω_ξ 78
$\omega_{\xi,\eta}$ 61
$\omega(\pi;\xi)$ 37
$\omega(\pi;\xi,\eta)$ 36
$\Omega(A)$ 16

$\partial_e K$ 232
$\bigvee_{i\in I} e_i$ 290
$\bigwedge_{i\in I} e_i$ 290
$\sum_{i\in I}^{\oplus} \{\mathfrak{M}_i, \mathfrak{H}_i\}$ 73
$\sum_{i\in I}^{\oplus} \{\pi_i, \mathfrak{H}_i\}$ 41
$\int_\Gamma^\oplus \mathfrak{H}(\gamma)\, d\mu(\gamma)$ 272
$\int_\Gamma^\oplus \mathfrak{M}(\gamma)_*\, d\mu(\gamma)$ 285
$\int_\Gamma^\oplus \mathfrak{M}(\gamma)\, d\mu(\gamma)$ 273
$\int_\Gamma^\oplus \{\mathfrak{M}(\gamma), \mathfrak{H}(\gamma)\}\, d\mu(\gamma)$ 274
$\int_\Gamma^\oplus x(\gamma)\, d\mu(\gamma)$ 272
$\int_\Gamma^\oplus \xi(\gamma)\, d\mu(\gamma)$ 272
$\int_\Gamma^\oplus \pi_\gamma\, d\mu(\gamma)$ 278
$\int_\Gamma \varphi(\gamma)\, d\mu(\gamma)$ 285

Subject Index

A^*-algebras 42
Abelian Banach algebras 13–16
Abelian projections 296
Absolute value of h 20
Absolute value of φ 142
Absolute value of x 24
Absolutely continuous spectrum 66
Absolutely invariant projections 363
Action of a group 363
  ergodic 363
Action of G on A 46
Adjoint cross-norm 190
Adjoint functional of f 36
Affine functions 158
α-homogeneous central projections 299
Amplification isomorphism 184
Approximate identity of A 26
Arens–Mackey topology 153
*-algebra of operators on $\mathfrak{H}$ 72
  nondegenerate 72
*-homomorphisms 21
*-operation 2
Asymmetric Riesz decomposition theorem 29
Atomic von Neumann algebras 155
Atomic representations 176
  universal atomic representation 176
$\mathscr{A}$-valued trace 298

Banach algebras 2
  abelian 13–16
  involutive 2
  semisimple 16
  unital 3
Barycenter of μ 233
B^*-algebra 55
Borel spaces 375
Boundary measures 235
Boundary set of f 234

Calkin algebra on $\mathfrak{H}$ 55
Canonical center valued trace 314
Canonical unitary involution of $L^2(\mathscr{M}, \tau)$ 337
C^*-algebra of operators 72
C^*-algebras 2
  dual 157
  monotone closed 137
  primitive 178
  weakly compact 157
C^*-homomorphisms 188
C^*-subalgebra of A generated by E 18
C^*-subalgebras 18
  unital 18
Cauchy–Schwarz inequality 37
Cayley transform 81
Center valued trace 298, 314

Center valued trace *(cont.)*
canonical 314
Centrally orthogonal projections 292
Central positive linear functionals 343
Central (representing) measure of φ 247
Central support of a projection 223
$C(\infty)$-direct sums 157
Closed projections 168
Commutant of $\pi(A)$ 53
Complementary split faces 159
Completely positive linear maps 194, 200
Components of $\{\pi, \mathfrak{H}\}$ 41
Concave functions 158
Conditional expectation of $\mathcal{M}$ onto $\mathcal{N}$ with respect to τ 332
Constant field of Hilbert spaces 272
Convex functions 158, 231
Cosine of a projection 308
Coupling constants 339
Coupling functions 339
Covariance C^*-algebra of $\{\Omega, G\}$ 46
Covariance von Neumann algebra of $(\mathcal{A}, G, \alpha)$ 364
Covariant representations 46
Covariant systems 363
Crossed products 364
Cross-norms 188
adjoint 190
greatest 189
injective 189
projective 189
Cyclic projections 293
Cyclic representation of A induced by ω 41
Cyclic subsets 77
Cyclic vectors 41

Decomposable operators 258
Definition ideal of a trace 319
Diagonal algebras 259, 273
Diagonal operators 259, 273
Dimension theory 289
Direct integral of the bounded measurable operator field 273
Direct integral of measurable fields of Hilbert spaces 272
Direct integral of von Neumann algebras 274
Direct integral of representations 278
Direct sums 41
Direct sums of von Neumann algebras 73
Disintegration 275
Double commutation theorem 74
Dual C^*-algebras 157

Effros Borel structure 264
Enveloping C^*-algebra of A 42
Equivalence of projections 290
Ergodic actions 363
Ergodic states 47
Essential space of π 36
Exponential function 12
Extreme boundary of K 232
Extremely disconnected spaces 104
Extreme points 47

Faces 159
complementary split 159
Factorial states 248
Factors 72
Faithful linear functionals 36
Faithful representations 36
Faithful semifinite normal traces 309
Faithful traces on a von Neumann algebra 309
Final projections 290
Finite von Neumann algebras 296
Finite projections 296
Finite traces of a von Neumann algebra 309
Fredholm operators 55
Free automorphisms 363
Fundamental sequence of μ-measurable vector fields 270

G-abelian actions 252
Gelfand–Naimark–Segal construction 41
Gelfand representation of A 16
Generalized center valued traces 330
General linear group 4
Greatest cross-norm 189
Group C^*-algebra of G 45
Group measure space construction 374
Group von Neumann algebra of G 364
Groups of infinite conjugacy class 368

Hereditary subcones 146
Hermitian elements 17
Hermitian linear functionals 36
Hermitian sesquilinear forms 59
Hull of $\mathfrak{m}$ 33
Hyperstonean spaces 107

ICC-groups 368
Identity modulo $\mathfrak{m}$ 13
Imaginary part of x 17
Index group of A 12
Induced von Neumann algebra of $\mathscr{M}'$ on $\mathfrak{K}$ 76
Induction of $\mathscr{M}'$ onto $\mathscr{M}'_e$ 76
Infinite von Neumann algebras 296
Infinite projections 296
Initial projections 290
Injective C^*-cross-norm 207
Injective C^*-tensor products 207
Injective cross-norm 189
Injective tensor products of Banach spaces 189
Integral representations of states 230–253
Invariant subsets 123
Invariant subspaces of $\mathscr{M}_*$ 123
Invertible elements 4
Involutions 2
  unitary 337
Involutive Banach algebras 2
Irreducible representations 41
Isomorphic von Neumann algebras 72

Joint spectrum of $(a_1, \ldots, a_n)$ 80
Jordan decomposition of a finite Radon measure 120
Jordan homomorphisms 187
Jordan product 187

Kadison transitivity theorem 92
Kaplansky density theorem 100
Kernel of Γ 33

(Left) group von Neumann algebra of G 364
Left invariant subsets 123
Left invariant subspaces 123
Left kernel of ω 37
Left multiplier of A 169
Left regular representation 45
Left support of x 291
Left support projections 140
$\mathscr{L}(\mathfrak{H})$-valued measurable functions 258
Liftings 259
Logarithmic function 12
L^1 group algebra of G 6
Lower envelope of f 232
Lower semicontinuous functions 104, 158
Lusin spaces 378

Matrix units 185
Maximal abelian von Neumann algebras 104
Maximal regular ideals 14
Measurable fields of
  Hilbert spaces 269
  von Neumann algebras 273
  representations 278
Measurable operator fields 272
Measurable operator valued functions 258
Measurable vector fields 270
Minimal projections 51
Modular lattices 303
Monotone closed C^*-algebras 137
Multiplicity-free representations 246
Multiplicity of π 53
Multiplier algebra of A 169
Multiplier of A 169
μ-measurable E-valued functions 253

Negative part of h 20
von Neumann algebras 72
  atomic 155
  covariance von Neumann algebra of $(\mathscr{A}, G, \alpha)$ 364
  finite 296
  infinite 296
  isomorphic 72
  (left) group von Neumann algebra of G 364
  maximal abelian 104
  of finite rank 155
  properly infinite 296
  purely infinite 296
  semifinite 297
  σ-finite 78
  spatially isomorphic 72
  type I 296
  type I_α 302
  type II 296
  type III 296
  type II_∞ 296
  type II_1 296
  type I_n 302
Noncommutative Egoroff's theorem 85
Noncommutative Lusin's theorem 87
Nondegenerate *-algebra of operators 72

Nondegenerate representations 36
Normal elements of involutive algebras 17
Normalizers 373
Normal linear functionals 127, 134, 137
Normal measures 106
Normal part of ω 127
Normal representations 128, 136
Normal traces on a von Neumann algebra 309
n-positive linear maps 194, 200
Nuclear operators 63

One parameter groups 57
Open projections 168
Operator-monotone functions 93
Operators affiliated with $\mathscr{M}$ 229
Operators of Hilbert–Schmidt class 66, 191
Operators of trace class 63
Orthogonal measures 240

Phillips lemma 179
Polar decomposition of φ 142
Polarization identity 59
Polish spaces 264, 375
Positive cones 23
Positive elements 24
Positive linear functionals 36
  central 343
  tracial 343
Positive part of h 20
Positive sesquilinear forms 59
Predual of a W^*-algebra 134
Predual of $\mathscr{M}$ 127
Primary states 248
Primitive C^*-algebras 178
Principal component of $G(A)$ 12
Projection lattices 290
Projections 17
  absolutely invariant 363
  abelian 296
  α-homogeneous 299
  centrally orthogonal 292
  closed 168
  cyclic 293
  finite 296
  infinite 296
  minimal 51
  of norm one 131
  open 168
  properly infinite 296
  purely infinite 296

Projective C^*-cross-norm 206
Projective C^*-tensor products 206
Projective cross-norm 189
Projective tensor products of Banach spaces 189
Properly infinite von Neumann algebras 296
Properly infinite projections 296
Proper representations 36
Pseudoconcentrated measures 246
Pure linear functionals 43
Purely infinite von Neumann algebras 296
Purely infinite projections 296

Quasi-equivalent representations 125
Quasi-spectrum of x 7
Quasi-state space of A 161
Quotient algebras of C^*-algebras 31

Radical of A 16
Real orthogonality 223
Real part of x 17
Real subspaces 223
Reduced von Neumann algebra of M on $\mathfrak{K}$ 76
Regular elements 4
Regular ideals 13
Regular subalgebras 373
Representations 35
  atomic 176
  covariant 46
  faithful 36
  irreducible 41
  multiplicity-free 246
  nondegenerate 36
  normal 128, 136
  proper 36
  quasi-equivalent 125
  tensor products of 222
  topologically irreducible 41
  unitarily equivalent 36
  universal 122
Representation space 35
Representing measure of a point 233
Resolvant of x 6
Restricted group C^*-algebra of G 45
Restriction of π to A_i 204
Resultant of μ 233
Right invariant subsets 123
Right invariant subspaces 123
Right kernel of ω 37
Right multiplier of A 169

Right support of x 291
Right support projections 140

Self-adjoint elements 17
Self-adjoint linear functionals 36
Semifinite von Neumann algebras 297
Semifinite traces on a von Neumann algebra 309
Semiregular subalgebras 373
Semisimple Banach algebras 16
Separating subsets 77
Sesquilinear forms 59
  hermitian 59
  positive 59
  symmetric 59
σ-finite von Neumann algebras 78
σ-strong* (operator) topology 68
σ-strong (operator) topology 68
σ-weak (operator) topology 67
Simplices 251
Sine of a projection 308
Singular functionals 127
Singular part of ω 127
Souslin sets 377
Souslin spaces 377
Spatial isomorphisms 72
Spatially isomorphic von Neumann algebras 72
Spectral mapping theorem 11
Spectral radius of x 7
Spectrum of A 16
Spectrum of x 6
Standard Borel spaces 264
States 36
  ergodic 47
  factorial 248
  primary 248
  tracial 343
State spaces 166
Stonean spaces 104
Stone–Čech compactification of Γ 21
Strictly convex functions 235
Strong* (operator) topology 68
Strong (operator) topology 68
Subrepresentations 41
Support of a
  left invariant subspace 124
  positive linear functional 140
  representation 126
  trace 315
Support projections of 134
  left invariant subspaces 124
  representations 126
Symmetric sesquilinear forms 59

Tensor product representation 222
Tensor products of
  Banach spaces 188
  C^*-algebras 203
  Hilbert spaces 182
  von Neumann algebras 183
  operators 182
  W^*-algebras 220
Thick ideals 169
Topological transformation groups 6
Topologically irreducible representations 41
Traces on a von Neumann algebra 309
  faithful 309
  finite 309
  normal 309
  semifinite 309
Tracial positive linear functionals 343
Transitivity theorem 92
Type I von Neumann algebras 296
Type I_α von Neumann algebras 302
Type II von Neumann algebras 296
Type III von Neumann algebras 296
Type II_∞ von Neumann algebras 296
Type II_1 von Neumann algebras 296
Type I_n von Neumann algebras 302

Uniform (operator) topology 68
Unital Banach algebras 3
Unitarily equivalent representations 36
Unitary elements 17
Unitary group of A 17
Unitary implementation of an action 371
Unitary involutions 337
Universal atomic representation of a C^*-algebra 176
Universal enveloping von Neumann algebra of A 122
Universally measurable elements 173
Universal representations 122
Up–down theorem 96
Up–down–up theorem 97
Upper envelope of f 232
Upper semicontinuous regularization of f 106

W^*-algebras 130
W^*-tensor products 221
Weakly compact C^*-algebras 157
Weak (operator) topology 68

C*-Algebras and W*-Algebras
by **S. Sakai**
(Ergebnisse der Mathematik und ihrer Grenzgebiete, Volume 60)

". . .this book is an excellent and comprehensive survey of the theory of von Neumann algcbras. It includes all the fundamental results of the subject, and is a valuable reference for both the beginner and the expert."

—*Mathematical Reviews*

1971/xii, 256 pp./Cloth
ISBN 0-387-05347-6

Tomita's Theory of Modular Hilbert Algebras and its Applications
By **M. Takesaki**
(Lecture Notes in Mathematics, Volume 128)

1970/ii, 123 pp./Paper
ISBN 0-387-04917-7

Duality for Crossed Products of von Neumann Algebras
By **Y. Nakagami** and **M. Takesaki**
(Lecture Notes in Mathematics, Volume 731)

1979/IX, 139 pp./Paper
ISBN 0-387-09522-5

An Invitation to C*-Algebras
By **W. Arveson,** Department of Mathematics, University of California, Berkeley
(Graduate Texts in Mathematics, Volume 39)

A concise introduction to C*-algebras and their representations on Hilbert spaces, accessible to students of mathematics and physics with some knowledge of functional analysis and measure theory. Multiplicity theory for normal operators, C*-algebras of compact operators, the structure of type I C*-algebras, standard and analytic Borel spaces including Souslin's theorem and a discussion of cross sections, the spectrum of a C*-algebra, and classification theory for representations of separable type I C*-algebras are discussed in detail.

1976/x, 110 pp./Cloth
ISBN 0-387-90176-0

Classical Banach Spaces I
Sequence Spaces
by **J. Lindenstrauss** and **L. Tzafriri**
(Ergebnisse der Mathematik und ihrer Grenzgebiete, Volume 92)

The first part of a multi-volume work entitled *Classical Banach Spaces,* this volume is devoted to the study of sequence spaces within the framework of the isomorphic theory of Banach spaces, Main emphasis is on detailed analysis of the structure of the classical sequence spaces c_0 and l_p, $1 \leqslant p \leqslant \infty$ and closely related spaces. In addition, the book contains central results concerning the approximation property and bases in general spaces. Other topics discussed are the structure theory of spaces with symmetric bases, Orlicz sequence spaces, and applications of operator theory to the study of Banach spaces. The book focuses on important recent results and current research directions as well as a concise presentation of the standard material required for understanding future volumes. A basic knowledge of elementary functional analysis is assumed.

1977/xiii/188 pp./Cloth
ISBN 0-387-08072-4

Classical Banach Spaces II
Function Spaces
by **J. Lindenstrauss** and **L. Tzafriri**
(Ergebnisse der Mathematik und ihrer Grenzgebiete, Volume 97)

Devoted to the substantial progress made in recent studies of Banach lattices, this second part of *Classical Banach Spaces* concentrates on the notions of p-convexity, p-concavity, and their variants. More extensive use is made of ideas and results from probability theory, and brief discussions are presented where needed. Basic definitions, examples, and results found useful in rearrangement invariant spaces, Boyd indices, the Haar and the trigonometric systems, and relevant results on complemented subspaces are covered. Applications of the Poisson process to r.i. function spaces and interpolation spaces and their applications are demonstrated as well. Beyond an average understanding of functional analysis and measure theory, only basic knowledge of the material presented in Volume I is assumed.

1978/256 pp./Cloth
ISBN 0-387-08888-1